Quantum Mechanics

2nd edition

30th May, 2007
W. y. Y.

We work with leading authors to develop the
strongest educational materials in physics,
bringing cutting-edge thinking and best learning
practice to a global market.

Under a range of well-known imprints, including
Prentice Hall, we craft high quality print and
electronic publications which help readers to
understand and apply their content,
whether studying or at work.

To find out more about the complete range of our
publishing please visit us on the World Wide Web at:
www.pearsoned.co.uk

Quantum Mechanics

2nd edition

B. H. Bransden and C. J. Joachain

Harlow, England • London • New York • Boston • San Francisco • Toronto • Sydney • Singapore • Hong Kong
Tokyo • Seoul • Taipei • New Delhi • Cape Town • Madrid • Mexico City • Amsterdam • Munich • Paris • Milan

Pearson Education Limited
Edinburgh Gate
Harlow
Essex CM20 2JE
England

and Associated Companies throughout the world

Visit us on the World Wide Web at:
www.pearsoned.co.uk

First published under the Longman Scientific & Technical imprint 1989
Second edition 2000

© Pearson Education Limited 1989, 2000

The rights of B. H. Bransden and C. J. Joachain to be identified as the authors of this Work have been asserted by them in accordance with the Copyright, Designs and Patents Act 1988.

All rights reserved. No part of this publication may be reproduced, stored in a retrieval system, or transmitted in any form or by any means, electronic, mechanical, photocopying, recording or otherwise, without either the prior written permission of the publisher or a licence permitting restricted copying in the United Kingdom issued by the Copyright Licensing Agency Ltd, 90 Tottenham Court Road, London WIT 4LP.

ISBN 0582-35691-1

British Library Cataloguing-in-Publication Data
A catalogue record for this book can be obtained from the British Library

Library of Congress Cataloging-in-Publication Data
Bransden, B. H., 1926-
  Quantum mechanics / B.H. Bransden and C.J. Joachain.– 2nd ed.
    p. cm.
  Rev. ed. of: Introduction to quantum mechanics. 1989.
  Includes bibliographical references and index.
  ISBN 0-582-35691-1
  1. Quantum theory. I. Joachain, C.J. (Charles Jean) II. Bransden, B. H., 1926- Introduction to quantum mechanics. III. Title.
QC174.12.B74 2000
530.12–dc21   99-055742

10 9 8 7 6 5 4
07 06 05

Typeset in Times at 10pt by 56.

Produced by Pearson Education Asia Pte Ltd.,
Printed in Great Britain by Henry Ling Limited, at the Dorset Press, Dorchester, DT1 1HD.

Contents

Preface to the Second Edition   xi
Preface to the First Edition   xiii
Acknowledgements   xiv

1 The origins of quantum theory   1
1.1 Black body radiation   2
1.2 The photoelectric effect   12
1.3 The Compton effect   16
1.4 Atomic spectra and the Bohr model of the hydrogen atom   19
1.5 The Stern–Gerlach experiment. Angular momentum and spin   33
1.6 De Broglie's hypothesis. Wave properties of matter and the genesis of quantum mechanics   38
Problems   45

2 The wave function and the uncertainty principle   51
2.1 Wave–particle duality   52
2.2 The interpretation of the wave function   56
2.3 Wave functions for particles having a definite momentum   58
2.4 Wave packets   60
2.5 The Heisenberg uncertainty principle   69
Problems   76

3 The Schrödinger equation   81
3.1 The time-dependent Schrödinger equation   82
3.2 Conservation of probability   86
3.3 Expectation values and operators   90
3.4 Transition from quantum mechanics to classical mechanics. The Ehrenfest theorem   97
3.5 The time-independent Schrödinger equation. Stationary states   100
3.6 Energy quantisation   104
3.7 Properties of the energy eigenfunctions   115

3.8	General solution of the time-dependent Schrödinger equation for a time-independent potential	120
3.9	The Schrödinger equation in momentum space	124
	Problems	128

4 One-dimensional examples — 133

4.1	General formulae	133
4.2	The free particle	134
4.3	The potential step	141
4.4	The potential barrier	150
4.5	The infinite square well	156
4.6	The square well	163
4.7	The linear harmonic oscillator	170
4.8	The periodic potential	182
	Problems	189

5 The formalism of quantum mechanics — 193

5.1	The state of a system	193
5.2	Dynamical variables and operators	198
5.3	Expansions in eigenfunctions	203
5.4	Commuting observables, compatibility and the Heisenberg uncertainty relations	210
5.5	Unitary transformations	216
5.6	Matrix representations of wave functions and operators	220
5.7	The Schrödinger equation and the time evolution of a system	231
5.8	The Schrödinger and Heisenberg pictures	238
5.9	Path integrals	240
5.10	Symmetry principles and conservation laws	245
5.11	The classical limit	256
	Problems	260

6 Angular momentum — 265

6.1	Orbital angular momentum	266
6.2	Orbital angular momentum and spatial rotations	270
6.3	The eigenvalues and eigenfunctions of L^2 and L_z	275
6.4	Particle on a sphere and the rigid rotator	289
6.5	General angular momentum. The spectrum of J^2 and J_z	292

6.6	Matrix representations of angular momentum operators	296
6.7	Spin angular momentum	299
6.8	Spin one-half	303
6.9	Total angular momentum	311
6.10	The addition of angular momenta	315
	Problems	323

7 The Schrödinger equation in three dimensions — 327

7.1	Separation of the Schrödinger equation in Cartesian coordinates	328
7.2	Central potentials. Separation of the Schrödinger equation in spherical polar coordinates	336
7.3	The free particle	341
7.4	The three-dimensional square well potential	347
7.5	The hydrogenic atom	351
7.6	The three-dimensional isotropic oscillator	367
	Problems	370

8 Approximation methods for stationary problems — 375

8.1	Time-independent perturbation theory for a non-degenerate energy level	375
8.2	Time-independent perturbation theory for a degenerate energy level	386
8.3	The variational method	399
8.4	The WKB approximation	408
	Problems	427

9 Approximation methods for time-dependent problems — 431

9.1	Time-dependent perturbation theory. General features	431
9.2	Time-independent perturbation	435
9.3	Periodic perturbation	443
9.4	The adiabatic approximation	447
9.5	The sudden approximation	458
	Problems	466

10 Several- and many-particle systems — 469

10.1	Introduction	469
10.2	Systems of identical particles	472
10.3	Spin-1/2 particles in a box. The Fermi gas	478

10.4	Two-electron atoms	485
10.5	Many-electron atoms	492
10.6	Molecules	498
10.7	Nuclear systems	506
	Problems	511

11 The interaction of quantum systems with radiation — 515

11.1	The electromagnetic field and its interaction with one-electron atoms	516
11.2	Perturbation theory for harmonic perturbations and transition rates	522
11.3	Spontaneous emission	527
11.4	Selection rules for electric dipole transitions	533
11.5	Lifetimes, line intensities, widths and shapes	538
11.6	The spin of the photon and helicity	544
11.7	Photoionisation	545
11.8	Photodisintegration	550
	Problems	555

12 The interaction of quantum systems with external electric and magnetic fields — 557

12.1	The Stark effect	557
12.2	Interaction of particles with magnetic fields	563
12.3	One-electron atoms in external magnetic fields	574
12.4	Magnetic resonance	576
	Problems	585

13 Quantum collision theory — 587

13.1	Scattering experiments and cross-sections	588
13.2	Potential scattering. General features	592
13.3	The method of partial waves	595
13.4	Applications of the partial-wave method	599
13.5	The integral equation of potential scattering	608
13.6	The Born approximation	615
13.7	Collisions between identical particles	620
13.8	Collisions involving composite systems	627
	Problems	635

14 Quantum statistics — 641
- 14.1 The density matrix — 642
- 14.2 The density matrix for a spin-1/2 system. Polarisation — 645
- 14.3 The equation of motion of the density matrix — 654
- 14.4 Quantum mechanical ensembles — 655
- 14.5 Applications to single-particle systems — 661
- 14.6 Systems of non-interacting particles — 663
- 14.7 The photon gas and Planck's law — 667
- 14.8 The ideal gas — 668
- Problems — 676

15 Relativistic quantum mechanics — 679
- 15.1 The Klein–Gordon equation — 679
- 15.2 The Dirac equation — 684
- 15.3 Covariant formulation of the Dirac theory — 690
- 15.4 Plane wave solutions of the Dirac equation — 696
- 15.5 Solutions of the Dirac equation for a central potential — 702
- 15.6 Non-relativistic limit of the Dirac equation — 711
- 15.7 Negative-energy states. Hole theory — 715
- Problems — 717

16 Further applications of quantum mechanics — 719
- 16.1 The van der Waals interaction — 719
- 16.2 Electrons in solids — 723
- 16.3 Masers and lasers — 735
- 16.4 The decay of K-mesons — 746
- 16.5 Positronium and charmonium — 753

17 Measurement and interpretation — 759
- 17.1 Hidden variables? — 759
- 17.2 The Einstein–Podolsky–Rosen paradox — 760
- 17.3 Bell's theorem — 762
- 17.4 The problem of measurement — 766
- 17.5 Time evolution of a system. Discrete or continuous? — 772

A Fourier integrals and the Dirac delta function — 775
- A.1 Fourier series — 775

	A.2	Fourier transforms	776
	B	**WKB connection formulae**	**783**
		References	**787**
		Table of fundamental constants	**789**
		Table of conversion factors	**791**
		Index	**793**

Preface to the Second Edition

The purpose of this book remains as outlined in the preface to the first edition: to provide a core text in quantum mechanics for students in physics at undergraduate level. It has not been found necessary to make major alterations to the contents of the book. However, we have taken advantage of the opportunity provided by the preparation of a new edition to make a number of minor improvements throughout the text, to introduce some new topics and to include a new chapter on relativistic quantum mechanics. This inclusion stems from a reconsideration of our earlier decision to exclude this material. We believe that a significant number of core courses now include an introduction to relativistic quantum mechanics; this is the subject of the new chapter (Chapter 15). Among the other important changes are the inclusion of the Feynman path integral approach to quantum mechanics (Chapter 5), a discussion of the Berry phase (Chapter 9) with applications (Chapters 10 and 12), an account of the Aharonov–Bohm effect (Chapter 12) and a discussion of quantum jumps (Chapter 17). We have also included the integral equation of potential scattering in our treatment of quantum collision theory (Chapter 13) and have given a more extended discussion of Bose–Einstein condensation in Chapter 14.

It is a pleasure to acknowledge the many helpful comments made to us by colleagues who have used the first edition of this book. Their remarks have been of great benefit to us in preparing this new edition. One of us (CJJ) would like to thank Professor H. Walther for his hospitality at the Max-Planck-Institut für Quantenoptik in Garching, where part of this work was carried out. We also wish to thank Mrs R. Lareppe for her expert and careful typing of the manuscript.

B. H. Bransden, Durham
C. J. Joachain, Brussels
August 1999

Preface to the First Edition

The study of quantum mechanics and its applications pervades much of the modern undergraduate course in physics. Virtually all undergraduates are expected to become familiar with the principles of non-relativistic quantum mechanics, with a variety of approximation methods and with the application of these methods to simple systems occurring in atomic, nuclear and solid state physics. This core material is the subject of this book. We have firmly in mind students of physics, rather than of mathematics, whose mathematical equipment is limited, particularly at the beginning of their studies. Relativistic quantum theory, the application of group theoretical methods and many-body techniques are usually taught in the form of optional courses and we have made no attempt to cover more advanced material of this nature. Although a fairly large number of examples drawn from atomic, nuclear and solid state physics are given in the text, we assume that the reader will be following separate systematic courses on those subjects, and only as much detail as necessary to illustrate the theory is given here.

Following an introductory chapter in which the evidence that led to the development of quantum theory is reviewed, we develop the concept of a wave function and its interpretation, and discuss Heisenberg's uncertainty relations. Chapter 3 is devoted to the Schrödinger equation and in the next chapter a variety of applications to one-dimensional problems is discussed. The next three chapters deal with the formal development of the theory, the properties of angular momenta and the application of Schrödinger's wave mechanics to simple three-dimensional systems.

Chapters 8 and 9 deal with approximation methods for time-independent and time-dependent problems, respectively, and these are followed by six chapters in which the theory is illustrated through application to a range of specific systems of fundamental importance. These include atoms, molecules, nuclei and their interaction with static and radiative electromagnetic fields, the elements of collision theory and quantum statistics. Finally, in Chapter 17, we discuss briefly some of the difficulties that arise in the interpretation of quantum theory. Problem sets are provided covering all the most important topics, which will help the student monitor his understanding of the theory.

We wish to thank our colleagues and students for numerous helpful discussions and suggestions. Particular thanks are due to Professor A. Aspect, Dr P. Francken, Dr R. M. Potvliege, Dr P. Castoldi and Dr J. M. Frère. It is also a pleasure to thank Miss P. Carse, Mrs E. Péan and Mrs M. Leclercq for their patient and careful typing of the manuscript, and Mrs H. Joachain-Bukowinski and Mr C. Depraetere for preparing a large number of the diagrams.

B. H. Bransden, Durham
C. J. Joachain, Brussels
November 1988

Acknowledgements

We are indebted to the following for permission to reproduce copyright material: Oxford University Press for figs. 17.1, 17.2 & 17.3 adapted from figs. 1.4, 1.1 & 1.8, pp. 2–12, from an article by A. Aspect and P. Grangier in *Quantum Concepts in Space and Time* edited by Penrose and Isham (1986), the *American Journal of Physics* and A. Tonomura *et al.* for fig. 2.3, *Physics World* and L. Kouwenhoven and C. Marcus for fig. 16.7 and the *Journal of Optical Communications* and Th. Sauter *et al.* for fig. 17.5.

1 The origins of quantum theory

1.1 Black body radiation 2
1.2 The photoelectric effect 12
1.3 The Compton effect 16
1.4 Atomic spectra and the Bohr model of the hydrogen atom 19
1.5 The Stern–Gerlach experiment. Angular momentum and spin 33
1.6 De Broglie's hypothesis. Wave properties of matter and the genesis of quantum mechanics 38
Problems 45

Until the end of the nineteenth century, classical physics appeared to be sufficient to explain all physical phenomena. The universe was conceived as containing matter, consisting of particles obeying Newton's laws of motion and radiation (waves) following Maxwell's equations of electromagnetism. The theory of special relativity, formulated by A. Einstein in 1905 on the basis of a critical analysis of the notions of space and time, generalised classical physics to include the region of high velocities. In the theory of special relativity the velocity c of light plays a fundamental role: it is the upper limit of the velocity of any material particle. Newtonian mechanics is an accurate approximation to relativistic mechanics only in the 'non-relativistic' regime, that is when all relevant particle velocities are small with respect to c. It should be noted that Einstein's theory of relativity does not modify the clear distinction between matter and radiation which is at the root of classical physics. Indeed, all pre-quantum physics, non-relativistic or relativistic, is now often referred to as classical physics.

During the late nineteenth century and the first quarter of the twentieth, however, experimental evidence accumulated which required new concepts radically different from those of classical physics. In this chapter we shall discuss some of the key experiments which prompted the introduction of these new concepts: *the quantisation of physical quantities* such as energy and angular momentum, *the particle properties of radiation* and *the wave properties of matter*. We shall see that they are directly related to the existence of a universal constant, called *Planck's constant h*. Thus, just as the velocity c of light plays a central role in relativity, so does Planck's constant in *quantum physics*. Because Planck's constant is very small when measured in 'macroscopic' units (such as SI units), quantum physics essentially deals with phenomena at the atomic and subatomic levels. As we shall see in this chapter, the new ideas were first introduced in a more or less *ad hoc* fashion. They evolved later to become part of a new theory, *quantum mechanics*, which we will begin to study in Chapter 2.

1.1 Black body radiation

We start by considering the problem which led to the birth of quantum physics, namely the formulation of the *black body radiation law*. It is a matter of common experience that the surface of a hot body emits energy in the form of electromagnetic radiation. In fact, this emission occurs at any temperature greater than absolute zero, the emitted radiation being continuously distributed over all wavelengths. The distribution in wavelength, or *spectral distribution* depends on temperature. At low temperature (below about 500 °C), most of the emitted energy is concentrated at relatively long wavelengths, such as those corresponding to infrared radiation. As the temperature increases, a larger fraction of the energy is radiated at lower wavelengths. For example, at temperatures between 500 and 600 °C, a large enough fraction of the emitted energy has wavelengths within the visible spectrum, so that the body 'glows', and at 3000 °C the spectral distribution has shifted sufficiently to the lower wavelengths for the body to appear 'white hot'. Not only does the spectral distribution change with temperature, but the total power (energy per unit time) radiated increases as the body becomes hotter.

When radiation falls on the surface of a body some is reflected and some is absorbed. For example, dark bodies absorb most of the radiation falling on them, while light-coloured bodies reflect most of it. The *absorption coefficient* of a material surface at a given wavelength is defined as the fraction of the radiant energy, incident on the surface, which is absorbed at that wavelength. Now, if a body is in thermal equilibrium with its surroundings, and therefore is at constant temperature, it must emit and absorb the same amount of radiant energy per unit time, for otherwise its temperature would rise or fall. The radiation emitted or absorbed under these circumstances is known as *thermal radiation*.

A *black body* is defined as a body which absorbs all the radiant energy falling upon it. In other words its absorption coefficient is equal to unity at all wavelengths. Thermal radiation absorbed or emitted by a black body is called *black body radiation* and is of special importance. Indeed, G. R. Kirchhoff proved in 1859 by using general thermodynamical arguments that, for any wavelength, the ratio of the *emissive power* or spectral emittance (defined as the power emitted per unit area at a given wavelength) to the absorption coefficient is the same for all bodies at the same temperature, and is equal to the emissive power of a black body at that temperature. This relation is known as *Kirchhoff's law*. Since the maximum value of the absorption coefficient is unity and corresponds to a black body, it follows from Kirchhoff's law that the black body is not only the most efficient absorber, but is also the most efficient emitter of electromagnetic energy. Moreover, it is clear from Kirchhoff's law that the emissive power of a black body does not depend on the nature of the body. Hence black body radiation has 'universal' properties and is therefore of particular interest.

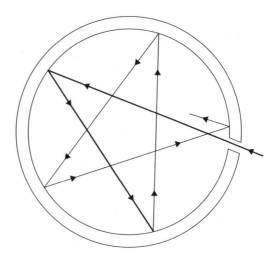

good approximation to a black body. A cavity kept at a constant temperature with blackened interior walls is connected to the outside by a small hole. To an outside observer this small hole appears like a black body surface because any radiation incident from outside on the hole will be almost completely absorbed after multiple reflections on the inner surface of the cavity. Because the cavity is in thermal equilibrium, the radiation inside it can be closely identified with black body radiation, and the hole also emits like a black body.

A perfect black body is of course an idealisation, but it can be very closely approximated in the following way. Consider a cavity kept at a constant temperature, whose interior walls are blackened (see Fig. 1.1). To an outside observer, a small hole made in the wall of such a cavity behaves like a black body surface. The reason is that any radiation incident from the outside upon the hole will pass through it and will almost completely be absorbed in multiple reflections inside the cavity, so that the hole has an effective absorption coefficient close to unity. Since the cavity is in thermal equilibrium, the radiation within it and that escaping from the small opening can thus be closely identified with the thermal radiation from a black body. It should be noted that the hole appears black only at low temperatures, where most of the energy is emitted at wavelengths longer than those corresponding to visible light.

Let us denote by R the *total emissive power* (or *total emittance*) of a black body, that is the total power emitted per unit area of the black body. In 1879 J. Stefan found an empirical relation between the quantity R and the absolute temperature T of a black body

$$R(T) = \sigma T^4 \tag{1.1}$$

where $\sigma = 5.67 \times 10^{-8}$ W m^{-2} K^{-4} is a constant known as Stefan's constant. (Throughout this book we use SI units; the symbol W denotes a watt and K refers to a degree Kelvin). In 1884, L. Boltzmann deduced the relation (1.1) from thermodynamics; it is now called the *Stefan–Boltzmann law*.

We now consider the spectral distribution of black body radiation. We denote by

$R(\lambda, T)$ the emissive power or spectral emittance of a black body, so that $R(\lambda, T)d\lambda$ is the power emitted per unit area from a black body at the absolute temperature T, corresponding to radiation with wavelengths between λ and $\lambda + d\lambda$. The total emissive power $R(T)$ is of course the integral of $R(\lambda, T)$ over all wavelengths,

$$R(T) = \int_0^\infty R(\lambda, T)d\lambda \tag{1.2}$$

and by the Stefan–Boltzmann law $R(T) = \sigma T^4$. Since R depends only on the temperature, it follows that the spectral emittance $R(\lambda, T)$ is a 'universal' function, in agreement with the conclusions drawn previously from Kirchhoff's law.

The first accurate measurements of $R(\lambda, T)$ were made by O. Lummer and E. Pringsheim in 1899. The observed spectral emittance $R(\lambda, T)$ is shown plotted against λ, for a number of different temperatures, in Fig. 1.2. We see that, for fixed λ, $R(\lambda, T)$ increases with increasing T. At each temperature, there is a wavelength $\lambda_{\max}$ for which $R(\lambda, T)$ has its maximum value; this wavelength varies inversely with the temperature:

$$\lambda_{\max} T = b \tag{1.3}$$

a result which is known as Wien's displacement law. The constant b which appears in (1.3) is called the *Wien displacement constant* and has the value $b = 2.898 \times 10^{-3}$ m K.

We have seen above that if a small hole is made in a cavity whose walls are uniformly heated to a given temperature, this hole will emit black body radiation, and that the radiation inside the cavity is also that of a black body. Using the second law of thermodynamics, Kirchhoff proved that the flux of radiation in the cavity is the same in all directions, so that the radiation is isotropic. He also showed that the radiation is homogeneous, namely the same at every point inside the cavity, and that it is identical in all cavities at the same temperature. Furthermore, all these statements hold at each wavelength.

Instead of using the spectral emittance $R(\lambda, T)$, it is convenient to specify the spectrum of black body radiation inside the cavity in terms of a quantity $\rho(\lambda, T)$ which is called the (wavelength) *spectral distribution function* or (wavelength) *monochromatic energy density*. It is defined so that $\rho(\lambda, T)d\lambda$ is the energy density (that is, the energy per unit volume) of the radiation in the wavelength interval $(\lambda, \lambda+d\lambda)$, at the absolute temperature T. As we expect on physical grounds, $\rho(\lambda, T)$ is proportional to $R(\lambda, T)$, and it can be shown[1] that the proportionality constant is $4/c$, where c is the velocity of light *in vacuo*

$$\rho(\lambda, T) = \frac{4}{c} R(\lambda, T). \tag{1.4}$$

Hence, measurements of the spectral emittance $R(\lambda, T)$ also determine the spectral distribution function $\rho(\lambda, T)$.

---

[1] See, for example, Richtmyer *et al.* (1969).

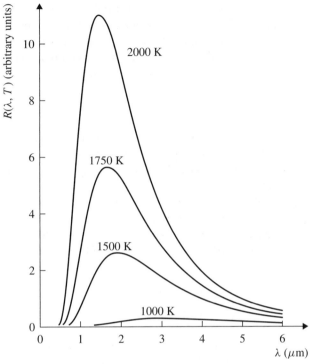

Figure 1.2 Spectral distribution of black body radiation. The spectral emittance $R(\lambda, T)$ is plotted as a function of the wavelength λ for different absolute temperatures.

Using general thermodynamical arguments, W. Wien showed in 1893 that the function $\rho(\lambda, T)$ had to be of the form

$$\rho(\lambda, T) = \lambda^{-5} f(\lambda T) \tag{1.5}$$

where $f(\lambda T)$ is a function of the single variable λT, which cannot be determined from thermodynamics. It is a simple matter to show (Problem 1.3) that Wien's law (1.5) includes the Stefan–Boltzmann law (1.1) as well as Wien's displacement law (1.3). Of course, the values of the Stefan constant σ and of the Wien displacement constant b cannot be obtained until the function $f(\lambda T)$ is known.

In order to determine the function $f(\lambda T)$ – and hence $\rho(\lambda, T)$ – one must go beyond thermodynamical reasoning and use a more detailed theoretical model. After some attempts by Wien, Lord Rayleigh and J. Jeans derived a spectral distribution function $\rho(\lambda, T)$ from the laws of classical physics in the following way. First, from electromagnetic theory, it follows that the thermal radiation within a cavity must exist in the form of standing electromagnetic waves. The number of such waves – or in other words the number of modes of oscillation of the electromagnetic field in the cavity – per unit volume, with wavelengths within the interval λ to $\lambda + d\lambda$, can be shown[1] to be $(8\pi/\lambda^4)d\lambda$, so that $n(\lambda) = 8\pi/\lambda^4$ is the number of modes per unit volume and per unit wavelength range. This number is independent of the size and shape of a sufficiently large cavity. Now, if $\bar{\varepsilon}$ denotes the average energy in the mode with wavelength λ, the spectral distribution function $\rho(\lambda, T)$ is simply the product of $n(\lambda)$ and $\bar{\varepsilon}$, and hence may be written as

$$\rho(\lambda, T) = \frac{8\pi}{\lambda^4}\bar{\varepsilon} \tag{1.6}$$

Rayleigh and Jeans then suggested that the standing waves of electromagnetic radiation are caused by the constant absorption and emission of radiation by atoms in the wall of the cavity, these atoms acting as electric dipoles, that is linear harmonic oscillators of frequency $\nu = c/\lambda$. The energy, ε, of each of these classical oscillators can take any value between 0 and ∞. However, since the system is in thermal equilibrium, the average energy $\bar{\varepsilon}$ of an assemblage of these oscillators can be obtained from classical statistical mechanics by weighting each value of ε with the Boltzmann probability distribution factor $\exp(-\varepsilon/kT)$, where k is Boltzmann's constant. Setting $\beta = 1/kT$, we have

$$\bar{\varepsilon} = \frac{\int_0^\infty \varepsilon \exp(-\beta\varepsilon)d\varepsilon}{\int_0^\infty \exp(-\beta\varepsilon)d\varepsilon}$$

$$= -\frac{d}{d\beta}\log\left[\int_0^\infty \exp(-\beta\varepsilon)d\varepsilon\right] = \frac{1}{\beta} = kT. \tag{1.7}$$

This result is in agreement with the classical law of *equipartition of energy*, according to which the average energy per degree of freedom of a dynamical system in equilibrium is equal to $kT/2$. In the present case the linear harmonic oscillators must be assigned $kT/2$ for the contribution to the average energy coming from their kinetic energy, plus another contribution $kT/2$ arising from their potential energy.

Inserting the value (1.7) of $\bar{\varepsilon}$ into (1.6) gives the Rayleigh–Jeans spectral distribution law

$$\rho(\lambda, T) = \frac{8\pi}{\lambda^4}kT \tag{1.8}$$

from which, using (1.5), we see that $f(\lambda T) = 8\pi k(\lambda T)$.

In the limit of long wavelengths, the Rayleigh–Jeans result (1.8) approaches the experimental values, as shown in Fig. 1.3. However, as can be seen from this figure, $\rho(\lambda, T)$ does not exhibit the observed maximum, and diverges as $\lambda \to 0$. This behaviour at short wavelengths is known as the 'ultraviolet catastrophe'. As a consequence, the total energy per unit volume

$$\rho_{\text{tot}}(T) = \int_0^\infty \rho(\lambda, T)d\lambda \tag{1.9}$$

is seen to be infinite, which is clearly incorrect.

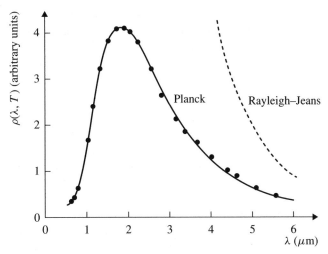

Figure 1.3 Comparison of the Rayleigh–Jeans and Planck spectral distribution laws with experiment at 1600 K. The dots represent experimental points.

Planck's quantum theory

No solution to these difficulties can be found using classical physics. However, in December 1900, M. Planck presented a new form of the black body radiation spectral distribution, based on a revolutionary hypothesis. He postulated that the energy of an oscillator of a given frequency ν cannot take arbitrary values between zero and infinity, but can only take on the discrete values $n\varepsilon_0$, where n is a positive integer or zero, and ε_0 is a finite amount, or *quantum*, of energy, which may depend on the frequency ν. In this case the average energy of an assemblage of oscillators of frequency ν, in thermal equilibrium, is given by

$$\bar{\varepsilon} = \frac{\sum_{n=0}^{\infty} n\varepsilon_0 \exp(-\beta n\varepsilon_0)}{\sum_{n=0}^{\infty} \exp(-\beta n\varepsilon_0)} = -\frac{d}{d\beta}\left[\log \sum_{n=0}^{\infty} \exp(-\beta n\varepsilon_0)\right]$$

$$= -\frac{d}{d\beta}\left[\log\left(\frac{1}{1-\exp(-\beta\varepsilon_0)}\right)\right] = \frac{\varepsilon_0}{\exp(\beta\varepsilon_0)-1} \quad (1.10)$$

where we have assumed, as did Planck, that the Boltzmann probability distribution factor can still be used. Substituting the new value (1.10) of $\bar{\varepsilon}$ into (1.6), we find that

$$\rho(\lambda, T) = \frac{8\pi}{\lambda^4} \frac{\varepsilon_0}{\exp(\varepsilon_0/kT) - 1}. \quad (1.11)$$

In order to satisfy Wien's law (1.5), ε_0 must be taken to be proportional to the frequency ν

$$\varepsilon_0 = h\nu = hc/\lambda \quad (1.12)$$

where h is a fundamental physical constant, called *Planck's constant*. The *Planck spectral distribution law* for $\rho(\lambda, T)$ is thus given by

$$\rho(\lambda, T) = \frac{8\pi hc}{\lambda^5} \frac{1}{\exp(hc/\lambda kT) - 1} \tag{1.13}$$

and we see from (1.5) that in Planck's theory the function $f(\lambda T)$ is given by $f(\lambda T) = 8\pi hc[\exp(hc/\lambda kT) - 1]^{-1}$.

By expanding the denominator in the Planck expression (1.13), it is easy to show (Problem 1.4) that at long wavelengths $\rho(\lambda, T) \to 8\pi kT/\lambda^4$, in agreement with the Rayleigh–Jeans formula (1.8). On the other hand, for short wavelengths, the presence of $\exp(hc/\lambda kT)$ in the denominator of the Planck radiation law (1.13) ensures that $\rho \to 0$ as $\lambda \to 0$. The physical reason for this behaviour is clear. At long wavelengths the quantity $\varepsilon_0 = hc/\lambda$ is small with respect to kT or, in other words, the quantum steps are small with respect to thermal energies; as a result the quantum states are almost continuously distributed, and the classical equipartition law is essentially unaffected. On the contrary, at short wavelengths, the available quantum states are widely separated in energy in comparison to thermal energies, and can be reached only by the absorption of high-energy quanta, a relatively rare phenomenon.

The value of λ for which the Planck spectral distribution (1.13) is a maximum can be evaluated (Problem 1.5), and it is found that

$$\lambda_{\max} T = \frac{hc}{4.965k} = b \tag{1.14}$$

where b is Wien's displacement constant. Moreover, in Planck's theory the total energy density is finite and we find from (1.9) and (1.13) that (Problem 1.6)

$$\rho_{\text{tot}} = aT^4, \qquad a = \frac{8\pi^5}{15} \frac{k^4}{h^3 c^3}. \tag{1.15}$$

Since ρ_{tot} is related to the total emissive power R by $\rho_{\text{tot}} = 4R/c$, where R is given by the Stefan–Boltzmann law (1.1), we see that Stefan's constant σ is given by

$$\sigma = \frac{2\pi^5}{15} \frac{k^4}{h^3 c^2}. \tag{1.16}$$

Equations (1.14) and (1.16) relate b and σ to the three fundamental constants c, k and h. In 1900, the velocity of light, c, was known accurately, and the experimental values of σ and b were also known. Using this data, Planck calculated both the values of k and h, which he found to be $k = 1.346 \times 10^{-23}$ J K^{-1} and $h = 6.55 \times 10^{-34}$ J s. (The symbol J denotes a joule and s a second.) This was not only the most accurate value of Boltzmann's constant k available at the time, but also, most importantly, the first calculation of Planck's constant h. Using his values of k and h, Planck obtained very good agreement with the experimental data for the spectral distribution of black body radiation over the entire range of wavelengths (see Fig. 1.3). The modern value of k is given by[2] $k = 1.38066 \times 10^{-23}$ J K^{-1} and that of h is

$$h = 6.62618 \times 10^{-34} \text{ J s.} \tag{1.17}$$

We remark that the physical dimensions of h are those of [energy] × [time] = [length] × [momentum]. These dimensions are those of a physical quantity called *action*, and consequently Planck's constant h is also known as the *fundamental quantum of action*. As seen from (1.17), the numerical value of h, when expressed in 'macroscopic units', such as SI units, is very small, which is in agreement with the statement made at the beginning of this chapter. We therefore expect that if, for a physical system, every variable having the dimension of action is very large when measured in units of h, then quantum effects will be negligible and the laws of classical physics will be sufficiently accurate.

As an illustration, let us consider a *macroscopic* linear harmonic oscillator of mass 10^{-2} kg, maximum velocity $v_{\max} = 10^{-1}$ m s^{-1} and maximum amplitude $x_0 = 10^{-2}$ m. The frequency of this oscillator is $\nu = v_{\max}/(2\pi x_0) \simeq 1.6$ Hz, its period is $T = \nu^{-1} \simeq 0.63$ s and its energy is given by $E = mv_{\max}^2/2 = 5 \times 10^{-5}$ J. The product of the energy times the period has the dimensions of an action, with the value 3.14×10^{-5} J s, which is about 5×10^{28} times larger than h! We also see that at the frequency $\nu = 1.6$ Hz of this oscillator, the quantum of energy $\varepsilon_0 = h\nu \simeq 10^{-33}$ J. Hence the ratio $\varepsilon_0/E \simeq 2 \times 10^{-29}$ is utterly negligible, and quantum effects can be neglected in this case. On the contrary, for high-frequency electromagnetic waves in black body radiation quantum effects are very important, as we have seen above. In the remaining sections of this chapter we shall discuss several examples of physical phenomena occurring in microphysics, where quantum effects are also of crucial importance.

The idea of quantisation of energy, in which the energy of a system can only take certain discrete values, was totally at variance with classical physics, and Planck's theory was not accepted readily. It should be noted in this respect that some aspects of Planck's derivation of the black body radiation law (1.13) are not completely sound. A revised proof of Planck's black body radiation law, based on modern quantum theory, will be given in Chapter 14. However, Planck's fundamental postulate about the quantisation of energy is correct, and it was not long before the quantum concept was used to explain other phenomena. In particular, as we shall see in Section 1.2, A. Einstein in 1905 was able to understand the photoelectric effect by extending Planck's idea of the quantisation of energy. In Planck's theory, the oscillators representing the source of the electromagnetic field could only vibrate with energies given by $n\varepsilon_0 = nh\nu (n = 0, 1, 2, \ldots)$. In contrast, Einstein assumed that the electromagnetic field itself was quantised and that light consists of corpuscles, called *light quanta* or *photons*, each photon travelling with the velocity c of light and having an energy

$$E = h\nu = hc/\lambda. \tag{1.18}$$

---

[2] See the Table of Fundamental Constants at the end of the book, p. 789.

According to Einstein, black body radiation may thus be considered as a photon gas in thermal equilibrium with the atoms in the walls of the cavity; it is this idea which will be developed in Chapter 14 to re-derive Planck's radiation law (1.13) from quantum statistical mechanics. We remark that since each photon has an energy hc/λ, the number dN of black body radiation photons per unit volume, with wavelengths between λ and $\lambda + d\lambda$, at absolute temperature T, is

$$dN = \frac{\rho(\lambda, T) d\lambda}{hc/\lambda} = \frac{8\pi}{\lambda^4} \frac{d\lambda}{\exp(hc/\lambda kT) - 1}. \quad (1.19)$$

The total number of black body photons per unit volume, at absolute temperature T, can be obtained by integrating (1.19) over all wavelengths. The result is (Problem 1.9)

$$N = 2.029 \times 10^7 T^3 \text{ photons m}^{-3}. \quad (1.20)$$

The average energy $\bar{E}$ of a black body photon at absolute temperature T is then readily deduced by dividing the total energy density ρ_{tot}, given by (1.15), by the total number N of photons per unit volume. We find in this way that

$$\bar{E} = 3.73 \times 10^{-23} T \text{ J}. \quad (1.21)$$

In quantum physics it is particularly convenient to specify energies in units of *electronvolts* (eV) or multiples of them. The electronvolt is defined to be the energy acquired by an electron passing through a potential difference of one volt. Since the charge of the electron has the absolute value $e = 1.60 \times 10^{-19}$ C, we see that[3]

$$\begin{aligned} 1 \text{ eV} &= 1.60 \times 10^{-19} \text{ C V} \\ &= 1.60 \times 10^{-19} \text{ J}. \end{aligned} \quad (1.22)$$

Hence the average energy of a black body photon is given in eV by

$$\bar{E} = \frac{3.73 \times 10^{-23}}{1.60 \times 10^{-19}} T \text{ eV} = 2.33 \times 10^{-4} T \text{ eV}. \quad (1.23)$$

The 3 K cosmic black body radiation

Before leaving the subject of black body radiation, we briefly discuss a particularly fascinating example of it. In 1964 A. A. Penzias and R. W. Wilson, using a radio-telescope, detected at a wavelength $\lambda = 7.35$ cm an isotropic radio 'noise' of cosmic origin, whose intensity corresponded to an effective temperature of about 3 K. This 'noise' was soon interpreted as due to black body radiation at an absolute temperature of approximately 3 K, which fills the universe uniformly and hence is incident on the Earth with equal intensity from all directions. Measurements of the intensity of this radiation at other wavelengths confirmed that its spectral distribution is given by Planck's radiation law (1.13), for a temperature of about 3 K.

The presence of this cosmic radiation provides strong evidence for the *big bang* theory of the origin of the universe, according to which the expansion of the universe

---

[3] Conversion factors between various units are given in a table at the end of this volume (p. 791).

began from a state of enormous density and temperature, that is an extremely hot fireball of particles and radiation. To see why this is the case, let us analyse black body radiation in an expanding universe. Assuming that the size of the universe has grown by a factor S, the wavelengths will be increased by the same factor because of the Doppler shift, so that the new values of the wavelengths are given by $\lambda' = S\lambda$. Now, after this expansion, the energy density $\rho(\lambda', T)\mathrm{d}\lambda'$ in the wavelength interval $(\lambda', \lambda' + \mathrm{d}\lambda')$ at absolute temperature T is smaller than the former energy density $\rho(\lambda, T)\mathrm{d}\lambda$ by a factor S^{-4}. Indeed, the volume of the universe having increased by a factor S^3, the number of photons per unit volume has dropped by this factor (we assume that no photons are created or destroyed). Moreover, since $\lambda' = S\lambda$, the energy hc/λ' of a photon at the wavelength λ' is smaller than that of a photon at wavelength λ by a factor S. Hence

$$\rho(\lambda', T)\mathrm{d}\lambda' = S^{-4}\rho(\lambda, T)\mathrm{d}\lambda = \frac{8\pi hc}{S^4 \lambda^5} \frac{\mathrm{d}\lambda}{\exp(hc/\lambda kT) - 1} \tag{1.24}$$

and therefore, since $\mathrm{d}\lambda' = S\mathrm{d}\lambda$, we have

$$\rho(\lambda', T) = \frac{8\pi hc}{(\lambda')^5} \frac{1}{\exp(hcS/\lambda' kT) - 1}. \tag{1.25}$$

Upon comparison with (1.13) we see that if we replace T by a new temperature

$$T' = \frac{T}{S} \tag{1.26}$$

the expression (1.25) is identical with the Planck radiation law for the energy density $\rho(\lambda', T')$. Thus black body radiation in an expanding universe can still be described by the Planck formula, in which the temperature T' decreases according to (1.26) as the universe expands. The cosmic radiation at $T' = 3$ K which is now observed is therefore 'fossil' radiation, cooled by expansion, originating from an epoch when the universe was smaller and hotter than at the present time. It is estimated[4] that this radiation comes from an epoch when the universe was about 1 million years old and was flooded with radiation at a temperature $T \simeq 3000$ K. Using (1.26) we see that since that time the universe has expanded by a factor $S \simeq 1000$.

Further applications of Planck's quantisation postulate

We have seen above that the quantisation postulate introduced in 1900 by Planck was successful in explaining the black body radiation problem. We have also mentioned that the quantum idea was used by Einstein in 1905 to understand the photoelectric effect in terms of light quanta or photons; we shall return to this subject in Section 1.2. In 1907, Einstein also used the Planck formula (1.10) for the average energy of an oscillator to study the variation of the specific heats of solids with temperature, a problem which could not be solved by using classical physics. Einstein's results were improved in 1912 by P. Debye, and the excellent agreement between the Debye theory

---

[4] For an excellent account of this subject, see Weinberg (1977).

and experiment provided additional support for the existence of energy quanta. As we shall see in the last four sections of this chapter, the quantum concept also proved to be essential in understanding the Compton effect, the existence of atomic line spectra and the Stern–Gerlach experiment, and it played a central role in predicting the wave properties of matter, thereby giving birth to the new quantum mechanics.

1.2 The photoelectric effect

In 1887 H. Hertz performed the celebrated experiments in which he produced and detected electromagnetic waves, thus confirming Maxwell's theory. Ironically enough, in the course of the same experiments he also discovered a phenomenon which ultimately led to the description of light in terms of corpuscles: the *photons*. Specifically, Hertz observed that ultraviolet light falling on metallic electrodes facilitated the passage of a spark. Further work by W. Hallwachs, M. Stoletov, P. Lenard and others showed that charged particles are ejected from metallic surfaces irradiated by high-frequency electromagnetic waves. This phenomenon is called the *photoelectric effect*. In 1900, Lenard measured the charge-to-mass ratio of the charged particles by performing experiments similar to those which had led J. J. Thomson to discover the electron[5], and in this way he was able to identify the charged particles as electrons.

In his experiments to establish the mechanism of the photoelectric effect. Lenard used an apparatus shown in schematic form in Fig. 1.4. In an evacuated glass tube, ultraviolet light incident on a polished metal cathode C (called a photocathode) liberates electrons. If some of these electrons strike the anode A, there is a current I in the external circuit. Lenard studied this current as a function of the potential difference V between the surface and the anode. The variation of the photoelectric current I with V is shown in Fig. 1.5. When V is positive, the electrons are attracted towards the anode. As V is increased the current I increases until it saturates when V is large enough so that all the emitted electrons reach the anode. Lenard also observed that if V is reversed, so that the cathode becomes positive with respect to the anode, there is a definite negative voltage $-V_0$ at which the photoelectric current ceases, implying that the emission of electrons from the cathode stops (see Fig. 1.5). From this result it follows that the photoelectrons are emitted with velocities up to a maximum $v_{\max}$ and that the voltage $-V_0$ is just sufficient to repel the fastest photoelectrons (having the maximum kinetic energy $T_{\max} = mv_{\max}^2/2$) so that

$$eV_0 = \tfrac{1}{2}mv_{\max}^2. \tag{1.27}$$

The potential V_0 is called the *stopping potential*. The fact that not all the photoelectrons have the same kinetic energy is readily explained: the electrons having the maximum kinetic energy $T_{\max}$ are emitted from the surface of the photocathode, while those having a lower energy originate from inside the photocathode, and thus lose energy in reaching the surface.

---

[5] See Bransden and Joachain (1983).

1.2 The photoelectric effect

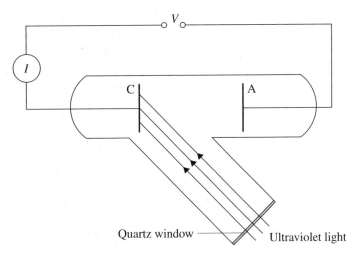

Figure 1.4 Schematic drawing of Lenard's apparatus for investigating the photoelectric effect.

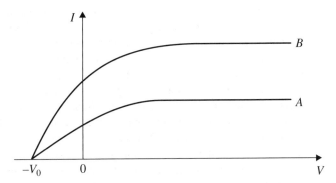

Figure 1.5 Variation of the photoelectric current I with the potential difference V between the cathode and the anode, for two values $A < B$ of the intensity of the light incident on the cathode. No current is observed when V is less than $-V_0$; the stopping potential V_0 is found to be independent of the light intensity.

Lenard found that the photoelectric current I is proportional to the intensity of the incident light (see Fig. 1.5). This result can be understood by using classical electromagnetic theory, which predicts that the number of electrons emitted per unit time should be proportional to the intensity of the incident light. However, the following important features exhibited by the experimental data cannot be explained in terms of the classical electromagnetic theory:

(1) There is a minimum, or *threshold* frequency ν_t of the radiation, characteristic of the surface, below which no emission of electrons takes place, no matter what the intensity of the incident radiation, or for how long it falls on the surface.

According to the classical wave theory, the photoelectric effect should occur for any frequency of the incident radiation, provided that the radiation intensity is large enough to give the energy required for ejecting the photoelectrons.

(2) The stopping potential V_0 and, hence, the maximum kinetic energy $T_{\max} = mv_{\max}^2/2$ of the photoelectrons are found to depend *linearly* on the frequency of the radiation and to be independent of its intensity. According to classical electromagnetic theory, the maximum kinetic energy of the emitted electrons should increase with the energy density (or intensity) of the incident radiation, independently of the frequency.

(3) Electron emission takes place immediately the light shines on the surface, with *no detectable time delay*. Now, in the classical wave theory of light, the light energy is spread uniformly over the wave front. To eject an electron from an atom described by classical mechanics, enough energy would have to be concentrated over an area of atomic dimensions, and to achieve such a concentration would require a certain time delay. Experiments can be arranged for which the predicted time delay is minutes, or even hours, and yet no detectable time lag is actually observed.

In 1905, Einstein offered an explanation of these seemingly strange aspects of the photoelectric effect, based on a generalisation of Planck's postulate of the quantisation of energy. In order to explain the spectral distribution of black body radiation, Planck had assumed that the processes of absorption and emission of radiation by matter do not take place continuously, but in finite quanta of energy $h\nu$. Einstein went further and advanced the idea that these quantum properties were inherent in the nature of electromagnetic radiation itself, so that light consists of quanta (corpuscles) called *photons*[6], each photon having an energy $E = h\nu = hc/\lambda$ (see (1.18)). The photons are sufficiently localised so that the whole quantum of energy can be absorbed by a single atom of the cathode at one time. Thus, when a photon falls on a metallic surface, its entire energy $h\nu$ is used to eject an electron from an atom. However, since electrons do not normally escape from surfaces, a certain minimum energy W is required for the ejected electron to leave the surface. This minimum energy, which depends on the metal, is called the *work function*. It follows that the maximum kinetic energy of a photoelectron is given in terms of the frequency ν by the linear relation

$$\tfrac{1}{2}mv_{\max}^2 = h\nu - W \tag{1.28}$$

which is called Einstein's equation. The threshold frequency ν_t is determined by the work function since in this case $v_{\max} = 0$, so that

$$h\nu_t = W. \tag{1.29}$$

The number of electrons emerging from the metal surface per unit time is proportional to the number of photons striking the surface per unit time, but the intensity of the

---

[6] In his 1905 paper entitled 'On a heuristic point of view concerning the creation and conversion of light', Einstein used the word *quantum of light*. The word *photon* was introduced by G. N. Lewis in 1926.

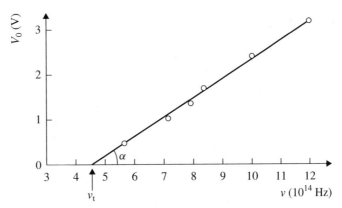

Figure 1.6 Millikan's results (circles) for the stopping potential V_0 as a function of the frequency ν. The data fall on a straight line, of slope $\tan \alpha = h/e$.

radiation is also proportional to the number of photons falling on a certain area per unit time, since each photon carries a fixed energy $h\nu$. It follows that the photoelectric current is proportional to the intensity of the radiation and that all the experimental observations are explained by Einstein's theory.

A series of very accurate measurements carried out between 1914 and 1916 by R. A. Millikan provided further confirmation of Einstein's theory. Combining (1.27) and (1.28), we see that the stopping potential V_0 satisfies the equation

$$V_0 = \frac{h}{e}\nu - \frac{W}{e}. \tag{1.30}$$

Millikan measured, for a given surface, V_0 as a function of ν. As seen from Fig. 1.6, his results indeed fell on a straight line, of slope h/e. Knowing the absolute charge e of the electron from his earlier 'oil-drop' experiments, Millikan obtained for h the value 6.56×10^{-34} J s, which agreed very well with Planck's result $h = 6.55 \times 10^{-34}$ J s determined from the black body spectral distribution. It is interesting that Millikan was able to use visible, rather than ultraviolet light for his photoelectric experiments using surfaces of lithium, sodium and potassium which have small values of the work function W.

Although the photoelectric effect provides compelling evidence for a *corpuscular* theory of light, it must not be forgotten that the existence of diffraction and interference phenomena demonstrate that light also exhibits a *wave* behaviour. This dual aspect of electromagnetic radiation is incompatible with classical physics. As we shall see below, the wave–particle duality is a general characteristic of all physical quantities, and the paradoxes resulting from this situation can only be resolved by using the new concepts embodied in quantum theory.

1.3 The Compton effect

The corpuscular nature of electromagnetic radiation was exhibited in a spectacular way in a quite different experiment performed in 1923 by A. H. Compton, in which a beam of X-rays was scattered through a block of material. X-rays had been discovered by W. K. Röntgen in 1895 and were known to be electromagnetic radiation of high frequency. The scattering of X-rays by various substances was first studied by C. G. Barkla in 1909, who interpreted his results with the help of J. J. Thomson's classical theory, developed around 1900. According to this theory, the oscillating electric field of the radiation acts on the electrons contained in the atoms of the target material. This interaction forces the atomic electrons to vibrate with the same frequency as the incident radiation. The oscillating electrons, in turn, radiate electromagnetic waves of the same frequency. The net effect is that the incident radiation is scattered with no change in wavelength, and this is called *Thomson scattering*. In general, Barkla found that the scattered intensity predicted by Thomson's theory agreed well with his experimental data. However, he found that some of his results were anomalous, particularly in the region of 'hard' X-rays, which correspond to shorter wavelengths. At the time of Barkla's work, it was not possible to measure the wavelengths of X-rays, and a further advance could not be made until M. von Laue in 1912, and later W. L. Bragg had shown that the wavelengths could be determined by studying the diffraction of X-rays by crystals. The experiment of Compton, which we shall now describe, was only possible because a precise determination of X-ray wavelengths could be made using a crystal spectrometer.

The experimental arrangement used by Compton is sketched in Fig. 1.7. He irradiated a graphite target with a nearly monochromatic beam of X-rays, of wavelength λ_0. He then measured the intensity of the scattered radiation as a function of wavelength. His results, illustrated in Fig. 1.8, showed that although part of the scattered radiation had the same wavelength λ_0 as the incident radiation, there was also a second component of wavelength λ_1, where $\lambda_1 > \lambda_0$. This phenomenon, called the *Compton effect*, could not be explained by the classical Thomson model. The shift in wavelength between the incident and scattered radiation, the *Compton shift* $\Delta\lambda = \lambda_1 - \lambda_0$, was found to vary with the angle of scattering (see Fig. 1.8) and to be proportional to $\sin^2(\theta/2)$, where θ is the angle between the incident and scattered beams. Further investigation showed $\Delta\lambda$ to be independent of both λ_0 and of the material used as the scatterer, and that the value of the constant of proportionality was 0.048×10^{-10} m.

In order to understand the origin of the wavelength shift $\Delta\lambda$, Compton suggested that the modified line at wavelength λ_1 could be attributed to X-ray photons scattered by loosely bound electrons in the atoms of the target. In fact, it is a good approximation to treat such electrons as *free*, since their binding energies are small compared with the energy of an X-ray photon; this explains why the results are independent of the nature of the material used for the target.

Let us then consider the scattering of an X-ray photon by a free electron, which can

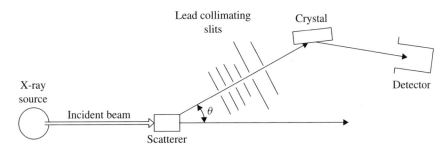

Figure 1.7 Schematic diagram of Compton's apparatus.

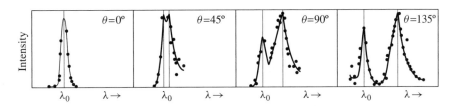

Figure 1.8 Compton's data for the scattering of X-rays by graphite.

be taken to be initially at rest. After the collision, the electron recoils (see Fig. 1.9) and since its velocity is not always small compared with the velocity of light, c, it is necessary to use *relativistic kinematics*. The relevant formulae will be quoted without derivation[7]. In particular, the total energy, E, of a particle having a rest mass m and moving with a velocity $\mathbf{v}$ is given by

$$E = \frac{mc^2}{(1 - v^2/c^2)^{1/2}}. \tag{1.31}$$

The kinetic energy T of the particle is defined as the difference between its total energy E and its rest mass energy mc^2, so that

$$T = E - mc^2. \tag{1.32}$$

The corresponding momentum of the particle is

$$\mathbf{p} = \frac{m\mathbf{v}}{(1 - v^2/c^2)^{1/2}} \tag{1.33}$$

and from (1.31) and (1.33) we see that the energy and momentum are related by

$$E^2 = m^2c^4 + p^2c^2. \tag{1.34}$$

---

[7] For a discussion of the theory of special relativity, see for example the text by Taylor and Wheeler (1966).

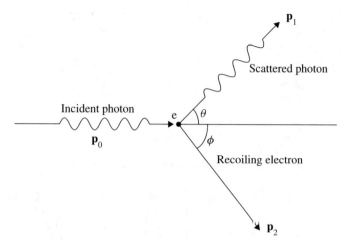

Figure 1.9 A photon of momentum $\mathbf{p}_0$ is incident on a free electron e at rest. After the collision, the photon has a momentum $\mathbf{p}_1$ while the electron recoils with a momentum $\mathbf{p}_2$.

Since the velocity of a photon is c and its energy $E = h\nu = hc/\lambda$ is finite, we see from (1.31) that we must take the mass of a photon to be zero, in which case we observe from (1.34) that the magnitude of its momentum is

$$p = E/c = h/\lambda. \tag{1.35}$$

Let us now apply these formulae to the situation depicted in Fig. 1.9, where a photon of energy $E_0 = hc/\lambda_0$ and momentum $\mathbf{p}_0$ (with $p_0 = E_0/c = h/\lambda_0$) collides with an electron of rest mass m initially at rest. After the collision, the photon has an energy $E_1 = hc/\lambda_1$ and a momentum $\mathbf{p}_1$ (with $p_1 = E_1/c = h/\lambda_1$) in a direction making an angle θ with the direction of incidence, while the electron recoils with a momentum $\mathbf{p}_2$ making an angle ϕ with the incident direction. Conservation of momentum yields $\mathbf{p}_0 = \mathbf{p}_1 + \mathbf{p}_2$ or, in other words

$$p_0 = p_1 \cos\theta + p_2 \cos\phi \tag{1.36a}$$
$$0 = p_1 \sin\theta - p_2 \sin\phi \tag{1.36b}$$

from which we find that

$$p_2^2 = p_0^2 + p_1^2 - 2p_0 p_1 \cos\theta. \tag{1.37}$$

Conservation of energy yields the relation

$$E_0 + mc^2 = E_1 + (m^2 c^4 + p_2^2 c^2)^{1/2} \tag{1.38}$$

and therefore, if we denote by T_2 the kinetic energy of the electron after the collision, we have

$$\begin{aligned} T_2 &= (m^2 c^4 + p_2^2 c^2)^{1/2} - mc^2 \\ &= E_0 - E_1 = c(p_0 - p_1) \end{aligned} \tag{1.39}$$

so that

$$p_2^2 = (p_0 - p_1)^2 + 2mc(p_0 - p_1). \tag{1.40}$$

Combining (1.40) with (1.37) we then find that

$$\begin{aligned} mc(p_0 - p_1) &= p_0 p_1 (1 - \cos\theta) \\ &= 2 p_0 p_1 \sin^2(\theta/2). \end{aligned} \tag{1.41}$$

Multiplying both sides of (1.41) by $h/(mcp_0 p_1)$ and using the fact that $\lambda_0 = h/p_0$ and $\lambda_1 = h/p_1$, we finally obtain

$$\Delta\lambda = \lambda_1 - \lambda_0 = 2\lambda_C \sin^2(\theta/2) \tag{1.42}$$

where the constant λ_C is given by

$$\lambda_C = \frac{h}{mc} \tag{1.43}$$

and is called the *Compton wavelength* of the electron. Equation (1.42) is known as the *Compton equation*. The calculated value of $(2\lambda_C)$ is 0.048 52 Å[8], and this agrees very well with the experimental value which is 0.048 Å.

The existence of the unmodified component of the scattered radiation, which has the same wavelength λ_0 as the incident radiation, can be explained by assuming that it results from scattering by electrons so tightly bound that the entire atom recoils. In this case, the mass to be used in (1.43) is M, the mass of the entire atom, and since $M \gg m$, the Compton shift $\Delta\lambda$ is negligible. For the same reason, there is no Compton shift for light in the visible region because the photon energy in this case is not large compared with the binding energy of even the loosely bound electrons. In contrast, for very energetic γ-rays only the shifted line is observed, since the photon energies are large compared with the binding energies of even the tightly bound electrons.

The recoil electrons predicted by Compton's theory were observed in 1923 by W. Bothe and also by C. T. R. Wilson. A little later, in 1925, W. Bothe and H. Geiger demonstrated that the scattered photon and the recoiling electron appear simultaneously. Finally, in 1927, A. A. Bless measured the energy of the ejected electrons, which he found to be in agreement with the prediction of Compton's theory.

1.4 Atomic spectra and the Bohr model of the hydrogen atom

Isaac Newton was the first to resolve white light into separate colours by dispersion with a prism. However, it was not until 1752 that T. Melvill showed that light from

---

[8] The Ångström unit of length, abbreviated as Å, is such that 1 Å = 10^{-10} m.

an incandescent gas is composed of a number of *discrete* wavelengths[9], now called *emission lines* because of the corresponding lines appearing on a photographic plate. Such *emission line spectra* are produced in particular when an electric discharge passes through a gas, or when a volatile salt is put into a flame, and the emitted light is dispersed by a prism. It was subsequently discovered that atoms also exhibit *absorption line spectra* when they are exposed to light having a continuous spectrum. For example, if white light is passed through an absorbing layer of an element and is then analysed with a spectrograph, it is found that the photographic plate is darkened everywhere, except for a number of unexposed lines. These lines must therefore correspond to certain *discrete* wavelengths missing from the continuous background, which have been absorbed by the atoms of the layer. In a crucial experiment performed in 1859, G. R. Kirchhoff showed that for a given element the wavelengths of the absorption lines coincide with those of the corresponding emission lines. He also understood that each element has its own *characteristic* line spectrum. This fact is of great importance, since it is the basis of chemical analysis by spectroscopic methods; it is also used in astrophysics to determine the presence of particular elements in the Sun, in the stars and in interstellar space.

A major discovery in the search for *regularities* in the line spectra of atoms was made in 1885 by J. Balmer, who studied the spectrum of atomic hydrogen. As seen from Fig. 1.10, in the visible and near ultraviolet regions this spectrum consists of a series of lines (denoted by $H_\alpha, H_\beta, H_\gamma, \ldots$), now called the *Balmer series*; these lines come closer together as the wavelength decreases, and reach a limit at a wavelength $\lambda = 3646$ Å. There is an apparent regularity in this spectrum, and Balmer observed that the wavelengths of the nine lines known at the time satisfied the simple formula

$$\lambda = C \frac{n^2}{n^2 - 4} \qquad (1.44)$$

where C is a constant equal to 3646 Å, and n is an integer taking on the values $3, 4, 5, \ldots$, for the lines $H_\alpha, H_\beta, H_\gamma, \ldots$ respectively. In 1889, J. R. Rydberg found that the lines of the Balmer series could be described in a more useful way in terms of the *wave number* $\tilde{\nu} = 1/\lambda = \nu/c$. According to Rydberg, the wave numbers of the Balmer lines are given by

$$\tilde{\nu} = R_H \left(\frac{1}{2^2} - \frac{1}{n^2} \right) \qquad (1.45)$$

where $n = 3, 4, 5, \ldots$, and R_H is a constant, called the *Rydberg constant* for atomic hydrogen. Its value determined from spectroscopic measurements is

$$R_H = 109\,677.58 \text{ cm}^{-1}. \qquad (1.46)$$

---

[9] This is in contrast to the continuous spectrum of electromagnetic radiation emitted from the surface of a hot solid, which we discussed in Section 1.1. Indeed, line spectra are emitted by atoms in rarefied gases, while in a solid there is a very large number of densely packed vibrating atoms, so that neighbouring spectral lines overlap, and as a result a continuous spectrum is emitted.

1.4 Atomic spectra and the Bohr model

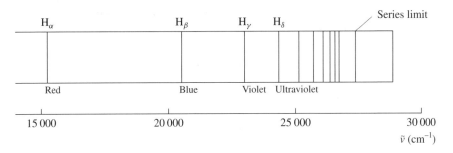

Figure 1.10 The Balmer series of atomic hydrogen.

Once it was written in the form (1.45), the Balmer–Rydberg formula could be easily generalised and applied to other series of atomic hydrogen spectral lines discovered subsequently in the ultraviolet and infrared regions. The generalised Balmer–Rydberg formula reads

$$\tilde{\nu}_{ab} = R_H \left(\frac{1}{n_a^2} - \frac{1}{n_b^2} \right); \quad \begin{aligned} n_a &= 1, 2, \ldots \\ n_b &= 2, 3, \ldots \end{aligned} \quad (1.47)$$

where $\tilde{\nu}_{ab}$ is the wave number of either an emission or an absorption line, and n_a and n_b are positive integers with $n_b > n_a$. A particular series of lines is obtained by setting n_a to be a fixed positive integer and letting n_b take on the values $n_a + 1, n_a + 2, \ldots$. The series are given different names after their discoverers. In particular, the series with $n_a = 1$ is known as the Lyman series and lies in the ultraviolet part of the spectrum; that with $n_a = 2$ is the Balmer series discussed above; the series with $n_a = 3, 4$ and 5 lie in the infrared region and are called, respectively, the Paschen, Brackett and Pfund series. Within each series the lines are labelled $\alpha, \beta, \gamma, \ldots$, in order of increasing wave number, as illustrated in Fig. 1.10 for the case of the Balmer series.

As seen from (1.47), the wave number $\tilde{\nu}_{ab}$ of a line in the atomic hydrogen spectrum can be expressed in the form

$$\tilde{\nu}_{ab} = T_a - T_b \quad (1.48)$$

that is as the difference of two *spectral terms*

$$T_a = R_H/n_a^2, \quad T_b = R_H/n_b^2. \quad (1.49)$$

For other atoms than hydrogen the formula (1.48) can still be applied, although the spectral terms T_a and T_b usually have a more complicated form than (1.49). Thus, the wave number $\tilde{\nu}_{ab}$ of *any* line emitted or absorbed by an atom can be expressed as a difference of two spectral terms T_a and T_b. An atomic spectrum is therefore characterised by a *set* of spectral terms, called the *term system* of the atom. This generalisation of the Balmer–Rydberg formula was obtained in 1908 by W. Ritz. As a consequence, if the wave numbers of three spectral lines are associated with three terms as

$$\tilde{v}_{ij} = T_i - T_j, \qquad \tilde{v}_{jk} = T_j - T_k, \qquad \tilde{v}_{ik} = T_i - T_k \tag{1.50}$$

we have

$$\tilde{v}_{ik} = (T_i - T_j) + (T_j - T_k) = \tilde{v}_{ij} + \tilde{v}_{jk} \tag{1.51}$$

which is an example of the *Ritz combination principle*.

The existence of atomic line spectra, which exhibit the regularities discussed above, cannot be explained by models of atomic structure based on classical physics. As we shall shortly see, N. Bohr was able in 1913 to explain the spectrum of the hydrogen atom by introducing the quantum concept into the physics of atoms. In formulating his new theory, Bohr used a picture of the atom evolved from the work of E. Rutherford, H. W. Geiger and E. Marsden, which we now discuss briefly.

The nuclear atom

In a series of experiments Rutherford, Geiger and Marsden studied the scattering of alpha particles (doubly ionised helium atoms) by atoms of thin metallic foils[10]. To interpret the results of these experiments, Rutherford postulated in 1911 that all the positive charge and almost all the mass of an atom is concentrated in a positively charged *nucleus* of very small dimension ($\simeq 10^{-14}$ m) compared with the dimension of the atom as a whole ($\simeq 10^{-10}$ m). The predictions of Rutherford's theory of alpha-particle scattering, based on this *nuclear atom* model, were fully confirmed in 1913 by further experiments performed by Geiger and Marsden.

The excellent agreement between the experimental results and the conclusions reached by Rutherford was interpreted as establishing the correctness of the concept of the nuclear atom. However, there were still important difficulties with the nuclear atom model, due to the fact that there exists no stable arrangement of positive and negative point charges at rest. Therefore, one must consider a *planetary model* of the atom, in which the negatively charged electrons move in orbits in the Coulomb field of the positively charged nucleus. Now, a particle moving on a curved trajectory is accelerating, and an accelerated charged particle can be shown from electromagnetic theory to radiate, thus losing energy. In fact, the laws of classical physics, applied to the Rutherford planetary atom, imply that in a time of the order of 10^{-10} s all the energy of the revolving electrons would be radiated away and the electrons would collapse into the nucleus. This is clearly contrary to experiment and is another piece of evidence that the classical laws of motion must be modified on the atomic scale.

---

[10] A discussion of Rutherford scattering may be found in Bransden and Joachain (1983). See also Section 13.6.

Bohr's model of the hydrogen atom

A major step forward was taken by N. Bohr in 1913. Combining the concepts of Rutherford's nuclear atom, Planck's quanta and Einstein's photons, he was able to explain the observed spectrum of atomic hydrogen.

Bohr assumed, as in the Rutherford model, that an electron in an atom moves in an orbit about the nucleus under the influence of the electrostatic attraction of the nucleus. Circular or elliptical orbits are allowed by classical mechanics, and Bohr elected to consider circular orbits for simplicity. He then postulated that instead of the infinity of orbits which are possible in classical mechanics, only a certain set of stable orbits, which he called *stationary states* are allowed. As a result, atoms can only exist in certain allowed energy levels, with internal energies $E_a, E_b, E_c, \ldots$. Bohr further postulated that an electron in a stable orbit does not radiate electromagnetic energy, and that radiation can only take place when a transition is made between the allowed energy levels. To obtain the frequency of the radiation, he made use of the idea that the energy of electromagnetic radiation is quantised and is carried by photons, each photon associated with the frequency ν carrying an energy $h\nu$. Thus, if a photon of frequency ν is absorbed by an atom, conservation of energy requires that[11]

$$h\nu = E_b - E_a \tag{1.52}$$

where E_a and E_b are the internal energies of the atom in the initial and final states, with $E_b > E_a$. Similarly, if the atom passes from a state of energy E_b to another state of lower energy E_a, the frequency of the emitted photon must be given by the *Bohr frequency relation* (1.52).

We note that because of the existence of an energy–frequency relationship, we can use frequency (or wave number) units of energy where convenient. For example, using (1.52) and the fact that the frequency ν corresponds to a wave number $\tilde{\nu} = \nu/c$, we find that one electronvolt of energy can be converted in hertz or in inverse centimetres as

$$\begin{aligned} 1 \text{ eV} &\equiv 2.417\,97 \times 10^{14} \text{ Hz} \\ &\equiv 8065.48 \text{ cm}^{-1}. \end{aligned} \tag{1.53}$$

Other conversions of units are given in the table at the end of the book. We also remark that the Bohr frequency relation (1.52) implies that the *terms* of spectroscopy can be interpreted as being the energies of the various allowed energy levels of the atom.

For the case of one-electron (also called hydrogenic) atoms[12], Bohr was able to modify the classical planetary model to obtain the quantisation of energy levels. To achieve this aim, he made the additional assumption that the magnitude of the orbital angular momentum of the electron moving in a circular orbit around the nucleus can

---

[11] We neglect here small recoil effects, which will be considered below.

[12] The word 'atom' denotes here a neutral atom (such as the hydrogen atom H) as well as an ion (such as He^+, Li^{2+} and so on).

only take one of the values $L = nh/2\pi = n\hbar$, where the *quantum number* n is a positive integer, $n = 1, 2, 3, \ldots$, and the commonly occurring quantity $h/2\pi$ is conventionally denoted by $\hbar$. The allowed energy levels of the bound system made up of a nucleus and an electron can then be determined in the following way.

Let us consider an electron moving with a non-relativistic velocity v in a circular orbit of radius r around the nucleus. We shall make the approximation (which we shall remove later) that the nucleus is infinitely heavy compared with the electron, and is therefore at rest. The Coulomb attractive force acting on the electron, due to its electrostatic interaction with the nucleus of charge Ze, can be equated with the electron mass m times the centripetal acceleration (v^2/r), giving

$$\frac{Ze^2}{(4\pi\varepsilon_0)r^2} = \frac{mv^2}{r}, \tag{1.54}$$

where ε_0 is the permittivity of free space. A second equation is obtained from Bohr's postulate that the magnitude of the orbital angular momentum of the electron is quantised:

$$L = mvr = n\hbar, \qquad n = 1, 2, 3, \ldots. \tag{1.55}$$

From (1.54) and (1.55), we obtain the allowed values of v and r

$$v = \frac{Ze^2}{(4\pi\varepsilon_0)\hbar n} \tag{1.56}$$

and

$$r = \frac{(4\pi\varepsilon_0)\hbar^2 n^2}{Zme^2}. \tag{1.57}$$

The total energy of the electron is the sum of its kinetic energy and its potential energy. The kinetic energy, T, is given by

$$T = \frac{1}{2}mv^2 = \frac{m}{2\hbar^2}\left(\frac{Ze^2}{4\pi\varepsilon_0}\right)^2 \frac{1}{n^2} \tag{1.58}$$

where we have used the result (1.56). Choosing the zero of the energy scale in such a way that the electron has a total energy $E = 0$ when it is at rest and completely separated from the nucleus ($r = \infty$), the potential energy of the electron is given by $V = -Ze^2/(4\pi\varepsilon_0)r$. Hence, using (1.57) we have

$$V = -\frac{Ze^2}{(4\pi\varepsilon_0)r} = -\frac{m}{\hbar^2}\left(\frac{Ze^2}{4\pi\varepsilon_0}\right)^2 \frac{1}{n^2}. \tag{1.59}$$

The allowed values obtained by Bohr for the total energy $T + V$ of the bound electron are thus given by

$$E_n = -\frac{m}{2\hbar^2}\left(\frac{Ze^2}{4\pi\varepsilon_0}\right)^2 \frac{1}{n^2}, \qquad n = 1, 2, 3, \ldots. \tag{1.60}$$

Since n may take on all integral values from 1 to $+\infty$, the *energy spectrum* corresponding to the bound states of the one-electron atom contains an infinite number of *discrete* energy levels. The subscript n in E_n reminds us that this *quantisation of the energy levels* is due to the presence of the quantum number n, is called the *principal quantum number* to distinguish it from other quantum numbers we shall meet later. The state with the lowest energy is known as the *ground state*; we see from (1.60) that it corresponds to the value $n = 1$. The states corresponding to the values $n = 2, 3, \ldots$, are called *excited states* of the atom, since their energies are greater that the energy of the ground state; the energies of the excited states are seen to converge to the value zero as $n \to \infty$. A commonly used notation is one in which an electron in an energy level with $n = 1, 2, 3, \ldots$, is said to be in the K, L, M, ..., shell, respectively.

It is convenient to represent graphically the energy spectrum of an atom by means of an *energy level diagram*, in which the ordinate gives the energy, and the energy levels or terms are drawn as horizontal lines. Fig. 1.11 shows the energy level diagram of atomic hydrogen, drawn according to the prediction (1.60) of the Bohr model. The energy scale on the right is in electronvolts; on the left another scale, increasing from top to bottom, gives the wave number $\tilde{\nu}$ in cm^{-1}.

Using the Bohr frequency relation (1.52), the frequencies of the spectral lines corresponding to transitions between two energy levels E_a and E_b are

$$\nu_{ab} = \frac{m}{4\pi \hbar^3} \left(\frac{Ze^2}{4\pi \varepsilon_0} \right)^2 \left(\frac{1}{n_a^2} - \frac{1}{n_b^2} \right) \tag{1.61}$$

where n_a and n_b are positive integers and $n_b > n_a$; the corresponding wave numbers are given by $\tilde{\nu}_{ab} = \nu_{ab}/c$. For the case of atomic hydrogen, where $Z = 1$, we have therefore

$$\tilde{\nu}_{ab} = R(\infty) \left(\frac{1}{n_a^2} - \frac{1}{n_b^2} \right) \tag{1.62}$$

where the constant $R(\infty)$ is given by

$$R(\infty) = \frac{m}{4\pi c \hbar^3} \left(\frac{e^2}{4\pi \varepsilon_0} \right)^2 \tag{1.63}$$

and we have written $R(\infty)$ to recall that we are using the infinite nuclear mass approximation. The theoretical result (1.62) is seen to have exactly the same form as the empirical Balmer–Rydberg formula (1.47). Moreover, Bohr's theory also gives the value of $R(\infty)$ in terms of fundamental constants, and this value can be compared with the experimental Rydberg constant R_H appearing in (1.47). Evaluating $R(\infty)$ from (1.63) one finds that $R(\infty) = 109\,737$ cm^{-1}, in good (but not perfect) agreement with the experimental value of R_H given by (1.46).

Returning to equation (1.60), we see that the Bohr energy levels of a one-electron atom may be written as

$$E_n = -I_\mathrm{P}/n^2, \qquad n = 1, 2, 3, \ldots \tag{1.64}$$

where

$$I_\mathrm{P} = \frac{m}{2\hbar^2} \left(\frac{Ze^2}{4\pi \varepsilon_0} \right)^2 = hcR(\infty)Z^2 = 13.6\ Z^2\ \mathrm{eV}. \tag{1.65}$$

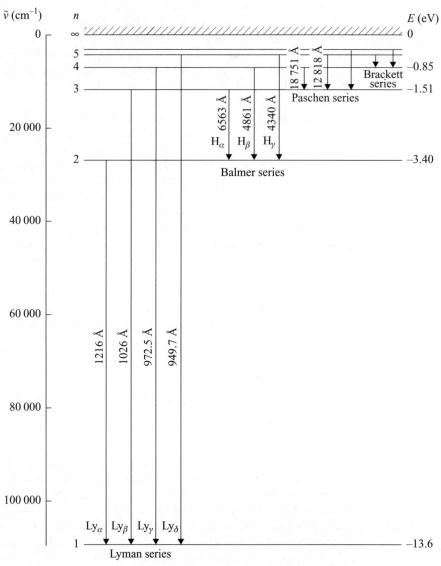

Figure 1.11 Energy level diagram of the hydrogen atom. The energy scale on the right is in electronvolts; on the left another scale, increasing from top to bottom, gives the wave number $\tilde{\nu}$ in cm^{-1}. The principal quantum number n corresponding to each energy level is also indicated. A spectral line, resulting from the transition of the atom from one energy level to another is represented by a vertical line joining the two energy levels. The numbers against the lines indicate the corresponding wavelengths in ångströms (1 Å = 10^{-10} m). For clarity, only transitions between lower-lying levels are shown.

Because the zero of the energy scale has been chosen so that it corresponds to an electron at rest, infinitely far from the nucleus, the (positive) amount of work required to remove the electron from its nth orbit to infinity (that is, to *ionise* the atom) is just $-E_n = I_P/n^2$. The energy needed to remove one electron from an atom in its ground state is called the *ionisation potential* of the atom. Since the ground state of one-electron atoms corresponds to the value $n = 1$, we see that the quantity I_P given by (1.65) is the value of the ionisation potential of hydrogenic atoms predicted by the Bohr theory (in the infinite nuclear mass approximation). The ionisation of atoms may occur in several ways, for example in collision processes or when a photon of sufficiently high frequency is absorbed by the atom. The latter phenomenon is known as *photoionisation* and is the process responsible for the photoelectric effect.

It is interesting to note that if the atom is initially in a bound state of (negative) energy E_n, and then absorbs an energy greater than the *binding energy* $|E_n|$, it will be ionised and the ejected electron will have a *positive energy*. States of positive energy therefore correspond to the situation in which the nucleus–electron system is *unbound*. It is clear that these states are relevant in *collision* processes between electrons and nuclei. In such a collision event, an electron having a given initial kinetic energy and coming from a very large distance is scattered by a nucleus and after being deflected recedes to an infinite separation without forming a bound system with the nucleus. Since the initial kinetic energy of the electron may be chosen arbitrarily, it is obvious that unbound states of the nucleus–electron system may have any positive energy value. Therefore, joining the bound state spectrum of one-electron atoms at the value $E = 0$, there is a *continuous* energy spectrum extending from $E = 0$ to $E = +\infty$, which is indicated in Fig. 1.11 by hatchings. Transitions may thus occur not only between two bound states of the discrete energy spectrum (bound–bound transitions), but also between a state of the discrete spectrum and one of the continuous spectrum (bound–free transitions), or between two states of the continuous spectrum (free–free transitions).

Let us now evaluate a few key quantities which appear in Bohr's theory. The radius of the orbit of the electron in the ground state of the hydrogen atom is known as the (first) Bohr radius of hydrogen and is denoted by a_0. It is obtained by setting $Z = n = 1$ in (1.57) so that

$$a_0 = \frac{(4\pi\varepsilon_0)\hbar^2}{me^2} = 5.29 \times 10^{-11} \text{ m}. \tag{1.66}$$

The velocity v_0 of the electron in the first Bohr orbit of the hydrogen atom is seen from (1.56) to be given by

$$v_0 = \frac{e^2}{(4\pi\varepsilon_0)\hbar} = \alpha c \tag{1.67}$$

where we have introduced the dimensionless constant

$$\alpha = \frac{e^2}{(4\pi\varepsilon_0)\hbar c} \tag{1.68}$$

which is known as the *fine-structure constant* and has the value $\alpha \simeq 1/137$. Thus $v_0 \simeq c/137$ and it is reasonable to treat the electron of an hydrogen atom in a non-relativistic way, as done by Bohr. We also remark in this context that the bound-state energies (1.60) of one-electron atoms may be rewritten as

$$E_n = -\frac{1}{2}mc^2 \frac{(Z\alpha)^2}{n^2}. \tag{1.69}$$

Thus, provided Z is not too large, the binding energies $|E_n|$ are small with respect to mc^2, the rest energy of the electron; this again indicates that a non-relativistic treatment of the problem is a sensible approximation.

Finite nuclear mass

Although the approximation in which the nucleus is assumed to be infinitely heavy is good enough for many purposes, an improvement can be made by taking into account the fact that the mass of the nucleus is finite. If we consider a bound state of the nucleus–electron system in the Bohr model, then both the electron of mass m and the nucleus of mass M rotate about the centre of mass of the system. Since no forces external to the atom are present, the centre of mass will either be at rest or in uniform motion. As we are not interested in the motion of the atom as a whole, we shall take the centre of mass to be at rest and to be the origin of our coordinate system.

The magnitude of the total orbital angular momentum of the system is readily shown to be[13]

$$L = \mu v r \tag{1.70}$$

where r is the distance between the nucleus and the electron, v is the velocity of the electron with respect to the nucleus and

$$\mu = \frac{mM}{m+M} \tag{1.71}$$

is the *reduced mass*. Bohr's quantisation condition for the orbital angular momentum thus becomes

$$L = \mu v r = n\hbar, \qquad n = 1, 2, 3, \ldots \tag{1.72}$$

which is the same as (1.55), but with μ replacing m.

Since the centre of mass is at rest in our coordinate system, the kinetic energy of the system is given by $T = \mu v^2/2$. The potential energy V, due to the electrostatic attraction between the nucleus and the electron, is of course still given by $V = -Ze^2/(4\pi\varepsilon_0)r$, since it does not depend on the masses of the particles. A

---

[13] See, for example, Bransden and Joachain (1983).

reasoning entirely similar to that made above in the infinite nuclear mass case then yields for the allowed energy levels of the bound nucleus–electron system

$$E_n = -\frac{\mu}{2\hbar^2}\left(\frac{Ze^2}{4\pi\varepsilon_0}\right)^2 \frac{1}{n^2} \tag{1.73}$$

which is the result (1.60) with m replaced by μ. Similarly, the allowed values of r are given by (1.57) with again μ replacing m,

$$r = \frac{(4\pi\varepsilon_0)\hbar^2 n^2}{Z\mu e^2} = \frac{n^2}{Z}\frac{m}{\mu}a_0 = \frac{n^2}{Z}a_\mu \tag{1.74}$$

where $a_\mu = (m/\mu)a_0$ is the *modified Bohr radius*.

As a consequence of the finite mass correction, the calculated value of the Rydberg constant becomes

$$R(M) = \frac{\mu}{m}R(\infty) = \frac{1}{1+(m/M)}R(\infty). \tag{1.75}$$

For atomic hydrogen the nuclear mass M is equal to M_P, the proton mass, and the corresponding theoretical value of $R(M_P)$ is 109 681 cm^{-1}, which agrees with the experimental value (1.46) to better than four parts in 10^5.

Because of the nuclear mass effect all the frequencies of the spectral lines of one-electron atoms are reduced by the factor $\mu/m = (1 + m/M)^{-1}$ with respect to their values calculated in the infinite nuclear mass approximation. As a result, there is an *isotopic shift* between the spectral lines of different isotopes[14]. For example, there is such a shift between the spectral lines of ordinary hydrogen (proton + electron) and those of its heavy isotope, deuterium. The nucleus of the deuterium atom, called the *deuteron*, contains a proton and a neutron, and hence has $Z = 1$ and a mass $M \simeq 2M_P$. The ratio of frequencies of corresponding lines in deuterium and 'ordinary' hydrogen is 1.000 27, which is easily detectable. In fact, it is through this isotopic shift that deuterium was discovered.

Another consequence of the fact that the mass of the nucleus is finite is that the *recoil* of the atom must be taken into account when using the Bohr frequency relation (1.52). This is readily done by using momentum and energy conservation. Let us consider the emission or absorption of a photon of energy $h\nu$ by an atom initially at rest. After the transition, the atom recoils with a momentum **P** whose magnitude P is equal to the magnitude $h\nu/c$ of the photon momentum, and with a kinetic energy $P^2/2M = (h\nu)^2/2Mc^2$, where M is the mass of the atom (which is essentially equal to the mass of the nucleus). If ν_0 denotes the frequency of the transition uncorrected for the recoil effect, conservation of energy yields for the fractional change in the frequency due to recoil (Problem 1.21)

$$\frac{\Delta\nu_R}{\nu} = \pm\frac{h\nu}{2Mc^2} \tag{1.76}$$

---

[14] We recall that isotopes are atoms of the same chemical element having different atomic masses. The nuclei of different isotopes contain the same number Z of protons, and hence have the same atomic number Z; they differ by their number of neutrons, N, so that they have different mass numbers $A = N + Z$.

where $\Delta\nu_R = \nu - \nu_0$, the plus sign corresponding to the absorption and the minus sign to the emission of a photon. Now, $2Mc^2$ is at least of the order of 10^9 eV (for a proton, $M_P c^2 \simeq 9.4 \times 10^8$ eV), so that for atomic transitions involving photons having energies $h\nu$ of a few electronvolts we find that $\Delta\nu_R/\nu \simeq 10^{-9}$, which is extremely small. Nevertheless, equation (1.76) does seem to imply that the atomic absorption spectrum of an atom is not completely identical to its emission spectrum, so that photons emitted by an atom could not be absorbed by another atom of the same kind, in contradiction with experiment. This apparent paradox will be resolved in Chapter 2, when we discuss the Heisenberg uncertainty relation for time and energy.

Limitation of the Bohr model

Although the Bohr model is successful in predicting the energy levels of one-electron atoms, and the idea of energy quantisation in atoms is correct, the model is unsatisfactory in many respects. Indeed, Bohr's theory is a hybrid one combining principles taken over from classical theory with new postulates breaking sharply with classical physics. For example, the stability problem is bypassed by postulating that an electron in one of the allowed stationary orbits does not radiate electromagnetic energy, and the quantisation of angular momentum is introduced in an *ad hoc* way. The hypothesis that only circular orbits are allowed is also arbitrary. Moreover, despite the fact that it provided an explanation of the regularities in X-ray spectra observed by Moseley in 1913[15], the Bohr model cannot be generalised to deal with atomic systems containing two or more electrons. Among other objections are the lack of a reliable method to calculate the rate of transitions between the different energy levels when radiation is emitted or absorbed, and the inability to handle unbound systems. In later work, W. Wilson and A. Sommerfeld showed how to remove the restriction to circular orbits, and Sommerfeld also obtained relativistic corrections to the Bohr model. However, the other objections still persisted and the theory – now called the *old quantum theory* – was eventually superseded by the quantum mechanics developed by E. Schrödinger, W. Heisenberg, P. A. M. Dirac and others, following the ideas of L. de Broglie.

The Bohr correspondence principle

In 1923, Bohr formulated a heuristic principle which had already inspired the development of the old quantum theory, and which proved to be of great help in the early development of quantum mechanics. This principle, known as the *correspondence principle*, states that *quantum theory results must tend asymptotically to those obtained from classical physics in the limit of large quantum numbers*. In other words, classical physics results are 'macroscopically correct' and may be considered as limiting cases of quantum mechanical results when the quantum discontinuities may be neglected.

As an illustration of the correspondence principle, let us consider the Bohr model for one-electron atoms, and compute the frequency of the radiation emitted in a

---

[15] See Bransden and Joachain (1983).

transition from the energy level E_n to that immediately below, E_{n-1}, when n is large. Assuming for simplicity that the nuclear mass is infinite, we find from (1.61) that this frequency is given by

$$\nu = \frac{m}{4\pi\hbar^3}\left(\frac{Ze^2}{4\pi\varepsilon_0}\right)^2\left[\frac{1}{(n-1)^2}-\frac{1}{n^2}\right]. \tag{1.77}$$

Now, when $n \gg 1$, one has $(n-1)^{-2} - n^{-2} \simeq 2n^{-3}$ and therefore, for large n,

$$\nu \simeq \frac{m}{2\pi\hbar^3}\left(\frac{Ze^2}{4\pi\varepsilon_0}\right)^2\frac{1}{n^3}. \tag{1.78}$$

On the other hand, in classical physics, an electron moving in a circular orbit of radius r with a velocity v would be expected to radiate with the frequency of its rotational motion, namely $\nu_{\text{cl}} = v/(2\pi r)$. Using the expressions (1.56) and (1.57) for v and r, we find that

$$\nu_{\text{cl}} = \frac{v}{2\pi r} = \frac{m}{2\pi\hbar^3}\left(\frac{Ze^2}{4\pi\varepsilon_0}\right)^2\frac{1}{n^3} \tag{1.79}$$

in agreement with (1.78).

The Franck and Hertz experiment

As we have seen above, the bound-state energies of atoms are quantised, so that only certain discrete energy values are allowed. This *energy quantisation of atoms* was confirmed directly by an experiment performed in 1914 by J. Franck and G. Hertz. A schematic diagram of the apparatus is shown in Fig. 1.12. In an evacuated tube, electrons are ejected from a heated cathode C and accelerated toward a wire grid G, maintained at a positive potential V_1 with respect to the cathode. The electrons accelerated by the potential V_1 attain a kinetic energy $mv^2/2 = eV_1$. Some of them pass through the grid and are collected by a plate P, thus causing a current I to flow in the collector circuit. The collector P is at a slightly lower potential V_2 than the grid, $V_2 = V_1 - \Delta V$, where $\Delta V \ll V_1$. The small retarding potential ΔV between the grid G and the collector P has the effect of reducing the kinetic energy of the electrons slightly, but not enough to stop them being collected.

The tube is now filled with mercury vapour. The electrons collide with the atoms of mercury, and if the collisions are elastic, so that there is no transfer of energy from the electrons to the internal structure of the atoms, the current I will be unaffected by the introduction of the gas. This follows because mercury atoms are too heavy to gain appreciable kinetic energy when struck by electrons. The electrons are deflected but retain the same kinetic energy. In contrast, if an electron makes an inelastic collision with a mercury atom in which it loses an energy E, exciting the mercury atom to a level of greater internal energy, then its final kinetic energy will be $mv_1^2/2 = (eV_1 - E)$. If eV_1 is equal to E, or is only a little larger, the retarding potential ΔV will be sufficient to prevent this electron from reaching the collecting plate, and hence from contributing to the current I.

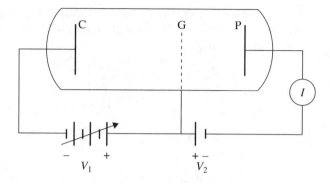

Figure 1.12 Schematic diagram of the Franck and Hertz experiment.

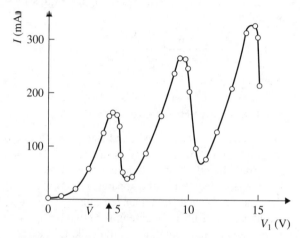

Figure 1.13 The variation of the current I as a function of the accelerating voltage V_1 in the Franck and Hertz experiment.

The experiment is carried out by gradually increasing V_1 from zero and measuring the current I as a function of V_1. The result obtained is illustrated in Fig. 1.13. The current I is seen to fall sharply at a potential $\bar{V}$, which for mercury is equal to 4.9 V. The results can be interpreted by supposing that the first excited level of the mercury atom has an energy about 4.9 eV higher than the ground state level. Thus, when the colliding electrons have an energy $eV_1 < 4.9$ eV, they cannot excite the mercury atoms to a level of higher internal energy, so that the collisions are elastic. On the contrary, when eV_1 reaches the value 4.9 eV, a large number of the colliding electrons excite mercury atoms to this level, losing their energy in the process; as a result the current I is sharply reduced.

If the voltage V_1 is further increased, the observed current I is seen to exhibit additional increases followed by sharp dips. Some of these dips are due to the excitation of higher discrete levels of the mercury atoms. Other dips arise from the fact that some colliding electrons have sufficient energy to excite two or more atoms to the first excited state, so that they lose 4.9 eV of energy more than once; such dips corresponding to multiple excitations of the first excited level are seen every 4.9 eV.

We have noted before (see (1.53)) that an energy of 1 eV corresponds to a wave number of 8065 cm^{-1}, so that if the above interpretation of the Franck and Hertz experiment is correct, a line would be expected in the mercury spectrum corresponding to a transition from the first excited state at 4.9 eV to the ground state, with a wave number of 4.9×8065 cm$^{-1} \simeq 39\,500$ cm^{-1}. Franck and Hertz were indeed able to verify the existence of such a line, and to show that radiation of this wave number was only emitted from the mercury vapour when V_1 exceeded 4.9 eV.

The Franck–Hertz experiment, and corresponding experiments using other gases and vapours, provide excellent confirmation of the discrete nature of bound-state energy levels of atoms. It can also be verified that when sufficient energy is available to ionise an atom, the energy of the ejected electron can take any positive value. Therefore, *the energy level spectrum of an atom consists of two parts: discrete negative energies corresponding to bound states and a continuum of positive energies corresponding to unbound (ionised) states.*

1.5 The Stern–Gerlach experiment. Angular momentum and spin

We shall now discuss another experiment of fundamental importance, carried out by O. Stern and W. Gerlach in 1922, to measure the magnetic dipole moments of atoms. The results demonstrated, once more, the inability of classical mechanics to describe atomic phenomena and confirmed the necessity of a quantum theory of angular momentum, which had been suggested by Bohr's model.

Let us first understand how an atom comes to possess a magnetic moment. In the Bohr model of a hydrogenic atom, an electron occupies a circular orbit, rotating with an orbital angular momentum **L**. A moving charge is equivalent to an electric current, so that an electron moving in a closed orbit forms a current loop, and this in turn creates a magnetic dipole[16]. In fact, whatever model of atomic structure we make, the electrons can be expected to possess angular momentum and accordingly atoms possess magnetic moments.

A circulating current of magnitude I enclosing a small plane area dA gives rise to a magnetic dipole moment

$$\mathcal{M} = I \, d\mathbf{A} \tag{1.80}$$

where the direction of $\mathcal{M}$ is along the normal to the plane of the current loop, as shown in Fig. 1.14. When the current I is due to an electron moving with a velocity v in a circle of radius r, it is given by

$$I = \frac{ev}{2\pi r}. \tag{1.81}$$

---

[16] See, for example, Duffin (1968).

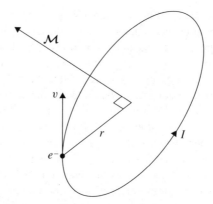

Figure 1.14 The magnetic dipole moment due to a current loop.

Since the area enclosed is πr^2, we have $\mathcal{M} = evr/2 = eL/2m$, and as the direction of the current is opposite to the direction of rotation of the electron,

$$\mathcal{M} = -\frac{e}{2m}\mathbf{L} \tag{1.82}$$

where $\mathbf{L} = \mathbf{r} \times \mathbf{p}$ is the orbital angular momentum of the electron, $\mathbf{p} = m\mathbf{v}$ being its (linear) momentum. The Bohr quantisation rule (1.55) suggests that $\hbar$ is a natural unit of angular momentum, so that we shall rewrite equation (1.82) as

$$\mathcal{M} = -\mu_B(\mathbf{L}/\hbar) \tag{1.83}$$

where

$$\mu_B = \frac{e\hbar}{2m}. \tag{1.84}$$

Because $(L/\hbar)$ is dimensionless, μ_B has the dimensions of a magnetic moment. It is known as the *Bohr magneton* and has the numerical value

$$\mu_B = 9.27 \times 10^{-24} \text{ J T}^{-1} \tag{1.85}$$

where J denotes a joule and T a tesla.

Interaction with a magnetic field

If an atom with a magnetic moment $\mathcal{M}$ is placed in a magnetic field $\mathcal{B}$, the (potential) energy of interaction is (see Duffin, 1968)

$$W = -\mathcal{M} \cdot \mathcal{B}. \tag{1.86}$$

The system experiences a torque Γ, where

$$\Gamma = \mathcal{M} \times \mathcal{B} \tag{1.87}$$

1.5 The Stern–Gerlach experiment

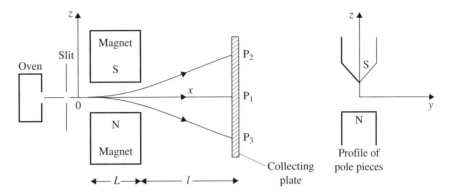

Figure 1.15 The Stern–Gerlach apparatus.

and a net force **F**, where

$$\mathbf{F} = -\nabla W. \tag{1.88}$$

Combining (1.86) with (1.88), we see that the components of **F** are

$$F_x = \mathcal{M} \cdot \frac{\partial \mathcal{B}}{\partial x}, \qquad F_y = \mathcal{M} \cdot \frac{\partial \mathcal{B}}{\partial y}, \qquad F_z = \mathcal{M} \cdot \frac{\partial \mathcal{B}}{\partial z}. \tag{1.89}$$

If the magnetic field is uniform, no net force is experienced by a magnetic dipole, which precesses with a constant angular frequency. For an orbiting electron this angular frequency is

$$\omega_L = \frac{\mu_B}{\hbar} \mathcal{B}. \tag{1.90}$$

It is called the *Larmor angular frequency*. On the other hand, in an inhomogeneous magnetic field, an atom experiences a net force proportional to the magnitude of the magnetic moment.

The Stern–Gerlach experiment

In 1921, Stern suggested that magnetic moments of atoms could be measured by detecting the deflection of an atomic beam by such an inhomogeneous field. The experiment was carried out a year later by Stern and Gerlach. The apparatus is shown in schematic form in Fig. 1.15.

The first experiments were made using atoms of silver. A sample of silver metal is vaporised in an oven and a fraction of the atoms, emerging from a small hole, is collimated by a system of slits so that it enters the magnetic field region as a narrow and nearly parallel atomic beam. This beam is then passed between the poles of a magnet shaped to produce an inhomogeneous field, as shown in Fig. 1.15. Finally, the beam is detected by allowing it to fall on a cool plate. The density of the deposit is proportional to the intensity of the beam and to the length of time for which the

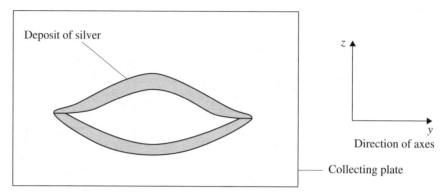

Figure 1.16 Results of the Stern–Gerlach experiment for silver. The atomic beam is split into two components having nearly equal and opposite deflections along the z-axis. Only in the centre is the field gradient sufficient to cause the splitting. The pattern is smeared because of the range of velocities of atoms in the beam. The shape of the upper line results from the greater inhomogeneity of the magnetic field near the upper-pole face.

beam falls on the plate. An important requirement for performing such an atomic beam experiment is a good vacuum, which allows the atoms of the beam to travel through the apparatus with a negligible probability of colliding with a molecule of the residual gas.

Taking the shape of the magnets as shown in Fig. 1.15, the force on each atom is given from (1.89) by

$$F_x = \mathcal{M}_z \frac{\partial \mathcal{B}_z}{\partial x}, \qquad F_y = \mathcal{M}_z \frac{\partial \mathcal{B}_z}{\partial y}, \qquad F_z = \mathcal{M}_z \frac{\partial \mathcal{B}_z}{\partial z}. \tag{1.91}$$

The magnet is symmetrical about the xz plane, and the beam is confined to this plane. It follows that $\partial \mathcal{B}_z/\partial y = 0$. Also, apart from edge effects, $\partial \mathcal{B}_z/\partial x = 0$, so that the only force on the atoms in the beam is in the z direction.

In the incident beam, the direction of the magnetic moment $\mathcal{M}$ of the atoms is completely at random, and in the z direction it would be expected that every value of $\mathcal{M}_z$ such that $-\mathcal{M} \leqslant \mathcal{M}_z \leqslant \mathcal{M}$ would occur, with the consequence that the deposit on the collecting plate would be spread continuously over a region symmetrically disposed about the point of no deflection. The surprising result that Stern and Gerlach obtained in their experiments with silver atoms was that *two distinct* and *separate* lines of spots were formed on the plate (see Fig. 1.16), symmetrically about the point of no deflection. Similar results were found for atoms of copper and gold, and in later work for sodium, potassium, caesium and hydrogen.

The quantisation of the component of the magnetic moment along the direction defined by the magnetic field is termed *space quantisation*. This implies that the component of the angular momentum in a certain direction is quantised so that it can only take certain values. In general, for each type of atom, the values of $\mathcal{M}_z$ will range from $(\mathcal{M}_z)_{\max}$ to $-(\mathcal{M}_z)_{\max}$ and, correspondingly, L_z will range from $-(L_z)_{\max}$ to $(L_z)_{\max}$. If we denote the observed multiplicity of values of $\mathcal{M}_z$ (and hence L_z) by α, we can try to interpret α and to deduce the allowed values of L by using the Bohr model. Indeed, the Bohr quantisation of angular momentum suggests that orbital angular momentum only occurs in integral units of $\hbar$. We may postulate that the magnitude of orbital angular momentum can only take values $L = l\hbar$, where l is a positive integer or zero. Thus the maximum value of L_z is $+l\hbar$, and its minimum value is $-l\hbar$. If L_z is also quantised in the form

$$L_z = m\hbar \tag{1.92}$$

where m is a positive or negative integer or zero, then m must take on the values $-l, -l+1, \ldots, l-1, l$, and the multiplicity α must be equal to $(2l+1)$. The number m is known as a *magnetic quantum number*. In fact, as we shall see in Chapter 6, the result (1.92) turns out to be correct in quantum mechanics, but the possible values of the total orbital angular momentum will be shown to be of the form $L = \sqrt{l(l+1)}\hbar$ with $l = 0, 1, 2, \ldots$, rather than of the form $L = l\hbar$ suggested by the Bohr model. However the results of Stern and Gerlach for silver do not fit with this scheme, since the multiplicity of values of the z component of the angular momentum for silver is $\alpha = 2$. This implies that $(2l+1) = 2$, giving $l = \frac{1}{2}$, which is a non-integral value.

Electron spin

The explanation of this result for silver came in 1925, when S. Goudsmit and G. E. Uhlenbeck analysed the splitting of spectral lines occurring when atoms are placed in a magnetic field (the Zeeman effect). They showed that this splitting could be explained if electrons, in addition to the magnetic moment produced by orbital motion, were assumed to possess an *intrinsic magnetic moment* $\mathcal{M}_s$, where the component of $\mathcal{M}_s$ in a given direction can take only the two values $\pm \mathcal{M}_s$. One can postulate that this intrinsic magnetic moment is due to an *intrinsic angular momentum*, or *spin* of the electron, which we denote by **S**. By analogy with (1.83), we then have

$$\mathcal{M}_s = -g_s \mu_B (\mathbf{S}/\hbar) \tag{1.93}$$

where g_s is the *spin gyromagnetic ratio*. If we introduce a *spin quantum number s*, analogous to l, so that the multiplicity of the spin component in a given direction is $(2s+1)$, then for an electron we must have $s = \frac{1}{2}$, and the possible values of the component S_z of the electron spin **S** in the z direction are $\pm\hbar/2$; the magnitude S of the electron spin will be shown in Chapter 6 to be given by $S = \sqrt{s(s+1)}\hbar = \sqrt{3/4}\hbar$.

Since the magnetic moment of an electron is due partly to its orbital angular momentum and partly to its spin angular momentum, we can use equations (1.83) and (1.93) to write this (total) magnetic moment as

$$\mathcal{M} = -\mu_B(\mathbf{L} + g_s\mathbf{S})/\hbar. \tag{1.94}$$

Accurate measurements have shown that the electron spin gyromagnetic ratio g_s is very nearly equal to 2.

The discovery of the spin of the electron is of fundamental importance. In fact, it is now known that all particles can be assigned an intrinsic angular momentum (spin), and therefore a spin quantum number s. In some cases, such as the pion (π-meson), $s = 0$, but in others, such as the electron, the proton and the neutron one has $s = 1/2$, and for other 'elementary particles' the spin quantum number may be $s = 1, s = \frac{3}{2}, \ldots$. The theory of the spin angular momentum will be discussed in Chapter 6.

Total angular momentum

The total angular momentum of an atom is obtained by adding vectorially all angular momenta of the particles it contains. Because the mass of the nucleus is much larger than the electron mass, the angular momentum of the nucleus leads to very small hyperfine effects which we shall neglect here. We therefore consider only the angular momenta of the electrons in the atom. By adding vectorially all the orbital and spin angular momenta of these electrons, we obtain the total electronic angular momentum, which we denote by **J**. A measurement of the component of **J** in a given direction (which we call the z direction) can only yield $(2j + 1)$ possible values, given by $m_j \hbar$, where the *magnetic quantum number m_j* can only take the values $-j, -j + 1, \ldots, j - 1, j$. Thus j is the maximum possible value of a component of **J**, measured in units of $\hbar$, in any given direction. It is found that j can take integral or half-odd-integral values only; $j = 0, \frac{1}{2}, 1, \frac{3}{2}, \ldots$. For an angular momentum whose component in a given direction has the multiplicity $(2j + 1)$, a measurement of the magnitude of the angular momentum produces the value $\sqrt{j(j+1)}\hbar$. Thus, in a Stern–Gerlach experiment, a beam of atoms with angular momentum of magnitude $\sqrt{j(j+1)}\hbar$ will produce $(2j+1)$ lines of spots on the detecting screen, symmetrically disposed about the point of zero deflection. The Stern–Gerlach results for silver atoms can then be explained if we assume that the orbital angular momentum of a silver atom is zero, but its spin angular momentum is equal to the spin of an electron, so that in this case $j = s = \frac{1}{2}$ and *two* lines of spots will appear on the screen. Similar experiments carried out with other atoms have confirmed the central features which emerge from the Stern–Gerlach experiment: *the quantisation of the z component of angular momentum* and *the existence of the electron spin*[17].

1.6 De Broglie's hypothesis. Wave properties of matter and the genesis of quantum mechanics

We have seen before that in addition to its classical wave properties, electromagnetic radiation also exhibits particle characteristics, as in the photoelectric effect and in the Compton effect. This dual nature of electromagnetic radiation, which is one of

---

[17] It should be noted that experiments of the Stern–Gerlach type cannot be performed directly on beams of electrons because of the deflection of the beam due to the Lorentz force.

the most striking features due to the appearance of quanta in physics, was generally recognised, although not understood, around 1923. On the other hand, at that time matter was believed to be purely of a corpuscular nature.

In 1923–24, L. de Broglie made a great unifying, bold hypothesis that *material particles might also possess wave-like properties*, so that, like radiation, they would exhibit a dual nature. In this way the wave–particle duality would be a universal characteristic of nature. The energy of a photon is given by $E = h\nu$, where ν is the frequency of the radiation; the magnitude of the photon momentum is $p = h\nu/c = h/\lambda$, where λ is the wavelength. For free material particles, de Broglie, using arguments based on the properties of wave packets[18], proposed that the associated matter waves also have a frequency ν and a wavelength λ, related respectively to the energy E and to the magnitude p of the momentum of the particle by

$$\nu = \frac{E}{h} \tag{1.95}$$

and

$$\lambda = \frac{h}{p}. \tag{1.96}$$

The wavelength $\lambda = h/p$ is known as the *de Broglie wavelength of a particle*. We remark that since $\lambda\nu = c$ for a photon, only one of the two relations (1.95)–(1.96) is required to obtain both its wavelength and frequency from its particle properties of energy and momentum. In contrast, for a material (massive) particle, one needs both relations (1.95) and (1.96) to obtain the associated frequency ν and wavelength λ.

The de Broglie idea immediately gives a *qualitative* explanation of the quantum condition (1.55) used in the Bohr model of one-electron atoms. Indeed, let us suppose that an electron in a hydrogenic atom moves in a circular orbit of radius r, with velocity v. If this is to be a stable stationary state, the wave associated with the electron must be a standing wave, and a whole number of wavelengths must fit into the circumference $2\pi r$. Thus

$$n\lambda = 2\pi r \qquad n = 1, 2, 3, \ldots. \tag{1.97}$$

Since $\lambda = h/p$ and $L = rp$, we immediately find the condition

$$L = nh/2\pi = n\hbar \tag{1.98}$$

which is identical with (1.55). Later, in 1925 and 1926, these qualitative ideas were incorporated into the systematic theory of quantum mechanics developed by E. Schrödinger, which will be discussed in the next chapter.

De Broglie wrote the relations (1.95) and (1.96) for the general case of a relativistic particle. For a free particle of mass m moving at a non-relativistic speed v, one has $E \simeq mc^2 + mv^2/2$. Now, in the non-relativistic approximation, the energy $mv^2/2$ of a free particle is defined so that it does not contain the rest energy mc^2. This amounts to a modification of the zero of the energy scale, which has the effect of adding a

---

[18] Wave packets will be discussed in Chapter 2.

constant frequency $\nu_0 = mc^2/h$ to the right of (1.95). We shall see in Chapter 2 that this addition has no observable effect. Since $p = mv$ for a non-relativistic particle, we also remark that the de Broglie wavelength is given in the non-relativistic approximation by

$$\lambda = \frac{h}{mv}. \qquad (1.99)$$

We now turn to the question of the experimental confirmation of de Broglie's hypothesis. We first recall that in order to observe the interference and diffraction effects which are characteristic of the wave properties of light, some geometric parameters of the instrument (such as apertures or slits) must have dimensions comparable to the wavelength of the light. If, on the contrary, this wavelength is much smaller than all the relevant dimensions of the optical instruments, we are in the domain of geometrical optics, where interference and diffraction effects are negligible. By analogy, we expect that in order to detect the presence of matter waves, an appropriate 'grating' having dimensions comparable to the de Broglie wavelength $\lambda = h/p$ of the particle is required.

Let us estimate the value of the de Broglie wavelength associated with a particle. For a macroscopic particle of mass $m = 10^{-3}$ kg, moving at a speed $v = 1$ m s^{-1}, we find from (1.99) that the associated de Broglie wavelength $\lambda \simeq 6.6 \times 10^{-31}$ m = 6.6×10^{-21} Å, which is orders of magnitude smaller than any existing aperture. On the other hand, let us consider a non-relativistic electron, of mass $m = 9.1 \times 10^{-31}$ kg, which has been accelerated by a potential difference V_0, so that it has a kinetic energy $mv^2/2 = eV_0$. Using (1.99), we obtain for the associated de Broglie wavelength

$$\lambda = \frac{h}{(2meV_0)^{1/2}} = \frac{12.3}{[V_0(\text{Volts})]^{1/2}} \text{Å} \qquad (1.100)$$

where we have used the fact that 1 eV = 1.6×10^{-19} J. Thus, the de Broglie wavelengths associated with electrons of energy 1, 10, 100 and 1000 eV are, respectively, 12.3, 3.89, 1.23 and 0.39 Å, comparable to the wavelength of X-rays. Hence, in macroscopic situations, such as in J. J. Thomson's experiments, where beams of electrons move under the influence of electric and magnetic fields, the de Broglie electron wavelengths are exceedingly small compared with the dimensions of any obstacles or slits in the apparatus, so that no interference or diffraction effects can be observed. However, the spacing of atoms in a crystal lattice is of the order of an ångström and therefore, just as in the case of X-rays, a crystal can be used as a grating to observe the diffraction and interference effects due to the electron matter waves. Experiments of this type were performed in 1927 by C. J. Davisson and L. H. Germer, and independently by G. P. Thomson.

In the Davisson–Germer experiment, the *reflection* of electrons from the face of a crystal was investigated. Electrons from a heated filament were accelerated through a potential difference V_0 and emerged from the 'electron gun' with kinetic energy eV_0. This beam of monoenergetic electrons was directed to strike at normal incidence the surface of a single nickel crystal, and the number $N(\theta)$ of electrons scattered at an

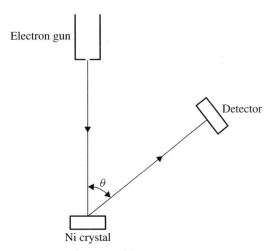

Figure 1.17 Schematic diagram of the Davisson–Germer experiment. Electrons strike at normal incidence the surface of a single nickel crystal. The number of electrons scattered at an angle θ to the incident direction is measured by means of a detector.

angle θ to the incident direction was measured by means of a detector (see Fig. 1.17). The data obtained by Davisson and Germer for 54 eV electrons are shown in Fig. 1.18. The scattered intensity is seen to fall from a maximum at $\theta = 0°$ to a minimum near 35°, and then to rise to a peak near 50°. The strong scattering at $\theta = 0°$ is expected from either a particle or a wave theory, but the peak at 50° can only be explained by constructive interference of the waves scattered by the regular crystal lattice.

Referring to Fig. 1.19, the Bragg condition for constructive interference is

$$n\lambda = 2d \sin \theta_B \tag{1.101}$$

where d is the spacing of the Bragg planes and n is an integer. If D denotes the spacing of the atoms in the crystal, we see from Fig. 1.19 that $d = D \sin \alpha$, with $\alpha = \pi/2 - \theta_B$. Moreover, the scattering angle $\theta = 2\alpha$, so that the Bragg condition (1.101) becomes

$$n\lambda = D \sin \theta. \tag{1.102}$$

Experiments in which X-rays were diffracted established that for a nickel crystal the atomic spacing is $D = 2.15$ Å. Assuming that the peak at $\theta = 50°$ corresponds to first-order diffraction ($n = 1$) we see from (1.102) that the experimental electron wavelength is given by $\lambda = (2.15 \sin 50°)$ Å $= 1.65$ Å. On the other hand, the de Broglie wavelength of a 54 eV electron is 1.67 Å, which agrees with the value of 1.65 Å, within the experimental error. By varying the voltage V_0, measurements were also performed at other incident electron energies, which confirmed the variation of λ with momentum as predicted by the formula (1.100), derived from the de Broglie relation (1.96). Higher-order maxima, corresponding to $n > 1$ in (1.102), were also observed and found to be in agreement with the theoretical predictions.

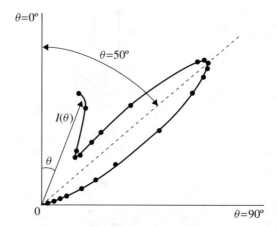

Figure 1.18 Polar plot of the scattered intensity as a function of the scattering angle θ for 54 eV electrons in the Davisson–Germer experiment. At each angle the intensity $I(\theta)$ is given by the distance of the point from the origin. A maximum is observed at $\theta = 50°$, which can only be explained by constructive interference of the waves scattered by the crystal lattice.

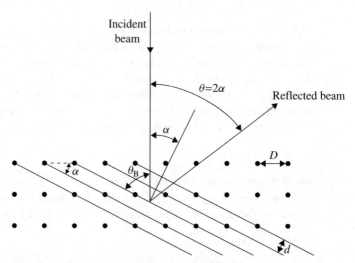

Figure 1.19 The scattering of electron waves by a crystal. Constructive interference occurs when the Bragg condition $n\lambda = 2d \sin \theta_B$ is satisfied.

In the experiment of G. P. Thomson, the *transmission* of electrons through a thin foil of polycrystalline material was analysed. A beam of monoenergetic electrons was directed towards the foil, and after passing through it the scattered electrons struck a photographic plate (see Fig. 1.20). This method is analogous to the Debye–Scherrer method used in the study of X-ray diffraction. Because the foil consists of many small randomly oriented microcrystals, 'classical' electrons behaving only as particles would yield a blurred image. However, the result obtained by G. P. Thomson

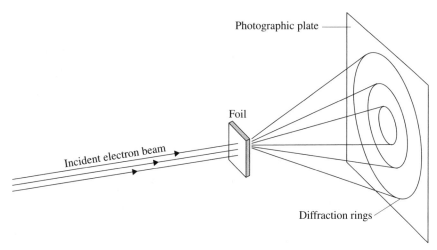

Figure 1.20 The experimental arrangement of G. P. Thomson for observing electron diffraction through a thin film of polycrystalline material.

was similar to the X-ray Debye–Scherrer diffraction pattern, which consists of a series of concentric rings. In the same way, when an electron beam passes through a single crystal, Laue spot patterns are observed, as in the case of the diffraction of X-rays.

The wave behaviour of electrons has also been demonstrated by observing the diffraction of electrons from edges and slits. For example, in 1961 C. Jönsson performed an electron diffraction experiment analogous to T. Young's famous double slit experiment, which in 1803 gave conclusive evidence of the wave properties of light. The principle of the experiment is illustrated in Fig. 1.21. Jönsson used 40 keV electrons, having a de Broglie wavelength $\lambda = 5 \times 10^{-2}$ Å. The slits, formed in copper foil, were very small (about 0.5×10^{-6} m wide), the slit separation being typically $d \simeq 2 \times 10^{-6}$ m. The interference fringes were observed on a screen at a distance $D = 0.35$ m from the slits. The spacing s of adjacent fringes is given by $s \simeq D\lambda/d \simeq 10^{-6}$ m. Because this spacing is very small, the fringes were magnified by placing electrostatic lenses between the slits and the screen.

According to de Broglie, not only electrons but *all material particles* possess wave-like characteristics. This universality of matter waves has been confirmed by a number of experiments. In all cases the measured wavelength was found to agree with the de Broglie formula (1.96). For example, in 1931 I. Esterman, R. Frisch and O. Stern observed the diffraction of helium atoms and hydrogen molecules by a crystal. Later, the diffraction of neutrons by crystals was also shown to occur. More recently, an atomic interferometer based on a Young double-slit arrangement has been demonstrated in 1991 by O. Carnal and J. Mlynek. In this apparatus, a beam of helium atoms from a gas reservoir G was passed through a source slit of width 2×10^{-6} m in a thin gold foil (see Fig. 1.22). The beam impinged on a screen S_1 having two slits, each of width 10^{-6} m, separated by a distance $d = 8 \times 10^{-6}$ m.

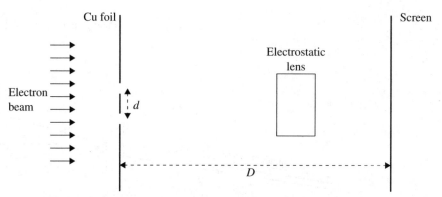

Figure 1.21 Schematic diagram illustrating the Jönsson double slit electron diffraction experiment.

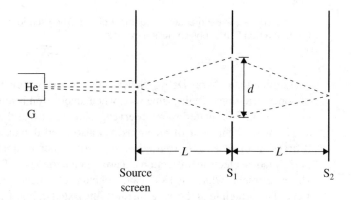

Figure 1.22 Schematic diagram of the atomic interferometer of Carnal and Mlynek.

The interference pattern was then observed on a detector screen S_2. The distance L between screens was fixed to be 64 cm. The mean velocity of the atoms, and hence their mean de Broglie wavelength, could be adjusted by changing the temperature of the gas reservoir. The experiments were performed with atoms with mean de Broglie wavelengths of $\lambda = 0.56$ Å and $\lambda = 1.03$ Å, corresponding to reservoir temperatures $T = 295$ K and $T = 83$ K, respectively.

It is worth stressing that the wave nature of matter is directly related to the finiteness of the Planck constant h. If h were zero, then the de Broglie wavelength $\lambda = h/p$ of a material particle would also vanish, and the particle would obey the laws of classical mechanics. It is precisely because Planck's constant is 'small' (when measured in units appropriate for the description of macroscopic phenomena) that the wave behaviour of matter is not apparent on the macroscopic scale. Thus, just as *geometrical optics is the short-wavelength limit of wave optics, classical mechanics can be considered as the short-wavelength limit of wave* (or *quantum*) *mechanics*.

The requirement that classical mechanics is contained in quantum mechanics in the limiting case $\lambda \to 0$ (i.e. $h \to 0$) is in accordance with the *correspondence principle*, which we discussed in Section 1.4.

Problems

1.1 Calculate the total emissive power of a black body at the following temperatures: (a) 1 K; (b) 300 K (room temperature); (c) 2000 K; (d) 4000 K. Find in each case the wavelength λ_{max} at which the spectral emittance $R(\lambda, T)$ has its maximum value.

1.2 (a) Calculate the surface temperature T of the Sun (in K) by assuming that it is a spherical black body with a radius of 7×10^8 m. The intensity of the solar radiation at the surface of the Earth is 1.4×10^3 W m^{-2} and the distance between the Sun and the Earth is 1.5×10^{11} m.

(b) Using the value of T obtained in (a), determine the wavelength λ_{max} at which the spectral emittance $R(\lambda, T)$ of the Sun (considered as a perfect black body) has its maximum value.

1.3 Using Wien's law (1.5), show the following:

(a) If the spectral distribution function of black body radiation, $\rho(\lambda, T)$, is known at one temperature, then it can be obtained at any temperature (so that a single curve can be used to represent black body radiation at all temperatures).

(b) The total emissive power is given by $R = \sigma T^4$ (the Stefan–Boltzmann law), where σ is a constant.

(c) The wavelength λ_{max} at which $\rho(\lambda, T)$ – or $R(\lambda, T)$ – has its maximum is such that $\lambda_{max} T = b$ (Wien's displacement law), where b is a constant.

1.4 Obtain the Rayleigh–Jeans spectral distribution law (1.8) as the long-wavelength limit of the Planck spectral distribution law (1.13).

1.5 Using Planck's spectral distribution law (1.13) for $\rho(\lambda, T)$, prove that

$$\lambda_{max} T = \frac{hc}{4.965k}$$

where λ_{max} is the wavelength at which $\rho(\lambda, T)$ has its maximum value for a given absolute temperature T. From this result and the values of h, c and k given in the table of fundamental constants at the end of the book (p. 789), obtain the constant b which occurs in Wien's displacement law (1.3).

(*Hint*: To solve the equation $d\rho/d\lambda = 0$, set $x = hc/\lambda kT$ and show that x must satisfy the equation $x = 5(1 - e^{-x})$. Then, upon writing $x = 5 - \varepsilon$ show that $\varepsilon \simeq 0.035$ so that $x = 4.965$ is an approximate solution.)

1.6 Using Planck's spectral distribution law (1.13) for $\rho(\lambda, T)$, prove that the total energy density ρ_{tot} is given by $\rho_{tot} = aT^4$, where $a = 8\pi^5 k^4/15h^3c^3$.

$$\left(\text{Hint:} \int_0^\infty \frac{x^3}{e^x - 1} dx = \frac{\pi^4}{15}. \right)$$

1.7 Consider the following three stars:

(1) a red star (such as Barnard's star) having a surface temperature of 3000 K;
(2) a yellow star (such as the Sun) having a surface temperature of 6000 K; and
(3) a blue-white star (such as Vega) having a surface temperature of 10 000 K.

Assuming that these stars radiate like black bodies, obtain for each of them:

(a) the total emissive power;
(b) the wavelength λ_{max} at which their spectral emittance $R(\lambda, T)$ peaks;
(c) the fraction of their energy which is radiated in the visible range of the electromagnetic spectrum (4000 Å $\leq \lambda \leq$ 7000 Å, where 1 Å $= 10^{-10}$ m).

(*Hint*: Use the integral given in Problem 1.6 and integrate Planck's spectral distribution numerically between the wavelengths $\lambda_1 = 4000$ Å and $\lambda_2 = 7000$ Å.)

1.8 What is the energy of:

(a) a γ-ray photon of wavelength $\lambda = 10^{-13}$ m?
(b) a photon of visible light, of wavelength $\lambda = 5 \times 10^{-7}$ m?
(c) a radio-wave photon of wavelength $\lambda = 10$ m?

1.9 Consider black body radiation at absolute temperature T. Show that:

(a) The number of photons per unit volume is $N = 2.029 \times 10^7 \, T^3$ photons m^{-3}.

$$\left(\text{Hint}: \int_0^\infty \frac{x^2}{e^x - 1} dx = 2.4041. \right)$$

(b) The average energy per photon is $\bar{E} = 3.73 \times 10^{-23} \, T$ J.

1.10 Consider an aluminium plate placed at a distance of 1 m from an ultraviolet source which radiates uniformly in all directions and has a power of 10^{-2} W. Suppose that an ejected photoelectron collects its energy from a classical atom modelled as a circular area of the plate whose radius is a typical atomic radius, $r \simeq 1$ Å. Assuming, as in classical physics that the energy is uniformly distributed over spherical wave fronts spreading out from the source, estimate the time required for a photoelectron to absorb the minimum amount of energy necessary to escape from the plate. The work function of aluminium is $W = 4.2$ eV.

1.11 The photoelectric work function W for lithium is 2.3 eV.

(a) Find the threshold frequency ν_t and the corresponding threshold wavelength λ_t.
(b) If ultraviolet light of wavelength $\lambda = 3000$ Å is incident on a lithium surface, calculate the maximum kinetic energy of the photoelectrons and the value of the stopping potential V_0.

1.12 A homogeneous light beam of wavelength $\lambda = 3000$ Å and intensity 5×10^{-2} W m^{-2} falls on a sodium surface. The photoelectric work function W for sodium is 2.3 eV. Calculate:

(a) the average number of electrons emitted per m² and per second; (you may assume that each photon striking the surface ejects an electron); and
(b) the maximum kinetic energy of the photoelectrons.

1.13 In a photoelectric experiment in which monochromatic light of wavelength λ falls on a potassium surface, it is found that the stopping potential is 1.91 V for $\lambda = 3000$ Å and 0.88 V for $\lambda = 4000$ Å. From these data, calculate:

(a) a value for Planck's constant, knowing that $e = 1.60 \times 10^{-19}$ C;
(b) the work function W for potassium; and
(c) the threshold frequency ν_t for potassium.

1.14 Consider the scattering of a photon by a free electron initially at rest. Referring to Fig. 1.9:

(a) prove that the angle ϕ at which the electron recoils is related to the angle θ of the scattered photon by

$$\tan \phi = \frac{\cot(\theta/2)}{1 + \lambda_C/\lambda_0} = \frac{\cot(\theta/2)}{1 + E_0/mc^2},$$

where $E_0 = hc/\lambda_0$ is the energy of the initial photon, m is the rest mass of the electron and $\lambda_C = h/mc$ is the Compton wavelength of the electron; and
(b) prove that the kinetic energy of the recoiling electron is given by

$$T_2 = E_0 \frac{(2\lambda_C/\lambda_0)\sin^2(\theta/2)}{1 + (2\lambda_C/\lambda_0)\sin^2(\theta/2)} = E_0 \frac{(2E_0/mc^2)\sin^2(\theta/2)}{1 + (2E_0/mc^2)\sin^2(\theta/2)}.$$

1.15 An X-ray photon of wavelength $\lambda_0 = 1$ Å is incident on a free electron which is initially at rest. The photon is scattered at an angle $\theta = 30°$ from the incident direction.

(a) Calculate the Compton shift $\Delta\lambda$.
(b) What is the angle ϕ (measured from the incident photon direction) at which the electron recoils?
(c) What is the kinetic energy T_2 of the recoiling electron?
(d) What fraction of its original energy does the photon lose?

1.16 Consider the Compton scattering of a photon of wavelength λ_0 by a free electron moving with a momentum of magnitude P in the same direction as that of the incident photon.

(a) Show that in this case the Compton equation (1.42) becomes

$$\Delta\lambda = 2\lambda_0 \frac{(p_0 + P)c}{E - Pc} \sin^2\left(\frac{\theta}{2}\right)$$

where $p_0 = h/\lambda_0$ is the magnitude of the incident photon momentum, θ is the photon scattering angle and $E = (m^2c^4 + P^2c^2)^{1/2}$ is the initial electron energy.

(b) What is the maximum value of the electron momentum after the collision? Compare with the case $P = 0$ considered in the text.
(c) Show that if the free electron initially moves with a momentum of magnitude P in a direction opposite to that of the incident photon, the Compton shift becomes

$$\Delta\lambda = 2\lambda_0 \frac{(p_0 - P)c}{E + Pc} \sin^2\left(\frac{\theta}{2}\right).$$

1.17 Consider a photon of energy $E_0 = 2$ eV which is scattered through an angle $\theta = \pi$ from an electron having a kinetic energy $T = 10$ GeV (1 GeV $= 10^9$ eV) and moving initially in a direction opposite to that of the photon. What is the energy E_1 of the scattered photon?
(*Hint*: Use the result (c) of Problem 1.16.)

1.18 Using the infinite nuclear mass approximation in the Bohr model, calculate:

(a) the ionisation potential (in eV), and
(b) the wavelengths of the first three lines of the Lyman, Balmer and Paschen series,

for H, He$^+$ ($Z = 2$), Li^{2+} ($Z = 3$) and C^{5+} ($Z = 6$).

1.19 Consider a one-electron atom (or ion), the nucleus of which contains A nucleons (Z protons and $N = A - Z$ neutrons). The mass of that nucleus is given approximately by $M \simeq AM_p$, where $M_p \simeq 1.67 \times 10^{-27}$ kg is the proton mass. Using this value of M, obtain the relative correction $\Delta E/E$ to the Bohr energy levels due to the finite nuclear mass for the case of atomic hydrogen ($A = 1, Z = 1$), deuterium ($A = 2, Z = 1$), tritium ($A = 3, Z = 1$) and ^{4}He$^+$($A = 4, Z = 2$).

1.20 Calculate the difference in wavelengths $\Delta\lambda$ between the Balmer H$_\alpha$ lines in atomic hydrogen and deuterium.

1.21 Show that the fractional change in the frequency of a photon absorbed or emitted by an atom initially at rest is

$$\frac{\nu - \nu_0}{\nu} = \pm \frac{h\nu}{2Mc^2}$$

where M is the mass of the atom and ν_0 is the frequency of the transition uncorrected for the recoil of the atom. In the above equation the plus sign corresponds to the absorption and the minus sign to the emission of a photon.

1.22 In an experiment of the Franck–Hertz type, atomic hydrogen is bombarded by electrons. It is found that the current I is sharply reduced when the potential V_1 has the values 10.2 V and 12.1 V, and three different spectral lines are observed.

(a) To which spectral series do these lines belong? What are the quantum numbers of the initial and final states in the transition corresponding to each spectral line?
(b) What is the wavelength of each of these spectral lines?

1.23 A beam of silver atoms, for which $\mathcal{M}_z = \pm\mu_B$, passes through an inhomogeneous magnetic field, as in the Stern–Gerlach apparatus shown schematically in Fig. 1.15. The field gradient is $\partial B_z/\partial z = 10^3$ T m^{-1}, the length of the pole piece is $L = 0.1$ m, the distance to the screen is $l = 1$ m and the temperature of the oven is $T = 600$ K. Assuming that the velocity of the silver atoms is equal to the root-mean-square velocity of $(3kT/M)^{1/2}$ (where k is the Boltzmann constant and M is the mass of a silver atom) calculate the maximum separation ($P_2 P_3$ in Fig. 1.15) of the two beams on the screen.

1.24 At what energy (in eV) will the non-relativistic expression (1.99) for the de Broglie wavelength lead to a 1% relative error for (a) an electron and (b) a proton?

1.25 Calculate the de Broglie wavelength of

(a) a mass of 1 g moving at a velocity of 1 m s^{-1},
(b) a free electron having a kinetic energy of 200 eV,
(c) a free electron having a kinetic energy of 50 GeV (1 GeV = 10^9 eV), and
(d) a free proton having a kinetic energy of 200 eV.

(*Hint*: Use the relativistic formula in (c).)

1.26 What is the kinetic energy of

(a) an electron,
(b) an alpha particle (doubly ionised helium atom), and
(c) a neutron,

having a de Broglie wavelength of 1 Å?

1.27 Thermal neutrons are neutrons in thermal equilibrium with matter at a given temperature; they have a kinetic energy of $3kT/2$, where k is the Boltzmann constant and T is the absolute temperature. What is the energy (in eV) and the de Broglie wavelength (in Å) of a thermal neutron at room temperature ($T = 300$ K)?

1.28 Consider electrons striking at normal incidence the surface of a nickel crystal, and undergoing first-order Bragg diffraction at a Bragg angle $\theta_B = 30°$. Knowing that the atomic separation in a nickel crystal is $D = 2.15$ Å calculate

(a) the de Broglie wavelength of the electrons, and
(b) the kinetic energy of the electrons.

1.29 The spacing of the Bragg planes in a NaCl crystal is $d = 2.82$ Å. Calculate the angular positions of the first- and second-order diffraction maximum for 100 eV electrons incident on the crystal surface at the Bragg angle θ_B.

1.30 A crystal has a set of Bragg planes separated by a distance $d = 1.1$ Å. What is the highest-order Bragg diffraction (i.e. the highest value of n in equation (1.101)) for incident neutrons having a kinetic energy of 0.5 eV?

(*Hint*: The maximum Bragg angle that can be attained is 90°.)

2 The wave function and the uncertainty principle

2.1 Wave–particle duality   52
2.2 The interpretation of the wave function   56
2.3 Wave functions for particles having a definite momentum   58
2.4 Wave packets   60
2.5 The Heisenberg uncertainty principle   69
    Problems   76

In the previous chapter we discussed a number of phenomena which cannot be explained within the framework of classical physics. As a result of this experimental evidence, revolutionary concepts had to be introduced such as those of *quantisation* and of *wave–particle duality*.

Attempts to construct a theoretical structure which incorporates these concepts in a satisfactory way met first with great difficulties, until a new theory, called *quantum mechanics*, was elaborated between the years 1925 and 1930. Quantum mechanics provides a consistent description of matter on the microscopic scale, and can be considered as one of the greatest intellectual achievements of the twentieth century. Two equivalent formulations of the theory were proposed at nearly the same time. The first, known as *matrix mechanics*, was developed in the years 1925 and 1926 by W. Heisenberg, M. Born and P. Jordan. In this approach, only physically observable quantities appear, and to each physical quantity the theory associates a matrix. These matrices obey a non-commutative algebra, which is the essential difference between matrix mechanics and classical mechanics. The second form of quantum mechanics, called *wave mechanics*, was proposed in 1925 by E. Schrödinger, following the ideas put forward in 1923 by L. de Broglie about matter waves. The equivalence of matrix mechanics and wave mechanics was proved in 1926 by Schrödinger. In fact, both matrix mechanics and wave mechanics are particular forms of a general formulation of quantum mechanics, which was developed by P. A. M. Dirac in 1930.

The general formulation of quantum mechanics, as well as matrix mechanics, requires a certain amount of abstract mathematics, and hence we shall defer discussion of it until Chapter 5. Wave mechanics, on the other hand, is more suitable for a first contact with quantum theory, and this is the approach which we shall use in most of this book. In this chapter we shall discuss the fundamental ideas underlying quantum mechanics in their simplest form. We begin with an analysis of wave–particle duality, introducing the notion of a wave function and its probabilistic interpretation. We then go on to construct wave functions corresponding to particles having a well-defined momentum, and to obtain localised wave functions by superposing plane waves into wave packets. Following this, we discuss the Heisenberg uncertainty relations which impose limits on the accuracy of simultaneous measurements of pairs of 'complementary' variables, such as position and momentum, or time and energy.

2.1 Wave–particle duality

In Chapter 1 we discussed several experiments which demonstrate conclusively that material particles possess wave-like properties, exhibiting interference and diffraction effects. On the other hand, we saw that electromagnetic radiation, which had been known for a long time to exhibit wave properties, can also show a particle-like behaviour. We shall now analyse this wave–particle duality in more detail.

Let us consider an idealised experiment in which monoenergetic particles, for example electrons, emitted by a source are directed on to a screen S_1 containing two slits A and B (see Fig. 2.1). At some distance beyond this screen a second screen S_2 is placed, incorporating detectors which can record each electron striking the screen S_2 at a given point. On detection, every electron exhibits purely particle-like properties, that is its mass and charge are localised, being never spread over more than one detector at a given time. In contrast, if after some time the total number of electrons arriving at the screen S_2 is plotted as a function of position, a diffraction pattern is found, which is characteristic of waves (see Fig. 2.2(a)). Thus, in a single experiment, both the particle and wave aspects of the electron are exhibited. A realisation of this ideal experiment by Jönsson has been described in Chapter 1. A similar experiment

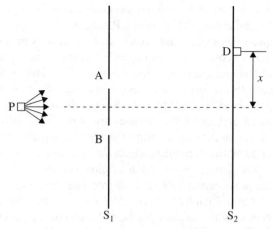

Figure 2.1 The two-slit experiment. Monoenergetic particles from a source at P fall on a screen S_1 containing slits at A and B. Detectors are placed on a second screen S_2 to record the number of particles arriving at each point. A particle detector D is indicated at a position x.

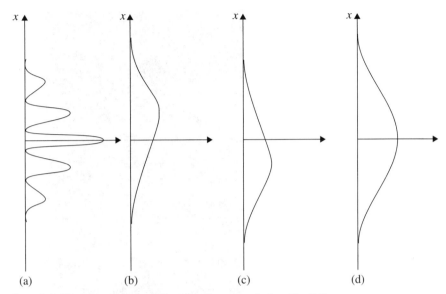

Figure 2.2 The intensity at a position x on the screen S_2 (see Fig. 2.1).
(a) A diffraction pattern characteristic of waves.
(b) The intensity with slit A open and B closed.
(c) The intensity with slit B open and A closed.
(d) An intensity distribution characteristic of classical particles when A and B are both open.

can be carried out with light. In this case the light can be detected by the photoelectric effect, showing its particle (photon) aspect, while the recorded intensity displays the diffraction pattern characteristic of the wave theory of light.

It might at first be supposed that the diffraction pattern is in some way due to interference between electrons (or photons) passing through the two slits. However, if the incident intensity is reduced until at any instant there is no more than one particle between the source and the detecting screen, the interference pattern is still accumulated. This was demonstrated originally in 1909 by G. I. Taylor, who photographed the diffraction pattern formed by the shadow of a needle, using a very weak source such that the exposures lasted for months. It can be concluded that interference does not occur between photons, but is a property of a single photon. This was confirmed in more recent experiments performed in 1989 by A. Aspect, P. Grangier and G. Roger. Material particles such as electrons exhibit a similar behaviour, as was demonstrated directly in 1989 by A. Tonomura, J. Endo, T. Matsuda, T. Kawasaki and H. Ezawa in two-slit experiments in which the accumulation of the interference pattern due to incoming single electrons was observed (see Fig. 2.3). It should be noted that if one slit is closed in a two-slit experiment, the diffraction pattern does not appear, so we may infer that when both slits are open the particle is not localised before it is detected, and hence must be considered as having passed through both slits!

54 ■ The wave function and the uncertainty principle

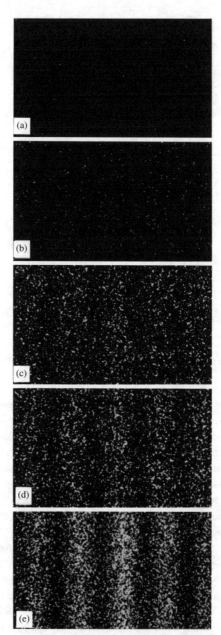

Figure 2.3 Buildup of an interference pattern by accumulating single electrons in the two-slit experiment of Tonomura *et al.*
(a) Number of electrons = 10; (b) number of electrons = 100;
(c) number of electrons = 3000; (d) number of electrons = 20 000; and
(e) number of electrons = 70 000.
(By courtesy of A. Tonomura.)

Another way of expressing these facts is to say that while in transit the electron or photon behaves like a wave, manifesting its particle-like property only on detection. This is of course in complete contradiction to the classical viewpoint, which would lead us to suppose that each particle being indivisible must pass either through one slit or the other. Let us put this to the test by detecting the particles as they pass through the slits. We can now record the particles which have passed through slit A and entered the detectors on screen S_2. Since all these particles passed through slit A, slit B might as well have been closed, in which case the intensity distribution will not show diffraction but will be as illustrated in Fig. 2.2(b). Similarly, if slit B is open and A is closed, the intensity distribution is that shown in Fig. 2.2(c). If we add the intensity distributions of Figs. 2.2(b) and 2.2(c) we obtain the intensity pattern shown in Fig. 2.2(d), which is very different from the diffraction pattern obtained in the absence of any knowledge about which slit the particles passed through (see Fig. 2.2(a)). Hence, if the particle nature of an electron, a photon, etc., is established by monitoring its trajectory, it cannot simultaneously behave like a wave. The wave and particle aspects of electrons, photons, etc., are *complementary*.

Let us now return to the case in which there is only one particle at a time transiting the apparatus and both slits are open. The place on the screen S_2 at which a given particle will be detected cannot be predicted, because if it could be predicted this would be equivalent to determining the trajectory, which we have seen would eliminate the diffraction pattern. What is predictable is the intensity distribution which builds up after a large number of individual events have occured. This suggests that for an individual particle the process is of a *statistical* nature, so that one can only determine the probability P that a particle will hit the screen S_2 at a certain point. By probability in this context we mean the number of times that an event occurs divided by the total number of events. The intensity of the pattern formed on the screen S_2, in the present case, is then proportional to the probability P.

In the classical theory of light, the intensity of light at each point is determined by the square of the amplitude of a wave. For example, in Young's two-slit experiment, the light intensity on the recording screen is given by the square of the amplitude of the wave formed by the *superposition* of the secondary waves arising from each slit. This classical wave theory cannot of course be used as it stands because it does not account for the particle aspect of light. However, it suggests, by analogy, that in quantum mechanics a *wave function* or *state function* $\Psi(x, y, z, t)$ can be introduced, which plays the role of a *probability amplitude*. We shall see later that in general the wave function Ψ is a *complex* quantity. We then expect that the probability $P(x, y, z, t)$ of finding the particle at a particular point within a volume V about the point with coordinates (x, y, z) at time t is proportional to $|\Psi|^2$:

$$P(x, y, z, t) \propto |\Psi(x, y, z, t)|^2. \tag{2.1}$$

Since probabilities are *real positive* numbers, we have associated in (2.1) the probability P with the square of the modulus of the wave function Ψ.

Let Ψ_A be the wave function at a particular point on the screen S_2 corresponding to waves spreading from the slit A. Similarly, let Ψ_B be the wave function at the same point corresponding to waves spreading from the slit B. The two intensity distributions of Fig. 2.2(b) and (c) corresponding to experiments performed with only one open slit are determined respectively by the probability distributions

$$P_A \propto |\Psi_A|^2, \qquad P_B \propto |\Psi_B|^2. \tag{2.2}$$

On the other hand, when both slits are open, the wave function Ψ is taken to be the sum of the two contributions Ψ_A and Ψ_B,

$$\Psi = \Psi_A + \Psi_B. \tag{2.3}$$

The corresponding probability distribution

$$P \propto |\Psi_A + \Psi_B|^2 \tag{2.4}$$

then determines the intensity pattern illustrated in Fig. 2.2(a). It is important to notice that in (2.4) the probability amplitudes Ψ_A and Ψ_B have been added and not the probabilities P_A and P_B. If the latter were the case there would be no possibility of obtaining interference patterns characteristic of a wave theory.

2.2 The interpretation of the wave function

In analysing the two-slit experiment, we introduced the concept that the probability of finding a particle at a given location is proportional to the square of the modulus of the wave function associated with the particle. This concept may be restated more precisely in the form of a fundamental assumption made by M. Born in 1926, which can be formulated in the following way. Let us imagine a very large number of identical, independent systems, each of them consisting of a single particle moving under the influence of some given external force. All these systems are identically prepared, and this ensemble is assumed to be described by a single wave function $\Psi(x, y, z, t)$ which contains all the information that can be known about them. It is then postulated that if measurements of the position of the particle are made on each of the systems, the probability (that is the statistical frequency) of finding the particle within the volume element $d\mathbf{r} \equiv dx\,dy\,dz$ about the point $\mathbf{r} \equiv (x, y, z)$ at the time t is

$$P(\mathbf{r}, t)d\mathbf{r} = |\Psi(\mathbf{r}, t)|^2 d\mathbf{r} \tag{2.5}$$

so that

$$P(\mathbf{r}, t) = |\Psi(\mathbf{r}, t)|^2 = \Psi^*(\mathbf{r}, t)\Psi(\mathbf{r}, t) \tag{2.6}$$

is the *position probability density*. Thus we see that the interpretation of the wave function introduced by Born is a statistical one. For convenience, we shall often speak of the wave function associated with a particular system, but it must always be understood that this is shorthand for the wave function associated with an ensemble of identical and identically prepared systems, as required by the statistical nature of the theory.

Since the probability of finding the particle somewhere must be unity, we deduce from (2.5) that the wave function $\Psi(\mathbf{r}, t)$ should be *normalised to unity*, so that

$$\int |\Psi(\mathbf{r}, t)|^2 d\mathbf{r} = 1 \tag{2.7}$$

where the integral extends over all space. A wave function for which the integral on the left of (2.7) is finite is said to be *square integrable*: such a wave function can always be normalised to unity by multiplying it by an appropriate complex constant.

It is important to notice that since $|\Psi(\mathbf{r}, t)|^2$ is the physically significant quantity, two wave functions which differ from each other by a constant multiplicative factor of modulus one (that is, a constant phase factor of the form $\exp(i\alpha)$, where α is a real number) are equivalent, and satisfy the same normalisation condition.

The superposition principle

As we have seen in the previous section, in order to account for interference effects, it must be possible to *superpose* wave functions. This means that if one possible state of an ensemble of identical systems is described by a wave function Ψ_1 and another state of this ensemble by a wave function Ψ_2, then any *linear combination*

$$\Psi = c_1 \Psi_1 + c_2 \Psi_2 \tag{2.8}$$

where c_1 and c_2 are complex constants, is also a wave function describing a possible state of the ensemble.

Let us write the (complex) wave functions Ψ_1 and Ψ_2 in the form

$$\Psi_1 = |\Psi_1| e^{i\alpha_1}, \qquad \Psi_2 = |\Psi_2| e^{i\alpha_2}. \tag{2.9}$$

Using (2.8), we find that the square of the modulus of Ψ is given by

$$|\Psi|^2 = |c_1 \Psi_1|^2 + |c_2 \Psi_2|^2 + 2\mathrm{Re}\{c_1 c_2^* |\Psi_1||\Psi_2| \exp[i(\alpha_1 - \alpha_2)]\} \tag{2.10}$$

so that, in general, $|\Psi|^2 \neq |c_1 \Psi_1|^2 + |c_2 \Psi_2|^2$, in keeping with the discussion of Section 2.1. It is worth stressing that although the quantity $|\Psi|^2$ is unaffected if Ψ is multiplied by an *overall* phase factor $\exp(i\alpha)$ (where α is a real constant) it does depend on the *relative* phase $(\alpha_1 - \alpha_2)$ of Ψ_1 and Ψ_2 through the third term on the right of (2.10), which is an *interference* term.

Finally, we emphasise that unlike classical waves (such as sound waves or water waves) the wave function $\Psi(\mathbf{r}, t)$ is an *abstract* quantity, the interpretation of which is of a *statistical* nature. This wave function is assumed to provide a *complete* description of the dynamical state of an ensemble. Indeed, we shall see later that the knowledge of the wave function enables one to predict for each dynamical variable (position, momentum, energy and so on) a *statistical distribution* of values obtained in measurements.

2.3 Wave functions for particles having a definite momentum

In this section we begin to investigate how wave functions can be found, considering the simple case of *free* particles. The experiments exhibiting the corpuscular nature of the electromagnetic radiation, which we discussed in Chapter 1, require that with the electromagnetic field one associates a particle, the photon, whose energy E and magnitude p of momentum are related to the frequency ν and wavelength λ of the electromagnetic radiation by

$$E = h\nu, \qquad p = \frac{h}{\lambda}. \tag{2.11}$$

On the other hand, we have seen in Section 1.6 that de Broglie was led to associate matter waves with particles in such a way that the frequency ν and the wavelength λ of the wave were linked with the particle energy E and the magnitude p of its momentum by the same relations (2.11). The de Broglie relation $\lambda = h/p$ was confirmed by the results of a number of experiments exhibiting the wave nature of matter. Following de Broglie, we shall assume that the relations (2.11) hold for all types of particles and field quanta. Introducing the angular frequency $\omega = 2\pi\nu$, the wave number $k = 2\pi/\lambda$ and the reduced Planck constant $\hbar = h/2\pi$, we may write the relations (2.11) in the more symmetric form

$$E = \hbar\omega, \qquad p = \hbar k. \tag{2.12}$$

Let us consider a *free* particle of mass m, moving along the x-axis with a definite momentum $\mathbf{p} = p_x \hat{\mathbf{x}}$, where $\hat{\mathbf{x}}$ is a unit vector along the x-axis, and a corresponding energy E. Assuming that $p_x > 0$, so that the particle moves in the positive x-direction, we associate with this particle a wave travelling in the same direction with a fixed wave number k. Such a wave is a *plane wave* and can be written as

$$\Psi(x, t) = A \exp\{i[kx - \omega(k)t]\} \tag{2.13a}$$

where A is a constant. This plane wave has a wavelength $\lambda = 2\pi/k$ and an angular frequency ω. Since from (2.12) $k = p/\hbar$ (with $p = p_x$) and $\omega = E/\hbar$, the wave function (2.13a) can be expressed as

$$\Psi(x, t) = A \exp\{i[p_x x - E(p_x)t]/\hbar\}, \tag{2.13b}$$

In writing (2.13a,b) we have taken $\omega(k)$ and $E(p_x)$ as functions to be specified later. We note that the wave function (2.13) satisfies the two relations

$$-i\hbar \frac{\partial}{\partial x} \Psi = p_x \Psi \tag{2.14}$$

and

$$i\hbar \frac{\partial}{\partial t} \Psi = E\Psi \tag{2.15}$$

the significance of which will emerge shortly.

2.3 Wave functions for particles having a definite momentum

This one-dimensional treatment is easily extended to three dimensions. To a free particle of mass m, having a well-defined momentum $\mathbf{p}$ and an energy E, we now associate a plane wave

$$\begin{aligned}\Psi(\mathbf{r}, t) &= A \exp\{i[\mathbf{k}\cdot\mathbf{r} - \omega(k)t]\} \\ &= A \exp\{i[\mathbf{p}\cdot\mathbf{r} - E(p)t]/\hbar\}\end{aligned} \qquad (2.16)$$

where the propagation vector (or wave vector) $\mathbf{k}$ is related to the momentum $\mathbf{p}$ by

$$\mathbf{p} = \hbar\mathbf{k} \qquad (2.17)$$

with

$$k = |\mathbf{k}| = \frac{|\mathbf{p}|}{\hbar} = \frac{2\pi}{\lambda} \qquad (2.18)$$

and the angular frequency ω is related to the energy by $\omega = E/\hbar$. Again, the functions $\omega(k)$ and $E(p)$ will be specified later. The equation (2.15) remains unchanged for the plane wave (2.16), while (2.14) is now replaced by its obvious generalisation

$$-i\hbar\nabla\Psi = \mathbf{p}\Psi \qquad (2.19)$$

where ∇ is the gradient operator, having Cartesian components $(\partial/\partial x, \partial/\partial y, \partial/\partial z)$. The relations (2.15) and (2.19) show that for a free particle the energy and momentum can be represented by the differential operators

$$E_{\mathrm{op}} = i\hbar\frac{\partial}{\partial t}, \qquad \mathbf{p}_{\mathrm{op}} = -i\hbar\nabla \qquad (2.20)$$

acting on the wave function Ψ. It is a *postulate* of wave mechanics that when the particle is not free the dynamical variables E and $\mathbf{p}$ are still represented by these differential operators.

According to the discussion of Section 2.2, wave functions should be normalised to unity if the probability interpretation is to be maintained. For one-dimensional systems, the normalisation condition (2.7) reduces to

$$\int_{-\infty}^{+\infty} |\Psi(x, t)|^2 \mathrm{d}x = 1. \qquad (2.21)$$

However, the plane wave (2.13) does not satisfy this requirement, since the integral on the left of (2.21) is given in this case by

$$\int_{-\infty}^{+\infty} |\Psi(x, t)|^2 \mathrm{d}x = |A|^2 \int_{-\infty}^{+\infty} \mathrm{d}x \qquad (2.22)$$

and hence does not exist. Similarly, the three-dimensional plane wave (2.16) cannot be normalised according to (2.7). There are two ways out of this difficulty. The first is to give up the concept of *absolute* probabilities when dealing with wave functions such as (2.13) or (2.16) which are not square integrable. Instead, $|\Psi(\mathbf{r}, t)|^2 \mathrm{d}\mathbf{r}$ is then interpreted as the *relative* probability of finding the particle at time t in a volume element $\mathrm{d}\mathbf{r}$ centred about $\mathbf{r}$, so that the ratio $|\Psi(\mathbf{r}_1, t)|^2/|\Psi(\mathbf{r}_2, t)|^2$ gives the probability of finding the particle within a volume element centred around $\mathbf{r} = \mathbf{r}_1$, compared with that of finding it within the same volume element at $\mathbf{r} = \mathbf{r}_2$. For the particular case of the plane wave (2.16), we see that $|\Psi|^2 = |A|^2$, so that there is an equal chance of finding the particle at any point. The plane wave (2.16) therefore describes the *idealised* situation of a free particle having a perfectly well-defined momentum, but which is completely '*delocalised*'. This suggests a second way out of the difficulty, which is to give up the requirement that the free particle should have a precisely defined momentum, and to superpose plane waves corresponding to different momenta to form a localised *wave packet*, which can be normalised to unity. It is to this question that we now turn our attention.

2.4 Wave packets

We have seen in the preceding section that plane waves such as (2.13) or (2.16) associated with free particles having a definite momentum are completely delocalised. To describe a particle which is confined to a certain spatial region, a *wave packet* can be formed by superposing plane waves of different wave numbers. Of course, in this case the momentum no longer has a precise value, but we shall construct a wave packet which 'represents' a particle having fairly precise values of both momentum and position.

Let us begin by considering the one-dimensional case. In order to describe a free particle confined to a region of the x-axis, we superpose plane waves of the form (2.13), where we now allow $p_x = \hbar k$ to be either positive or negative. The most general superposition of this kind is then given by the integral

$$\Psi(x,t) = (2\pi\hbar)^{-1/2} \int_{-\infty}^{+\infty} e^{i[p_x x - E(p_x)t]/\hbar} \phi(p_x) dp_x \tag{2.23}$$

where the factor $(2\pi\hbar)^{-1/2}$ in front of the integral has been chosen for future convenience. The function $\phi(p_x)$ is the amplitude of the plane wave corresponding to the momentum p_x. In general it is a complex function, but it is sufficient for our present purposes to discuss only the case for which $\phi(p_x)$ is real.

Let us assume that $\phi(p_x)$ is sharply peaked about some value $p_x = p_0$, falling rapidly to zero outside an interval $(p_0 - \Delta p_x, p_0 + \Delta p_x)$. Writing (2.23) in the form

$$\Psi(x,t) = (2\pi\hbar)^{-1/2} \int_{-\infty}^{+\infty} e^{i\beta(p_x)/\hbar} \phi(p_x) dp_x \tag{2.24}$$

where

$$\beta(p_x) = p_x x - E(p_x) t \tag{2.25}$$

we see that $|\Psi(x,t)|$ is largest when $\beta(p_x)$ is nearly constant in the vicinity of $p_x = p_0$. Indeed, if $\beta(p_x)$ were varying significantly over the interval $(p_0 - \Delta p_x, p_0 + \Delta p_x)$, the factor $\exp[i\beta(p_x)/\hbar]$ would oscillate rapidly, so that the value of the integral on the right of (2.24) would be small. Thus $|\Psi(x,t)|$ will only be significant in a limited region, its maximum value occurring when the *stationary phase condition*

$$\left[\frac{d\beta(p_x)}{dp_x}\right]_{p_x=p_0} = 0 \tag{2.26}$$

is satisfied. This condition determines the *centre of the wave packet*, which upon using (2.25) is seen to travel according to the law

$$x = v_g t \tag{2.27}$$

where

$$v_g = \left[\frac{dE(p_x)}{dp_x}\right]_{p_x=p_0}. \tag{2.28}$$

It follows from (2.27) that the centre of the wave packet moves with the constant velocity v_g, which is known as the *group velocity* of the packet. From (2.28) and the fact that $E = \hbar\omega$, and $p_x = \hbar k$, we see that the group velocity can also be written as

$$v_g = \left[\frac{d\omega(k)}{dk}\right]_{k=k_0} \tag{2.29}$$

with $k_0 = p_0/\hbar$. We remark that this velocity is, in general, different from the *phase velocity* v_{ph}, which is the velocity of propagation of the individual plane waves (2.13) and is given for a particular plane wave $A \exp[i(k_0 x - \omega(k_0)t)]$ by

$$v_{ph} = \frac{\omega(k_0)}{k_0} = \frac{E(p_0)}{p_0}. \tag{2.30}$$

In the macroscopic limit the motion of a particle must be governed by the laws of classical mechanics, in accordance with the correspondence principle (see Section 1.4). In this limit the extension of the wave packet is negligible so that the group velocity v_g can be identified with the classical velocity $v = p_0/m$ of the particle

$$v_g = v = \frac{p_0}{m}. \tag{2.31}$$

Combining this result with (2.28) allows us to determine the functional dependence of $E(p_x)$ on p_x. We have

$$\frac{dE(p_x)}{dp_x} = \frac{p_x}{m} \tag{2.32}$$

so that $E(p_x) = p_x^2/2m +$ constant. We may set the constant of integration equal to zero because the zero of energy can be chosen arbitrarily, only *energy differences* being of physical interest. Hence we have

$$E(p_x) = \frac{p_x^2}{2m}. \tag{2.33}$$

It should be noted that since $E = h\nu$, the absolute value of the frequency has no physical significance in quantum mechanics, in contrast with classical wave theory (e.g. sound waves). We remark that since in our case $E(p_0) = p_0^2/2m$, the phase velocity (2.30) is given by

$$v_{\text{ph}} = \frac{p_0^2/2m}{p_0} = \frac{p_0}{2m} = \frac{v_g}{2}. \tag{2.34}$$

Let us return to the expression (2.23) of the wave packet and express $E(p_x) = p_x^2/2m$ in the form

$$\begin{aligned} E(p_x) &= \frac{p_0^2}{2m} + \frac{p_0}{m}(p_x - p_0) + \frac{(p_x - p_0)^2}{2m} \\ &= E(p_0) + v_g(p_x - p_0) + \frac{(p_x - p_0)^2}{2m}. \end{aligned} \tag{2.35}$$

Since the function $\phi(p_x)$ in (2.23) is negligible except in the interval $(p_0 - \Delta p_x, p_0 + \Delta p_x)$ we can neglect the third term on the right of (2.35), provided t is small enough so that

$$\frac{1}{2m\hbar}(\Delta p_x)^2 t \ll 1. \tag{2.36}$$

Indeed, if the condition (2.36) is satisfied, the quantity $\exp[-i(p_x - p_0)^2 t/2m\hbar]$ which occurs in the integrand on the right of (2.23) is approximately equal to unity. Making this approximation, equation (2.23) reduces to

$$\Psi(x, t) = e^{i[p_0 x - E(p_0)t]/\hbar} F(x, t) \tag{2.37}$$

where

$$F(x, t) = (2\pi\hbar)^{-1/2} \int_{-\infty}^{+\infty} e^{i(p_x - p_0)(x - v_g t)/\hbar} \phi(p_x) dp_x. \tag{2.38}$$

The wave packet (2.37) is the product of a plane wave of wavelength $\lambda_0 = h/|p_0|$ and angular frequency $\omega_0 = E(p_0)/\hbar$ times a *modulating* amplitude or *envelope* function $F(x, t)$ such that $|\Psi(x, t)|^2 = |F(x, t)|^2$. Since

$$F(x, t = 0) = F(x + v_g t, t) \tag{2.39}$$

this envelope function propagates *without change of shape* with the group velocity v_g (see Fig. 2.4). It should be borne in mind that this is only true for times t satisfying the condition (2.36); at later times the shape of the wave packet will change as it propagates.

Fourier transforms and momentum space wave function

Looking back at the wave packet (2.23), defining $\psi(x) \equiv \Psi(x, t = 0)$ and using the results of Appendix A, we see that the functions

$$\psi(x) = (2\pi\hbar)^{-1/2} \int e^{ip_x x/\hbar} \phi(p_x) dp_x \tag{2.40}$$

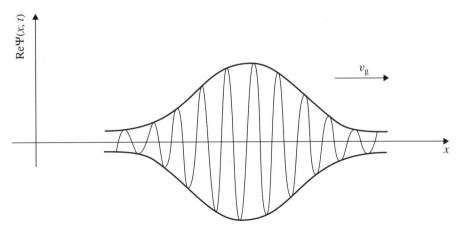

Figure 2.4 The function $\text{Re}\Psi(x, t)$ for a wave packet propagating along the x-axis, with a group velocity v_g.

and

$$\phi(p_x) = (2\pi\hbar)^{-1/2} \int e^{-ip_x x/\hbar} \psi(x) dx \quad (2.41)$$

are *Fourier transforms* of each other. More generally, at time t, we can introduce a function $\Phi(p_x, t)$ such that

$$\Psi(x, t) = (2\pi\hbar)^{-1/2} \int_{-\infty}^{+\infty} e^{ip_x x/\hbar} \Phi(p_x, t) dp_x \quad (2.42)$$

and

$$\Phi(p_x, t) = (2\pi\hbar)^{-1/2} \int_{-\infty}^{+\infty} e^{-ip_x x/\hbar} \Psi(x, t) dx \quad (2.43)$$

are also mutual Fourier transforms. The function $\Phi(p_x, t)$ is called the *wave function in momentum space*, and we see that $\phi(p_x) = \Phi(p_x, t = 0)$. The definition of the momentum space wave function given by (2.43) is completely general and holds for all types of wave function $\Psi(x, t)$, including the free particle wave packets which we have been considering.

From Parseval's theorem (see equation (A.43) of Appendix A), we infer that if the wave function $\phi(p_x)$ is normalised to unity in the sense that

$$\int_{-\infty}^{+\infty} |\phi(p_x)|^2 dp_x = 1 \quad (2.44)$$

then the wave function $\psi(x) \equiv \Psi(x, t = 0)$ given by (2.40) is also normalised to unity. Moreover, once $\Psi(x, t)$ is normalised to unity at $t = 0$, it remains normalised to unity at all times. Indeed,

$$\int_{-\infty}^{+\infty} \Psi^*(x,t)\Psi(x,t)\mathrm{d}x = (2\pi\hbar)^{-1} \int_{-\infty}^{+\infty} \mathrm{d}x \int_{-\infty}^{+\infty} \mathrm{d}p_x \int_{-\infty}^{+\infty} \mathrm{d}p'_x \mathrm{e}^{\mathrm{i}[(p_x-p'_x)x]/\hbar}$$
$$\times \mathrm{e}^{-\mathrm{i}[E(p_x)-E(p'_x)]t/\hbar} \phi^*(p'_x)\phi(p_x)$$
$$= \int_{-\infty}^{+\infty} \mathrm{d}p_x \int_{-\infty}^{+\infty} \mathrm{d}p'_x \delta(p_x - p'_x) \mathrm{e}^{-\mathrm{i}[E(p_x)-E(p'_x)]t/\hbar} \phi^*(p'_x)\phi(p_x)$$
$$= \int_{-\infty}^{+\infty} \phi^*(p_x)\phi(p_x)\mathrm{d}p_x$$
$$= 1 \quad (2.45)$$

where in the second line we have introduced the Dirac delta function $\delta(p_x - p'_x)$ such that (see (A.18))

$$(2\pi\hbar)^{-1} \int_{-\infty}^{+\infty} \mathrm{e}^{\mathrm{i}[(p_x-p'_x)x]/\hbar}\mathrm{d}x = (2\pi)^{-1} \int_{-\infty}^{+\infty} \mathrm{e}^{\mathrm{i}[(p_x-p'_x)x']}\mathrm{d}x'$$
$$= \delta(p_x - p'_x) \quad (2.46)$$

and used the property (A.26) in the third line. The result (2.45) expresses the *conservation of probability*, which is clearly a requirement of the theory. We note that if $\Psi(x,t)$ is normalised to unity, so is also the momentum space wave function $\Phi(p_x,t)$.

Gaussian wave packet

As an example, we shall now study the particular case in which the function $\phi(p_x)$ is a Gaussian function peaked about the value p_0

$$\phi(p_x) = C \exp\left[-\frac{(p_x - p_0)^2}{2(\Delta p_x)^2}\right] \quad (2.47)$$

where Δp_x, which we call the width of the distribution in p_x, is a constant such that $|\phi(p_x)|^2$ drops to $1/\mathrm{e}$ of its maximum value at $p_x = p_0 \pm \Delta p_x$ (see Fig. 2.5(a)). The constant C in (2.47) is a *normalisation constant* which we shall choose in such a way that the normalisation condition (2.44) is satisfied. Using the known result

$$\int_{-\infty}^{+\infty} \mathrm{e}^{-\alpha u^2}\mathrm{e}^{-\beta u}\mathrm{d}u = \left(\frac{\pi}{\alpha}\right)^{1/2}\mathrm{e}^{\beta^2/4\alpha} \quad (2.48)$$

with $u = p_x - p_0$, $\alpha = (\Delta p_x)^{-2}$ and $\beta = 0$, we have

$$\int_{-\infty}^{+\infty} |\phi(p_x)|^2 \mathrm{d}p_x = |C|^2 \pi^{1/2} \Delta p_x. \quad (2.49)$$

The normalisation condition (2.44) is therefore fulfilled by taking

$$|C|^2 = \pi^{-1/2}(\Delta p_x)^{-1}. \quad (2.50)$$

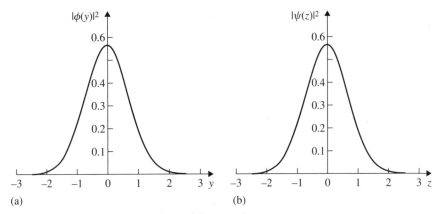

Figure 2.5 (a) The function $|\phi(y)|^2 = \pi^{-1/2}\exp(-y^2)$, where $y = (p_x - p_0)/\Delta p_x$.
(b) The function $|\psi(z)|^2 = \pi^{-1/2}\exp(-z^2)$, where $z = (\Delta p_x/\hbar)x$.

The constant C is determined by (2.50) apart from a phase factor of unit modulus, which can be set equal to one, so that we can take C to be given by

$$C = \pi^{-1/4}(\Delta p_x)^{-1/2}. \tag{2.51}$$

Substituting $\phi(p_x)$ given by (2.47) into (2.40) and using (2.48) and (2.51), we find that

$$\psi(x) \equiv \Psi(x, t=0) = \pi^{-1/4}\hbar^{-1/2}(\Delta p_x)^{1/2}e^{ip_0 x/\hbar}e^{-(\Delta p_x)^2 x^2/2\hbar^2}. \tag{2.52}$$

Apart from the phase factor $\exp(ip_0 x/\hbar)$, this function is again a Gaussian. We remark that $|\psi(x)|^2$ has a maximum at $x = 0$ and falls to $1/e$ of its maximum value at $x = \pm\Delta x$, where $\Delta x = \hbar/\Delta p_x$ is the width of the distribution in the x variable (see Fig. 2.5(b)). Given the above definitions of the 'widths' Δx and Δp_x, we see that for a Gaussian wave packet $\Delta x \Delta p_x = \hbar$. Thus if we decrease Δp_x so that the wave function in momentum space, $\phi(p_x)$, is more sharply peaked about $p_x = p_0$, then Δx will increase and $\psi(x)$ becomes increasingly 'delocalised'. Conversely, if Δp_x is increased, so that $\phi(p_x)$ is 'delocalised' in momentum space, then $\psi(x)$ will become more strongly localised about $x = 0$. We shall return shortly to this important property, which is of a general nature. We note from (2.40) and (A.26) that in the limit in which $\phi(p_x)$ is the delta function $\delta(p_x - p_0)$, the wave function $\psi(x)$ becomes the plane wave $(2\pi\hbar)^{-1/2}\exp(ip_0 x/\hbar)$, which is completely delocalised.

Let us now examine how the Gaussian wave packet evolves in time. Using (2.23), (2.33), (2.47), (2.51) and (2.48) we find that

$$\Psi(x, t) = \pi^{-1/4}\left[\frac{\Delta p_x/\hbar}{1 + i(\Delta p_x)^2 t/m\hbar}\right]^{1/2}$$
$$\times \exp\left[\frac{ip_0 x/\hbar - (\Delta p_x/\hbar)^2 x^2/2 - ip_0^2 t/2m\hbar}{1 + i(\Delta p_x)^2 t/m\hbar}\right] \tag{2.53}$$

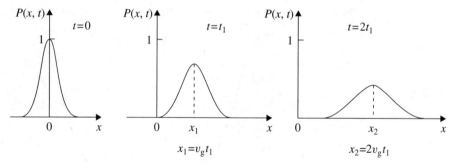

Figure 2.6 The position probability density $P(x,t) = |\Psi(x,t)|^2$ for a Gaussian wave packet at times $t = 0$, $t = t_1$ and $t = 2t_1$, plotted in arbitrary units.

and the corresponding position probability density is

$$P(x,t) = |\Psi(x,t)|^2 = \pi^{-1/2} \frac{\Delta p_x/\hbar}{[1 + (\Delta p_x)^4 t^2/m^2\hbar^2]^{1/2}}$$

$$\times \exp\left[-\frac{(\Delta p_x/\hbar)^2 (x - v_g t)^2}{1 + (\Delta p_x)^4 t^2/m^2\hbar^2}\right] \quad (2.54)$$

where we recall that $v_g = p_0/m$ is the group velocity of the packet. It is clear from (2.54) that the centre of the wave packet moves uniformly with the velocity v_g. The width of the packet, defined so that $P(x,t)$ falls to $1/e$ of its maximum value at the points $x - v_g t = \pm \Delta x$, is given by

$$\Delta x(t) = \frac{\hbar}{\Delta p_x}\left[1 + \frac{(\Delta p_x)^4}{m^2\hbar^2} t^2\right]^{1/2} \quad (2.55)$$

and hence increases with time. However, if the time is sufficiently small so that

$$t \ll t_1 = \frac{m\hbar}{(\Delta p_x)^2} \quad (2.56)$$

the second term in brackets in (2.55) is very small and the wave packet propagates without changing its width appreciably. This is in accordance with our general discussion (see (2.36)–(2.39)). The spreading of the probability density (2.54) is illustrated in Fig. 2.6, where $P(x,t)$ is shown for times $t = 0$, $t = t_1$ and $t = 2t_1$.

To take a particular case, consider a Gaussian wave packet associated with an electron which at time $t = 0$ is localised to within a distance 10^{-10} m characteristic of atomic dimensions, so that $\Delta p_x = \hbar/\Delta x \simeq 10^{-24}$ kg m s^{-1}. According to (2.55) the wave packet will have spread to twice its size at time $t = \sqrt{3}t_1 \simeq 10^{-16}$ s (see Problem 2.7). On the other hand, for a macroscopic object having a mass of 1 g, whose position is initially defined within an accuracy $\Delta x \simeq 10^{-6}$ m, we find that the width of the packet doubles after a time $t \simeq 10^{19}$ s, which is larger than the estimated age of the universe.

A word of caution should be said about the interpretation of these results. Let us suppose that we have a wave packet representing an electron, which is spread so that the width of the packet is, for example, $\Delta x = 1$ km at a given time. If an electron detector is placed at a particular position at that time, it will record the presence or absence of the 'complete' electron, since when the electron manifests itself in the detection process it is indivisible. Before the electron is detected the wave function determines the *probability* that the electron will be found at a certain place, at a given time. As soon as the electron has been detected, its location is of course known to within a precision $\Delta x' \ll \Delta x$, so that a new wave function must describe the situation. This change of the wave function upon measurement is called the '*collapse of the wave packet*'. A more careful analysis of this measurement problem can be based on the study of the combined wave function of the measured system and the measuring apparatus. Using this approach, it will be shown in Chapter 17 that the idea of an instantaneous 'collapse' of 'reduction' of the wave function on measurement can be avoided.

Wave packets in three dimensions

Our discussions of one-dimensional free-particle wave packets can easily be extended to three dimensions. By superposing plane waves of the form (2.16) we obtain the wave packet

$$\Psi(\mathbf{r}, t) = (2\pi\hbar)^{-3/2} \int_{-\infty}^{+\infty} dp_x \int_{-\infty}^{+\infty} dp_y \int_{-\infty}^{+\infty} dp_z e^{i[\mathbf{p}\cdot\mathbf{r} - E(p)t]/\hbar} \phi(\mathbf{p})$$

$$= (2\pi\hbar)^{-3/2} \int e^{i[\mathbf{p}\cdot\mathbf{r} - E(p)t]/\hbar} \phi(\mathbf{p}) d\mathbf{p} \tag{2.57}$$

where $d\mathbf{p} = dp_x dp_y dp_z$ is the volume element in momentum space. Writing $\psi(\mathbf{r}) \equiv \Psi(\mathbf{r}, t = 0)$ we see from Appendix A that

$$\psi(\mathbf{r}) = (2\pi\hbar)^{-3/2} \int e^{i\mathbf{p}\cdot\mathbf{r}/\hbar} \phi(\mathbf{p}) d\mathbf{p} \tag{2.58}$$

and

$$\phi(\mathbf{p}) = (2\pi\hbar)^{-3/2} \int e^{-i\mathbf{p}\cdot\mathbf{r}/\hbar} \psi(\mathbf{r}) d\mathbf{r} \tag{2.59}$$

are three-dimensional *Fourier transforms* of each other. Paralleling our discussion of the one-dimensional case, a *momentum space wave function* $\Phi(\mathbf{p}, t)$ can be introduced as the Fourier transform of the wave function $\Psi(\mathbf{r}, t)$, so that

$$\Psi(\mathbf{r}, t) = (2\pi\hbar)^{-3/2} \int e^{i\mathbf{p}\cdot\mathbf{r}/\hbar} \Phi(\mathbf{p}, t) d\mathbf{p} \tag{2.60}$$

and

$$\Phi(\mathbf{p}, t) = (2\pi\hbar)^{-3/2} \int e^{-i\mathbf{p}\cdot\mathbf{r}/\hbar} \Psi(\mathbf{r}, t) d\mathbf{r} \tag{2.61}$$

and we note that $\phi(\mathbf{p}) = \Phi(\mathbf{p}, t = 0)$. As in the one-dimensional case (see (2.43)) this definition of the momentum space wave function is completely general, and holds for all types of wave functions $\Psi(\mathbf{r}, t)$. If $\phi(\mathbf{p})$ is normalised to unity, then $\Psi(\mathbf{r}, t)$ is also normalised to unity at all times, i.e. satisfies (2.7). The momentum space wave function $\Phi(\mathbf{p}, t)$ will then also be normalised to unity, satisfying

$$\int |\Phi(\mathbf{p}, t)|^2 d\mathbf{p} = 1. \tag{2.62}$$

As a result, the quantity

$$\Pi(\mathbf{p}, t) d\mathbf{p} = |\Phi(\mathbf{p}, t)|^2 d\mathbf{p} = \Phi^*(\mathbf{p}, t) \Phi(\mathbf{p}, t) d\mathbf{p} \tag{2.63}$$

is the probability at time t that the momentum of the particle lies within the momentum space volume element $d\mathbf{p} \equiv dp_x dp_y dp_z$ about the point $\mathbf{p} \equiv (p_x, p_y, p_z)$.

A three-dimensional wave packet associated with a free particle having fairly well determined values of both position coordinates (x, y, z) and momentum coordinates (p_x, p_y, p_z) can be constructed analogously to the one-dimensional case. Assuming that the function $\phi(\mathbf{p})$ in (2.57) is peaked about $\mathbf{p} = \mathbf{p}_0$, and setting

$$\beta(\mathbf{p}) = \mathbf{p} \cdot \mathbf{r} - E(p)t \tag{2.64}$$

one finds (Problem 2.8) that the centre of the wave packet, defined by the condition[1]

$$[\nabla_\mathbf{p} \beta(\mathbf{p})]_{\mathbf{p}=\mathbf{p}_0} = 0 \tag{2.65}$$

travels with a uniform motion according to the law

$$\mathbf{r} = \mathbf{v}_g t \tag{2.66}$$

where

$$\mathbf{v}_g = [\nabla_\mathbf{p} E(p)]_{\mathbf{p}=\mathbf{p}_0} \tag{2.67}$$

is the *group velocity* of the wave packet. Equations (2.66) and (2.67) are the generalisations of equations (2.27) and (2.28), respectively. In the classical limit the group velocity $\mathbf{v}_g$ must be equal to the velocity $\mathbf{v} = \mathbf{p}_0/m$ of the particle, from which we find that the functional relation between E and p is given by

$$E(p) = \frac{p^2}{2m} \tag{2.68}$$

apart from an additive constant which can be chosen to be zero.

---

[1] In equation (2.65), $\nabla_\mathbf{p} \beta(\mathbf{p})$ is a vector having Cartesian components $\partial \beta(p_x, p_y, p_z)/\partial p_x$, $\partial \beta(p_x, p_y, p_z)/\partial p_y$ and $\partial \beta(p_x, p_y, p_z)/\partial p_z$.

Wave packets in a slowly varying potential

The general idea we have developed for the motion of a free-particle wave packet in the classical limit can be extended to describe the motion of a particle in a potential $V(\mathbf{r})$ provided that the potential does not vary appreciably over a distance comparable to the de Broglie wavelength of the particle. In this case the centre of the wave packet travels along the trajectory followed by a classical particle moving in the potential $V(\mathbf{r})$. As the centre of the wave packet moves along this trajectory, the de Broglie wavelength changes slowly, being determined by the relation

$$\lambda = \frac{h}{p} = \frac{h}{[2m(E - V(\mathbf{r}))]^{1/2}} \tag{2.69}$$

where $p = [2m(E - V(\mathbf{r}))]^{1/2}$ is the classical local momentum of the particle.

2.5 The Heisenberg uncertainty principle

We have shown in the case of a one-dimensional Gaussian wave packet that the 'width' Δx of the distribution in the position variable x is linked with the 'width' Δp_x of the distribution in the momentum p_x by the relation $\Delta x \Delta p_x \gtrsim \hbar$. In fact, it is a general property of Fourier transforms that 'widths' in position and momentum satisfy the relation

$$\Delta x \Delta p_x \gtrsim \hbar \tag{2.70}$$

where the sign $\gtrsim$ means 'greater than or of the order of'. In the context of quantum mechanics this is called the *Heisenberg uncertainty relation for position and momentum*, according to which a state cannot be prepared in which both the position and momentum of a particle can be defined simultaneously to arbitrary accuracy. In fact, the product of the uncertainty Δx in the precision with which the position can be defined with the uncertainty Δp_x in the precision with which the momentum can be defined, cannot be made smaller than a quantity of order $\hbar$. At this point we have not given a precise definition of the uncertainties Δx and Δp_x, but this will be done later, in Chapter 5.

The relation (2.70) is easily generalised to three dimensions by using the properties of three-dimensional Fourier transforms. The three-dimensional form of the Heisenberg uncertainty relations for position and momentum is

$$\Delta x \Delta p_x \gtrsim \hbar, \qquad \Delta y \Delta p_y \gtrsim \hbar, \qquad \Delta z \Delta p_z \gtrsim \hbar. \tag{2.71}$$

It should be noted that there is no relation between the uncertainty in one Cartesian component of the position vector of a particle, for example Δx, and the uncertainty in a *different* Cartesian component of the momentum, for example Δp_y. The only restrictions are on the 'complementary' pairs: $\Delta x, \Delta p_x$; $\Delta y, \Delta p_y$; and $\Delta z, \Delta p_z$.

It is worth stressing that the Heisenberg uncertainty relations do not place any restriction on the precision with which a position measurement of a particle can be made. However, once such a measurement has been made and the particle is known to be confined to some region of extent Δa, the ensemble of such particles is subsequently described by a wave function also of extent Δa. Subsequent measurements of the momentum on each of the identical systems composing the ensemble will then produce a range of values with a spread of order $\hbar/\Delta a$.

Suppose now that before the position measurement was made the particle was in a state of definite momentum. We see then that the act of measuring the position forces the system into a state in which the momentum is no longer known exactly but has become uncertain by an amount $\Delta p \gtrsim \hbar/\Delta a$. Thus the Heisenberg uncertainty relations do not refer to the period before the state was prepared, but only to the current situation.

Similarly, it can be seen that the Heisenberg relations do not restrict the precision with which a momentum measurement can be made, but once the system is known to have a momentum defined to within a precision Δp, subsequent measurements of the position must produce results spread over a range $\gtrsim \hbar/\Delta p$. The uncertainty Δp can be made arbitrarily small, but the more precisely the momentum of the system is known the greater the range of results that will be obtained in a series of position measurements on an ensemble of such systems.

Finally, it is hardly necessary to point out that the inherent limitations on measurement imposed by the Heisenberg relations have nothing to do with the 'experimental errors' that occur in actual measurements.

The γ-ray microscope

As a first illustration of the Heisenberg uncertainty relations, let us consider a conceptual ('*gedanken*') experiment first discussed by Heisenberg, which attempts to measure the position of a particle as accurately as possible. This experiment consists of illuminating the particle and observing the image through a microscope (see Fig. 2.7). If λ is the wavelength of the incident radiation, then the x component of the particle position can be determined to a precision Δx given by the resolving power of the microscope, namely

$$\Delta x = \frac{\lambda}{\sin \theta} \qquad (2.72)$$

where θ is the half-angle subtended at the particle P by the lens L. This resolving power can be made very small by using radiation of short wavelength, such as γ rays. Now, for any measurement to be possible at least one photon must enter the microscope after scattering from the particle. This Compton scattering imparts to the particle a recoil momentum of the order of magnitude of the photon momentum $p_\gamma = h/\lambda$. However, this recoil momentum cannot be exactly known, since the direction of the scattered photon is only determined to within the angle θ. Thus there is an uncertainty in the recoil momentum of the particle in the x direction by an amount

$$\Delta p_x \simeq \frac{h}{\lambda} \sin \theta \qquad (2.73)$$

2.5 The Heisenberg uncertainty principle

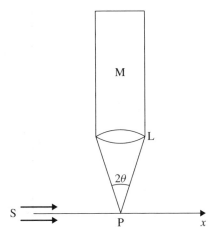

Figure 2.7 The Heisenberg γ-ray microscope. Photons from a source S are scattered into a microscope M from a particle located at P. The resolving power of the microscope is $\lambda/\sin\theta$, where λ is the photon wavelength and 2θ is the angle subtended by the lens L at the particle.

and upon combining (2.73) with (2.72) we see that after the observation one always has

$$\Delta x \Delta p_x \simeq h \tag{2.74}$$

which is consistent with the Heisenberg uncertainty relation (2.70).

The two-slit experiment

As a second example, let us return to the two-slit experiment discussed in Section 2.1, and analyse it in the light of the uncertainty principle. Suppose we attempt to discover which slit a particle (for example an electron) goes through. This might be done by placing a microscope behind one of the slits (see Fig. 2.8). In order to discriminate between particles which have passed through slit A or slit B the resolving power of the microscope must be less than the distance d between the slits. From (2.72) we see that the wavelength of the light illuminating the particle must be smaller than d. The corresponding photon momentum $p_\gamma = h/\lambda$ must therefore be larger than h/d. The interaction of a photon having this momentum with the electron will make the electron momentum p uncertain by an amount Δp which is of the order of p_γ, and hence will be such that

$$d\Delta p \gtrsim h \tag{2.75}$$

in agreement with the uncertainty relations. Consequently, the direction of motion of the particle also becomes uncertain by an amount $\Delta\theta$, where (see Fig. 2.8)

$$\Delta\theta \simeq \frac{\Delta p}{p} \simeq \frac{h}{pd} \simeq \frac{\lambda_e}{d} \tag{2.76}$$

72 ■ The wave function and the uncertainty principle

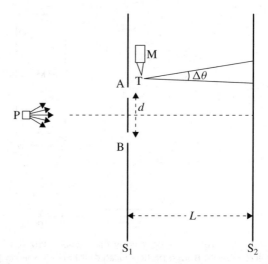

Figure 2.8 In the two-slit experiment of Fig. 2.1 a microscope M is placed behind the slit A and the area behind the slit is illuminated. A particle is located at the point T, but as a result of the photon–particle interaction the direction of motion of the particle becomes uncertain by $\Delta\theta$.

where $\lambda_e = h/p$ is the de Broglie wavelength of the electron. This in turn produces an uncertainty $\lambda_e L/d$ in the position of the particle on the screen, which is comparable to the distance between successive maxima of the interference pattern. Thus, if an attempt is made to determine through which slit the particles pass, the interference pattern is destroyed.

Stability of atoms

The position–momentum uncertainty relations (2.71) can be used to obtain an estimate of the binding energy of the hydrogen atom in its ground state. To this end, let us write the total classical energy of the electron in the field of the proton (assumed to be infinitely heavy) as

$$E = \frac{p^2}{2m} - \frac{e^2}{(4\pi\varepsilon_0)r}. \tag{2.77}$$

The first term on the right-hand side is the kinetic energy, which is positive, while the second term represents the negative potential energy. Classically, there is no lower limit to the value of the total energy E because r, the radius of the electronic orbit, can be made as small as one pleases. This is not the case in quantum mechanics. If p is interpreted as the average momentum of the electron in a given state, it follows that the momentum is determined within a range Δp of the order of p. From the uncertainty relations (2.71), this in turn implies that the smallest value of the uncertainty Δr of the position of the electron is of the order of $\hbar/p$. Assuming that the average value of the radius r is of the order of Δr, we then have $rp \simeq \hbar$, from which

$$E \simeq \frac{\hbar^2}{2mr^2} - \frac{e^2}{(4\pi\varepsilon_0)r}. \tag{2.78}$$

As a result, there is a *minimum* value of the total energy E at $r \simeq r_0$ such that $dE/dr = 0$, namely

$$r_0 \simeq \frac{(4\pi\varepsilon_0)\hbar^2}{me^2} = a_0 \tag{2.79}$$

where a_0 is the first Bohr radius (1.66). Substituting $r = a_0$ in (2.78) we find that the ground-state energy E_0 is given approximately by

$$E_0 \simeq -\frac{e^2}{(4\pi\varepsilon_0)2a_0}. \tag{2.80}$$

The fact that this estimate of E_0 coincides with the actual ground-state energy of atomic hydrogen must not be taken too seriously, as the argument is clearly qualitative. However, the important point is that this reasoning shows that there is a *minimum* value of the total energy E of an atom, which is *the lowest value of E compatible with the uncertainty principle*. We can therefore understand in this way why atoms are stable.

The uncertainty relation for time and energy

A time–energy uncertainty relation analogous to the position–momentum uncertainty relations (2.71) can be obtained in the following way. Let $\Psi(t) \equiv \Psi(\mathbf{r}_0, t)$ be a wave function, at a fixed point $\mathbf{r} = \mathbf{r}_0$, associated with a single particle state. We consider the case such that $\Psi(t)$ is a pulse or 'time packet', which is negligible except in a time interval Δt. This time packet can be expressed as a superposition of monochromatic waves of angular frequency ω by the Fourier integral (see Appendix A)

$$\Psi(t) = (2\pi)^{-1/2} \int_{-\infty}^{+\infty} G(\omega) e^{-i\omega t} d\omega \tag{2.81}$$

where the function $G(\omega)$ is given by

$$G(\omega) = (2\pi)^{-1/2} \int_{-\infty}^{+\infty} \Psi(t) e^{i\omega t} dt. \tag{2.82}$$

As $\Psi(t)$ takes only significant values for a duration Δt, it follows from the general properties of Fourier transforms that $G(\omega)$ is only significant for a range of angular frequencies such that

$$\Delta\omega \Delta t \gtrsim 1. \tag{2.83}$$

Since $E = \hbar\omega$, the width of the distribution in energy, ΔE, satisfies the time–energy uncertainty relation

$$\Delta E \Delta t \gtrsim \hbar. \tag{2.84}$$

The interpretation of this relationship is somewhat different from that of the position–momentum uncertainty relations (2.71) because the time t is a parameter, and not a dynamical variable like x or p_x. The relation (2.84) implies that if the dynamical state exists only for a time of order Δt, then the energy of the state cannot be defined to a precision better than $\hbar/\Delta t$. In other words, if we consider an ensemble of identically prepared systems described by the wave function Ψ, then the measurement of the energy on each member of the ensemble will produce a range of values spread over an interval ΔE of extent greater than or of the order of $\hbar/\Delta t$.

Energy width and natural lifetime of excited states of atoms

As an example, let us consider an atomic transition from an excited state of energy E_b, in which a photon is emitted (see Fig. 2.9(a)). The time of emission of the photon from a particular atom cannot be predicted. However, as we shall see in Chapter 11, quantum mechanics does predict the *probability* that the transition takes place at a given time. This in turn determines the *average duration* (over a large number of atoms) of the excited state b, called the *lifetime* τ_b of that state. Clearly, the wave function Ψ describing the atomic system in the state b must be negligible outside a time interval Δt of the order of τ_b. From the uncertainty relation (2.84) it follows that the energy of the excited state b is not sharply defined, but is uncertain at least by an amount

$$\Delta E_b = \frac{\hbar}{\tau_b} \qquad (2.85)$$

known as the *natural energy width* of the state b. We see from (2.85) that the shorter the lifetime of an excited state, the larger the natural energy width of that state. On the other hand, the ground state a is stable, since in this state the system cannot make a transition to a state of lower energy. The lifetime of the ground state is therefore infinite ($\tau_a = \infty$) and the ground state energy E_a is perfectly sharp ($\Delta E_a = 0$).

Because of the natural energy width ΔE_b of the excited state b, the energy emitted (or absorbed) in a transition between the atomic states a and b is not sharply defined. Instead, the photons emitted (or absorbed) in transitions between the states a and b have an energy distribution of natural width ΔE_b about the value $E_b - E_a$. As a result, the spectral line associated with this transition does not have the sharply defined frequency $\nu_{ab} = (E_b - E_a)/h$, but has a distribution of frequencies of width at least as great as $\Delta \nu = \Delta E_b/h$ about the value ν_{ab}. The quantity $\Delta \nu$ is called the *natural linewidth* of that spectral line. If the transition occurs between two excited states a and b (see Fig. 2.9(b)), the photons emitted or absorbed in the transition will have an energy distribution of natural width

$$\Delta E_{ab} = \hbar \left(\frac{1}{\tau_a} + \frac{1}{\tau_b} \right) \qquad (2.86)$$

where τ_a is the lifetime of the lower state a. The natural linewidth of the spectral line is then $\Delta \nu_{ab} = \Delta E_{ab}/h$.

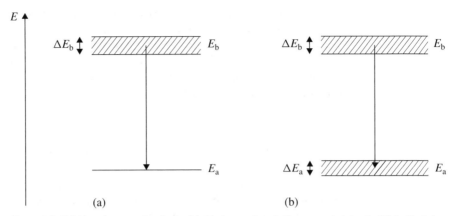

Figure 2.9 Widths of energy levels. In (a) the lower state is the ground state. In (b) both states are excited states.

The preceding discussion concerning the relation between the lifetime and the natural energy width of excited states is not confined to atomic states, but is equally applicable to other quantum systems such as molecules or nuclei.

Natural energy width and resonance absorption of radiation

In Chapter 1, we showed that because of the *recoil* of the atom the wavelength of an emission line is slightly different from that of the corresponding absorption line, with the consequence that radiation emitted by one atom could not be absorbed by another atom of the same kind (see (1.76)). This is contrary to what is observed. The paradox can be resolved by recognising that the energy levels of excited states have a finite width, so that an atom can emit or absorb photons within a range of energies determined by the natural energy widths of the levels concerned. Thus, if the change in frequency due to the recoil is smaller than the natural linewidth associated with the transition, an atom will be able to absorb radiation emitted by another atom of the same kind. Now, as seen from (1.76), the largest change of frequency due to recoil occurs for atoms of the smallest mass–hydrogen atoms. For a typical atomic transition involving photons having an energy $h\nu \simeq 2$ eV the change of frequency due to recoil is therefore of the order of or less than

$$|\Delta \nu_R| = \frac{(h\nu)^2}{2hM_pc^2} \simeq 5 \times 10^5 \text{ Hz}. \qquad (2.87)$$

On the other hand, a typical atomic lifetime is about 10^{-8} s, corresponding to a natural linewidth of 1.5×10^7 Hz. Since this value is considerably larger than $|\Delta \nu_R|$ resonance absorption can indeed occur.

In contrast, for transitions occurring between energy levels of nuclei the change of frequency $|\Delta \nu_R|$ due to recoil often exceeds the natural linewidth, with the consequence that the γ-ray photons emitted by one nucleus cannot in general be absorbed

The uncertainty principle and complementarity

The uncertainty relations for position and momentum and for energy and time are particular examples of the *uncertainty principle*, formulated by W. Heisenberg in 1927[2]. According to this principle, it is impossible to prepare states in which both members of certain pairs of variables, called *complementary variables*, have values determined with arbitrary precision. If one of the variables of a complementary pair is measured, information is inevitably lost about the other. We have studied in some detail the complementary pairs (x, p_x) and (E, t), but as we shall see in Chapter 5 other pairs of complementary variables exist which satisfy uncertainty relations.

The Heisenberg uncertainty principle is clearly a consequence of wave–particle duality, and reflects the fact that quantum mechanics, although a complete theory, provides a less detailed description of a physical system than does classical physics. In his *complementarity principle*, introduced in 1928, Bohr described this situation by stating that the wave and particle aspects of physical systems are complementary, both aspects being needed for a complete description of nature.

Problems

2.1 Consider the wave packet $\psi(x) \equiv \Psi(x, t = 0)$ given by

$$\psi(x) = C e^{ip_0 x/\hbar} e^{-|x|/(2\Delta x)}$$

where C is a normalisation constant.

(a) Normalise $\psi(x)$ to unity.
(b) Obtain the corresponding momentum space wave function $\phi(p_x)$ and verify that it is normalised to unity according to (2.44).
(c) Suggest a reasonable definition of the width Δp_x of the momentum distribution and show that $\Delta x \Delta p_x \gtrsim \hbar$.

2.2 Consider the momentum space wave function

$$\phi(p_x) = \begin{cases} 0, & |p_x - p_0| > \gamma \\ C, & |p_x - p_0| \leqslant \gamma \end{cases}$$

where C and γ are constants.

---

[2] A detailed discussion of the uncertainty principle can be found in Heisenberg's book *The Physical Principles of the Quantum Theory* (1949).

(a) Normalise $\phi(p_x)$ to unity according to (2.44).
(b) Find the corresponding wave function $\psi(x)$ in configuration space, and verify that $\psi(x)$ is normalised to unity.

$$\left(\text{Hint: Use the result } \int_{-\infty}^{+\infty} \frac{\sin^2 x}{x^2} dx = \pi.\right)$$

(c) Using reasonable definitions of the widths Δx and Δp_x, show that

$$\Delta x \Delta p_x \gtrsim \hbar.$$

2.3 The family of functions $\delta_L(x)$ is defined by

$$\delta_L(x) = (2\pi)^{-1} \int_{-L}^{+L} e^{ikx} dk.$$

Evaluate the integral and show that $\delta_L(x)$ behaves like the Dirac delta function $\delta(x)$ as $L \to \infty$.

2.4 Let $\delta_\varepsilon(x)$ be a family of functions defined by

$$\delta_\varepsilon(x) = (2\pi)^{-1} \int_{-\infty}^{+\infty} e^{ikx} e^{-\varepsilon k^2} dk, \qquad \varepsilon > 0.$$

Evaluate the integral and show that $\delta_\varepsilon(x)$ has the properties of the Dirac delta function $\delta(x)$ as $\varepsilon \to 0$.

2.5 Verify that the Dirac delta function can be represented in the following ways

$$\delta(x) = \lim_{\varepsilon \to 0^+} \frac{\varepsilon/\pi}{x^2 + \varepsilon^2} \tag{1}$$

$$= \lim_{\varepsilon \to 0^+} \frac{\sin(x/\varepsilon)}{\pi x} \tag{2}$$

$$= \lim_{\varepsilon \to 0^+} \frac{\varepsilon}{\pi x^2}[1 - \cos(x/\varepsilon)] \tag{3}$$

$$= \lim_{\varepsilon \to 0} \frac{\theta(x+\varepsilon) - \theta(x)}{\varepsilon} \tag{4}$$

where in equation (4) $\theta(x)$ is the step function

$$\theta(x) = \begin{cases} 1, & x > 0 \\ 0, & x < 0. \end{cases}$$

2.6 (a) By multiplying both sides of the following equations by a differentiable function $f(x)$, and integrating over x, verify the following equations:

$$\delta(x) = \delta(-x)$$

$$\delta'(x) = -\delta'(-x), \qquad \delta'(x) = \frac{d}{dx}\delta(x)$$

$$x\delta(x) = 0$$

$$x\delta'(x) = -\delta(x)$$

$$\delta(ax) = \frac{1}{|a|}\delta(x), \qquad a \neq 0$$

$$\delta(x^2 - a^2) = \frac{1}{2|a|}[\delta(x-a) + \delta(x+a)].$$

(b) Prove the following relations

$$\int \delta(a-x)\delta(x-b)dx = \delta(a-b)$$

$$f(x)\delta(x-a) = f(a)\delta(x-a).$$

2.7 (a) A Gaussian wave packet is associated with an electron localised at time $t = 0$ to within a distance of 10^{-10} m. Show that this wave packet will have spread to twice its size after a time $t \simeq 10^{-16}$ s.

(b) Consider now a Gaussian wave packet associated with a proton localised at time $t = 0$ to within a distance of 10^{-14} m. Find the time after which this wave packet will have doubled in size.

2.8 Let $\Psi(\mathbf{r}, t)$ be a three-dimensional wave packet defined by (2.57). Show that if the function $\phi(\mathbf{p})$ is peaked about $\mathbf{p} = \mathbf{p}_0$, the centre of the wave packet, defined by the condition (2.65), travels uniformly according to the law $\mathbf{r} = \mathbf{v}_g t$, where the group velocity $\mathbf{v}_g$ is given by (2.67).

2.9 A monoenergetic beam of electrons falls on a screen containing a slit of width d. The coordinates of the electrons in the x direction (taken along the screen) are therefore known at the moment after passing the slit with the accuracy $\Delta x = d$. Now, because of the wave nature of the electron, diffraction phenomena are produced by the slit. The emergent beam has a finite angle of divergence θ which, according to diffraction theory is such that

$$\sin\theta = \frac{\lambda}{d}$$

where λ is the de Broglie wavelength of the electrons. Show that after the electrons have passed through the slit the component of their momentum along the x-axis becomes uncertain by an amount $\Delta p_x \gtrsim \hbar/d$.

2.10 An electron is confined within a region of atomic dimensions, of the order of 10^{-10} m. Find the uncertainty in its momentum. Repeat the calculation for a proton confined to a region of nuclear dimensions, of the order of 10^{-14} m.

2.11 Using the uncertainty principle, find the minimum value (in MeV) of the kinetic energy of a nucleon confined within a nucleus of radius $R = 5 \times 10^{-15}$ m.

2.12 Calculate the relative frequency spread $\Delta \nu / \nu$ for a nanosecond (10^{-9} s) pulse from a CO_2 laser, in which the nominal photon energy is $h\nu = 0.112$ eV.

2.13 A beam of monoenergetic electrons is used to raise atoms to an excited state in a Franck–Hertz experiment. If this excited state has a lifetime of 10^{-9} s, calculate the spread in energy of the inelastically scattered electrons.

3 The Schrödinger equation

3.1 The time-dependent Schrödinger equation   82
3.2 Conservation of probability   86
3.3 Expectation values and operators   90
3.4 Transition from quantum mechanics to classical mechanics. The Ehrenfest theorem   97
3.5 The time-independent Schrödinger equation. Stationary states   100
3.6 Energy quantisation   104
3.7 Properties of the energy eigenfunctions   115
3.8 General solution of the time-dependent Schrödinger equation for a time-independent potential   120
3.9 The Schrödinger equation in momentum space   124
Problems   128

In Chapter 2 we introduced the concept of the wave function to describe the properties of a physical system, and we saw that in the position representation the wave function Ψ of a particle is a function of its space coordinates $\mathbf{r}$ and of the time t. Clearly, in order to make progress we need a method of determining wave functions systematically. In this chapter we shall show that the wave function $\Psi(\mathbf{r}, t)$ can be calculated from a partial differential equation called the *Schrödinger wave equation*. We shall discuss some general properties of its solutions, examine how expectation values of physical quantities can be calculated from the knowledge of $\Psi(\mathbf{r}, t)$, and analyse the transition from quantum mechanics to classical mechanics. We shall also see how the Schrödinger theory accounts for the quantisation of energy levels, and we shall conclude the chapter with a brief discussion of the Schrödinger equation in momentum space.

In searching for an equation to be satisfied by the wave function Ψ, we shall be guided by the following principles. First, the equation should be *linear* and *homogeneous* so that the *superposition principle* holds. That is, if Ψ_1 and Ψ_2 are solutions of the wave equation for a given system, any linear combination $c_1\Psi_1 + c_2\Psi_2$ (where c_1 and c_2 are constants) must also be a solution. Secondly, the results obtained by using the equation should agree with those of classical mechanics in macroscopic situations, in accord with the *correspondence principle*. Finally, in order to satisfy the hypothesis that the evolution of the system is entirely determined once the wave function is known at a particular time, the equation should be of first order in the time derivative $\partial/\partial t$. If the equation contained the second derivative $\partial^2 \Psi/\partial t^2$, it would be necessary to specify both Ψ and $\partial \Psi/\partial t$ at a certain time to obtain a unique solution, in contradiction to the hypothesis that only a knowledge of Ψ is required.

3.1 The time-dependent Schrödinger equation

Let us begin by considering the one-dimensional, non-relativistic motion of a free particle of mass m, having a well-defined momentum $\mathbf{p} = p_x \hat{\mathbf{x}}$ (where $\hat{\mathbf{x}}$ is the unit vector along the x-axis) of magnitude $p = |p_x|$ and an energy E. Assuming that the particle is travelling in the positive x direction (so that $p_x = p$) we have seen in Chapter 2 that this particle is described by a monochromatic plane wave of wave number $k = p_x/\hbar$ and angular frequency $\omega = E/\hbar$, namely

$$\Psi(x,t) = A e^{i(kx-\omega t)} = A e^{i(p_x x - Et)/\hbar} \tag{3.1}$$

where A is a constant. The angular frequency ω is connected with the wave number by the relation $\omega = \hbar k^2/2m$, which is equivalent to the classical relation

$$E = \frac{p_x^2}{2m} \tag{3.2}$$

connecting the momentum and energy of the particle. Now, if we differentiate (3.1) with respect to time, we have

$$\frac{\partial \Psi}{\partial t} = -\frac{iE}{\hbar} \Psi. \tag{3.3}$$

On the other hand, differentiating (3.1) twice with respect to x, we find that

$$\frac{\partial^2 \Psi}{\partial x^2} = -\frac{p_x^2}{\hbar^2} \Psi. \tag{3.4}$$

Hence, using (3.2), we see that the plane wave (3.1) satisfies the partial differential equation

$$i\hbar \frac{\partial}{\partial t} \Psi(x,t) = -\frac{\hbar^2}{2m} \frac{\partial^2}{\partial x^2} \Psi(x,t). \tag{3.5}$$

More generally, since the equation (3.5) is *linear* and *homogeneous*, it will be satisfied by a linear superposition of plane waves (3.1). For example, the wave packet (2.23)

$$\Psi(x,t) = (2\pi\hbar)^{-1/2} \int_{-\infty}^{+\infty} e^{i[p_x x - E(p_x)t]/\hbar} \phi(p_x) \mathrm{d}p_x \tag{3.6}$$

associated with a 'localised' free particle moving in one dimension, is also a solution of the equation (3.5), since

$$i\hbar \frac{\partial}{\partial t}\Psi(x,t) = (2\pi\hbar)^{-1/2} \int_{-\infty}^{+\infty} E(p_x) e^{i[p_x x - E(p_x)t]/\hbar} \phi(p_x) dp_x$$

$$= (2\pi\hbar)^{-1/2} \int_{-\infty}^{+\infty} \frac{p_x^2}{2m} e^{i[p_x x - E(p_x)t]/\hbar} \phi(p_x) dp_x$$

$$= -\frac{\hbar^2}{2m} \frac{\partial^2}{\partial x^2} \Psi(x,t). \tag{3.7}$$

The wave equation (3.5) is known as the *time-dependent Schrödinger equation* for the motion of a *free* particle in *one dimension*. We remark that this equation satisfies the requirements that we formulated at the beginning of this chapter. Indeed, we have already seen that it is a *linear* and *homogeneous* equation for the wave function Ψ. Furthermore, remembering that in wave mechanics the total energy E is represented by the operator $E_{op} = i\hbar \partial/\partial t$, and the component p_x of the momentum by the operator $(p_x)_{op} = -i\hbar \partial/\partial x$ (see (2.20)), we note that equation (3.5) may be written in the form

$$E_{op}\Psi(x,t) = \frac{1}{2m}[(p_x)_{op}]^2 \Psi(x,t) \tag{3.8}$$

which is the quantum-mechanical 'translation' of the classical relation (3.2). This analogy with classical mechanics is in agreement with the *correspondence principle*. We also remark that the Schrödinger wave equation (3.5) is of *first order* in the time derivative $\partial/\partial t$. Hence, if the wave function $\Psi(x,t)$ is given at a certain time t_0, it is determined at all other times by this equation.

The generalisation of these considerations to free-particle motion in three dimensions is straightforward. A free particle of mass m having a well-defined momentum $\mathbf{p}$ and an energy E is now described by a plane wave

$$\Psi(\mathbf{r},t) = A e^{i(\mathbf{k}\cdot\mathbf{r} - \omega t)}$$
$$= A e^{i(\mathbf{p}\cdot\mathbf{r} - Et)/\hbar} \tag{3.9}$$

characterised by the wave vector $\mathbf{k} = \mathbf{p}/\hbar$ and the angular frequency $\omega = E/\hbar$. Again we have $\omega = \hbar k^2/2m$, which is equivalent to the classical relation

$$E = \frac{\mathbf{p}^2}{2m} \tag{3.10}$$

between the momentum and energy of the particle. It is then readily verified that the plane wave (3.9) satisfies the partial differential equation

$$i\hbar \frac{\partial}{\partial t}\Psi(\mathbf{r},t) = -\frac{\hbar^2}{2m}\nabla^2 \Psi(\mathbf{r},t) \tag{3.11}$$

where

$$\nabla^2 \equiv \frac{\partial^2}{\partial x^2} + \frac{\partial^2}{\partial y^2} + \frac{\partial^2}{\partial z^2} \tag{3.12}$$

is the Laplacian operator. The wave equation (3.11), which is the direct generalisation of (3.5), is the *three-dimensional time-dependent Schrödinger equation for a free particle*. As in the one-dimensional case, it is a *linear* and *homogeneous* equation which is therefore satisfied by arbitrary linear superpositions of plane waves (3.9), in particular by wave packets of the form (2.57) associated with 'localised' free particles. Equation (3.11) is also clearly of first order in the time derivative $\partial/\partial t$. Finally, using the fact that in wave mechanics the total energy E and the momentum $\mathbf{p}$ are represented by the differential operators (see (2.20))

$$E_{\text{op}} = i\hbar \frac{\partial}{\partial t}, \qquad \mathbf{p}_{\text{op}} = -i\hbar \nabla \tag{3.13}$$

we observe that the free-particle Schrödinger equation (3.11) may also be written in the form

$$E_{\text{op}} \Psi(\mathbf{r}, t) = \frac{1}{2m} (\mathbf{p}_{\text{op}})^2 \Psi(\mathbf{r}, t) \tag{3.14}$$

in formal analogy with the classical equation (3.10). Note that the quantity $\mathbf{p}^2/2m$ is represented by the operator[1]

$$T = \frac{1}{2m} (\mathbf{p}_{\text{op}})^2 = -\frac{\hbar^2}{2m} \nabla^2 \tag{3.15}$$

which is called the kinetic energy operator of the particle.

Let us now try to generalise the free-particle Schrödinger equation (3.11) to the case of a particle moving in a field of force. We shall assume that the force $F(\mathbf{r}, t)$ acting on the particle is derivable from a potential

$$\mathbf{F}(\mathbf{r}, t) = -\nabla V(\mathbf{r}, t) \tag{3.16}$$

so that, for a classical particle, the total energy E is given by the sum of its kinetic energy $\mathbf{p}^2/2m$ and its potential energy $V(\mathbf{r}, t)$

$$E = \frac{\mathbf{p}^2}{2m} + V(\mathbf{r}, t). \tag{3.17}$$

Since the potential energy V does not depend on $\mathbf{p}$ or E, the above discussion of the free-particle case suggests using (3.13) to write

$$E_{\text{op}} \Psi(\mathbf{r}, t) = \left[\frac{1}{2m} (\mathbf{p}_{\text{op}})^2 + V(\mathbf{r}, t) \right] \Psi(\mathbf{r}, t) \tag{3.18}$$

so that the generalisation of the free-particle Schrödinger equation (3.11) reads

$$i\hbar \frac{\partial}{\partial t} \Psi(\mathbf{r}, t) = \left[-\frac{\hbar^2}{2m} \nabla^2 + V(\mathbf{r}, t) \right] \Psi(\mathbf{r}, t). \tag{3.19}$$

This is the celebrated *time-dependent Schrödinger wave equation* for a particle moving in a potential, which was proposed by E. Schrödinger in 1926. It is the

---

[1] In what follows we shall drop the subscript 'op' which denotes operators, unless there is a possibility of confusion with a quantity (number, vector, etc.) denoted by the same symbol.

basic equation of non-relativistic quantum mechanics. We want to emphasise that we have not formally *derived* this equation, but have only made it appear plausible. The Schrödinger equation, like Newton's laws, cannot be proved to be true. Its best possible justification comes from an exhaustive comparison of the predictions based on this equation with experiment. It is the successful application of the Schrödinger equation to many problems which demonstrated its correctness in non-relativistic quantum mechanics.

The operator appearing inside the brackets on the right of the Schrödinger equation (3.19) will be seen shortly to play a very important role. It is called the Hamiltonian operator of the particle and is denoted by H. Using (3.15), we have

$$H = -\frac{\hbar^2}{2m}\nabla^2 + V$$
$$= \frac{1}{2m}(\mathbf{p}_{op})^2 + V = T + V \tag{3.20}$$

and the time-dependent Schrödinger equation (3.19) may therefore be rewritten in the form

$$i\hbar\frac{\partial}{\partial t}\Psi(\mathbf{r},t) = H\Psi(\mathbf{r},t). \tag{3.21}$$

Now, in classical mechanics the total energy E of a system, when expressed in terms of the coordinates of the system, of their respective momenta and of the time, is called the (classical) Hamiltonian H_{cl} of this system. In particular, for the case of a classical particle moving in a potential, we have from (3.17)

$$E = H_{cl}(\mathbf{r},\mathbf{p},t) \tag{3.22}$$

where

$$H_{cl}(\mathbf{r},\mathbf{p},t) = \frac{\mathbf{p}^2}{2m} + V(\mathbf{r},t). \tag{3.23}$$

Upon comparison of (3.20) and (3.23), we see that the quantum mechanical Hamiltonian operator H is obtained from the classical Hamiltonian H_{cl} by performing the substitution $\mathbf{p} \to \mathbf{p}_{op} = -i\hbar\nabla$, namely

$$H \equiv H_{cl}(\mathbf{r},-i\hbar\nabla,t). \tag{3.24}$$

As a result, the time-dependent Schrödinger equation (3.21) may be obtained by starting from the classical equation (3.22), making the substitutions

$$E \to E_{op} = i\hbar\frac{\partial}{\partial t}, \qquad \mathbf{p} \to \mathbf{p}_{op} = -i\hbar\nabla \tag{3.25}$$

and applying the operators E_{op} and $H \equiv H_{cl}(\mathbf{r},-i\hbar\nabla,t)$ on both sides of the equation to the wave function $\Psi(\mathbf{r},t)$. This makes even more apparent the analogy with classical mechanics required by the correspondence principle. Later in this chapter we shall discuss in more detail the classical limit of the Schrödinger equation.

As in the case of the free particle, the time-dependent Schrödinger equation (3.19) for a particle moving in a potential is *linear* and *homogeneous*. Therefore, if $\Psi_1(\mathbf{r}, t)$ and $\Psi_2(\mathbf{r}, t)$ are distinct solutions of (3.19)

$$\Psi(\mathbf{r}, t) = c_1 \Psi_1(\mathbf{r}, t) + c_2 \Psi_2(\mathbf{r}, t) \tag{3.26}$$

is also a solution, where c_1 and c_2 are complex constants. More generally, an arbitrary linear superposition of solutions of the time-dependent Schrödinger equation (3.19) is also a solution of the equation, in agreement with the superposition principle. We also emphasise that the time-dependent Schrödinger equation (3.19) is of *first order in the time derivative* $\partial/\partial t$, so that once the initial value of the wave function is given at some time t_0, namely $\Psi(\mathbf{r}, t_0)$, its value at all other times can be found by solving the equation. The basic mathematical problem is therefore to obtain a solution $\Psi(\mathbf{r}, t)$ of the Schrödinger equation (3.19) which satisfies given initial conditions.

Continuity conditions

Provided the potential $V(\mathbf{r}, t)$ is a continuous function of each of the Cartesian coordinates x, y and z, it is clear from the Schrödinger equation (3.19) that each of $\Psi(\mathbf{r}, t)$, $\partial \Psi/\partial t$ and $\nabla \Psi$ is also a continuous function of x, y and z. If $V(\mathbf{r}, t)$ exhibits finite discontinuities (jumps) as a function of x, y and z, then from the Schrödinger equation we see that $\nabla^2 \Psi$ exhibits corresponding finite jumps as a function of x, y and z. From this it follows that $\nabla \Psi$ must be continuous as a function of x, y and z, for otherwise $\nabla^2 \Psi$ would become infinite at the points where $\nabla \Psi$ changed discontinuously. In turn, since $\nabla \Psi$ is continuous, Ψ and $\partial \Psi/\partial t$ must also be continuous as functions of x, y and z.

Turning now to the time dependence, we see from (3.19) that if $V(\mathbf{r}, t)$ is a continuous function of t, so will be $\Psi(\mathbf{r}, t)$ and $\partial \Psi/\partial t$. However, if $V(\mathbf{r}, t)$ exhibits finite discontinuities as a function of t, so will $\partial \Psi/\partial t$, while $\Psi(\mathbf{r}, t)$ remains a continuous function of t.

3.2 Conservation of probability

As we have explained in Chapter 2, the wave function associated with a particle in quantum mechanics has a *statistical interpretation*. According to Born's postulate (see Section 2.2), if a particle is described by a wave function $\Psi(\mathbf{r}, t)$ normalised to unity the probability of finding the particle at time t within the volume element $\mathrm{d}\mathbf{r} \equiv \mathrm{d}x\mathrm{d}y\mathrm{d}z$ about the point $\mathbf{r} \equiv (x, y, z)$ is

$$P(\mathbf{r}, t)\mathrm{d}\mathbf{r} = |\Psi(\mathbf{r}, t)|^2 \mathrm{d}\mathbf{r} \tag{3.27}$$

so that $P(\mathbf{r}, t) = |\Psi(\mathbf{r}, t)|^2 = \Psi^*(\mathbf{r}, t)\Psi(\mathbf{r}, t)$ is the *position probability density*. Assuming that the particle is confined within a given volume V_0 (for example, a room), we have seen in Chapter 2 that *at any arbitrary time t* the wave function $\Psi(\mathbf{r}, t)$ can be chosen to satisfy the *normalisation condition*

$$\int |\Psi(\mathbf{r}, t)|^2 d\mathbf{r} = 1 \tag{3.28}$$

where the integral extends over all space. This condition simply expresses the fact that the probability of finding the particle somewhere at time t is unity. We noted in Chapter 2 that wave functions for which the integral on the left of (3.28) exists are said to be *square integrable*, and that some wave functions cannot be normalised in this way. For example, in the case of plane waves such as (3.9) the normalisation integral on the left of (3.28) diverges. However, we recognised that this difficulty is due to the fact that a plane wave represents the idealised physical situation of a free particle which has a perfectly well-defined momentum and hence is totally 'delocalised'. If the requirement that the particle should have a completely well-defined momentum is given up, we have seen in Chapter 2 that localised wave packets can be constructed, which can be normalised to unity. For the time being we shall only consider square integrable wave functions Ψ which satisfy the normalisation condition (3.28).

Let us now ask what happens as time changes. The interpretation of $|\Psi(\mathbf{r}, t)|^2$ as a position probability density clearly requires that the probability of finding the particle somewhere must remain unity as time varies, which is to say that *probability is conserved*. In other words, once $\Psi(\mathbf{r}, t)$ is normalised according to (3.28) at a given time t, it must remain so at all times. The normalisation integral on the left of (3.28) must therefore be independent of time, namely

$$\frac{\partial}{\partial t} \int P(\mathbf{r}, t) d\mathbf{r} = 0 \tag{3.29}$$

where $P(\mathbf{r}, t)$ is given by (3.27) and we recall that the integral extends over all space.

In order to verify that this relation is indeed satisfied in Schrödinger's theory, let us start by considering the probability of finding the particle at time t within a finite volume V, namely

$$\int_V P(\mathbf{r}, t) d\mathbf{r} = \int_V |\Psi(\mathbf{r}, t)|^2 d\mathbf{r} \tag{3.30}$$

where the integrals extend over the volume V. The rate of change of this probability is

$$\frac{\partial}{\partial t} \int_V P(\mathbf{r}, t) d\mathbf{r} = \int_V \left[\Psi^* \left(\frac{\partial \Psi}{\partial t} \right) + \left(\frac{\partial \Psi^*}{\partial t} \right) \Psi \right] d\mathbf{r}. \tag{3.31}$$

Now, the time dependence of the wave function $\Psi(\mathbf{r}, t)$ is not arbitrary but is governed by the time-dependent Schrödinger equation (3.19). Using this equation and its complex conjugate

$$-i\hbar \frac{\partial}{\partial t} \Psi^*(\mathbf{r}, t) = \left[-\frac{\hbar^2}{2m} \nabla^2 + V(\mathbf{r}, t) \right] \Psi^*(\mathbf{r}, t) \tag{3.32}$$

where we have used the fact that $V(\mathbf{r}, t)$ is a *real* quantity, we find that

$$\frac{\partial}{\partial t} \int_V P(\mathbf{r}, t) \mathrm{d}\mathbf{r} = \frac{\mathrm{i}\hbar}{2m} \int_V [\Psi^*(\nabla^2 \Psi) - (\nabla^2 \Psi^*)\Psi] \mathrm{d}\mathbf{r}$$

$$= \frac{\mathrm{i}\hbar}{2m} \int_V \nabla \cdot [\Psi^*(\nabla \Psi) - (\nabla \Psi^*)\Psi] \mathrm{d}\mathbf{r}$$

$$= -\int_V \nabla \cdot \mathbf{j} \, \mathrm{d}\mathbf{r}. \tag{3.33}$$

In the last line we have introduced the vector

$$\mathbf{j}(\mathbf{r}, t) = \frac{\hbar}{2m\mathrm{i}} [\Psi^*(\nabla \Psi) - (\nabla \Psi^*)\Psi] \tag{3.34}$$

whose physical significance will be discussed shortly. Using Green's theorem[2], we can convert the volume integral on the right of (3.33) into an integral over the surface S bounding the volume V

$$\frac{\partial}{\partial t} \int_V P(\mathbf{r}, t) \mathrm{d}\mathbf{r} = -\int_S \mathbf{j} \cdot \mathrm{d}\mathbf{S} \tag{3.35}$$

where $\mathrm{d}\mathbf{S}$ is a vector whose magnitude is equal to an element $\mathrm{d}S$ of the surface S, and whose direction is that of the outward normal to $\mathrm{d}S$.

The above relations are valid for any finite volume V. In order to study the time rate of change of the normalisation integral (3.28), we must extend the volume to the entire space. The surface S in (3.35) then recedes to infinity. Since Ψ is square integrable it vanishes at large distances[3] so that the surface integral in (3.35) is equal to zero, and the condition (3.29) is satisfied. It is worth noting that the proof of (3.29) depends on the facts that the Schrödinger equation (3.19) is of first order in the time derivative $\partial/\partial t$ (this allowed us to eliminate the time derivatives on the right of (3.31)) and that the potential energy $V(\mathbf{r}, t)$ is *real*.

Probability conservation and the Hermiticity of the Hamiltonian

The condition (3.29) expressing the conservation in time of the normalisation of the wave function can also be formulated in terms of the Hamiltonian operator H. Indeed,

---

[2] Green's theorem (also called the divergence theorem of Gauss) states that the surface integral of the component of a vector $\mathbf{A}$ along the outward normal taken over a closed surface S is equal to the integral of the divergence of $\mathbf{A}$ taken over the volume V enclosed by the surface. That is

$$\int_V \nabla \cdot \mathbf{A} \, \mathrm{d}\mathbf{r} = \int_S \mathbf{A} \cdot \mathrm{d}\mathbf{S}$$

where $\mathrm{d}\mathbf{S} = \hat{\mathbf{n}} \mathrm{d}S$, $\hat{\mathbf{n}}$ being the unit vector along the outward normal and $\mathrm{d}S$ being an element of the surface S.

[3] If the particle is confined within a volume V_0, then Ψ vanishes outside that volume, and hence over a surface bounding any volume that contains V_0.

using the form (3.21) of the time-dependent Schrödinger equation, and its complex conjugate

$$-i\hbar \frac{\partial}{\partial t}\Psi^*(\mathbf{r},t) = [H\Psi(\mathbf{r},t)]^* \tag{3.36}$$

we can write the left-hand side of (3.29) as

$$\frac{\partial}{\partial t}\int P(\mathbf{r},t)d\mathbf{r} = \frac{\partial}{\partial t}\int |\Psi(\mathbf{r},t)|^2 d\mathbf{r}$$

$$= \int \left[\Psi^*\left(\frac{\partial \Psi}{\partial t}\right) + \left(\frac{\partial \Psi^*}{\partial t}\right)\Psi\right]d\mathbf{r}$$

$$= (i\hbar)^{-1}\int [\Psi^*(H\Psi) - (H\Psi)^*\Psi]d\mathbf{r} \tag{3.37}$$

so that the condition (3.29) becomes

$$\int \Psi^*(H\Psi)d\mathbf{r} = \int (H\Psi)^*\Psi d\mathbf{r}. \tag{3.38}$$

This condition must hold for all square integrable wave functions $\Psi(\mathbf{r},t)$, and hence is a restriction on the Hamiltonian operator H. Operators which satisfy the condition (3.38) for all functions Ψ of the function space in which they act are called *Hermitian operators*. We have thus shown that the requirement of probability conservation implies that the Hamiltonian operator H which appears in the Schrödinger equation (3.21) must be Hermitian when acting on square integrable wave functions. Of course, since the condition (3.29) expressing the conservation in time of the normalisation of $\Psi(\mathbf{r},t)$ has been shown above to follow from the Schrödinger equation (3.19), it follows that the Hamiltonian (3.20) of a particle in a real potential $V(\mathbf{r},t)$ is a Hermitian operator.

Probability current density

Let us now return to equation (3.35). Since the rate of change of the probability of finding the particle in the volume V is equal to the probability flux passing through the surface S bounding V, the vector **j** given by (3.34) can be interpreted as a *probability current density*. The equation

$$\frac{\partial}{\partial t}P(\mathbf{r},t) + \boldsymbol{\nabla}\cdot\mathbf{j}(\mathbf{r},t) = 0 \tag{3.39}$$

which follows from (3.33) is analogous to the *continuity equation* expressing charge conservation in electrodynamics or the conservation of matter in hydrodynamics. It has the familiar form associated with the conservation law for a classical fluid of density P and current density **j** in a medium where there are no sources or sinks. Note that the probability current density (3.34) may also be written as

$$\mathbf{j}(\mathbf{r},t) = \text{Re}\left\{\Psi^* \frac{\hbar}{im}\boldsymbol{\nabla}\Psi\right\}. \tag{3.40}$$

Since the operator $(\hbar/im)\nabla$ represents the quantity $\mathbf{p}/m$ (that is, the velocity $\mathbf{v}$ of the particle) we see that $\mathbf{j}$ corresponds to the product of the velocity and the density. Thus it is reasonable to interpret $\mathbf{j}$ as a probability current density. We also remark from (3.40) that $\mathbf{j}$ vanishes if the wave function Ψ is real. In order to describe situations in which the probability current is non-zero it is therefore necessary to use complex wave functions. Finally, we remark that since the wave function Ψ and its gradient $\nabla\Psi$ are continuous functions of $\mathbf{r}$, the probability current density $\mathbf{j}(\mathbf{r}, t)$ as well as the probability density $P(\mathbf{r}, t)$ have no discontinuous changes as $\mathbf{r}$ varies.

3.3 Expectation values and operators

Let $\Psi(\mathbf{r}, t)$ be the wave function of a particle, normalised to unity. Given the probabilistic interpretation attached to Ψ, we shall now show how information concerning the behaviour of the particle can be extracted from the knowledge of Ψ. First of all, since $P(\mathbf{r}, t)\mathrm{d}\mathbf{r} = \Psi^*(\mathbf{r}, t)\Psi(\mathbf{r}, t)\mathrm{d}\mathbf{r}$ is the probability of finding the particle in the volume element $\mathrm{d}\mathbf{r}$ about the point $\mathbf{r}$ at the time t, the *expectation* value (or average value) of the position vector $\mathbf{r}$ of the particle, which we shall write $\langle \mathbf{r} \rangle$, is given by

$$\langle \mathbf{r} \rangle = \int \mathbf{r} P(\mathbf{r}, t)\mathrm{d}\mathbf{r}$$

$$= \int \Psi^*(\mathbf{r}, t)\mathbf{r}\Psi(\mathbf{r}, t)\mathrm{d}\mathbf{r}. \tag{3.41}$$

The order of the factors in the integrand is obviously immaterial; the one we have adopted in the second line of (3.41) has been chosen for future convenience.

As a result of the interpretation of $P(\mathbf{r}, t)$ as the position probability density, the physical meaning to be attributed to the expectation value $\langle \mathbf{r} \rangle$ is the following: it is the average value of the measurements of $\mathbf{r}$ performed on a very large number of equivalent, identically prepared independent systems represented by the wave function Ψ. We remark that equation (3.41) is equivalent to the three equations

$$\langle x \rangle = \int \Psi^*(\mathbf{r}, t)x\Psi(\mathbf{r}, t)\mathrm{d}\mathbf{r} \tag{3.42a}$$

$$\langle y \rangle = \int \Psi^*(\mathbf{r}, t)y\Psi(\mathbf{r}, t)\mathrm{d}\mathbf{r} \tag{3.42b}$$

$$\langle z \rangle = \int \Psi^*(\mathbf{r}, t)z\Psi(\mathbf{r}, t)\mathrm{d}\mathbf{r} \tag{3.42c}$$

It should be noted that the expectation values are functions only of the time, since the space coordinates have been integrated out. We also remark that the quantities $\langle x \rangle$, $\langle y \rangle$ and $\langle z \rangle$ are obviously real, so that the three components x, y and z, considered as operators acting on the wave function $\Psi(\mathbf{r}, t)$ on the right of equations (3.42) are Hermitian (see (3.38)).

More generally, the expectation value of an arbitrary function $f(\mathbf{r}, t) \equiv f(x, y, z, t)$ of the coordinates of the particle and of the time is given by

$$\langle f(\mathbf{r}, t) \rangle = \int f(\mathbf{r}, t) P(\mathbf{r}, t) d\mathbf{r}$$

$$= \int \Psi^*(\mathbf{r}, t) f(\mathbf{r}, t) \Psi(\mathbf{r}, t) d\mathbf{r} \tag{3.43}$$

provided, of course, that the integral exists. The order of the factors in the integrand has again been chosen for future convenience. Note that $f(\mathbf{r}, t)$, considered as an operator acting on $\Psi(\mathbf{r}, t)$, is Hermitian if the function f is real. As an example, the expectation value of the potential energy is

$$\langle V(\mathbf{r}, t) \rangle = \int \Psi^*(\mathbf{r}, t) V(\mathbf{r}, t) \Psi(\mathbf{r}, t) d\mathbf{r}. \tag{3.44}$$

In order to calculate the expectation value of the momentum $\mathbf{p}$ of the particle, we recall that if $\Phi(\mathbf{p}, t)$ is the wave function of the particle in momentum space, normalised to unity, then $\Pi(\mathbf{p}, t) d\mathbf{p} = \Phi^*(\mathbf{p}, t) \Phi(\mathbf{p}, t) d\mathbf{p}$ is the probability of finding at time t the momentum of the particle within the volume element $d\mathbf{p} \equiv dp_x dp_y dp_z$ about the point $\mathbf{p} \equiv (p_x, p_y, p_z)$ in momentum space (see (2.63)). The expectation value of $\mathbf{p}$ is therefore given by

$$\langle \mathbf{p} \rangle = \int \mathbf{p} \Pi(\mathbf{p}, t) d\mathbf{p}$$

$$= \int \Phi^*(\mathbf{p}, t) \mathbf{p} \Phi(\mathbf{p}, t) d\mathbf{p} \tag{3.45}$$

an equation which is equivalent to the three equations

$$\langle p_x \rangle = \int \Phi^*(\mathbf{p}, t) p_x \Phi(\mathbf{p}, t) d\mathbf{p} \tag{3.46a}$$

$$\langle p_y \rangle = \int \Phi^*(\mathbf{p}, t) p_y \Phi(\mathbf{p}, t) d\mathbf{p} \tag{3.46b}$$

$$\langle p_z \rangle = \int \Phi^*(\mathbf{p}, t) p_z \Phi(\mathbf{p}, t) d\mathbf{p}. \tag{3.46c}$$

More generally, the expectation value of an arbitrary function $g(\mathbf{p}, t) \equiv g(p_x, p_y, p_z, t)$ is

$$\langle g(\mathbf{p}, t) \rangle = \int g(\mathbf{p}, t) \Pi(\mathbf{p}, t) d\mathbf{p}$$

$$= \int \Phi^*(\mathbf{p}, t) g(\mathbf{p}, t) \Phi(\mathbf{p}, t) d\mathbf{p}. \tag{3.47}$$

For example, the expectation value of the kinetic energy is

$$\left\langle \frac{\mathbf{p}^2}{2m} \right\rangle = \int \Phi^*(\mathbf{p}, t) \frac{\mathbf{p}^2}{2m} \Phi(\mathbf{p}, t) d\mathbf{p} \tag{3.48}$$

We shall now obtain another form of the expectation value of **p** which involves the wave function in configuration space, $\Psi(\mathbf{r}, t)$, rather than the wave function in momentum space, $\Phi(\mathbf{p}, t)$. To this end, we consider first the expectation value of p_x, given by (3.46a). Using (2.61) to express $\Phi(\mathbf{p}, t)$ in terms of $\Psi(\mathbf{r}, t)$ and writing in a similar way

$$\Phi^*(\mathbf{p}, t) = (2\pi\hbar)^{-3/2} \int e^{i\mathbf{p}\cdot\mathbf{r}'/\hbar} \Psi^*(\mathbf{r}', t) d\mathbf{r}' \tag{3.49}$$

we obtain upon substitution in (3.46a)

$$\langle p_x \rangle = (2\pi\hbar)^{-3} \int d\mathbf{p} \int d\mathbf{r} \int d\mathbf{r}' e^{i\mathbf{p}\cdot\mathbf{r}'/\hbar} \Psi^*(\mathbf{r}', t) p_x e^{-i\mathbf{p}\cdot\mathbf{r}/\hbar} \Psi(\mathbf{r}, t). \tag{3.50}$$

Now we observe that

$$p_x e^{-i\mathbf{p}\cdot\mathbf{r}/\hbar} = p_x e^{-i(p_x x + p_y y + p_z z)/\hbar}$$
$$= i\hbar \frac{\partial}{\partial x} e^{-i\mathbf{p}\cdot\mathbf{r}/\hbar} \tag{3.51}$$

so that we may write (3.50) in the form

$$\langle p_x \rangle = (2\pi\hbar)^{-3} \int d\mathbf{p} \int d\mathbf{r} \int d\mathbf{r}' e^{i\mathbf{p}\cdot\mathbf{r}'/\hbar} \Psi^*(\mathbf{r}', t) \left(i\hbar \frac{\partial}{\partial x} e^{-i\mathbf{p}\cdot\mathbf{r}/\hbar} \right) \Psi(\mathbf{r}, t). \tag{3.52}$$

Let us now integrate by parts with respect to x. The integrated part, which is proportional to the value of the wave function $\Psi(\mathbf{r}, t)$ at $|x| = \infty$, vanishes since $\Psi(\mathbf{r}, t)$ is a normalisable wave function which is equal to zero in the limit $|x| \to \infty$. We therefore have

$$\langle p_x \rangle = (2\pi\hbar)^{-3} \int d\mathbf{p} \int d\mathbf{r} \int d\mathbf{r}' \Psi^*(\mathbf{r}', t) e^{i\mathbf{p}\cdot(\mathbf{r}'-\mathbf{r})/\hbar} \left[-i\hbar \frac{\partial}{\partial x} \Psi(\mathbf{r}, t) \right]. \tag{3.53}$$

Now, from equation (A.50) of Appendix A we have

$$\delta(\mathbf{r}' - \mathbf{r}) = (2\pi\hbar)^{-3} \int e^{i\mathbf{p}\cdot(\mathbf{r}'-\mathbf{r})/\hbar} d\mathbf{p} \tag{3.54}$$

where δ is the Dirac delta function. The **p** integration in (3.53) is thus readily performed to yield

$$\langle p_x \rangle = \int d\mathbf{r} \int d\mathbf{r}' \Psi^*(\mathbf{r}', t) \delta(\mathbf{r}' - \mathbf{r}) \left[-i\hbar \frac{\partial}{\partial x} \Psi(\mathbf{r}, t) \right] \tag{3.55}$$

and from equation (A.51) of Appendix A, we can perform the $\mathbf{r}'$ integration to obtain

$$\langle p_x \rangle = \int \Psi^*(\mathbf{r}, t) \left(-i\hbar \frac{\partial}{\partial x} \right) \Psi(\mathbf{r}, t) d\mathbf{r}. \tag{3.56a}$$

Proceeding in a similar way with $\langle p_y \rangle$ and $\langle p_z \rangle$, we find that

$$\langle p_y \rangle = \int \Psi^*(\mathbf{r}, t) \left(-i\hbar \frac{\partial}{\partial y} \right) \Psi(\mathbf{r}, t) d\mathbf{r} \tag{3.56b}$$

and
$$\langle p_z \rangle = \int \Psi^*(\mathbf{r},t)\left(-i\hbar\frac{\partial}{\partial z}\right)\Psi(\mathbf{r},t)d\mathbf{r}. \tag{3.56c}$$

The three equations (3.56) are equivalent to the single equation

$$\langle \mathbf{p} \rangle = \int \Psi^*(\mathbf{r},t)(-i\hbar\nabla)\Psi(\mathbf{r},t)d\mathbf{r}. \tag{3.57}$$

We also remark that equations (3.46) imply that the expectation values $\langle p_x \rangle$, $\langle p_y \rangle$ and $\langle p_z \rangle$ are real, so that the operators $(-i\hbar\partial/\partial x)$, $(-i\hbar\partial/\partial y)$ and $(-i\hbar\partial/\partial z)$ acting on the wave function $\Psi(\mathbf{r},t)$ are Hermitian.

The result (3.57) can be generalised to more complicated functions. For example, if n is a positive integer, we have by using a straightforward generalisation of the above method (Problem 3.3)

$$\langle p_x^n \rangle = \int \Psi^*(\mathbf{r},t)\left(-i\hbar\frac{\partial}{\partial x}\right)^n \Psi(\mathbf{r},t)d\mathbf{r}. \tag{3.58}$$

More generally, if $g(p_x, p_y, p_z, t)$ is a polynomial or an absolutely convergent series in p_x, p_y, p_z, one has

$$\langle g(p_x,p_y,p_z,t) \rangle = \int \Psi^*(\mathbf{r},t) g\left(-i\hbar\frac{\partial}{\partial x},-i\hbar\frac{\partial}{\partial y},-i\hbar\frac{\partial}{\partial z},t\right)\Psi(\mathbf{r},t)d\mathbf{r} \tag{3.59}$$

or, using a more compact notation

$$\langle g(\mathbf{p},t) \rangle = \int \Psi^*(\mathbf{r},t) g(-i\hbar\nabla,t)\Psi(\mathbf{r},t)d\mathbf{r} \tag{3.60}$$

and we note that the operator $g(-i\hbar\nabla, t)$ is Hermitian if the function g is real.

Looking back at the expectation values (3.41)–(3.44) and (3.56)–(3.60) obtained from the wave function $\Psi(\mathbf{r},t)$ in configuration space, we see that they can all be written in the form

$$\langle A \rangle = \int \Psi^*(\mathbf{r},t) A \Psi(\mathbf{r},t)d\mathbf{r} \tag{3.61}$$

where A is the operator associated with the quantity whose average value is to be evaluated. If we are dealing with a function $f(\mathbf{r},t)$ of the coordinates of the particles and of the time, then the action of A on Ψ consists of multiplying the wave function Ψ by the function $f(\mathbf{r},t)$. If the quantity whose expectation value is to be calculated is a function $g(\mathbf{p},t)$, the operator A is obtained by making in $g(\mathbf{p},t)$ the substitution $\mathbf{p} \to -i\hbar\nabla$; we note that this is precisely the substitution which allowed us to obtain the Schrödinger equation (see (3.13)).

It is important to remark that the operators which we have considered thus far are either multipliers or derivatives, and hence belong to the category of *linear* operators. An operator A is said to be linear if its action on any two functions Ψ_1 and Ψ_2 is such that

$$A(c_1\Psi_1 + c_2\Psi_2) = c_1(A\Psi_1) + c_2(A\Psi_2) \tag{3.62}$$

where c_1 and c_2 are arbitrary complex numbers. Most of the operators relevant in quantum mechanics are linear, and in what follows, unless otherwise stated, the term 'operator' will mean a linear operator.

Until now we have considered quantities which are *either* functions $f(\mathbf{r}, t)$ of $\mathbf{r}$ and t *or* functions $g(\mathbf{p}, t)$ of $\mathbf{p}$ and t. However, some dynamical variables can be functions of $\mathbf{r}$, $\mathbf{p}$ and t, and thus can contain $\mathbf{r}$ and $\mathbf{p}$ simultaneously. This is the case, for example, for the total energy $E = \mathbf{p}^2/2m + V(\mathbf{r}, t)$ of a particle moving in a potential, to which is associated the Hamiltonian operator (3.20), namely $H = -(\hbar^2/2m)\nabla^2 + V(\mathbf{r}, t)$. In order to find a method for evaluating the average value of E, we shall in accordance with the correspondence principle, require that

$$\langle E \rangle = \left\langle \frac{\mathbf{p}^2}{2m} \right\rangle + \langle V \rangle \tag{3.63}$$

so that we obtain the correct averages in the classical limit (see (3.22)–(3.23)). Remembering the substitution rules (3.25) we may also write the relation (3.63) as

$$\left\langle i\hbar \frac{\partial}{\partial t} \right\rangle = \left\langle -\frac{\hbar^2}{2m}\nabla^2 \right\rangle + \langle V \rangle \tag{3.64}$$

and we see that the above equation is consistent with the Schrödinger equation (3.19) provided that the expectation value is defined in the general case with the operator acting on the right on Ψ, and multiplied on the left by Ψ^*. Thus we have

$$\langle E \rangle = \int \Psi^*(\mathbf{r}, t)\left(i\hbar\frac{\partial}{\partial t}\right)\Psi(\mathbf{r}, t)\mathrm{d}\mathbf{r}$$

$$= \int \Psi^*(\mathbf{r}, t)H\Psi(\mathbf{r}, t)\mathrm{d}\mathbf{r} = \int \Psi^*(\mathbf{r}, t)\left[-\frac{\hbar^2}{2m}\nabla^2 + V(\mathbf{r}, t)\right]\Psi(\mathbf{r}, t)\mathrm{d}\mathbf{r}$$

$$= \langle H \rangle. \tag{3.65}$$

Generalising the above results, we are therefore led to postulate that if the dynamical state of a particle is described by the configuration space wave function $\Psi(\mathbf{r}, t)$, normalised to unity, the expectation value of a dynamical variable must be calculated as follows.

(1) One first associates with the dynamical variable $\mathcal{A} \equiv A(\mathbf{r}, \mathbf{p}, t)$ representing a physical quantity, the linear operator

$$A(\mathbf{r}, -i\hbar\nabla, t) \tag{3.66}$$

obtained by performing the substitution $\mathbf{p} \to -i\hbar\nabla$ wherever the momentum $\mathbf{p}$ occurs. This rule, however, needs some qualifications, which will be discussed shortly.

(2) One then calculates the required expectation value from the expression

$$\langle A \rangle = \int \Psi^*(\mathbf{r}, t) A(\mathbf{r}, -i\hbar\nabla, t)\Psi(\mathbf{r}, t)\mathrm{d}\mathbf{r}. \tag{3.67}$$

3.3 Expectation values and operators

Table 3.1 Physical quantities and corresponding operators acting in configuration space.

Physical quantity	Operator
Position coordinate x	x
Position vector $\mathbf{r}$	$\mathbf{r}$
x component of momentum p_x	$-i\hbar \frac{\partial}{\partial x}$
Momentum $\mathbf{p}$	$-i\hbar \nabla$
Kinetic energy $T = \frac{\mathbf{p}^2}{2m}$	$-\frac{\hbar^2}{2m}\nabla^2$
Potential energy $V(\mathbf{r}, t)$	$V(\mathbf{r}, t)$
Total energy $\frac{\mathbf{p}^2}{2m} + V(\mathbf{r}, t)$	$H = -\frac{\hbar^2}{2m}\nabla^2 + V(\mathbf{r}, t)$

If the wave function $\Psi(\mathbf{r}, t)$ is not normalised to unity, the expression (3.67) must of course be replaced by

$$\langle A \rangle = \frac{\int \Psi^*(\mathbf{r}, t) A(\mathbf{r}, -i\hbar \nabla, t) \Psi(\mathbf{r}, t) d\mathbf{r}}{\int \Psi^*(\mathbf{r}, t) \Psi(\mathbf{r}, t) d\mathbf{r}}. \tag{3.68}$$

The operators A associated with physical quantities $\mathcal{A}$ are subject to an important restriction, which arises in the following way. The results of measurements of $\mathcal{A}$, and hence the expectation value $\langle A \rangle$, must obviously be *real* quantities. As a consequence, for any wave function Ψ, the condition

$$\int \Psi^* A \Psi d\mathbf{r} = \int (A\Psi)^* \Psi d\mathbf{r} \tag{3.69}$$

has to be satisfied, which means that the operator A associated with the dynamical quantity $\mathcal{A}$ must be *Hermitian*. A few important (linear) Hermitian operators acting in configuration space are listed in Table 3.1, together with the physical quantities to which they correspond.

As we have already seen, the requirement of the reality of expectation values – and hence the Hermitian character of the operators A associated with dynamical quantities – is automatically satisfied for real functions $f(\mathbf{r}, t)$ or $g(\mathbf{p}, t)$. However, for functions which depend on *both* $\mathbf{r}$ and $\mathbf{p}$, this is not necessarily the case. Consider, for example, a particle moving in one dimension, described by the wave function $\Psi(x, t)$, normalised to unity, and suppose that we wish to calculate the expectation value of the quantity xp_x. Following the rule stated above, we write

$$\langle xp_x \rangle = \int_{-\infty}^{+\infty} \Psi^*(x, t) x \left(-i\hbar \frac{\partial}{\partial x} \right) \Psi(x, t) dx. \tag{3.70}$$

Integrating by parts and dropping the arguments of Ψ and Ψ^* to simplify the notation, we have

$$\langle xp_x \rangle = [-i\hbar x \Psi^* \Psi]_{-\infty}^{+\infty} + i\hbar \int_{-\infty}^{+\infty} \Psi \frac{\partial}{\partial x}(x\Psi^*) dx. \tag{3.71}$$

The integrated part vanishes since $|\Psi| \to 0$ when $|x| \to \infty$. Thus

$$\langle xp_x \rangle = i\hbar \int_{-\infty}^{+\infty} \Psi x \frac{\partial \Psi^*}{\partial x} dx + i\hbar \int_{-\infty}^{+\infty} \Psi^* \Psi dx$$
$$= \langle xp_x \rangle^* + i\hbar \tag{3.72}$$

where $\langle xp_x \rangle^*$ denotes the complex conjugate of $\langle xp_x \rangle$. Since the second term on the right of (3.72) is non-zero, it is clear that $\langle xp_x \rangle$ is *not* a real quantity, and therefore that the operator $xp_x \equiv x(-i\hbar \partial/\partial x)$ is not Hermitian. Similarly, it is readily checked (Problem 3.4) that

$$\langle p_x x \rangle = \langle p_x x \rangle^* - i\hbar \tag{3.73}$$

so that $\langle p_x x \rangle$ is not real and the operator $p_x x \equiv (-i\hbar \partial/\partial x)x$ is not Hermitian. However, a glance at equations (3.72) and (3.73) shows that

$$\left\langle \frac{xp_x + p_x x}{2} \right\rangle = \left\langle \frac{xp_x + p_x x}{2} \right\rangle^* \tag{3.74}$$

so that the operator $(xp_x + p_x x)/2$, obtained by taking the *mean* of the possible orders in which x and p_x appear, is Hermitian.

It is important to stress that in contrast to classical mechanics, where all quantities obey the rules of ordinary algebra, we are dealing in quantum mechanics with *operators*, which in general *do not commute* with each other. That is, if A and B are two operators, the product AB is not necessarily equal to the product BA. The *commutator* of two operators A and B is defined as the difference $AB - BA$ and is denoted by the symbol $[A, B]$:

$$[A, B] = AB - BA. \tag{3.75}$$

If their commutator vanishes, the two operators A and B commute: $AB = BA$.

As an example of operators which do not commute, let us consider the two operators x and $p_x = -i\hbar \partial/\partial x$. For any wave function $\Psi(\mathbf{r}, t)$, we have

$$[x, p_x]\Psi = (xp_x - p_x x)\Psi$$
$$= -i\hbar \left[x \frac{\partial \Psi}{\partial x} - \frac{\partial}{\partial x}(x\Psi) \right]$$
$$= i\hbar \Psi \tag{3.76}$$

so that we may write the relation $[x, p_x] = i\hbar$. More generally,

$$[x, p_x] = [y, p_y] = [z, p_z] = i\hbar. \tag{3.77}$$

As a consequence of the non-commutativity of operators, different quantum mechanical operators may correspond to equivalent classical quantities. For example, to the three equivalent classical expressions

$$xp_x, \; p_x x, \; \tfrac{1}{2}(xp_x + p_x x) \tag{3.78}$$

correspond the three different operators

$$x\left(-i\hbar\frac{\partial}{\partial x}\right), \left(-i\hbar\frac{\partial}{\partial x}\right)x, \frac{1}{2}\left[x\left(-i\hbar\frac{\partial}{\partial x}\right) + \left(-i\hbar\frac{\partial}{\partial x}\right)x\right] \quad (3.79)$$

and we have seen above that only the last one is Hermitian.

We can now give a more precise meaning to the rule (3.66) which associates a linear operator A with a dynamical quantity. As stated previously, we must perform the substitution $\mathbf{p} \to -i\hbar\nabla$ wherever the momentum $\mathbf{p}$ appears. However, when necessary, we must also remove the ambiguity in the order of the factors of x and p_x (and similarly for y, p_y and z, p_z). To this end we shall require that the resulting operator A be Hermitian. In the case of the products xp_x and p_xx we have seen that this 'Hermitisation' can be accomplished by taking the mean between the two possible orders of x and p_x. This result can be generalised (Problem 3.5) and, in practice, the following procedure can be used:

(1) The function is ordered in such a way that all the factors involving x occur together, as do all the factors involving p_x (and similarly for y, p_y and z, p_z).
(2) One then replaces $x^k p_x^l$ by $(x^k p_x^l + p_x^l x^k)/2$, which guarantees that the operator A will be Hermitian. Although this procedure is still not completely unambiguous[4], it will be fully adequate to handle all the cases studied in this book.

We conclude this section with the following remark. Although we started our discussion of expectation values by giving equal emphasis to the configuration-space wave function $\Psi(\mathbf{r}, t)$ and the momentum-space wave function $\Phi(\mathbf{p}, t)$ (compare for example equations (3.41) and (3.45)), we continued our treatment by using exclusively the configuration-space wave function $\Psi(\mathbf{r}, t)$. The reason is that the Schrödinger equation and its interpretation have been developed earlier in this chapter in the more familiar configuration space. At the end of this chapter we shall restore the symmetry between the treatments in configuration and momentum space by discussing the Schrödinger equation, as well as operators and expectation values, in momentum space.

3.4 Transition from quantum mechanics to classical mechanics. The Ehrenfest theorem

According to the correspondence principle, we expect that the motion of a wave packet should agree with that of the corresponding classical particle whenever the distances and momenta involved in describing the motion of the particle are so large that the uncertainty principle may be ignored. In order to investigate this point, we

---
[4] For a detailed discussion of this point, see Bohm (1951), Chapter 9.

shall prove a theorem which is due to P. Ehrenfest (1927). It states that Newton's fundamental equations of classical dynamics, written in the form[5]

$$\frac{d\mathbf{r}}{dt} = \frac{\mathbf{p}}{m} \qquad (3.80a)$$

and

$$\frac{d\mathbf{p}}{dt} = -\nabla V \qquad (3.80b)$$

are exactly satisfied by the *expectation (average) values* of the corresponding operators in quantum mechanics, these expectation values being calculated according to equation (3.67), where $\Psi(\mathbf{r}, t)$ is a square-integrable solution of the Schrödinger equation (3.19), normalised to unity. It is worth stressing that well-defined trajectories do not exist in quantum mechanics, so that equations such as (3.80) which give the values of $d\mathbf{r}/dt$ and $d\mathbf{p}/dt$ along a classical path cannot be written. However, it is possible to study the *time rates of change of the expectation values* $\langle \mathbf{r} \rangle$ and $\langle \mathbf{p} \rangle$.

Let us consider first the expectation value of x, which is given by (3.42a). The time rate of change of $\langle x \rangle$ is

$$\frac{d}{dt}\langle x \rangle = \frac{d}{dt} \int \Psi^*(\mathbf{r}, t) x \Psi(\mathbf{r}, t) d\mathbf{r}$$

$$= \int \Psi^*(\mathbf{r}, t) x \frac{\partial \Psi(\mathbf{r}, t)}{\partial t} d\mathbf{r} + \int \frac{\partial \Psi^*(\mathbf{r}, t)}{\partial t} x \Psi(\mathbf{r}, t) d\mathbf{r}. \qquad (3.81)$$

This equation can be transformed by using the Schrödinger equation (3.21) and its complex conjugate (3.36). Dropping the arguments of Ψ and Ψ^* for notational simplicity, we have in this way

$$\frac{d}{dt}\langle x \rangle = (i\hbar)^{-1} \left[\int \Psi^* x (H\Psi) d\mathbf{r} - \int (H\Psi)^* x \Psi d\mathbf{r} \right]$$

$$= (i\hbar)^{-1} \left[\int \Psi^* x \left(-\frac{\hbar^2}{2m} \nabla^2 \Psi + V\Psi \right) d\mathbf{r} \right.$$

$$\left. - \int \left(-\frac{\hbar^2}{2m} \nabla^2 \Psi^* + V\Psi^* \right) x \Psi d\mathbf{r} \right]. \qquad (3.82)$$

The terms involving the potential V cancel out, so that

$$\frac{d}{dt}\langle x \rangle = \frac{i\hbar}{2m} \int [\Psi^* x (\nabla^2 \Psi) - (\nabla^2 \Psi^*) x \Psi] d\mathbf{r}. \qquad (3.83)$$

---

[5] Note that equations (3.80) indeed reduce to the familiar Newtonian equations

$$m \frac{d^2 \mathbf{r}}{dt^2} = \frac{d\mathbf{p}}{dt} = \mathbf{F}$$

where the force $\mathbf{F}$ is assumed to derive from a potential, $\mathbf{F} = -\nabla V$.

Let us consider the second contribution to the integral. Using Green's first identity[6] and remembering that the volume V is the entire space, we obtain

$$\int (\nabla^2 \Psi^*) x \Psi \, d\mathbf{r} = \int_S x \Psi (\nabla \Psi^*) . d\mathbf{S} - \int (\nabla \Psi^*) . \nabla (x \Psi) d\mathbf{r}. \tag{3.84}$$

The first integral on the right is over the infinite bounding surface S, and hence is equal to zero because the wave function Ψ vanishes at large distances. Consequently, we have

$$\int (\nabla^2 \Psi^*) x \Psi \, d\mathbf{r} = - \int (\nabla \Psi^*) . \nabla (x \Psi) d\mathbf{r}. \tag{3.85}$$

Using again Green's first identity, we have

$$- \int (\nabla \Psi^*) . \nabla (x \Psi) d\mathbf{r} = - \int_S \Psi^* \nabla (x \Psi) . d\mathbf{S} + \int \Psi^* \nabla^2 (x \Psi) d\mathbf{r}. \tag{3.86}$$

Again the surface integral vanishes, so that

$$\int (\nabla^2 \Psi^*) x \Psi \, d\mathbf{r} = \int \Psi^* \nabla^2 (x \Psi) d\mathbf{r}. \tag{3.87}$$

Putting this result back into the equation (3.83) for $d\langle x \rangle / dt$, we find that

$$\frac{d}{dt} \langle x \rangle = \frac{i\hbar}{2m} \int \Psi^* [x \nabla^2 \Psi - \nabla^2 (x \Psi)] d\mathbf{r}$$

$$= -\frac{i\hbar}{m} \int \Psi^* \frac{\partial \Psi}{\partial x} d\mathbf{r}. \tag{3.88}$$

On the other hand, the expectation value of the x-component of the momentum is given by (3.56a), so that we have

$$\frac{d}{dt} \langle x \rangle = \frac{\langle p_x \rangle}{m} \tag{3.89}$$

which is the quantum counterpart of the x-component of the classical equation (3.80a).

Let us now calculate the time rate of change of $\langle p_x \rangle$. We have from (3.56a)

$$\frac{d}{dt} \langle p_x \rangle = -i\hbar \frac{d}{dt} \int \Psi^* \frac{\partial \Psi}{\partial x} d\mathbf{r}$$

$$= -i\hbar \left[\int \Psi^* \frac{\partial}{\partial x} \frac{\partial \Psi}{\partial t} d\mathbf{r} + \int \frac{\partial \Psi^*}{\partial t} \frac{\partial \Psi}{\partial x} d\mathbf{r} \right]. \tag{3.90}$$

---

[6] Green's first identity states that if $u(x, y, z)$ and $v(x, y, z)$ are scalar functions of position with continuous derivatives of at least the second order, then

$$\int_V [u(\nabla^2 v) + (\nabla u).(\nabla v)] d\mathbf{r} = \int_S u(\nabla v).d\mathbf{S}$$

where V is a volume bounded by the closed surface S. To obtain (3.84) set $u = x \Psi$ and $v = \Psi^*$.

Replacing, respectively, $\partial\Psi/\partial t$ and $\partial\Psi^*/\partial t$ according to the Schrödinger equation (3.21) and its complex conjugate (3.36), we obtain

$$\frac{d}{dt}\langle p_x \rangle = -\int \Psi^* \frac{\partial}{\partial x}\left(-\frac{\hbar^2}{2m}\nabla^2\Psi + V\Psi\right)d\mathbf{r} + \int \left(-\frac{\hbar^2}{2m}\nabla^2\Psi^* + V\Psi^*\right)\frac{\partial\Psi}{\partial x}d\mathbf{r}$$

$$= \frac{\hbar^2}{2m}\int \left[\Psi^*\left(\nabla^2 \frac{\partial\Psi}{\partial x}\right) - (\nabla^2\Psi^*)\frac{\partial\Psi}{\partial x}\right]d\mathbf{r}$$

$$- \int \Psi^*\left[\frac{\partial}{\partial x}(V\Psi) - V\frac{\partial\Psi}{\partial x}\right]d\mathbf{r}. \tag{3.91}$$

Assuming that $\partial\Psi/\partial x$, as well as Ψ, vanishes at large distances, the first integral on the right of (3.91) is equal to zero by Green's second identity[7], in which $u = \Psi^*$ and $v = \partial\Psi/\partial x$. The second integral on the right of (3.91) is just

$$-\int \Psi^*\left[\frac{\partial}{\partial x}(V\Psi) - V\frac{\partial\Psi}{\partial x}\right]d\mathbf{r} = -\int \Psi^* \frac{\partial V}{\partial x}\Psi d\mathbf{r}$$

$$= -\left\langle \frac{\partial V}{\partial x}\right\rangle \tag{3.92}$$

so that

$$\frac{d}{dt}\langle p_x \rangle = -\left\langle \frac{\partial V}{\partial x}\right\rangle \tag{3.93}$$

which is the quantum counterpart of the x-component of the classical equation (3.80b). Equations (3.89) and (3.93), together with similar ones for the y- and z-components, constitute the mathematical formulation of the Ehrenfest theorem.

3.5 The time-independent Schrödinger equation. Stationary states

Let us now consider the particular case such that the potential energy V of the particle does not depend on the time. The Hamiltonian operator $H = -(\hbar^2/2m)\nabla^2 + V(\mathbf{r})$ is then also time-independent, and the time-dependent Schrödinger equation (3.19) simplifies considerably. Indeed, we shall show below that it admits particular solutions of the form

$$\Psi(\mathbf{r}, t) = \psi(\mathbf{r})f(t) \tag{3.94}$$

which are products of functions of $\mathbf{r}$ and t separately; we shall also see that the general solution of (3.19) can be expressed as a sum of such 'separable' solutions.

---

[7] Green's second identity states that if $u(x, y, z)$ and $v(x, y, z)$ are scalar functions of position with continuous derivatives of at least the second order, then

$$\int_V [u(\nabla^2 v) - v(\nabla^2 u)]d\mathbf{r} = \int_S [u(\nabla v) - v(\nabla u)] \cdot d\mathbf{S}$$

where V is a volume bounded by the closed surface S.

3.5 The time-independent Schrödinger equation. Stationary states

In order to prove that particular solutions of the Schrödinger equation (3.19) can be written in the product form (3.94) when V is independent of t, we apply to that equation the method of *separation of variables*. Substituting (3.94) into (3.19), we have

$$i\hbar \psi(\mathbf{r}) \frac{\mathrm{d}f(t)}{\mathrm{d}t} = \left[-\frac{\hbar^2}{2m} \nabla^2 \psi(\mathbf{r}) + V(\mathbf{r}) \psi(\mathbf{r}) \right] f(t). \tag{3.95}$$

Dividing both sides of this equation by $\Psi(\mathbf{r}, t) = \psi(\mathbf{r}) f(t)$, we find that

$$i\hbar \frac{1}{f(t)} \frac{\mathrm{d}f(t)}{\mathrm{d}t} = \frac{1}{\psi(\mathbf{r})} \left[-\frac{\hbar^2}{2m} \nabla^2 \psi(\mathbf{r}) + V(\mathbf{r}) \psi(\mathbf{r}) \right]. \tag{3.96}$$

Since the left-hand side depends only on t and the right-hand side only on $\mathbf{r}$, both sides must be equal to a constant. As this constant has the dimensions of an energy, we shall denote it tentatively by E. We therefore obtain the two equations

$$i\hbar \frac{\mathrm{d}}{\mathrm{d}t} f(t) = E f(t) \tag{3.97}$$

and

$$\left[-\frac{\hbar^2}{2m} \nabla^2 + V(\mathbf{r}) \right] \psi(\mathbf{r}) = E \psi(\mathbf{r}). \tag{3.98}$$

The first equation can be immediately integrated to give

$$f(t) = C \exp(-iEt/\hbar) \tag{3.99}$$

where C is an arbitrary constant. Because the solution (3.94) is in product form and equations (3.97) and (3.98) are homogeneous, there is no loss of generality in taking $C = 1$ and writing the particular solution (3.94) in the form

$$\Psi(\mathbf{r}, t) = \psi(\mathbf{r}) \exp(-iEt/\hbar). \tag{3.100}$$

Equation (3.98), which must be satisfied by the function $\psi(\mathbf{r})$, is called the *time-independent Schrödinger equation*. In contrast to the time-dependent Schrödinger equation (3.19), which describes the time development of the wave function $\Psi(\mathbf{r}, t)$, we shall see shortly that equation (3.98) is an *eigenvalue equation*. Before proceeding with the analysis of (3.98), it is therefore useful to review a few basic features of eigenvalue equations.

Eigenvalue equations are equations of the type

$$A \psi_n = a_n \psi_n \tag{3.101}$$

where A is an operator and a_n is a number. A solution ψ_n of such an equation is called an *eigenfunction* corresponding to the *eigenvalue* a_n of the operator A. We see that the eigenvalue equation (3.101) states that the operator A, acting on certain functions ψ_n (the eigenfunctions) will give back these functions multiplied by constants a_n (the eigenvalues). In order that this eigenvalue problem be well defined, the conditions of regularity and the boundary conditions to be satisfied by the functions ψ_n must be specified. If more than one linearly independent eigenfunction corresponds to the same eigenvalue a_n, this eigenvalue is said to be *degenerate*; the *degree of degeneracy* is defined as the number of linearly independent eigenfunctions corresponding to that eigenvalue.

Let us now return to the time-independent Schrödinger equation (3.98). Remembering that the Hamiltonian operator of the particle is given by $H = -(\hbar^2/2m)\nabla^2 + V(\mathbf{r})$, we may write this equation in the form of the eigenvalue equation

$$H\psi(\mathbf{r}) = E\psi(\mathbf{r}) \tag{3.102}$$

where it is understood that the eigenfunction $\psi(\mathbf{r})$ corresponds to the eigenvalue E. When we wish to emphasise this fact we shall add a subscript E to the eigenfunction $\psi(\mathbf{r})$, thus writing $\psi_E(\mathbf{r}) \equiv \psi(\mathbf{r})$. We remark that since H is assumed to be time-independent, a 'separable' solution (3.100) of the Schrödinger equation (3.19) is also an eigenfunction of H corresponding to the eigenvalue E, and hence satisfies the equation

$$H\Psi = E\Psi. \tag{3.103}$$

This last conclusion can also be reached by acting on the wave function (3.100) with the total energy operator $E_{\text{op}} = i\hbar\partial/\partial t$. We obtain in this way

$$i\hbar\frac{\partial}{\partial t}\Psi = E\Psi \tag{3.104}$$

which shows that the 'separable' wave function (3.100) is an eigenfunction of the energy operator $i\hbar\partial/\partial t$ with eigenvalue E. Using (3.104) and the fact that (3.100) is a solution of (3.21) when H is time-independent we retrieve the result (3.103), as expected.

The above discussion strongly suggests that the wave function (3.100) corresponds to a state in which the total energy has the precise (numerical) value E. To show that this is the case, we first note that E is *real*. Indeed, for any 'separable' wave function of the form (3.100) the position probability density is given by

$$\begin{aligned} P(\mathbf{r}, t) &= \Psi^*(\mathbf{r}, t)\Psi(\mathbf{r}, t) \\ &= \psi^*(\mathbf{r})\psi(\mathbf{r}) \exp\left[-\frac{i}{\hbar}(E - E^*)t\right]. \end{aligned} \tag{3.105}$$

Assuming that Ψ is square integrable, we have from probability conservation, as expressed by equation (3.29)

$$(E - E^*)\int \Psi^*(\mathbf{r}, t)\Psi(\mathbf{r}, t)\,\mathrm{d}\mathbf{r} = 0 \tag{3.106}$$

where the integration is over all space. Hence $E = E^*$ and E is real. Now we have seen in Section 3.2 that the probability conservation equation (3.29) implies that the Hamiltonian operator H is Hermitian. We may therefore say that the eigenvalues E of the time-independent Schrödinger equation (3.102) are real because the Hamiltonian H is Hermitian. More generally, we shall see in Chapter 5 that the eigenvalues of *any* Hermitian operator are always real.

3.5 The time-independent Schrödinger equation. Stationary states

Let us now calculate the expectation value of the total energy in a state described by the wave function (3.100). Using (3.65), (3.21) and (3.103), we have

$$\langle E \rangle = \int \Psi^*(\mathbf{r}, t) \left(i\hbar \frac{\partial}{\partial t} \right) \Psi(\mathbf{r}, t) d\mathbf{r} = \int \Psi^*(\mathbf{r}, t) H \Psi(\mathbf{r}, t) d\mathbf{r}$$
$$= E \tag{3.107}$$

where we have assumed that the wave function (3.100) is normalised to unity. From (3.64) and (3.65) we may also write the above relation as

$$E = \langle H \rangle = \langle T \rangle + \langle V \rangle. \tag{3.108}$$

Thus the number E is the expectation value of the total energy in the state (3.100). More generally, we see that for a state (3.100) normalised to unity we have

$$\langle E^n \rangle = \int \Psi^*(\mathbf{r}, t) H^n \Psi(\mathbf{r}, t) d\mathbf{r} = E^n. \tag{3.109}$$

Hence, if $f(E)$ is a function of the energy which can be expanded in an absolutely convergent power series in E,

$$f(E) = \sum_n a_n E^n \tag{3.110}$$

its expectation value, in a state (3.100) normalised to unity, is given by

$$\langle f(E) \rangle = \sum_n a_n \langle E^n \rangle = \sum_n a_n E^n = f(E) \tag{3.111}$$

so that the wave function (3.100) describes a state in which the total energy has the definite numerical value E. In other words, a measurement of the energy on any member of an ensemble of identically prepared systems described by the wave function (3.100) will produce the same numerical value E. For this reason the eigenvalues E appearing in the time-independent Schrödinger equation (3.102) are called *energy eigenvalues*, the corresponding eigenfunctions $\psi_E(\mathbf{r}) \equiv \psi(\mathbf{r})$ of the Hamiltonian operator H being the *energy eigenfunctions*. Since $\psi(\mathbf{r})$ and the 'separable' state (3.100) differ only by the time-dependent phase factor $\exp(-iEt/\hbar)$, the spatial factor $\psi(\mathbf{r})$ is called a *time-independent wave function*.

Stationary states

The states (3.100) corresponding to a precise value E of the total energy have interesting properties. First of all, since E is real, we see from (3.105) that the position probability density corresponding to these states is given by

$$P(\mathbf{r}) = \psi^*(\mathbf{r}) \psi(\mathbf{r}) = |\psi(\mathbf{r})|^2 \tag{3.112}$$

and hence is constant in time. For this reason the states (3.100) are called *stationary states*. We also note from (3.34) and (3.100) that the probability current density corresponding to stationary states reads

$$\mathbf{j}(\mathbf{r}) = \frac{\hbar}{2mi}\{\psi^*(\mathbf{r})[\nabla\psi(\mathbf{r})] - [\nabla\psi^*(\mathbf{r})]\psi(\mathbf{r})\} \quad (3.113)$$

and is also constant in time. Moreover, the continuity equation (3.39) reduces to

$$\nabla\cdot\mathbf{j}(\mathbf{r}) = 0. \quad (3.114)$$

An example of a stationary state is the plane wave (3.9), which can be written in the 'separable' form (3.100), with $\psi(\mathbf{r}) = A\exp(i\mathbf{p}\cdot\mathbf{r}/\hbar)$. It is a stationary state describing a free particle of well-defined momentum $\mathbf{p}$ and definite energy $E = \mathbf{p}^2/2m$.

It is important to note that for stationary states (3.100) the expectation value of any operator A is independent of the time t, provided that A itself does not depend explicitly on t. Indeed, for such an operator we have, from (3.68) and (3.100),

$$\langle A \rangle = \frac{\int \Psi^*(\mathbf{r},t) A(\mathbf{r},-i\hbar\nabla) \Psi(\mathbf{r},t) d\mathbf{r}}{\int \Psi^*(\mathbf{r},t) \Psi(\mathbf{r},t) d\mathbf{r}}$$

$$= \frac{\int \psi^*(\mathbf{r}) A(\mathbf{r},-i\hbar\nabla) \psi(\mathbf{r}) d\mathbf{r}}{\int \psi^*(\mathbf{r}) \psi(\mathbf{r}) d\mathbf{r}} \quad (3.115)$$

and we see that $\langle A \rangle$ is time-independent. In particular, since the time-independent operators $\mathbf{r}$ and $-i\hbar\nabla$ are associated respectively with the basic dynamical variables $\mathbf{r}$ and $\mathbf{p}$, the expectation values of $\mathbf{r}$ and $\mathbf{p}_{op} = -i\hbar\nabla$ are independent of time for stationary states.

It is apparent from the time-independent Schrödinger equation (3.102) that if ψ is a solution of that equation corresponding to the eigenvalue E, all multiples of ψ are also solutions of (3.102) corresponding to that eigenvalue. Since wave functions differing by a constant multiplicative factor describe the same physical state, we shall not regard two solutions of (3.102) to be distinct if they differ only by a constant factor. Note that since $\Psi^*(\mathbf{r},t)\Psi(\mathbf{r},t) = \psi^*(\mathbf{r})\psi(\mathbf{r})$ for a stationary state (3.100), it follows that if this state is normalised to unity, then

$$\int \psi^*(\mathbf{r}) \psi(\mathbf{r}) d\mathbf{r} = 1 \quad (3.116)$$

so that the energy eigenfunctions $\psi(\mathbf{r})$ are also normalised to unity. Clearly, in this case two solutions of (3.102) which are multiples of each other can differ only by a constant 'phase factor' of unit modulus, having the form $\exp(i\alpha)$, where α is a real constant.

3.6 Energy quantisation

We have seen above that the time-independent Schrödinger equation (3.102) is an *eigenvalue equation*. We shall now show that *physically acceptable* solutions of that equation exist only for certain values of the total energy E.

To see how this comes about, let us consider the simple case of a particle moving in one dimension in a potential $V(x)$. The stationary states (3.100) then read

$$\Psi(x, t) = \psi(x) \exp(-iEt/\hbar) \tag{3.117}$$

and the energy eigenfunctions $\psi(x)$ are solutions of the time-independent Schrödinger equation

$$H\psi(x) \equiv -\frac{\hbar^2}{2m} \frac{d^2 \psi(x)}{dx^2} + V(x)\psi(x) = E\psi(x) \tag{3.118}$$

which we shall rewrite for convenience as

$$\frac{d^2 \psi(x)}{dx^2} = \frac{2m}{\hbar^2}[V(x) - E]\psi(x). \tag{3.119}$$

Since this is a second-order linear differential equation, it always has two linearly independent solutions, for *any* E. Provided $V(x)$ is finite everywhere, we see from (3.119) that the second derivative $d^2\psi(x)/dx^2$ is also finite, so that $d\psi(x)/dx$, and, therefore, $\psi(x)$ are *continuous* for all x, in agreement with the general continuity properties of the wave function noted at the end of Section 3.1. Moreover, in order to maintain the probabilistic interpretation of the wave function discussed in Chapter 2 and in Section 3.2, we shall impose the *physical* requirement that a solution $\psi(x)$ of equation (3.119) must be *finite* and single-valued everywhere. The eigenfunctions $\psi(x)$ we are looking for are therefore solutions of equation (3.119), such that each solution is finite, continuous and has a continuous derivative over the entire interval $(-\infty, +\infty)$. It is worth noting that since both E and $V(x)$ are *real*, if $\psi(x)$ is an eigenfunction of equation (3.119), so is also its complex conjugate $\psi^*(x)$. Consequently, the real part $[\psi(x) + \psi^*(x)]/2$ and the imaginary part $[\psi(x) - \psi^*(x)]/2i$ of an eigenfunction $\psi(x)$ are also solutions of (3.119). Because both $[\psi(x) + \psi^*(x)]/2$ and $[\psi(x) - \psi^*(x)]/2i$ are *real* functions, we only need to know the real eigenfunctions of (3.119) to construct all the eigenfunctions corresponding to a given eigenvalue. In the remaining part of this section we shall take advantage of this fact which allows us to restrict our attention to the real solutions of (3.119), without loss of generality. In particular, this will permit us to plot real solutions $\psi(x)$ in the usual way, instead of representing for instance the real and imaginary parts of a complex $\psi(x)$.

As an example, let us study the case of a potential $V(x)$ having the form shown in Fig. 3.1. This potential is equal to the value V_- for $x \to -\infty$. As x increases, $V(x)$ decreases and attains a minimum value $V_{min} = V(x_0)$ at $x = x_0$. For values of x larger than x_0 the potential $V(x)$ is assumed to increase until it eventually becomes equal to the value V_+ when $x \to +\infty$. Four cases must be distinguished, depending on the value of the energy E (see Fig. 3.1).

Case 1: $E < V_{min}$

In this case the quantity $V(x) - E$ is always positive, so that from (3.119) it is apparent that the second derivative of the wave function, $d^2\psi(x)/dx^2$, has always the same

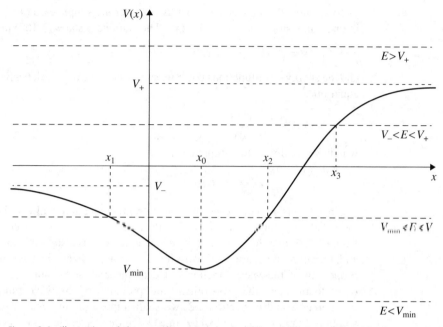

Figure 3.1 Illustration of the one-dimensional potential well $V(x)$ considered in the text. For $x \to -\infty$ the potential $V(x)$ tends to V_- and for $x \to +\infty$ it tends to V_+. At $x = x_0$ it attains the minimum value $V_{\min}$. Also shown are four values of the total energy E (represented by horizontal lines) corresponding to the four cases which can occur. The points x_1 and x_2 are the classical turning points corresponding to the case $V_{\min} < E < V_-$, and x_3 is the classical turning point corresponding to the case $V_- < E < V_+$.

sign as $\psi(x)$. Now the behaviour of a function $\psi(x)$ near a point $x = \bar{x}$, where $\psi(x)$ and $d^2\psi(x)/dx^2$ have the same sign is easily predicted and is illustrated in Fig. 3.2. If $\psi(\bar{x}) > 0$, then $\psi(x)$ will be concave upwards in the vicinity of $x = \bar{x}$ (see Fig. 3.2(a)); if $\psi(\bar{x}) < 0$ then $\psi(x)$ will be concave downwards in this vicinity (see Fig. 3.2(b)); if $\psi(\bar{x}) = 0$ then $\psi(x)$ will 'escape' away from the x-axis on both sides of the point $x = \bar{x}$ (see Fig. 3.2(c)). Since $V(x) - E > 0$ for *all* x, it is clear that a solution $\psi(x)$ of (3.119) which remains finite everywhere cannot be found. Indeed, $|\psi(x)|$ grows without limit either (a) as $x \to +\infty$ *and* $x \to -\infty$ or (b) as $x \to +\infty$ *or* $x \to -\infty$. The best we can do is to select from among the two linearly independent solutions of (3.119) a function $\psi(x)$ which approaches the x-axis asymptotically either on the left side (see Fig. 3.3(a)) or on the right side (see Fig. 3.3(b)), but then this solution will necessarily 'blow up' on the other side. We therefore conclude that there is *no physically acceptable solution* of equation (3.119) when $E < V(x)$ for all x. We also remark that because the kinetic energy $E - V(x)$ is everywhere negative, no classical motion would be possible in this case.

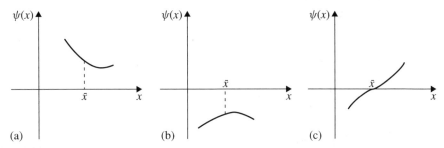

Figure 3.2 The local behaviour of a real solution $\psi(x)$ of equation (3.119) near a point $x = \bar{x}$, when $E < V_{\min}$, for the three cases (a) $\psi(\bar{x}) > 0$, (b) $\psi(\bar{x}) < 0$ and (c) $\psi(\bar{x}) = 0$.

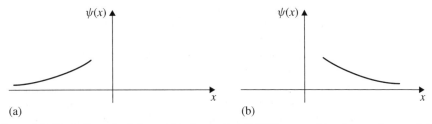

Figure 3.3 Illustration of solutions $\psi(x)$ of equation (3.119) approaching the x-axis asymptotically (a) on the left side and (b) on the right side, for the case $E < V_{\min}$. In both cases it is assumed that $\psi(x) > 0$.

Case 2: $V_{\min} < E < V_-$

Referring to Fig. 3.1, we see that $E - V(x)$ vanishes at the two points $x = x_1$ and $x = x_2$. The points for which $E = V(x)$ are called the classical *turning points* because they are the limits of the motion of a classical particle of energy E. Indeed, the regions for which the kinetic energy $E - V(x)$ is negative are inaccessible to a classical particle. At a turning point the kinetic energy is zero, so that a classical particle would stop and turn around at this point. In the present case the motion of a classical particle would therefore be confined to the interval $x_1 \leqslant x \leqslant x_2$.

Quantum mechanically, we shall now show that physically admissible solutions only exist for certain *discrete* values of the energy E. To see this, let us examine the behaviour of the solutions $\psi(x)$ of (3.119) in the 'internal' region $x_1 < x < x_2$ and in the two 'external' regions $x < x_1$ and $x > x_2$. In the internal region, where $E - V(x) > 0$, $d^2\psi(x)/dx^2$ is of opposite sign to $\psi(x)$. As a result, if $\bar{x}$ is a point belonging to that region, and if $\psi(\bar{x}) > 0$, then $\psi(x)$ will be concave downwards in the vicinity of $x = \bar{x}$ (see Fig. 3.4(a)); if $\psi(\bar{x}) < 0$ then $\psi(x)$ will be concave upwards in this vicinity (see Fig. 3.4(b)); if $\psi(\bar{x}) = 0$, then $\psi(x)$ will turn towards the x-axis on both sides of $\bar{x}$ (see Fig. 3.4(c)). Thus, when $x_1 < x < x_2$, $\psi(x)$ is always concave towards the x-axis and hence exhibits an *oscillatory* behaviour. It is worth noting that $\psi(x)$ might have several zeros in the internal region. The general solution of (3.119) in the internal region is a linear combination of two linearly independent

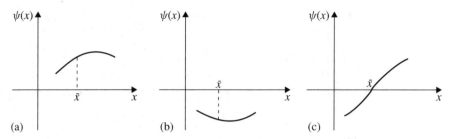

Figure 3.4 The local behaviour of a real solution $\psi(x)$ of equation (3.119) near a point $x = \bar{x}$ belonging to the internal region where $E - V(x) > 0$ for the three cases:
(a) $\psi(\bar{x}) > 0$, (b) $\psi(\bar{x}) < 0$, and (c) $\psi(\bar{x}) = 0$.

solutions, both of which are oscillatory.

In the external region $x < x_1$, the second derivative $d^2\psi(x)/dx^2$ has the same sign as $\psi(x)$, and therefore two linearly independent solutions of (3.119) can be found, one which tends to zero as $x \to -\infty$, the other one being such that $|\psi(x)|$ increases without limit as $x \to -\infty$. The general solution of (3.119) in the external region $x < x_1$ is a linear combination of these two particular solutions. Likewise, in the external region $x > x_2$, two particular linearly independent solutions of (3.119) can be obtained: one which tends to zero as $x \to +\infty$ and the other one such that $|\psi(x)|$ 'blows up' as $x \to +\infty$. Again, the general solution of (3.119) in the region $x > x_2$ is a linear combination of these two particular solutions.

Now, the only physically admissible solution in the external region $x < x_1$ is the one tending to zero as $x \to -\infty$, and this solution is *unique*, apart from an irrelevant multiplicative constant. The continuity of $\psi(x)$ and $d\psi(x)/dx$ at $x = x_1$ then provides two conditions[8] which determine a unique linear combination of the two independent oscillatory solutions in the internal region $x_1 < x < x_2$. Similarly, the continuity of $\psi(x)$ and $d\psi(x)/dx$ at $x = x_2$ provides two conditions which determine a *unique* linear combination of the two independent solutions (one tending to zero and the other one blowing up as $x \to \infty$) in the external region $x > x_2$. Thus, in general, for a given value of E such that $V_{\min} < E < V_-$, the unique solution $\psi(x)$ of (3.119) which tends to zero as $x \to -\infty$ will contain in the region $x > x_2$ a component such that $|\psi(x)|$ increases without limit as $x \to +\infty$ (see Fig. 3.5(a) and 3.5(b)); it is therefore physically inadmissible. However, since the value of $d^2\psi(x)/dx^2$ and, hence, the curvature of the solution $\psi(x)$ depend on the energy E, there may be certain discrete value of E for which the solution in the internal region will join smoothly to the solutions in the external regions which tend to zero as $x \to -\infty$ and $x \to +\infty$ (see Fig. 3.5(c)). Such solutions, which are finite, continuous and having a continuous derivative everywhere are *eigenfunctions* of (3.119). Since the probability density $P(x) = |\psi(x)|^2$ corresponding to these physically admissible solutions decreases to

---

[8] The two conditions that $\psi(x)$ and $d\psi(x)/dx$ be continuous at a given point are often referred to as the conditions for 'smooth' matching (or joining) at that point.

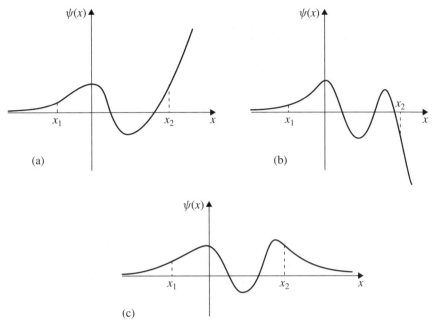

Figure 3.5 Diagrams (a), (b) and (c) show the behaviour of the solution of equation (3.119) at energies just below, just above, and at a bound state energy, respectively. All three solutions decrease to zero as $x \to -\infty$. It is seen that in cases (a) and (b) the solutions diverge as $x \to +\infty$, while in case (c) the solution tends asymptotically to zero as $x \to +\infty$.

zero as $|x| \to \infty$, these eigenfunctions are said to represent *bound states*, and the corresponding *discrete* energies are called *bound-state energies*. We have therefore obtained a fundamental result: the *quantisation of the bound-state energies*, and we see that the determination of the quantised energies appears in Schrödinger's wave mechanics as an eigenvalue problem. The number of discrete energy levels depends on the nature of the potential $V(x)$; it can be finite or infinite.

It is important to remark that in contrast to classical mechanics, where in the case $V_{\min} < E < V_-$ the motion is strictly confined to the internal region (between the turning points), a quantum mechanical bound-state wave function $\psi(x)$ extends outside the internal region and vanishes only in the limits $x \to \pm\infty$. Thus, although the position probability density $P(x) = |\psi(x)|^2$ decreases as one goes deeper and deeper into the classically forbidden regions, it is nevertheless non-vanishing at finite distances, and becomes equal to zero only in the limits $x \to \pm\infty$.

It is apparent from the above discussion that apart from a multiplicative constant the physically admissible wave function corresponding to an allowed energy value is *unique*, so that *in one dimension the bound-state (discrete) energy eigenvalues are non-degenerate*. (In more than one dimension they may be degenerate.) This property can also be proved as follows. Let us suppose for a moment that the contrary is true, and let $\psi_1(x)$ and $\psi_2(x)$ be two linearly independent eigenfunctions of (3.119) corresponding to the same energy eigenvalue E. Since both $\psi_1(x)$ and $\psi_2(x)$ satisfy the equation (3.119), we have

$$\frac{\psi_1''}{\psi_1} = \frac{\psi_2''}{\psi_2} = \frac{2m}{\hbar^2}(V - E) \tag{3.120}$$

where we have used the notation $\psi_1'' \equiv d^2\psi_1/dx^2$ and $\psi_2'' \equiv d^2\psi_2/dx^2$. Using (3.120) we see that

$$\psi_1\psi_2'' - \psi_2\psi_1'' = (\psi_1\psi_2')' - (\psi_2\psi_1')' = 0 \tag{3.121}$$

where the prime denotes a derivative with respect to x. Integrating this relation, we find that

$$\psi_1\psi_2' - \psi_2\psi_1' = \text{constant.} \tag{3.122}$$

We recognise on the left-hand side of this equation the *Wronskian*[9] of the two solutions ψ_1 and ψ_2. Since we are dealing with bound states, both ψ_1 and ψ_2 vanish at infinity, so that the constant in (3.122) must be equal to zero. Hence the Wronskian of ψ_1 and ψ_2 vanishes over the entire interval $(-\infty, +\infty)$ and

$$\frac{\psi_1'}{\psi_1} = \frac{\psi_2'}{\psi_2}. \tag{3.123}$$

---

[9] The *Wronskian* (determinant) of the two functions $f(x)$ and $g(x)$ is defined as

$$W[f, g] = \begin{vmatrix} f & g \\ f' & g' \end{vmatrix} = fg' - gf'.$$

If $W[f, g] = 0$ at a point x_0 of the x-axis, then $f'/f = g'/g$ at that point, so that the two functions f and g have the same logarithmic derivative at the point x_0. If $W[f, g]$ vanishes over the entire interval $(-\infty, +\infty)$ the two functions f and g are multiples of each other. Suppose now that $f(x)$ and $g(x)$ are, respectively, solutions of the equations

$$f'' + F(x)f = 0 \tag{1}$$
$$g'' + G(x)g = 0 \tag{2}$$

in an interval (a, b) where the functions $F(x)$ and $G(x)$ are piecewise continuous. Multiplying (1) by g, (2) by f and subtracting term by term, we obtain

$$fg'' - gf'' = (F - G)fg. \tag{3}$$

The left-hand side of this equation is just the derivative of the Wronskian $W[f, g]$. Upon integration over the interval (a, b), we therefore find that the overall variation of the Wronskian in the interval (a, b) is given by

$$W[f, g]_a^b = \int_a^b [F(x) - G(x)]f(x)g(x)dx. \tag{4}$$

This property is known as the *Wronskian theorem*. As a consequence of this theorem, we see that if ψ_1 and ψ_2 are two solutions of the Schrödinger equation (3.119) corresponding to the *same* energy eigenvalue E, their Wronskian is independent of x and is therefore a constant:

$$W[\psi_1, \psi_2] = \text{constant.} \tag{5}$$

Integrating this equation once more, we find that $\psi_1 = c\psi_2$, where c is a constant; this result is in contradiction to the assumed linear independence of the two eigenfunctions.

Case 3: $V_- < E < V_+$

Looking again at Fig. 3.1, we note that in this energy interval there is only one classical turning point, at $x = x_3$, no classical motion being allowed in the region $x > x_3$ for which the kinetic energy $E - V(x)$ is negative. A classical particle moving to the right would thus be reflected at the turning point. Note that classically the particle is *unbound* since it can move in an infinite region of the x-axis.

Quantum mechanically, we see that for $x < x_3$ there are two linearly independent solutions of (3.119) both of which are oscillatory, while for $x > x_3$ two linearly independent solutions can be found, one which tends to zero as $x \to +\infty$ and the other which is unbounded when $x \to +\infty$. The only physically admissible solution in the region $x > x_3$ is the one tending to zero as $x \to +\infty$, and this solution is *unique* (except for a multiplicative constant). The continuity of $\psi(x)$ and $\mathrm{d}\psi(x)/\mathrm{d}x$ at $x = x_3$ then yields two conditions which determine a *unique* linear combination of the two independent oscillatory solutions in the region $x < x_3$. A typical eigenfunction obtained in this way is illustrated in Fig. 3.6(a). It is important to remark that this 'smooth matching' at $x = x_3$ can be made for *every* energy E in the interval $V_- < E < V_+$. We therefore conclude that in this energy interval the allowed energies form a *continuum* and are *non-degenerate*.

We also note that in the present case the physically acceptable solutions of (3.119) tend to zero as $x \to +\infty$, but are non-zero and finite as $x \to -\infty$, so that the particle can be found 'at infinity'. Wave functions which are non-zero and finite at infinity are said to correspond to *unbound* (or *scattering*) states. Such wave functions are clearly not square integrable, and hence cannot be normalised by imposing the condition (3.28). We shall return to this point in the next section.

Case 4: $E > V_+$

In this case the kinetic energy $E - V(x)$ is everywhere positive. Classically, the particle is *unbound* and there are two independent states of motion, one in which the particle moves from left to right, and the other one in which it moves from right to left. Since $\mathrm{d}^2\psi(x)/\mathrm{d}x^2$ is opposite in sign to $\psi(x)$ for all x in this energy region, *two* linearly independent *oscillatory* solutions of (3.119) exist which are physically admissible for every energy. Thus when $E > V_+$ the energy eigenvalues of (3.119) form a continuum and are doubly degenerate. The corresponding eigenfunctions $\psi(x)$ are non-zero and finite as $x \to +\infty$ and $x \to -\infty$; they describe *unbound* (*scattering*) states. A typical eigenfunction of this kind is displayed in Fig. 3.6(b).

To summarise, we see that the eigenfunctions (that is the physically admissible solutions) of the time-independent Schrödinger equation (3.119) correspond to *bound* or *unbound* (*scattering*) states according to whether they vanish or are merely finite at infinity. The values of the total energy E for which the time-independent Schrödinger

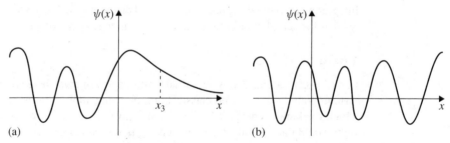

Figure 3.6 Typical eigenfunctions of the Schrödinger equation (3.119): (a) for the case $V_- < E < V_+$, and (b) for the case $E > V_+$.

equation has physically admissible solutions constitute the *energy spectrum*. For a particle moving in a bounded potential $V(x)$ this energy spectrum may contain a set of *discrete* energy levels corresponding to the *bound states*; in addition, it contains a *continuous* range of energies extending to $E = +\infty$, which corresponds to the *unbound states* (see Fig. 3.7). The bound-state energies must belong to a region in which the total energy E is such that the classical motion would be bound. Moreover, the energies of the bound states are *quantised*, since only a discrete set of them is allowed. As we have already seen in Section 1.4, the lowest discrete energy level is called the *ground-state energy* of the system, and all discrete higher energy levels are said to correspond to *excited* states. The energies of the unbound states are such that the particle would classically also be unbound; since these energies form a continuum there is no energy quantisation for unbound states.

Similar conclusions can be reached for the case of a particle of mass m moving in three dimensions in a potential $V(\mathbf{r})$. To be physically acceptable, a solution $\psi(\mathbf{r})$ of the time-independent Schrödinger equation (3.98) and its partial derivatives of the first order must be *continuous, single-valued* and *bounded* over all space. According to whether they vanish or are merely finite at infinity, the eigenfunctions of (3.98) are said to correspond to *bound states* or to *unbound (scattering)* states. It can be shown that if $V(\mathbf{r})$ is finite as $r \to \infty$ in any direction, the energy spectrum may contain a certain number of *discrete* (quantised) energy levels, corresponding to *bound states*; in addition it contains a *continuous* range of energies corresponding to unbound states. The bound states are analogous to the closed orbits of classical mechanics; their energies cannot exceed the minimum value V_- that $V(\infty)$ has in any direction; their number is finite (including zero) or infinite depending on the form of V. The scattering states are analogous to the open orbits of classical mechanics; their energies extend continuously from V_- to $+\infty$. A number of examples will be analysed in Chapters 4 and 7 to illustrate, respectively, the one-dimensional and three-dimensional situations.

It is interesting to note that if the potential energy V is changed by a constant (finite) amount, the eigenfunctions of the time-independent Schrödinger equation (3.98) are unchanged, while the eigenenergies are shifted by that constant (Problem 3.8). Indeed, the addition of a constant to the potential energy has no physical effect (the force

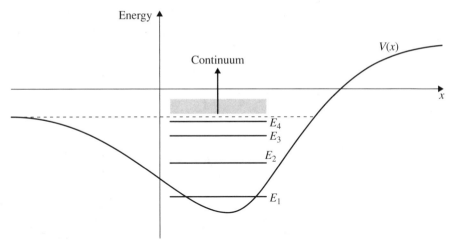

Figure 3.7 A typical energy spectrum for a one-dimensional potential well, containing both discrete and continuum energy levels.

corresponding to a constant potential energy being equal to zero) and corresponds simply to a change in the origin of the energy scale. We may therefore choose this constant as we please. For example, if $V(\mathbf{r})$ is bounded as $r \to \infty$, it is convenient to choose V_- (the smallest value of $V(\infty)$ in any direction) to be zero. In this case the discrete bound-state energies (if any) will be negative and the continuous spectrum will extend from $E = 0$ to $E = +\infty$.

Infinite potential energy

Until now we have considered potentials which are *finite* everywhere; this clearly includes potentials which exhibit discontinuities of the first kind (finite jumps). However, when the potential energy V is infinite[10] anywhere, the boundary conditions to be satisfied by the solutions of the time-independent Schrödinger equation must be modified. Indeed, because a particle cannot have an infinite potential energy, it cannot penetrate in a region of space where $V = \infty$. Consequently, its wave function $\Psi(\mathbf{r}, t)$ – and hence also the time-independent wave function $\psi(\mathbf{r})$ – must *vanish* everywhere in that region. The continuity of the wave function then requires that it also vanishes on the border of that region. We remark that since V exhibits a discontinuity of the second kind (infinite jump) across that border, $\nabla^2 \psi$ will also exhibit such a discontinuity, so that $\nabla \psi$ will be *discontinuous* (and hence undetermined) there. In particular, a boundary surface at which there is an infinite potential step acts like a perfectly rigid, impenetrable 'wall' where the wave function of the particle vanishes and its gradient is undetermined.

---

[10] Potentials which have discontinuities (finite or infinite) are of course mathematical idealisations of actual physical potentials.

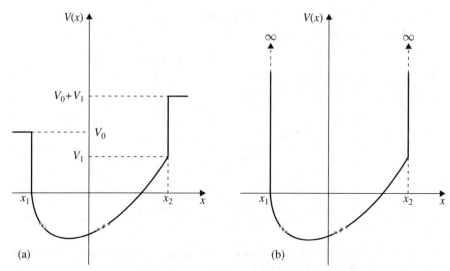

Figure 3.8 Illustration of the two potential wells described in the text: (a) $V(x)$ jumps upwards at $x = x_1$ and $x = x_2$ by the finite amount $V_0 > 0$, (b) corresponds to the case $V_0 \to \infty$.

The foregoing considerations will be illustrated in Section 4.3, where we shall study in detail the one-dimensional potential step. Here, we analyse briefly another example, which will be useful shortly in connection with the 'normalisation' of unbound wave functions. We consider a one-dimensional potential well such as that represented in Fig. 3.8(a). We assume that $V(x)$ is finite for $x_1 < x < x_2$. At $x = x_1$ it 'jumps' upwards from zero to the finite value V_0, and remains equal to V_0 for $x < x_1$. Similarly at $x = x_2$, the potential 'jumps' by the positive amount V_0 from the value V_1 (which for definiteness is taken to be positive) and is equal to $V_1 + V_0$ for $x > x_2$. According to the analysis made earlier in this section, when the total energy E is smaller than V_0, there may be a certain number of discrete bound states, while for $E > V_0$ the allowed energies form a continuum. Thus, as V_0 increases the continuum part of the spectrum will shrink and in the limit $V_0 \to +\infty$ (see Fig. 3.8(b)) the energy spectrum will be *entirely discrete*, the allowed energy levels being determined by the boundary conditions

$$\psi(x_1) = \psi(x_2) = 0 \tag{3.124}$$

which express the fact that the energy eigenfunctions must vanish at the two impenetrable walls of this one-dimensional box. The explicit determination of the energy levels and eigenfunctions will be done in Section 4.5 for the particular case of an infinite square well.

The above results are readily generalised to three dimensions. If we enclose a three-dimensional system in a box with impenetrable walls, its energy spectrum will be *entirely discrete*, the allowed energies being obtained from the requirement that the wave function vanishes at the walls. The energy levels and eigenfunctions of a 'free' particle constrained to move inside a three-dimensional box will be obtained in this way in Section 7.1.

Finally, we remark that as the walls of the box recede towards infinity, we expect on physical grounds that the energy spectrum will resemble more and more closely that corresponding to the potential without walls. This fact will be verified explicitly on the examples studied in Sections 4.5 and 7.1.

3.7 Properties of the energy eigenfunctions

In this section we shall obtain some important results concerning the energy eigenfunctions, that is the physically acceptable solutions of the time-independent Schrödinger equation (3.102). It will be convenient here to use the more explicit notation $\psi_E(\mathbf{r})$ to denote an energy eigenfunction corresponding to the eigenvalue E.

First of all, let us examine the question of the *normalisation* of the energy eigenfunctions. We have seen in the preceding section that there are two kinds of eigenfunctions of (3.102): the bound states, which vanish at infinity and correspond to discrete energies, and the unbound (scattering) states, which are only finite at infinity, and whose corresponding energies form a continuum. The bound-state eigenfunctions are square integrable, and hence can readily be normalised to unity in the familiar fashion (see (3.116)). However, the unbound states cannot be normalised in that way. As a result, one cannot define absolute probabilities and expectation values of physical quantities in such unbound states. A way of avoiding this difficulty, which we shall adopt here, is to enclose the system in a box with impenetrable walls, so that the entire eigenvalue spectrum becomes discrete. Since all the eigenfunctions now correspond to bound states, they can be normalised according to

$$\int_V \psi_E^*(\mathbf{r})\psi_E(\mathbf{r})d\mathbf{r} = 1 \tag{3.125}$$

where V denotes the volume of the box. We shall choose the box to be very large (macroscopic) so that its effect on the energy spectrum is essentially negligible[11]. From now on all the integrals in this section will be assumed to be performed over the volume of a very large box.

We shall now demonstrate that two energy eigenfunctions ψ_E and $\psi_{E'}$ corresponding to *unequal* eigenvalues E and E' are *orthogonal*, namely

$$\int \psi_{E'}^*(\mathbf{r})\psi_E(\mathbf{r})d\mathbf{r} = 0, \qquad E \neq E'. \tag{3.126}$$

The proof is very simple. First, since $H\psi_E = E\psi_E$ we have, upon premultiplying both sides by $\psi_{E'}^*$

$$\psi_{E'}^*(H\psi_E) = E\psi_{E'}^*\psi_E \tag{3.127}$$

---

[11] Alternative procedures to 'normalise' unbound states will be discussed in Sections 4.2 and 5.3.

where we have omitted the arguments of the functions, for notational simplicity. Now, because $H\psi_{E'} = E'\psi_{E'}$, we can also write

$$(H\psi_{E'})^* = E'\psi_{E'}^* \tag{3.128}$$

where we have used the fact that E' is real. Multiplying both sides of (3.128) on the right by ψ_E, we obtain

$$(H\psi_{E'})^*\psi_E = E'\psi_{E'}^*\psi_E. \tag{3.129}$$

Subtracting (3.129) from (3.127) and integrating over the volume V, we obtain

$$(E - E')\int \psi_{E'}^*\psi_E d\mathbf{r} = \int [\psi_{E'}^\dagger(H\psi_E) - (H\psi_{E'})^*\psi_E]d\mathbf{r}$$
$$= 0 \tag{3.130}$$

where in the last step we have used the fact that H is Hermitian. Since we have assumed that $E \neq E'$ we see that (3.126) follows from (3.130).

Combining the normalisation relation (3.125) and the orthogonality relation (3.126), we can say that the energy eigenfunctions ψ_E satisfy the *orthonormality* relations

$$\int \psi_{E'}^*(\mathbf{r})\psi_E(\mathbf{r})d\mathbf{r} = \delta_{EE'} \tag{3.131}$$

where $\delta_{EE'}$ is the Kronecker δ-symbol, such that

$$\delta_{EE'} = \begin{cases} 1 & \text{if } E = E' \\ 0 & \text{if } E \neq E' \end{cases}. \tag{3.132}$$

We have just seen that eigenfunctions corresponding to *different* energy eigenvalues are orthogonal. Let us now see what happens when an energy eigenvalue is degenerate, so that there is more than one linearly independent eigenfunction of (3.102) corresponding to that eigenvalue. Assuming that the degeneracy of E is of order α, we denote by $\psi_{E1}, \psi_{E2}, \ldots, \psi_{E\alpha}$ a set of α linearly independent eigenfunctions corresponding to that energy and normalised according to (3.125). The orthogonality proof given above, which made use of the fact that $E \neq E'$, is clearly not applicable in the present case. Indeed, in general, the eigenfunctions $\psi_{Er}(r = 1, 2, \ldots, \alpha)$ belonging to a degenerate eigenvalue E are not orthogonal. However, given a set of α linearly independent, normalisable functions, it is always possible to construct a new set of α mutually *orthogonal* functions by using the *Schmidt orthogonalisation*

3.7 Properties of the energy eigenfunctions

procedure. Starting from the function ψ_{E1}, we form the sequence

$$\phi_{E1} = \psi_{E1} \tag{3.133a}$$

$$\phi_{E2} = a_{21}\phi_{E1} + \psi_{E2} \tag{3.133b}$$

$$\phi_{E3} = a_{31}\phi_{E1} + a_{32}\phi_{E2} + \psi_{E3} \tag{3.133c}$$

$$\vdots$$

$$\phi_{E\alpha} = a_{\alpha 1}\phi_{E1} + a_{\alpha 2}\phi_{E2} + \cdots + a_{\alpha,\alpha-1}\phi_{E\alpha-1} + \psi_{E\alpha} \tag{3.133d}$$

where a_{ij} denote the coefficients which we shall determine in such a way that each function of the sequence be orthogonal to all the preceding ones.

First, from the orthogonality requirement

$$\int \phi_{E1}^* \phi_{E2} d\mathbf{r} = 0 \tag{3.134}$$

we deduce by using (3.133a) and (3.133b) that

$$a_{21} \int \phi_{E1}^* \phi_{E1} d\mathbf{r} + \int \phi_{E1}^* \phi_{E2} d\mathbf{r} = 0 \tag{3.135}$$

and since $\phi_{E1}(=\psi_{E1})$ is normalised to unity, we have

$$a_{21} = -\int \phi_{E1}^* \psi_{E2} d\mathbf{r}. \tag{3.136}$$

With the coefficient a_{21} determined in this way we obtain from (3.133b) a function

$$\phi_{E2} = \left[-\int \phi_{E1}^* \psi_{E2} d\mathbf{r}\right] \phi_{E1} + \psi_{E2} \tag{3.137}$$

orthogonal to ϕ_{E1}. Note that since the (normalised) functions ϕ_{E1} and ψ_{E2} are linearly independent it follows from the Schwarz inequality that $|a_{21}| < 1$. Consequently, using (3.133b), (3.136) and the fact that the functions ϕ_{E1} and ϕ_{E2} are orthogonal, it is seen that the normalisation integral for ϕ_{E2} never vanishes, so that the function ϕ_{E2} is normalisable.

Next, from the two orthogonality conditions

$$\int \phi_{E1}^* \phi_{E3} d\mathbf{r} = 0 \tag{3.138a}$$

and

$$\int \phi_{E2}^* \phi_{E3} d\mathbf{r} = 0 \tag{3.138b}$$

one deduces in a similar way that

$$a_{31} + \int \phi_{E1}^* \psi_{E3} d\mathbf{r} = 0 \tag{3.139a}$$

and

$$a_{32} \int \phi_{E2}^* \phi_{E2} d\mathbf{r} + \int \phi_{E2}^* \psi_{E3} d\mathbf{r} = 0 \tag{3.139b}$$

where we have used (3.134). From (3.133c) and (3.139) we obtain in this fashion a function

$$\phi_{E3} = \left[-\int \phi_{E1}^* \psi_{E3} d\mathbf{r}\right]\phi_{E1} + \left[-\frac{\int \phi_{E2}^* \psi_{E3} d\mathbf{r}}{\int \phi_{E2}^* \phi_{E2} d\mathbf{r}}\right]\phi_{E2} + \psi_{E3} \tag{3.140}$$

which is orthogonal to ϕ_{E1} and ϕ_{E2}, and normalisable, since ϕ_{E1}, ϕ_{E2} and ψ_{E3} are linearly independent.

This process can be continued until we arrive at the function

$$\phi_{E\alpha} = -\sum_{i=1}^{\alpha-1} \frac{\int \phi_{Ei}^* \psi_{E\alpha} d\mathbf{r}}{\int \phi_{Ei}^* \phi_{Ei} d\mathbf{r}} \phi_{Ei} + \psi_{E\alpha} \tag{3.141}$$

which is orthogonal to the functions $\phi_{E1}, \phi_{E2}, \ldots, \phi_{E,\alpha-1}$ and is normalisable. The set of orthogonal functions $\phi_{Er}(r = 1, 2, \ldots, \alpha)$ being determined in this way[12], we can construct the functions

$$\chi_{Er} = \frac{\phi_{Er}}{\left[\int \phi_{Er}^* \phi_{Er} d\mathbf{r}\right]^{1/2}}, \quad r = 1, 2, \ldots, \alpha \tag{3.142}$$

which constitute a set of *mutually orthogonal* functions, *normalised to unity*. These functions satisfy, therefore, the orthonormality relations

$$\int \chi_{Es}^*(\mathbf{r}) \chi_{Er}(\mathbf{r}) d\mathbf{r} = \delta_{rs} \quad (r, s = 1, 2, \ldots, \alpha). \tag{3.143}$$

To summarise, we see that although the α linearly independent eigenfunctions $\psi_{Er}(r = 1, 2, \ldots, \alpha)$ corresponding to the eigenenergy E need not be mutually orthogonal, it is always possible to construct from linear combinations of them a new set of α eigenfunctions which are orthonormal. From now on we shall assume that this orthogonalisation procedure has already been carried out for each degenerate energy eigenvalue. Since, in addition, the eigenfunctions corresponding to different energy eigenvalues are orthogonal, it follows that *all* the energy eigenfunctions can be assumed to satisfy the orthonormality relations

$$\int \psi_{E's}^*(\mathbf{r}) \psi_{Er}(\mathbf{r}) d\mathbf{r} = \delta_{EE'} \delta_{rs} \tag{3.144}$$

and hence form an *orthonormal set*.

We shall postulate that the energy spectrum obtained by solving the time-independent Schrödinger equation (3.102) represents all the physically

---

[12] Note that this set is by no means unique, since we could for example have started the sequence of functions ϕ_{Er} by identifying ϕ_{E1} with ψ_{E2}, or with a linear combination of ψ_{E1} and ψ_{E2}. Thus there are infinitely many possible sets of orthogonal functions ϕ_{Er}.

realisable energies of the system. This implies that the set of energy eigenfunctions is *complete*, in the sense that any physically admissible wave function can be expressed as a superposition of them. The *general* solution $\Psi(\mathbf{r}, t)$ of the time-dependent Schrödinger equation (3.21) can therefore be expanded in terms of the energy eigenfunctions as

$$\Psi(\mathbf{r}, t) = \sum_E C_E(t) \psi_E(\mathbf{r}) \tag{3.145}$$

where the expansion coefficients $C_E(t)$ depend on the time. In writing the above equation, we have dropped the degeneracy index for notational simplicity, it being understood that the summation includes α times an energy eigenvalue whose degeneracy is of order α.

The formula (3.145) is analogous to the familiar expression giving the expansion of a vector in terms of a complete set of orthonormal basis vectors in a vector space. The complete set of orthonormal functions ψ_E plays the role of a set of basis vectors, and the wave function Ψ may be thought of as a 'vector' whose 'components' along the basis 'vectors' ψ_E are the coefficients C_E. This analogy with ordinary vector spaces has a profound significance, as we shall see in Chapter 5.

We remark that since the 'vector' Ψ evolves in time according to the time-dependent Schrödinger equation (3.21), its components C_E must also be functions of the time. If $\Psi(\mathbf{r}, t)$ is given at a particular time, its 'components' are readily obtained in the following way. Multiplying (3.145) on the left by $\psi_{E'}^*(\mathbf{r})$, integrating over the volume V of our (large) box and making use of the orthonormality relations satisfied by the energy eigenfunctions, we find that

$$\int \psi_{E'}^*(\mathbf{r}) \Psi(\mathbf{r}, t) d\mathbf{r} = \sum_E C_E(t) \int \psi_{E'}^*(\mathbf{r}) \psi_E(\mathbf{r}) d\mathbf{r}$$

$$= \sum_E C_E(t) \delta_{EE'}$$

$$= C_{E'}(t). \tag{3.146}$$

We conclude this section by mentioning an interesting property of the energy eigenfunctions corresponding to *bound states* of *one-dimensional* systems. Since we are dealing here exclusively with bound states there is no need to enclose the system in a finite box. Assuming that the potential energy $V(x)$ is finite everywhere, we have seen in the previous section that the bound state eigenfunctions vanish when $x \to \pm\infty$. Now, if bound-state energies exist and if they are placed in order of increasing magnitude ($E_1 < E_2 < E_3 < \cdots$), then the corresponding eigenfunctions $\psi_1(x), \psi_2(x), \psi_3(x), \ldots$, will occur in increasing number of their zeros, the nth eigenfunction $\psi_n(x)$ having $(n-1)$ zeros for finite values of x. Moreover, between two consecutive zeros of $\psi_n(x)$, the following eigenfunctions will have at least one zero. This is known as the *oscillation theorem* (see Problem 3.9). If the particle is enclosed in a one-dimensional box with impenetrable walls such that $V(x_1) = V(x_2) = +\infty$, then its motion is confined to the interval $x_1 < x < x_2$ and the $(n-1)$ zeros of $\psi_n(x)$ referred to in the oscillation theorem will occur *within* that interval. In addition, of course, $\psi_n(x)$ must vanish in this case at the walls (i.e. at $x = x_1$ and $x = x_2$) so that the total number of its zeros will be $n + 1$.

3.8 General solution of the time-dependent Schrödinger equation for a time-independent potential

In the preceding section we have seen that the general solution $\Psi(\mathbf{r}, t)$ of the time-dependent Schrödinger equation (3.21) can be expanded in terms of the energy eigenfunctions $\psi_E(\mathbf{r})$ according to (3.145). We shall now determine the expansion coefficients $C_E(t)$ for the case of a particle moving in a *time-independent* potential $V(\mathbf{r})$. To this end, we substitute (3.145) in the Schrödinger wave equation (3.21), and use the facts that the Hamiltonian H is time-independent and that $H\psi_E = E\psi_E$. Thus

$$i\hbar \frac{\partial}{\partial t} \sum_E C_E(t)\psi_E(\mathbf{r}) = H \sum_E C_E(t)\psi_E(\mathbf{r})$$

$$= \sum_E C_E(t) H \psi_E(\mathbf{r})$$

$$= \sum_E C_E(t) E \psi_E(\mathbf{r}). \tag{3.147}$$

Premultiplying both sides of this equation by $\psi_{E'}^*(\mathbf{r})$ and integrating over the volume V of the box we get

$$i\hbar \frac{\partial}{\partial t} \sum_E C_E(t) \int \psi_{E'}^*(\mathbf{r})\psi_E(\mathbf{r})d\mathbf{r} = \sum_E C_E(t) E \int \psi_{E'}^*(\mathbf{r})\psi_E(\mathbf{r})d\mathbf{r}. \tag{3.148}$$

Now, since the energy eigenfunctions are orthonormal, this equation reduces to

$$i\hbar \frac{d}{dt} C_E(t) = E C_E(t) \tag{3.149}$$

where we have written d/dt instead of $\partial/\partial t$ since only the time variable appears in (3.149). This equation is readily integrated to yield

$$C_E(t) = C_E(t_0) \exp[-iE(t - t_0)/\hbar]. \tag{3.150}$$

Using (3.145) and the above result, we thus find that the *general* solution of the time-dependent Schrödinger equation (3.21) may be written for a *time-independent potential* as

$$\Psi(\mathbf{r}, t) = \sum_E C_E(t_0) \exp[-iE(t - t_0)/\hbar] \psi_E(\mathbf{r}) \tag{3.151}$$

or

$$\Psi(\mathbf{r},t) = \sum_E c_E \psi_E(\mathbf{r}) \exp(-iEt/\hbar) \tag{3.152}$$

where the constants c_E are given by

$$c_E = C_E(t_0) \exp(iEt_0/\hbar). \tag{3.153}$$

Remembering that in the case of a time-independent potential the *stationary states* $\psi_E(\mathbf{r}) \exp(-iEt/\hbar)$ are particular solutions of the time-dependent Schrödinger equation, we see that the general solution (3.152) is an *arbitrary linear superposition of stationary states*. This is to be expected since according to the discussion of Section 3.1 an arbitrary linear superposition of solutions of the time-dependent Schrödinger equation must also be a solution of that equation, as required by the superposition principle.

The constants c_E can be evaluated from the knowledge of the wave function Ψ at any particular time t_0. Indeed, using (3.153) and (3.146), we have

$$c_E = \exp(iEt_0/\hbar) \int \psi_E^*(\mathbf{r}) \Psi(\mathbf{r},t_0) d\mathbf{r} \tag{3.154}$$

so that the general solution (3.152) may be written as

$$\Psi(\mathbf{r},t) = \sum_E \left[\int \psi_E^*(\mathbf{r}') \Psi(\mathbf{r}',t_0) d\mathbf{r}' \right] \exp[-iE(t-t_0)/\hbar] \psi_E(\mathbf{r}). \tag{3.155}$$

This expression gives $\Psi(\mathbf{r},t)$ at any time, once it is known at the time $t = t_0$. If we define the function

$$K(\mathbf{r},t;\mathbf{r}',t_0) = \sum_E \psi_E^*(\mathbf{r}') \psi_E(\mathbf{r}) \exp[-iE(t-t_0)/\hbar] \tag{3.156}$$

we see that we may recast (3.155) in the simpler form

$$\Psi(\mathbf{r},t) = \int K(\mathbf{r},t;\mathbf{r}',t_0) \Psi(\mathbf{r}',t_0) d\mathbf{r}'. \tag{3.157}$$

This shows that the 'propagation' of the wave function from the time t_0 to the time t is controlled by the function $K(\mathbf{r},t;\mathbf{r}',t_0)$, which is therefore called a *propagator*.

Let us now return to the form (3.152) of our general solution. In what follows we shall also need the complex conjugate of that solution

$$\Psi^*(\mathbf{r},t) = \sum_{E'} c_{E'}^* \psi_{E'}^*(\mathbf{r}) \exp(iE't/\hbar) \tag{3.158}$$

where we have used a different (dummy) summation symbol for further convenience. If we require that the general solution (3.152) be normalised to unity, we see from (3.152) and (3.158) that we must have

$$\int \Psi^*(\mathbf{r},t)\Psi(\mathbf{r},t)\mathrm{d}\mathbf{r} = \sum_E\sum_{E'} c_{E'}^* c_E \exp[-\mathrm{i}(E-E')t/\hbar] \int \psi_{E'}^*(\mathbf{r})\psi_E(\mathbf{r})\mathrm{d}\mathbf{r}$$

$$= 1 \qquad (3.159)$$

where the integration is over the volume of the box. Now, because the energy eigenfunctions are orthonormal, we have

$$\sum_E\sum_{E'} c_{E'}^* c_E \exp[-\mathrm{i}(E-E')t/\hbar] \int \psi_{E'}^*(\mathbf{r})\psi_E(\mathbf{r})\mathrm{d}\mathbf{r}$$

$$= \sum_E\sum_{E'} c_{E'}^* c_F \exp[-\mathrm{i}(E-E')t/\hbar]\delta_{FF'}$$

$$= \sum_E |c_E|^2 \qquad (3.160)$$

and therefore the normalisation condition (3.159) reduces to

$$\sum_E |c_E|^2 = 1. \qquad (3.161)$$

It is important to note that in contrast to the case of stationary states, the position probability density corresponding to a superposition of stationary states is *time-dependent*. Indeed, using (3.152) and (3.158), this probability density is given by

$$P(\mathbf{r},t) = \Psi^*(\mathbf{r},t)\Psi(\mathbf{r},t)$$

$$= \sum_E\sum_{E'} c_{E'}^* c_E \exp[-\mathrm{i}(E-E')t/\hbar]\psi_{E'}^*(\mathbf{r})\psi_E(\mathbf{r})$$

$$= \sum_E |c_E|^2|\psi_E(\mathbf{r})|^2 + \sum_E\sum_{\substack{E'\\ E\neq E'}} c_{E'}^* c_E \exp[-\mathrm{i}(E-E')t/\hbar]\psi_{E'}^*(\mathbf{r})\psi_E(\mathbf{r})$$

$$(3.162)$$

where in the last line we have separated the contributions to $P(\mathbf{r},t)$ arising from the diagonal ($E = E'$) and non-diagonal ($E \neq E'$) terms in the double summation on E and E'. We see that the non-diagonal terms are responsible for the time-dependence of $P(\mathbf{r},t)$.

Since a wave function $\Psi(\mathbf{r},t)$ which is a superposition of stationary states leads to a time-dependent position probability density, it follows that the expectation values of physical quantities in a state described by such a wave function depend in general on the time. An exception occurs for the total energy, whose expectation value in a state described by the general wave function (3.152) is

$$\langle E \rangle = \langle H \rangle = \int \Psi^*(\mathbf{r},t) H \Psi(\mathbf{r},t) d\mathbf{r}$$

$$= \sum_E \sum_{E'} c_{E'}^* c_E \exp[-i(E-E')t/\hbar] \int \psi_{E'}^*(\mathbf{r}) H \psi_E(\mathbf{r}) d\mathbf{r}$$

$$= \sum_E \sum_{E'} c_{E'}^* c_E \exp[-i(E-E')t/\hbar] \, E \int \psi_{E'}^*(\mathbf{r}) \psi_E(\mathbf{r}) d\mathbf{r}$$

$$= \sum_E \sum_{E'} c_{E'}^* c_E \exp[-i(E-E')t/\hbar] \, E \, \delta_{EE'}$$

$$= \sum_E |c_E|^2 E \tag{3.163}$$

where we have used the fact that $H\psi_E = E\psi_E$ and the orthonormality of the energy eigenfunctions. As seen from (3.163) the average value of the total energy is time-independent, as expected.

It is important to remember that the summation on E in all the previous equations also includes implicitly a sum over a degeneracy index for degenerate energy eigenvalues. If we write this index explicitly, the general solution (3.152) reads

$$\Psi(\mathbf{r},t) = \sum_E \sum_r c_{Er} \psi_{Er}(\mathbf{r}) \exp(-iEt/\hbar). \tag{3.164}$$

Assuming (without loss of generality) that the energy eigenfunctions satisfy the orthonormality relations (3.144), the normalisation condition becomes

$$\sum_E \sum_r |c_{Er}|^2 = 1 \tag{3.165}$$

and equation (3.163) is replaced by

$$\langle E \rangle = \sum_E \sum_r |c_{Er}|^2 E. \tag{3.166}$$

Since we have assumed that the energy spectrum obtained by solving the time-independent Schrödinger equation (3.102) represents all the physically realisable energies of the system, it follows that the energy eigenvalues E are the only possible results of precise measurements of the total energy of the particle. The above relations then suggest that we interpret the quantity

$$P(E) = |c_E|^2 \tag{3.167}$$

as the *probability that a measurement of the total energy will yield the value E*, provided that this energy value is *non-degenerate*; the coefficient c_E is the corresponding *probability amplitude.* If the energy eigenvalue E is α times degenerate, it follows from (3.165) and (3.166) that the probability of obtaining this value is

$$P(E) = \sum_{r=1}^{\alpha} |c_{Er}|^2. \tag{3.168}$$

Equation (3.166) gives the average of the values found in a large number of precise measurements of the total energy, performed on equivalent, independent systems described by the same wave function (3.164). It should be noted that in general $\langle E \rangle$ will not be equal to one of the energy eigenvalues. However, if the system is in a particular stationary state corresponding to a given *non-degenerate* energy eigenvalue $\bar{E}$ then its wave function reduces to

$$\Psi(\mathbf{r}, t) = \psi_{\bar{E}}(\mathbf{r}) \exp(-i\bar{E}t/\hbar) \tag{3.169}$$

so that $c_{\bar{E}} = 1$ and all the other coefficients vanish. Consequently $\langle E \rangle = \bar{E}$ and, upon measurement of the total energy, we are certain to obtain the value $\bar{E}$. If the energy eigenvalue $\bar{E}$ is α times *degenerate* and the system is in an arbitrary superposition of states corresponding to that energy eigenvalue, then its wave function is given by

$$\Psi(\mathbf{r}, t) = \sum_{r=1}^{\alpha} c_{\bar{E}r} \psi_{\bar{E}r}(\mathbf{r}) \exp(-i\bar{E}t/\hbar). \tag{3.170}$$

Upon comparison with (3.164) we see that all the coefficients c_{Er} vanish for $E \neq \bar{E}$. Hence, using (3.166) and (3.165) we have $\langle E \rangle = \bar{E}$, and a measurement of the total energy will certainly give the value $\bar{E}$.

Until now we have enclosed the system in a large box of volume V, so that the entire energy spectrum is discrete. If we remove that constraint, the energy spectrum will in general contain a discrete and a continuous part. The general solution of the time-dependent Schrödinger equation for a time-independent potential will then still be given by the expression (3.152) (or more explicitly by (3.164)), provided the summation over the allowed energies E also includes an *integration* over the continuous part of the energy spectrum. In what follows the summation symbol on E will always be understood to have that meaning, when necessary. The normalisation of general wave functions (3.152) (or (3.164)) which contain unbound states will be discussed in Chapter 5.

3.9 The Schrödinger equation in momentum space

In Chapter 2 we defined the momentum space wave function $\Phi(\mathbf{p}, t)$ of a particle as the Fourier transform of its configuration space wave function $\Psi(\mathbf{r}, t)$, and we saw that $\Phi(\mathbf{p}, t)$ plays in momentum space the role assigned to the wave function $\Psi(\mathbf{r}, t)$ in configuration space. In particular, assuming that $\Phi(\mathbf{p}, t)$ is normalised to unity, $\Pi(\mathbf{p}, t)d\mathbf{p} = |\Phi(\mathbf{p}, t)|^2 d\mathbf{p}$ is the probability of finding at time t the momentum of the particle within the volume element

3.9 The Schrödinger equation in momentum space

$d\mathbf{p} \equiv dp_x dp_y dp_z$ about the point $\mathbf{p}$ in momentum space. In this section we shall further pursue the analogy between the formulations of quantum mechanics in configuration and momentum space.

Since the configuration space wave function $\Psi(\mathbf{r}, t)$ satisfies the time-dependent Schrödinger equation (3.19), we can readily obtain the corresponding time-dependent Schrödinger equation for the momentum space wave function $\Phi(\mathbf{p}, t)$. Remembering that $H = \mathbf{p}^2/2m + V(\mathbf{r}, t)$, we can transform (3.19) in momentum space by writing first

$$i\hbar \frac{\partial}{\partial t} \left[(2\pi\hbar)^{-3/2} \int e^{-i\mathbf{p}\cdot\mathbf{r}/\hbar} \Psi(\mathbf{r}, t) d\mathbf{r} \right]$$

$$= (2\pi\hbar)^{-3/2} \int e^{-i\mathbf{p}\cdot\mathbf{r}/\hbar} \left[\frac{\mathbf{p}^2}{2m} + V(\mathbf{r}, t) \right] \Psi(\mathbf{r}, t) d\mathbf{r} \qquad (3.171)$$

which, upon using (2.61), becomes

$$i\hbar \frac{\partial}{\partial t} \Phi(\mathbf{p}, t) = \frac{\mathbf{p}^2}{2m} \Phi(\mathbf{p}, t) + (2\pi\hbar)^{-3/2} \int e^{-i\mathbf{p}\cdot\mathbf{r}/\hbar} V(\mathbf{r}, t) \Psi(\mathbf{r}, t) d\mathbf{r}. \qquad (3.172)$$

The second term on the right-hand side can be transformed by using the convolution theorem for Fourier transforms (see equations (A.53)–(A.54) of Appendix A). In this way (3.172) becomes

$$i\hbar \frac{\partial}{\partial t} \Phi(\mathbf{p}, t) = \frac{\mathbf{p}^2}{2m} \Phi(\mathbf{p}, t) + \int \tilde{V}(\mathbf{p} - \mathbf{p}', t) \Phi(\mathbf{p}', t) d\mathbf{p}' \qquad (3.173)$$

where

$$\tilde{V}(\mathbf{p} - \mathbf{p}', t) = (2\pi\hbar)^{-3} \int e^{-i(\mathbf{p}-\mathbf{p}')\cdot\mathbf{r}/\hbar} V(\mathbf{r}, t) d\mathbf{r}. \qquad (3.174)$$

Equation (3.173) is the time-dependent Schrödinger equation for the momentum space wave function. Because of the integral on the right this equation is generally more complicated than the corresponding time-dependent Schrödinger equation (3.19) in configuration space. This explains why the configuration space Schrödinger equation is used in most applications.

Expectation values

In Section 3.3 we used the momentum space wave function to obtain in a natural way the expectation values of the momentum $\mathbf{p}$ (see (3.45)) and of arbitrary functions $g(\mathbf{p}, t)$ of $\mathbf{p}$ and t (see (3.47)). We shall now generalise these results in order to be able to calculate the expectation values of other dynamical variables in terms of $\Phi(\mathbf{p}, t)$.

Let us consider first the expectation value of x. Starting from (3.42a) and expressing $\Psi(\mathbf{r}, t)$ and $\Psi^*(\mathbf{r}, t)$ in terms of $\Phi(\mathbf{p}, t)$ and $\Phi^*(\mathbf{p}, t)$, we obtain by following an argument similar to that used after (3.50) (Problem 3.13)

$$\langle x \rangle = \int \Phi^*(\mathbf{p}, t) \left(i\hbar \frac{\partial}{\partial p_x} \right) \Phi(\mathbf{p}, t) d\mathbf{p} \qquad (3.175a)$$

where the wave function $\Psi(\mathbf{r}, t)$ – and hence $\Phi(\mathbf{p}, t)$ – is normalised to unity. Similarly, we have

$$\langle y \rangle = \int \Phi^*(\mathbf{p}, t) \left(i\hbar \frac{\partial}{\partial p_y} \right) \Phi(\mathbf{p}, t) d\mathbf{p} \tag{3.175b}$$

and

$$\langle z \rangle = \int \Phi^*(\mathbf{p}, t) \left(i\hbar \frac{\partial}{\partial p_z} \right) \Phi(\mathbf{p}, t) d\mathbf{p}. \tag{3.175c}$$

The above equations imply that in momentum space the dynamical variables x, y and z are represented respectively by the differential operators $i\hbar\partial/\partial p_x$, $i\hbar\partial/\partial p_y$ and $i\hbar\partial/\partial p_z$. We remark that since the expectation values $\langle x \rangle$, $\langle y \rangle$ and $\langle z \rangle$ are real these operators acting on $\Phi(\mathbf{p}, t)$ are Hermitian. The three equations (3.175) can be summarised by writing

$$\langle \mathbf{r} \rangle = \int \Phi^*(\mathbf{p}, t)(i\hbar \nabla_\mathbf{p}) \Phi(\mathbf{p}, t) d\mathbf{p} \tag{3.176}$$

so that in momentum space the dynamical variable $\mathbf{r}$ is represented by the operator

$$\mathbf{r}_{op} = i\hbar \nabla_\mathbf{p} \tag{3.177}$$

where the gradient must be taken with respect to $\mathbf{p}$. On the other hand, it is clear from (3.45) that the action of $\mathbf{p}$ on $\Phi(\mathbf{p}, t)$ consists of multiplying the wave function $\Phi(\mathbf{p}, t)$ by $\mathbf{p}$. More generally, we see from (3.47) that the action of a function $g(\mathbf{p}, t)$ on $\Phi(\mathbf{p}, t)$ consists of multiplying $\Phi(\mathbf{p}, t)$ by $g(\mathbf{p}, t)$.

Following a reasoning similar to that used in Section 3.3 we can therefore formulate the following rules for obtaining the expectation value of a dynamical variable in a state specified by the momentum space wave function $\Phi(\mathbf{p}, t)$ normalised to unity.

(1) One associates with the dynamical variable $\mathcal{A} \equiv A(\mathbf{r}, \mathbf{p}, t)$ the linear operator

$$A(i\hbar \nabla_\mathbf{p}, \mathbf{p}, t) \tag{3.178}$$

obtained by performing the substitution $\mathbf{r} \rightarrow i\hbar \nabla_\mathbf{p}$ wherever the position vector $\mathbf{r}$ occurs. In addition, we must require that the operator A be Hermitian. A few important Hermitian operators acting in momentum space are listed in Table 3.2, together with the corresponding physical quantities. This Table should be compared with Table 3.1, where this correspondence was given for operators acting in configuration space.

(2) One then calculates the required expectation value from the expression

$$\langle A \rangle = \int \Phi^*(\mathbf{p}, t) A(i\hbar \nabla_\mathbf{p}, \mathbf{p}, t) \Phi(\mathbf{p}, t) d\mathbf{p}. \tag{3.179}$$

If the wave function $\Phi(\mathbf{p}, t)$ is not normalised to unity, the above expression must be replaced by

$$\langle A \rangle = \frac{\int \Phi^*(\mathbf{p}, t) A(i\hbar \nabla_\mathbf{p}, \mathbf{p}, t) \Phi(\mathbf{p}, t) d\mathbf{p}}{\int \Phi^*(\mathbf{p}, t) \Phi(\mathbf{p}, t) d\mathbf{p}}. \tag{3.180}$$

3.9 The Schrödinger equation in momentum space

Table 3.2 Physical quantities and corresponding operators acting in momentum space.

Physical quantity	Operator
Position coordinate x	$i\hbar \frac{\partial}{\partial p_x}$
Position vector $\mathbf{r}$	$i\hbar \nabla_{\mathbf{p}}$
x-component of momentum p_x	p_x
Momentum $\mathbf{p}$	$\mathbf{p}$
Kinetic energy $T = \frac{\mathbf{p}^2}{2m}$	$\frac{\mathbf{p}^2}{2m}$
Potential energy $V(\mathbf{r}, t)$	$V(i\hbar \nabla_{\mathbf{p}}, t)$
Total energy $\frac{\mathbf{p}^2}{2m} + V(\mathbf{r}, t)$	$H = \frac{\mathbf{p}^2}{2m} + V(i\hbar \nabla_{\mathbf{p}}, t)$

Stationary states

If the potential energy V is time-independent, the time-dependent Schrödinger equation (3.173) admits particular solutions of the separable form

$$\Phi(\mathbf{p}, t) = \phi(\mathbf{p}) \exp(-iEt/\hbar) \tag{3.181}$$

corresponding to precise values of the total energy E. The probability density in momentum space corresponding to these states is

$$\Pi(\mathbf{p}) = |\phi(\mathbf{p})|^2 = \phi^*(\mathbf{p})\phi(\mathbf{p}) \tag{3.182}$$

and hence is constant in time. The separable solutions (3.181) are clearly the Fourier transforms of the stationary states $\psi(\mathbf{r}) \exp(-iEt/\hbar)$ given by (3.100), so that the time-independent wave functions $\psi(\mathbf{r})$ and $\phi(\mathbf{p})$ are related by the Fourier transformation

$$\phi(\mathbf{p}) = (2\pi\hbar)^{-3/2} \int e^{-i\mathbf{p}\cdot\mathbf{r}/\hbar} \psi(\mathbf{r}) d\mathbf{r}, \tag{3.183a}$$

$$\psi(\mathbf{r}) = (2\pi\hbar)^{-3/2} \int e^{i\mathbf{p}\cdot\mathbf{r}/\hbar} \phi(\mathbf{p}) d\mathbf{p}. \tag{3.183b}$$

Upon substitution of (3.181) into the Schrödinger equation (3.173), we find that the wave function $\phi(\mathbf{p})$ must satisfy the time-independent Schrödinger eigenvalue equation

$$\frac{\mathbf{p}^2}{2m}\phi(\mathbf{p}) + \int \tilde{V}(\mathbf{p} - \mathbf{p}')\phi(\mathbf{p}')d\mathbf{p}' = E\phi(\mathbf{p}) \tag{3.184}$$

where

$$\tilde{V}(\mathbf{p} - \mathbf{p}') = (2\pi\hbar)^{-3} \int e^{-i(\mathbf{p}-\mathbf{p}')\cdot\mathbf{r}/\hbar} V(\mathbf{r}) d\mathbf{r}. \tag{3.185}$$

The equation (3.184) is an integral equation, and hence is generally more difficult to solve than the corresponding time-independent Schrödinger equation (3.98) in configuration space.

Problems

3.1 In obtaining equations (3.33) and (3.39) we have assumed that the potential $V(\mathbf{r}, t)$ is a real quantity.

(a) Prove that if $V(\mathbf{r}, t)$ is complex the continuity equation (3.39) becomes

$$\frac{\partial}{\partial t} P(\mathbf{r}, t) + \boldsymbol{\nabla} \cdot \mathbf{j}(\mathbf{r}, t) = \frac{2}{\hbar} [\text{Im } V(\mathbf{r}, t)] P(\mathbf{r}, t)$$

so that the addition of an imaginary part of the potential describes the presence of sources if Im $V > 0$ or sinks (absorbers) if Im $V < 0$. Show that if the wave function $\Psi(\mathbf{r}, t)$ is square integrable

$$\frac{\partial}{\partial t} \int P(\mathbf{r}, t) \mathrm{d}\mathbf{r} = \frac{2}{\hbar} \int [\text{Im } V(\mathbf{r}, t)] P(\mathbf{r}, t) \mathrm{d}\mathbf{r}.$$

(b) Assuming that Im V is time-independent and that $\Psi(\mathbf{r}, t)$ is normalised to unity at $t = 0$, prove that if $P(\mathbf{r}, t)$ is expanded about $t = 0$ as

$$P(\mathbf{r}, t) = P_0(\mathbf{r}) + P_1(\mathbf{r}) t + \cdots$$

then

$$\int P(\mathbf{r}, t) \mathrm{d}\mathbf{r} = 1 + Ct + \cdots$$

Find an expression for C in terms of $P_0(\mathbf{r})$ and Im $V(\mathbf{r})$, and verify that if Im $V(\mathbf{r}) < 0$ the quantity $\int P(\mathbf{r}, t) \mathrm{d}\mathbf{r}$ decreases with increasing time, so that absorption occurs.

3.2 Prove that the operator $p_x = -i\hbar \partial/\partial x$ is Hermitian.
(*Hint*: consider the expectation value

$$\langle p_x \rangle = \int \Psi^*(\mathbf{r}, t) \left[-i\hbar \frac{\partial}{\partial x} \right] \Psi(\mathbf{r}, t) \mathrm{d}\mathbf{r}$$

and integrate by parts.)

3.3 Starting from the expression

$$\langle p_x^n \rangle = \int \Phi^*(\mathbf{p}, t) p_x^n \Phi(\mathbf{p}, t) \mathrm{d}\mathbf{p}$$

where n is a positive integer, prove equation (3.58).

3.4 Prove equation (3.73).

3.5 Show that the operator $x^k p_x^l$ is not Hermitian (where k and l are positive integers), but that the combination $(x^k p_x^l + p_x^l x^k)/2$ is Hermitian.

3.6 Show that, for a general one-dimensional free-particle wave packet

$$\Psi(x,t) = (2\pi\hbar)^{-1/2} \int_{-\infty}^{+\infty} \exp[i(p_x x - p_x^2 t/2m)/\hbar]\phi(p_x)\mathrm{d}p_x$$

(a) the expectation value $\langle p_x \rangle$ of the momentum does not change with time, and
(b) the expectation value $\langle x \rangle$ of the position coordinate satisfies the equation

$$\langle x \rangle = \langle x \rangle_{t=t_0} + \frac{\langle p_x \rangle}{m}(t - t_0)$$

in agreement with the correspondence principle.

(*Hints*: Part (a) may be proved by obtaining the result

$$\langle p_x \rangle = \int_{-\infty}^{+\infty} \phi^*(p_x) p_x \phi(p_x)\mathrm{d}p_x.$$

For part (b), use the fact that

$$\frac{\partial}{\partial p_x}\exp[i(p_x x - p_x^2 t/2m)/\hbar] = \frac{i}{\hbar}(x - p_x t/m)\exp[i(p_x x - p_x^2 t/2m)/\hbar]$$

to show that

$$\langle x \rangle = \int_{-\infty}^{+\infty} \phi^*(p_x)\left[i\hbar\frac{\partial}{\partial p_x} + \frac{p_x}{m}t\right]\phi(p_x)\mathrm{d}p_x.)$$

3.7 Show that if $\Psi(\mathbf{r},t)$ is a square integrable wave function normalised to unity, then

$$\frac{\mathrm{d}}{\mathrm{d}t}\langle x^2 \rangle = \frac{1}{m}[\langle xp_x \rangle + \langle p_x x \rangle]$$

where the expectation values are defined according to (3.67).

3.8 Prove that if a constant V_0 is added to the potential energy $V(\mathbf{r})$

(a) the energy eigenvalues of the time-independent Schrödinger equation (3.98) are shifted from E to $E + V_0$; and
(b) the corresponding eigenfunctions $\psi(\mathbf{r})$ remain unchanged.

3.9 Consider a one-dimensional system with bound-state energies placed in order of increasing magnitude ($E_1 < E_2 < E_3 < \cdots$) and let $\psi_1(x), \psi_2(x), \psi_3(x), \ldots$, be the corresponding energy eigenfunctions. Prove that between two consecutive zeros of $\psi_n(x)$, the eigenfunction $\psi_{n+1}(x)$ must possess at least one zero. Note that the bound states of one-dimensional systems are non-degenerate, and that the energy eigenfunctions $\psi_n(x)$ can be taken to be real.

(*Hint*: From the Wronskian theorem (p. 110) and the Schrödinger eigenvalue equations satisfied by ψ_n and ψ_{n+1}, obtain first the relationship

$$[\psi_n' \psi_{n+1} - \psi_{n+1}' \psi_n]_a^b = \frac{2m}{\hbar^2}(E_{n+1} - E_n) \int_a^b \psi_n \psi_{n+1} dx.$$

Then complete the proof by taking a and b to be two consecutive zeros of ψ_n.)

3.10 Let $H = \mathbf{p}^2/2m + V(\mathbf{r})$ be a time-independent Hamiltonian. Prove that the propagator $K(\mathbf{r}, t; \mathbf{r}', t_0)$ defined by (3.156) satisfies the time-dependent Schrödinger equation

$$\left[i\hbar \frac{\partial}{\partial t} - H\right] K(\mathbf{r}, t; \mathbf{r}', t_0) = 0$$

subject to the condition

$$K(\mathbf{r}, t_0; \mathbf{r}', t_0) = \delta(\mathbf{r} - \mathbf{r}').$$

3.11 Show that the general one-dimensional free-particle wave packet $\Psi(x, t)$ of Problem 3.6 can be expressed in terms of its initial value $\Psi(x, t_0)$ as

$$\Psi(x, t) = \int_{-\infty}^{+\infty} K(x, t; x', t_0) \Psi(x', t_0) dx'$$

where the propagator $K(x, t; x', t_0)$ is given by

$$K(x, t; x', t_0) = \left[\frac{m}{2\pi i\hbar(t - t_0)}\right]^{1/2} \exp\left[\frac{im(x - x')^2}{2\hbar(t - t_0)}\right]$$

(*Hint*: Use the result (2.48), which also holds if α is purely imaginary, provided it is regarded as a limit. That is, if $\alpha = a + ib$, $a > 0$, it is the limit in which $a \to 0$.)

3.12 Let E_n denote the bound-state energy eigenvalues of a one-dimensional system and let $\psi_n(x)$ denote the corresponding energy eigenfunctions. Let $\Psi(x, t)$ be the wave function of the system, normalised to unity, and suppose that at $t = 0$ it is given by

$$\Psi(x, t = 0) = \frac{1}{\sqrt{2}} e^{i\alpha_1} \psi_1(x) + \frac{1}{\sqrt{3}} e^{i\alpha_2} \psi_2(x) + \frac{1}{\sqrt{6}} e^{i\alpha_3} \psi_3(x)$$

where the α_i are real constants.

(a) Write down the wave function $\Psi(x, t)$ at time t.
(b) Find the probability that at time t a measurement of the energy of the system gives the value E_2.
(c) Does $\langle x \rangle$ vary with time? Does $\langle p_x \rangle$ vary with time? Does $E = \langle H \rangle$ vary with time?

3.13 Prove equations (3.175).

3.14 Prove that the time-dependent Schrödinger equation (3.173) for the momentum space wave function can be written in the form

$$i\hbar \frac{\partial}{\partial t} \Phi(\mathbf{p}, t) = \frac{\mathbf{p}^2}{2m} \Phi(\mathbf{p}, t) + V(i\hbar \nabla_{\mathbf{p}}, t) \Phi(\mathbf{p}, t)$$

provided $V(\mathbf{r}, t)$ is an analytic function of x, y, z.
(*Hint*: Show first that

$$V(\mathbf{r}, t) = \exp\left(\frac{i}{\hbar} \mathbf{p} \cdot \mathbf{r}\right) V(i\hbar \nabla_{\mathbf{p}}, t) \exp\left(-\frac{i}{\hbar} \mathbf{p} \cdot \mathbf{r}\right).)$$

4 One-dimensional examples

4.1 General formulae 133
4.2 The free particle 134
4.3 The potential step 141
4.4 The potential barrier 150
4.5 The infinite square well 156
4.6 The square well 163
4.7 The linear harmonic oscillator 170
4.8 The periodic potential 182
Problems 189

In this chapter we shall study a few simple one-dimensional quantum mechanical problems in order to acquire some practice in dealing with the Schrödinger equation, and to analyse various interesting phenomena predicted by the quantum theory. One-dimensional problems are of interest not only because several physical situations are effectively one-dimensional, but also because a number of more complicated problems can be reduced to the solution of equations similar to the one-dimensional Schrödinger equation, as we shall see later in this book.

4.1 General formulae

We consider a particle of mass m moving on the x-axis in a time-independent potential $V(x)$. The time-dependent Schrödinger equation corresponding to this one-dimensional motion is

$$i\hbar \frac{\partial}{\partial t}\Psi(x,t) = \left[-\frac{\hbar^2}{2m}\frac{\partial^2}{\partial x^2} + V(x)\right]\Psi(x,t). \tag{4.1}$$

As we have shown in Chapter 3, the fact that the potential is time-independent implies that we can look for stationary-state solutions of (4.1) having the form

$$\Psi(x,t) = \psi(x)\exp(-iEt/\hbar) \tag{4.2}$$

where E is the energy of the stationary state. The general solution of (4.1) is an arbitrary superposition of stationary states (4.2). The time-independent wave function

$\psi(x)$ is a solution of the time-independent Schrödinger equation

$$-\frac{\hbar^2}{2m}\frac{d^2\psi(x)}{dx^2} + V(x)\psi(x) = E\psi(x) \tag{4.3}$$

which, in the present one-dimensional case, is just an ordinary differential equation. It is this equation which we shall solve below for several particular forms of the potential $V(x)$.

We also recall that the position probability density associated with a wave function $\Psi(x,t)$ is given by

$$P(x,t) = |\Psi(x,t)|^2 \tag{4.4}$$

and that the corresponding probability current density (or probability flux) is

$$j(x,t) = \frac{\hbar}{2im}\left[\Psi^*(x,t)\frac{\partial \Psi(x,t)}{\partial x} - \Psi(x,t)\frac{\partial \Psi^*(x,t)}{\partial x}\right]. \tag{4.5}$$

For stationary states (4.2) both P and j are, of course, time-independent. In particular, since $|\Psi(x,t)|^2 = |\psi(x)|^2$, the position probability density is given by

$$P(x) = |\psi(x)|^2. \tag{4.6}$$

Moreover, for one-dimensional stationary states (4.2) the equation (3.39) expressing probability conservation reduces to $dj/dx = 0$, so that the probability current density

$$j = \frac{\hbar}{2im}\left[\psi^*(x)\frac{d\psi(x)}{dx} - \psi(x)\frac{d\psi^*(x)}{dx}\right] \tag{4.7}$$

is also independent of x.

4.2 The free particle

Let us start by considering the simplest case, namely that of a potential which is constant: $V(x) = V_0$. The force acting on the particle, $F(x) = -dV/dx$, then vanishes so that the particle is free. Without loss of generality we can take the constant V_0 to be zero, since the addition of a constant amount to the potential energy merely shifts the energy eigenvalues of the Schrödinger equation (4.3) by that amount and leaves the eigenfunctions $\psi(x)$ unchanged (see Problem 3.8).

We must therefore solve the time-independent Schrödinger equation

$$-\frac{\hbar^2}{2m}\frac{d^2\psi(x)}{dx^2} = E\psi(x). \tag{4.8}$$

Writing

$$k = \left(\frac{2m}{\hbar^2}E\right)^{1/2} \tag{4.9}$$

we see that two linearly independent solutions of (4.8) are $\exp(ikx)$ and $\exp(-ikx)$ or, equivalently, the pair of real solutions $\sin kx$ and $\cos kx$. The general solution of (4.8) is therefore the linear combination

$$\psi(x) = A e^{ikx} + B e^{-ikx} \tag{4.10}$$

where A and B are arbitrary constants. It is clear that for a solution to be physically acceptable the quantity k cannot have an imaginary part, because if it did $\psi(x)$ would increase exponentially at one of the limits $x = -\infty$ or $x = +\infty$, and possibly at both limits. Since $E = \hbar^2 k^2/2m$, we must therefore have $E \geqslant 0$, which confirms the statement made in Section 3.6 that the energy cannot remain lower than the potential (here $V = 0$) over the entire interval $(-\infty, +\infty)$. Because any non-negative value of E is allowed, the energy spectrum is *continuous*, extending from $E = 0$ to $E = +\infty$. This, of course, is not surprising, since E is the kinetic energy of the free particle. Note that each positive energy eigenvalue is *doubly degenerate* because there are two linearly independent eigenfunctions, $\exp(ikx)$ and $\exp(-ikx)$, corresponding to the same positive energy $E = \hbar^2 k^2/2m$. We also remark that the basic solutions of (4.8) may be written in the form

$$\psi_{k_x}(x) = C \exp(ik_x x) \tag{4.11a}$$

or

$$\psi_{p_x}(x) = C \exp(ip_x x/\hbar) \tag{4.11b}$$

where C is a constant, $k_x = \pm k$ and $p_x = \hbar k_x$ is the momentum of the particle, of magnitude $p = |p_x| = \hbar k$.

Momentum eigenfunctions

The functions (4.11) are not only eigenfunctions of the free-particle Hamiltonian $p_{\text{op}}^2/2m = -(\hbar^2/2m)\partial^2/\partial x^2$ corresponding to the energy $E = p^2/2m = \hbar^2 k^2/2m$, they are also *momentum eigenfunctions*, that is eigenfunctions of the momentum operator $p_{\text{op}} = -i\hbar \partial/\partial x$. Indeed, the eigenvalue equation for the momentum operator reads

$$-i\hbar \frac{\partial}{\partial x} \psi_{p_x}(x) = p_x \psi_{p_x}(x) \tag{4.12}$$

and we see immediately that the functions (4.11) are solutions of this equation. The eigenvalues $p_x = \hbar k_x$ must be real, so that the eigenfunctions (4.11) remain finite as x tends to $+\infty$ or $-\infty$. Since this is the only constraint on p_x, the spectrum of the operator p_{op} is *continuous*, extending from $-\infty$ to $+\infty$. Note that the two functions $\exp(ikx)$ and $\exp(-ikx)$, which are eigenfunctions of the free-particle Hamiltonian corresponding to the same energy $E = \hbar^2 k^2/2m$, are eigenfunctions of the momentum operator p_{op} corresponding to the two different eigenvalues $p_x = \hbar k$ and $p_x = -\hbar k$, respectively. Finally, we remark that according to the Fourier analysis of Section 2.4, any physically admissible wave function $\Psi(x, t)$ can be expressed as a superposition of the momentum eigenfunctions (4.11).

Physical interpretation of the free-particle solutions

Let us now return to the general solution (4.10) of the free-particle Schrödinger equation (4.8). Substituting (4.10) into (4.2), we find that a stationary state of energy $E > 0$ for a free particle has the general form

$$\Psi(x, t) = (Ae^{ikx} + Be^{-ikx})e^{-iEt/\hbar}$$
$$= Ae^{i(kx-\omega t)} + Be^{-i(kx+\omega t)} \quad (4.13)$$

where $\omega = E/\hbar$ is the angular frequency.

In order to interpret the wave function (4.13) physically, let us consider some particular cases. If we set $B = 0$ in (4.13), the resulting wave function is the *plane wave*

$$\Psi(x, t) = Ae^{i(kx-\omega t)} \quad (4.14)$$

which, as we have seen in Chapter 2, is associated with a free particle of mass m moving along the x-axis in the positive direction with a definite momentum of magnitude $p = \hbar k$ (so that $p_x = \hbar k$) and a corresponding energy $E = p^2/2m = \hbar^2 k^2/2m$. This plane wave has an angular frequency $\omega = E/\hbar = \hbar k^2/2m$ and a wave number $k = p/\hbar = 2\pi/\lambda$, where λ is the de Broglie wavelength of the particle. It represents a vibration *travelling* in the positive x-direction with a phase velocity $v_{\text{ph}} = \omega/k = \hbar k/2m$. We recall that the particle velocity $v = p/m = \hbar k/m$ is *not* equal to the phase velocity v_{ph}, but is equal to the group velocity $v_g = d\omega/dk$ of the wave packet formed by superposing plane waves of different wave numbers.

The position probability density corresponding to the plane wave (4.14) is

$$P = |\Psi(x, t)|^2 = |A|^2. \quad (4.15)$$

It is not only independent of time (as for any stationary state), but is also independent of the variable x, so that the position of the particle on the x-axis is completely unknown. This is in accordance with the Heisenberg uncertainty relation (2.70) which implies that a particle moving along the x-axis with a perfectly well-defined momentum ($\Delta p_x = 0$) cannot be localised at all along this axis ($\Delta x = \infty$). We may thus think of the plane wave (4.14) as representing the idealised situation of a particle of perfectly known momentum, moving somewhere in a beam of infinite length.

Using (4.7) with $\psi(x) = A\exp(ikx)$, we find that the probability current density corresponding to the plane wave (4.14) is given by

$$j = \frac{\hbar}{2im}(A^* e^{-ikx} A i k e^{ikx} - Ae^{ikx} A^*(-ik)e^{-ikx})$$
$$= \frac{\hbar k}{m}|A|^2 = \frac{p}{m}|A|^2 = v|A|^2 \quad (4.16)$$

and is independent of x and t, in agreement with the general result obtained at the end of Section 4.1 for one-dimensional stationary states. It is also interesting to note that by using (4.15) we can rewrite (4.16) in the form $j = vP$. This relation is the same as the familiar one between flux, velocity and density in classical hydrodynamics. In particular, if the position probability density P is equal to unity (i.e. to one particle per unit length), the probability per unit time that the particle will cross the point x in the direction of increasing x is equal to its velocity v, as we expect.

Another particular case of (4.13) is obtained by setting $A = 0$. This gives the plane wave

$$\Psi(x, t) = B e^{-i(kx+\omega t)}. \tag{4.17}$$

The corresponding position probability density is

$$P = |\Psi(x, t)|^2 = |B|^2 \tag{4.18}$$

and the probability current density, obtained from (4.7) with $\psi(x) = B\exp(-ikx)$, is

$$j = -\frac{\hbar k}{m}|B|^2 = -\frac{p}{m}|B|^2 = -v|B|^2. \tag{4.19}$$

Thus the plane wave (4.17) corresponds to a vibration of wave number $k = p/\hbar$ and angular frequency $\omega = E/\hbar = \hbar k^2/2m$ travelling in the *negative* x-direction with a phase velocity $v_{\text{ph}} = \omega/k = \hbar k/2m$. It is associated with a free particle moving along the x-axis in the negative direction with a well-defined momentum of magnitude $p = mv = \hbar k$ (so that $p_x = -\hbar k$) and a given energy $E = p^2/2m = \hbar^2 k^2/2m$, but whose position on the x-axis is completely unknown.

Two other interesting particular cases of (4.13) are obtained by choosing $A = B$ or $A = -B$. If we first set $A = B$, so that we add two plane waves travelling in opposite directions with equal amplitudes, the wave function (4.13) becomes

$$\begin{aligned}\Psi(x, t) &= A(e^{ikx} + e^{-ikx})e^{-i\omega t}\\ &= C\cos kx\, e^{-i\omega t}\end{aligned} \tag{4.20}$$

where $C = 2A$. In contrast to the travelling waves (4.14) and (4.17), this is a *standing wave* whose nodes are fixed in space at values

$$x_n = \pm\left(\frac{\pi}{2} + n\pi\right)\Big/ k, \quad n = 0, 1, 2, \ldots \tag{4.21}$$

for which $\cos kx$ vanishes. The position probability density associated with the wave function (4.20) is

$$P(x) = |C|^2 \cos^2 kx \tag{4.22}$$

and from (4.7), with $\psi(x) = C\cos kx$, the corresponding probability flux is given by

$$\begin{aligned}j &= \frac{\hbar}{2im}(-C^*\cos kx\, Ck\sin kx + C\cos kx\, C^* k\sin kx)\\ &= 0.\end{aligned} \tag{4.23}$$

More generally, the probability flux (4.7) is seen to vanish for any function $\psi(x)$ which is real, apart from a multiplicative constant. In the present case, the result (4.23) is readily understood, since the probability flux $v|A|^2$ associated with the plane wave $A\exp[i(kx - \omega t)]$ is cancelled by the probability flux $-v|A|^2$ corresponding to the plane wave $A\exp[-i(kx + \omega t)]$; the net probability per unit time that the particle will cross a point x is thus equal to zero. We can therefore associate the standing wave (4.20) with a free particle moving along the x-axis with a momentum whose magnitude $p = \hbar k$ is known precisely, but whose direction is unknown.

It is also worth noting that the position probability density (4.22) vanishes at the points x_n given by (4.21), so that a free particle in the state described by the wave function (4.20) can never be found at these points. This is clearly an interference effect between the two travelling plane waves from which the standing wave (4.20) was constructed; it is entirely due to the wave properties associated with the particle.

If we set $A = -B$ in (4.13), we obtain another standing wave, namely

$$\Psi(x,t) = A(e^{ikx} - e^{-ikx})e^{-i\omega t}$$
$$= D\sin kx\, e^{-i\omega t}, \qquad D = 2iA \qquad (4.24)$$

with properties similar to those discussed above for the standing wave (4.20). In particular, the probability flux associated with the wave function (4.24) is equal to zero, and its position probability density

$$P(x) = |D|^2 \sin^2 kx \qquad (4.25)$$

vanishes at the values $x_n = \pm n\pi/k\,(n = 0, 1, 2, \ldots)$ for which $\sin kx = 0$.

Let us now return to the general free-particle stationary state (4.13), which is the superposition of two plane waves travelling in opposite directions along the x-axis, both with phase velocity $v_{\text{ph}} = \omega/k = \hbar k/2m$ and with amplitudes A and B, respectively. The position probability density corresponding to the wave function (4.13) is readily obtained by substituting (4.10) into (4.6); it is given by

$$P(x) = |A|^2 + |B|^2 + (AB^*e^{2ikx} + A^*Be^{-2ikx}) \qquad (4.26)$$

and exhibits interference effects of the two plane waves. Using (4.7) and (4.10) we also find that the probability current density associated with the wave function (4.13) is given by

$$j = v[|A|^2 - |B|^2] \qquad (4.27)$$

which, as we expect, is just the sum of the probability current densities (4.16) and (4.19).

'Normalisation' of the free-particle wave function

Since the integral

$$I = \int_{-\infty}^{+\infty} |Ae^{ikx} + Be^{-ikx}|^2 dx \qquad (4.28)$$

is infinite for all values of A and B (except, of course $A = B = 0$) the free-particle wave functions (4.10) and momentum eigenfunctions cannot be made to satisfy the normalisation condition

$$\int_{-\infty}^{+\infty} |\psi(x)|^2 dx = 1. \tag{4.29}$$

As we pointed out in Chapter 2, for wave functions of this kind which are not square integrable we can only speak of *relative probabilities*. This will be sufficient for treating applications in this chapter involving such eigenfunctions, corresponding to continuous eigenvalues, which remain finite at great distances. Nevertheless, it is often desirable to be able to treat these eigenfunctions on the same footing as the square integrable eigenfunctions corresponding to discrete eigenvalues. There are several ways of doing this. For example, we may recognise that the momentum of a particle is never known exactly, and work with wave packets. However, this approach is cumbersome, and would prevent us from using simple and convenient wave functions such as the plane wave $\exp(ikx)$, which is the idealisation of a very broad, square integrable wave packet.

An alternative way is to 'normalise' such wave functions in a sense different from that of (4.29). In particular, we may enclose the particle in a box (in the present case a 'one-dimensional' box of length L) at the walls of which the wave function of the particle must obey boundary conditions. In order to illustrate this *box normalisation* procedure, let us consider the momentum eigenfunctions (i.e. the plane waves) given by (4.11a). In this case, it is convenient to require that these wave functions satisfy *periodic boundary conditions* at the walls, so that $\psi_{k_x}(x + L) = \psi_{k_x}(x)$. Because of this periodicity condition, k_x is no longer an arbitrary real quantity; it is restricted to the values

$$k_x = \frac{2\pi}{L} n, \qquad n = 0, \pm 1, \pm 2, \ldots . \tag{4.30}$$

As a result, the spectrum of energy eigenvalues of the Schrödinger equation (4.8) becomes discrete:

$$E_n = \frac{\hbar^2 k_x^2}{2m} = \frac{2\pi^2 \hbar^2}{mL^2} n^2 \tag{4.31}$$

each eigenvalue (except $E = 0$) being doubly degenerate. Note that as L increases, the spacing of the successive energy levels decreases, so that for a macroscopic 'box' the spectrum is essentially continuous. The momentum eigenfunctions (4.11a) can now be normalised by requiring that, in the basic 'box' $(0, L)$ or $(-L/2, +L/2)$ of side L,

$$\int_{-L/2}^{+L/2} |\psi_{k_x}(x)|^2 dx = 1 \tag{4.32}$$

so that $|C|^2 = L^{-1}$. Choosing the arbitrary phase of the normalisation constant C to be zero, we have $C = L^{-1/2}$, and the normalised momentum eigenfunctions (4.11a)

are given by
$$\psi_{k_x}(x) = L^{-1/2} \exp(ik_x x). \tag{4.33}$$

We note that these eigenfunctions are *orthonormal* since
$$\int_{-L/2}^{+L/2} \psi^*_{k'_x}(x)\psi_{k_x}(x)\mathrm{d}x = L^{-1}\int_{-L/2}^{+L/2} \exp[i(k_x - k'_x)x]\mathrm{d}x$$
$$= \delta_{k_x k'_x} \tag{4.34}$$

where we have used the periodicity condition expressed by (4.30). We also remark that the momentum eigenfunctions (4.11) cannot be normalised by choosing a box with impenetrable (rigid) walls, since these eigenfunctions never vanish anywhere, and therefore cannot exist within such a box.

Instead of imposing a box normalisation with periodic boundary conditions, we can also use the Dirac delta function (discussed in Appendix A) to set up a *delta function normalisation* similar to that given by (4.34), but which allows the momentum eigenfunctions (4.11a) to retain their form over the entire x-axis, for all real values of k_x. Indeed, using equation (A.18) of Appendix A, we have
$$\int_{-\infty}^{+\infty} \exp[i(k_x - k'_x)x]\mathrm{d}x = 2\pi\delta(k_x - k'_x). \tag{4.35}$$

Taking the arbitrary phase of the normalisation constant C in (4.11a) to be zero, we see that if we choose $C = (2\pi)^{-1/2}$, the momentum eigenfunctions
$$\psi_{k_x}(x) = (2\pi)^{-1/2}\exp(ik_x x) \tag{4.36}$$
satisfy the orthonormality relation
$$\int_{-\infty}^{+\infty} \psi^*_{k'_x}(x)\psi_{k_x}(x)\mathrm{d}x = \delta(k_x - k'_x). \tag{4.37}$$

We also see with the help of (A.18) that the wave functions (4.36) satisfy the *closure relation*
$$\int_{-\infty}^{+\infty} \psi^*_{k_x}(x')\psi_{k_x}(x)\mathrm{d}k_x = \delta(x - x'). \tag{4.38}$$

A similar relation can also be established in the limit of large L for the momentum eigenfunctions (4.33) satisfying the box normalisation with periodic boundary conditions (Problem 4.1).

The normalisation condition expressed by (4.37) is often referred to as *k-normalisation*. Other delta-function normalisations for wave functions belonging to the continuous spectrum may be used. For example, the momentum eigenfunctions
$$\psi_{p_x}(x) = (2\pi\hbar)^{-1/2}\exp(ip_x x/\hbar) \tag{4.39}$$
are such that
$$\int_{-\infty}^{+\infty} \psi^*_{p'_x}(x)\psi_{p_x}(x)\mathrm{d}x = \delta(p_x - p'_x) \tag{4.40}$$

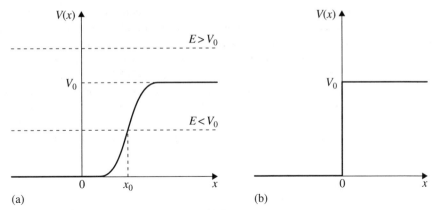

Figure 4.1 (a) A smoothly varying, infinitely wide potential barrier $V(x)$, tending to zero as one goes to the left, and to the constant value $V_0 > 0$ as one goes to the right. Two values of the total energy of the particle are indicated by dashed lines, corresponding to the cases $E < V_0$ and $E > V_0$. When $E < V_0$ there is a classical turning point at $x = x_0$, where $E = V(x_0)$. (b) The step potential (4.41).

and this normalisation is called *p-normalisation*. Another frequently used normalisation, called the energy normalisation, is the subject of Problem 4.2.

4.3 The potential step

Our next example is that of a particle moving in a potential $V(x)$ which has the form of an infinitely wide potential barrier (see Fig. 4.1(a)). As we go to the left, the potential tends to the constant value zero, while as we go to the right it tends to the constant value $V_0 > 0$. We have also indicated in Fig. 4.1(a) two values of the total energy E of the particle. According to classical mechanics a particle incident from the left with an energy $E < V_0$ would always be *reflected* at the *turning point* x_0, where its kinetic energy vanishes. The region to the right of x_0 is therefore classically forbidden for the particle. On the other hand, following classical mechanics, a particle incident from the left with an energy $E > V_0$ would never be reflected, but would always be *transmitted*.

Let us now study this problem by using quantum mechanics. Since the potential is time-independent the motion of a particle of energy E is described by the wave function $\Psi(x, t) = \psi(x) \exp(-iEt/\hbar)$, where $\psi(x)$ is a solution of the time-independent Schrödinger equation (4.3). To simplify our task we shall replace the smoothly varying potential of Fig. 4.1(a) by the *step potential* shown in Fig. 4.1(b). Note that we have chosen the turning point x_0 to be at the origin, so that

$$V(x) = \begin{cases} 0, & x < 0 \\ V_0, & x > 0. \end{cases} \tag{4.41}$$

It is clear that this potential step is an idealisation of potentials of the form shown in Fig. 4.1(a). It is the simplest example of a *piecewise constant* potential which is pieced together from constant portions in adjacent ranges of the x-axis. Other examples of piecewise constant potentials will be considered later in this chapter.

Before solving the time-independent Schrödinger equation (4.3) for the potential step (4.41), we recall from our discussion of Section 3.6 that the total energy E can never be lower than the absolute minimum of the potential. Thus for the potential step (4.41) there is no acceptable solution of the Schrödinger equation (4.3) when $E < 0$. In what follows we shall study successively the two cases $0 < E < V_0$, and $E > V_0$.

Case 1: $0 < E < V_0$

Since the step potential (4.41) is such that $V(x) = 0$ for $x < 0$ and $V(x) = V_0 > E$ for $x > 0$, the Schrödinger equation (4.3) becomes, in this case,

$$\frac{d^2\psi(x)}{dx^2} + k^2\psi(x) = 0, \qquad k = \left(\frac{2m}{\hbar^2}E\right)^{1/2}, \qquad x < 0 \tag{4.42a}$$

and

$$\frac{d^2\psi(x)}{dx^2} - \kappa^2\psi(x) = 0, \qquad \kappa = \left[\frac{2m}{\hbar^2}(V_0 - E)\right]^{1/2}, \qquad x > 0. \tag{4.42b}$$

Equation (4.42a), valid for $x < 0$, describes the motion of a free particle of wave number k. Its general solution is

$$\psi(x) = Ae^{ikx} + Be^{-ikx}, \qquad x < 0 \tag{4.43a}$$

where A and B are arbitrary constants. The second equation (4.42b), valid for $x > 0$, has the general solution

$$\psi(x) = Ce^{\kappa x} + De^{-\kappa x}, \qquad x > 0 \tag{4.43b}$$

where C and D are arbitrary constants.

It follows from the discussion of Section 3.6 that in order to be an acceptable solution of the Schrödinger equation (4.3) the eigenfunction $\psi(x)$ and its first derivative $d\psi(x)/dx$ must be *finite* and *continuous* for all values of x. Now, if $C \neq 0$ it is clear that the eigenfunction $\psi(x)$ given by (4.43b) for $x > 0$ will not remain finite in the limit $x \to +\infty$, since $\exp(\kappa x)$ diverges in that limit. To prevent this unacceptable behaviour, we must choose the constant $C = 0$, so that for $x > 0$ the eigenfunction $\psi(x)$ is given by $\psi(x) = D\exp(-\kappa x)$. Thus we have

$$\psi(x) = \begin{cases} Ae^{ikx} + Be^{-ikx}, & x < 0 \\ De^{-\kappa x}, & x > 0 \end{cases} \qquad \begin{matrix}(4.44a)\\(4.44b)\end{matrix}$$

and we note that, since one of the two linearly independent solutions for $x > 0$ has been excluded, the energy eigenvalues for $0 < E < V_0$ are non-degenerate.

Next, let us consider the point of discontinuity of the potential at $x = 0$. We recall from our discussion of Section 3.1 that the existence of discontinuities of the first kind (i.e. finite jumps) in the potential does not modify the conditions of continuity imposed on the function $\psi(x)$ and on its first derivative $d\psi(x)/dx$. Indeed, at a finite discontinuity of the potential it follows from the Schrödinger equation (4.3) that the *second* derivative $d^2\psi(x)/dx^2$ will exhibit a finite jump, but the integral of $d^2\psi(x)/dx^2$ remains continuous, so that $d\psi(x)/dx$, and hence $\psi(x)$, are continuous everywhere. This means that at $x = 0$ the solutions (4.44a) and (4.44b) must join 'smoothly' in such a way that $\psi(x)$ and $d\psi(x)/dx$ are continuous. The condition that $\psi(x)$ be continuous at $x = 0$ gives the relation

$$A + B = D \tag{4.45a}$$

while the requirement that $d\psi(x)/dx$ be continuous at $x = 0$ yields

$$ik(A - B) = -\kappa D. \tag{4.45b}$$

From equations (4.45) we deduce that

$$A = \frac{1 + i\kappa/k}{2}D, \qquad B = \frac{1 - i\kappa/k}{2}D \tag{4.46}$$

so that the solution $\psi(x)$ of the Schrödinger equation (4.42) is given by equations (4.44), with A and B obtained in terms of D from (4.46). Note that since the oscillatory part (4.44a) of $\psi(x)$ can be joined smoothly to its exponential part (4.44b) for all values of E between 0 and V_0, the spectrum is *continuous*. The remaining constant D can be determined by 'normalising' the eigenfunction $\psi(x)$, but this will not be required in what follows. We also remark that since the quantity

$$\frac{B}{A} = \frac{1 - i\kappa/k}{1 + i\kappa/k} = \frac{1 - i(V_0/E - 1)^{1/2}}{1 + i(V_0/E - 1)^{1/2}} \tag{4.47}$$

is of modulus one it may be written in the form

$$\frac{B}{A} = e^{i\alpha}, \qquad \alpha = 2\tan^{-1}\left[-\left(\frac{V_0}{E} - 1\right)^{1/2}\right] \tag{4.48}$$

while

$$\frac{D}{A} = \frac{2}{1 + i\kappa/k} = \frac{2}{1 + i(V_0/E - 1)^{1/2}} = 1 + e^{i\alpha}. \tag{4.49}$$

Hence the eigenfunction (4.44) can be rewritten in terms of the constant A as

$$\psi(x) = \begin{cases} 2Ae^{i\alpha/2}\cos(kx - \frac{\alpha}{2}), & x < 0 \\ 2Ae^{i\alpha/2}\cos\frac{\alpha}{2}e^{-\kappa x}, & x > 0 \end{cases} \tag{4.50a}$$
$$\tag{4.50b}$$

and is seen to be real, apart from a multiplicative constant. As an example, the eigenfunction (4.50) with $A = 2^{-1}\exp(-i\alpha/2)$ is shown in Fig. 4.2 for the case $\alpha = -\pi/3$, which corresponds to $E = 0.75V_0$.

We may readily interpret the solution $\psi(x)$ as follows. Let us first consider the region $x < 0$ and return to equation (4.44a). We have seen in Section 4.2 that

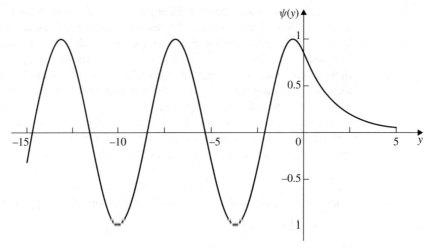

Figure 4.2 The eigenfunction ψ given by equations (4.50) with $A = 2^{-1}\exp(-i\alpha/2)$, as a function of the variable $y = kx$, for the case $\alpha = -\pi/3$, corresponding to an incident particle energy $E = 0.75 V_0$.

the functions $\exp(ikx)$ and $\exp(-ikx)$, when multiplied by the factor $\exp(-iEt/\hbar)$, represent plane waves moving towards the right and towards the left, respectively. Moreover, the probability current density corresponding to the wave function (4.44a) is given (see (4.27)) by

$$j = v[|A|^2 - |B|^2], \qquad x < 0 \qquad (4.51)$$

where $v = \hbar k/m$ is the magnitude of the velocity of the particle. The first term $A\exp(ikx)$ of $\psi(x)$ in the region $x < 0$ therefore corresponds to a plane wave of amplitude A *incident* from the left on the potential step, while the second term $B\exp(-ikx)$ corresponds to a *reflected* plane wave of amplitude B. The *reflection coefficient* R is defined as the ratio of the intensity of the reflected probability current density $(v|B|^2)$ to the intensity of the incident probability current density $(v|A|^2)$. Thus

$$R = \frac{v|B|^2}{v|A|^2} = \frac{|B|^2}{|A|^2} \qquad (4.52)$$

and we see that the quantity R is also equal to the ratio of the intensity of the reflected wave to that of the incident wave. Note that R is independent of the absolute normalisation of the wave function $\psi(x)$. Now, from (4.48) we have $|A|^2 = |B|^2$, so that $R = 1$ and the reflection is *total*. This result is in agreement with the prediction of classical mechanics, according to which all particles with total energy $E < V_0$ are reflected by the potential step. However, we see from (4.6) and (4.50a) that to the left of the barrier the position probability density

$$P(x) = 4|A|^2 \cos^2\left(kx - \frac{\alpha}{2}\right), \qquad x < 0 \tag{4.53}$$

exhibits an oscillatory behaviour, which is a purely quantum mechanical interference effect between the incident and reflected waves.

Since $|A|^2 = |B|^2$, the probability current density (4.51) vanishes for $x < 0$. By substituting the expression of $\psi(x)$ given by (4.44b) on the right of equation (4.7), it is also readily verified that $j = 0$ for $x > 0$. Hence there is no net current anywhere and therefore no net momentum in the state (4.44). This result could have been predicted directly by looking at the form (4.50) of the wave function. Indeed, as we already remarked in Section 4.2, the probability current density (4.7) vanishes for any function $\psi(x)$ which is real except for a multiplicative constant.

Let us now examine in more detail the eigenfunction $\psi(x)$ in the region $x > 0$, as given by (4.44b). The interesting feature is that this function *does not vanish* in this region. Thus, using (4.6) and (4.44b), we find that the position probability density is given to the right of the potential step by

$$P(x) = |D|^2 e^{-2\kappa x}, \qquad x > 0. \tag{4.54}$$

Although this expression decreases rapidly with increasing x, *there is a finite probability of finding the particle in the classically inaccessible region $x > 0$*. This striking non-classical phenomenon of *barrier penetration* is of great importance in the understanding of various phenomena in quantum physics, and will be illustrated in the following sections. In the present case, however, the wave-like property of penetration into the classically excluded region cannot be demonstrated experimentally. Indeed, in order to observe the particle in the region $x > 0$, it must be localised within a distance of order $\Delta x \simeq \kappa^{-1}$, for which the probability (4.54) is appreciable. As a result, its momentum would be uncertain by an amount

$$\Delta p_x \gtrsim \frac{\hbar}{\Delta x} \simeq \hbar\kappa = [2m(V_0 - E)]^{1/2} \tag{4.55}$$

and its energy would be uncertain by a quantity

$$\Delta E = \frac{(\Delta p_x)^2}{2m} \gtrsim V_0 - E \tag{4.56}$$

so that it would no longer be possible to state with certainty that the total energy E of the particle is less than V_0.

It is interesting to examine the limiting case of an infinitely high potential barrier, so that $V_0 \to \infty$ (the energy E being kept constant). Then $\kappa \to \infty$ and we see from (4.44b) that $\psi(x) \to 0$ in the classically forbidden region $x > 0$. From (4.47)–(4.49) we also deduce that

$$\lim_{V_0 \to \infty} \frac{B}{A} = -1, \qquad \lim_{V_0 \to \infty} \frac{D}{A} = 0 \tag{4.57}$$

so that as $V_0 \to \infty$ we have $B = -A$ and $D = 0$. The wave function (4.44) therefore becomes, in this case,

$$\psi(x) = \begin{cases} A(e^{ikx} - e^{-ikx}), & x < 0 \\ 0, & x > 0. \end{cases} \quad (4.58)$$

Thus, for an infinitely high barrier the wave function vanishes at the barrier (for $x = 0$) and beyond it (for $x > 0$), in agreement with the discussion of Section 3.6. We also remark that at the point $x = 0$, where the potential makes an infinite jump, the slope of the wave function (4.58) changes suddenly from the finite value $2ikA$ to zero. Note that this discontinuity of the slope is not in contradiction with the condition of 'smooth' joining (which requires both $\psi(x)$ and $d\psi(x)/dx$ to be continuous) which only holds when the potential is either continuous or exhibits *finite* jumps.

Case 2: $E > V_0$

Let us now return to finite potential steps and consider the case for which the total energy E exceeds V_0. The Schrödinger equation (4.3) then becomes

$$\frac{d^2\psi(x)}{dx^2} + k^2\psi(x) = 0, \qquad k = \left(\frac{2m}{\hbar^2}E\right)^{1/2}, \qquad x < 0 \quad (4.59a)$$

$$\frac{d^2\psi(x)}{dx^2} + k'^2\psi(x) = 0, \qquad k' = \left[\frac{2m}{\hbar^2}(E - V_0)\right]^{1/2}, \qquad x > 0 \quad (4.59b)$$

and is readily solved to give

$$\psi(x) = \begin{cases} Ae^{ikx} + Be^{-ikx}, & x < 0 & (4.60a) \\ Ce^{ik'x} + De^{-ik'x}, & x > 0 & (4.60b) \end{cases}$$

where A, B, C and D are integration constants. We see that in both regions $x < 0$ and $x > 0$ the wave function (4.60) has an oscillatory character. Moreover, it is doubly degenerate since it consists of a linear combination of two linearly independent solutions corresponding to the same energy.

In order to make progress in the interpretation of the wave function (4.60), we must specify the physical situation which we want to study. Clearly, the particle can be incident on the step either from the left or from the right. We shall consider here the case when the particle is incident from the left. We must then discard the second term $D \exp(-ik'x)$ on the right of (4.60b), since this term corresponds to a reflected wave travelling in the direction of decreasing x in the region $x > 0$, and there is nothing for x large and positive which can cause such a reflection. Hence we must set $D = 0$, so that the wave function (4.60) becomes

$$\psi(x) = \begin{cases} Ae^{ikx} + Be^{-ikx}, & x < 0 & (4.61a) \\ Ce^{ik'x}, & x > 0. & (4.61b) \end{cases}$$

We see that it consists of an *incident* wave (of amplitude A) and a *reflected* wave (of amplitude B), both of wave number k, in the region $x < 0$, and of a *transmitted* wave (of amplitude C) of wave number k' in the region $x > 0$. Thus, in contrast with classical mechanics which says that a particle with energy $E > V_0$ will always pass the potential step, quantum mechanics predicts that the incident wave will be partly transmitted and partly reflected.

The constants A, B and C in (4.61) can be related by joining 'smoothly' the solutions (4.61a) and (4.61b), so that $\psi(x)$ and $d\psi(x)/dx$ are continuous at $x = 0$. Continuity of $\psi(x)$ at $x = 0$ requires that

$$A + B = C \tag{4.62a}$$

and continuity of $d\psi(x)/dx$ at $x = 0$ yields the further condition

$$k(A - B) = k'C. \tag{4.62b}$$

These two equations are readily solved to give

$$\frac{B}{A} = \frac{k - k'}{k + k'} \tag{4.63}$$

and

$$\frac{C}{A} = \frac{2k}{k + k'}. \tag{4.64}$$

Since the two solutions (4.61a) and (4.61b) can be joined smoothly in this way for all values of $E > V_0$, the spectrum is again *continuous*.

Let us now consider the probability current density j associated with the wave function (4.61). Substituting this wave function in (4.7), we find that

$$j = v[|A|^2 - |B|^2], \quad x < 0 \tag{4.65a}$$

$$= v'|C|^2, \quad x > 0 \tag{4.65b}$$

where $v = \hbar k/m$ and $v' = \hbar k'/m$ are the magnitudes of the particle velocity in the regions $x < 0$ and $x > 0$, respectively. Using (4.63) and (4.64) we have

$$\frac{|B|^2}{|A|^2} + \frac{v'}{v}\frac{|C|^2}{|A|^2} = 1 \tag{4.66}$$

so that $v'|C|^2 = v[|A|^2 - |B|^2]$ and the probability current density j has the same constant value everywhere, as it should for any one-dimensional stationary state. Note, however, that in contrast to the case $0 < E < V_0$ studied above, this constant value is not zero in the present situation.

The *reflection coefficient* R has been defined in (4.52) to be the ratio of the intensity of the reflected probability current density to that of the incident probability current density. Using (4.63), we therefore have

$$R = \frac{|B|^2}{|A|^2} = \frac{(k - k')^2}{(k + k')^2} = \frac{[1 - (1 - V_0/E)^{1/2}]^2}{[1 + (1 - V_0/E)^{1/2}]^2}, \quad E > V_0. \tag{4.67}$$

The reflection coefficient is plotted in Fig. 4.3 as a function of the ratio E/V_0. We have also included in this graph the case $E/V_0 < 1$ studied above, for which $R = 1$.

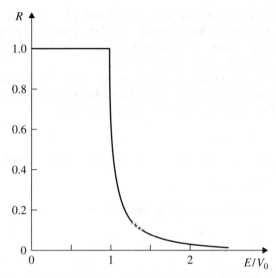

Figure 4.3 The reflection coefficient R for particles incident on the step potential (4.41), as a function of the ratio E/V_0.

Note that $R = 1$ when $E = V_0$. As E/V_0 continues to increase beyond unity, R decreases monotonically and tends to zero when $E/V_0 \to \infty$.

We can also define the *transmission coefficient* T as the ratio of the intensity of the transmitted probability current density to that of the incident probability current density. Thus T, like R, is independent of the absolute normalisation of the wave function $\psi(x)$. Using (4.64) we obtain

$$T = \frac{v'|C|^2}{v|A|^2} = \frac{4kk'}{(k+k')^2} = \frac{4(1-V_0/E)^{1/2}}{[1+(1-V_0/E)^{1/2}]^2}, \qquad E > V_0 \qquad (4.68)$$

and we see that both R and T depend only on the ratio V_0/E. Moreover, we deduce from (4.66) that

$$R + T = 1. \qquad (4.69)$$

We have seen above that the wave function describing the motion of the particle is partially reflected and partially transmitted at the potential step. Similarly, the probability flux incident upon the potential step is split into a reflected flux and a transmitted flux, the total probability flux being conserved by virtue of (4.69). The particle itself is of course not split at the potential step, since particles are always observed as complete entities. In any particular event the particle will either be reflected or transmitted, the probability of reflection being R and that of transmission being T for a large number of events.

The fact that potential steps can reflect particles for which $E > V_0$ (i.e. having enough energy to be classically transmitted) is clearly due to the *wave-like* properties of particles embodied in the quantum mechanical formalism. It is therefore not surprising that this behaviour is analogous to a phenomenon occurring in classical *wave optics*, namely the reflection of light at a boundary where the index of refraction changes abruptly. Indeed, let us consider the regions to the left and to the right of the potential step as two 'optical media'. In the first medium ($x < 0$) the de Broglie wavelength of the particle is $\lambda = 2\pi/k$, and in the second medium ($x > 0$) it is given by $\lambda' = 2\pi/k'$. The index of refraction n of medium 2 relative to medium 1 is defined in the usual way as the ratio of the two wavelengths

$$n = \frac{\lambda}{\lambda'} = \frac{k'}{k} = \left(1 - \frac{V_0}{E}\right)^{1/2} \tag{4.70}$$

and in terms of this index of refraction we have

$$R = \left(\frac{1-n}{1+n}\right)^2 \tag{4.71}$$

which is precisely the result obtained in wave optics.

Until now we have analysed the situation for which the particle is incident from the left, but it is also instructive to consider the case when the particle is incident on the potential step from the right. We must then set $A = 0$ in the general wave function (4.60), since the term $A \exp(ikx)$ now corresponds to a reflected wave in the region $x < 0$, and there is no reflector for $x \to -\infty$. Following the method given above, it is straightforward to obtain the reflection and transmission coefficients for this case (Problem 4.3). It is found that they are the same as those given by (4.67) and (4.68) for a particle incident from the left, a result which could have been anticipated since both R and T remain unchanged if k and k' are exchanged. This clearly shows that partial reflection is not due to an increase of $V(x)$ in the direction of motion of the particle, but results from an *abrupt change* in the potential (and hence in the de Broglie wavelength of the particle), just as optical reflection occurs at a boundary where the index of refraction changes abruptly. Thus, if we return to the smoothly varying potential shown in Fig. 4.1(a), we can predict that partial reflection of particles with energy $E > V_0$ will only occur significantly if the potential $V(x)$ changes by an appreciable amount over a short distance. This can be proved by solving explicitly the Schrödinger equation (4.3) for potentials of the form shown in Fig. 4.1(a). It is found that partial reflection for $E > V_0$ occurs if the potential $V(x)$ varies significantly over a distance of the order of the de Broglie wavelength λ of the incident particle. On the other hand, reflection is negligible when the potential varies slowly over that de Broglie wavelength. This, of course, is the case in the classical limit for which any realistic potential $V(x)$ of the form shown in Fig. 4.1(a) changes by a negligible amount over the de Broglie wavelength, which tends to zero in that limit.

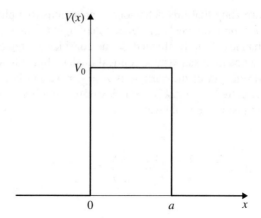

Figure 4.4 A rectangular potential barrier of height V_0 and thickness a.

4.4 The potential barrier

We now proceed to study the motion of a particle in a potential $V(x)$ which has the form of the rectangular barrier shown in Fig. 4.4. Thus

$$V(x) = \begin{cases} 0, & x < 0 \\ V_0, & 0 < x < a \\ 0, & x > a \end{cases} \tag{4.72}$$

with $V_0 > 0$. According to classical mechanics, a particle of total energy E incident upon this barrier would always be reflected if $E < V_0$ and it would be transmitted if $E > V_0$. In view of what we learned in the preceding section, we expect that quantum mechanics will lead to a different prediction. Indeed, as we shall now see, the quantum mechanical treatment of this problem leads to the conclusion that both reflection and transmission happen with finite probability for most (positive) values of the particle energy.

We begin by remarking that, as in the case of the potential step, there is no admissible solution of the Schrödinger equation (4.3) for the potential (4.72) when $E < 0$. We shall thus assume that $E > 0$. Note that the problem we want to analyse belongs to the category of one-dimensional *scattering problems*, in which a particle of energy $E > 0$ interacts with a potential in a given interval, and is free outside this interval. In the present case the particle interacts with the potential in the interval $(0, a)$. In the external regions $x < 0$ and $x > a$ the particle is free, and the general solution of the Schrödinger equation (4.3) is given by

$$\psi(x) = \begin{cases} Ae^{ikx} + Be^{-ikx}, & x < 0 \tag{4.73a} \\ Ce^{ikx} + De^{-ikx}, & x > a \tag{4.73b} \end{cases}$$

where A, B, C, D are constants and $k = (2mE/\hbar^2)^{1/2}$. It is interesting to remark that solutions of the form (4.73) are valid in the force-free regions that are external to the interval in which *any* scattering potential acts.

The particle can obviously be incident upon the barrier either from the left or from the right. We shall study here the case when it is incident on the barrier from the left. Since there is nothing at large positive values of x to cause a reflection, we must set $D = 0$ in (4.73b). The wave function is therefore given in the external regions by

$$\psi(x) = \begin{cases} A e^{ikx} + B e^{-ikx}, & x < 0 \\ C e^{ikx}, & x > a. \end{cases} \quad \begin{matrix} \text{(4.74a)} \\ \text{(4.74b)} \end{matrix}$$

We see that in the region $x < 0$ the wave function consists of an *incident wave* of amplitude A and a *reflected wave* of amplitude B, both of wave number k. In the region $x > a$ it is a pure *transmitted* wave of amplitude C and wave number k. This interpretation is readily checked by substituting the wave function (4.74) into (4.7). The resulting probability current density (which is constant since we are dealing with one-dimensional stationary states) is given by

$$j = \begin{cases} v[|A|^2 - |B|^2], & x < 0 \\ v|C|^2, & x > a \end{cases} \quad \begin{matrix} \text{(4.75a)} \\ \text{(4.75b)} \end{matrix}$$

where $v = \hbar k/m$ is the magnitude of the particle velocity. The quantities $v|A|^2$, $v|B|^2$ and $v|C|^2$ are, respectively, the intensities of the incident, reflected and transmitted probability current densities. Following the definitions of Section 4.3, a *reflection coefficient* R and a *transmission coefficient* T can be introduced, which are independent of the absolute normalisation of the wave function $\psi(x)$. They are given by

$$R = \frac{|B|^2}{|A|^2}, \qquad T = \frac{|C|^2}{|A|^2}. \quad \text{(4.76)}$$

Note that in writing the expression for T we have used the fact that the velocity v of the particle is the same in both regions $x < 0$ and $x > a$.

Until now we have studied the solution $\psi(x)$ of the Schrödinger equation in the external regions $x < 0$ and $x > a$. The nature of the solution in the internal region $0 < x < a$ depends on whether $E < V_0$ or $E > V_0$. We shall analyse these two cases in succession.

Case 1: $E < V_0$

Remembering that $V(x) = V_0$ for $0 < x < a$, and setting $\kappa = [2m(V_0 - E)/\hbar^2]^{1/2}$ as in (4.42b), the solution of the Schrödinger equation in the internal region is given by

$$\psi(x) = F e^{\kappa x} + G e^{-\kappa x}, \qquad 0 < x < a. \quad \text{(4.77)}$$

The five constants A, B, C, F and G can be related by requiring that $\psi(x)$ and $d\psi(x)/dx$ be continuous at both points $x = 0$ and $x = a$ where the potential is discontinuous. Thus at $x = 0$ we have, from (4.74a) and (4.77),

$$A + B = F + G \tag{4.78a}$$
$$ik(A - B) = \kappa(F - G) \tag{4.78b}$$

while at $x = a$ we find from (4.74b) and (4.77) that

$$Ce^{ika} = Fe^{\kappa a} + Ge^{-\kappa a} \tag{4.79a}$$
$$ikCe^{ika} = \kappa(Fe^{\kappa a} - Ge^{-\kappa a}). \tag{4.79b}$$

Eliminating F and G and solving for the ratios B/A and C/A, we obtain

$$\frac{B}{A} = \frac{(k^2 + \kappa^2)(e^{2\kappa a} - 1)}{e^{2\kappa a}(k + i\kappa)^2 - (k - i\kappa)^2} \tag{4.80a}$$

and

$$\frac{C}{A} = \frac{4ik\kappa e^{-ika}e^{\kappa a}}{e^{2\kappa a}(k + i\kappa)^2 - (k - i\kappa)^2} \tag{4.80b}$$

so that the reflection and transmission coefficients are given by

$$R = \frac{|B|^2}{|A|^2} = \left[1 + \frac{4k^2\kappa^2}{(k^2 + \kappa^2)^2 \sinh^2(\kappa a)}\right]^{-1} = \left[1 + \frac{4E(V_0 - E)}{V_0^2 \sinh^2(\kappa a)}\right]^{-1} \tag{4.81a}$$

and

$$T = \frac{|C|^2}{|A|^2} = \left[1 + \frac{(k^2 + \kappa^2)^2 \sinh^2(\kappa a)}{4k^2\kappa^2}\right]^{-1} = \left[1 + \frac{V_0^2 \sinh^2(\kappa a)}{4E(V_0 - E)}\right]^{-1}. \tag{4.81b}$$

It is also readily verified that $R + T = 1$, which expresses conservation of the probability flux.

The above equations lead to the remarkable prediction that a particle has a certain probability of 'leaking through' a potential barrier which is completely 'opaque' from the point of view of classical mechanics. This phenomenon, which is one of the most striking features of quantum mechanics, is known as *barrier penetration* or the *tunnel effect*, and is quite common in atomic, nuclear and solid state physics. For example, the tunnel effect is fundamental in explaining the emission of alpha particles by radioactive nuclei, a process which we shall study in Chapter 8 after learning how to calculate the transmission coefficient for more general potential barriers than the rectangular one considered here. In Fig. 4.5 we illustrate schematically the tunnel effect by showing the modulus square of the wave function, $|\psi(x)|^2$, for a rectangular potential barrier (4.72) such that $mV_0a^2/\hbar^2 = 0.25$ and an incident particle energy $E = 0.75V_0$. Of particular interest is the fact that in the region $x > a$, one has $|\psi(x)|^2 = |C|^2$, which is non-zero. We also note the attenuation of $|\psi(x)|^2$ inside the barrier, and the fact that since $R < 1$ we have $|B| < |A|$, so that the wave function $\psi(x)$ – and hence $|\psi(x)|^2$ – never goes through zero in the region $x < 0$.

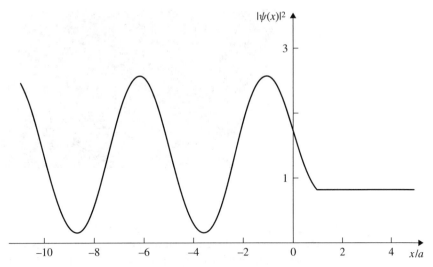

Figure 4.5 The modulus square of the wave function, $|\psi(x)|^2$, for the case of a rectangular barrier such that $mV_0a^2/\hbar^2 = 0.25$. The incident particle energy is $E = 0.75V_0$. The coefficient A in (4.73a) has been taken to be $A = 1$.

Going back to equation (4.81b), we see that $T \to 0$ in the limit $E \to 0$, and that T is a monotonically increasing function of E. Also, when the particle energy approaches the top of the barrier (from below), we have

$$\lim_{E \to V_0} T = \left(1 + \frac{mV_0a^2}{2\hbar^2}\right)^{-1}. \tag{4.82}$$

Note that the dimensionless number $mV_0a^2/\hbar^2$ may be considered as a measure of the 'opacity' of the barrier. In the classical limit this opacity becomes very large, and the transmission coefficient T remains vanishingly small for all energies E such that $0 \leqslant E \leqslant V_0$. Finally, if $\kappa a \gg 1$ we can write $\sinh(\kappa a) \simeq 2^{-1}\exp(\kappa a)$, so that the transmission coefficient is given approximately by

$$T \simeq \frac{16E(V_0 - E)}{V_0^2} e^{-2\kappa a} \tag{4.83}$$

and is very small.

The formula (4.83) has an important application in the *scanning tunnelling microscope*. A sharp metal needle is brought very close to a metal. The energy of the electrons within the needle and the surface is less than in the gap between them, so that a potential barrier exists through which the electrons can tunnel. If an electrical voltage is applied between the needle and the surface, a current will flow, whose magnitude depends on the transmission coefficient (4.83), in which the width a of the barrier is the height of the needle above the surface. Because of the exponential dependence of the transmission coefficient T on a, the current is a sensitive measure

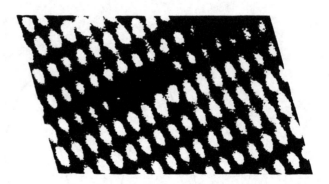

Figure 4.6 A scanning tunnelling microscope image of the surface of a silicon crystal.

of the height of the needle above the surface. This height can be adjusted as the needle scans the surface so that the current remains constant. In this way contour maps of surfaces have been obtained with an accuracy better than 10^{-11} m in the direction perpendicular to the surface and with a resolution of 0.5×10^{-9} m in directions parallel to the surface. In Fig. 4.6 a constant-current image of the surface of a silicon crystal is shown. A periodic array is seen corresponding to surface unit cells. The black area near the centre is caused by a small defect. Further applications of tunnelling are discussed in Section 8.4.

Case 2: $E > V_0$

The solution of the Schrödinger equation in the internal region is now given by

$$\psi(x) = F e^{ik'x} + G e^{-ik'x} \qquad 0 < x < a \qquad (4.84)$$

where $k' = [2m(E - V_0)/\hbar^2]^{1/2}$ as in (4.59b).

The conditions that $\psi(x)$ and $d\psi(x)/dx$ be continuous at the two points $x = 0$ and $x = a$ provide four relations between the five constants A, B, C, F and G. Thus, after elimination of F and G, we can solve again for the ratios B/A and C/A. Actually, it is easier to obtain these ratios directly by replacing κ by ik' in the equations (4.80). We find in this way that the reflection and transmission coefficients are now given by

$$R = \frac{|B|^2}{|A|^2} = \left[1 + \frac{4k^2 k'^2}{(k^2 - k'^2)^2 \sin^2(k'a)}\right]^{-1} = \left[1 + \frac{4E(E - V_0)}{V_0^2 \sin^2(k'a)}\right]^{-1} \qquad (4.85a)$$

$$T = \frac{|C|^2}{|A|^2} = \left[1 + \frac{(k^2 - k'^2)^2 \sin^2(k'a)}{4k^2 k'^2}\right]^{-1} = \left[1 + \frac{V_0^2 \sin^2(k'a)}{4E(E - V_0)}\right]^{-1} \qquad (4.85b)$$

and we see again that $R + T = 1$. The most important point about these equations is that the transmission coefficient T is in general less than unity, in contradiction to the classical prediction that the particle should always pass the potential barrier when $E > V_0$. In fact, we see from (4.85b) that we only have $T = 1$ when $k'a = \pi, 2\pi, 3\pi, \ldots$, namely whenever the thickness a of the barrier is equal to a half-integral or integral

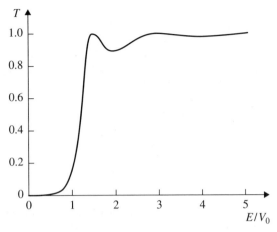

Figure 4.7 The transmission coefficient for a rectangular potential barrier such that $mV_0a^2/\hbar^2 = 10$, as a function of the ratio E/V_0.

number of de Broglie wavelengths $\lambda' = 2\pi/k'$ in that region. This effect results from destructive interference between reflections at the two points $x = 0$ and $x = a$. We also remark that when E tends to V_0 (this time from above), T joins smoothly to the value given by (4.82), while T becomes asymptotically equal to unity when E is large compared to V_0.

In Fig. 4.7 we show the transmission coefficient T, obtained from equations (4.81b) and (4.85b), as a function of E/V_0, for a barrier such that $mV_0a^2/\hbar^2 = 10$. Note that even when E is somewhat smaller than V_0 the transmission coefficient T is not negligible, which shows the importance of the tunnel effect in this case.

Since the phenomena of barrier penetration are characteristic of waves, we should not be surprised that they have a classical analogue in wave optics. Consider for example two optical media 1 and 2 (such as glass and air) having refractive indices n_1 and n_2, with $n_1 > n_2$, and suppose that a plane wave is incident from the optical medium 1 on the plane interface of the two media (see Fig. 4.8(a)). If the angle of incidence θ is larger than the critical angle θ_c, and if the optical medium 2 extends to infinity below the interface, the wave will be *totally reflected*, as shown in Fig. 4.8(a). However, despite the fact that the wave does not propagate in the optical medium 2, classical electromagnetic theory shows that the electric field does penetrate in that medium, its amplitude decreasing exponentially as one goes further away from the interface. This situation is therefore analogous to the potential step problem with $E < V_0$, for which we saw in the previous section that the reflection is total ($R = 1$), but the wave function $\psi(x)$ does penetrate in the classically forbidden region ($x > 0$) and decreases exponentially in that region (see (4.44b)).

Consider now the situation depicted in Fig. 4.8(b), where the optical medium 2 is reduced to a thin layer separating two optical media of type 1. In this case, a wave incident at an angle $\theta > \theta_c$ will be *partly reflected* and *partly transmitted*.

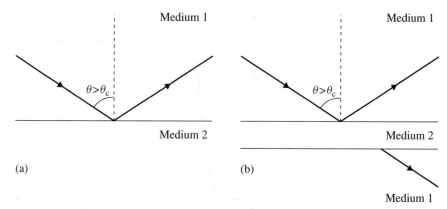

Figure 4.8 (a) Illustration of the total reflection of a plane electromagnetic wave at the plane interface of two media having different refractive indices. (b) Illustration of frustrated total reflection.

This phenomenon, called *frustrated total reflection*, can be fully understood by using classical electromagnetic theory, and is the analogue of the quantum mechanical tunnel effect. Just as the reduction of the optical medium 2 to a layer allows the transmission of the electromagnetic wave to occur, so does the reduction of a step potential (having infinite thickness) to a barrier potential (of finite thickness) permit the transmission of particles of energy $E < V_0$ in the quantum mechanical problem.

4.5 The infinite square well

Thus far we have only considered examples for which the Schrödinger eigenvalue equation (4.3) has solutions for every positive value of the energy, so that the energy spectrum forms a *continuum*. This is the case whenever the classical motion is not confined to a given region of space. We shall now study the opposite case of a particle of mass m which is *bound* so that the corresponding classical motion is periodic. In contrast with classical mechanics, solutions of the Schrödinger equation (4.3) then exist only for certain *discrete* values of the energy, as we have seen in Section 3.6. In this section we shall analyse a simple example such that the energy spectrum consists only of discrete bound states. This is the infinite square well potential, corresponding to the motion of a particle constrained by impenetrable walls to move in a region of width L, where the potential energy is constant. Taking this constant to be zero, and defining $a = L/2$, the potential energy for this problem is

$$V(x) = \begin{cases} 0, & -a < x < a \\ \infty, & |x| > a \end{cases} \tag{4.86}$$

where we have chosen the origin of the x-axis to be at the centre of the well. This potential is illustrated in Fig. 4.9.

4.5 The infinite square well

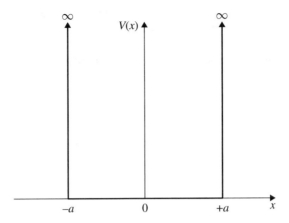

Figure 4.9 The infinite square well potential.

As the potential energy is infinite at $x = \pm a$, the probability of finding the particle outside the well is zero. The wave function $\psi(x)$ must therefore vanish for $|x| > a$, and we only need to solve the Schrödinger equation (4.3) inside the well. Moreover, since the wave function must be continuous, $\psi(x)$ must vanish at the constraining walls, namely

$$\psi(x) = 0 \quad \text{at} \quad x = \pm a. \tag{4.87}$$

We shall see shortly that this *boundary condition* leads to the quantisation of the energy. Note that the derivative of the wave function, $d\psi(x)/dx$, cannot vanish at $x = \pm a$, since then the eigenfunction $\psi(x)$ would be the trivial solution $\psi = 0$. Thus $d\psi(x)/dx$ will be discontinuous at the points $x = \pm a$, where the potential makes infinite jumps. As we have already remarked in Section 3.6, discontinuities of $d\psi(x)/dx$ are allowed at the points where the potential exhibits infinite discontinuities.

The Schrödinger equation for $|x| < a$ is simply

$$-\frac{\hbar^2}{2m}\frac{d^2\psi(x)}{dx^2} = E\psi(x). \tag{4.88}$$

Setting $k = (2mE/\hbar^2)^{1/2}$, the general solution of (4.88) is a linear combination of the two linearly independent solutions $\exp(ikx)$ and $\exp(-ikx)$ or, equivalently, of the pair of real solutions $\sin kx$ and $\cos kx$. In the present problem it is convenient to use the real solutions, so that we shall write the general solution in the form

$$\psi(x) = A\cos kx + B\sin kx, \quad k = \left(\frac{2m}{\hbar^2}E\right)^{1/2}. \tag{4.89}$$

Let us now apply the boundary conditions (4.87). We find in this way that the following two conditions must be obeyed

$$A\cos ka = 0, \quad B\sin ka = 0. \tag{4.90}$$

It is clear that we cannot allow both A and B to vanish, since this would yield the physically uninteresting trivial solution $\psi(x) = 0$. Also, we cannot make both $\cos ka$ and $\sin ka$ to vanish for a given value of k. Hence, there are *two* possible classes of solutions. For the *first class* $B = 0$ and $\cos ka = 0$, so that the only allowed values of k are

$$k_n = \frac{n\pi}{2a} = \frac{n\pi}{L} \tag{4.91}$$

with $n = 1, 3, 5, \ldots$. The corresponding eigenfunctions $\psi_n(x) = A_n \cos k_n x$ can be normalised so that

$$\int_{-a}^{+a} \psi_n^*(x)\psi_n(x)\mathrm{d}x = 1 \tag{4.92}$$

from which the normalisation constants are found (within an arbitrary multiplicative factor of modulus one) to be given by $A_n = a^{-1/2}$. The normalised eigenfunctions of the first class can therefore be written as

$$\psi_n(x) = \frac{1}{\sqrt{a}} \cos \frac{n\pi}{2a} x, \qquad n = 1, 3, 5, \ldots. \tag{4.93}$$

In the same way, we find that for the *second class* of solutions $A = 0$ and $\sin ka = 0$, so that the allowed values of k are given by (4.91), with $n = 2, 4, 6, \ldots$. The normalised eigenfunctions of the second class are therefore

$$\psi_n(x) = \frac{1}{\sqrt{a}} \sin \frac{n\pi}{2a} x, \qquad n = 2, 4, 6, \ldots. \tag{4.94}$$

For both classes of eigenfunctions it is unnecessary to consider negative values of n, since these lead to solutions which are not linearly independent of those corresponding to positive n.

Taking into account the two classes of solutions, we find that the values of k are quantised, being given by $k_n = n\pi/L$, with $n = 1, 2, 3, \ldots$. The corresponding de Broglie wavelengths are $\lambda_n = 2\pi/k_n = 2L/n$, so that eigenfunctions are only obtained if a half-integral or integral number of de Broglie wavelengths can fit into the box. In contrast with the classical result, according to which a particle can move in the potential (4.86) with any positive energy, we see that in the quantum mechanical problem the energy is *quantised*, the *energy eigenvalues* being

$$E_n = \frac{\hbar^2 k_n^2}{2m} = \frac{\hbar^2}{8m} \frac{\pi^2 n^2}{a^2} = \frac{\hbar^2}{2m} \frac{\pi^2 n^2}{L^2}, \qquad n = 1, 2, 3, \ldots \tag{4.95}$$

so that the energy spectrum consists of an *infinite number* of *discrete* energy levels corresponding to *bound states*. Note that there is just one eigenfunction for each level, so that the energy levels are *non-degenerate*. We also see from (4.93) and (4.94) that the nth eigenfunction has $(n - 1)$ nodes within the potential, and that the (real)

4.5 The infinite square well

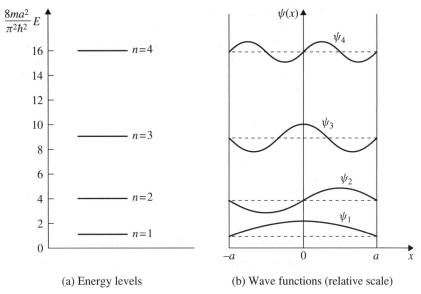

(a) Energy levels   (b) Wave functions (relative scale)

Figure 4.10 (a) The first four energy eigenvalues and (b) the corresponding eigenfunctions of the infinite square well. In (b) the x-axis for the eigenfunction ψ_n is drawn at a height corresponding to the energy E_n.

eigenfunctions $\psi_n(x)$ and $\psi_m(x)$ corresponding to different eigenvalues E_n and E_m are orthogonal:

$$\int_{-a}^{+a} \psi_m^*(x)\psi_n(x)\mathrm{d}x = 0, \qquad n \neq m. \tag{4.96}$$

All these results are in agreement with the discussion of Sections 3.6 and 3.7. The first few energy eigenvalues and corresponding eigenfunctions are shown in Fig. 4.10. It is worth noting that the lowest energy or *zero-point* energy is $E_{n=1} = \hbar^2\pi^2/8ma^2$, so that there is no state of zero energy. This is in agreement with the requirements of the uncertainty principle. Indeed, the position uncertainty is roughly given by $\Delta x \simeq a$. The corresponding momentum uncertainty is therefore $\Delta p_x \gtrsim \hbar/a$, leading to a minimum kinetic energy of order $\hbar^2/ma^2$, in qualitative agreement with the value of E_1.

Parity

There is an important difference between the two classes of eigenfunction which we have obtained. That is, the eigenfunctions (4.93) belonging to the first class are such that $\psi_n(-x) = \psi_n(x)$ and are therefore *even* functions of x, while the eigenfunctions (4.94) of the second class are such that $\psi_n(-x) = -\psi_n(x)$ and hence are *odd*. This division of the eigenfunctions $\psi_n(x)$ into eigenfunctions having a definite *parity* (even for the first class, odd for the second class) is a direct consequence of the fact that the potential is *symmetric* about $x = 0$, i.e. is an even function, $V(-x) = V(x)$, as we shall now show.

For this purpose let us study the behaviour of the Schrödinger eigenvalue equation (4.3) under the operation of *reflection through the origin*, $x \to -x$, which is also called the *parity* operation. If the potential is symmetric, the Hamiltonian $H = -(\hbar^2/2m)d^2/dx^2 + V(x)$ does not change when x is replaced by $-x$: it is *invariant* under the parity operation. Thus, if we change the sign of x in the Schrödinger equation (4.3), we have

$$-\frac{\hbar^2}{2m}\frac{d^2\psi(-x)}{dx^2} + V(x)\psi(-x) = E\psi(-x) \tag{4.97}$$

so that both $\psi(x)$ and $\psi(-x)$ are solutions of the same equation, with the same eigenvalue E. Two cases may arise.

Case 1: The eigenvalue E is non-degenerate

The two eigenfunctions $\psi(x)$ and $\psi(-x)$ can then differ only by a multiplicative constant

$$\psi(-x) = \alpha\psi(x). \tag{4.98}$$

Changing the sign of x in this equation yields

$$\psi(x) = \alpha\psi(-x) \tag{4.99}$$

and by combining these two equations we find that $\psi(x) = \alpha^2\psi(x)$. Hence $\alpha^2 = 1$ so that $\alpha = \pm 1$ and

$$\psi(-x) = \pm\psi(x) \tag{4.100}$$

which shows that the eigenfunctions $\psi(x)$ have a *definite parity*, being either even or odd for the parity operation $x \to -x$.

In particular, since bound states in one dimension are non-degenerate, every one-dimensional bound-state wave function in a symmetric potential must be either even or odd. Moreover, even functions clearly have an even number (including zero) of nodes, and odd functions have an odd number of nodes. As a consequence, if the energy levels are ordered by increasing energy values, the corresponding eigenfunctions are alternately even or odd, with the ground-state wave function being always an even function. As seen from (4.93) and (4.94), the results we have obtained above for the infinite square well are in agreement with these conclusions.

Case 2: The eigenvalue E is degenerate

In this case more than one linearly independent eigenfunction corresponds to the eigenvalue E, and these eigenfunctions need not have a definite parity. However, it is always possible to construct linear combinations of these eigenfunctions, such that each has definite parity. Indeed, let us assume that the eigenfunction $\psi(x)$ does not have a definite parity. We may then write

$$\psi(x) = \psi_+(x) + \psi_-(x) \tag{4.101}$$

where

$$\psi_+(x) = \tfrac{1}{2}[\psi(x) + \psi(-x)] \tag{4.102a}$$

obviously has even parity, while

$$\psi_-(x) = \tfrac{1}{2}[\psi(x) - \psi(-x)] \tag{4.102b}$$

is odd. Substituting (4.101) into the Schrödinger equation (4.3) we have

$$\left[-\frac{\hbar^2}{2m}\frac{d^2}{dx^2} + V(x) - E\right]\psi_+(x) + \left[-\frac{\hbar^2}{2m}\frac{d^2}{dx^2} + V(x) - E\right]\psi_-(x) = 0. \tag{4.103}$$

Changing x to $-x$ and using the fact that $V(-x) = V(x)$, $\psi_+(-x) = \psi_+(x)$ and $\psi_-(-x) = -\psi_-(x)$, we find that

$$\left[-\frac{\hbar^2}{2m}\frac{d^2}{dx^2} + V(x) - E\right]\psi_+(x) - \left[-\frac{\hbar^2}{2m}\frac{d^2}{dx^2} + V(x) - E\right]\psi_-(x) = 0. \tag{4.104}$$

Therefore, upon adding and subtracting (4.103) and (4.104), we see that $\psi_+(x)$ and $\psi_-(x)$ are separately solutions of the Schrödinger equation (4.3), corresponding to the same eigenvalue E. This completes the proof that *for a symmetric* (even) *potential the eigenfunctions $\psi(x)$ of the one-dimensional Schrödinger equation (4.3) can always be chosen to have definite parity*, without loss of generality.

The fact that eigenfunctions of the Schrödinger equation (4.3) can always be chosen to be even or odd when $V(-x) = V(x)$ often simplifies the calculations. In particular, we only have to obtain these eigenfunctions for positive values of x, and we know that odd functions vanish at the origin, while even functions must have zero slope at $x = 0$. This is apparent in the case of the infinite square well, and will again be illustrated below in our study of the finite square well and the linear harmonic oscillator.

Wave function regeneration

The general solution of the time-dependent Schrödinger equation (4.1) for a particle in the infinite square well (4.86) is given by

$$\Psi(x, t) = \sum_{n=1}^{\infty} c_n \psi_n(x) \exp(-iE_n t/\hbar) \tag{4.105}$$

where the energy eigenvalues E_n are given by (4.95), and the corresponding eigenfunctions are given by (4.93) for n odd and (4.94) for n even. The constants c_n can be determined from the knowledge of the wave function $\Psi(x, t)$ at any particular time t_0. Following the procedure leading to (3.154), and setting $t_0 = 0$, we have

$$c_n = \int_{-a}^{+a} \psi_n^*(x) \Psi(x, t = 0) dx \tag{4.106}$$

where we have used the normalisation condition (4.92) and the orthogonality relation (4.96).

As the time passes, the form of the wave packet $\Psi(x, t)$ changes. However, we shall now show that at the time $t = T$, where $T = 2\pi\hbar/E_1$, the wave packet is regenerated, so that

$$\Psi(x, t = T) = \Psi(x, t = 0). \tag{4.107}$$

Indeed, from (4.105), and since $E_n = n^2 E_1$, we have

$$\Psi(x, t = T) = \sum_{n=1}^{\infty} c_n \psi_n(x) \exp\left(-in^2 E_1 \frac{2\pi}{E_1}\right)$$

$$= \sum_{n=1}^{\infty} c_n \psi_n(x) \exp(-i2\pi n^2). \tag{4.108}$$

Since $2n^2$ is an even integer, it follows that $\exp(-i2\pi n^2) = 1$, and hence

$$\Psi(x, t = T) = \sum_{n=1}^{\infty} c_n \psi_n(x) = \Psi(x, t = 0). \tag{4.109}$$

By repeating this argument, we see that the wave function is completely regenerated at times sT, where $s = 1, 2, 3, \ldots$ is a positive integer.

Another interesting feature of the infinite square well wave function is that

$$\Psi(x, t = (2s - 1)T/2) = -\Psi(-x, t = 0) \tag{4.110}$$

which means that the wave function at times $(2s - 1)T/2$ is a reflection of that at $t = 0$ through the origin. To prove this, we note that

$$\Psi(x, t = (2s - 1)T/2) = \sum_{n=1}^{\infty} c_n \psi_n(x) \exp[-i\pi n^2(2s - 1)]. \tag{4.111}$$

When n is odd, n^2 is odd and $\exp[-i\pi n^2(2s - 1)] = -1$. On the other hand, when n is even, n^2 is even and $\exp[-i\pi n^2(2s - 1)] = 1$. Thus

$$\Psi(x, t = (2s - 1)T/2) = \sum_{n=1}^{\infty} c_n (-1)^n \psi_n(x). \tag{4.112}$$

Since $\psi_n(x) = \psi_n(-x)$ when n is odd, and $\psi_n(x) = -\psi_n(-x)$ when n is even, the result (4.110) follows.

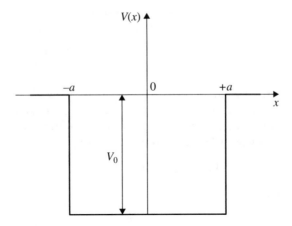

Figure 4.11 The one-dimensional square well potential (4.113) of depth V_0 and range a.

4.6 The square well

Let us now consider the square (or rectangular) well potential such that

$$V(x) = \begin{cases} -V_0, & |x| < a \\ 0, & |x| > a \end{cases} \tag{4.113}$$

where the positive constant V_0 is the depth of the well and a is its range. This potential is illustrated in Fig. 4.11. As in the case of the infinite square well studied in the preceding section, we have chosen the origin of the x-axis at the centre of the well, so that the potential is symmetric about $x = 0$. Two cases must be distinguished, corresponding respectively to positive or negative values of the energy E. When $E > 0$ the particle is *unconfined*; the corresponding *scattering* problem will be analysed later. The case $E < 0$ corresponds to a particle which is *confined* and hence is in a *bound state*; it is this situation which we shall study first.

Case 1: $E < 0$

Since the energy E cannot be lower than the absolute minimum of the potential, we must have $-V_0 \leqslant E < 0$. Inside the well the Schrödinger equation (4.3) reads

$$\frac{\mathrm{d}^2 \psi(x)}{\mathrm{d}x^2} + \alpha^2 \psi(x) = 0, \quad \alpha = \left[\frac{2m}{\hbar^2}(V_0 + E)\right]^{1/2} = \left[\frac{2m}{\hbar^2}(V_0 - |E|)\right]^{1/2},$$
$$|x| < a \tag{4.114a}$$

and outside the well it can be written as

$$\frac{d^2\psi(x)}{dx^2} - \beta^2\psi(x) = 0, \quad \beta = \left(-\frac{2m}{\hbar^2}E\right)^{1/2} = \left(\frac{2m}{\hbar^2}|E|\right)^{1/2}, \quad |x| > a \tag{4.114b}$$

where we have introduced the *binding energy* $|E| = -E$ of the particle.

The potential being an even function of x, we know that the solutions have definite parity, and hence are determined by positive values of x. The *even* solutions of equations (4.114) are given for $x > 0$, by

$$\psi(x) = A\cos\alpha x, \quad 0 < x < a \tag{4.115a}$$

and

$$\psi(x) = Ce^{-\beta x}, \quad x > a. \tag{4.115b}$$

In writing this last equation we have taken into account the fact that the wave function cannot contain a term like $D\exp(\beta x)$, which would become infinite when $x \to +\infty$; we must therefore set $D = 0$. The requirements that ψ and $d\psi/dx$ be continuous at $x = a$ yield the two equations

$$A\cos\alpha a = Ce^{-\beta a} \tag{4.116a}$$

$$-\alpha A\sin\alpha a = -\beta Ce^{-\beta a} \tag{4.116b}$$

from which we deduce that

$$\alpha\tan\alpha a = \beta. \tag{4.117}$$

Note that instead of requiring both ψ and $d\psi/dx$ to be continuous at $x = a$, we may simply ask that the *logarithmic derivative* of the wave function, $\psi^{-1}(d\psi/dx)$, be *continuous* at $x = a$. Indeed, this condition yields directly equation (4.117), since the normalisation constants A and C disappear by taking the ratio $\psi^{-1}(d\psi/dx)$. The requirement of continuity for the logarithmic derivative is therefore a simple way of expressing the conditions of smooth joining.

The *odd* solutions of equations (4.114) are given for positive x by

$$\psi(x) = B\sin\alpha x, \quad 0 < x < a \tag{4.118a}$$

and

$$\psi(x) = Ce^{-\beta x}, \quad x > a. \tag{4.118b}$$

By requiring that $\psi^{-1}(d\psi/dx)$ be continuous at $x = a$ we obtain the equation

$$\alpha\cot\alpha a = -\beta. \tag{4.119}$$

The *energy levels* of the bound states are found by solving the transcendental equations (4.117) and (4.119), either numerically or graphically. We shall use here a simple graphical procedure to obtain the allowed values of the energy. We first introduce the dimensionless quantities $\xi = \alpha a$ and $\eta = \beta a$, so that the equations to be solved are

$$\xi \tan \xi = \eta \quad \text{(for even states)} \tag{4.120a}$$

$$\xi \cot \xi = -\eta \quad \text{(for odd states).} \tag{4.120b}$$

Note that both ξ and η must be positive, and such that

$$\xi^2 + \eta^2 = \gamma^2 \tag{4.121}$$

where $\gamma = (2mV_0a^2/\hbar^2)^{1/2}$. The energy levels may therefore be found by determining the points of intersection of the circle $\xi^2 + \eta^2 = \gamma^2$, of known radius γ, with the curve $\eta = \xi \tan \xi$ (for even states), or with the curve $\eta = -\xi \cot \xi$ (for odd states). This is illustrated in Fig. 4.12(a) for even states and in Fig. 4.12(b) for odd states. Several conclusions can be deduced from these figures by inspection. First, the bound-state energy levels are seen to be *non-degenerate*, as expected since we are dealing with a one-dimensional problem. Second, their number is *finite* and depends on the dimensionless parameter γ, which may be called the 'strength parameter' of the potential. Note that for a fixed value of the mass m of the particle, γ depends on the parameters of the square well through the combination V_0a^2. In particular, we see from Fig. 4.12(a) that if $0 < \gamma \leqslant \pi$ there is one even bound state, while if $\pi < \gamma \leqslant 2\pi$ there are two such bound states. More generally, it is clear that there will be N_e even-parity bound states if $(N_e - 1)\pi < \gamma \leqslant N_e\pi$. On the other hand, we note from Fig. 4.12(b) that if $0 < \gamma \leqslant \pi/2$, there is no bound state of odd parity, while if $\pi/2 < \gamma \leqslant 3\pi/2$, there is one such bound state. More generally, there are N_0 odd parity bound states if $(N_0 - 1/2)\pi < \gamma \leqslant (N_0 + 1/2)\pi$. This brings us to our third conclusion, namely that as the strength parameter γ increases, energy levels corresponding respectively to even and odd solutions appear successively (see Fig. 4.13), the total number of bound states being equal to N, if $(N - 1)\pi/2 < \gamma \leqslant N\pi/2$. Hence the bound-state spectrum consists of *alternating even and odd states*, the ground state being always even, the next state odd, and so on.

Having solved the equations (4.120) to determine the bound-state energy levels, we can readily obtain explicitly the corresponding normalised eigenfunctions from equations (4.115) (for even states) and (4.118) (for odd states). As an example, we illustrate in Fig. 4.14 the four energy eigenfunctions of a square well for which $\gamma = 5$. The corresponding eigenenergies are found to be given respectively by $E_1 = -0.93V_0$, $E_2 = -0.73V_0$, $E_3 = -0.41V_0$ and $E_4 = -0.04V_0$. The eigenfunctions $\psi_1(x)$ and $\psi_3(x)$ are seen to be even, while $\psi_2(x)$ and $\psi_4(x)$ are odd, as expected from the foregoing discussion. Moreover, the nth eigenfunction has $(n - 1)$ nodes, in agreement with the oscillation theorem (see Section 3.7). It is worth stressing that the eigenfunctions extend into the classically forbidden region $|x| > a$. In fact, we see from (4.115b) and (4.118b) that in this region the eigenfunctions fall off like $\exp(-\beta|x|)$, so that they extend to a distance given roughly by $\beta^{-1} = \hbar/(2m|E|)^{1/2}$ which increases as the binding energy $|E|$ of the particle decreases.

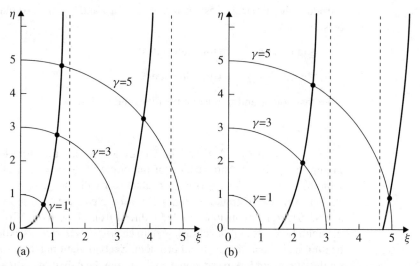

Figure 4.12 Graphical determination of the energy levels for a one-dimensional square well potential (a) for even states and (b) for odd states. The energy levels are obtained by finding the points of intersection of the circle $\xi^2 + \eta^2 = \gamma^2$, of known radius $\gamma = (2mV_0 a^2/\hbar^2)^{1/2}$ with the curve $\eta = \xi \tan \xi$ (for even states), or with the curve $\eta = -\xi \cot \xi$ (for odd states). The graphical solution of equations (4.120) is illustrated for three values of γ. When $\gamma = 1$ there is one (even) bound state, when $\gamma = 3$ there are two bound states (one even, one odd) and when $\gamma = 5$ there are four bound states (two even, two odd). The asymptotes are indicated by vertical dashed lines.

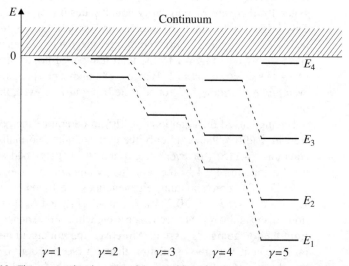

Figure 4.13 The energy level spectra for one-dimensional square wells of various depths V_0. The value of the strength parameter $\gamma = (2mV_0 a^2/\hbar^2)^{1/2}$ corresponding to each energy level spectrum is indicated. The energy levels E_1 and E_3 correspond to even solutions, while E_2 and E_4 correspond to odd solutions.

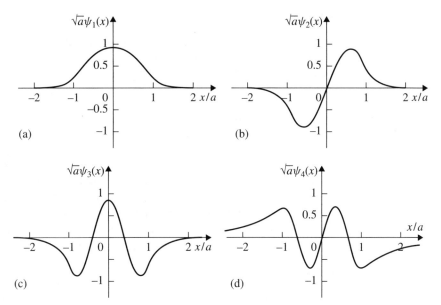

Figure 4.14 The four energy eigenfunctions of a one-dimensional square well potential such that $\gamma = 5$.

Finally, let us examine the limiting case of an infinitely deep square well, for which $V_0 \to \infty$ (so that $\gamma \to \infty$). In order to make contact with the results of Section 4.5, it is convenient to start from a slightly different finite square well, such that $V(x) = 0$ for $|x| < a$ and $V(x) = V_0$ for $|x| > a$. This corresponds to making an upward shift of the zero of the energies by an amount V_0. As we know (Problem 3.8), such a shift has the effect of adding the constant V_0 to all the energy eigenvalues, and leaves the corresponding eigenfunctions unchanged. Thus the bound states occur for $0 \leqslant E < V_0$; they are still determined by the transcendental equations (4.117) (for even states) and (4.119) (for odd states), where now $\alpha = (2mE/\hbar^2)^{1/2}$ and $\beta = [2m(V_0 - E)/\hbar^2]^{1/2}$. When $V_0 \to \infty$, it is apparent that the roots of equations (4.120) will be given by

$$\xi_n = n\frac{\pi}{2}, \qquad n = 1, 2, \ldots \tag{4.122}$$

and since $\xi = \alpha a = (2mE/\hbar^2)^{1/2}a$, the corresponding energy eigenvalues are

$$E_n = \frac{\hbar^2}{8m}\frac{\pi^2 n^2}{a^2} \tag{4.123}$$

in agreement with the result (4.95). Moreover, since $\beta \to \infty$ when $V_0 \to \infty$, the eigenfunctions (4.115) and (4.118) then vanish for $|x| \geqslant a$, as we found in Section 4.5 for the infinite square well.

Case 2: $E > 0$

Let us return to the square well (4.113), and consider the case of positive energies, so that the particle is not bound. Assuming that the particle is incident upon the well from the left, the solution of the Schrödinger equation (4.3) in the *external regions* $x < -a$ and $x > a$ is given by

$$\psi(x) = \begin{cases} Ae^{ikx} + Be^{-ikx}, & x < -a \\ Ce^{ikx}, & x > a \end{cases} \quad \begin{matrix} \text{(4.124a)} \\ \text{(4.124b)} \end{matrix}$$

where $k = (2mE/\hbar^2)^{1/2}$, A, B, C are constants, and we have used the fact that there is no reflector at large positive values of x (so that there is no term of the form $\exp(-ikx)$ in (4.124b)). In the region $x < -a$ the wave function is seen to consist of an *incident wave* of amplitude A and a *reflected wave* of amplitude B, while in the region $x > a$ it is a pure *transmitted wave* of amplitude C. In the *internal region* $-a < x < a$, the solution of the Schrödinger equation is

$$\psi(x) = Fe^{i\alpha x} + Ge^{-i\alpha x} \tag{4.125}$$

where $\alpha = [2m(V_0 + E)/\hbar^2]^{1/2}$.

Looking back at equations (4.74) and (4.84), we see that the present problem of scattering by a square well is closely related to the scattering by a potential barrier (with $E > V_0$) which was studied in Section 4.4. We shall therefore only outline the solution here, leaving the details of the calculations as an exercise (Problem 4.9). It is worth noting that in one-dimensional scattering problems there is a lack of symmetry between the external regions to the left and to the right of the potential, since the particle is assumed to be incident on the potential in a given direction. Hence in this case there is no advantage in dealing with solutions having a definite parity.

The five constants A, B, C, F and G appearing in (4.124) and (4.125) can be related by requiring that the wave function $\psi(x)$ and its derivative $d\psi(x)/dx$ be continuous at $x = \pm a$. This smooth joining of the external and internal solutions can be done for any positive value of E, so that the spectrum is *continuous* for $E > 0$. After eliminating F and G, one can solve for the ratios B/A and C/A and obtain the *reflection coefficient* $R = |B/A|^2$ and the *transmission coefficient* $T = |C/A|^2$. These are given by

$$R = \left[1 + \frac{4k^2\alpha^2}{(k^2 - \alpha^2)^2 \sin^2(\alpha L)}\right]^{-1} = \left[1 + \frac{4E(V_0 + E)}{V_0^2 \sin^2(\alpha L)}\right]^{-1} \tag{4.126a}$$

and

$$T = \left[1 + \frac{(k^2 - \alpha^2)^2 \sin^2(\alpha L)}{4k^2\alpha^2}\right]^{-1} = \left[1 + \frac{V_0^2 \sin^2(\alpha L)}{4E(V_0 + E)}\right]^{-1} \tag{4.126b}$$

where we have written $L = 2a$. We have, of course, $R + T = 1$. We also remark that the above results may be obtained from the corresponding ones given in (4.85) for the potential barrier by making the obvious substitutions $V_0 \to -V_0$, $k' \to \alpha$ and

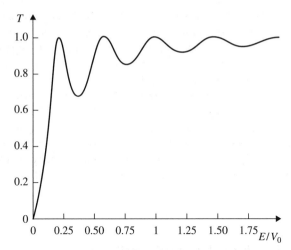

Figure 4.15 The transmission coefficient for a square well potential such that $\gamma = 10$, as a function of the ratio E/V_0.

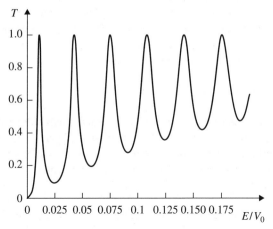

Figure 4.16 The transmission coefficient for a strong square well potential such that $\gamma = 100$, as a function of the ratio E/V_0.

$a \to L$ (we recall that the thickness of the potential barrier considered in Section 4.4 was denoted by a, while here the corresponding quantity is called L).

An important point about the results (4.126) is again that the transmission coefficient T is in general less than unity, in contradiction to the classical prediction according to which the particle should always be transmitted. The behaviour of T as a function of the energy E is readily deduced from (4.126b). We see that $T = 0$ when $E = 0$. As E increases, T rises and then fluctuates between its maximum value ($T = 1$) reached when $\alpha L = n\pi$ ($n = 1, 2, \ldots$) and minima near $\alpha L = (2n+1)\pi/2$. Note that, as in the case of the potential barrier, perfect transmission occurs when the thickness L is equal to an integral or half-integral number of de Broglie wavelengths $(2\pi/\alpha)$ in the internal region. Finally, when E is large compared with V_0, the transmission coefficient becomes asymptotically equal to unity. This is illustrated in Fig. 4.15, where T is plotted as a function of E/V_0 for a square well whose strength parameter is $\gamma = 10$.

The case of a particle of low energy scattered by a very strong well is of particular interest. Using (4.126b) we see that the transmission coefficient is then small almost everywhere, except for a series of *sharp maxima* (where $T = 1$), which are reached when the condition $\alpha L = n\pi$ is obeyed. Such pronounced peaks in the transmission coefficient are said to represent *resonances*; they are clearly typical of the wave-like behaviour of the particle, and are illustrated in Fig. 4.16 for a square well such that $\gamma = 100$.

4.7 The linear harmonic oscillator

We shall now study the one-dimensional motion of a particle of mass m which is attracted to a fixed centre by a force proportional to the displacement from that centre. Thus, choosing the origin as the centre of force, the restoring force is given by $F = -kx$ (Hooke's law), where k is the force constant. This force can thus be represented by the potential energy

$$V(x) = \tfrac{1}{2}kx^2 \tag{4.127}$$

which is shown in Fig. 4.17(a). Such a parabolic potential is of great importance in quantum physics as well as in classical physics, since it can be used to approximate an arbitrary continuous potential $W(x)$ in the vicinity of a stable equilibrium position at $x = a$ (see Fig. 4.17(b)). Indeed, if we expand $W(x)$ in a Taylor series about $x = a$, we have

$$W(x) = W(a) + (x-a)W'(a) + \tfrac{1}{2}(x-a)^2 W''(a) + \cdots \tag{4.128a}$$

with

$$W'(a) = \left(\frac{dW(x)}{dx}\right)_{x=a}, \quad W''(a) = \left(\frac{d^2W(x)}{dx^2}\right)_{x=a}. \tag{4.128b}$$

Because $W(x)$ has a minimum at $x = a$ we have $W'(a) = 0$ and $W''(a) > 0$. Choosing a as the origin of coordinates and $W(a)$ as the origin of the energy scale, we see that the harmonic oscillator potential (4.127) (with $k = W''(a)$) is the first approximation to $W(x)$. The linear harmonic oscillator is therefore the prototype for systems in which there exist small vibrations about a point of stable equilibrium. This will be illustrated in Chapter 10 for the case of the vibrational motion of nuclei in molecules.

4.7 The linear harmonic oscillator   171

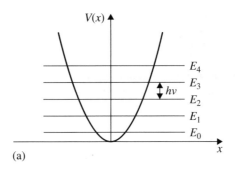

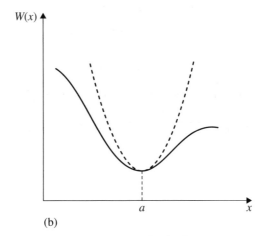

Figure 4.17 (a) The parabolic potential well $V(x) = \frac{1}{2}kx^2$. This is the potential of the linear harmonic oscillator. Also shown are the first few energy eigenvalues (4.151). (b) A continuous potential well $W(x)$, represented by the solid line can be approximated in the vicinity of a stable equilibrium position at $x = a$ by a linear harmonic oscillator potential, shown as the dashed line.

As the potential energy for a linear harmonic oscillator is given by (4.127), the corresponding Hamiltonian operator is

$$H = -\frac{\hbar^2}{2m}\frac{d^2}{dx^2} + \frac{1}{2}kx^2 \tag{4.129}$$

and the Schrödinger eigenvalue equation reads

$$-\frac{\hbar^2}{2m}\frac{d^2\psi(x)}{dx^2} + \frac{1}{2}kx^2\psi(x) = E\psi(x). \tag{4.130}$$

Clearly, all eigenfunctions correspond to bound states of positive energy. It is convenient to rewrite (4.130) in terms of dimensionless quantities. To this end we first introduce the dimensionless eigenvalues

$$\lambda = \frac{2E}{\hbar\omega} \tag{4.131}$$

where
$$\omega = \left(\frac{k}{m}\right)^{1/2} \tag{4.132}$$
is the angular frequency of the corresponding classical oscillator. We shall also use the dimensionless variable
$$\xi = \alpha x \tag{4.133}$$
where
$$\alpha = \left(\frac{mk}{\hbar^2}\right)^{1/4} = \left(\frac{m\omega}{\hbar}\right)^{1/2}. \tag{4.134}$$
The Schrödinger equation (4.130) then becomes
$$\frac{d^2\psi(\xi)}{d\xi^2} + (\lambda - \xi^2)\psi(\xi) = 0. \tag{4.135}$$

Let us first analyse the behaviour of ψ in the asymptotic region $|\xi| \to \infty$. For any finite value of the total energy E the quantity λ becomes negligible with respect to ξ^2 in the limit $|\xi| \to \infty$, so that in this limit equation (4.135) reduces to
$$\left(\frac{d^2}{d\xi^2} - \xi^2\right)\psi(\xi) = 0. \tag{4.136}$$
For large enough $|\xi|$ it is readily verified that the functions
$$\psi(\xi) = \xi^p e^{\pm \xi^2/2} \tag{4.137}$$
satisfy the equation (4.136) so far as the leading terms (which are of order $\xi^2\psi$) are concerned, when p has any finite value. Because the wave function ψ must be bounded everywhere, including at $\xi = \pm\infty$, the physically acceptable solution must contain only the minus sign in the exponent. The asymptotic analysis therefore suggests looking for solutions to equation (4.135) which are valid for all ξ having the form
$$\psi(\xi) = e^{-\xi^2/2} H(\xi) \tag{4.138}$$
where $H(\xi)$ are functions which must not affect the asymptotic behaviour of ψ. Substituting (4.138) into (4.135) we obtain for $H(\xi)$ the differential equation
$$\frac{d^2H}{d\xi^2} - 2\xi \frac{dH}{d\xi} + (\lambda - 1)H = 0 \tag{4.139}$$
which is called the *Hermite equation*.

In order to solve this equation, let us expand $H(\xi)$ in a power series in ξ. Since the harmonic oscillator potential (4.127) is such that $V(-x) = V(x)$, we know from our discussion of Section 4.5 that the eigenfunctions $\psi(x)$ of the Schrödinger equation (4.130) must have a definite parity. We shall therefore consider separately the even and odd states.

Even states

For these states we have $\psi(-\xi) = \psi(\xi)$ and also, therefore, $H(-\xi) = H(\xi)$, so that we can write for $H(\xi)$ the power series

$$H(\xi) = \sum_{k=0}^{\infty} c_k \xi^{2k}, \qquad c_0 \neq 0 \tag{4.140}$$

which contains only *even* powers of ξ. Substituting (4.140) into the Hermite equation (4.139), we find that

$$\sum_{k=0}^{\infty} [2k(2k-1)c_k \xi^{2(k-1)} + (\lambda - 1 - 4k)c_k \xi^{2k}] = 0 \tag{4.141}$$

or

$$\sum_{k=0}^{\infty} [2(k+1)(2k+1)c_{k+1} + (\lambda - 1 - 4k)c_k]\xi^{2k} = 0. \tag{4.142}$$

This equation will be satisfied provided the coefficient of each power of ξ separately vanishes, so that we obtain the *recursion relation*

$$c_{k+1} = \frac{4k+1-\lambda}{2(k+1)(2k+1)} c_k. \tag{4.143}$$

Thus, given $c_0 \neq 0$, all the coefficients c_k can be determined successively by using the above equation. We have therefore obtained a series representation of the even solution (4.140) of the Hermite equation.

If this series does not terminate, we see from (4.143) that for large k

$$\frac{c_{k+1}}{c_k} \sim \frac{1}{k}. \tag{4.144}$$

This ratio is the same as that of the series for $\xi^{2p} \exp(\xi^2)$, where p has a finite value. Using (4.138), we then find that the wave function $\psi(\xi)$ has an asymptotic behaviour of the form

$$\psi(\xi) \underset{|\xi| \to \infty}{\sim} \xi^{2p} e^{\xi^2/2} \tag{4.145}$$

which is obviously unacceptable. The only way to avoid this divergence of $\psi(\xi)$ at large $|\xi|$ is to require that the series (4.140) *terminates*, which means that $H(\xi)$ must be a *polynomial* in the variable ξ^2. Let the highest power of ξ^2 appearing in this polynomial be ξ^{2N}, where $N = 0, 1, 2, \ldots$, is a positive integer or zero. Thus in (4.140) we have $c_N \neq 0$, while the coefficient c_{N+1} must vanish. Using the recursion relation (4.143) we see that this will happen if and only if λ takes on the discrete values

$$\lambda = 4N + 1, \qquad N = 0, 1, 2, \ldots. \tag{4.146}$$

To each value $N = 0, 1, 2, \ldots$, of N will then correspond an even function $H(\xi)$ which is a polynomial of order $2N$ in ξ, and an even, physically acceptable, wave function $\psi(\xi)$ which is given by (4.138).

Odd states

In this case we have $\psi(-\xi) = -\psi(\xi)$, and hence $H(-\xi) = -H(\xi)$. Thus we begin by writing for $H(\xi)$ the power series

$$H(\xi) = \sum_{k=0}^{\infty} d_k \xi^{2k+1}, \qquad d_0 \neq 0 \tag{4.147}$$

which contains only *odd* powers of ξ. Substituting (4.147) into the Hermite equation (4.139), we obtain for the coefficients d_k the recursion relation

$$d_{k+1} = \frac{4k + 3 - \lambda}{2(k+1)(2k+3)} d_k. \tag{4.148}$$

For large k we have $d_{k+1}/d_k \sim k^{-1}$, so that the wave function $\psi(\xi)$, given by (4.138), will again diverge at large $|\xi|$ unless the series (4.147) for $H(\xi)$ terminates. Let the highest power of ξ in (4.147) be ξ^{2N+1}, where $N = 0, 1, 2, \ldots$. Since $d_N \neq 0$, while d_{N+1} is required to vanish, we see from the recursion relation (4.148) that λ must take one of the discrete values

$$\lambda = 4N + 3, \qquad N = 0, 1, 2, \ldots . \tag{4.149}$$

To each value $N = 0, 1, 2, \ldots$, of N will then correspond an odd function $H(\xi)$ which is a polynomial of order $2N + 1$ in ξ, and an odd, physically acceptable wave function $\psi(\xi)$ given by (4.138).

Energy levels

Putting together the results which we have obtained for the even and odd cases, we see from (4.146) and (4.149) that the eigenvalue λ must take on one of the discrete values

$$\lambda = 2n + 1, \qquad n = 0, 1, 2, \ldots \tag{4.150}$$

where the quantum number n is a positive integer or zero. Using (4.131) we therefore find that the energy spectrum of the linear harmonic oscillator is given by

$$E_n = \left(n + \tfrac{1}{2}\right)\hbar\omega = \left(n + \tfrac{1}{2}\right)h\nu,$$
$$n = 0, 1, 2, \ldots \tag{4.151}$$

where $\nu = \omega/2\pi$ is the frequency of the corresponding classical oscillator.

In contrast with classical mechanics, which predicts that the energy E of a linear harmonic oscillator can have any value, we see from (4.151) that its quantum mechanical energy spectrum consists of an infinite sequence of *discrete* levels (see Fig. 4.17(a)). For any finite eigenvalue (4.151) the particle is *bound*. The energy levels (4.151) are equally spaced and are similar to those discovered in 1900 by Planck for the radiation field modes (see Section 1.1). This is due to the fact that a decomposition of the electromagnetic field into normal modes is essentially a decomposition into uncoupled harmonic oscillators. We notice, however, that according to (4.151) the linear harmonic oscillator even in its lowest state ($n = 0$), has the energy $\hbar\omega/2$. The finite value $\hbar\omega/2$ of the ground-state energy level, which is called the *zero-point energy* of the linear harmonic oscillator, is clearly also a quantum phenomenon. As in the case of the infinite square well discussed in Section 4.5, the existence of this zero-point energy is directly related to the uncertainty principle (see Problem 4.12). In classical mechanics the lowest possible energy of the oscillator would of course be zero, corresponding to the particle being at rest at the origin, but in quantum mechanics this is forbidden by the uncertainty relation (2.70). We also remark that the eigenvalues (4.151) are *non-degenerate*, since for each value of the quantum number n there exists only one eigenfunction (apart from an arbitrary multiplicative constant); this is in agreement with the observation, already made, that one-dimensional bound states are non-degenerate.

Hermite polynomials

Let us now return to the wave functions $\psi(\xi)$. Using (4.138) and collecting our results for both even and odd cases, we see that the physically acceptable solutions of equation (4.135), corresponding to the eigenvalues (4.150), are given by

$$\psi_n(\xi) = e^{-\xi^2/2} H_n(\xi) \tag{4.152}$$

where the functions $H_n(\xi)$ are polynomials of order n. Both $\psi_n(\xi)$ and $H_n(\xi)$ have the parity of n. Moreover, the polynomials $H_n(\xi)$ satisfy the Hermite equation (4.139) with $\lambda = 2n + 1$, namely

$$\frac{d^2 H_n}{d\xi^2} - 2\xi \frac{dH_n}{d\xi} + 2n H_n = 0. \tag{4.153}$$

The polynomials $H_n(\xi)$ are called *Hermite polynomials*. It is clear from the foregoing discussion that they are uniquely defined, except for an arbitrary multiplicative constant. This constant is traditionally chosen so that the highest power of ξ appears with the coefficient 2^n in $H_n(\xi)$. This is consistent with the following definition of the Hermite polynomials:

$$H_n(\xi) = (-1)^n e^{\xi^2} \frac{d^n e^{-\xi^2}}{d\xi^n} \tag{4.154a}$$

$$= e^{\xi^2/2} \left(\xi - \frac{d}{d\xi} \right)^n e^{-\xi^2/2}. \tag{4.154b}$$

The first few Hermite polynomials, obtained from (4.154), are

$$\begin{aligned}
H_0(\xi) &= 1 \\
H_1(\xi) &= 2\xi \\
H_2(\xi) &= 4\xi^2 - 2 \\
H_3(\xi) &= 8\xi^3 - 12\xi \\
H_4(\xi) &= 16\xi^4 - 48\xi^2 + 12 \\
H_5(\xi) &= 32\xi^5 - 160\xi^3 + 120\xi.
\end{aligned} \tag{4.155}$$

Note that the definition (4.154) implies that $H_n(\xi)$ has n real zeros.

Another definition of the Hermite polynomials $H_n(\xi)$, which is equivalent to (4.154) involves the use of a *generating function* $G(\xi, s)$. That is

$$G(\xi, s) = e^{-s^2+2s\xi} \tag{4.156a}$$

$$= \sum_{n=0}^{\infty} \frac{H_n(\xi)}{n!} s^n. \tag{4.156b}$$

These relations mean that if the function $\exp(-s^2+2s\xi)$ is expanded in a power series in s, the coefficients of successive powers of s are just $1/n!$ times the Hermite polynomials $H_n(\xi)$. Using equations (4.156) it is straightforward to prove (Problem 4.13) that the Hermite polynomials satisfy the recursion relations

$$H_{n+1}(\xi) - 2\xi H_n(\xi) + 2n H_{n-1}(\xi) = 0 \tag{4.157}$$

and

$$\frac{dH_n(\xi)}{d\xi} = 2n H_{n-1}(\xi). \tag{4.158}$$

The lowest-order differential equation for H_n which can be constructed from these two recursion relations is then readily seen to be the equation (4.153) satisfied by the Hermite polynomials. Moreover, the equivalence of the two definitions (4.154) and (4.156) of the Hermite polynomials can be proved by using both expressions for $G(\xi, s)$ given in (4.156), differentiating n times with respect to s, and then letting s tend to zero (Problem 4.14).

The wave functions for the linear harmonic oscillator

Using (4.152), we see that to each of the discrete values E_n of the energy, given by (4.151), there corresponds one, and only one, physically acceptable eigenfunction, namely

$$\psi_n(x) = N_n e^{-\alpha^2 x^2/2} H_n(\alpha x) \tag{4.159}$$

where we have returned to our original variable x. Both $H_n(\alpha x)$ and $\psi_n(x)$ have the parity of n and have n real zeros. The quantity N_n, written on the right of (4.159) is a constant which (apart from an arbitrary phase factor) can be determined by requiring that the wave function (4.159) be normalised to unity. That is

$$\int_{-\infty}^{+\infty} |\psi_n(x)|^2 dx = \frac{|N_n|^2}{\alpha} \int_{-\infty}^{+\infty} e^{-\xi^2} H_n^2(\xi) d\xi = 1. \tag{4.160}$$

In order to evaluate the integral on the right of (4.160), we consider the generating function $G(\xi, s)$ given by (4.156) as well as the second generating function

$$G(\xi, t) = e^{-t^2+2t\xi}$$

$$= \sum_{m=0}^{\infty} \frac{H_m(\xi)}{m!} t^m. \tag{4.161}$$

Using (4.156) and (4.161), we may then write

$$\int_{-\infty}^{+\infty} e^{-\xi^2} G(\xi, s) G(\xi, t) d\xi = \sum_{n=0}^{\infty} \sum_{m=0}^{\infty} \frac{s^n t^m}{n! m!} \int_{-\infty}^{+\infty} e^{-\xi^2} H_n(\xi) H_m(\xi) d\xi. \quad (4.162)$$

Since

$$\int_{-\infty}^{+\infty} e^{-x^2} dx = \sqrt{\pi} \quad (4.163)$$

the integral on the left-hand side of (4.162) is simply

$$\int_{-\infty}^{+\infty} e^{-\xi^2} e^{-s^2+2s\xi} e^{-t^2+2t\xi} d\xi = e^{2st} \int_{-\infty}^{+\infty} e^{-(\xi-s-t)^2} d(\xi - s - t)$$

$$= \sqrt{\pi} e^{2st}$$

$$= \sqrt{\pi} \sum_{n=0}^{\infty} \frac{(2st)^n}{n!}. \quad (4.164)$$

Equating the coefficients of equal powers of s and t on the right-hand sides of (4.162) and (4.164), we find that

$$\int_{-\infty}^{+\infty} e^{-\xi^2} H_n^2(\xi) d\xi = \sqrt{\pi} 2^n n! \quad (4.165)$$

and

$$\int_{-\infty}^{+\infty} e^{-\xi^2} H_n(\xi) H_m(\xi) d\xi = 0, \qquad n \neq m. \quad (4.166)$$

From (4.160) and (4.165) we see that apart from an arbitrary complex multiplicative factor of modulus one the normalisation constant N_n is given by

$$N_n = \left(\frac{\alpha}{\sqrt{\pi} 2^n n!} \right)^{1/2} \quad (4.167)$$

so that the normalised linear harmonic oscillator eigenfunctions are given by

$$\psi_n(x) = \left(\frac{\alpha}{\sqrt{\pi} 2^n n!} \right)^{1/2} e^{-\alpha^2 x^2/2} H_n(\alpha x). \quad (4.168)$$

Moreover, the result (4.166) implies that

$$\int_{-\infty}^{+\infty} \psi_n^*(x) \psi_m(x) dx = 0, \qquad n \neq m \quad (4.169)$$

so that the (real) harmonic oscillator wave functions $\psi_n(x)$ and $\psi_m(x)$ are *orthogonal* if $n \neq m$, in agreement with the fact that they correspond to non-degenerate energy eigenvalues. We may of course summarise equations (4.160) and (4.169) by writing that the set of normalised eigenfunctions (4.168) satisfies the relations

$$\int_{-\infty}^{+\infty} \psi_n^*(x) \psi_m(x) dx = \delta_{nm} \quad (4.170)$$

and hence is *orthonormal*.

It is worth noting at this point that other integrals involving harmonic oscillator wave functions can also be evaluated by using the generating function (4.156) of the Hermite polynomials. For example, let us consider the integral

$$x_{nm} = \int_{-\infty}^{+\infty} \psi_n^*(x) x \psi_m(x) dx = \frac{N_n N_m}{\alpha^2} \int_{-\infty}^{+\infty} e^{-\xi^2} \xi H_n(\xi) H_m(\xi) d\xi \qquad (4.171)$$

which we shall need further in this book. Using again the two generating functions $G(\xi, s)$ and $G(\xi, t)$, we now look at the quantity

$$\int_{-\infty}^{+\infty} e^{-\xi^2} \xi G(\xi, s) G(\xi, t) d\xi = \sum_{n=0}^{\infty} \sum_{m=0}^{\infty} \frac{s^n t^m}{n! m!} \int_{-\infty}^{+\infty} e^{-\xi^2} \xi H_n(\xi) H_m(\xi) d\xi. \qquad (4.172)$$

Using the result (4.163), one finds that the integral on the left-hand side is given by

$$\int_{-\infty}^{+\infty} e^{-\xi^2} \xi e^{-s^2 + 2s\xi} e^{-t^2 + 2t\xi} d\xi = \sqrt{\pi}(s+t)e^{2st}$$

$$= \sqrt{\pi} \sum_{n=0}^{\infty} \frac{2^n}{n!} (s^{n+1} t^n + s^n t^{n+1}). \qquad (4.173)$$

Upon comparison of the coefficients of equal powers of s and t on the right-hand sides of (4.172) and (4.173), and using (4.167), we see that

$$x_{nm} = \begin{cases} 0, & m \neq n \pm 1 \\ \frac{1}{\alpha}\left(\frac{n+1}{2}\right)^{1/2}, & m = n+1 \\ \frac{1}{\alpha}\left(\frac{n}{2}\right)^{1/2}, & m = n-1. \end{cases} \qquad (4.174)$$

This result may also be obtained (Problem 4.15) by using the recurrence relation (4.157) together with the orthonormality relation (4.170) and the result (4.167). We remark in particular that for any harmonic oscillator wave function ψ_n the expectation value of x vanishes:

$$\langle x \rangle = x_{nn} = \int_{-\infty}^{+\infty} \psi_n^*(x) x \psi_n(x) dx = 0. \qquad (4.175)$$

This could have been anticipated on general symmetry grounds. Indeed, the harmonic oscillator wave functions have a *definite* parity (even when n is even, odd when n is odd), so that $|\psi_n(x)|^2$ is an *even* function of x. The integrand $x|\psi_n(x)|^2$ appearing in (4.175) is therefore an *odd* function of x, so that its integral taken from $-\infty$ to $+\infty$ vanishes.

Comparison with classical theory

The eigenfunctions $\psi_n(x)$ of the linear harmonic oscillator corresponding to the four lowest eigenvalues ($n = 0$ to 3) are plotted on the left of Fig. 4.18. On the right are displayed the corresponding position probability densities $|\psi_n|^2$, together with the limits of motion for a classical oscillator having the corresponding energy E_n, and the classical probability density P_c for such an oscillator. According to classical mechanics, the position of the particle is given by $x = x_0 \sin \omega t$ (where x_0 is the amplitude of oscillation), its speed is $v = \omega x_0 \cos \omega t$, and its energy is $E = m\omega^2 x_0^2/2$. The classical motion takes place between the turning points, such that $E = V(x)$, located at $\pm x_0 = \pm(2E/m\omega^2)^{1/2}$. The probability $P_c(x)dx$ that the classical particle will be found in the interval dx in a random observation is equal to the fraction of the total time spent by the particle in that interval. If $T = 2\pi/\omega$ denotes the period of oscillation, we therefore have

$$P_c(x)dx = \frac{1}{T}\frac{2dx}{v} = \frac{dx}{\pi(x_0^2 - x^2)^{1/2}}. \tag{4.176}$$

As expected, the classical probability density $P_c(x)$ is largest in the vicinity of the turning points $\pm x_0$, where the speed of the classical particle vanishes. In terms of the reduced variable $\xi = \alpha x$ the classical turning points are located at $\pm \xi_0 = \pm \alpha x_0 = \pm \lambda^{1/2}$ (where we recall that $\lambda = 2E/\hbar\omega$), and the classical probability density is

$$P_c(\xi) = \frac{1}{\pi(\xi_0^2 - \xi^2)^{1/2}}. \tag{4.177}$$

It is clear from Fig. 4.18 that for low values of the quantum number n the quantum mechanical position probability densities $|\psi_n|^2$ are very different from the corresponding densities P_c for the classical oscillator. For example, in the case of the ground state ($n = 0$), the quantum mechanical probability density $|\psi_0|^2$ has its maximum at $x = 0$, while classically the particle is most likely to be found at the end-points of its motion, as we have seen above. As predicted on general grounds (see Section 3.6) the wave functions ψ_n curve towards the axis and have n zeros in the classically allowed region of motion. Outside that region the wave functions curve away from the axis and decrease rapidly, but there is nevertheless a finite probability of finding the particle outside the classically permitted region. As n increases the agreement between the classical and quantum mechanical probability densities improves. This is in accordance with Bohr's *correspondence principle*, and is further illustrated in Fig. 4.19, where the quantum mechanical position probability density $|\psi_n|^2$ for $n = 20$ is plotted together with the probability density P_c of the corresponding classical oscillator, having a total energy $E_{n=20} = (41/2)\hbar\omega$. Apart from the rapid fluctuations of the quantum mechanical curve, the general agreement between the classical and quantum results is seen to be quite good.

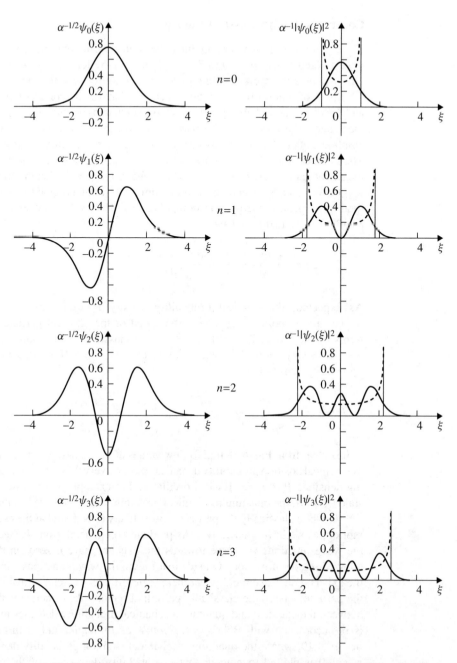

Figure 4.18 The first four energy eigenfunctions $\psi_n(\xi)$, $n = 0, 1, 2, 3$, of the linear harmonic oscillator are plotted on the left. On the right are shown the corresponding position probability densities $|\psi_n(\xi)|^2$ (solid curves), together with the limits of motion for a classical oscillator having the energy E_n, and the classical probability density $P_c(\xi)$ for such an oscillator (dashed curve).

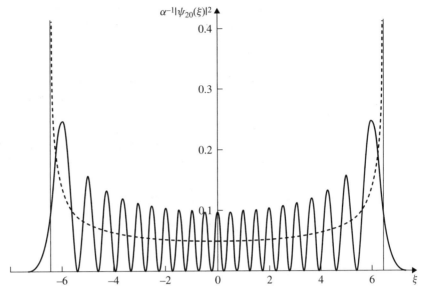

Figure 4.19 Comparison of the quantum mechanical position probability density for the state $n = 20$ of a linear harmonic oscillator (solid curve) with the probability density of the corresponding classical oscillator (dashed curve), having a total energy $E_{n=20} = (41/2)\hbar\omega$.

It is also interesting to evaluate the expectation value of the potential energy in the state ψ_n, namely

$$\langle V \rangle = \int_{-\infty}^{+\infty} \psi_n^*(x) \frac{1}{2} k x^2 \psi_n(x) dx$$

$$= \frac{1}{2} k \langle x^2 \rangle \tag{4.178}$$

where

$$\langle x^2 \rangle = \int_{-\infty}^{+\infty} \psi_n^*(x) x^2 \psi_n(x) dx. \tag{4.179}$$

The integral on the right of this equation can be evaluated by using the generating function (4.156), or by making use of the recursion relation (4.157) in conjunction with the orthonormality relation (4.170). The result is (Problem 4.16)

$$\langle x^2 \rangle = \frac{2n+1}{2\alpha^2} = \left(n + \frac{1}{2}\right) \frac{\hbar}{m\omega} \tag{4.180}$$

so that, using (4.132), (4.178) and (4.151), we have

$$\langle V \rangle = \frac{1}{2}\left(n + \frac{1}{2}\right)\hbar\omega = \frac{1}{2} E_n. \tag{4.181}$$

From this result we can also deduce the average value of the kinetic energy operator $T = p_x^2/2m = -(\hbar^2/2m)\mathrm{d}^2/\mathrm{d}x^2$ in the state ψ_n. That is

$$\langle T \rangle = E_n - \langle V \rangle = \tfrac{1}{2} E_n. \tag{4.182}$$

Thus, for any harmonic oscillator eigenstate ψ_n, the average potential and kinetic energies are each equal to one-half of the total energy, as in the case of the classical oscillator.

We have seen above that $\langle x \rangle = 0$ for any harmonic oscillator eigenfunction ψ_n. By using the relation (4.158) together with the orthonormality relation (4.170), it is readily proved (Problem 4.16) that

$$\langle p_x \rangle = \int \psi_n^*(x) \left(-i\hbar \frac{\mathrm{d}}{\mathrm{d}x} \right) \psi_n(x) \mathrm{d}x = 0 \tag{4.183}$$

and from (4.182) we also deduce that

$$\langle p_x^2 \rangle = 2m\langle T \rangle = mE_n = \left(n + \tfrac{1}{2}\right) m\hbar\omega. \tag{4.184}$$

4.8 The periodic potential

As our final example of one-dimensional problems, we shall consider the motion of a particle in a periodic potential of period l, so that

$$V(x+l) = V(x). \tag{4.185}$$

As an illustration we show in Fig. 4.20 a periodic potential with rectangular sections, called the *Kronig–Penney* potential, which can be used as a model of the interaction to which an electron is subjected in a crystal lattice consisting of a regular array of single atoms separated by the distance l.

Bloch waves

We shall first deduce some general consequences due to the periodicity of the potential (4.185). Although a real crystal is of course of finite length, we shall assume, as a useful idealisation, that equation (4.185) is true for all values of x, over the entire x-axis. Thus, if $\psi(x)$ is a solution of the Schrödinger equation (4.3), corresponding to the energy E, so also is $\psi(x+l)$. In addition, because the Schrödinger equation (4.3) is a second-order linear equation, any solution $\psi(x)$ may be represented as a linear combination of two linearly independent solutions $\psi_1(x)$ and $\psi_2(x)$

$$\psi(x) = c_1 \psi_1(x) + c_2 \psi_2(x). \tag{4.186}$$

Now $\psi_1(x+l)$ and $\psi_2(x+l)$ are also solutions of (4.3), and hence may be represented as linear combinations of $\psi_1(x)$ and $\psi_2(x)$

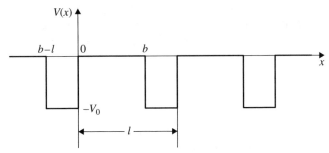

Figure 4.20 A periodic potential with rectangular sections, called the *Kronig–Penney* potential. The period has length l. The wells represent schematically the attraction exerted by the atoms (ions) of the crystal lattice on the electrons.

$$\psi_1(x+l) = a_{11}\psi_1(x) + a_{12}\psi_2(x)$$
$$\psi_2(x+l) = a_{21}\psi_1(x) + a_{22}\psi_2(x). \tag{4.187}$$

Thus, using (4.186), (4.187) and the fact that $\psi(x+l)$ is also a solution, we may write

$$\psi(x+l) = (c_1 a_{11} + c_2 a_{21})\psi_1(x) + (c_1 a_{12} + c_2 a_{22})\psi_2(x)$$
$$= d_1\psi_1(x) + d_2\psi_2(x). \tag{4.188}$$

The relation between the coefficients (c_1, c_2) and (d_1, d_2) clearly involves the matrix multiplication

$$\begin{pmatrix} d_1 \\ d_2 \end{pmatrix} = \begin{pmatrix} a_{11} & a_{21} \\ a_{12} & a_{22} \end{pmatrix} \begin{pmatrix} c_1 \\ c_2 \end{pmatrix}. \tag{4.189}$$

Let us see what happens if we diagonalise the 2×2 matrix in this equation. We must then solve the equation

$$\begin{vmatrix} a_{11} - \lambda & a_{21} \\ a_{12} & a_{22} - \lambda \end{vmatrix} = 0 \tag{4.190}$$

which is a quadratic equation for λ, having two solutions λ_1 and λ_2. If (c_1, c_2) is an eigenvector corresponding to one of the eigenvalues $\lambda = \lambda_i (i = 1, 2)$, we have $d_1 = \lambda c_1$ and $d_2 = \lambda c_2$. Thus, among the solutions, $\psi(x)$, there are two having the property

$$\psi(x+l) = \lambda \psi(x) \tag{4.191}$$

where λ is a constant factor. This result is known as *Floquet's theorem*. Of course, we see immediately from (4.191) that

$$\psi(x+nl) = \lambda^n \psi(x), \qquad n = 0, \pm 1, \pm 2, \ldots. \tag{4.192}$$

Now, let ψ_{λ_1} and ψ_{λ_2} be two solutions of the Schrödinger equation (4.3) corresponding to the energy E, which satisfy (4.191) and correspond respectively to the eigenvalues λ_1 and λ_2 of equation (4.190). Let ψ'_{λ_1} and ψ'_{λ_2} denote their respective derivatives. If

$$W = \begin{vmatrix} \psi_{\lambda_1} & \psi_{\lambda_2} \\ \psi'_{\lambda_1} & \psi'_{\lambda_2} \end{vmatrix} \tag{4.193}$$

denotes the Wronskian determinant of ψ_{λ_1} and ψ_{λ_2}, we have from (4.191)

$$W(x+l) = \lambda_1 \lambda_2 W(x). \tag{4.194}$$

Since the Wronskian of two solutions of the Schrödinger equation (4.3) corresponding to the same energy eigenvalue E is a constant (see footnote[9] of Chapter 3, p. 110) it follows that

$$\lambda_1 \lambda_2 = 1. \tag{4.195}$$

Let us now return to (4.191). If $|\lambda| > 1$, it is clear from (4.192) that ψ will grow above all limits when $x \to +\infty$, and decrease below all limits if $x \to -\infty$. The opposite will happen if $|\lambda| < 1$. Hence, physically admissible solutions of the Schrödinger equation exist only if $|\lambda| = 1$. Taking into account the result (4.195), we may therefore write the two quantities λ_1 and λ_2 in the form

$$\lambda_1 = e^{iKl}, \qquad \lambda_2 = e^{-iKl} \tag{4.196}$$

where K is a real number. Since $\exp(i2\pi n) = 1$, it is clear that a complete set of wave functions can be obtained by restricting the values of K to the interval

$$-\frac{\pi}{l} \leqslant K \leqslant \frac{\pi}{l}. \tag{4.197}$$

Therefore, all physically admissible solutions must satisfy the relation

$$\psi(x + nl) = e^{inKl} \psi(x), \qquad n = 0, \pm 1, \pm 2, \ldots \tag{4.198}$$

which is known as the *Bloch condition*. If we write the solution $\psi(x)$ in the form

$$\psi(x) = e^{iKx} u_K(x) \tag{4.199}$$

it follows from (4.198) that

$$u_K(x + l) = u_K(x). \tag{4.200}$$

This result is known as *Bloch's theorem*. We see that the *Bloch wave function* (4.199) represents a travelling wave of wavelength $2\pi/K$, whose amplitude $u_K(x)$ is periodic, with the same period l as that of the crystal lattice.

Energy bands

Let us now discuss the energy spectrum. For simplicity, we shall focus our attention on the Kronig–Penney potential (see Fig. 4.20) which exhibits the basic features of the interaction felt by an electron in a crystal and leads to a readily solvable problem. We choose the zero of the energy scale to coincide with the top of the wells, and consider successively the two cases for which the electron energy E is such that $-V_0 < E < 0$ and $E > 0$.

Case 1: $-V_0 < E < 0$

The Schrödinger equation is

$$\frac{d^2\psi(x)}{dx^2} + \alpha^2\psi(x) = 0 \qquad \text{inside the wells} \tag{4.201a}$$

and

$$\frac{d^2\psi(x)}{dx^2} - \beta^2\psi(x) = 0 \qquad \text{between the wells} \tag{4.201b}$$

where (see (4.114))

$$\alpha = \left[\frac{2m}{\hbar^2}(V_0 + E)\right]^{1/2}, \qquad \beta = \left(-\frac{2m}{\hbar^2}E\right)^{1/2}. \tag{4.202}$$

Let us solve the equations (4.201) for a given 'cell', for example that extending from $b - l$ to b. We have

$$\psi(x) = Ae^{i\alpha x} + Be^{-i\alpha x}, \qquad b - l < x < 0 \tag{4.203a}$$

$$\psi(x) = Ce^{\beta x} + De^{-\beta x}, \qquad 0 < x < b. \tag{4.203b}$$

Now, according to the Bloch theorem, the solution $\psi(x)$ of the Schrödinger equation for a periodic potential has the general form (4.199), where the amplitude $u_K(x)$ has the same period l as that of the crystal lattice. In other words, the electron does not 'belong' to any one atom (ion) of the lattice, but has an equal probability of being found in the neighbourhood of any of them. Using (4.199) and (4.203), we find in the 'cell' $(b - l, b)$

$$u_K(x) = Ae^{i(\alpha-K)x} + Be^{-i(\alpha+K)x}, \qquad b - l < x < 0 \tag{4.204a}$$

$$u_K(x) = Ce^{(\beta-iK)x} + De^{-(\beta+iK)x}, \qquad 0 < x < b \tag{4.204b}$$

and since $u_K(x)$ is periodic, the equations (4.204) determine $u_K(x)$ – and hence also $\psi(x)$ – for all values of x. For example, if we want to write the function $u_K(x)$ in the region $b < x < l$ we use the fact that $u_K(x) = u_K(x - l)$, and upon replacing x by $x - l$ in (4.204a), we have

$$u_K(x) = Ae^{i(\alpha-K)(x-l)} + Be^{-i(\alpha+K)(x-l)}, \qquad b < x < l. \tag{4.205}$$

In order to be a physically acceptable solution of the Schrödinger equation, the wave function $\psi(x)$ and its first derivative $d\psi(x)/dx$ – and hence $u_K(x)$ and $du_K(x)/dx$

– must be continuous at both edges of each potential well. The periodicity of $u_K(x)$ guarantees that if $u_K(x)$ and $du_K(x)/dx$ are continuous at the edges of just *one* well, they will be continuous at the edges of *every* well. Now, from (4.204) we see that by requiring $u_K(x)$ and $du_K(x)/dx$ to be continuous at $x = 0$ we obtain the two relations

$$A + B = C + D, \tag{4.206a}$$

$$i(\alpha - K)A - i(\alpha + K)B = (\beta - iK)C - (\beta + iK)D \tag{4.206b}$$

while the continuity conditions applied at $x = b$ to the functions (4.204b) and (4.205) yield the two equations

$$Ae^{-i(\alpha-K)c} + Be^{i(\alpha+K)c} = Ce^{(\beta-iK)b} + De^{-(\beta+iK)b} \tag{4.206c}$$

$$i(\alpha - K)Ae^{-i(\alpha-K)c} - i(\alpha + K)Be^{i(\alpha+K)c} = (\beta - iK)Ce^{(\beta-iK)b}$$
$$-(\beta + iK)De^{-(\beta+iK)b} \tag{4.206d}$$

where $c = l - b$.

The equations (4.206) constitute a system of four homogeneous equations for the four unknown quantities A, B, C and D. In order for this system to have a non-trivial solution the determinant of the coefficients of A, B, C and D must vanish. This results in the condition

$$\cos \alpha c \cosh \beta b - \frac{\alpha^2 - \beta^2}{2\alpha\beta} \sin \alpha c \sinh \beta b = \cos Kl \tag{4.207}$$

which is an implicit equation for the energy E.

Case 2: $E > 0$

The modifications required to treat the case $E > 0$ are very simple. We see from (4.202) that when the energy E is positive the quantity β becomes imaginary. We shall therefore set $\beta = ik$, with $k = (2mE/\hbar^2)^{1/2}$. Since no assumption about the reality of β has been made in deriving (4.207), we may at once rewrite this equation for the case $E > 0$ as

$$\cos \alpha c \cos kb - \frac{\alpha^2 + k^2}{2\alpha k} \sin \alpha c \sin kb = \cos Kl \tag{4.208}$$

and we see that this is again an implicit equation for the energy E.

We can summarise both equations (4.207) and (4.208) by writing

$$F(E) = \cos Kl \tag{4.209}$$

where the function $F(E)$ represents the left-hand sides of (4.207) and (4.208). In writing the equation (4.209) we have used the fact that the two left-hand sides join smoothly at $E = 0$, so that only one function $F(E)$ is required for the full energy range $E > -V_0$. The function $F(E)$ is shown in Fig. 4.21 for typical values of b, l and V_0. The remarkable feature which emerges from this graph is that the equation (4.209) cannot be satisfied for certain ranges of values of E. Indeed, since K is real, we have

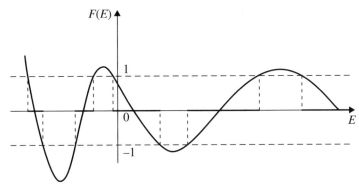

Figure 4.21 Plot of $F(E)$, which represents the left-hand sides of equations (4.207) and (4.208), as a function of the energy E. The heavy lines on the abscissa show the allowed values of E, corresponding to conduction bands; they are separated by forbidden energy gaps.

$|\cos Kl| \leqslant 1$, so that the values of E for which $|F(E)| > 1$ are inaccessible. As a result, the allowed values of E fall into bands satisfying the condition $|F(E)| \leqslant 1$, which are indicated by heavy lines in Fig. 4.21. These bands are known as the *conduction bands* of the lattice; they are separated by *forbidden energy gaps*.

It can be proved (Problem 4.20) that if the periodic distance l is increased without changing c or V_0 (i.e. if the separation between the wells is increased) the energy bands for $-V_0 < E < 0$ become narrower, and contract in the limit $l \to \infty$ into the discrete energy levels of an isolated potential well. This behaviour, which is illustrated in Fig. 4.22, is easy to understand. Indeed, as the spacing l between atoms is increased, the motion of an electron in one of the atoms of the crystal will be less and less affected by the presence of the other atoms, so that each atom will behave as if it were isolated. We also see from Fig. 4.21 that the bands corresponding to the lowest-lying levels are the narrowest; this is due to the fact that the electrons which are most tightly bound to the atoms are less likely to be perturbed by the presence of the other atoms. Note that the electrons in the low-lying bands (with energies $E < 0$) can only go from one atom to the other by 'tunnelling' through the potential barriers between the wells.

Until now we have not discussed the *boundary conditions* which must be satisfied by the wave function $\psi(x)$ at each end of the linear chain of atoms (i.e. at the 'surface' of the one-dimensional crystal we are considering). We could obviously ask that $\psi = 0$ at both ends of the chain so that the electrons cannot escape from the crystal. Unfortunately, this requirement, which leads to standing wave solutions of the Schrödinger equation, is not easy to implement. In particular, we see that the Bloch wave functions (4.199), which are travelling waves, do not satisfy such boundary conditions, so that superpositions of Bloch waves would be necessary to construct the required standing waves. For this reason, we shall disregard the (small) surface effects and adopt *periodic boundary conditions*, such that the wave function ψ is required to take the same value at both ends of the chain. This means that an

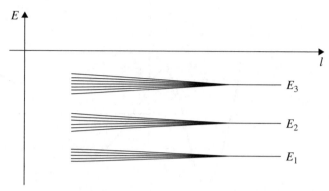

Figure 4.22 When the separation between the wells is increased, the energy bands for $-V_0 < E < 0$ shrink and in the limit $l \to \infty$ contract into the discrete energy levels of an isolated potential well.

electron leaving the crystal at one end simultaneously re-enters it at the corresponding point on the opposite face with its momentum unchanged.

An equivalent way of formulating the periodic boundary conditions is to consider that the one-dimensional lattice has the form of a *closed loop* containing N atoms (where $N \simeq 10^{23}$ is of the order of Avogadro's number). In order for the wave function to be single-valued, we must therefore have

$$\psi(x + Nl) = \psi(x) \tag{4.210}$$

and from (4.198) we see that this requirement is equivalent to the condition

$$e^{iNKl} = 1. \tag{4.211}$$

As a result, the possible values of K are discrete and are given by

$$K = \frac{2\pi n}{Nl}, \qquad n = 0, \pm 1, \pm 2, \ldots, \tag{4.212}$$

Moreover, we see from (4.199), (4.207) and (4.208) that the wave function $\psi(x)$, as well as the corresponding energy E, are unchanged if K is increased or decreased by an integral multiple of $2\pi/l$. We may therefore, without any loss of generality, assume that K is confined within a given interval of length $2\pi/l$, for example the interval (4.197). From (4.197) and (4.212) we immediately deduce that there are N allowed values of K. As K takes on each of these possible values, the energy is equal to every one of its corresponding allowed values in each band. Hence every conduction band contains N allowed energy levels. This result is readily explained if one remembers that in the limit $l \to \infty$ each band contracts into a single level which is N-fold degenerate, since the electron can be bound to any one of the N atoms. For finite values of l this degeneracy is removed and each discrete atomic level spreads into a band of N energy levels.

Problems

4.1 Show that the momentum eigenfunctions

$$\psi_{k_x}(x) = L^{-1/2} \exp(ik_x x)$$

satisfying the periodic boundary conditions $k_x = 2\pi n/L (n = 0, \pm 1, \pm 2, \ldots)$ obey in the limit $L \to \infty$ the closure relation

$$\int_{-\infty}^{+\infty} \psi_{k_x}^*(x') \psi_{k_x}(x) dk_x = \delta(x - x').$$

(*Hint*: Convert the sum $\sum_{n=-\infty}^{+\infty}$ to the integral

$$\int_{-\infty}^{+\infty} dn = (L/2\pi) \int_{-\infty}^{+\infty} dk_x.)$$

4.2 Consider the momentum eigenfunctions

$$\psi_E(x) = c(E) \exp(ikx).$$

Determine $c(E)$ so that $\psi_E(x)$ is 'normalised' on the energy scale according to

$$\int_{-\infty}^{+\infty} \psi_{E'}^*(x) \psi_E(x) dx = \delta(E - E').$$

(*Hint*: Use equation (A.32) of Appendix A to obtain the relation

$$\delta(\sqrt{E} - \sqrt{E'}) = 2\sqrt{E}\delta(E - E'), \qquad E \neq 0.)$$

4.3 Consider a particle incident from the right on the potential step (4.41). Prove that the reflection and transmission coefficients are the same as when the particle is incident on the potential step from the left.

4.4 Consider an electron of energy E incident on the potential step (4.41), where $V_0 = 10$ eV. Calculate the reflection coefficient R and the transmission coefficient T (a) when $E = 5$ eV, (b) when $E = 15$ eV and (c) when $E = 10$ eV.

4.5 Determine the reflection coefficient R and transmission coefficient T:

(a) of an electron of energy $E = 1$ eV for a rectangular barrier potential (4.72) such that $V_0 = 2$ eV and $a = 1$ Å; and
(b) of a proton of the same energy for the same potential barrier.

4.6 Consider a particle of mass m moving in one dimension in an infinite square well of width L, such that the origin 0 has been chosen to be the left corner of the well. Show that the energy eigenvalues are given by (4.95), while the corresponding normalised eigenfunctions are

$$\psi_n(x) = \left(\frac{2}{L}\right)^{1/2} \sin\left(\frac{n\pi}{L}x\right), \qquad n = 1, 2, 3, \ldots$$

and compare these eigenfunctions with those obtained in the text (see (4.93) and (4.94)).

4.7 A particle of mass m moves in one dimension in the infinite square well (4.86). Suppose that at time $t = 0$ its wave function is

$$\Psi(x, t = 0) = A(a^2 - x^2)$$

where A is a normalisation constant.

(a) Find the probability P_n of obtaining the value E_n of the particle energy, where E_n is one of the energy eigenvalues.
(b) Determine the expectation value $\langle E \rangle$ of the energy.

4.8 Solve numerically the equations (4.120) to obtain the energy levels and corresponding normalised eigenfunctions of a one-dimensional square well (4.113) such that $\gamma = (2mV_0a^2/\hbar^2)^{1/2} = 5$.

4.9 Derive the results (4.126) for the reflection and transmission coefficients corresponding to a one-dimensional square well.

4.10 Consider a particle of mass m moving in one dimension in the potential well

$$V(x) = \begin{cases} 0, & 0 \leqslant |x| < a \\ V_0, & a < |x| < b \\ \infty, & |x| > b \end{cases}$$

where V_0, a and b are positive quantities and $b > a$. The energy E of the particle is such that $0 < E < V_0$.

(a) The eigenfunctions of the Schrödinger equation (4.3) for this potential can be assumed to be either even or odd. Why?
(b) Determine the normalised even eigenfunctions $\psi_n^+(x)$ and write down an expression allowing the corresponding energy eigenvalues E_n^+ to be obtained.
(c) Repeat the calculation of (b) for the normalised odd eigenfunctions $\psi_n^-(x)$ and the corresponding energy levels E_n^-.
(d) Show that when $a = b$ one retrieves the eigenfunctions and energy eigenvalues of a particle in the infinite square well (4.86).

4.11 Consider the one-dimensional motion of a particle of mass m in the double potential well

$$V(x) = \begin{cases} V_0, & 0 \leqslant |x| < b \\ 0, & b < |x| < c \\ \infty, & |x| > c \end{cases}$$

where V_0, b and c are positive quantities and $b < c$. The energy E of the particle is such that $0 < E < V_0$.

(a) Let $\psi_n^+(x)$ and $\psi_n^-(x)$ denote respectively the normalised even and odd eigenfunctions of the Schrödinger equation (4.3) for this potential, and let E_n^+ and E_n^- be the corresponding energy eigenvalues. Determine the functions $\psi_n^+(x)$ and $\psi_n^-(x)$ and write down the energy quantisation conditions giving E_n^+ and E_n^-.

(b) Show that for each doublet (E_n^+, E_n^-) the lowest energy level is E_n^+.

(c) Prove that when $V_0 \to \infty$ (so that an infinite barrier separates the two wells), each doublet (E_n^+, E_n^-) merges into a single (twofold degenerate) energy level E_n, the energy values $E_n = (\hbar^2/2m)(\pi^2 n^2/L^2)$, $n = 1, 2, 3, \ldots$, being those of the infinite square well (see (4.95)) with $L = c - b$.

4.12 By using the uncertainty relation (2.70) obtain an estimate of the zero-point energy of a linear harmonic oscillator.

4.13 Prove equations (4.157) and (4.158) by using the generating function (4.156) of the Hermite polynomials.

4.14 Prove that the definition (4.154a) of the Hermite polynomials is equivalent to the definition by means of the generating function given in equations (4.156).

4.15 Obtain the result (4.174) for x_{nm} by using the recurrence relation (4.157) of the Hermite polynomials, the orthonormality relation (4.170), and the result (4.167).

4.16 Prove equations (4.180) and (4.183).

4.17 Show that $f(\xi) = \xi \exp(\xi^2/2)$ is an eigenfunction of the linear harmonic oscillator equation (4.135) corresponding to the eigenvalue $\lambda = -3$. Is this eigenfunction physically acceptable? Explain.

4.18 Consider the motion of a particle of mass m in a one-dimensional potential $V(x) = \lambda \delta(x)$, where $\delta(x)$ is the Dirac delta function.

(a) Assuming that $\lambda > 0$, obtain the reflection coefficient R and the transmission coefficient T.

(Hint: For a potential of the form $V(x) = \lambda \delta(x)$, the integration of the Schrödinger eigenvalue equation (4.3) from $-\varepsilon$ to $+\varepsilon$ gives

$$\left.\frac{d\psi}{dx}\right|_{x=+\varepsilon} - \left.\frac{d\psi}{dx}\right|_{x=-\varepsilon} = \frac{2m}{\hbar^2}\lambda \int_{-\varepsilon}^{+\varepsilon} \delta(x)\psi(x)dx$$

$$= \frac{2m}{\hbar^2}\lambda \psi(x=0).)$$

(b) Consider now the case for which $\lambda < 0$. Show that there is only one bound state and find its energy as a function of $|\lambda|$. Using (4.120), verify that your result agrees with that obtained for the square well in the limit $V_0 \to \infty$, $a \to 0$, in such a way that $2aV_0$ is held fixed and equal to $|\lambda|$.

4.19 Consider the motion of a particle of mass m in the one-dimensional periodic potential

$$V(x) = \frac{V_0}{l} \sum_{n=-\infty}^{+\infty} \delta(x - nl)$$

where δ denotes the Dirac delta function and l is the periodic distance. Prove that for such a potential equation (4.209) reads

$$\cos kl + \frac{mV_0}{\hbar^2} \frac{\sin kl}{kl} = \cos Kl$$

where $k = (2mE/\hbar^2)^{1/2}$, and show that the band edges occur when $Kl = n\pi$, $n = 0, \pm 1, \pm 2, \ldots$.

4.20 Prove that, in the Kronig–Penney model, if the distance l is increased without changing c or V_0, the energy bands for $-V_0 < E < 0$ become narrower, and contract in the limit $l \to \infty$ into the discrete energy levels of an isolated potential well.

5 The formalism of quantum mechanics

5.1 The state of a system   193
5.2 Dynamical variables and operators   198
5.3 Expansions in eigenfunctions   203
5.4 Commuting observables, compatibility and the Heisenberg uncertainty relations   210
5.5 Unitary transformations   216
5.6 Matrix representations of wave functions and operators   220
5.7 The Schrödinger equation and the time evolution of a system   231
5.8 The Schrödinger and Heisenberg pictures   238
5.9 Path integrals   240
5.10 Symmetry principles and conservation laws   245
5.11 The classical limit   256
Problems   260

The basic ideas of quantum mechanics were introduced in Chapters 2 and 3 by using a formulation of the theory known as *wave mechanics*. These ideas were applied in Chapter 4 to some simple one-dimensional problems. In this chapter we shall present the principles of quantum mechanics in a more general way, as a set of postulates justified by the experimental verification of their consequences. In particular, we shall see that there exist alternative ways of formulating quantum mechanics which are equivalent to wave mechanics. We shall also in the course of this chapter develop the mathematical tools necessary to make progress in the theory.

5.1 The state of a system

In classical physics, the dynamical state of a system is determined at each instant of time by the knowledge of the physical quantities–dynamical variables–such as the position vectors and the momenta of the particles which constitute the system. It is postulated that all the dynamical variables associated with the system may, in principle, be measured simultaneously with infinite precision.

The situation is very different in quantum physics, because of the central role played by the *measurement* process. Indeed, when a given dynamical variable is measured, the dynamical state of the system is in general modified in an unpredictable way, and we have seen in Chapter 2 that according to the Heisenberg uncertainty relations this sets a limit to the precision with which 'complementary' dynamical variables can be measured simultaneously. One is therefore led to abandon the classical assumption that all the dynamical variables of a system have well-defined values at every instant. Instead, quantum mechanics only predicts the number n of times that a particular result will be obtained when a large number N of identical, independent, identically prepared physical systems (called a *statistical ensemble* or, for short, an *ensemble*) are subjected to a measurement process. In other words, quantum mechanics predicts the *statistical frequency* n/N or *probability* of an event. In order to make these predictions, a set of postulates are required, which we shall now formulate.

> **Postulate 1**
>
> To an ensemble of physical systems one can, in certain cases, associate a *wave function* or *state function* which contains all the information that can be known about the ensemble. This function is in general complex; it may be multiplied by an arbitrary complex number without altering its physical significance.

Let us make the following two remarks about this first postulate:

(1) The words 'in certain cases' mean that some ensembles cannot be described by a single state function (determined apart from a complex multiplicative constant). Such ensembles will not be considered here; they will be discussed in Chapter 14.

(2) Although in principle one should always use ensembles, it is common practice to speak of the wave function associated with a particular system, and we shall often do so for convenience. However, as stated in Section 2.2, it should always be understood that this is shorthand for the wave function associated with an ensemble.

Let us first consider a physical system consisting of a single particle in a given potential $V(\mathbf{r}, t)$. We assume that the particle is structureless, i.e. has no internal degrees of freedom such as spin[1]. Expressed in configuration space, the state function associated with an ensemble of such systems is a wave function $\Psi(\mathbf{r}, t)$ depending on the position vector $\mathbf{r}$ of the particle and on the time t. As we saw in Section 2.2, a wave function $\Psi(\mathbf{r}, t)$ is said to be *square integrable* if its normalisation integral

$$I = \int |\Psi(\mathbf{r}, t)|^2 d\mathbf{r} \tag{5.1}$$

is finite, where the integration is over all space. Since, according to Postulate 1, two wave functions Ψ and $c\Psi$ which differ by an arbitrary complex multiplicative

---

[1] The case of particles with spin will be discussed in Chapter 6.

constant c represent the same state, it is convenient when working with square integrable functions to choose this constant so that the wave function is normalised to unity:

$$\int |\Psi(\mathbf{r}, t)|^2 d\mathbf{r} = 1. \tag{5.2}$$

The quantity

$$P(\mathbf{r}, t) = |\Psi(\mathbf{r}, t)|^2 \tag{5.3}$$

can then be interpreted as a position probability density, as explained in Section 2.2. We note that the wave functions Ψ and $\exp(i\alpha)\Psi$ which differ by the phase factor $\exp(i\alpha)$–where α is a real number–not only describe the same state, but also have the same normalisation. We also recall that some wave functions, such as plane waves, are not square integrable, in which case other 'normalisation' procedures can be used, for example enclosing the system in a large box, or using the Dirac delta function 'normalisation'.

The generalisation of these considerations to the case of a physical system containing N structureless particles is straightforward. The state function associated with an ensemble of such systems, expressed in configuration space, is a wave function $\Psi(\mathbf{r}_1, \ldots, \mathbf{r}_N, t)$ which depends on the position vectors $\mathbf{r}_1, \ldots, \mathbf{r}_N$ of the particles and on the time t. This wave function is said to be square integrable if the normalisation integral

$$I = \int |\Psi(\mathbf{r}_1, \ldots, \mathbf{r}_N, t)|^2 d\mathbf{r}_1 \ldots d\mathbf{r}_N \tag{5.4}$$

is finite. Again, since Ψ and $c\Psi$ (where c is a complex multiplicative constant) describe the same state, we can choose c so that the wave function is normalised to unity:

$$\int |\Psi(\mathbf{r}_1, \ldots, \mathbf{r}_N, t)|^2 d\mathbf{r}_1 \ldots d\mathbf{r}_N = 1 \tag{5.5}$$

in which case a constant phase factor of the form $\exp(i\alpha)$ is still left undetermined. When the wave function is normalised to unity, the quantity

$$P(\mathbf{r}_1, \ldots, \mathbf{r}_N, t) = |\Psi(\mathbf{r}_1, \ldots, \mathbf{r}_N, t)|^2 \tag{5.6}$$

can be interpreted as a position probability density, in the sense that $P(\mathbf{r}_1, \ldots, \mathbf{r}_N, t) d\mathbf{r}_1 \ldots d\mathbf{r}_N$ is the probability of finding at time t particle 1 in the volume element $d\mathbf{r}_1$ about $\mathbf{r}_1$, particle 2 in the volume element $d\mathbf{r}_2$ about $\mathbf{r}_2$, and so on. We remark that

$$P_1(\mathbf{r}_1, t) \equiv \int P(\mathbf{r}_1, \mathbf{r}_2, \ldots, \mathbf{r}_N, t) d\mathbf{r}_2 \ldots d\mathbf{r}_N \tag{5.7}$$

is the position probability density of particle 1 at the point $\mathbf{r}_1$ at time t, independently of the positions of the other particles. Similar position probability densities can clearly be introduced for the other particles.

> **Postulate 2**
> The superposition principle.

According to the *superposition principle*, which we have already discussed in Chapters 2 and 3, the dynamical states of a quantum system are linearly superposable. Thus, if the state function Ψ_1 is associated with one possible state of a statistical ensemble of physical systems, and the state function Ψ_2 with another state of this ensemble, then any linear combination

$$\Psi = c_1\Psi_1 + c_2\Psi_2 \tag{5.8}$$

where c_1 and c_2 are complex constants, is also a state function associated with a possible state of the ensemble. As we already stressed in Section 2.2, the *relative* phase of Ψ_1 and Ψ_2 in (5.8) is important, since it does affect the physical quantity $|\Psi|^2$ (see (2.10)).

Momentum space wave functions

Instead of using the configuration space wave function $\Psi(\mathbf{r}, t)$ to describe the state of an ensemble of one-particle systems, we could as well use the corresponding *momentum space* wave function $\Phi(\mathbf{p}, t)$, which is the Fourier transform of $\Psi(\mathbf{r}, t)$ (see (2.61)). We saw in Section 2.4 that if the configuration space wave function $\Psi(\mathbf{r}, t)$ is normalised to unity, the corresponding momentum space wave function $\Phi(\mathbf{p}, t)$ will also be normalised to unity in momentum space:

$$\int |\Phi(\mathbf{p}, t)|^2 d\mathbf{p} = 1 \tag{5.9}$$

so that the quantity

$$\Pi(\mathbf{p}, t) = |\Phi(\mathbf{p}, t)|^2 \tag{5.10}$$

can be interpreted as the probability density in momentum space for finding the momentum of the particle in the volume element $d\mathbf{p}$ about $\mathbf{p}$. Likewise the state function associated with an ensemble of N-particle physical systems can be 'represented' by a momentum space wave function $\Phi(\mathbf{p}_1, \ldots, \mathbf{p}_N, t)$ which is the Fourier transform of the configuration space wave function $\Psi(\mathbf{r}_1, \ldots, \mathbf{r}_N, t)$:

$$\Phi(\mathbf{p}_1, \ldots, \mathbf{p}_N, t) = (2\pi\hbar)^{-3N/2} \int \exp\left[-\frac{i}{\hbar}(\mathbf{p}_1 \cdot \mathbf{r}_1 + \cdots + \mathbf{p}_N \cdot \mathbf{r}_N)\right]$$
$$\times \Psi(\mathbf{r}_1, \ldots, \mathbf{r}_N, t) d\mathbf{r}_1 \ldots d\mathbf{r}_N. \tag{5.11}$$

If $\Psi(\mathbf{r}_1, \ldots, \mathbf{r}_N, t)$ is normalised to unity according to (5.5), then by a direct generalisation of the Parseval theorem discussed in Appendix A, we have in momentum space

$$\int |\Phi(\mathbf{p}_1, \ldots, \mathbf{p}_N, t)|^2 d\mathbf{p}_1 \ldots d\mathbf{p}_N = 1 \tag{5.12}$$

so that the quantity

$$\Pi(\mathbf{p}_1, \ldots, \mathbf{p}_N, t) = |\Phi(\mathbf{p}_1, \ldots, \mathbf{p}_N, t)|^2 \tag{5.13}$$

is the probability density in momentum space for finding the momentum of particle 1 in the volume element $d\mathbf{p}_1$ about $\mathbf{p}_1$, the momentum of particle 2 in the volume element $d\mathbf{p}_2$ about $\mathbf{p}_2$, and so on.

The fact that both the wave function Ψ in configuration space and the wave function Φ in momentum space 'represent' the same state suggests that a more abstract quantity can be introduced of which Ψ and Φ are two explicit representations. We shall return to this point in Section 5.6.

Dirac bracket notation

A very convenient notation, due to Dirac, will now be introduced. The scalar product of two square integrable functions $\Psi_1(\mathbf{r})$ and $\Psi_2(\mathbf{r})$ is denoted by the symbol $\langle \Psi_1|\Psi_2 \rangle$, namely

$$\langle \Psi_1|\Psi_2 \rangle \equiv \int \Psi_1^*(\mathbf{r})\Psi_2(\mathbf{r})d\mathbf{r} \tag{5.14a}$$

and the generalisation to functions $\Psi_1(\mathbf{r}_1, \ldots, \mathbf{r}_N)$ and $\Psi_2(\mathbf{r}_1, \ldots, \mathbf{r}_N)$ is straightforward. That is,

$$\langle \Psi_1|\Psi_2 \rangle \equiv \int \Psi_1^*(\mathbf{r}_1, \ldots, \mathbf{r}_N)\Psi_2(\mathbf{r}_1, \ldots, \mathbf{r}_N)d\mathbf{r}_1 \ldots d\mathbf{r}_N. \tag{5.14b}$$

The symbol $|\Psi_2\rangle$ is known as a *ket* while $\langle\Psi_1|$ is known as a *bra*. From the definition (5.14) we have

$$\langle \Psi_1|\Psi_2 \rangle = \langle \Psi_2|\Psi_1 \rangle^*. \tag{5.15a}$$

Moreover, if c is a complex number and Ψ_3 a third function, we also have

$$\langle \Psi_1|c\Psi_2 \rangle = c\langle \Psi_1|\Psi_2 \rangle, \tag{5.15b}$$

$$\langle c\Psi_1|\Psi_2 \rangle = c^*\langle \Psi_1|\Psi_2 \rangle, \tag{5.15c}$$

$$\langle \Psi_3|\Psi_1 + \Psi_2 \rangle = \langle \Psi_3|\Psi_1 \rangle + \langle \Psi_3|\Psi_2 \rangle. \tag{5.15d}$$

Two functions Ψ_1 and Ψ_2 are said to be orthogonal if their scalar product vanishes:

$$\langle \Psi_1|\Psi_2 \rangle = 0. \tag{5.16}$$

Using the Dirac bracket notation, we see that the normalisation condition (5.5) can be written compactly as

$$\langle \Psi|\Psi \rangle = 1. \tag{5.17}$$

We remark that this notation can be used just as well for momentum space wave functions. For example, the normalisation condition (5.12) can be written as $\langle \Phi | \Phi \rangle = 1$.

5.2 Dynamical variables and operators

We have seen in Chapter 2 and 3 that with each dynamical variable is associated an operator acting on the wave function. Following this idea, one can formulate the next postulate as follows.

> **Postulate 3**
> With every dynamical variable is associated a linear operator.

We recall that an operator A is linear if it has the property (see (3.62))

$$A(c_1\Psi_1 + c_2\Psi_2) = c_1(A\Psi_1) + c_2(A\Psi_2) \tag{5.18}$$

where Ψ_1 and Ψ_2 are two functions and c_1 and c_2 are complex constants.

The rules for associating a linear operator with a dynamical variable have already been discussed in Sections 3.3 and 3.9 for the case of one-particle systems. The generalisation of these rules to systems of N particles is as follows. If the dynamical state of the system is described by a wave function $\Psi(\mathbf{r}_1, \ldots, \mathbf{r}_N, t)$ in configuration space, one associates with the dynamical variable $\mathcal{A} = A(\mathbf{r}_1, \ldots, \mathbf{r}_N, \mathbf{p}_1, \ldots, \mathbf{p}_N, t)$ the linear operator

$$A(\mathbf{r}_1, \ldots, \mathbf{r}_N, -i\hbar\nabla_1, \ldots, -i\hbar\nabla_N, t) \tag{5.19}$$

obtained by performing the *substitution* $\mathbf{p}_i \to -i\hbar\nabla_i (i = 1, 2, \ldots, N)$ whenever the momentum $\mathbf{p}_i$ occurs. If, on the other hand, the dynamical state of the system is described by a wave function $\Phi(\mathbf{p}_1, \ldots, \mathbf{p}_N, t)$ in momentum space, one associates with the dynamical variable $\mathcal{A} = A(\mathbf{r}_1, \ldots, \mathbf{r}_N, \mathbf{p}_1, \ldots, \mathbf{p}_N, t)$ the linear operator

$$A(i\hbar\nabla_{\mathbf{p}_1}, \ldots, i\hbar\nabla_{\mathbf{p}_N}, \mathbf{p}_1, \ldots, \mathbf{p}_N, t) \tag{5.20}$$

obtained by performing the *substitution* $\mathbf{r}_i \to i\hbar\nabla_{\mathbf{p}_i} (i = 1, 2, \ldots, N)$ whenever the position vector $\mathbf{r}_i$ occurs.

It is worth noting that these substitution rules only apply to dynamical variables expressed in Cartesian coordinates. We shall also see in Chapter 6 that quantum mechanics uses operators such as the spin which have no classical analogue, and which therefore cannot be obtained from the substitution rules.

Eigenvalues and eigenfunctions

We have seen in Section 3.5 that a function ψ_n is said to be an *eigenfunction* of an operator A if the result of operating on ψ_n with A is to multiply ψ_n by a number a_n called an eigenvalue:

$$A\psi_n = a_n\psi_n. \tag{5.21}$$

Postulate 4
The only result of a precise measurement of the dynamical variable $\mathcal{A}$ is one of the eigenvalues a_n of the linear operator A associated with $\mathcal{A}$.

The totality of the eigenvalues of an operator A is called the *spectrum of A*. Since the results of measurements are real numbers it follows that the spectrum of any operator representing a dynamical variable must be real. In some cases the spectrum of an operator consists only of discrete eigenvalues, in others it consists of a continuous range of eigenvalues, or a mixture of both.

Hermitian operators

We shall now discuss the consequences of the requirement that the allowed class of linear operators representing dynamical variables must have real eigenvalues. The operators required are called *Hermitian operators*, and are defined by the condition

$$\langle X|(A\Psi)\rangle = \langle (AX)|\Psi\rangle \tag{5.22}$$

where Ψ and X are square-integrable functions. We note that if $\Psi = X$,

$$\langle \Psi|(A\Psi)\rangle = \langle (A\Psi)|\Psi\rangle \tag{5.23}$$

which for a one-particle system is identical to (3.69). Although it appears that (5.22) is a more general definition of Hermiticity than (5.23), the two definitions are in fact equivalent (see Problem 5.2). The matrix element $\langle X|(A\Psi)\rangle$ is usually written in the Dirac notation as

$$\langle X|(A\Psi)\rangle \equiv \langle X|A|\Psi\rangle \tag{5.24}$$

and from (5.23) we see that if A is Hermitian the matrix element $\langle \Psi|A|\Psi\rangle$ is real.

If ψ_n is an eigenfunction of the operator A corresponding to the eigenvalue a_n, then from (5.21) we have

$$\langle \psi_n|A|\psi_n\rangle = a_n\langle \psi_n|\psi_n\rangle. \tag{5.25}$$

In addition, since

$$(A\psi_n)^* = a_n^*\psi_n^* \tag{5.26}$$

we also have

$$\langle (A\psi_n)|\psi_n\rangle = a_n^*\langle\psi_n|\psi_n\rangle. \tag{5.27}$$

If A is Hermitian, the left-hand sides of (5.25) and (5.27) are equal, and hence $a_n = a_n^*$. Therefore a sufficient (but not a necessary) condition for an operator to have real eigenvalues is for it to be Hermitian. From now on the linear operator A associated with a dynamical variable $\mathcal{A}$ will be taken to be Hermitian. If the operator A obtained by using the substitution rules (5.19) and (5.20) is not Hermitian, the Hermitisation procedure discussed in Section 3.3 must be carried out.

If the wave function of a system is one of the eigenfunctions ψ_n of the operator A, corresponding to the eigenvalue a_n, then a measurement of the dynamical variable $\mathcal{A}$ will certainly produce the result a_n. In this case we shall say that the system is in an eigenstate of A characterised by the eigenvalue a_n. On the other hand, if the wave function is not an eigenfunction of A, then in a measurement of A any one of the results $a_1, a_2, \ldots$, can be obtained. Which result will be obtained is impossible to predict. However, as we shall see later, the *probability* of obtaining a particular result a_n can be obtained.

Postulate 5

If a series of measurements is made of the dynamical variable A on an ensemble of systems[2], described by the wave function Ψ, the expectation or average value of this dynamical variable is

$$\langle A\rangle = \frac{\langle\Psi|A|\Psi\rangle}{\langle\Psi|\Psi\rangle}. \tag{5.28}$$

Since A is a Hermitian operator it follows that $\langle A\rangle$ is real. If the wave function Ψ is normalised to unity, we have $\langle\Psi|\Psi\rangle = 1$ and

$$\langle A\rangle = \langle\Psi|A|\Psi\rangle. \tag{5.29}$$

It is important to stress that $\langle A\rangle$ does not represent the average of a classical statistical distribution of the dynamical variable A between the systems being measured. Each systems is identical and is in the *same* state described by the wave function Ψ. The actual value of A obtained in an experiment on a single system is inherently unpredictable (unless Ψ is an eigenfunction of A). Since Ψ contains the maximum possible information about the system, there is no way of specifying the state more completely in a way which would allow the value of A to be predicted.

Some measurements are not immediately repeatable, for example if the energy of a particle is measured by noting the length of the track it makes while slowing down in a photographic plate. In contrast, measurements of the component in a certain direction of the magnetic moment of an atom in a Stern–Gerlach experiment (see Chapter 1) can be repeated immediately by passing the beam through a second apparatus. In such a case, it is reasonable to expect that if a particular result a_n is obtained in the first measurement, the same result will be obtained if the measurement is repeated immediately. Since the result of the second measurement can be predicted with certainty, we deduce that after the first measurement the state of the system is described by the eigenfunction ψ_n of A belonging to the eigenvalue a_n. Hence, in this case the act of measurement has a 'filtering' effect so that whatever the state of the system before the measurement, it is certainly in an eigenstate of the measured quantity immediately afterwards.

Adjoint operator

It is useful to introduce an operator $A^\dagger$, called the *adjoint* or *Hermitian conjugate* of a linear operator A by the relation

$$\langle X|A^\dagger|\Psi\rangle = \langle(AX)|\Psi\rangle$$
$$= \langle\Psi|A|X\rangle^* \tag{5.30}$$

where Ψ and X are any pair of square-integrable functions.

If we define a bra $\langle\Phi|$ by the relation

$$\langle\Phi| = \langle X|A^\dagger \tag{5.31}$$

where the operator $A^\dagger$ acts to the left on the bra $\langle X|$, then it follows from (5.30) that the kets $|\Phi\rangle$ and $|X\rangle$ are related by

$$|\Phi\rangle = A|X\rangle \tag{5.32}$$

A linear operator A satisfying

$$A = A^\dagger \tag{5.33}$$

is said to be *self-adjoint*, and from (5.22) we see that a self-adjoint operator is Hermitian. The adjoint of an operator is the operator generalisation of the complex conjugate of a complex number, and a self-adjoint operator is the operator generalisation of a real number. It is important to note that the operator $A^\dagger$ is not, in general, equal to the operator A^* (obtained by replacing every i appearing in A by $-i$). For example, the configuration space operator $p_x = -i\hbar\partial/\partial x$ is Hermitian, so that $p_x^\dagger = p_x$, while $p_x^* = i\hbar\partial/\partial x = -p_x$. Hence $p_x^\dagger \neq p_x^*$.

Three important properties of the adjoint operator (see Problem 5.4) are

$$(cA)^\dagger = c^* A^\dagger \tag{5.34}$$

where c is a complex number,

$$(A+B)^\dagger = A^\dagger + B^\dagger \tag{5.35}$$

and

$$(AB)^\dagger = B^\dagger A^\dagger. \tag{5.36}$$

Functions of operators

If a function $f(z)$ can be expanded in a power series,

$$f(z) = \sum_{i=0}^{\infty} c_i z^i \tag{5.37}$$

then the operator function $f(A)$ can be defined as

$$f(A) = \sum_{i=0}^{\infty} c_i A^i. \tag{5.38}$$

Consequently, if ψ_n is one of the eigenfunctions of A, corresponding to the eigenvalue a_n, $A^i \psi_n = (a_n)^i \psi_n$ and thus

$$f(A)\psi_n = f(a_n)\psi_n. \tag{5.39}$$

The adjoint operator to $f(A)$ can be obtained as follows. Using (5.34)–(5.36) and (5.38), we see that

$$[f(A)]^\dagger = \sum_{i=0}^{\infty} c_i^* (A^i)^\dagger = \sum_{i=0}^{\infty} c_i^* (A^\dagger)^i$$

$$= f^*(A^\dagger). \tag{5.40}$$

In particular, if A is a self-adjoint operator, we have

$$[f(A)]^\dagger = f^*(A). \tag{5.41}$$

Inverse and unitary operators

The unit operator I is the operator that leaves any function Ψ unchanged

$$I\Psi = \Psi. \tag{5.42}$$

If, given an operator A, there exists another operator B such that

$$BA = AB = I \tag{5.43}$$

then B is said to be the *inverse* of A and one writes

$$B = A^{-1}. \tag{5.44}$$

A linear operator U is said to be *unitary* if

$$U^{-1} = U^\dagger \tag{5.45a}$$

or

$$UU^\dagger = U^\dagger U = I. \tag{5.45b}$$

Such an operator can be expressed in the form

$$U = e^{iA} \tag{5.46}$$

where A is a Hermitian operator. Indeed, using (5.41), we see that

$$U^{\dagger} = (e^{iA})^{\dagger} = e^{-iA} \tag{5.47}$$

from which (5.45b) follows.

Projection operators

An operator Λ is said to be *idempotent* if

$$\Lambda^2 = \Lambda. \tag{5.48}$$

If, in addition, Λ is Hermitian, it is called a *projection operator*.

Any function Ψ can be expressed in terms of two orthogonal functions Φ and X by means of a projection operator. This can be seen as follows. We first write

$$\Psi = \Phi + X \tag{5.49}$$

with $\Phi = \Lambda \Psi$ and $X = (I - \Lambda)\Psi$. Now

$$\begin{aligned}\langle \Phi | X \rangle &= \langle \Lambda \Psi | (I - \Lambda) \Psi \rangle \\ &= \langle \Psi | \Lambda - \Lambda^2 | \Psi \rangle \\ &= 0 \end{aligned} \tag{5.50}$$

where in the second line we have used the fact that Λ is Hermitian and in the third line we have used (5.48). Note that $I - \Lambda$ is also a projection operator, since it is Hermitian, and

$$\begin{aligned}(I - \Lambda)^2 &= I - 2\Lambda + \Lambda^2 \\ &= I - \Lambda. \end{aligned} \tag{5.51}$$

5.3 Expansions in eigenfunctions

We shall now study in more detail the properties of the solutions of eigenvalue equations such as (5.21), generalising the discussion of Section 3.7. We assume that in (5.21) the operator A is a linear, Hermitian operator representing a dynamical variable, so that the eigenvalues a_n are real. We first consider the case where all the eigenfunctions ψ_n are square integrable and hence can be taken to be normalised to unity:

$$\langle \psi_n | \psi_n \rangle = 1. \tag{5.52}$$

Orthogonality

If ψ_i and ψ_j are two eigenfunctions of A corresponding to *different* eigenvalues a_i and a_j, then

$$A\psi_i = a_i\psi_i \tag{5.53a}$$

and

$$A\psi_j = a_j\psi_j. \tag{5.53b}$$

Hence

$$\begin{aligned}(a_i - a_j)\langle\psi_i|\psi_j\rangle &= \langle a_i\psi_i|\psi_j\rangle - \langle\psi_i|a_j\psi_j\rangle \\ &= \langle(A\psi_i)|\psi_j\rangle - \langle\psi_i|(A\psi_j)\rangle \\ &= 0 \end{aligned} \tag{5.54}$$

where we have used the fact that A is Hermitian. Since $a_i \neq a_j$ it follows from (5.54) that

$$\langle\psi_i|\psi_j\rangle = 0, \qquad i \neq j \tag{5.55}$$

so that eigenfunctions belonging to different eigenvalues are *orthogonal*.

Degeneracy

We have seen in Section 3.5 that an eigenvalue a_n is said to be *degenerate* if there is more than one linearly independent eigenfunction belonging to that eigenvalue, the *degree* of degeneracy being the number of such linearly independent eigenfunctions.

Suppose that α is the degree of degeneracy of the eigenvalue a_n, so that the corresponding eigenfunctions can be labelled ψ_{nr} (with $r = 1, 2, \ldots, \alpha$), and

$$A\psi_{nr} = a_n\psi_{nr}, \qquad r = 1, 2, \ldots, \alpha. \tag{5.56}$$

By using the Schmidt orthogonalisation procedure discussed in Section 3.7, it is always possible to arrange that all the eigenfunctions ψ_{nr} are mutually orthogonal and each of them can be normalised to unity. Since the eigenfunctions belonging to different eigenvalues are mutually orthogonal, there is no loss of generality in writing that all the eigenfunctions satisfy the *orthonormality relations*

$$\langle\psi_{ir}|\psi_{js}\rangle = \delta_{ij}\delta_{rs} \tag{5.57}$$

where δ_{mn} is the Kronecker delta symbol such that

$$\delta_{mn} = \begin{cases} 1, & m = n \\ 0, & m \neq n. \end{cases} \tag{5.58}$$

5.3 Expansions in eigenfunctions

To avoid cumbersome subscripts, unless it is necessary to distinguish the eigenfunctions belonging to a degenerate eigenvalue explicitly, we shall label all the orthonormal eigenfunctions of the operator A by a single index and write

$$\langle \psi_m | \psi_n \rangle = \delta_{mn}. \tag{5.59}$$

Postulate 6
A wave function representing any dynamical state can be expressed as a linear combination of the eigenfunctions of A, where A is the operator associated with a dynamical variable.

For the case of purely discrete eigenvalues, which we shall consider first, we have

$$\Psi = \sum_n c_n \psi_n. \tag{5.60}$$

The number of eigenfunctions in the set $\{\psi_n\}$ is in some cases finite, and in others infinite. Since *all* wave functions can be expanded in the set of eigenfunctions $\{\psi_n\}$, the set is said to be *complete*. The completeness of the set of eigenfunctions of some operators can be proved, but in general it must be postulated, as stated above. It should be noted that not all Hermitian operators possess a complete set of eigenfunctions; those that do are called *observables*.

The coefficients of the expansion (5.60) can be found by using the orthonormality relation (5.59). Taking the scalar product (see (5.14)) of both sides of (5.60) with the eigenfunction ψ_m, we obtain

$$\begin{aligned}
\langle \psi_m | \Psi \rangle &= \sum_n c_n \langle \psi_m | \psi_n \rangle \\
&= \sum_n c_n \delta_{mn} \\
&= c_m.
\end{aligned} \tag{5.61}$$

More explicitly, for the case of a one-particle system, we have

$$\begin{aligned}
\Psi(\mathbf{r}, t) &= \sum_n \left[\int \psi_n^*(\mathbf{r}') \Psi(\mathbf{r}', t) d\mathbf{r}' \right] \psi_n(\mathbf{r}) \\
&= \int \left[\sum_n \psi_n^*(\mathbf{r}') \psi_n(\mathbf{r}) \right] \Psi(\mathbf{r}', t) d\mathbf{r}'
\end{aligned} \tag{5.62}$$

and hence

$$\sum_n \psi_n^*(\mathbf{r}') \psi_n(\mathbf{r}) = \delta(\mathbf{r} - \mathbf{r}') \tag{5.63a}$$

where we have used the property (A.51) of the Dirac delta function. The relation (5.63a) is called the *closure* relation; it expresses the completeness of the set of functions $\{\psi_n\}$. The obvious generalisation of this relation for systems of N particles is

$$\sum_n \psi_n^*(\mathbf{r}_1', \ldots, \mathbf{r}_N') \psi_n(\mathbf{r}_1, \ldots, \mathbf{r}_N) = \delta(\mathbf{r}_1 - \mathbf{r}_1') \ldots \delta(\mathbf{r}_N - \mathbf{r}_N'). \tag{5.63b}$$

Using the closure relation (5.63a) we can write the scalar product of two wave functions as

$$\begin{aligned}
\langle X|\Psi\rangle &= \int X^*(\mathbf{r}, t) \Psi(\mathbf{r}, t) \mathrm{d}\mathbf{r} \\
&= \int X^*(\mathbf{r}, t) \delta(\mathbf{r} - \mathbf{r}') \Psi(\mathbf{r}', t) \mathrm{d}\mathbf{r} \mathrm{d}\mathbf{r}' \\
&= \sum_n \int X^*(\mathbf{r}, t) \psi_n(\mathbf{r}) \mathrm{d}\mathbf{r} \int \psi_n^*(\mathbf{r}') \Psi(\mathbf{r}', t) \mathrm{d}\mathbf{r}' \\
&= \sum_n \langle X|\psi_n\rangle \langle \psi_n|\Psi\rangle
\end{aligned} \tag{5.64}$$

where we have considered the one-particle case for ease of notation. We see from (5.64) that in the Dirac notation the closure relation can be written in the compact form

$$\sum_n |\psi_n\rangle \langle \psi_n| = I \tag{5.65}$$

where I is the unit operator.

Probability amplitudes

In the state described by a wave function Ψ, normalised to unity, the expectation value of an observable A is given by (5.29). Expanding Ψ in the complete orthonormal set of eigenfunctions $\{\psi_n\}$ of A according to (5.60), and expanding Ψ^* similarly, we have

$$\begin{aligned}
\langle A\rangle &= \langle \Psi|A|\Psi\rangle \\
&= \sum_m \sum_n c_m^* c_n \langle \psi_m|A|\psi_n\rangle \\
&= \sum_m \sum_n c_m^* c_n a_n \langle \psi_m|\psi_n\rangle \\
&= \sum_n |c_n|^2 a_n
\end{aligned} \tag{5.66}$$

where the orthonormality relation (5.59) has been used to obtain the last line. Since Ψ is normalised to unity, $\langle\Psi|\Psi\rangle = 1$, we also have

$$\sum_n |c_n|^2 = 1. \tag{5.67}$$

In view of the fact that the possible results of measurements of A are the eigenvalues a_n, and since the average value obtained in series of measurements of a large number of identically prepared systems, all in the same state described by Ψ, is the expectation value $\langle A \rangle$, it is reasonable, following M. Born, to interpret the quantity

$$P_n = |c_n|^2 = |\langle\psi_n|\Psi\rangle|^2 \tag{5.68}$$

as the probability that in a given measurement the particular value a_n will be obtained. The condition (5.67) is thus seen to express the fact that the probability of obtaining some result is unity. The coefficients $c_n = \langle\psi_n|\Psi\rangle$ are called *probability amplitudes*.

In obtaining the above results we have implicitly included the degeneracy index in the summations. If a particular eigenvalue a_n is α times degenerate, and $\psi_{nr}(r = 1, \ldots, \alpha)$ are the corresponding orthonormal eigenfunctions, we have instead of (5.60) the explicit expansion

$$\Psi = \sum_n \sum_{r=1}^{\alpha} c_{nr} \psi_{nr} \tag{5.69}$$

with

$$c_{nr} = \langle\psi_{nr}|\Psi\rangle. \tag{5.70}$$

A simple reworking of (5.66) then yields

$$\langle A \rangle = \sum_n \sum_{r=1}^{\alpha} |c_{nr}|^2 a_n \tag{5.71}$$

so that the probability of obtaining upon measurement of A the degenerate eigenvalue a_n is

$$P_n = \sum_{r=1}^{\alpha} |c_{nr}|^2 = \sum_{r=1}^{\alpha} |\langle\psi_{nr}|\Psi\rangle|^2. \tag{5.72}$$

After a measurement leading to the value a_n, the system is described by the (unnormalised) wave function

$$\Psi_n = \sum_{r=1}^{\alpha} c_{nr} \psi_{nr} \tag{5.73}$$

and if the measurement is immediately repeated the value a_n will be obtained with certainty.

The continuous spectrum

So far in this chapter, observables which possess only a spectrum of discrete eigenvalues have been considered. Our discussion is therefore not sufficiently general, as we have already encountered observables for which the spectrum is purely continuous, or for which the spectrum possesses both a discrete and a continuum part. For example, the momentum operator p_x belongs to the former category (see Section 4.2) and the Hamiltonian operator for a particle in an attractive one-dimensional square well belongs to the latter category (see Section 4.6). Thus, in general, in order to expand an arbitrary wave function in the complete set of eigenfunctions of an operator, eigenfunctions corresponding to continuous as well as discrete eigenvalues are required. As we have seen in Chapters 3 and 4, the eigenfunctions corresponding to the continuous spectrum are not normalisable in the usual sense. One way of avoiding this difficulty is to render the spectrum purely discrete by enclosing the system in a large box with walls which are either impenetrable, or where periodic boundary conditions are imposed. In that case, the various expansion and closure formulae remain the same as for the discrete case. The physical results obtained from the theory are independent of the size of the normalisation box, provided that it is large enough. The alternative is to 'normalise' the eigenfunctions corresponding to the continuous eigenvalues in terms of the Dirac delta functions, as we did for the plane waves in Section 4.2.

Consider an observable A with a spectrum containing both discrete eigenvalues a_n and a continuous range of eigenvalues which we denote by a. The corresponding eigenfunctions are ψ_n and ψ_a, respectively. For the sake of simplicity we shall not display the degeneracy indices. Thus we have

$$A\psi_n = a_n\psi_n, \qquad A\psi_a = a\psi_a. \tag{5.74}$$

The continuous eigenvalues must clearly be real, as are the discrete ones. The discrete eigenfunctions will be taken to be orthonormal, as before.

According to the 'expansion' Postulate 6, an arbitray wave function Ψ must be expandable in the complete set $\{\psi_n, \psi_a\}$,

$$\Psi = \sum_n c_n\psi_n + \int c(a)\psi_a \mathrm{d}a \tag{5.75}$$

where the integral runs over the whole range of values of a. Taking the case in which Ψ is normalised to unity, the expectation value of the observable A is given from (5.29) and (5.75) by

$$\langle A \rangle = \langle \Psi | A | \Psi \rangle$$

$$= \sum_m \sum_n c_m^* c_n \langle \psi_m | A | \psi_n \rangle + \sum_m \int da\, c_m^* c(a) \langle \psi_m | A | \psi_a \rangle$$

$$+ \sum_n \int da'\, c^*(a') c_n \langle \psi_{a'} | A | \psi_n \rangle + \int da \int da'\, c^*(a') c(a) \langle \psi_{a'} | A | \psi_a \rangle$$

$$= \sum_m \sum_n c_m^* c_n a_n \langle \psi_m | \psi_n \rangle + \sum_m \int da\, c_m^* c(a) a \langle \psi_m | \psi_a \rangle$$

$$+ \sum_n \int da'\, c^*(a') c_n a_n \langle \psi_{a'} | \psi_n \rangle$$

$$+ \int da \int da'\, c^*(a') c(a) a \langle \psi_{a'} | \psi_a \rangle \tag{5.76}$$

where in writing the last two lines we have used (5.74).

Now, in order to maintain the interpretation of the coefficients c_n and $c(a)$ as probability amplitudes, we demand that the generalisation of the basic results (5.66) should be, in the present case,

$$\langle A \rangle = \sum_n |c_n|^2 a_n + \int |c(a)|^2 a \, da. \tag{5.77}$$

Comparing (5.76) with (5.77) and recollecting that $\langle \psi_m | \psi_n \rangle = \delta_{mn}$, we deduce the following:

(1) All the eigenfunctions belonging to the continuum spectrum must be orthogonal to all those belonging to the discrete spectrum:

$$\langle \psi_m | \psi_a \rangle = 0. \tag{5.78}$$

(2) The eigenfunctions belonging to the continuum spectrum must satisfy the orthonormality condition

$$\langle \psi_{a'} | \psi_a \rangle = \delta(a - a'). \tag{5.79}$$

Using these results, together with (5.75), we then find that the coefficients c_n and $c(a)$ are given, respectively, by

$$c_n = \langle \psi_n | \Psi \rangle, \qquad c(a) = \langle \psi_a | \Psi \rangle. \tag{5.80}$$

The closure relation (5.63a) for a one-particle system now reads

$$\sum_n \psi_n^*(\mathbf{r}') \psi_n(\mathbf{r}) + \int \psi_a^*(\mathbf{r}') \psi_a(\mathbf{r}) da = \delta(\mathbf{r} - \mathbf{r}') \tag{5.81a}$$

and the corresponding result for a N-particle system is

$$\sum_n \psi_n^*(\mathbf{r}_1', \ldots, \mathbf{r}_N')\psi_n(\mathbf{r}_1, \ldots, \mathbf{r}_N) + \int \psi_a^*(\mathbf{r}_1', \ldots, \mathbf{r}_N')\psi_a(\mathbf{r}_1, \ldots, \mathbf{r}_N)\mathrm{d}a$$

$$= \delta(\mathbf{r}_1 - \mathbf{r}_1')\ldots\delta(\mathbf{r}_N - \mathbf{r}_N'). \tag{5.81b}$$

If the box normalisation is employed (so that the entire spectrum becomes discrete) and we denote by ψ_i the normalised eigenfunction corresponding to the discrete eigenvalue $a_i (i = 1, 2, \ldots)$, the connection between the eigenfunction ψ_i and the continuum eigenfunction ψ_a can be established as follows. Assuming the normalisation box to be very large, the eigenvalues a_i corresponding to the continuous spectrum are densely distributed, and for these eigenvalues the index i can be treated as a continuous variable. Hence, setting $i \equiv i(a)$ and introducing a *density of states*

$$\rho(a) = \frac{\mathrm{d}i}{\mathrm{d}a} \tag{5.82}$$

which is equal to the number of discrete states within a unit range of a, we have

$$\sum_i c_i \psi_i \to \int c_i \psi_i \mathrm{d}i = \int \rho(a) c_i \psi_i \mathrm{d}a. \tag{5.83}$$

Requiring that

$$\int \rho(a) c_i \psi_i \mathrm{d}a = \int c(a) \psi_a \mathrm{d}a \tag{5.84}$$

and also (from the closure relation and with $\sum_i \to \int \rho(a)\mathrm{d}a$)

$$\int \rho(a) \psi_i^*(\mathbf{r}')\psi_i(\mathbf{r})\mathrm{d}a = \int \psi_a^*(\mathbf{r}')\psi_a(\mathbf{r})\mathrm{d}a \tag{5.85}$$

one can make the identifications

$$\psi_a = [\rho(a)]^{1/2}\psi_i \tag{5.86a}$$

and

$$c(a) = [\rho(a)]^{1/2}c_i \tag{5.86b}$$

which allow one to 'translate' formulae written using the 'box normalisation' into those written using the delta function 'normalisation'.

5.4 Commuting observables, compatibility and the Heisenberg uncertainty relations

We saw in Chapter 3 that the *commutator* of two operators A and B is defined as

$$[A, B] = AB - BA. \tag{5.87}$$

If the commutator vanishes when acting on *any* wave function, the two operators A and B are said to commute, $AB = BA$. The operators representing the 'canonically conjugate' position and momentum variables of a particle satisfy the fundamental commutation relations (see (3.77))

$$[x, p_x] = [y, p_y] = [z, p_z] = i\hbar \tag{5.88}$$

with all other pairs of operators (for example y and p_x) commuting. If a system contains several particles with position vectors $\mathbf{r}_i$ and momenta $\mathbf{p}_i$ ($i = 1, 2, \ldots, N$), it is clear that *all operators referring to one particle commute with all those referring to another*. The only pairs out of the operators (x_1, y_1, z_1), $(x_2, y_2, z_2), \ldots,$ and (p_{1x}, p_{1y}, p_{1z}), $(p_{2x}, p_{2y}, p_{2z}), \ldots,$ which do not commute are the canonically conjugate pairs (x_i, p_{ix}), (y_i, p_{iy}) and (z_i, p_{iz}) for which, in conformity with (5.88)

$$[x_i, p_{ix}] = [y_i, p_{iy}] = [z_i, p_{iz}] = i\hbar \qquad (i = 1, 2, \ldots, N). \tag{5.89}$$

Commuting observables

Let us suppose that A and B are two observables. If there exists a complete set of functions ψ_n such that each function is simultaneously an eigenfunction of A and of B, the observables A and B are said to be *compatible*. If the eigenvalues of A and B corresponding to the eigenfunction ψ_n are denoted by a_n and b_n, respectively, then

$$A\psi_n = a_n\psi_n, \qquad B\psi_n = b_n\psi_n. \tag{5.90}$$

In a state described by ψ_n, a measurement of A must produce the precise result a_n and a measurement of B the precise result b_n, with no limit on the precision with which A and B can be measured simultaneously. Examples of such compatible observables are the Cartesian components x, y and z of the position vector $\mathbf{r}$ of a particle, and others are the Cartesian components p_x, p_y and p_z of its momentum $\mathbf{p}$. In contrast, x and p_x are not compatible, since, by the uncertainty relations, both of these quantities cannot be measured simultaneously to arbitrary precision. More generally, several observables $A, B, C, \ldots,$ are said to be compatible if they possess a common set of eigenfunctions. In this case, all these observables can be measured simultaneously to arbitrary precision.

If A and B are two compatible observables, and ψ_n is a common eigenfunction, we have

$$\begin{aligned} AB\psi_n &= a_n b_n \psi_n \\ &= b_n a_n \psi_n \\ &= BA\psi_n. \end{aligned} \tag{5.91}$$

Since any wave function Ψ can be expanded in the complete set of eigenfunctions ψ_n according to (5.60)[3], we find, using (5.91),

$$(AB - BA)\Psi = \sum_n c_n (AB - BA)\psi_n = 0 \tag{5.92}$$

so that

$$[A, B] = 0 \tag{5.93}$$

and two compatible observables commute.

The converse of this result, that two operators which commute possess a complete set of common eigenfunctions, will now be proved. First, consider the case for which A has *non-degenerate* eigenvalues a_n. Then if A and B commute,

$$A(B\psi_n) = BA\psi_n = a_n(B\psi_n) \tag{5.94}$$

and therefore $(B\psi_n)$ is an eigenfunction of A belonging to the eigenvalue a_n. Since a_n is non-degenerate, $(B\psi_n)$ can only differ from ψ_n by a multiplicative constant which we call b_n

$$B\psi_n = b_n \psi_n. \tag{5.95}$$

Thus we see that ψ_n is simultaneously an eigenfunction of the operators A and B belonging to the eigenvalues a_n and b_n, respectively.

Now consider the case in which a_n is a degenerate eigenvalue of A, of degree α, with corresponding linearly independent eigenfunctions ψ_{nr} ($r = 1, 2, \ldots, \alpha$). Since A and B commute, $(B\psi_{nr})$ is an eigenfunction of A belonging to the degenerate eigenvalue a_n. It follows that $(B\psi_{nr})$ can be expanded in terms of the α linearly independent functions $\psi_{n1}, \psi_{n2}, \ldots, \psi_{n\alpha}$.

$$B\psi_{nr} = \sum_{s=1}^{\alpha} c_{rs} \psi_{ns} \tag{5.96}$$

where c_{rs} are the expansion coefficients. Let us form a linear combination of the functions ψ_{nr} with α constants d_r, so that

$$B \sum_{r=1}^{\alpha} d_r \psi_{nr} = \sum_{r=1}^{\alpha} \sum_{s=1}^{\alpha} d_r c_{rs} \psi_{ns}. \tag{5.97}$$

Therefore, $\sum_r d_r \psi_{nr}$ is an eigenfunction of B belonging to an eigenvalue b_n provided that

$$\sum_{r=1}^{\alpha} d_r c_{rs} = b_n d_s, \qquad s = 1, 2, \ldots, \alpha. \tag{5.98}$$

---

[3] For notational simplicity we only treat here the case of a discrete spectrum.

This is a system of α homogeneous linear equations for the α constants d_r. This system has a non-trivial solution if

$$\det |c_{rs} - b_n \delta_{rs}| = 0 \qquad (5.99)$$

where det means the determinant. This is an equation of order α for b_n, having α roots. Corresponding to each root, $b_n = b_n^{(k)}$, where $k = 1, 2, \ldots, \alpha$, is a solution $d_r^{(k)}$ of (5.98) and we see by construction that

$$\phi_n^{(k)} = \sum_{r=1}^{\alpha} d_r^{(k)} \psi_{nr} \qquad (5.100)$$

is simultaneously an eigenfunction of A, belonging to the eigenvalue a_n, and of B, belonging to the eigenvalue $b_n^{(k)}$. The eigenvalue a_n together with the eigenvalue $b_n^{(k)}$ completely specify a particular simultaneous eigenfunction $\phi_n^{(k)}$ of A and B, so that when both operators are considered together the degeneracy is removed.

The foregoing analysis can be extended to show that if $A, B, C, \ldots$, are a set of commuting observables, then a complete set of simultaneous eigenfunctions of these observables exists. The largest set of commuting observables which can be found (for a given system) is called a *complete set of commuting observables*. In this case the eigenvalues $a_n, b_n, c_n \ldots$, completely specify a simultaneous eigenfunction ψ_n of $A, B, C, \ldots$ (apart from a multiplicative constant), so that the degeneracy is completely removed.

Commutator algebra

The following relations satisfied by commutators are useful and are readily proved (see Problem 5.6)

$$[A, B] = -[B, A] \qquad (5.101\text{a})$$

$$[A, B + C] = [A, B] + [A, C] \qquad (5.101\text{b})$$

$$[A, BC] = [A, B]C + B[A, C] \qquad (5.101\text{c})$$

$$[A, [B, C]] + [B, [C, A]] + [C, [A, B]] = 0 \qquad (5.101\text{d})$$

The Heisenberg uncertainty relations

In Section 2.5 we discussed the Heisenberg uncertainty principle, and found the 'order of magnitude' uncertainty relations (2.71) for the position and momentum of a particle. We shall now obtain a precise form of these uncertainty relations by adopting an accurate definition of the uncertainties $\Delta x, \Delta p_x, \ldots$, which appear in these relations. At the same time, we shall derive a more general expression of the Heisenberg uncertainty relations, which is valid for two canonically conjugate observables A and B, such that $[A, B] = i\hbar$.

Let us consider two observables A and B. Let $\langle A \rangle \equiv \langle \Psi | A | \Psi \rangle$ be the expectation value of A in a given state Ψ (normalised to unity) and let $\langle B \rangle \equiv \langle \Psi | B | \Psi \rangle$ be the expectation value of B in the state Ψ. We define the uncertainty ΔA to be

$$\Delta A = [\langle (A - \langle A \rangle)^2 \rangle]^{1/2} \tag{5.102}$$

so that

$$(\Delta A)^2 = \langle (A - \langle A \rangle)^2 \rangle = \langle A^2 \rangle - \langle A \rangle^2 \tag{5.103}$$

is the mean-square deviation about the expectation value $\langle A \rangle$. Similarly, we define the uncertainty ΔB to be

$$\Delta B = [\langle (B - \langle B \rangle)^2 \rangle]^{1/2}. \tag{5.104}$$

We shall now prove that

$$\Delta A \Delta B \geqslant \frac{1}{2} |\langle [A, B] \rangle|. \tag{5.105}$$

To this end, we first introduce the linear Hermitian operators

$$\bar{A} = A - \langle A \rangle, \qquad \bar{B} = B - \langle B \rangle \tag{5.106}$$

which are such that their expectation values vanish. In terms of these operators, we have

$$(\Delta A)^2 = \langle \bar{A}^2 \rangle, \qquad (\Delta B)^2 = \langle \bar{B}^2 \rangle \tag{5.107}$$

and we also note that

$$[\bar{A}, \bar{B}] = [A - \langle A \rangle, B - \langle B \rangle] = [A, B]. \tag{5.108}$$

Next, we consider the linear (but not Hermitian) operator

$$C = \bar{A} + i\lambda \bar{B} \tag{5.109}$$

where λ is a real constant. The adjoint of C is the operator $C^\dagger = \bar{A} - i\lambda \bar{B}$ and we note that the expectation value of $CC^\dagger$ is real and non-negative, since

$$\langle CC^\dagger \rangle = \langle \Psi | CC^\dagger | \Psi \rangle = \langle C^\dagger \Psi | C^\dagger \Psi \rangle \geqslant 0. \tag{5.110}$$

From (5.109) and (5.110) it follows that the expectation value

$$\langle (\bar{A} + i\lambda \bar{B})(\bar{A} - i\lambda \bar{B}) \rangle = \langle \bar{A}^2 + \lambda^2 \bar{B}^2 - i\lambda [\bar{A}, \bar{B}] \rangle \tag{5.111}$$

is real and non-negative. Using (5.111), (5.107) and (5.108), we see that the function

$$\begin{aligned} f(\lambda) &= \langle \bar{A}^2 \rangle + \lambda^2 \langle \bar{B}^2 \rangle - i\lambda \langle [\bar{A}, \bar{B}] \rangle \\ &= (\Delta A)^2 + \lambda^2 (\Delta B)^2 - i\lambda \langle [A, B] \rangle \end{aligned} \tag{5.112}$$

is also real and non-negative, which implies that $\langle [A, B] \rangle$ is purely imaginary. Now, the function $f(\lambda)$ has a minimum for

$$\lambda_0 = \frac{i}{2} \frac{\langle [A, B] \rangle}{(\Delta B)^2} \tag{5.113}$$

and the value of $f(\lambda)$ at the minimum is

$$f(\lambda_0) = (\Delta A)^2 + \frac{1}{4}\frac{(\langle[A, B]\rangle)^2}{(\Delta B)^2}. \tag{5.114}$$

Since this value is non-negative, we must have

$$(\Delta A)^2(\Delta B)^2 \geqslant -\frac{1}{4}(\langle[A, B]\rangle)^2 \tag{5.115}$$

and the property (5.105) follows by remembering that $\langle[A, B]\rangle$ is purely imaginary.

For two observables which are canonically conjugate, so that $[A, B] = i\hbar$ we have $\langle[A, B]\rangle = i\hbar$ and we therefore deduce from (5.115) that

$$\Delta A \Delta B \geqslant \frac{\hbar}{2}. \tag{5.116}$$

In particular, for the pairs of canonically conjugate variables (x, p_x), (y, p_y) and (z, p_z), we can state the position–momentum uncertainty relations in the precise form

$$\Delta x \Delta p_x \geqslant \frac{\hbar}{2}, \qquad \Delta y \Delta p_y \geqslant \frac{\hbar}{2}, \qquad \Delta z \Delta p_z \geqslant \frac{\hbar}{2} \tag{5.117}$$

with

$$\Delta x = [\langle(x - \langle x\rangle)^2\rangle]^{1/2}, \qquad \Delta p_x = [\langle(p_x - \langle p_x\rangle)^2\rangle]^{1/2} \tag{5.118}$$

and similar definitions for Δy, Δp_y, Δz and Δp_z.

The minimum uncertainty wave packet

It is clear that the equality sign in (5.105), giving the minimum uncertainty product, holds when $\lambda = \lambda_0$ and $C^\dagger \Psi = 0$ (see (5.110) and (5.114)) so that

$$(\bar{A} - i\lambda_0 \bar{B})\Psi = 0. \tag{5.119}$$

This equation can be used to find the wave function Ψ such that the product $\Delta A \Delta B$ takes its minimum value.

As an example, let us consider the motion of a particle in one dimension. Setting $A = x$ and $B = p_x$, the minimum value of the uncertainty product $\Delta x \Delta p_x$ at a definite instant of time (say $t = 0$) is given by

$$\Delta x \Delta p_x = \frac{\hbar}{2}. \tag{5.120}$$

Writing $\psi(x) \equiv \Psi(x, t = 0)$, and using (5.119) with $\bar{A} = x - \langle x\rangle$, $\bar{B} = p_x - \langle p_x\rangle$ and $\lambda_0 = i\langle[x, p_x]\rangle/[2(\Delta p_x)^2] = -\hbar/[2(\Delta p_x)^2]$, we find that

$$\left(-i\hbar\frac{d}{dx} - \langle p_x\rangle\right)\psi(x) = \frac{2i(\Delta p_x)^2}{\hbar}(x - \langle x\rangle)\psi(x). \tag{5.121}$$

This first order differential equation is readily integrated to give the *minimum uncertainty wave function*

$$\psi(x) = C \exp\left(\frac{i}{\hbar}\langle p_x \rangle x\right) \exp\left[-\frac{(\Delta p_x)^2 (x - \langle x \rangle)^2}{\hbar^2}\right] \tag{5.122}$$

where C is a normalisation constant. This wave function is seen to be a Gaussian wave packet.

5.5 Unitary transformations

We shall now show that by acting with a unitary operator on the wave function describing a state of a system, one obtains a new wave function which provides a completely equivalent description of this state. The application of a unitary operator to every wave function associated with a system is called a unitary transformation.

Let Ψ and X be two functions and let A be a linear, Hermitian operator such that

$$A\Psi = X. \tag{5.123}$$

Let us apply the unitary transformation U, so that

$$\Psi' = U\Psi, \qquad X' = UX. \tag{5.124}$$

Writing

$$A'\Psi' = X' \tag{5.125}$$

we have

$$A'U\Psi = UX = UA\Psi \tag{5.126}$$

so that

$$A'U = UA. \tag{5.127}$$

Since $UU^\dagger = U^\dagger U = I$ (see (5.45b)), it follows at once from (5.127) that

$$A' = UAU^\dagger, \qquad A = U^\dagger A' U. \tag{5.128}$$

A number of results will now be proved.

(1) If A is Hermitian, then A' is also Hermitian.
   From (5.128) we have

$$A'^\dagger = (UAU^\dagger)^\dagger = UA^\dagger U^\dagger \tag{5.129}$$

and since $A = A^\dagger$, we find that

$$A'^\dagger = UAU^\dagger = A'. \tag{5.130}$$

(2) Operator equations remain unchanged in form.
Consider for example the operator equation

$$A = c_1 B + c_2 C D \tag{5.131}$$

where c_1 and c_2 are complex constants and B, C, D are operators. Using again the fact that $UU^\dagger = U^\dagger U = I$, we have

$$UAU^\dagger = c_1 U B U^\dagger + c_2 U C U^\dagger U D U^\dagger \tag{5.132}$$

or

$$A' = c_1 B' + c_2 C' D' \tag{5.133}$$

where A', B', C' and D' are the transforms of A, B, C and D, respectively. We also note that if A and B are two operators such that $[A, B] = c$, where c is a complex number, and A' and B' are their transforms, then

$$[A, B] = [A', B'] = c. \tag{5.134}$$

It follows that the fundamental commutation relations (5.88) remain unchanged under unitary transformations.

(3) The eigenvalues of A' are the same as those of A.
Indeed, we can rewrite the eigenvalue equation (5.21) as

$$AU^\dagger U \psi_n = a_n U^\dagger U \psi_n. \tag{5.135}$$

Operating from the left throughout with U, we obtain

$$(UAU^\dagger)(U\psi_n) = a_n (UU^\dagger)(U\psi_n) \tag{5.136}$$

so that

$$A' \psi'_n = a_n \psi'_n \tag{5.137}$$

with $\psi'_n = U \psi_n$. Hence $A' = UAU^\dagger$ has the same eigenvalues as A.

(4) The quantities $\langle X|A|\Psi \rangle$ are unchanged by a unitary transformation.
The proof is as follows:

$$\begin{aligned} \langle X|A|\Psi \rangle &= \langle X|U^\dagger U A U^\dagger U|\Psi \rangle \\ &= \langle (UX)|UAU^\dagger|(U\Psi) \rangle \\ &= \langle X'|A'|\Psi' \rangle \end{aligned} \tag{5.138}$$

where in the second line we have used (5.30). A corollary of (5.138) is that expectation values remain unchanged under unitary transformations:

$$\langle \Psi|A|\Psi \rangle = \langle \Psi'|A'|\Psi' \rangle. \tag{5.139}$$

By choosing $A = I$ in (5.138), we also see that

$$\langle X|\Psi \rangle = \langle X'|\Psi' \rangle \tag{5.140}$$

so that the scalar product is invariant under unitary transformations. Consequently, the normalisation is also preserved in such transformations, since

$$\langle \Psi | \Psi \rangle = \langle \Psi' | \Psi' \rangle. \tag{5.141}$$

From the above results we see that physical quantities such as eigenvalues and expectation values can be equally well calculated from the transformed wave functions $\Psi', X', \ldots$, and operators $A', B', \ldots$, as from the original ones. This raises the possibility that a dynamical problem may be easier to solve by finding a suitable unitary transformation leading to a new set of wave functions and operators. It is also worth noting that each different unitary transformation leads to a new form or 'representation' of the operators associated with the basic dynamical variables x, y, z and p_x, p_y, p_z, while the fundamental commutation relations (5.88) remain invariant. This suggests that an alternative formulation of quantum mechanics can be developed by considering the commutation relations (5.88) as one of the basic postulates of the theory[4].

As an example, consider a particle moving in one dimension, for which the wave function in the position representation is $\Psi(x, t)$, the position and momentum operators being given in that representation by $x_{op} = x$ and $(p_x)_{op} = -i\hbar \partial/\partial x$, respectively. The Fourier transform (2.43) of $\Psi(x, t)$ defines the momentum space wave function $\Phi(p_x, t)$, and this transformation is unitary. Indeed, we can write

$$\Phi(p_x, t) = U\Psi(x, t)$$

$$= (2\pi\hbar)^{-1/2} \int_{-\infty}^{+\infty} e^{-ip_x x/\hbar} \Psi(x, t) dx, \tag{5.142}$$

the inverse transformation (2.42) being

$$\Psi(x, t) = U^{-1}\Phi(p_x, t)$$

$$= (2\pi\hbar)^{-1/2} \int_{-\infty}^{+\infty} e^{ip_x x/\hbar} \Phi(p_x, t) dp_x$$

$$= U^\dagger \Phi(p_x, t). \tag{5.143}$$

Thus

$$U^\dagger U \Psi(x, t) = U^\dagger \Phi(p_x, t) = \Psi(x, t) \tag{5.144a}$$

and also

$$UU^\dagger \Phi(p_x, t) = U\Psi(x, t) = \Phi(p_x, t) \tag{5.144b}$$

---

[4] See Dirac (1958).

from which we see that $U^\dagger U = UU^\dagger = I$, so that the integral operator U defined by (5.142) is unitary. By Parseval's theorem (see equation (A.43) of Appendix A)

$$\int_{-\infty}^{+\infty} |\Psi(x,t)|^2 dx = \int_{-\infty}^{+\infty} |\Phi(p_x,t)|^2 dp_x \tag{5.145}$$

which is just a particular case of (5.141).

As we saw in Table 3.2, in the momentum representation the position and momentum operators are given by $x_{\text{op}} = i\hbar \partial/\partial p_x$ and $(p_x)_{\text{op}} = p_x$, respectively. It is straightforward to verify that in both the position and momentum representations the commutation relation $[x_{\text{op}}, (p_x)_{\text{op}}] = i\hbar$ is satisfied, which illustrates the invariance of the basic commutation relations (5.88) under unitary transformations.

Infinitesimal unitary transformations

If the unitary operator U is very close to the unit operator I, the unitary transformation is said to be infinitesimal. We may then write

$$U = I + i\varepsilon F \tag{5.146}$$

where ε is a real, arbitrary small parameter, and F is an operator which must be *Hermitian*. Indeed, from (5.45) and (5.146) we deduce that to first order in ε

$$I = U^\dagger U = (I - i\varepsilon F^\dagger)(I + i\varepsilon F) \simeq I - i\varepsilon F^\dagger + i\varepsilon F \tag{5.147}$$

from which $\varepsilon(F - F^\dagger) = 0$ and

$$F = F^\dagger. \tag{5.148}$$

The operator F is called the *generator* of the infinitesimal unitary transformation. We remark that if an infinitesimal unitary transformation is performed, the transformed wave functions are given by

$$\Psi' \equiv \Psi + \delta\Psi = (I + i\varepsilon F)\Psi \tag{5.149}$$

so that

$$\delta\Psi = i\varepsilon F\Psi \tag{5.150}$$

while from (5.128), (5.146) and (5.148) the transformed operators are

$$\begin{aligned} A' \equiv A + \delta A &= (I + i\varepsilon F)A(I - i\varepsilon F) \\ &= A + i\varepsilon FA - i\varepsilon AF + \mathcal{O}(\varepsilon^2) \\ &= A + i\varepsilon[F, A] + \mathcal{O}(\varepsilon^2) \end{aligned} \tag{5.151}$$

and therefore, to first order in ε,

$$\delta A = i\varepsilon[F, A]. \tag{5.152}$$

5.6 Matrix representations of wave functions and operators

Let us consider a complete set of orthonormal functions $\{\psi_n\}$. To simplify the notation we shall assume that the index n is discrete. Any physical wave function Ψ can be expanded in terms of this set according to equation (5.60), where the coefficients c_n are determined by $c_n = \langle \psi_n | \Psi \rangle$ (see (5.61)). For a given set of functions $\{\psi_n\}$, the numbers c_n specify the wave function Ψ completely. The coefficients c_n are said to *represent* Ψ in the *basis* (or *representation*) $\{\psi_n\}$. We can think of the set of functions $\{\psi_n\}$ as being analogous to a set of orthogonal axes and the numbers c_n as being analogous to the components of a 'vector' Ψ along each of these axes.

The action of a linear and Hermitian operator A on a wave function produces another wave function

$$X = A\Psi. \tag{5.153}$$

The wave function X can also be expanded in terms of the basis $\{\psi_n\}$ as

$$X = \sum_m d_m \psi_m \tag{5.154}$$

where the coefficients of the expansion are given by $d_m = \langle \psi_m | X \rangle$. Using (5.153) and (5.60), we have

$$\begin{aligned} d_m &= \langle \psi_m | X \rangle \\ &= \langle \psi_m | A | \Psi \rangle \\ &= \sum_n \langle \psi_m | A | \psi_n \rangle c_n. \end{aligned} \tag{5.155}$$

The quantities

$$A_{mn} = \langle \psi_m | A | \psi_n \rangle \tag{5.156}$$

are called the *matrix elements* of the operator A in the basis $\{\psi_n\}$. The equation (5.155) can thus be written as

$$d_m = \sum_n A_{mn} c_n. \tag{5.157}$$

It relates the coefficients c_n which uniquely define Ψ to the coefficients d_m which uniquely define X, and is completely equivalent to the operator equation (5.153). Thus the set of matrix elements A_{mn} completely specifies the operator A within the basis $\{\psi_n\}$. We remark that equation (5.157) can be written as a *matrix equation*

$$\mathbf{d} = \mathbf{Ac} \tag{5.158}$$

or

$$\begin{pmatrix} d_1 \\ d_2 \\ \vdots \end{pmatrix} = \begin{pmatrix} A_{11} & A_{12} & \cdots \\ A_{21} & A_{22} & \cdots \\ \vdots & \vdots & \cdots \end{pmatrix} \begin{pmatrix} c_1 \\ c_2 \\ \vdots \end{pmatrix} \tag{5.159}$$

where $\mathbf{d}$ and $\mathbf{c}$ are column vectors and $\mathbf{A}$ is a square matrix. If the basis $\{\psi_n\}$ contains a finite number of functions, the matrix $\mathbf{A}$ is of finite dimension; otherwise $\mathbf{A}$ is of infinite dimension. Clearly there are as many different representations of wave functions and operators as there are basis sets of functions $\{\psi_n\}$.

Using (5.64) and the fact that $c_n = \langle \psi_n | \Psi \rangle$ and $d_n^* = \langle X | \psi_n \rangle$, the scalar product $\langle X | \Psi \rangle$ can be expressed as

$$\langle X | \Psi \rangle = \sum_n d_n^* c_n = \mathbf{d}^\dagger . \mathbf{c} \tag{5.160}$$

where $\mathbf{c}$ is a column vector with elements c_n and $\mathbf{d}^\dagger$ is a row vector with elements d_n^*. Thus (5.160) can be displayed as

$$\langle X | \Psi \rangle = (d_1^*\ d_2^* \cdots) \begin{pmatrix} c_1 \\ c_2 \\ \vdots \end{pmatrix}. \tag{5.161}$$

Matrix properties and definitions

Two matrices $\mathbf{A}$ and $\mathbf{B}$ can be added when they have the same number of rows and the same number of columns. Addition is commutative

$$\mathbf{A} + \mathbf{B} = \mathbf{B} + \mathbf{A} \tag{5.162}$$

and if $\mathbf{C} = \mathbf{A} + \mathbf{B}$ is the matrix sum, then

$$C_{mn} = A_{mn} + B_{mn}. \tag{5.163}$$

If $\mathbf{C} = \mathbf{AB}$ denotes the product of two matrices $\mathbf{A}$ and $\mathbf{B}$, then

$$C_{mn} = \sum_k A_{mk} B_{kn} \tag{5.164}$$

which requires that the number of rows of $\mathbf{B}$ must be equal to the number of columns of $\mathbf{A}$. It follows from (5.163) and (5.164) that multiplications is distributive

$$\mathbf{A}(\mathbf{B} + \mathbf{C}) = \mathbf{AB} + \mathbf{AC} \tag{5.165}$$

and associative

$$\mathbf{A}(\mathbf{BC}) = (\mathbf{AB})\mathbf{C}. \tag{5.166}$$

It is also apparent from (5.164) that, in general, $\mathbf{AB}$ is not equal to $\mathbf{BA}$. If $\mathbf{AB} = \mathbf{BA}$, the two matrices $\mathbf{A}$ and $\mathbf{B}$ are said to commute.

The inverse $\mathbf{A}^{-1}$ of a matrix $\mathbf{A}$ is such that

$$\mathbf{A}\mathbf{A}^{-1} = \mathbf{A}^{-1}\mathbf{A} = \mathbf{I} \tag{5.167}$$

where $\mathbf{I}$ is the unit matrix, having elements $I_{mn} = \delta_{mn}$. A matrix which possesses an inverse is said to be *non-singular*.

The matrix $\mathbf{A}^T$ is the transpose of $\mathbf{A}$ if

$$(A^T)_{mn} = A_{nm} \tag{5.168}$$

and $\mathbf{A}^\dagger$ is the adjoint of $\mathbf{A}$ if

$$(A^\dagger)_{mn} = A_{nm}^*. \tag{5.169}$$

A matrix is Hermitian if it is equal to its adjoint:

$$\mathbf{A} = \mathbf{A}^\dagger \tag{5.170}$$

so that

$$A_{mn} = A_{nm}^*. \tag{5.171}$$

If the inverse of a matrix is equal to its adjoint, this matrix is said to be unitary. Hence, for a unitary matrix $\mathbf{U}$,

$$\mathbf{U}^{-1} = \mathbf{U}^\dagger \tag{5.172a}$$

or

$$\mathbf{U}\mathbf{U}^\dagger = \mathbf{U}^\dagger\mathbf{U} = \mathbf{I}. \tag{5.172b}$$

The trace of a matrix is the sum of its diagonal elements:

$$\mathrm{Tr}\mathbf{A} = \sum_m A_{mm}. \tag{5.173}$$

Using these definitions, it is easy to prove (see Problem 5.10) that a Hermitian operator is represented by a Hermitian matrix, a unitary operator by a unitary matrix, the operator sum $A + B$ by the matrix sum $\mathbf{A} + \mathbf{B}$, and the operator product $C = AB$ by the matrix product $\mathbf{C} = \mathbf{AB}$, so that the algebraic properties of operators are mirrored by their matrix representations. For example, consider the operator product $C = AB$. We have

$$\begin{aligned} C_{mn} &= \langle \psi_m | AB | \psi_n \rangle \\ &= \sum_k \langle \psi_m | A | \psi_k \rangle \langle \psi_k | B | \psi_n \rangle \\ &= \sum_k A_{mk} B_{kn} \end{aligned} \tag{5.174}$$

where in the second line the closure relation (5.65) has been used. Looking back at (5.164), it follows that the matrix $\mathbf{C}$ is equal to the matrix product of $\mathbf{A}$ and $\mathbf{B}$.

A square matrix **A** is said to be *diagonal* if it has non-vanishing elements A_{mn} only when $m = n$. If the complete set of orthonormal eigenfunctions $\{\psi_n\}$ of an observable A is used as a basis, then the matrix **A** representing A in this basis is diagonal, the diagonal elements being the eigenvalues a_n of A. Indeed, since $A\psi_n = a_n \psi_n$, the matrix elements of A in the basis $\{\psi_n\}$ are

$$\begin{aligned} A_{mn} &= \langle \psi_m | A | \psi_n \rangle \\ &= a_n \langle \psi_m | \psi_n \rangle \\ &= a_n \delta_{mn}. \end{aligned} \quad (5.175)$$

Let **A** be the matrix representing the observable A in a given basis. The eigenvalue equation $A\psi_n = a_n\psi_n$ for the operator A becomes, in that representation, the matrix eigenvalue equation

$$\mathbf{A}\mathbf{u}_n = a_n \mathbf{u}_n \quad (5.176)$$

where $\mathbf{u}_n$ is called an *eigenvector* of the matrix **A**, belonging to the eigenvalue a_n. It is often convenient to use the Dirac ket notation and to write the eigenvector $\mathbf{u}_n$ belonging to the eigenvalue a_n as the *eigenket* $|a_n\rangle$. Correspondingly, the row vector $\mathbf{u}_n^\dagger$ will be written as the bra $\langle a_n|$. More generally, a simultaneous eigenvector of the matrices **A**, **B**, **C**, ..., corresponding to eigenvalues $a_n, b_m, c_j, \ldots$, will be denoted by $|a_n, b_m, c_j, \ldots\rangle$.

The eigenvalues a_n of the matrix equation (5.176) are real since **A** is Hermitian; they are solutions of the 'secular' equation

$$\det |\mathbf{A} - a_n \mathbf{I}| = 0. \quad (5.177)$$

The properties of the eigenfunctions ψ_n of A are mirrored by those of the eigenvectors $\mathbf{u}_n$ of **A**. In particular, two eigenvectors belonging to different eigenvalues are orthogonal (see (5.55)). Also, by using the Schmidt orthogonalisation procedure, it is possible to ensure that linearly independent eigenvectors $\mathbf{u}_{ir}$ belonging to degenerate eigenvalues a_i are mutually orthogonal. All eigenvectors can also be normalised to unity. Thus we have (compare with (5.57))

$$\mathbf{u}_{ir}^\dagger \cdot \mathbf{u}_{js} = \delta_{ij}\delta_{rs} \quad (5.178)$$

where the indices r and s refer to the degeneracy. Finally, we remark that in the basis $\{\psi_n\}$ of the eigenfunctions of A, in which the matrix **A** assumes the diagonal form (5.175), all the elements of the normalised eigenvector $\mathbf{u}_n$ are equal to zero, except the nth element, which is unity.

Change of representation and unitary transformations

We have already shown that different representations of quantum mechanical operators and wave functions can be obtained through unitary transformations. In the particular case of transformations between different matrix representations, the transforming operators are unitary matrices, as we shall now see. Suppose that $\{\psi_n\}$

and $\{\phi_m\}$ are two different orthonormal bases. Each member of the set $\{\psi_n\}$ can be expanded in the basis $\{\phi_m\}$ as

$$\psi_n = \sum_m U_{mn}\phi_m \qquad (5.179)$$

where the expansion coefficients U_{mn} are obtained by taking the scalar product of both sides of (5.179) with ϕ_m, and are given by

$$U_{mn} = \langle\phi_m|\psi_n\rangle. \qquad (5.180)$$

We shall now prove that the set of numbers U_{mn} are the elements of a *unitary* matrix. Indeed,

$$\begin{aligned}(UU^\dagger)_{mn} &= \sum_k U_{mk}(U^\dagger)_{kn} \\ &= \sum_k U_{mk}U_{nk}^* \\ &= \sum_k \langle\phi_m|\psi_k\rangle\langle\psi_k|\phi_n\rangle \\ &= \langle\phi_m|\phi_n\rangle = \delta_{mn} \end{aligned} \qquad (5.181a)$$

where we have used the closure relation (5.65). Similarly

$$(U^\dagger U)_{mn} = \delta_{mn} \qquad (5.181b)$$

so that $\mathbf{UU^\dagger = U^\dagger U = I}$, and $\mathbf{U}$ is a unitary matrix.

Suppose now that a wave function Ψ is represented in the basis $\{\psi_n\}$ by the coefficients c_n (forming a column vector $\mathbf{c}$) and in the basis $\{\phi_m\}$ by the coefficients c'_m (forming a column vector $\mathbf{c'}$). That is

$$\Psi = \sum_n c_n\psi_n = \sum_m c'_m\phi_m \qquad (5.182a)$$

with

$$c_n = \langle\psi_n|\Psi\rangle, \qquad c'_m = \langle\phi_m|\Psi\rangle. \qquad (5.182b)$$

Using the closure relation (5.65) and equation (5.180), we have

$$\begin{aligned}c'_m &= \langle\phi_m|\Psi\rangle \\ &= \sum_n \langle\phi_m|\psi_n\rangle\langle\psi_n|\Psi\rangle \\ &= \sum_n U_{mn}c_n,\end{aligned} \qquad (5.183)$$

a result which can be written in matrix form as

$$\mathbf{c' = Uc}. \qquad (5.184)$$

Similarly, if **A** is a matrix representing the operator A in the basis $\{\psi_n\}$ and **A**′ the matrix representing that operator in the basis $\{\phi_m\}$, then

$$\begin{aligned}
A'_{mn} &= \langle \phi_m | A | \phi_n \rangle \\
&= \sum_k \sum_l \langle \phi_m | \psi_k \rangle \langle \psi_k | A | \psi_l \rangle \langle \psi_l | \phi_n \rangle \\
&= \sum_k \sum_l U_{mk} A_{kl} (U^\dagger)_{ln}
\end{aligned} \qquad (5.185)$$

so that

$$\mathbf{A}' = \mathbf{U}\mathbf{A}\mathbf{U}^\dagger \quad \text{and} \quad \mathbf{A} = \mathbf{U}^\dagger \mathbf{A}' \mathbf{U}. \qquad (5.186)$$

We saw earlier that if two Hermitian operators are connected by a unitary transformation they have the same eigenvalues. This property is mirrored for Hermitian matrices, so that the problem of solving the matrix eigenvalue equationn (5.176) can be viewed as that of finding a unitary transformation which converts the Hermitian matrix **A** into a matrix $\mathbf{A}' = \mathbf{U}\mathbf{A}\mathbf{U}^\dagger$ which is diagonal. It is a fundamental theorem of matrix algebra that *any Hermitian matrix can be diagonalised by a unitary transformation*. Another important theorem states that in order that two Hermitian matrices **A** and **B** can be diagonalised by the same unitary transformation, it is necessary and sufficient that they commute (**AB** = **BA**).

Finally, an important property of unitary transformations is that they leave the trace of a matrix unaltered. Indeed, if the matrices **A** and **A**′ are connected by a unitary transformation, so that $\mathbf{A}' = \mathbf{U}\mathbf{A}\mathbf{U}^\dagger$, we have

$$\begin{aligned}
\operatorname{Tr} \mathbf{A}' &= \sum_m A'_{mm} \\
&= \sum_k \sum_l \sum_m U_{mk} A_{kl} (U^\dagger)_{lm} \\
&= \sum_k \sum_l \sum_m [(U^\dagger)_{lm} U_{mk}] A_{kl} \\
&= \sum_k \sum_l \delta_{lk} A_{kl} \\
&= \sum_k A_{kk} = \operatorname{Tr} \mathbf{A}.
\end{aligned} \qquad (5.187)$$

It follows from (5.187), and the fundamental diagonalisation theorem, that the trace of a Hermitian matrix is equal to the sum of its eigenvalues.

The state vector

By now we have seen that the wave functions and operators which appear in quantum mechanics can be represented in many ways, all connected by unitary transformations. This fact led Dirac to suggest that the state of a system could be described by a more abstract quantity called a *state vector*, denoted by a ket $|\Psi\rangle$. With each state vector is associated a conjugate quantity or bra $\langle\Psi|$ such that $\langle\Psi|\Psi\rangle$ is a real number representing the square of the 'length' or norm of $|\Psi\rangle$. The various representations we have studied correspond to specifying the 'components' of $|\Psi\rangle$ along different directions in the abstract space, just as ordinary three-dimensional vectors can be represented by components along any three independent directions. The linear Hermitian operators, corresponding to dynamical variables, act on the abstract state vectors, converting one into another. In Dirac's formulation of quantum theory the fundamental commutation relations play a central role, since these are independent of the choice of basis. Indeed, the superposition principle, together with the commutation relations for the basic operators (x, y, z) and (p_x, p_y, p_z) can be used to determine the physical quantities, such as expectation values or eigenvalues, without introducing a specific basis. In what follows we shall illustrate Dirac's approach by obtaining the energy levels of a linear harmonic oscillator.

The linear harmonic oscillator revisited

Let us consider a one-dimensional harmonic oscillator, the Hamiltonian of which is

$$H = \frac{p_x^2}{2m} + \frac{1}{2}kx^2 = \frac{p_x^2}{2m} + \frac{1}{2}m\omega^2 x^2 \qquad (5.188)$$

with $\omega = (k/m)^{1/2}$. In Section 4.7 we obtained the eigenvalues and eigenfunctions of this Hamiltonian by working in the position representation. We shall now find the eigenvalues of H by using a method due to Dirac, which does not make reference to any particular representation.

Let us introduce the operators

$$a_\pm = \frac{1}{\sqrt{2}}\left[\left(\frac{m\omega}{\hbar}\right)^{1/2} x \mp i\frac{p_x}{(m\hbar\omega)^{1/2}}\right]. \qquad (5.189)$$

Since x and p_x are Hermitian operators, a_+ and a_- are adjoints of each other: $a_+ = a_-^\dagger$ and $a_- = a_+^\dagger$. Using the basic commutation relation $[x, p_x] = i\hbar$, it is readily verified (Problem 5.12) that a_+ and a_- satisfy the commutation relation

$$[a_-, a_+] = 1. \qquad (5.190)$$

In terms of the operators a_+ and a_-, the Hamiltonian (5.188) can be rewritten in either of the forms

$$H = \frac{\hbar\omega}{2}(a_- a_+ + a_+ a_-) = \hbar\omega\left(a_- a_+ - \tfrac{1}{2}\right) = \hbar\omega\left(a_+ a_- + \tfrac{1}{2}\right) = \hbar\omega\left(N + \tfrac{1}{2}\right) \qquad (5.191)$$

where

$$N = a_+ a_-. \qquad (5.192)$$

Furthermore, we have (Problem 5.12)

$$[H, a_\pm] = \pm \hbar\omega a_\pm. \tag{5.193}$$

If $|E\rangle$ is an eigenvector (eigenket) of H belonging to the eigenvalue E, so that

$$H|E\rangle = E|E\rangle \tag{5.194}$$

we have from (5.193) that

$$\begin{aligned} Ha_\pm|E\rangle &= (a_\pm H \pm \hbar\omega a_\pm)|E\rangle \\ &= (E \pm \hbar\omega)a_\pm|E\rangle. \end{aligned} \tag{5.195}$$

It follows that the kets $a_\pm|E\rangle$ are also eigenvectors of H, belonging to the eigenvalues $E \pm \hbar\omega$. As a_+ raises and a_- lowers the value of E, the operators a_+ and a_- are called *raising* and *lowering* operators, respectively. Since H only contains the squares of the operators p_x and x, the expectation value of H in any state cannot be negative, and hence the eigenvalues of H must be non-negative. Let E_0 be the smallest of these eigenvalues, and $|E_0\rangle$ be the corresponding eigenket. We must have

$$a_-|E_0\rangle = 0 \tag{5.196}$$

for otherwise $a_-|E_0\rangle$ would be an eigenket corresponding to the eigenvalue $E_0 - \hbar\omega$, contrary to the hypothesis that E_0 is the lowest eigenvalue. Operating on (5.196) with $\hbar\omega a_+$, and using (5.191) and (5.192), we have

$$\hbar\omega a_+ a_-|E_0\rangle = \hbar\omega N|E_0\rangle = \left(H - \tfrac{1}{2}\hbar\omega\right)|E_0\rangle = 0 \tag{5.197}$$

from which we deduce that the lowest eigenvalue E_0 is given by $E_0 = \hbar\omega/2$. Using (5.195) we see that by operating repeatedly with a_+ on the eigenket $|E_0\rangle$, we obtain the sequence of (unnormalised) eigenkets

$$|E_0\rangle, a_+|E_0\rangle, a_+^2|E_0\rangle, \ldots, \tag{5.198}$$

the eigenket $a_+^n|E_0\rangle$ corresponding to the eigenvalue

$$E_n = \left(n + \tfrac{1}{2}\right)\hbar\omega, \qquad n = 0, 1, 2, \ldots \tag{5.199}$$

Let $|E_n\rangle$ be the normalised eigenket corresponding to the eigenvalue E_n and $|E_{n+1}\rangle$ be that corresponding to the eigenvalue E_{n+1}. From (5.198), we have

$$|E_{n+1}\rangle = C_{n+1} a_+ |E_n\rangle \tag{5.200}$$

where C_{n+1} is a normalisation coefficient. Since $\langle E_{n+1}|E_{n+1}\rangle = 1$ and $a_- = a_+^\dagger$, we find that

$$|C_{n+1}|^2 \langle E_n|a_- a_+|E_n\rangle = 1. \tag{5.201}$$

Now, since $a_- a_+ = (H/\hbar\omega) + 1/2$, and $H|E_n\rangle = E_n|E_n\rangle$, with $E_n = (n+1/2)\hbar\omega$, and remembering that $\langle E_n|E_n\rangle = 1$, we have

$$C_{n+1} = (n+1)^{-1/2} \tag{5.202}$$

where we have taken C_{n+1} to be real. From (5.200) and (5.201) we see that

$$a_+|E_n\rangle = (n+1)^{1/2}|E_{n+1}\rangle. \tag{5.203}$$

By starting from $n = 0$ and using repeatedly the relation (5.203), all the eigenvectors $|E_n\rangle$ can be obtained from $|E_0\rangle$:

$$|E_n\rangle = (n!)^{-1/2} a_+^n |E_0\rangle \tag{5.204}$$

Let $|E_{n-1}\rangle$ be the normalised eigenket corresponding to the eigenvalue E_{n-1}. From (5.200) and (5.202), we have

$$|E_n\rangle = C_n a_+ |E_{n-1}\rangle \tag{5.205}$$

with $C_n = n^{-1/2}$. Operating on both sides of this equation with a_-, we find that

$$a_-|E_n\rangle = n^{-1/2} a_- a_+ |E_{n-1}\rangle \tag{5.206}$$

Since $a_- a_+ = (H/\hbar\omega) + 1/2$ and $H|E_{n-1}\rangle = (n - 1/2)\hbar\omega|E_{n-1}\rangle$, we obtain

$$a_-|E_n\rangle = n^{1/2}|E_{n-1}\rangle \tag{5.207}$$

The operators a_+ and a_- can be used to calculate any property of the system. As an example, let us evaluate the expectation value of x^4 in the ground state $|E_0\rangle$ of the linear harmonic oscillator. From (5.189), we have

$$x = \left(\frac{\hbar}{2m\omega}\right)^{1/2} (a_+ + a_-) \tag{5.208}$$

and therefore

$$\langle E_0|x^4|E_0\rangle = \frac{\hbar^2}{4m^2\omega^2} \langle E_0| a_+^4 + a_+^3 a_- + a_+^2 a_- a_+ + a_+^2 a_-^2$$
$$+ a_+ a_- a_+^2 + a_+ a_- a_+ a_- + a_+ a_-^2 a_+ + a_+ a_-^3$$
$$+ a_- a_+^3 + a_- a_+^2 a_- + a_- a_+ a_- a_+ + a_- a_+ a_-^2$$
$$+ a_-^2 a_+^2 + a_-^2 a_+ a_- + a_-^3 a_+ + a_-^4 |E_0\rangle. \tag{5.209}$$

We now observe that since $a_-|E_0\rangle = 0$, and also $\langle E_0|a_+ = \langle E_0|a_-^\dagger = \langle a_- E_0| = 0$, (5.209) reduces to

$$\langle E_0|x^4|E_0\rangle = \frac{\hbar^2}{4m^2\omega^2} \langle E_0| a_- a_+^3 + a_- a_+ a_- a_+ + a_-^2 a_+^2 + a_-^3 a_+ |E_0\rangle. \tag{5.210}$$

Since we are evaluating a diagonal matrix element, we must also exclude all terms in which the number of operators a_+ is not equal to the number of operators a_-. Hence

$$\langle E_0|x^4|E_0\rangle = \frac{\hbar^2}{4m^2\omega^2} \langle E_0| a_- a_+ a_- a_+ + a_-^2 a_+^2 |E_0\rangle. \tag{5.211}$$

Using the relations (5.203) and (5.207), we have

$$\begin{aligned}
a_-a_+a_-a_+|E_0\rangle &= a_-a_+a_-|E_1\rangle \\
&= a_-a_+|E_0\rangle \\
&= a_-|E_1\rangle \\
&= |E_0\rangle.
\end{aligned} \tag{5.212}$$

Similarly,

$$\begin{aligned}
a_-^2 a_+^2 |E_0\rangle &= a_-^2 a_+|E_1\rangle \\
&= a_-^2 \sqrt{2}|E_2\rangle \\
&= a_- 2|E_1\rangle \\
&= 2|E_0\rangle
\end{aligned} \tag{5.213}$$

and hence

$$\langle E_0|x^4|E_0\rangle = \frac{3\hbar^2}{4m^2\omega^2} \tag{5.214}$$

Matrix representation in the $\{|E_n\rangle\}$ basis

In terms of the orthonormal set $\{|E_n\rangle\}$ with $n = 0, 1, 2, \ldots$, the matrix representation of H is a diagonal matrix with elements $E_n = (n + 1/2)\hbar\omega$, and that of the operator N is a diagonal matrix with elements n. That is

$$\mathbf{H} = \hbar\omega \begin{pmatrix} \frac{1}{2} & 0 & 0 & \cdots \\ 0 & \frac{3}{2} & 0 & \cdots \\ 0 & 0 & \frac{5}{2} & \\ \vdots & \vdots & \vdots & \vdots \end{pmatrix}, \quad \mathbf{N} = \begin{pmatrix} 0 & 0 & 0 & \cdots \\ 0 & 1 & 0 & \cdots \\ 0 & 0 & 2 & \\ \vdots & & & \vdots \end{pmatrix}. \tag{5.215}$$

From (5.200) and the orthogonality of eigenvectors corresponding to different eigenvalues, we have

$$\langle E_k|E_{n+1}\rangle = C_{n+1}\langle E_k|a_+|E_n\rangle = \delta_{k,n+1} \tag{5.216}$$

where the indices k and n can take the values $0, 1, 2, \ldots$. Hence, the matrix elements of a_+ in the $\{|E_n\rangle\}$ representation are, using (5.202),

$$(a_+)_{kn} = (n+1)^{1/2}\delta_{k,n+1} \tag{5.217}$$

and we see that $\mathbf{a}_+$ is a real matrix whose only non-zero elements are those of the diagonal immediately below the main diagonal. Moreover, since $\mathbf{a}_- = \mathbf{a}_+^\dagger$ and $\mathbf{a}_+$ is a real matrix, we find that the matrix elements of $\mathbf{a}_-$ are

$$(a_-)_{kn} = (k+1)^{1/2}\delta_{k+1,n} \tag{5.218}$$

so that $\mathbf{a}_-$ is a real matrix whose only non-zero elements lie on the diagonal immediately above the main diagonal. Explicitly,

$$\mathbf{a}_+ = \begin{pmatrix} 0 & 0 & 0 & \cdots \\ \sqrt{1} & 0 & 0 & \cdots \\ 0 & \sqrt{2} & 0 & \cdots \\ 0 & 0 & \sqrt{3} & \\ \vdots & \vdots & \vdots & \ddots \end{pmatrix}, \quad \mathbf{a}_- = \begin{pmatrix} 0 & \sqrt{1} & 0 & 0 & \cdots \\ 0 & 0 & \sqrt{2} & 0 & \cdots \\ 0 & 0 & 0 & \sqrt{3} & \cdots \\ \vdots & \vdots & \vdots & & \ddots \end{pmatrix}. \tag{5.219}$$

The matrix elements of x and p_x can be found immediately by using (5.189) (see Problem 5.14).

Transition from the $\{|E_n\rangle\}$ to the position representation

We shall now show how the linear harmonic oscillator eigenfunctions, given by (4.168) in the position representation, can be found by using the operators a_+ and a_-. We first recall that in the position representation the operator x is represented by ordinary multiplication by x, while the operator p_x is represented by $-i\hbar \partial/\partial x$ (with $\partial/\partial x \equiv d/dx$ in the present case). Hence, the operators $a_\pm$ introduced in (5.189) become

$$a_\pm = \frac{1}{\sqrt{2}} \left[\left(\frac{m\omega}{\hbar}\right)^{1/2} x \mp \frac{1}{(m\hbar\omega)^{1/2}} \frac{d}{dx} \right] \tag{5.220}$$

or, in terms of the variable $\xi = (m\omega/\hbar)^{1/2} x = \alpha x$

$$a_\pm = \frac{1}{\sqrt{2}} \left(\xi \mp \frac{d}{d\xi} \right). \tag{5.221}$$

Thus, in the position representation, the equation (5.196) becomes

$$\left(\xi + \frac{d}{d\xi} \right) \psi_0(\xi) = 0. \tag{5.222}$$

This equation has the solution

$$\psi_0(\xi) = N_0 e^{-\xi^2/2} \tag{5.223}$$

where N_0 is a constant. Hence

$$\psi_0(x) = N_0 e^{-\alpha^2 x^2/2}. \tag{5.224}$$

If N_0 is chosen to be real and such that $\psi_0(x)$ is normalised to unity, we have

$$N_0 = \left(\frac{\alpha}{\sqrt{\pi}} \right)^{1/2}. \tag{5.225}$$

All the other eigenfunctions can be found by using equation (5.204), which in the position representation becomes

$$\psi_n(\xi) = (n!)^{-1/2} \left[\frac{1}{\sqrt{2}} \left(\xi - \frac{d}{d\xi} \right) \right]^n \psi_0(\xi) \tag{5.226}$$

Using (4.154b), (5.223) and (5.225), we obtain

$$\psi_n(\xi) = \left(\frac{\alpha}{\sqrt{\pi}2^n n!}\right)^{1/2} e^{-\xi^2/2} H_n(\xi) \qquad (5.227)$$

where $H_n(\xi)$ is the Hermite polynomial of order n. Upon returning to the variable x, we see that (5.227) is identical to (4.168).

5.7 The Schrödinger equation and the time evolution of a system

The next postulate of quantum mechanics concerns the time evolution of a quantum system and may be formulated as follows.

> **Postulate 7**
> The time evolution of the wave function of a system is determined by the time-dependent Schrödinger equation
>
> $$i\hbar \frac{\partial}{\partial t} \Psi(t) = H \Psi(t) \qquad (5.228)$$
>
> where H is the Hamiltonian, or total energy operator of the system.

Assuming that the system has a classical analogue, its Hamiltonian operator can be obtained by applying the substitution rule (5.19) in the position representation, or (5.20) in the momentum representation, and carrying out the Hermitisation procedure described in Section 3.3, if necessary.

As a first example, let us consider a system of N spinless particles having masses m_i, position coordinates $\mathbf{r}_i$, momenta $\mathbf{p}_i$, and such that its potential energy $V(\mathbf{r}_1, \ldots, \mathbf{r}_N, t)$ depends only on the positions $\mathbf{r}_i$ and the time t. The (non-relativistic) Hamiltonian operator of this system is given by

$$H = \sum_{i=1}^{N} \frac{\mathbf{p}_i^2}{2m_i} + V(\mathbf{r}_1, \ldots, \mathbf{r}_N, t) \qquad (5.229)$$

where $\mathbf{p}_i = -i\hbar \nabla_i$ if one is working in the position representation.

As a second example, consider a particle of mass m and charge q moving in an electromagnetic field described by a vector potential $\mathbf{A}(\mathbf{r}, t)$ and a scalar potential $\phi(\mathbf{r}, t)$. Its (non-relativistic) classical Hamiltonian can be obtained (Goldstein 1980) by starting from the field free expression $E = \mathbf{p}^2/2m$ between the energy and the momentum of the particle, and making in it the substitutions

$$E \to E - q\phi, \quad \mathbf{p} \to \mathbf{p} - q\mathbf{A}. \qquad (5.230)$$

The resulting Hamiltonian

$$H = \frac{1}{2m}(\mathbf{p} - q\mathbf{A})^2 + q\phi \tag{5.231a}$$

becomes the (non-relativistic) quantum mechanical Hamiltonian operator by replacing the generalised momentum $\mathbf{p}$ by the corresponding momentum operator ($\mathbf{p}_{op} = -i\hbar \nabla$ in the position representation). Applying the 'Hermitisation' procedure discussed in Section 3.3 to the term $-(q/m)\mathbf{A}\cdot\mathbf{p}$, this Hamiltonian operator can be written in the form

$$H = \frac{\mathbf{p}^2}{2m} - \frac{q}{2m}(\mathbf{A}\cdot\mathbf{p} + \mathbf{p}\cdot\mathbf{A}) + \frac{q^2}{2m}\mathbf{A}^2 + q\phi \tag{5.231b}$$

where we have omitted the subscript on the momentum operator in order to simplify the notation.

The evolution operator

Since the Schrödinger equation (5.228) is a first-order differential equation in time, the state vector $\Psi(t)$ is determined for all t once it is specified at any given time t_0. We may therefore introduce an *evolution* operator $U(t, t_0)$ such that

$$\Psi(t) = U(t, t_0)\Psi(t_0) \tag{5.232}$$

with

$$U(t_0, t_0) = I. \tag{5.233}$$

Applying twice the definition (5.232), we also have

$$U(t, t_0) = U(t, t')U(t', t_0) \tag{5.234}$$

and

$$U^{-1}(t, t_0) = U(t_0, t) \tag{5.235}$$

so that the evolution operator exhibits the *group property*.

Substituting (5.232) into the Schrödinger equation (5.228), we see that the evolution operator $U(t, t_0)$ satisfies the equation

$$i\hbar \frac{\partial}{\partial t}U(t, t_0) = HU(t, t_0) \tag{5.236}$$

subject to the initial condition (5.233). We remark that the differential equation (5.236) together with the initial condition (5.233) can be replaced by the integral equation

$$U(t, t_0) = I - \frac{i}{\hbar}\int_{t_0}^{t} HU(t', t_0)dt'. \tag{5.237}$$

Conservation of probability requires that

$$\langle \Psi(t)|\Psi(t)\rangle = \langle \Psi(t_0)|\Psi(t_0)\rangle. \tag{5.238}$$

However, from (5.232)

$$\begin{aligned}\langle\Psi(t)|\Psi(t)\rangle &= \langle U(t,t_0)\Psi(t_0)|U(t,t_0)\Psi(t_0)\rangle \\ &= \langle\Psi(t_0)|U^\dagger(t,t_0)U(t,t_0)|\Psi(t_0)\rangle\end{aligned} \qquad (5.239)$$

from which

$$U^\dagger(t,t_0)U(t,t_0) = I. \qquad (5.240a)$$

Likewise, starting from $\langle\Psi(t_0)|\Psi(t_0)\rangle$, we obtain

$$U(t,t_0)U^\dagger(t,t_0) = I \qquad (5.240b)$$

so that from these two equations we conclude that $U(t,t_0)$ is a *unitary* operator.

The unitary character of the evolution operator is clearly connected to the Hermiticity of the Hamiltonian. We can display the connection by studying the change in the evolution operator induced after an arbitrary small time δt. We first have from (5.236)

$$i\hbar[U(t_0+\delta t, t_0) - U(t_0,t_0)] = HU(t_0+\delta t, t_0)\delta t. \qquad (5.241)$$

Hence, to first order in δt, and using the initial condition (5.233), we have

$$U(t_0+\delta t, t_0) = I - \frac{i}{\hbar}H\delta t. \qquad (5.242)$$

The Hamiltonian H is therefore the generator of an infinitesimal unitary transformation (see (5.146)) – in fact, an infinitesimal *time translation* – described by the evolution operator $U(t_0+\delta t, t_0)$.

Let us consider the particular case for which the Hamiltonian H is time-independent. A solution of (5.236) satisfying the initial condition (5.233) is then given by

$$U(t,t_0) = \exp\left[-\frac{i}{\hbar}H(t-t_0)\right] \qquad (5.243)$$

as is readily checked by recollecting that a function of an operator can be defined by its series expansion (see (5.38)) so that

$$\exp\left[-\frac{i}{\hbar}H(t-t_0)\right] = \sum_{n=0}^\infty \frac{1}{n!}\left(-\frac{i}{\hbar}\right)^n H^n(t-t_0)^n. \qquad (5.244)$$

Thus a formal solution of the time-dependent Schrödinger equation for a time-independent Hamiltonian H is given by

$$\Psi(t) = \exp\left[-\frac{i}{\hbar}H(t-t_0)\right]\Psi(t_0). \qquad (5.245)$$

As an example, suppose that we are dealing with the motion of a structureless particle in a time-independent potential $V(\mathbf{r})$. From (5.245) we may write the wave function $\Psi(\mathbf{r}, t)$ as

$$\Psi(\mathbf{r}, t) = \exp\left[-\frac{i}{\hbar}H(t - t_0)\right]\Psi(\mathbf{r}, t_0)$$

$$= \int \exp\left[-\frac{i}{\hbar}H(t - t_0)\right]\delta(\mathbf{r} - \mathbf{r}')\Psi(\mathbf{r}', t_0)\mathrm{d}\mathbf{r}'. \tag{5.246}$$

Now, according to the closure relation satisfied by the energy eigenfunctions, we have

$$\sum_E \psi_E^*(\mathbf{r}')\psi_E(\mathbf{r}) = \delta(\mathbf{r} - \mathbf{r}') \tag{5.247}$$

so that we can recast (5.246) in the form

$$\Psi(\mathbf{r}, t) = \sum_E \int \exp\left[-\frac{i}{\hbar}H(t - t_0)\right]\psi_E^*(\mathbf{r}')\psi_E(\mathbf{r})\Psi(\mathbf{r}', t_0)\mathrm{d}\mathbf{r}'. \tag{5.248}$$

Since $H\psi_E = E\psi_E$, it follows that

$$\Psi(\mathbf{r}, t) = \sum_E \left[\int \psi_E^*(\mathbf{r}')\Psi(\mathbf{r}', t_0)\mathrm{d}\mathbf{r}'\right]\exp[-iE(t - t_0)/\hbar]\psi_E(\mathbf{r}) \tag{5.249}$$

in agreement with the result (3.155) obtained in Chapter 3.

Time variation of expectation values

Let us consider an observable A. The expectation value $\langle A \rangle$ of this observable in the state Ψ, normalised to unity, is given by $\langle \Psi|A|\Psi \rangle$. The rate of change of this expectation value is therefore

$$\frac{\mathrm{d}}{\mathrm{d}t}\langle A \rangle = \frac{\mathrm{d}}{\mathrm{d}t}\langle \Psi|A|\Psi \rangle$$

$$= \left\langle \frac{\partial \Psi}{\partial t}\Big|A|\Psi \right\rangle + \left\langle \Psi\Big|\frac{\partial A}{\partial t}\Big|\Psi \right\rangle + \left\langle \Psi\Big|A\Big|\frac{\partial \Psi}{\partial t} \right\rangle$$

$$= -(i\hbar)^{-1}\langle H\Psi|A|\Psi \rangle + \left\langle \Psi\Big|\frac{\partial A}{\partial t}\Big|\Psi \right\rangle + (i\hbar)^{-1}\langle \Psi|AH|\Psi \rangle \tag{5.250}$$

where in writing the last line we have used the Schrödinger equation (5.228) and its complex conjugate. Since H is Hermitian, the first matrix element on the right of (5.250) may be written as $\langle \Psi|HA|\Psi \rangle$ so that

$$\frac{\mathrm{d}}{\mathrm{d}t}\langle A \rangle = (i\hbar)^{-1}\langle [A, H] \rangle + \left\langle \frac{\partial A}{\partial t} \right\rangle \tag{5.251}$$

where

$$\langle [A, H] \rangle = \langle \Psi|[A, H]|\Psi \rangle = \langle \Psi|AH - HA|\Psi \rangle \tag{5.252}$$

and

$$\left\langle \frac{\partial A}{\partial t} \right\rangle = \left\langle \Psi\Big|\frac{\partial A}{\partial t}\Big|\Psi \right\rangle. \tag{5.253}$$

In particular, if the operator A does not depend explicitly on time (that is, if $\partial A/\partial t = 0$) equation (5.251) reduces to

$$\frac{d}{dt}\langle A \rangle = (i\hbar)^{-1}\langle [A, H] \rangle. \tag{5.254}$$

Hence, if $\partial A/\partial t = 0$ and the operator A commutes with H, its expectation value does not vary in time, and we can say that *the observable A is a constant of the motion.*

Time-independent Hamiltonian

As an example of the preceding discussion, consider a system for which the Hamiltonian is *time-independent* ($\partial H/\partial t = 0$). Using (5.254) with $A = H$, we see that

$$\frac{d}{dt}\langle H \rangle = (i\hbar)^{-1}\langle [H, H] \rangle = 0 \tag{5.255}$$

so that the total energy is a constant of the motion. This is the analogue of energy conservation for conservative system in classical mechanics.

Let ψ_E be an eigenfunction of the time-independent Hamiltonian H corresponding to the eigenenergy E. For a *stationary state* $\Psi_E = \psi_E \exp(-iEt/\hbar)$ and a *time-independent* operator A it is clear that the expectation value $\langle \Psi_E | A | \Psi_E \rangle = \langle \psi_E | A | \psi_E \rangle$ does not depend on the time. Hence in this case (5.254) reduces to

$$\langle \psi_E | [A, H] | \psi_E \rangle = 0. \tag{5.256}$$

The virial theorem

We now apply the above result to the particular case of a particle of mass m moving in a potential $V(\mathbf{r})$, so that the corresponding Hamiltonian is

$$H = \frac{\mathbf{p}^2}{2m} + V(\mathbf{r}). \tag{5.257}$$

Moreover, we choose A to be the time-independent operator $\mathbf{r}.\mathbf{p}$. We then have from (5.256)

$$\langle \psi_E | [\mathbf{r}.\mathbf{p}, H] | \psi_E \rangle = 0. \tag{5.258}$$

Using the algebraic properties (5.101) of the commutators, together with the fundamental commutation relations (5.88) and the fact that $\mathbf{p} = -i\hbar \nabla$ in the position representation, we find that

$$\begin{aligned}
[\mathbf{r}.\mathbf{p}, H] &= \left[(xp_x + yp_y + zp_z), \frac{1}{2m}(p_x^2 + p_y^2 + p_z^2) + V(x, y, z) \right] \\
&= \frac{i\hbar}{m}(p_x^2 + p_y^2 + p_z^2) - i\hbar\left(x\frac{\partial V}{\partial x} + y\frac{\partial V}{\partial y} + z\frac{\partial V}{\partial z} \right) \\
&= 2i\hbar T - i\hbar(\mathbf{r}.\nabla V)
\end{aligned} \tag{5.259}$$

where $T = \mathbf{p}^2/2m = -(\hbar^2/2m)\nabla^2$ is the kinetic energy operator. From (5.258) and (5.259) we therefore deduce that, for a *stationary state*

$$2\langle T \rangle = \langle \mathbf{r} \cdot \nabla V \rangle \tag{5.260}$$

which is known as the *virial theorem*[5]. We remark that the result (5.260) may also be obtained by choosing the operator A to be $\mathbf{p} \cdot \mathbf{r}$ instead of $\mathbf{r} \cdot \mathbf{p}$. Indeed, the difference between $\mathbf{r} \cdot \mathbf{p}$ and $\mathbf{p} \cdot \mathbf{r}$ is a constant, and hence commutes with H. We also note that if the interaction potential V is spherically symmetric and proportional to r^s, one has for a stationary state

$$2\langle T \rangle = \left\langle r \frac{\partial V}{\partial r} \right\rangle$$
$$= s\langle V \rangle, \tag{5.261}$$

The generalisation of the virial theorem (5.260) to a system of N particles is the subject of Problem 5.16.

The Schrödinger equation for a two-body system

As an example of the time evolution of a system we shall consider the case of two particles, of masses m_1 and m_2, interacting via a time-independent potential $V(\mathbf{r}_1 - \mathbf{r}_2)$ which depends only upon the relative coordinate $\mathbf{r}_1 - \mathbf{r}_2$. The classical Hamiltonian of the system is therefore given by

$$H = \frac{p_1^2}{2m_1} + \frac{p_2^2}{2m_2} + V(\mathbf{r}_1 - \mathbf{r}_2). \tag{5.262}$$

Making the substitutions $\mathbf{p}_1 \to -i\hbar \nabla_{\mathbf{r}_1}$ and $\mathbf{p}_2 \to -i\hbar \nabla_{\mathbf{r}_2}$, we obtain the quantum mechanical Hamiltonian operator, and the corresponding time-dependent Schrödinger equation reads in configuration space

$$i\hbar \frac{\partial}{\partial t} \Psi(\mathbf{r}_1, \mathbf{r}_2, t) = \left[-\frac{\hbar^2}{2m_1} \nabla_{\mathbf{r}_1}^2 - \frac{\hbar^2}{2m_2} \nabla_{\mathbf{r}_2}^2 + V(\mathbf{r}_1 - \mathbf{r}_2) \right] \Psi(\mathbf{r}_1, \mathbf{r}_2, t). \tag{5.263}$$

---

[5] In classical mechanics the *virial* of a particle is defined as the quantity $-(1/2)\overline{\mathbf{F} \cdot \mathbf{r}}$, where $\mathbf{F}$ is the force acting on the particle and the bar denotes a *time average*. If the motion is periodic (or even if the motion is not periodic, but the coordinates and velocity of the particle remain finite) and $\bar{T}$ denotes the time average of the kinetic energy of the particle, one has

$$\bar{T} = -\frac{1}{2}\overline{\mathbf{F} \cdot \mathbf{r}}$$

and this relation is known as the *virial theorem*. If the force is derivable from a potential V the virial theorem becomes

$$2\bar{T} = \overline{\mathbf{r} \cdot \nabla V}$$

which is the classical analogue of (5.260).

This is a seven-dimensional partial-differential equation. However, just as in classical mechanics, the problem can be reduced by introducing the *relative coordinate*

$$\mathbf{r} = \mathbf{r}_1 - \mathbf{r}_2 \tag{5.264}$$

and the vector

$$\mathbf{R} = \frac{m_1 \mathbf{r}_1 + m_2 \mathbf{r}_2}{m_1 + m_2} \tag{5.265}$$

which determines the position of the *centre of mass* (CM) of the system. Changing variables from the coordinates $(\mathbf{r}_1, \mathbf{r}_2)$ to the new coordinates $(\mathbf{r}, \mathbf{R})$, we find after a straightforward calculation (Problem 5.18) that

$$-\frac{\hbar^2}{2m_1}\nabla_{\mathbf{r}_1}^2 - \frac{\hbar^2}{2m_2}\nabla_{\mathbf{r}_2}^2 = -\frac{\hbar^2}{2M}\nabla_{\mathbf{R}}^2 - \frac{\hbar^2}{2\mu}\nabla_{\mathbf{r}}^2 \tag{5.266}$$

where

$$M = m_1 + m_2 \tag{5.267}$$

is the *total mass* of the system and

$$\mu = \frac{m_1 m_2}{m_1 + m_2} \tag{5.268}$$

is the *reduced mass* of the two particles. The Schrödinger equation (5.263) therefore becomes

$$i\hbar \frac{\partial}{\partial t}\Psi(\mathbf{R}, \mathbf{r}, t) = \left[-\frac{\hbar^2}{2M}\nabla_{\mathbf{R}}^2 - \frac{\hbar^2}{2\mu}\nabla_{\mathbf{r}}^2 + V(\mathbf{r})\right]\Psi(\mathbf{R}, \mathbf{r}, t). \tag{5.269}$$

This equation can also be obtained by introducing the *relative momentum*

$$\mathbf{p} = \frac{m_2 \mathbf{p}_1 - m_1 \mathbf{p}_2}{m_1 + m_2} \tag{5.270}$$

together with the *total momentum*

$$\mathbf{P} = \mathbf{p}_1 + \mathbf{p}_2. \tag{5.271}$$

Since

$$\frac{\mathbf{p}_1^2}{2m_1} + \frac{\mathbf{p}_2^2}{2m_2} = \frac{\mathbf{P}^2}{2M} + \frac{\mathbf{p}^2}{2\mu} \tag{5.272}$$

the classical Hamiltonian (5.262) can be written as

$$H = \frac{\mathbf{P}^2}{2M} + \frac{\mathbf{p}^2}{2\mu} + V(\mathbf{r}). \tag{5.273}$$

Performing the substitutions $\mathbf{P} \to -i\hbar \nabla_{\mathbf{R}}$ and $\mathbf{p} \to -i\hbar \nabla_{\mathbf{r}}$ in (5.273), we then obtain the quantum mechanical Hamiltonian operator

$$H = -\frac{\hbar^2}{2M}\nabla_{\mathbf{R}}^2 - \frac{\hbar^2}{2\mu}\nabla_{\mathbf{r}}^2 + V(\mathbf{r}) \tag{5.274}$$

leading to the Schrödinger equation (5.269).

Two separations of the equation (5.269) can now be made. The time dependence can first be separated as in Section 3.5 since the potential $V(\mathbf{r})$ is independent of time. Second, the spatial part of the wave function $\Psi(\mathbf{R}, \mathbf{r}, t)$ can be separated into a product of functions of the centre of mass coordinate $\mathbf{R}$ and of the relative coordinate $\mathbf{r}$. Thus the Schrödinger equation (5.269) admits solutions of the form

$$\Psi(\mathbf{R}, \mathbf{r}, t) = \Phi(\mathbf{R})\psi(\mathbf{r}) \exp[-i(E_{CM} + E)t/\hbar] \tag{5.275}$$

where the functions $\Phi(\mathbf{R})$ and $\psi(\mathbf{r})$ satisfy, respectively, the equations

$$-\frac{\hbar^2}{2M}\nabla_{\mathbf{R}}^2 \Phi(\mathbf{R}) = E_{CM}\Phi(\mathbf{R}) \tag{5.276}$$

and

$$\left[-\frac{\hbar^2}{2\mu}\nabla_{\mathbf{r}}^2 + V(\mathbf{r})\right]\psi(\mathbf{r}) = E\psi(\mathbf{r}). \tag{5.277}$$

We see that the equation (5.276) is a time-independent Schrödinger equation describing the centre of mass as a free particle of mass M and energy E_{CM}. The second time-independent Schrödinger equation (5.277) describes the relative motion of the two particles; it is the same as the equation corresponding to the motion of a particle having the reduced mass μ in the potential $V(\mathbf{r})$. The total energy of the system is clearly

$$E_{\text{tot}} = E_{CM} + E. \tag{5.278}$$

We have therefore 'decoupled' the original two-body problem into two one-body problems: that of a free particle (the centre of mass) and that of a single particle of reduced mass μ in the potential $V(\mathbf{r})$. We remark that if we elect to work in the centre-of-mass system of the two particles, we need not be concerned with the motion of the centre of mass, the coordinates of which are eliminated.

5.8 The Schrödinger and Heisenberg pictures

Although there are many representations of wave functions and observables connected by unitary transformations, it is useful to distinguish certain classes of representations, called *pictures*, which differ in the way the time evolution of the system is treated.

The *Schrödinger picture*, which we have used until now, is one in which the operators (either in differential or matrix form) representing the position and momentum variables $\mathbf{r}_i$ and $\mathbf{p}_i$, are *time-independent*. The time evolution of a system is determined in this picture by a *time-dependent* wave function $\Psi(t)$ satisfying the Schrödinger equation (5.228). According to (5.232), the wave function $\Psi(t)$ is related to its value at time t_0 by performing the unitary transformation $\Psi(t) = U(t, t_0)\Psi(t_0)$, where $U(t, t_0)$ is the evolution operator. The time dependence of the expectation values of the basic time-independent operators $\mathbf{r}_i$ and $\mathbf{p}_i$ is given by (5.254). On the other hand, the rate of change of the expectation value of an operator which depends explicitly on time (such as a time-dependent potential $V(\mathbf{r}, t)$) is determined by equation (5.251).

The *Heisenberg picture* is obtained from the Schrödinger picture by applying to the Schrödinger wave function $\Psi(t)$ the unitary operator $U^\dagger(t, t_0) = U(t_0, t)$. The resulting wave function is the Heisenberg wave function (or state function) Ψ_H such that

$$\Psi_H = U^\dagger(t, t_0)\Psi(t) = U(t_0, t)\Psi(t) = \Psi(t_0). \tag{5.279}$$

Hence, in the Heisenberg picture, the wave function Ψ_H is *time-independent* and coincides at some particular fixed time t_0 with the Schrödinger wave function $\Psi(t_0)$. Using (5.279) and (5.128) we see that if A is an operator in the Schrödinger picture and A_H is the corresponding operator in the Heisenberg picture, we have

$$\begin{aligned} A_H(t) &= U^\dagger(t, t_0) A U(t, t_0) \\ &= U(t_0, t) A U^\dagger(t_0, t) \end{aligned} \tag{5.280}$$

and we note that $A_H(t)$ is time-dependent even if A does not depend on time.

The time variation of $A_H(t)$ can be determined as follows. Writing $U \equiv U(t_0, t)$, we first have from (5.280)

$$\frac{d}{dt} A_H(t) = \frac{\partial U}{\partial t} A U^\dagger + U \frac{\partial A}{\partial t} U^\dagger + U A \frac{\partial U^\dagger}{\partial t}. \tag{5.281}$$

Using (5.236) and the facts that H is Hermitian and U is unitary, we then find that

$$\begin{aligned} \frac{d}{dt} A_H(t) &= (i\hbar)^{-1}(-UHAU^\dagger + UAHU^\dagger) + U\frac{\partial A}{\partial t}U^\dagger \\ &= (i\hbar)^{-1}(-UHU^\dagger UAU^\dagger + UAU^\dagger UHU^\dagger) + U\frac{\partial A}{\partial t}U^\dagger. \end{aligned} \tag{5.282}$$

Defining the Heisenberg operators

$$H_H = UHU^\dagger \tag{5.283}$$

and

$$\left(\frac{\partial A}{\partial t}\right)_H = U\frac{\partial A}{\partial t}U^\dagger \tag{5.284}$$

we obtain

$$\frac{d}{dt} A_H(t) = (i\hbar)^{-1}[A_H, H_H] + \left(\frac{\partial A}{\partial t}\right)_H \tag{5.285}$$

which is the *Heisenberg equation of motion* for the operator A_H.

As an example, let us consider the case of the x-components of the position and momentum of a particle, x and p_x. Since these operators are time-independent in the Schrödinger picture ($\partial x/\partial t = \partial p_x/\partial t = 0$), we see from (5.284) and (5.285) that

$$\frac{dx_H}{dt} = (i\hbar)^{-1}[x_H, H_H] \tag{5.286a}$$

and
$$\frac{d(p_x)_H}{dt} = (i\hbar)^{-1}[(p_x)_H, H_H]. \tag{5.286b}$$

Using the fact that the basic commutation relations (5.88) are unchanged by unitary transformations, together with the results of Problem 5.7, one has

$$\frac{dx_H}{dt} = \frac{\partial H_H}{\partial (p_x)_H} \tag{5.287a}$$

and

$$\frac{d(p_x)_H}{dt} = -\frac{\partial H_H}{\partial x_H} \tag{5.287b}$$

which are formally identical to Hamilton's canonical equations in classical mechanics. Thus we see that the Heisenberg picture corresponds to a formulation of quantum dynamics which is formally close to classical mechanics.

It is clear that since the Schrödinger and Heisenberg pictures are related by a unitary transformation, all physical quantities such as eigenvalues and expectation values of observables are identical when calculated in either picture.

We also remark that if the Hamiltonian operator H in the Schrödinger picture is *time-independent*, the evolution operator in that picture is given by (5.243). The Heisenberg wave function Ψ_H is then related to the Schrödinger wave function by

$$\Psi_H \equiv \Psi(t_0) = \exp\left[\frac{i}{\hbar}H(t-t_0)\right]\Psi(t) \tag{5.288}$$

and we see from (5.283) that $H_H = H$.

It is sometimes useful to define other pictures in which a unitary transformation is applied to the Schrödinger wave function $\Psi(t)$ using not the full Hamiltonian H as in (5.288), but part of it. Such pictures are called *interaction pictures*.

5.9 Path integrals

As we saw in Section 3.1, the Schrödinger equation can be obtained from the classical Hamiltonian function by replacing the coordinates and momenta by the appropriate operators. An alternative formulation of quantum mechanics, developed by R. P. Feynman, is closely related to the classical Lagrangian function, and provides new insights into the structure of quantum mechanics and its classical limit. It has also provided a very useful framework in which a large variety of problems occurring in high-energy and statistical physics can be studied. To avoid a complicated notation Feynman's formulation, based on *path integrals*, will be discussed for the motion in one dimension of a particle of mass m moving in a time-independent potential $V(x)$. The classical Lagrangian function for this system is

$$L = T - V(x) \tag{5.289}$$

where T is the kinetic energy

$$T = \frac{1}{2}m\dot{x}^2 \tag{5.290}$$

and the dot denotes a derivative with respect to time. The equation of motion written in terms of L, is the *Euler–Lagrange* equation[6]

$$\frac{d}{dt}\left(\frac{\partial L}{\partial \dot{x}}\right) - \frac{\partial L}{\partial x} = 0 \tag{5.291}$$

and the solution of this second order equation, with boundary conditions $x(t_0) = a$, $\dot{x}(t_0) = b$, determines the classical path

$$x = x(t). \tag{5.292}$$

The equation of motion (5.291) follows from Hamilton's principle which states that the motion of the system from time t_0 to t_1 is such that the *action I* is stationary, where I is defined as

$$I(t_1, t_0) = \int_{t_0}^{t_1} L(\dot{x}(t), x(t)) dt, \tag{5.293}$$

and the integral is taken along the classical path (5.292).

The stationary condition means that if the line integral (5.293) is taken along a neighbouring path to (5.292) so that

$$x = x(t) + \epsilon \eta(t) \tag{5.294}$$

where ϵ is a small quantity and $\eta(t)$ is an arbitary function such that

$$\eta(t_0) = \eta(t_1) = 0, \tag{5.295}$$

then the change in I is of the order ϵ^2.

The Schrödinger wave function $\Psi(x, t_1)$ satisfies the equation (3.157), which for one-dimensional motion is

$$\Psi(x, t_1) = \int_{-\infty}^{\infty} K(x, t_1; x', t_0) \Psi(x', t_0) dx'. \tag{5.296}$$

Feynman showed that the propagator K could be expressed as

$$K(x, t_1; x', t_0) = \sum_p W_p \exp[iI_p(t_1, t_0)/\hbar] \tag{5.297}$$

where $I_p(t_1, t_0)$ is the classical action (5.293), in which the integration is taken along a path $x = x_p(t)$, the sum over p is over *all* paths $x_p(t)$ which connect $x(t_1)$ with $x(t_0)$, and W_p is a weighting factor (see Fig. 5.1). Since there is a continuum of such paths the sum over p represents a kind of integral called a path integral, which will be defined later.

---

[6] See for example Goldstein (1980) or Kibble and Berkshire (1996).

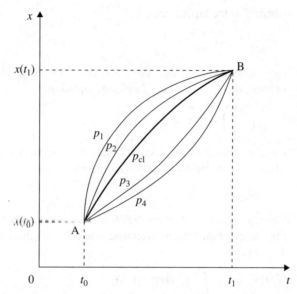

Figure 5.1 Different paths p_1, p_2 ... linking the end points A and B for motion in one dimension between times t_0 and t_1. The classical path is marked as p_{cl}.

For a particle with a wavelength which is short compared with the range of the potential, $I/\hbar$ is a very large quantity. The change in I between neighbouring paths is also large, so that the exponential in (5.297) oscillates rapidly with the consequence that the sum over paths averages to zero. The exceptional case is when I is stationary, so that the change in I is zero between neighbouring paths and a finite contribution to the sum results. This only occurs for the classical path p_{cl}, which is the only significant path for a macroscopic system. In contrast, for a microscopic particle, such as an electron moving in a potential of atomic dimensions, $I/\hbar$ is not large, so that many paths contribute to the propagator and the classical path is no longer dominant.

To derive equation (5.297), we write the propagator K in the form

$$K(x, t_1; x', t_0) = \exp\left(-\frac{i}{\hbar}H(t_1 - t_0)\right)\delta(x - x') \tag{5.298}$$

which is obtained by comparing (5.245) and (5.296). In the present case the Hamiltonian H is

$$H = \frac{p_x^2}{2m} + V(x). \tag{5.299}$$

The time interval $(t_1 - t_0)$ can be divided into N equal segments of width $\Delta t = (t_1 - t_0)/N$, so that the evolution operator $\exp[-iH(t_1 - t_0)/\hbar]$ can be expressed as a product of N terms:

$$\exp\left(-\frac{i}{\hbar}H(t_1-t_0)\right) = \exp\left(-\frac{i}{\hbar}HN\Delta t\right)$$

$$= \left[\exp\left(-\frac{i}{\hbar}H\Delta t\right)\right]^N$$

$$= \exp\left(-\frac{i}{\hbar}H\Delta t\right)\exp\left(-\frac{i}{\hbar}H\Delta t\right)\ldots\exp\left(-\frac{i}{\hbar}H\Delta t\right). \quad (5.300)$$

Using this expression in (5.298) and inserting a delta function between each factor of $\exp(-iH\Delta t/\hbar)$, the propagator K can be put in the form

$$K(x_N, t_1; x_0, t_0) = \int_{-\infty}^{\infty} dx_1 \ldots \int_{-\infty}^{\infty} dx_{N-1} \exp\left(-\frac{i}{\hbar}H\Delta t\right)\delta(x_N - x_{N-1})$$

$$\times \exp\left(-\frac{i}{\hbar}H\Delta t\right)\delta(x_{N-1} - x_{N-2})\ldots$$

$$\times \delta(x_2 - x_1)\exp\left(-\frac{i}{\hbar}H\Delta t\right)\delta(x_1 - x_0) \quad (5.301)$$

where we have set $x = x_N$ and $x' = x_0$ for notational convenience.

By taking N to be large so that Δt is small and $(\Delta t)^2 \ll \Delta t$, it can be seen using (5.244) that

$$\exp\left(-\frac{i}{\hbar}H\Delta t\right) \simeq \exp\left[-\frac{i}{\hbar}\frac{p_x^2}{2m}\Delta t\right]\exp\left(-\frac{i}{\hbar}V(x)\Delta t\right) \quad (5.302)$$

This relation is correct up to terms of order $(\Delta t)^2$ and is exact in the limit $N \to \infty$, that is $\Delta t \to 0$. Each of the delta functions in (5.301) can be expressed by the relation (A.18) of appendix A in the form

$$\delta(x_n - x_{n-1}) = (2\pi)^{-1}\int_{-\infty}^{\infty} dk \exp[ik(x_n - x_{n-1})], \quad n = 1, 2, \ldots N \quad (5.303)$$

Since the plane wave $\exp[ik(x_n - x_{n-1})]$ is an eigenfunction of the kinetic energy operator $p_x^2/2m$ belonging to the eigenvalue $\hbar^2 k^2/2m$, we have from (5.302) that in the limit of small Δt

$$\exp\left(-\frac{i}{\hbar}H\Delta t\right)\delta(x_n - x_{n-1})$$

$$= (2\pi)^{-1}\int_{-\infty}^{\infty} dk \exp\left(-i\frac{\hbar k^2}{2m}\Delta t + ik(x_n + x_{n-1})\right)$$

$$\times \exp\left(-\frac{i}{\hbar}V(x_{n-1})\Delta t\right). \quad (5.304)$$

Using the integral (2.48), which holds for imaginary values of α and β, we find with $\alpha = i\hbar\Delta t/(2m)$ and $\beta = i(x_n - x_{n-1})$ that

$$\exp\left(-\frac{i}{\hbar}H\Delta t\right)\delta(x_n - x_{n-1})$$
$$= \left(\frac{m}{2\pi i\hbar\Delta t}\right)^{1/2} \exp\left[\frac{im(x_n - x_{n-1})^2}{2\hbar\Delta t}\right] \exp\left[-\frac{i}{\hbar}V(x_{n-1})\Delta t\right]. \quad (5.305)$$

It should be noted that when $V = 0$, this expression reduces to the free particle propagator which expresses $\Psi(x_n, t_0+\Delta t)$ in terms of $\Psi(x_{n-1}, t_0)$ (see Problem 3.11).

From (5.301), (5.304) and (5.305) and taking the limit $N \to \infty$, we obtain an exact expression for the propagator K

$$K(x_N, t_1; x_0, t_0) = \lim_{N\to\infty}\left[\frac{m}{2\pi i\hbar\Delta t}\right]^{N/2} \int_{-\infty}^{\infty} dx_1 \ldots \int_{-\infty}^{\infty} dx_{N-1}$$
$$\times \exp\left\{\frac{i}{\hbar}\Delta t \sum_{n=1}^{N}\left[\frac{m(x_n - x_{n-1})^2}{2(\Delta t)^2} - V(x_{n-1})\right]\right\}. \quad (5.306)$$

The point x_n is the value of x at time $t_0 + n\Delta t$ and the set of points x_n, $n = 1, 2, 3, \ldots N$ defines a path between the end points (x_0, t_0) and (x_N, t_N) as shown in Fig. 5.2. The integrals over $x_1, x_2 \ldots x_{N-1}$ result in a sum over all paths and in the limit $N \to \infty$ define a path integral. In this limit

$$\lim_{\Delta t \to 0} \frac{(x_n - x_{n-1})^2}{(\Delta t)^2} \to (\dot{x}(t))^2 \quad (5.307)$$

and

$$\Delta t \sum_{n=1}^{N} \to \int dt \quad (5.308)$$

so that the exponent in (5.306) can be written in terms of the classical action

$$\lim_{N\to\infty} \frac{i}{\hbar}\Delta t \sum_{n=1}^{N}\left[\frac{m(x_n - x_{n-1})^2}{2(\Delta t)^2} - V(x_{n-1})\right]$$
$$= \frac{i}{\hbar}\int_{t_0}^{t_1}\left[\frac{1}{2}m(\dot{x}(t))^2 - V(x(t))\right]dt$$
$$= \frac{i}{\hbar}I(t_1, t_0). \quad (5.309)$$

If the infinitely dimensional integral in (5.306) is written symbolically as

$$\int \mathcal{D}(x(t)) = \lim_{N\to\infty}\left[\frac{m}{2\pi i\hbar\Delta t}\right]^{N/2} \int_{-\infty}^{\infty} dx_1 \ldots \int_{-\infty}^{\infty} dx_{N-1} \quad (5.310)$$

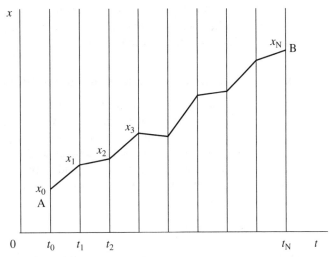

Figure 5.2 A path between $A(x_0, t_0)$ and $B(x_N, t_N)$ is defined by a series of points x_n at times $t_n = t_0 + n\Delta t$. By integrating over $x_1, x_2 \ldots x_{N-1}$ a sum over all paths is obtained.

the path integral form of K can be written as

$$K(x_N, t_1; x_0, t_0) = \int \mathcal{D}(x(t)) \exp\left(\frac{\mathrm{i}}{\hbar} I(t_1, t_0)\right). \tag{5.311}$$

Although it is convenient to express the sum over all paths by an integral symbol, it must not be forgotten that the path integral is defined as the limit $N \to \infty$ of the discretised expression (5.306).

In this discussion the propagator, and hence the wave function, have been expressed in path integral form by using the evolution operator (5.243) which was in turn derived from the Schrödinger equation. Alternatively it is possible to postulate the expression (5.311) where I is the classical action, and to develop quantum mechanics from it.

With the exception of a few potentials, including the harmonic oscillator, the propagator K cannot be evaluated exactly starting from (5.311), so that the path integral expression has a limited use in practical calculations. However, path integral methods can lead to an understanding of physical effects both in non-relativistic quantum mechanics and in its generalisation to relativistic quantum field theory (see for example Huang, 1998).

5.10 Symmetry principles and conservation laws

In this section we shall discuss the connection between symmetry principles and conservation laws. The symmetry operations with which we are concerned are transformations of the dynamical variables that leave the time-independent Hamiltonian operator of an *isolated system* invariant. We shall show that for each operation that leaves the Hamiltonian invariant there is a corresponding dynamical variable that is a constant of the motion and is conserved.

Spatial translations and conservation of momentum

Let us begin by analysing the connection between the translational properties of a system and the conservation of momentum. We shall make the basic assumption that the physical properties of an *isolated* physical system cannot be altered by a translation of the system by an arbitrary amount **a**. Such a translation has of course the same effect as one in which the system is undisturbed, but the origin of coordinates is displaced[7] by an amount $-\mathbf{a}$. Thus the new position vector of a point, $\mathbf{r}'$, is related to the old one, $\mathbf{r}$, by

$$\mathbf{r}' = T(\mathbf{a})\mathbf{r} \equiv \mathbf{r} + \mathbf{a} \tag{5.312}$$

where $T(\mathbf{a})$ is the operator effecting the transformation. The inverse translation $T^{-1}(\mathbf{a})$ is then defined by

$$\mathbf{r} = T^{-1}(\mathbf{a})\mathbf{r}' \equiv \mathbf{r}' - \mathbf{a}. \tag{5.313}$$

Let us now consider the simple case of a system consisting of a single structureless particle described at a given time by the spatial wave function $\psi(\mathbf{r})$. Let the system after being translated by an amount **a** be described by the wave function $\psi'(\mathbf{r})$. The two wave functions $\psi(\mathbf{r})$ and $\psi'(\mathbf{r})$ can be connected by an operator $U_T(\mathbf{a})$ defined so that

$$\psi'(\mathbf{r}) = U_T(\mathbf{a})\psi(\mathbf{r}). \tag{5.314}$$

Since this active translation of the system by the amount **a** is equivalent to shifting the coordinate origin by $-\mathbf{a}$, it is clear that

$$\psi'(\mathbf{r} + \mathbf{a}) = \psi'(T\mathbf{r})$$
$$= \psi(\mathbf{r}) \tag{5.315}$$

or, alternatively,

$$\psi'(\mathbf{r}) = U_T(\mathbf{a})\psi(\mathbf{r}) = \psi(T^{-1}\mathbf{r})$$
$$= \psi(\mathbf{r} - \mathbf{a}). \tag{5.316}$$

Because the physical properties of the system cannot be altered by this transformation, $U_T(\mathbf{a})$ must be a *unitary operator* (see Section 5.5). The explicit form of $U_T(\mathbf{a})$

---

[7] Transformations in which the system is moved relative to a particular coordinate system are said to be 'active' transformations. If, on the other hand, the system is considered to be fixed and the coordinate system is moved the transformations are said to be 'passive'.

can be determined by considering first the effect of an *infinitesimal translation* $\delta\mathbf{a}$ on the wave function. Using (5.316) and keeping terms to first order in $\delta\mathbf{a}$, we have

$$\psi'(\mathbf{r}) = \psi(\mathbf{r} - \delta\mathbf{a})$$
$$= \psi(\mathbf{r}) - \delta a_x \frac{\partial\psi(\mathbf{r})}{\partial x} - \delta a_y \frac{\partial\psi(\mathbf{r})}{\partial y} - \delta a_z \frac{\partial\psi(\mathbf{r})}{\partial z}$$
$$= (I - \delta\mathbf{a}\cdot\nabla)\psi(\mathbf{r}). \tag{5.317}$$

Comparing (5.317) with (5.314), we see that the infinitesimal unitary translation operator is

$$U_T(\delta\mathbf{a}) = I - \delta\mathbf{a}\cdot\nabla$$
$$= I - \frac{i}{\hbar}\delta\mathbf{a}\cdot\mathbf{p}_{op} \tag{5.318}$$

where $\mathbf{p}_{op} = -i\hbar\nabla$ is the momentum operator of the particle in the position representation. Thus we see that the momentum operator $\mathbf{p}_{op}$ is the generator of the infinitesimal translations.

A finite translation can be obtained by performing successive infinitesimal translations in steps of $\delta\mathbf{a}$. Let us write $\delta\mathbf{a} = \mathbf{a}/n$, where n is an integer, and take the limit $n \to \infty$. We then have

$$U_T(\mathbf{a}) = \lim_{n\to\infty}\left(I - \frac{i}{\hbar}\frac{\mathbf{a}\cdot\mathbf{p}_{op}}{n}\right)^n$$
$$= \exp\left(-\frac{i}{\hbar}\mathbf{a}\cdot\mathbf{p}_{op}\right). \tag{5.319}$$

If the wave function $\psi(\mathbf{r})$ is an eigenstate of the momentum operator $\mathbf{p}_{op}$ corresponding to an eigenvalue $\mathbf{p}$

$$\mathbf{p}_{op}\psi = \mathbf{p}\psi \tag{5.320}$$

it follows that

$$U_T(\mathbf{a})\psi = \exp\left(-\frac{i}{\hbar}\mathbf{a}\cdot\mathbf{p}\right)\psi. \tag{5.321}$$

Thus the action of $U_T(\mathbf{a})$ is to alter ψ by a phase factor, which clearly does not change the state of the system. Conversely, we see that if the state of a system is unaltered by a translation, it is an eigenstate of the momentum operator.

These results can be immediately generalised to a system of N particles. In this case, the infinitesimal unitary translation operator is given by

$$U_T(\delta\mathbf{a}) = I - \frac{i}{\hbar}\delta\mathbf{a}\cdot\mathbf{P}_{op} \tag{5.322}$$

where the generator of the transformation, $\mathbf{P}_{op}$, is the total momentum operator

$$\mathbf{P}_{op} = (\mathbf{p}_1)_{op} + (\mathbf{p}_2)_{op} + \cdots + (\mathbf{p}_N)_{op}. \tag{5.323}$$

Likewise, the unitary operator corresponding to a finite translation is given for an N-particle system by

$$U_T(\mathbf{a}) = \exp\left(-\frac{i}{\hbar}\mathbf{a}.\mathbf{P}_{op}\right). \tag{5.324}$$

Since the Hamiltonian H of an isolated system is invariant under any translation, it follows that

$$H' = U_T(\mathbf{a})HU_T^\dagger(\mathbf{a}) = H. \tag{5.325}$$

In particular, for an infinitesimal translation we have from (5.322), to order $\delta \mathbf{a}$

$$U_T(\delta\mathbf{a})HU_T^\dagger(\delta\mathbf{a}) = H - \frac{i}{\hbar}\delta\mathbf{a}.[\mathbf{P}_{op}, H] \tag{5.326}$$

and therefore, upon comparison of (5.326) with (5.325), we find that

$$[\mathbf{P}_{op}, H] = 0 \tag{5.327}$$

so that the total momentum is a constant of the motion (see (5.254)). Thus we see that *the conservation of the total momentum of an isolated system results from the invariance of its Hamiltonian under translations.*

Conservation laws and continuous symmetry transformations

The foregoing discussion can readily be generalised to any continuous symmetry transformation. Let the Hamiltonian H of an isolated system be invariant under a symmetry transformation S. If U_S is the unitary operator effecting this transformation, its action on the wave function Ψ of the system is given by

$$\Psi' = U_S\Psi. \tag{5.328}$$

Let A be an observable and A' its transform by the symmetry operation. Because the expectation value of measurements of A' made on the system in the dynamical state Ψ' must be equal to the expectation value of measurements of A made on the system in the dynamical state Ψ, we must have

$$\begin{aligned}\langle\Psi|A|\Psi\rangle &= \langle\Psi'|A'|\Psi'\rangle \\ &= \langle U_S\Psi|A'|U_S\Psi\rangle\end{aligned} \tag{5.329}$$

so that

$$A' = U_S A U_S^\dagger, \qquad A = U_S^\dagger A' U_S. \tag{5.330}$$

In particular, since the Hamiltonian H is invariant under the symmetry operation S,

$$H' = U_S H U_S^\dagger = H. \tag{5.331}$$

If the symmetry transformation is a continuous one, any operator U_S can be expressed as a product of operators $U_{\delta S}$ corresponding to infinitesimal symmetry transformations, such that

$$U_{\delta S}(\varepsilon) = I + i\varepsilon F_S \tag{5.332}$$

where ε is a real, arbitrarily small parameter and F_S, the generator of the infinitesimal unitary transformation, is an Hermitian operator. To first order in ε,

$$H' = U_{\delta S}(\varepsilon) H U_{\delta S}^{\dagger}(\varepsilon) = H + i\varepsilon[F_S, H]. \tag{5.333}$$

Comparing (5.333) with (5.331), we see that

$$[F_S, H] = 0 \tag{5.334}$$

and it follows from (5.254) that if F_S is time-independent its expectation value $\langle F_S \rangle$ will not vary in time, so that F_S is a *constant of the motion*.

Equation (5.334) generalises to any continuous symmetry transformation the result (5.327) derived for spatial translations. In Chapter 6 we shall examine the case of rotations, and shall show that *the conservation of the angular momentum of an isolated system is the consequence of the invariance of its Hamiltonian under rotations*. Here we consider briefly *time translations*, which, as we have seen in Section 5.7, are effected for a time-independent Hamiltonian by the evolution operator

$$U(t, t_0) = \exp\left[-\frac{i}{\hbar} H(t - t_0)\right]. \tag{5.335}$$

The generator of the corresponding infinitesimal transformation is the Hamiltonian H, and as H commutes with itself, it follows that if H is time-independent the energy is conserved. Thus, *the conservation of energy of an isolated system is a consequence of the invariance of its Hamiltonian with respect to time translations*.

Space reflection and parity conservation

Until now we have discussed continuous symmetry transformations. We now turn our attention to a *discrete* symmetry transformation: the *reflection* through the origin 0 of the coordinate system. This operation is also known as an *inversion* or *parity* operation. Corresponding to such an operation, there is a unitary operator, usually denoted by $\mathcal{P}$, and called the *parity* operator. It is such that, for a single particle (spatial) wave function $\psi(\mathbf{r})$,

$$\mathcal{P}\psi(\mathbf{r}) = \psi(-\mathbf{r}) \tag{5.336a}$$

and for several particles

$$\mathcal{P}\psi(\mathbf{r}_1, \mathbf{r}_2, \ldots, \mathbf{r}_N) = \psi(-\mathbf{r}_1, -\mathbf{r}_2, \ldots, -\mathbf{r}_N). \tag{5.336b}$$

The parity operator $\mathcal{P}$ is Hermitian ($\mathcal{P}^\dagger = \mathcal{P}$) since, for any two wave functions $\psi(\mathbf{r})$ and $\phi(\mathbf{r})$, we have

$$\int \phi^*(\mathbf{r})\mathcal{P}\psi(\mathbf{r})d\mathbf{r} = \int \phi^*(\mathbf{r})\psi(-\mathbf{r})d\mathbf{r}$$

$$= \int \phi^*(-\mathbf{r})\psi(\mathbf{r})d\mathbf{r}$$

$$= \int [\mathcal{P}\phi(\mathbf{r})]^*\psi(\mathbf{r})d\mathbf{r} \tag{5.337}$$

and the generalisation to several-particle wave functions is obvious.

It follows from the definition (5.336) that

$$\mathcal{P}^2 = I \tag{5.338}$$

so that the eigenvalues of $\mathcal{P}$ are $+1$ or -1, and the corresponding eigenstates are said to be *even* or *odd*. Thus, if $\psi_+(\mathbf{r})$ is an even eigenstate of $\mathcal{P}$ and $\psi_-(\mathbf{r})$ is an odd eigenstate, we have

$$\mathcal{P}\psi_+(\mathbf{r}) = \psi_+(-\mathbf{r}) = \psi_+(\mathbf{r}) \tag{5.339a}$$

and

$$\mathcal{P}\psi_-(\mathbf{r}) = \psi_-(-\mathbf{r}) = -\psi_-(\mathbf{r}). \tag{5.339b}$$

We note that

$$\int \psi_+^*(\mathbf{r})\psi_-(\mathbf{r})d\mathbf{r} = \int \psi_+^*(-\mathbf{r})\psi_-(-\mathbf{r})d\mathbf{r}$$

$$= -\int \psi_+^*(\mathbf{r})\psi_-(\mathbf{r})d\mathbf{r}. \tag{5.340}$$

Hence

$$\int \psi_+^*(\mathbf{r})\psi_-(\mathbf{r})d\mathbf{r} = 0 \tag{5.341}$$

so that the eigenstates $\psi_+(\mathbf{r})$ and $\psi_-(\mathbf{r})$ are *orthogonal*, in accordance with the fact that they belong to different eigenvalues of $\mathcal{P}$. They also form a complete set, since any function can be written as

$$\psi(\mathbf{r}) = \psi_+(\mathbf{r}) + \psi_-(\mathbf{r}) \tag{5.342}$$

where

$$\psi_+(\mathbf{r}) = \tfrac{1}{2}[\psi(\mathbf{r}) + \psi(-\mathbf{r})] \tag{5.343}$$

has even parity, while

$$\psi_-(\mathbf{r}) = \tfrac{1}{2}[\psi(\mathbf{r}) - \psi(-\mathbf{r})] \tag{5.344}$$

has odd parity. Again, the generalisation to systems of N particles is obvious.

5.10 Symmetry principles and conservation laws

The action of the parity operator $\mathcal{P}$ on the observables $\mathbf{r}$ and $\mathbf{p}_{op}$ is given by

$$\mathcal{P}\mathbf{r}\mathcal{P}^{\dagger} = -\mathbf{r} \tag{5.345}$$

and

$$\mathcal{P}\mathbf{p}_{op}\mathcal{P}^{\dagger} = -\mathbf{p}_{op}. \tag{5.346}$$

and we recall that $\mathcal{P}^{\dagger} = \mathcal{P}$.

The parity operation is equivalent to transforming a right-handed system of coordinates into a left-handed one. From our general discussion, we know that if the parity operator commutes with the Hamiltonian of the system, then parity is conserved and simultaneous eigenstates of the Hamiltonian and the parity operator can be formed. If the weak interaction (which is responsible, for example, for the beta-decay of nuclei) is disregarded, the parity operator commutes with the Hamiltonians of atomic and nuclear systems

$$[\mathcal{P}, H] = 0 \tag{5.347}$$

and parity is conserved.

Time-reversal invariance

A second important discrete transformation is one in which the time is reversed, $t \to -t$. Before we examine the effect of this transformation in quantum mechanics, it is instructive to recall how time-reversal invariance occurs in classical mechanics. For this purpose, we start from Newton's law of motion for a mass point,

$$m\frac{d^2\mathbf{r}}{dt^2} = \mathbf{F} \tag{5.348}$$

and assume that the force $\mathbf{F}$ only depends explicitly on the position coordinates. Then, because Newton's equations are of second order in t, we can associate to every solution $\mathbf{r}(t)$ of equation (5.348) another solution

$$\mathbf{r}'(t) = \mathbf{r}(-t). \tag{5.349}$$

The correspondence between the two solutions is illustrated by Fig. 5.3. We remark that the position of the particle at time t_0 in case (a) is the same as its position at time $-t_0$ in case (b), while the velocities (and hence the momenta) are reversed. Indeed,

$$\mathbf{v}'(t_0) = \left[\frac{d\mathbf{r}'(t)}{dt}\right]_{t=t_0} = -\left[\frac{d\mathbf{r}(-t)}{d(-t)}\right]_{t=t_0} = -\left[\frac{d\mathbf{r}(t')}{dt'}\right]_{t'=-t_0} = -\mathbf{v}(-t_0). \tag{5.350}$$

Let us now investigate the effect of the time-reversal transformation $t \to -t$ in quantum mechanics. We begin by considering the case of a spinless particle of mass m moving in a real time-independent potential $V(\mathbf{r})$. The corresponding time-dependent Schrödinger equation reads

$$i\hbar\frac{\partial}{\partial t}\Psi(\mathbf{r}, t) = \left[-\frac{\hbar^2}{2m}\nabla^2 + V(\mathbf{r})\right]\Psi(\mathbf{r}, t). \tag{5.351}$$

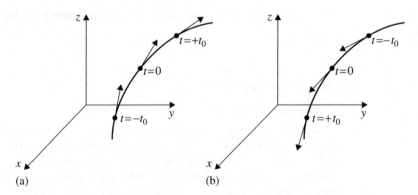

Figure 5.3 Two classical trajectories (a) and (b) which correspond by time reversal.

Changing t into $-t$, we obtain

$$-i\hbar \frac{\partial}{\partial t} \Psi(\mathbf{r}, -t) = \left[-\frac{\hbar^2}{2m} \nabla^2 + V(\mathbf{r}) \right] \Psi(\mathbf{r}, -t) \tag{5.352}$$

and we see that by taking the complex conjugate we can restore the form of the original equation (5.351):

$$i\hbar \frac{\partial}{\partial t} \Psi^*(\mathbf{r}, -t) = \left[-\frac{\hbar^2}{2m} \nabla^2 + V(\mathbf{r}) \right] \Psi^*(\mathbf{r}, -t). \tag{5.353}$$

It follows that if $\Psi(\mathbf{r}, t)$ is a solution of the time-dependent Schrödinger equation (5.351), so also is the 'time-reversed' wave function

$$\Psi'(\mathbf{r}, t) = \Psi^*(\mathbf{r}, -t). \tag{5.354}$$

It should be noted that this result depends on our choice of representation. For example, if we write (see (2.60))

$$\Psi(\mathbf{r}, t) = (2\pi\hbar)^{-3/2} \int e^{i\mathbf{p}\cdot\mathbf{r}/\hbar} \Phi(\mathbf{p}, t) d\mathbf{p} \tag{5.355}$$

and

$$\Psi'(\mathbf{r}, t) = (2\pi\hbar)^{-3/2} \int e^{i\mathbf{p}\cdot\mathbf{r}/\hbar} \Phi'(\mathbf{p}, t) d\mathbf{p} \tag{5.356}$$

we see from (5.354) that the momentum space wave functions $\Phi(\mathbf{p}, t)$ and $\Phi'(\mathbf{p}, t)$ are related by

$$\Phi'(\mathbf{p}, t) = \Phi^*(-\mathbf{p}, -t). \tag{5.357}$$

Thus in momentum space, as we change t into $-t$ we must not only take the complex conjugate of the wave function, but also change $\mathbf{p}$ into $-\mathbf{p}$. This last finding is consistent with the classical result (5.350).

5.10 Symmetry principles and conservation laws

Let us now consider a more general case, starting from the time-dependent Schrödinger equation

$$i\hbar \frac{\partial}{\partial t} \Psi(t) = H \Psi(t) \tag{5.358}$$

and assuming that the Hamiltonian H is time-independent. Changing t into $-t$, and taking the complex conjugate, we obtain

$$i\hbar \frac{\partial}{\partial t} \Psi^*(-t) = H^* \Psi^*(-t). \tag{5.359}$$

Let us first assume that the Hamiltonian H is real ($H^* = H$). Then, if $\Psi(t)$ is a solution of the time-dependent Schrödinger equation, so is also the time-reversed state vector

$$\Psi'(t) = \Psi^*(-t) = K\Psi(-t). \tag{5.360}$$

Here K denotes the operator of complex conjugation. We remark that this operator is anti-unitary[8], and such that $K^2 = I$ or $K = K^\dagger$.

In general, however, H is not real. In this case we suppose that there exists a unitary[9] operator U_τ, such that

$$U_\tau H^* U_\tau^\dagger = H. \tag{5.361}$$

This operator U_τ should obviously reduce to the unit operator I when H is real. Operating on both sides of (5.359) with U_τ and using (5.361), we find that

$$i\hbar \frac{\partial}{\partial t} U_\tau \Psi^*(-t) = H U_\tau \Psi^*(-t). \tag{5.362}$$

Hence, if $\Psi(t)$ is a solution of the time-dependent Schrödinger equation, so also is

$$\begin{aligned}\Psi'(t) &= U_\tau \Psi^*(-t) = U_\tau K \Psi(-t) \\ &= \mathcal{T} \Psi(-t)\end{aligned} \tag{5.363}$$

where

$$\mathcal{T} = U_\tau K \tag{5.364}$$

is the *time-reversal operator*. In analogy with our foregoing discussion we call $\Psi'(t)$ the time-reversed state vector corresponding to $\Psi(t)$. We remark that since K is anti-unitary and U_τ is unitary the time-reversal operator $\mathcal{T}$ is anti-unitary.

---

[8] An operator B is said to be *anti-linear* if

$$B(c_1 \Psi_1 + c_2 \Psi_2) = c_1^* (B\Psi_1) + c_2^* (B\Psi_2)$$

where Ψ_1 and Ψ_2 are two functions and c_1 and c_2 are complex constants. An *anti-unitary* operator V is anti-linear and, in addition, satisfies the relation

$$VV^\dagger = V^\dagger V = I.$$

[9] The condition that U_τ be a unitary operator is imposed by the requirement of preserving the normalisation of the state vector.

We have seen above that when the Hamiltonian H is real the operator U_τ reduces to the unit operator, so that in this case the time-reversal operator $\mathcal{T}$ is just the operator K of complex conjugation. When $H^* \neq H$ it is usually a simple matter to obtain the appropriate operator U_τ (and hence $\mathcal{T}$) by using the following considerations. First of all, we require that the position operator $\mathbf{r}$ be left unchanged under time reversal:

$$\mathbf{r}' = \mathcal{T}\mathbf{r}\mathcal{T}^\dagger = \mathbf{r}. \tag{5.365}$$

Hence, if we work in the position representation (where $\mathbf{r}$ is a purely real operator), we must have

$$\mathcal{T}\mathbf{r}\mathcal{T}^\dagger = U_\tau K \mathbf{r} K^\dagger U_\tau^\dagger = U_\tau \mathbf{r} U_\tau^\dagger = \mathbf{r} \tag{5.366}$$

so that U_τ must commute with $\mathbf{r}$. We also require that the momentum operator shall change sign under the time-reversal operation

$$\mathbf{p}'_{op} = \mathcal{T}\mathbf{p}_{op}\mathcal{T}^\dagger = -\mathbf{p}_{op}. \tag{5.367}$$

Since $\mathbf{p}_{op} = -i\hbar\nabla$ is a purely imaginary operator in the position representation, we have in that representation

$$\mathcal{T}\mathbf{p}_{op}\mathcal{T}^\dagger = U_\tau K(-i\hbar\nabla)K^\dagger U_\tau^\dagger = U_\tau(i\hbar\nabla)U_\tau^\dagger = -\mathbf{p}_{op} = i\hbar\nabla \tag{5.368}$$

which implies that U_τ must also commute with $\mathbf{p}_{op} = -i\hbar\nabla$.

Let us now return once more to the time-dependent Schrödinger equation (5.358). Changing t into $-t$ and inserting $\mathcal{T}^\dagger\mathcal{T}(=I)$ between H and Ψ, we obtain

$$-i\hbar\frac{\partial}{\partial t}\Psi(-t) = H\mathcal{T}^\dagger\mathcal{T}\Psi(-t) \tag{5.369}$$

and therefore

$$-\mathcal{T}i\hbar\frac{\partial}{\partial t}\Psi(-t) = \mathcal{T}H\mathcal{T}^\dagger\mathcal{T}\Psi(-t). \tag{5.370}$$

Now, using (5.363) and (5.364), we have

$$-U_\tau K i\hbar\frac{\partial}{\partial t}\Psi(-t) = -U_\tau(-i)\hbar\frac{\partial}{\partial t}\Psi^*(-t) = \mathcal{T}H\mathcal{T}^\dagger\Psi'(t) \tag{5.371}$$

or

$$i\hbar\frac{\partial}{\partial t}\Psi'(t) = \mathcal{T}H\mathcal{T}^\dagger\Psi'(t) \tag{5.372}$$

and we want this equation to be identical with the original Schrödinger equation (5.358). Hence, provided that a unitary operator U_τ satisfying (5.361) exists, we see that the requirement

$$\mathcal{T}H\mathcal{T}^\dagger = H \tag{5.373}$$

or

$$[\mathcal{T}, H] = 0 \tag{5.374}$$

is the necessary and sufficient condition for the time-reversal invariance of the Schrödinger equation. At one time it was supposed that all fundamental Hamiltonians satisfied the time-reversal invariance condition (5.374), but it is now known that the Hamiltonian corresponding to the weak interaction is not time-reversal invariant, as well as not being invariant under the parity operation.

Galilean transformations

Let us consider a frame of reference S specified by Cartesian axes OX, OY, OZ and a second frame S' with Cartesian axes $O'X, O'Y, O'Z$ such that the axes in the two frames are parallel, but the origin O' moves with a constant velocity $\mathbf{v}$ with respect to O. If $\mathbf{r}$ is the position vector of a particle in S at time t and $\mathbf{r}'$ is the position vector of the same particle at the same time in S', then

$$\mathbf{r}' = \mathbf{r} - \mathbf{v}t \tag{5.375}$$

where we have taken O and O' to coincide at the time $t = 0$. The transformation (5.375) is known as a *Galilean transformation*. In non-relativistic classical mechanics, if Newton's laws of motion are obeyed in a frame S (an *inertial frame*), then they also hold in all frames S' which can be obtained by a Galilean transformation from S.

We shall now examine how the Schrödinger equation for a single particle of mass m moving in a potential $V(\mathbf{r}, t)$ changes under the Galilean transformation (5.375). If $\Psi(\mathbf{r}, t)$ is the wave function in the frame S, it satisfies the Schrödinger equation

$$i\hbar \frac{\partial}{\partial t} \Psi(\mathbf{r}, t) = \left[-\frac{\hbar^2}{2m} \nabla_\mathbf{r}^2 + V(\mathbf{r}, t) \right] \Psi(\mathbf{r}, t) \tag{5.376}$$

To rewrite this equation in terms of $\mathbf{r}'$ we note that

$$\Psi(\mathbf{r}, t) = \Psi(\mathbf{r}' + \mathbf{v}t, t) \tag{5.377a}$$

$$\frac{\partial}{\partial t} \Psi(\mathbf{r}, t) = \frac{\partial}{\partial t} \Psi(\mathbf{r}' + \mathbf{v}t, t) - \mathbf{v} \cdot \nabla_{\mathbf{r}'} \Psi(\mathbf{r}' + \mathbf{v}t, t) \tag{5.377b}$$

$$\nabla_\mathbf{r} \Psi(\mathbf{r}, t) = \nabla_{\mathbf{r}'} \Psi(\mathbf{r}' + \mathbf{v}t, t) \tag{5.377c}$$

so that (5.376) can be written as

$$i\hbar \frac{\partial}{\partial t} \Psi(\mathbf{r}' + \mathbf{v}t, t) - i\hbar \mathbf{v} \cdot \nabla_{\mathbf{r}'} \Psi(\mathbf{r}' + \mathbf{v}t, t)$$
$$= \left(-\frac{\hbar^2}{2m} \nabla_{\mathbf{r}'}^2 + V(\mathbf{r}' + \mathbf{v}t, t) \right) \Psi(\mathbf{r}' + \mathbf{v}t, t) \tag{5.378}$$

This equation, although correct, is not in the form of a Schrödinger equation because of the term in $\mathbf{v} \cdot \nabla_{\mathbf{r}'}$ on the left-hand side. A wave function $\Psi'(\mathbf{r}', t)$ which satisfies a standard Schrödinger equation in the frame S' can be introduced by making the unitary transformation

$$\Psi(\mathbf{r}' + \mathbf{v}t, t) = \exp[i(m\mathbf{v} \cdot \mathbf{r}' + mv^2 t/2)/\hbar] \Psi'(\mathbf{r}', t) \tag{5.379}$$

The plane wave exponential factor takes account of the fact that a particle at rest in the frame S' has a momentum $m\mathbf{v}$ and a kinetic energy $mv^2/2$ when viewed from the frame S. Inserting (5.379) into (5.378) it is found that (Problem 5.21)

$$i\hbar \frac{\partial \Psi'(\mathbf{r}',t)}{\partial t} = \left(-\frac{\hbar^2}{2m}\nabla_{\mathbf{r}'}^2 + V'(\mathbf{r}',t)\right)\Psi'(\mathbf{r}',t) \tag{5.380}$$

where the potential in the S' frame, $V'(\mathbf{r}',t)$, is defined as

$$V'(\mathbf{r}',t) = V(\mathbf{r} - \mathbf{v}t, t) \tag{5.381}$$

It is seen that $\Psi'(\mathbf{r}',t)$ satisfies a normal Schrödinger equation. If the potential V is zero, so is the potential V', with the consequence that both Ψ and Ψ' satisfy the same Schrödinger equation for a free particle; that is, the free-particle Schrödinger equation is invariant under a Galilean transformation.

We shall now show that the Schrödinger equation for an isolated system of particles is also invariant under a Galilean transformation. To see this, consider first the Schrödinger equation for an isolated two-body system which is written in terms of the relative coordinate $\mathbf{r}$ and the centre of mass coordinate $\mathbf{R}$ in equation (5.269). There are no external interactions but the two particles interact by the potential $V(\mathbf{r})$. Under a Galilean transformation the relative coordinate $\mathbf{r}$ is unchanged, but the position $\mathbf{R}'$ of the centre of mass in S' is related to $\mathbf{R}$ by

$$\mathbf{R}' = \mathbf{R} - \mathbf{v}t \tag{5.382}$$

In the frame S' the wave function $\Psi(\mathbf{R}, \mathbf{r}, t)$ becomes $\Psi(\mathbf{R}' + \mathbf{v}t, \mathbf{r}, t)$ and introducing the unitary transformation

$$\Psi(\mathbf{R}' + \mathbf{v}t, t) = \exp[i(M\mathbf{v}\cdot\mathbf{R}' + Mv^2 t/2)\hbar]\Psi'(\mathbf{R}', \mathbf{r}, t) \tag{5.383}$$

it is found that $\Psi'(\mathbf{R}', \mathbf{r}, t)$ satisfies

$$i\hbar \frac{\partial \Psi'(\mathbf{R}', \mathbf{r}, t)}{\partial t} = \left[-\frac{\hbar^2}{2M}\nabla_{\mathbf{R}'}^2 - \frac{\hbar^2}{2\mu}\nabla_{\mathbf{r}}^2 + V(\mathbf{r})\right]\Psi'(\mathbf{R}', \mathbf{r}, t) \tag{5.384}$$

where we recall that $M = m_1 + m_2$ is the total mass of the system and $\mu = m_1 m_2/(m_1 + m_2)$ is the reduced mass of the two particles. The equation (5.384) is identical in form to (5.269) so that the two-body Schrödinger equation for an isolated system is invariant under a Galilean transformation. Clearly this argument can be repeated for a many-body system in which there are no external interactions and the interactions between particles only depend on internal coordinates $\mathbf{r}_{ij} = \mathbf{r}_i - \mathbf{r}_j$.

5.11 The classical limit

The transition from quantum mechanics to classical mechanics was discussed briefly in Section 3.4. In this section, we shall examine this question in more detail.

The Ehrenfest theorem

We begin by re-deriving the Ehrenfest theorem, using commutator algebra. Consider a particle of mass m moving in a potential $V(\mathbf{r}, t)$, so that its Hamiltonian is given by

$$H = \frac{\mathbf{p}^2}{2m} + V(\mathbf{r}, t) \tag{5.385}$$

Using equation (5.254) for the operator x, we find that

$$\frac{d}{dt}\langle x \rangle = (i\hbar)^{-1}\langle [x, H] \rangle$$

$$= (i\hbar)^{-1}\left\langle \left[x, \frac{\mathbf{p}^2}{2m} + V(\mathbf{r}, t)\right] \right\rangle \tag{5.386}$$

Since x commutes with $V(\mathbf{r}, t)$ and also with p_y and p_z, we have

$$\frac{d}{dt}\langle x \rangle = \frac{1}{2mi\hbar}\langle [x, p_x^2] \rangle$$

$$= \frac{1}{2mi\hbar}\langle [x, p_x]p_x + p_x[x, p_x] \rangle \tag{5.387}$$

But $[x, p_x] = i\hbar$ and therefore

$$\frac{d}{dt}\langle x \rangle = \frac{\langle p_x \rangle}{m} \tag{5.388}$$

Proceeding in the same way with the expectation values of y and z, we obtain

$$\frac{d}{dt}\langle \mathbf{r} \rangle = \frac{\langle \mathbf{p} \rangle}{m} \tag{5.389}$$

which is the first Ehrenfest relation.

Using now equation (5.254) for the operator p_x, we have

$$\frac{d}{dt}\langle p_x \rangle = (i\hbar)^{-1}\langle [p_x, H] \rangle$$

$$= (i\hbar)^{-1}\left\langle \left[p_x, \frac{\mathbf{p}^2}{2m} + V(\mathbf{r}, t)\right] \right\rangle$$

$$= -\left\langle \frac{\partial V}{\partial x} \right\rangle \tag{5.390}$$

where we have used the fact that $[p_x, \mathbf{p}^2] = 0$ together with the equation

$$[p_x, V] = -i\hbar \frac{\partial V}{\partial x} \tag{5.391}$$

Repeating the argument for p_y and p_z, we find that

$$\frac{d}{dt}\langle \mathbf{p} \rangle = -\langle \boldsymbol{\nabla} V(\mathbf{r}, t) \rangle \tag{5.392}$$

which is the second Ehrenfest relation.

The Ehrenfest relations do not imply that classical mechanics holds for the expectation (average) values of $\mathbf{r}$ and $\mathbf{p}$. For this to be true, we must have in addition that

$$\langle \nabla V(\mathbf{r}, t) \rangle = \nabla V(\langle \mathbf{r} \rangle, t). \tag{5.393}$$

This can only be approximately satisfied if the width of the position probability distribution is small compared with the distance over which the potential V varies appreciably.

To examine this point further, let us consider the simple case of one-dimensional motion in a potential $V(x)$. Expanding $dV(x)/dx \equiv V'(x)$ about the (time-dependent) expectation value $\langle x \rangle$, we have

$$V'(x) = V'(\langle x \rangle) + (x - \langle x \rangle)V''(\langle x \rangle) + \frac{1}{2}(x - \langle x \rangle)^2 V'''(\langle x \rangle) + \cdots \tag{5.394}$$

so that

$$\langle V'(x) \rangle = V'(\langle x \rangle) + \frac{1}{2}\langle (x - \langle x \rangle)^2 \rangle V'''(\langle x \rangle) + \cdots \tag{5.395}$$

where we have used the fact that $\langle x - \langle x \rangle \rangle = 0$. The quantity $-V'(\langle x \rangle)$ is just the classical force at the point $\langle x \rangle$. Thus, if only the first term is retained on the right of (5.395), we obtain

$$\frac{d}{dt}\langle p_x \rangle = -V'(\langle x \rangle), \tag{5.396}$$

which shows that the expectation values $\langle x \rangle$ and $\langle p_x \rangle$ obey Newton's law. However, the condition that the second and higher terms in the expansion (5.395) are small is a necessary, but not sufficient, condition for the motion to be classical. For example, in the case of the harmonic oscillator, where $V(x)$ is quadratic in x, the second and higher terms in (5.395) vanish, and yet the classical limit can only be recovered if the spacing between the discrete energy levels is small compared with the energy of the oscillator; this occurs in the limit of large quantum numbers.

The Hamilton–Jacobi equation

Let us consider the time-dependent Schrödinger equation for a particle of mass m in a potential $V(\mathbf{r}, t)$:

$$i\hbar \frac{\partial}{\partial t}\Psi(\mathbf{r}, t) = \left[-\frac{\hbar^2}{2m}\nabla^2 + V(\mathbf{r}, t) \right] \Psi(\mathbf{r}, t) \tag{5.397}$$

and look for a solution of the form

$$\Psi(\mathbf{r}, t) = \exp\left[\frac{i}{\hbar} W(\mathbf{r}, t) \right]. \tag{5.398}$$

Substituting (5.398) into (5.397), we find that the function W must satisfy the equation

$$\frac{\partial W}{\partial t} + \frac{1}{2m}(\nabla W)^2 + V - \frac{i\hbar}{2m}\nabla^2 W = 0 \tag{5.399}$$

which in the limit $\hbar \to 0$ reduces to

$$\frac{\partial W}{\partial t} + \frac{1}{2m}(\nabla W)^2 + V = 0. \tag{5.400}$$

This is the *Hamilton–Jacobi equation* for Hamilton's principal function W[10]. It can also be written as

$$\frac{\partial W}{\partial t} + H(\mathbf{r}, \mathbf{p}, t) = 0 \tag{5.401}$$

where H is the classical Hamiltonian, $\mathbf{r}$ and $\mathbf{p}$ are the classical position and (canonical conjugate) momentum respectively, and

$$\mathbf{p}(t) = \nabla W(\mathbf{r}, t). \tag{5.402}$$

Thus, for the one-body problem we are considering, the Hamilton–Jacobi equation is a first-order partial differential equation in the four variables x, y, z and t. Classical trajectories $\mathbf{r}(t)$, which depend on the initial values (at $t = t_0$) of the canonical variables $\mathbf{r}$ and $\mathbf{p}$, can be determined from the knowledge of the principal function $W(\mathbf{r}, t)$.

Let us now consider the case for which the potential $V(\mathbf{r})$ is time-independent, so that the Schrödinger equation (5.397) has stationary solutions of the form $\Psi(\mathbf{r}, t) = \psi(\mathbf{r}) \exp(-iEt/\hbar)$. In that case, we can write

$$W(\mathbf{r}, t) = S(\mathbf{r}) - Et \tag{5.403}$$

so that

$$\psi(\mathbf{r}) = \exp\left[\frac{i}{\hbar} S(\mathbf{r})\right]. \tag{5.404}$$

Substituting (5.403) into (5.399), we obtain for the function S the equation

$$\frac{1}{2m}(\nabla S)^2 - [E - V(\mathbf{r})] - \frac{i\hbar}{2m}\nabla^2 S = 0 \tag{5.405}$$

In the limit $\hbar \to 0$, this equation reduces to

$$\frac{1}{2m}(\nabla S)^2 = E - V(\mathbf{r}) \tag{5.406}$$

which is the equation determining Hamilton's characteristic function $S(\mathbf{r})$. For a time-independent potential $V(\mathbf{r})$, it follows from (5.402) that the classical trajectories $\mathbf{r}(t)$ are orthogonal to the surfaces of constant W. In 1834, Hamilton observed that in analogy with optics (where the rays of geometrical optics are the normals to wave fronts of constant phase), we may interpret classical mechanics as the 'geometrical optics' limit of a wave motion, the classical orbits being orthogonal to the wave fronts $W = $ constant. This idea was the starting point of the investigations of de Broglie

---

[10] It can be shown quite generally for any classical system that the numerical value of the action integral I introduced in Section 5.9 is the same as that of the corresponding Hamilton principal function W (see Goldstein 1980).

and Schrödinger, which led to the formulation of wave mechanics. In Section 8.4, we shall show how the semi-classical approximation method of Wentzel, Kramers and Brillouin can be based on these considerations.

Problems

5.1 The operators A_i ($i = 1, \ldots, 6$) are defined as follows

$$A_1 \psi(x) = [\psi(x)]^2 \qquad A_4 \psi(x) = x^2 \psi(x)$$

$$A_2 \psi(x) = \frac{d}{dx} \psi(x) \qquad A_5 \psi(x) = \sin[\psi(x)]$$

$$A_3 \psi(x) = \int_a^x \psi(x') dx' \qquad A_6 \psi(x) = \frac{d^2}{dx^2} \psi(x).$$

Which of the operators A_i are linear operators? Which are Hermitian?

5.2 Using the definition of Hermiticity (3.69) show, by expressing the wave function Ψ in the form $\Psi = c_1 \Psi_1 + c_2 \Psi_2$, where Ψ_1, Ψ_2 are a pair of square-integrable functions and c_1, c_2 are complex constants, that

$$\int \Psi_1^* A \Psi_2 d\mathbf{r} = \int (A \Psi_1)^* \Psi_2 d\mathbf{r}$$

in conformity with the general definition of Hermiticity (5.22).

5.3 If A and B are two Hermitian operators (a) show that AB and BA may not be Hermitian, but that $(AB + BA)$ and $i(AB - BA)$ are both Hermitian. (b) Prove that the expectation value of A^2 is always real and non-negative. (c) If $A^2 = 2$, find the eigenvalues of A.

5.4 Prove that (a) if A is a linear operator and c is a complex number then

$$(cA)^\dagger = c^* A^\dagger$$

(b) if A and B are linear operators then

$$(A + B)^\dagger = A^\dagger + B^\dagger, \quad \text{and} \quad (AB)^\dagger = B^\dagger A^\dagger.$$

5.5 Consider a free particle of mass m moving in one dimension so that its Hamiltonian is $H = p_x^2/2m$. Show that the eigenvalues of H are two-fold degenerate, and that the degeneracy can be removed by considering the simultaneous eigenfunctions of H and p_x.

5.6 Prove the relations (5.101) satisfied by commutators.

5.7 (a) Prove by induction that

$$[x^n, p_x] = i\hbar n x^{n-1}$$

and that
$$[x, p_x^n] = i\hbar n p_x^{n-1}$$
where n is a positive integer.

(b) Using these results show that if $f(x)$ can be expanded in a polynomial in x and $g(p_x)$ can be expanded in a polynomial in p_x, then
$$[f(x), p_x] = i\hbar df/dx$$
and
$$[x, g(p_x)] = i\hbar dg/dp_x.$$

5.8 If A and B are two operators such that $[A, B] = \lambda$, where λ is a complex number, and if μ is a second complex number, prove that
$$\exp[\mu(A + B)] = \exp(\mu A)\exp(\mu B)\exp(-\mu^2\lambda/2).$$

5.9 Show that for a one-dimensional harmonic oscillator the uncertainty product $\Delta x \Delta p_x$ is equal to $(n + \frac{1}{2})\hbar$ in the energy eigenstate $\psi_n(x)$, where Δx and Δp_x are defined by (5.118). Verify that the ground-state wave function has the form of a minimum uncertainty wave packet (5.122).

5.10 Prove that a Hermitian operator is represented by a Hermitian matrix, a unitary operator by a unitary matrix, and the operator sum $A + B$ by the matrix sum $\mathbf{A} + \mathbf{B}$.

5.11 The Hamiltonian operator H for a certain physical system is represented by the matrix
$$\mathbf{H} = \hbar\omega\begin{pmatrix} 1 & 0 & 0 \\ 0 & 2 & 0 \\ 0 & 0 & 2 \end{pmatrix}$$
while two other observables A and B are represented by the matrices
$$\mathbf{A} = \begin{pmatrix} 0 & \lambda & 0 \\ \lambda & 0 & 0 \\ 0 & 0 & 2\lambda \end{pmatrix}, \quad \mathbf{B} = \begin{pmatrix} 2\mu & 0 & 0 \\ 0 & 0 & \mu \\ 0 & \mu & 0 \end{pmatrix}$$
where λ and μ are real (non-zero) numbers.

(a) Find the eigenvalues and eigenvectors of $\mathbf{A}$ and $\mathbf{B}$.
(b) If the system is in a state described by the state vector
$$\mathbf{u} = c_1\mathbf{u}_1 + c_2\mathbf{u}_2 + c_3\mathbf{u}_3$$
where c_1, c_2 and c_3 are complex constants and
$$\mathbf{u}_1 = \begin{pmatrix} 1 \\ 0 \\ 0 \end{pmatrix}, \mathbf{u}_2 = \begin{pmatrix} 0 \\ 1 \\ 0 \end{pmatrix}, \mathbf{u}_3 = \begin{pmatrix} 0 \\ 0 \\ 1 \end{pmatrix}.$$

(i) find the relationship between c_1, c_2 and c_3 such that **u** is normalised to unity; and

(ii) find the expectation values of H, A and B.

(iii) What are the possible values of the energy that can be obtained in a measurement when the system is described by the state vector **u**? For each possible result find the wave function in the matrix representation immediately after the measurement.

5.12 Prove the commutation relations (5.190) and (5.193).

5.13 Calculate for a linear harmonic oscillator the matrix elements $(x^i)_{nm}$, where $i = 2, 3, 4$:

(a) by using matrix multiplication, starting from the expressions for x_{nm} given by (4.174), and

(b) by using the raising and lowering operators a_+ and a_-.

5.14 Find the matrix elements of x and p_x for the linear harmonic oscillator using $(a_+)_{kn}$ and $(a_-)_{kn}$ given respectively by (5.217) and (5.218).

5.15 Consider a particle of mass m moving in a potential $V(\mathbf{r}, t)$, so that its Hamiltonian is given by

$$H = \frac{\mathbf{p}^2}{2m} + V(\mathbf{r}, t).$$

Using the result (5.254) and commutator algebra, prove that

$$\frac{d}{dt}\langle x^2 \rangle = \frac{1}{m}(\langle xp_x \rangle + \langle p_x x \rangle).$$

5.16 Generalise the virial theorem (5.260) to a system of N particles, having a Hamiltonian of the form (5.229).

5.17 Consider a one-particle system, of which the Hamiltonian does not depend on x. Using (5.254), show that the x-component of the momentum is then a constant of the motion.

5.18 Prove equation (5.266).

5.19 Show that for a one-dimensional harmonic oscillator described in the Heisenberg picture by the Hamiltonian

$$H_H = \frac{1}{2m}p_H^2(t) + \frac{1}{2}k x_H^2(t)$$

where $p_H \equiv (p_x)_H$, the general solution of equations (5.287) is

$$x_H(t) = x_H(0)\cos\omega t + \frac{1}{m\omega}p_H(0)\sin\omega t,$$

$$p_H(t) = -m\omega x_H(0)\sin\omega t + p_H(0)\cos\omega t$$

where $\omega = (k/m)^{1/2}$.

Is the Hamiltonian H_H time-independent?

5.20 Consider the one-dimensional motion of a free particle and the three operators $H = p_x^2/2m$, p_x and $\mathcal{P}$ (the parity operator). Do these operators all mutually commute? If not, show that one can construct two different complete sets of commuting observables from these three operators.

5.21 Show that the wave function $\Psi'(\mathbf{r}', t)$ defined by the unitary transformation (5.379) satisfies the time-dependent Schrödinger equation (5.380).

6 Angular momentum

6.1 Orbital angular momentum 266
6.2 Orbital angular momentum and spatial rotations 270
6.3 The eigenvalues and eigenfunctions of $\mathbf{L}^2$ and L_z 275
6.4 Particle on a sphere and the rigid rotator 289
6.5 General angular momentum. The spectrum of $\mathbf{J}^2$ and J_z 292
6.6 Matrix representations of angular momentum operators 296
6.7 Spin angular momentum 299
6.8 Spin one-half 303
6.9 Total angular momentum 311
6.10 The addition of angular momenta 315
Problems 323

One of the most important concepts in classical mechanics is that of angular momentum. All isolated systems possess the fundamental invariance property that their total angular momentum is conserved. This property provides a powerful aid in studying a number of problems such as the dynamics of rigid bodies and the motion of planets. In this respect the angular momentum can be compared with the total energy and the total linear momentum, which are also conserved quantities for isolated systems.

The role of angular momentum in quantum mechanics is also of paramount importance. In this chapter, the necessary formalism for the description of angular momentum in quantum theory will be developed and some illustrative applications will be discussed. We shall see that the main differences from classical mechanics arise from the fact that in quantum physics the angular momentum is not an ordinary vector, but is a *vector operator*, whose three components do not commute with each other. We begin by considering the *orbital angular momentum*, and establish its relationship with rotations. Then, after analysing the general properties of angular momentum, we will study the important case of the *spin* (or intrinsic) angular momentum. Finally, we shall treat the problem of the addition of angular momenta.

6.1 Orbital angular momentum

Let us consider a particle of mass m. We denote by $\mathbf{p}$ its momentum and by $\mathbf{r}$ its position vector with respect to a fixed origin 0. In classical mechanics the orbital angular momentum of this particle with respect to 0 is defined as a vector $\mathbf{L}$, such that

$$\mathbf{L} = \mathbf{r} \times \mathbf{p} \tag{6.1}$$

This vector points in a direction at right angles to the plane containing $\mathbf{r}$ and $\mathbf{p}$, and is of magnitude $L = rp \sin \alpha$, where α is the angle between $\mathbf{r}$ and $\mathbf{p}$. The Cartesian components of $\mathbf{L}$ are

$$L_x = yp_z - zp_y \tag{6.2a}$$
$$L_y = zp_x - xp_z \tag{6.2b}$$
$$L_z = xp_y - yp_x. \tag{6.2c}$$

The corresponding operators representing L_x, L_y and L_z in quantum mechanics are obtained by replacing x, y, z and p_x, p_y, p_z by the appropriate operators obeying the fundamental commutation relations (5.88). Using the position representation of wave mechanics, such that $x_{\text{op}} = x$, $y_{\text{op}} = y$, $z_{\text{op}} = z$ and $(p_x)_{\text{op}} = -i\hbar \partial/\partial x$, $(p_y)_{\text{op}} = -i\hbar \partial/\partial y$, $(p_z)_{\text{op}} = -i\hbar \partial/\partial z$, we find that the operators L_x, L_y and L_z are given by

$$L_x = -i\hbar \left(y \frac{\partial}{\partial z} - z \frac{\partial}{\partial y} \right) \tag{6.3a}$$

$$L_y = -i\hbar \left(z \frac{\partial}{\partial x} - x \frac{\partial}{\partial z} \right) \tag{6.3b}$$

$$L_z = -i\hbar \left(x \frac{\partial}{\partial y} - y \frac{\partial}{\partial x} \right). \tag{6.3c}$$

In other words the orbital angular momentum is represented in the position representation of wave mechanics by the *vector operator*

$$\mathbf{L} = -i\hbar (\mathbf{r} \times \nabla) \tag{6.4}$$

whose Cartesian components are the differential operators given by (6.3). It is easy to verify (Problem 6.1) that $\mathbf{L}$ is Hermitian.

Using the rules (5.101) of commutator algebra and the basic commutation relations (5.88), one readily obtains the commutation relations satisfied by the operators L_x, L_y and L_z. For example, we have

$$[L_x, L_y] = [(yp_z - zp_y), (zp_x - xp_z)]$$
$$= [yp_z, zp_x] + [zp_y, xp_z] - [yp_z, xp_z] - [zp_y, zp_x]. \tag{6.5}$$

The first commutator on the right of this equation is

$$[yp_z, zp_x] = yp_z zp_x - zp_x yp_z. \tag{6.6}$$

Since y and p_x commute with each other and with z and p_z, we can write

$$[yp_z, zp_x] = yp_x[p_z, z]$$
$$= -i\hbar yp_x. \tag{6.7}$$

Similarly, the second commutator on the right of (6.5) becomes

$$[zp_y, xp_z] = zp_y xp_z - xp_z zp_y$$
$$= xp_y[z, p_z]$$
$$= i\hbar xp_y. \tag{6.8}$$

The third commutator on the right of (6.5) vanishes because y, x and p_z mutually commute, and so also does the last commutator, since z, p_y and p_x mutually commute. Thus we find that

$$[L_x, L_y] = i\hbar(xp_y - yp_x)$$
$$= i\hbar L_z. \tag{6.9a}$$

Similarly, one obtains

$$[L_y, L_z] = i\hbar L_x \tag{6.9b}$$

and

$$[L_z, L_x] = i\hbar L_y. \tag{6.9c}$$

The three commutation relations (6.9) are readily shown (Problem 6.2) to be equivalent to the vector commutation relation

$$\mathbf{L} \times \mathbf{L} = i\hbar \mathbf{L}. \tag{6.10}$$

Note that the operator character of $\mathbf{L}$ is clearly exhibited by this equation, since the vector product $\mathbf{L} \times \mathbf{L}$ would vanish if $\mathbf{L}$ were an ordinary vector.

It is apparent from equations (6.9) that the operators representing any two components of the orbital angular momentum do not commute. From the results of Section 5.4 it follows that they are not simultaneously measurable. As a result, it is, in general[1], impossible to assign simultaneously definite values for all components of the orbital angular momentum. Thus, if the system is in an eigenstate of one orbital angular momentum component (so that a measurement of that component will produce a definite value), it will in general not be in an eigenstate of either of the two other components.

The operator representing the square of the magnitude of the orbital angular momentum is defined as

$$\mathbf{L}^2 = L_x^2 + L_y^2 + L_z^2 \tag{6.11}$$

---

[1] The words 'in general' are required because we shall see shortly that functions which depend only on the magnitude r of the position vector $\mathbf{r}$ are simultaneous eigenstates of L_x, L_y and L_z with eigenvalue zero.

Using the rules of commutator algebra and the commutation relations (6.9), it is easy to show that $\mathbf{L}^2$ commutes with each of the three components of $\mathbf{L}$. For example, let us look at the commutator

$$[\mathbf{L}^2, L_x] = [L_x^2 + L_y^2 + L_z^2, L_x]. \tag{6.12}$$

Since the commutator of L_x with itself vanishes, we have $[L_x^2, L_x] = 0$ and

$$\begin{aligned}
[\mathbf{L}^2, L_x] &= [L_y^2 + L_z^2, L_x] \\
&= [L_y^2, L_x] + [L_z^2, L_x] \\
&= L_y[L_y, L_x] + [L_y, L_x]L_y + L_z[L_z, L_x] + [L_z, L_x]L_z \\
&= -i\hbar(L_y L_z + L_z L_y) + i\hbar(L_z L_y + L_y L_z) \\
&= 0. \tag{6.13a}
\end{aligned}$$

In the same way, we also find that

$$[\mathbf{L}^2, L_y] = [\mathbf{L}^2, L_z] = 0 \tag{6.13b}$$

and we may summarise the equations (6.13) by writing

$$[\mathbf{L}^2, \mathbf{L}] = 0. \tag{6.14}$$

It follows that simultaneous eigenfunctions of $\mathbf{L}^2$ and any one component of $\mathbf{L}$ can be found, so that both the magnitude of the orbital angular momentum and the value of any one of its components can be determined precisely. Furthermore, $\mathbf{L}^2$ and that component of $\mathbf{L}$ (say for example L_z) form a complete set for the specification of angular momentum states.

In obtaining the eigenvalues and eigenfunctions of $\mathbf{L}^2$ and of one of the components of $\mathbf{L}$, which we shall carry out in Section 6.3, it is convenient to express the orbital angular momentum operators in spherical polar coordinates (r, θ, ϕ) which are related to the Cartesian coordinates (x, y, z) of the vector $\mathbf{r}$ by

$$\begin{aligned}
x &= r \sin\theta \cos\phi \\
y &= r \sin\theta \sin\phi \\
z &= r \cos\theta
\end{aligned} \tag{6.15}$$

with $0 \leqslant r \leqslant \infty$, $0 \leqslant \theta \leqslant \pi$, $0 \leqslant \phi \leqslant 2\pi$ (see Fig. 6.1). After some algebra (Problem 6.3) it is found that

$$L_x = -i\hbar \left(-\sin\phi \frac{\partial}{\partial \theta} - \cot\theta \cos\phi \frac{\partial}{\partial \phi} \right) \tag{6.16a}$$

$$L_y = -i\hbar \left(\cos\phi \frac{\partial}{\partial \theta} - \cot\theta \sin\phi \frac{\partial}{\partial \phi} \right) \tag{6.16b}$$

$$L_z = -i\hbar \frac{\partial}{\partial \phi} \tag{6.16c}$$

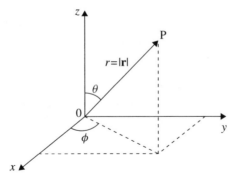

Figure 6.1 The spherical polar coordinates (r, θ, ϕ) of a point P. The position vector of P with respect to the origin is **r**.

and

$$\mathbf{L}^2 = -\hbar^2 \left[\frac{1}{\sin\theta} \frac{\partial}{\partial \theta} \left(\sin\theta \frac{\partial}{\partial \theta} \right) + \frac{1}{\sin^2\theta} \frac{\partial^2}{\partial \phi^2} \right]. \tag{6.17}$$

It should be noted that L_x, L_y, L_z and $\mathbf{L}^2$ are purely angular operators, which do not operate on the radial coordinate r. Therefore these operators commute with any function of r

$$[L_x, f(r)] = [L_y, f(r)] = [L_z, f(r)] = [\mathbf{L}^2, f(r)] = 0. \tag{6.18}$$

From the expressions (6.16) and (6.17) it is also clear that if a function is independent of θ and ϕ (and hence depends only on r), then it is simultaneously an eigenfunction of L_x, L_y, L_z and $\mathbf{L}^2$ corresponding in each case to the eigenvalue zero. This is the single exception to the rule that eigenfunctions of one component of $\mathbf{L}$ cannot be simultaneously eigenfunctions of either of the two other components.

Thus far we have considered the three Cartesian components of the orbital angular momentum operator $\mathbf{L}$, but in what follows we shall also need its component L_n along an arbitrary direction determined by the unit vector $\hat{\mathbf{n}} \equiv (n_x, n_y, n_z)$. For this purpose we use the following definition. If $\mathbf{V}$ is a vector operator which has Cartesian components V_x, V_y, V_z, the component V_n of $\mathbf{V}$ along the unit vector $\hat{\mathbf{n}}$ is

$$V_n \equiv \hat{\mathbf{n}} \cdot \mathbf{V} = n_x V_x + n_y V_y + n_z V_z. \tag{6.19}$$

In this way, one may define the three components V_u, V_v, V_w of $\mathbf{V}$ in any Cartesian frame whose axes are directed along the unit vectors $\hat{\mathbf{u}}$, $\hat{\mathbf{v}}$ and $\hat{\mathbf{w}}$ such that $\hat{\mathbf{w}} = \hat{\mathbf{u}} \times \hat{\mathbf{v}}$. In particular, the three components of $\mathbf{L}$ along $\hat{\mathbf{u}}$, $\hat{\mathbf{v}}$ and $\hat{\mathbf{w}}$ are given respectively by

$$L_u \equiv \hat{\mathbf{u}} \cdot \mathbf{L}, \qquad L_v \equiv \hat{\mathbf{v}} \cdot \mathbf{L}, \qquad L_w \equiv \hat{\mathbf{w}} \cdot \mathbf{L} \tag{6.20}$$

and it is readily verified (Problem 6.5) that the three commutation relations (6.9) for L_x, L_y and L_z imply analogous commutation relations for L_u, L_v and L_w, namely

$$[L_u, L_v] = i\hbar L_w, \qquad [L_v, L_w] = i\hbar L_u, \qquad [L_w, L_u] = i\hbar L_v. \tag{6.21}$$

Let us now consider an isolated system of N structureless particles. If $\mathbf{r}_i$ and $\mathbf{p}_i$ are respectively the position and momentum vectors of the ith particle, its orbital angular momentum is defined to be

$$\mathbf{L}_i = \mathbf{r}_i \times \mathbf{p}_i \tag{6.22}$$

so that in the position representation $\mathbf{L}_i$ is represented by the operator $-i\hbar(\mathbf{r}_i \times \boldsymbol{\nabla}_i)$ where $\boldsymbol{\nabla}_i$ is the gradient operator with respect to $\mathbf{r}_i$, having Cartesian components $(\partial/\partial x_i, \partial/\partial y_i, \partial/\partial z_i)$. The total orbital angular momentum $\mathbf{L}$ of the system is the vector sum of the orbital angular momenta of the N constituents:

$$\mathbf{L} = \sum_{i=1}^{N} \mathbf{L}_i. \tag{6.23}$$

Since the operators acting on different variables commute, all the components of $\mathbf{L}_i$ commute with all the components of $\mathbf{L}_k (i \neq k)$, and we have

$$\begin{aligned}
\mathbf{L} \times \mathbf{L} &= (\mathbf{L}_1 + \mathbf{L}_2 + \cdots + \mathbf{L}_N) \times (\mathbf{L}_1 + \mathbf{L}_2 + \cdots + \mathbf{L}_N) \\
&= \mathbf{L}_1 \times \mathbf{L}_1 + \mathbf{L}_2 \times \mathbf{L}_2 + \cdots + \mathbf{L}_N \times \mathbf{L}_N \\
&= i\hbar(\mathbf{L}_1 + \mathbf{L}_2 + \cdots + \mathbf{L}_N) = i\hbar \mathbf{L}
\end{aligned} \tag{6.24}$$

so that the components of the total orbital angular momentum satisfy the commutation relations

$$[L_x, L_y] = i\hbar L_z, \qquad [L_y, L_z] = i\hbar L_x, \qquad [L_z, L_x] = i\hbar L_y \tag{6.25}$$

as in the case of one particle.

6.2 Orbital angular momentum and spatial rotations

Considerable insight into the nature of the orbital angular momentum operator is obtained by analysing its connection with rotations in space. First of all, as in the case of the translations studied in Chapter 5, we must distinguish between the *passive* and *active* points of view in considering the effect of rotations on a physical system. According to the passive viewpoint, the coordinate axes are rotated while the system itself is untouched, while in the active viewpoint the axes are fixed and the system is rotated. The two viewpoints are equivalent in the sense that rotating the physical system about an axis or rotating the coordinate axes in the opposite direction amounts to the same thing. In what follows we shall adopt the active viewpoint and rotate the physical system.

Let us assume that we are dealing with an *isolated* system. Since space is intrinsically *isotropic* (i.e. all directions are equivalent), the physics of this system must be the same before and after a rotation is performed on it. Using the results of Section 5.10, it follows that under a rotation R the state vector Ψ describing the system must transform as

$$\Psi' = U_R \Psi \tag{6.26}$$

where U_R is a unitary operator (such that $U_R U_R^\dagger = U_R^\dagger U_R = I$). The normalisation is then preserved

$$\langle\Psi'|\Psi'\rangle = \langle U_R\Psi|U_R\Psi\rangle = \langle\Psi|U_R^\dagger U_R|\Psi\rangle = \langle\Psi|\Psi\rangle \tag{6.27}$$

and the probability amplitude $\langle X|\Psi\rangle$ that the system described by Ψ will be found in a state X is also unchanged by the rotation.

Consider now a dynamical variable (observable) to which corresponds an operator A, and let A' be the transform of A under the rotation R. Since the average value of A in a state Ψ must be equal to the average value of A' in the state $\Psi' = U_R\Psi$, we must have

$$\langle\Psi'|A'|\Psi'\rangle = \langle\Psi|U_R^\dagger A' U_R|\Psi\rangle = \langle\Psi|A|\Psi\rangle \tag{6.28}$$

and hence

$$A = U_R^\dagger A' U_R \quad \text{or} \quad A' = U_R A U_R^\dagger. \tag{6.29}$$

In particular, the Hamiltonian H of an isolated system must be *invariant* for spatial rotations. As a result, we have

$$H' = U_R H U_R^\dagger = H \tag{6.30}$$

so that $U_R H = H U_R$ and

$$[U_R, H] = 0. \tag{6.31}$$

In order to obtain an explicit form for the unitary rotation operator U_R, we begin by considering the case of a single structureless particle, whose wave function at a given time is $\psi(\mathbf{r})$. Under a rotation R the point $\mathbf{r}$ is moved to $\mathbf{r}' = R\mathbf{r}$. Except for a phase factor, which may be chosen equal to unity[2] the wave function ψ' describing the rotated state at $\mathbf{r}'$ must be equal to the original wave function ψ at the point $\mathbf{r}$, namely

$$\psi'(\mathbf{r}') = \psi(\mathbf{r}). \tag{6.32}$$

According to (6.26) we want to relate ψ and ψ' by a unitary operator U_R such that

$$\psi'(\mathbf{r}) = U_R \psi(\mathbf{r}). \tag{6.33}$$

Now, from (6.32) we have $\psi'(\mathbf{r}') = \psi(R^{-1}\mathbf{r}')$, and since this relation is valid for all points in space, it follows that $\psi'(\mathbf{r}) = \psi(R^{-1}\mathbf{r})$. We therefore obtain for the determination of U_R the relation

$$\psi'(\mathbf{r}) = U_R \psi(\mathbf{r}) = \psi(R^{-1}\mathbf{r}) \tag{6.34}$$

which specifies the one-to-one correspondence (6.26) in the position representation.

As a first example, let us consider an *infinitesimal rotation* by an angle $\delta\alpha$ in the positive (right-handed) sense about the z-axis, and let us denote by $U_z(\delta\alpha)$ the

---

[2] This point is discussed in detail by A. Messiah (1968), Chapter 15.

corresponding rotation operator U_R. To first order in $\delta\alpha$, the coordinates (x', y', z') of $\mathbf{r}' = R\mathbf{r}$ are given by

$$x' = x - y\delta\alpha$$
$$y' = x\delta\alpha + y$$
$$z' = z \qquad (6.35)$$

and those of $R^{-1}\mathbf{r}$ are obtained by changing $\delta\alpha$ into $-\delta\alpha$. Using (6.33), we have

$$\psi'(\mathbf{r}) = U_z(\delta\alpha)\psi(\mathbf{r}) = \psi(R^{-1}\mathbf{r}) = \psi(x + y\delta\alpha, y - x\delta\alpha, z) \qquad (6.36)$$

and retaining only first-order terms in $\delta\alpha$, we find that

$$U_z(\delta\alpha)\psi(\mathbf{r}) = \psi(x, y, z) + y\delta\alpha\frac{\partial\psi}{\partial x} - x\delta\alpha\frac{\partial\psi}{\partial y}$$

$$= \left[I - \delta\alpha\left(x\frac{\partial}{\partial y} - y\frac{\partial}{\partial x}\right)\right]\psi(\mathbf{r})$$

$$= \left[I - \frac{i}{\hbar}\delta\alpha L_z\right]\psi(\mathbf{r}) \qquad (6.37)$$

where we have used (6.3c). Hence we obtain the important relation

$$U_z(\delta\alpha) = I - \frac{i}{\hbar}\delta\alpha L_z \qquad (6.38)$$

between the infinitesimal rotation operator $U_z(\delta\alpha)$ and the z-component of the orbital angular momentum. Note that since $U_z(\delta\alpha)$ is an infinitesimal unitary transformation it follows from the discussion of Section 5.5 that L_z must be Hermitian.

Relations similar to (6.38) are readily derived for infinitesimal rotations about the x- and y-axes. As expected, they involve respectively the x- and y-components of $\mathbf{L}$,

$$U_x(\delta\alpha) = I - \frac{i}{\hbar}\delta\alpha L_x, \qquad U_y(\delta\alpha) = I - \frac{i}{\hbar}\delta\alpha L_y. \qquad (6.39)$$

More generally, let us consider an infinitesimal rotation by an angle $\delta\alpha$ in the positive sense about an axis oriented along an arbitrary unit vector $\hat{\mathbf{n}}$, so that

$$\mathbf{r}' = R\mathbf{r} = \mathbf{r} + \delta\alpha\hat{\mathbf{n}} \times \mathbf{r}. \qquad (6.40)$$

Denoting by $U_{\hat{\mathbf{n}}}(\delta\alpha)$ the corresponding infinitesimal rotation operator, we have

$$U_{\hat{\mathbf{n}}}(\delta\alpha)\psi(\mathbf{r}) = \psi(R^{-1}\mathbf{r}) = \psi(\mathbf{r} - \delta\alpha\hat{\mathbf{n}} \times \mathbf{r}). \qquad (6.41)$$

Keeping terms to first order in $\delta\alpha$, we find that

$$U_{\hat{\mathbf{n}}}(\delta\alpha)\psi(\mathbf{r}) = \psi(\mathbf{r}) - (\delta\alpha\hat{\mathbf{n}} \times \mathbf{r}).\nabla\psi(\mathbf{r})$$

$$= [I - \delta\alpha\hat{\mathbf{n}}.(\mathbf{r} \times \nabla)]\psi(\mathbf{r})$$

$$= \left[I - \frac{i}{\hbar}\delta\alpha\hat{\mathbf{n}}.\mathbf{L}\right]\psi(\mathbf{r}) \qquad (6.42)$$

where we have used the definition (6.4) of the operator **L**, and we recall that $\hat{\mathbf{n}} \cdot \mathbf{L}$ is the component of **L** along $\hat{\mathbf{n}}$. Hence

$$U_{\hat{\mathbf{n}}}(\delta\alpha) = I - \frac{i}{\hbar}\delta\alpha\,\hat{\mathbf{n}} \cdot \mathbf{L} \tag{6.43}$$

so that the orbital angular momentum operator **L** may be called the *generator of infinitesimal rotations*.

Except when they are performed about the same axis, rotations do not in general commute. Looking at (6.43), we therefore expect that the commutation relations (6.9) obeyed by the components of the orbital angular momentum operator could be re-derived by considering two successive infinitesimal rotations about different coordinate axes in opposite orders. To see this, let us perform an infinitesimal rotation R_x of angle $\delta\alpha$ about the x-axis, followed by an infinitesimal rotation R_y of angle $\delta\alpha$ about the y-axis. Let the corresponding infinitesimal rotation operators be $U_x(\delta\alpha)$ and $U_y(\delta\alpha)$, respectively. According to (6.41) we then have (remembering that the order of the operations reads from right to left),

$$\begin{aligned} U_y(\delta\alpha)U_x(\delta\alpha)\psi(\mathbf{r}) &= \psi[(R_y R_x)^{-1}\mathbf{r}] \\ &= \psi[R_x^{-1} R_y^{-1}\mathbf{r}] \\ &= \psi[R_x^{-1}(\mathbf{r} - \delta\alpha\,\hat{\mathbf{y}} \times \mathbf{r})] \\ &= \psi[\mathbf{r} - \delta\alpha\,\hat{\mathbf{y}} \times \mathbf{r} - \delta\alpha\,\hat{\mathbf{x}} \times (\mathbf{r} - \delta\alpha\,\hat{\mathbf{y}} \times \mathbf{r})] \end{aligned} \tag{6.44}$$

where $\hat{\mathbf{x}}$ and $\hat{\mathbf{y}}$ are unit vectors about the x- and y-axis, respectively. Similarly, if we perform first the infinitesimal rotation R_y about the y-axis, and then the infinitesimal rotation R_x about the x-axis, we obtain

$$U_x(\delta\alpha)U_y(\delta\alpha)\psi(\mathbf{r}) = \psi[\mathbf{r} - \delta\alpha\,\hat{\mathbf{x}} \times \mathbf{r} - \delta\alpha\,\hat{\mathbf{y}} \times (\mathbf{r} - \delta\alpha\,\hat{\mathbf{x}} \times \mathbf{r})]. \tag{6.45}$$

Subtracting (6.44) from (6.45), we find that the commutator of the two operators $U_x(\delta\alpha)$ and $U_y(\delta\alpha)$, acting on $\psi(\mathbf{r})$, gives

$$\begin{aligned}{} [U_x(\delta\alpha), U_y(\delta\alpha)]\psi(\mathbf{r}) &= \psi[\mathbf{r} - \delta\alpha(\hat{\mathbf{x}} + \hat{\mathbf{y}}) \times \mathbf{r} + (\delta\alpha)^2\hat{\mathbf{y}} \times (\hat{\mathbf{x}} \times \mathbf{r})] \\ &\quad - \psi[\mathbf{r} - \delta\alpha(\hat{\mathbf{y}} + \hat{\mathbf{x}}) \times \mathbf{r} + (\delta\alpha)^2\hat{\mathbf{x}} \times (\hat{\mathbf{y}} \times \mathbf{r})] \\ &= (\delta\alpha)^2[\hat{\mathbf{y}} \times (\hat{\mathbf{x}} \times \mathbf{r}) - \hat{\mathbf{x}} \times (\hat{\mathbf{y}} \times \mathbf{r})] \cdot \nabla \psi(\mathbf{r}) \\ &= (\delta\alpha)^2[y\hat{\mathbf{x}} - x\hat{\mathbf{y}}] \cdot \nabla \psi(\mathbf{r}) \\ &= -(\delta\alpha)^2(\hat{\mathbf{z}} \times \mathbf{r}) \cdot \nabla \psi(\mathbf{r}) \\ &= -(\delta\alpha)^2[\hat{\mathbf{z}} \cdot (\mathbf{r} \times \nabla)]\psi(\mathbf{r}) \\ &= -(\delta\alpha)^2\frac{i}{\hbar}L_z\psi(\mathbf{r}) \end{aligned} \tag{6.46}$$

where we have kept terms to leading order $(\delta\alpha)^2$ on the right-hand side. On the other hand, it follows from (6.39) that

$$[U_x(\delta\alpha), U_y(\delta\alpha)]\psi(\mathbf{r}) = \left[\left(I - \frac{i}{\hbar}\delta\alpha L_x\right)\left(I - \frac{i}{\hbar}\delta\alpha L_y\right)\right.$$
$$\left. - \left(I - \frac{i}{\hbar}\delta\alpha L_y\right)\left(I - \frac{i}{\hbar}\delta\alpha L_x\right)\right]\psi(\mathbf{r})$$
$$= -\frac{(\delta\alpha)^2}{\hbar^2}[L_x, L_y]\psi(\mathbf{r}) \qquad (6.47)$$

and therefore, upon comparison of (6.46) and (6.47) we find that $[L_x, L_y] = i\hbar L_z$, which is the commutation relation (6.9a). The commutation relations (6.9b) and (6.9c) may clearly be obtained in a similar way. The non-commutativity of different components of the orbital angular momentum operator $\mathbf{L}$ is therefore a direct consequence of the non-commutativity of spatial rotations about different axes. Notice that, to the extend that $(\delta\alpha)^2$ can be neglected, infinitesimal rotations do, in fact, commute.

Let us now consider rotations through *finite* angles. Any such finite rotation about an axis oriented along an arbitrary unit vector $\hat{\mathbf{n}}$ can be built up by a succession of infinitesimal rotations about that axis. If $U_{\hat{\mathbf{n}}}(\alpha)$ is the unitary operator corresponding to a finite rotation of angle α, we may therefore write

$$U_{\hat{\mathbf{n}}}(\alpha + \delta\alpha) = U_{\hat{\mathbf{n}}}(\delta\alpha)U_{\hat{\mathbf{n}}}(\alpha)$$
$$= \left(I - \frac{i}{\hbar}\delta\alpha\hat{\mathbf{n}}.\mathbf{L}\right)U_{\hat{\mathbf{n}}}(\alpha) \qquad (6.48)$$

where we have used (6.43). Thus

$$dU_{\hat{\mathbf{n}}}(\alpha) = U_{\hat{\mathbf{n}}}(\alpha + \delta\alpha) - U_{\hat{\mathbf{n}}}(\alpha) = \left(-\frac{i}{\hbar}\delta\alpha\hat{\mathbf{n}}.\mathbf{L}\right)U_{\hat{\mathbf{n}}}(\alpha) \qquad (6.49)$$

so that

$$\frac{dU_{\hat{\mathbf{n}}}(\alpha)}{d\alpha} = \left(-\frac{i}{\hbar}\hat{\mathbf{n}}.\mathbf{L}\right)U_{\hat{\mathbf{n}}}(\alpha) \qquad (6.50)$$

and upon integration, we find that

$$U_{\hat{\mathbf{n}}}(\alpha) = \exp\left(-\frac{i}{\hbar}\alpha\hat{\mathbf{n}}.\mathbf{L}\right) \qquad (6.51)$$

where the meaning of the operator $\exp(-i\alpha\hat{\mathbf{n}}.\mathbf{L}/\hbar)$ is (see (5.38))

$$\exp\left(-\frac{i}{\hbar}\alpha\hat{\mathbf{n}}.\mathbf{L}\right) = \sum_{n=0}^{\infty}\frac{1}{n!}\left(-\frac{i\alpha\hat{\mathbf{n}}.\mathbf{L}}{\hbar}\right)^n. \qquad (6.52)$$

The above considerations are readily extended to an isolated system of N structureless particles. In particular, replacing the vector coordinate $\mathbf{r}$ by the set of vector coordinates $(\mathbf{r}_1, \mathbf{r}_2, \ldots, \mathbf{r}_N)$ of the N particles, one finds (Problem 6.6) that the unitary rotation operators U_R are still given by equations (6.43) (for infinitesimal rotations) and (6.51) (for finite rotations), where **L** now denotes the *total* orbital angular momentum (6.23).

Let us now return to the important equation (6.31), which expresses the invariance of the Hamiltonian H of an isolated system for spatial rotations. Because any rotation can be expressed as a product of infinitesimal rotations, and since the total orbital angular momentum is the generator of infinitesimal rotations (see (6.43)), it follows that the invariance of H under rotations implies that, for an isolated system of structureless particles,

$$[\mathbf{L}, H] = 0 \tag{6.53}$$

so that the *total orbital angular momentum is conserved*. Conversely, if the Hamiltonian H of a system of structureless particles commutes with the total orbital angular momentum operator **L**, then H is invariant under rotations. It also follows from (6.53) and (6.14) that the operators H, $\mathbf{L}^2$ and L_z mutually commute. Thus, the eigenfunctions of H can be found among those common to $\mathbf{L}^2$ and L_z, so that the eigenvalue problem of H considerably simplifies.

6.3 The eigenvalues and eigenfunctions of $\mathbf{L}^2$ and L_z

Let us consider a single particle, for which the orbital angular momentum operator **L** is given by (6.4). In Section 6.1 we learned that the operator $\mathbf{L}^2$ commutes with any component of **L**, but that two components of **L** cannot accurately be known simultaneously. Thus simultaneous eigenfunction of $\mathbf{L}^2$ and of one of its components can be found, and we shall obtain them in this section, using spherical polar coordinates. Since we have seen in Section 6.1 that the components of **L** as well as $\mathbf{L}^2$ are purely angular operators, we only need to consider the polar angles (θ, ϕ). Moreover, because the expression (6.16c) for L_z is simpler than those for L_x and L_y in spherical polar coordinates, it is convenient to look for simultaneous eigenfunctions of $\mathbf{L}^2$ and of the component L_z.

Eigenfunctions and eigenvalues of L_z

We begin by finding the eigenvalues and eigenfunctions of the operator L_z alone. Denoting the eigenvalues of L_z by $m\hbar$, and the corresponding eigenfunctions by $\Phi_m(\phi)$, we have

$$L_z \Phi_m(\phi) = m\hbar \Phi_m(\phi) \tag{6.54}$$

or

$$-i\frac{\partial}{\partial \phi}\Phi_m(\phi) = m\Phi_m(\phi) \tag{6.55}$$

where we have used (6.16c). The solutions of this equation are

$$\Phi_m(\phi) = (2\pi)^{-1/2} e^{im\phi} \tag{6.56}$$

where $(2\pi)^{-1/2}$ is a normalisation constant. The equation (6.55) is then satisfied for any value of m. However, for the wave function to be *single-valued*[3], we must require that $\Phi_m(2\pi) = \Phi_m(0)$, or

$$e^{i2\pi m} = 1. \tag{6.57}$$

This relation is satisfied if m is restricted to positive or negative integers, or zero. Hence the eigenvalues of the operator L_z are equal to $m\hbar$, where

$$m = 0, \pm 1, \pm 2, \ldots \tag{6.58}$$

and a measurement of the z-component of the orbital angular momentum can only yield the values $0, \pm\hbar, \pm 2\hbar, \ldots$ Because the z-axis can be chosen along an arbitrary direction, we see that the component of the orbital angular momentum about any axis is quantised. The quantum number m is usually called the *magnetic quantum number*, a name which, as we shall see later, is justified by its prominent role in the study of charged particles in magnetic fields.

The eigenfunctions $\Phi_m(\phi)$ given by (6.56) are *orthonormal* over the range $0 \leqslant \phi \leqslant 2\pi$,

$$\int_0^{2\pi} \Phi_{m'}^*(\phi)\Phi_m(\phi)\mathrm{d}\phi = \delta_{mm'} \tag{6.59}$$

which means that they are normalised to unity and that eigenfunctions belonging to different eigenvalues ($m \neq m'$) are orthogonal. The eigenfunctions (6.56) also form a *complete set*, so that any function $f(\phi)$ defined on the interval $0 \leqslant \phi \leqslant 2\pi$ can be expanded as

$$f(\phi) = \sum_{m=-\infty}^{+\infty} a_m \Phi_m(\phi). \tag{6.60}$$

The coefficients a_m are found by multiplying both sides of (6.60) by $\Phi_{m'}^*(\phi)$, integrating over ϕ from 0 to 2π and using (6.59). This gives

$$a_m = \int_0^{2\pi} \Phi_m^*(\phi) f(\phi) \mathrm{d}\phi. \tag{6.61}$$

Inserting this expression of a_m (in which we change the name of the integration variable from ϕ to ϕ') into (6.60) and using the property (A.26) of the delta function, we see that the eigenfunctions $\Phi_m(\phi)$ satisfy the *closure relation*

$$\sum_{m=-\infty}^{+\infty} \Phi_m^*(\phi')\Phi_m(\phi) = \delta(\phi - \phi'). \tag{6.62}$$

---

[3] It should be stressed that the requirement for single-valuedness of the wave function only applies to its spatial part. As we shall see in Section 6.9, certain wave functions depending on internal degrees of freedom may be double-valued; in particular, this is the case for spin wave functions corresponding to particles having half-odd integer spin (see (6.268)).

Simultaneous eigenfunctions of L^2 and L_z

We now turn to the problem of obtaining the common eigenfunctions $Y_{lm}(\theta, \phi)$ of the operators L^2 and L_z, satisfying the two eigenvalue equations

$$L^2 Y_{lm}(\theta, \phi) = l(l+1)\hbar^2 Y_{lm}(\theta, \phi) \tag{6.63}$$

and

$$L_z Y_{lm}(\theta, \phi) = m\hbar Y_{lm}(\theta, \phi) \tag{6.64}$$

where the eigenvalues of L^2 have been written as $l(l+1)\hbar^2$. From our study of the eigenfunctions of L_z, we see that equation (6.64) implies that the functions $Y_{lm}(\theta, \phi)$ must be of the form

$$Y_{lm}(\theta, \phi) = \Theta_{lm}(\theta)\Phi_m(\phi) \tag{6.65}$$

where the functions $\Phi_m(\phi)$ are given by (6.56), with $m = 0, \pm 1, \pm 2, \ldots$

Let us now turn to equation (6.63). Using the expression (6.17) of L^2 in spherical polar coordinates, we have

$$\left[\frac{1}{\sin\theta}\frac{\partial}{\partial\theta}\left(\sin\theta\frac{\partial}{\partial\theta}\right) + \frac{1}{\sin^2\theta}\frac{\partial^2}{\partial\phi^2}\right] Y_{lm}(\theta, \phi) = -l(l+1)Y_{lm}(\theta, \phi). \tag{6.66}$$

Substituting (6.65) into (6.66) and making use of (6.56), we obtain for $\Theta_{lm}(\theta)$ the differential equation

$$\left[\frac{1}{\sin\theta}\frac{d}{d\theta}\left(\sin\theta\frac{d}{d\theta}\right) + \left\{l(l+1) - \frac{m^2}{\sin^2\theta}\right\}\right]\Theta_{lm}(\theta) = 0. \tag{6.67}$$

Our task is to find physically acceptable solutions of this equation over the range $0 \leqslant \theta \leqslant \pi$. For this purpose, it is convenient to introduce the new variable $w = \cos\theta$ and the new function

$$F_{lm}(w) = \Theta_{lm}(\theta) \tag{6.68}$$

so that (6.67) reads

$$\left[(1-w^2)\frac{d^2}{dw^2} - 2w\frac{d}{dw} + l(l+1) - \frac{m^2}{1-w^2}\right]F_{lm}(w) = 0 \tag{6.69}$$

with $-1 \leqslant w \leqslant 1$.

The case $m = 0$

It is easiest to start with the case $m = 0$, for which (6.69) reduces to

$$\left[(1-w^2)\frac{d^2}{dw^2} - 2w\frac{d}{dw} + l(l+1)\right]F_{l0}(w) = 0. \tag{6.70}$$

This differential equation is known as the *Legendre equation*. Since the point $w = 0$ is an ordinary (non-singular) point of this equation, we can try a power series solution of the form

$$F_{l0}(w) = \sum_{k=0}^{\infty} c_k w^k. \tag{6.71}$$

Inserting (6.71) into (6.70) and setting $\lambda = l(l+1)$, we find that

$$\sum_{k=0}^{\infty} [k(k-1)c_k w^{k-2} - k(k-1)c_k w^k - 2kc_k w^k + \lambda c_k w^k] = 0 \tag{6.72}$$

or

$$\sum_{k=0}^{\infty} [(k+1)(k+2)c_{k+2} + [\lambda - k(k+1)]c_k] w^k = 0. \tag{6.73}$$

This equation will be satisfied if the coefficient of each power of w vanishes. We therefore obtain for the coefficients c_k the *recursion relation*

$$c_{k+2} = \frac{k(k+1) - \lambda}{(k+1)(k+2)} c_k \tag{6.74}$$

so that the general power series solution of (6.70) is given by

$$F_{l0}(w) = c_0 \left[1 - \frac{\lambda}{2} w^2 - \frac{2.3 - \lambda}{3.4} \frac{\lambda}{2} w^4 + \cdots \right]$$
$$+ c_1 \left[w + \frac{1.2 - \lambda}{2.3} w^3 + \frac{3.4 - \lambda}{4.5} \frac{1.2 - \lambda}{2.3} w^5 + \cdots \right] \tag{6.75}$$

where c_0 and c_1 are arbitrary constants. Note that the first power series, which multiplies c_0 on the right of (6.75), contains only *even* powers of w, while the second power series (multiplying c_1) contains only *odd* powers of w. The occurrence of particular solutions which are either even functions of w (when $c_1 = 0$) or odd functions of w (when $c_0 = 0$) is to be expected, since equation (6.70) is invariant when $-w$ is substituted for w.

We see from (6.74) that if the series on the right of (6.75) do not terminate at some fixed values of k, then for large k we have $c_{k+2}/c_k \simeq k/(k+2)$, a behaviour which is similar to the divergent series Σk^{-1}, where k is even or odd. As a result, both series on the right of (6.75) diverge for $w^2 = 1$, i.e. for $w = \pm 1$, and in consequence do not yield an acceptable wave function[4]. Thus, in order to be satisfactory, the expression (6.75) for $F_{l0}(w)$ must contain only a finite number of terms. Using (6.74) and remembering that $\lambda = l(l+1)$, we see that by choosing l to be a positive integer to zero,

$$l = 0, 1, 2, \ldots \tag{6.76}$$

---

[4] This is not surprising, since $w = \pm 1$ are (regular) *singular* points of the Legendre equation (6.70). See, for example, Mathews and Walker (1973).

either the even or odd series in (6.75) can be terminated at the finite value $k = l$. The other series can be made to vanish by choosing $c_1 = 0$ (for even l) or $c_0 = 0$ (for odd l). The quantum number l, whose allowed values are given by (6.76), is called the *orbital angular momentum quantum number*.

The physically acceptable solutions of the Legendre equation (6.70), which exist only for $l = 0, 1, 2, \ldots$, are thus polynomials of degree l, called *Legendre polynomials*, and denoted by $P_l(w)$. From the above discussion it is clear that they are uniquely defined, apart from an arbitrary multiplicative constant. By convention this constant is chosen so that

$$P_l(1) = 1. \tag{6.77}$$

Since the Legendre polynomial $P_l(w)$ is in even or odd powers of w, depending on whether l is even or odd, we have

$$P_l(-w) = (-1)^l P_l(w). \tag{6.78}$$

An equivalent definition of the Legendre polynomials is

$$P_l(w) = (2)^{-l}(l!)^{-1} \frac{d^l}{dw^l}(w^2 - 1)^l \tag{6.79}$$

which is known as the Rodrigues formula. It is readily verified that the Legendre polynomials given by (6.79) satisfy the Legendre equation (6.70) and the relation (6.77). The Rodrigues formula can be used to prove various important properties of the Legendre polynomials. In particular, by performing integrations by parts, one finds that (Problem 6.7)

$$\int_{-1}^{+1} P_l(w) P_{l'}(w) dw = \frac{2}{2l+1} \delta_{ll'} \tag{6.80}$$

so that the Legendre polynomials are orthogonal for $l \neq l'$ over the range $-1 \leqslant w \leqslant 1$, but they are not normalised to unity. Of course, physically acceptable solutions of the Legendre equation (6.70) which are orthogonal for $l \neq l'$ and normalised to unity are easily constructed. From (6.80) we see that apart from a phase factor $\exp(i\alpha)$ of modulus one they are given by

$$F_{l0}(w) = \left(\frac{2l+1}{2}\right)^{1/2} P_l(w). \tag{6.81}$$

Another equivalent definition of the Legendre polynomials may be given in terms of a *generating function*, by

$$\begin{aligned} T(w, s) &= (1 - 2ws + s^2)^{-1/2} \\ &= \sum_{l=0}^{\infty} P_l(w) s^l, \qquad |s| < 1. \end{aligned} \tag{6.82}$$

By differentiating this generating function with respect to w and s one can show (Problem 6.8) that the Legendre polynomials satisfy the recurrence relations

$$(l+1)P_{l+1}(w) = (2l+1)wP_l(w) - lP_{l-1}(w) \tag{6.83a}$$

and

$$(1-w^2)\frac{dP_l}{dw} = -lwP_l(w) + lP_{l-1}(w). \tag{6.83b}$$

The lowest-order differential equation for P_l which can be constructed from these two recurrence relations is then seen to be the Legendre equation (6.70). Note that from (6.82) we immediately have $P_l(1) = 1$.

The Legendre polynomials $P_l(w)$ form a *complete set* on the interval $-1 \leqslant w \leqslant 1$, so that any function $f(w)$, defined on this interval, can be expanded as

$$f(w) = \sum_{l=0}^{\infty} a_l P_l(w) \tag{6.84}$$

and by using (6.80) we have

$$a_l = \frac{2l+1}{2} \int_{-1}^{+1} P_l(w) f(w) dw. \tag{6.85}$$

Inserting this expression of a_l into (6.84) and using (6.80), we also find that the Legendre polynomials satisfy the *closure relation*

$$\frac{1}{2} \sum_{l=0}^{\infty} (2l+1) P_l(w') P_l(w) = \delta(w-w'). \tag{6.86}$$

The first few Legendre polynomials are

$$P_0(w) = 1$$
$$P_1(w) = w$$
$$P_2(w) = \tfrac{1}{2}(3w^2 - 1)$$
$$P_3(w) = \tfrac{1}{2}(5w^3 - 3w)$$
$$P_4(w) = \tfrac{1}{8}(35w^4 - 30w^2 + 3)$$
$$P_5(w) = \tfrac{1}{8}(63w^5 - 70w^3 + 15w). \tag{6.87}$$

The general case

Having solved the Legendre equation (6.70) we shall now obtain the physically acceptable solutions of equation (6.69), for which m is not necessarily equal to zero. Since (6.69) is independent of the sign of m, the solutions of this equation can be characterised by l and $|m|$. Let us define the *associated Legendre function* $P_l^{|m|}(w)$ of degree l ($l = 0, 1, 2, \ldots$) and order $|m| \leqslant l$ by the relation

$$P_l^{|m|}(w) = (1-w^2)^{|m|/2} \frac{d^{|m|}}{dw^{|m|}} P_l(w), \qquad |m| = 0, 1, 2, \ldots \tag{6.88}$$

We note that for $m = 0$ we have $P_l^0(w) = P_l(w)$. Upon differentiating $|m|$ times the Legendre equation (6.70) satisfied by $P_l(w)$, we obtain

$$\left[(1-w^2)\frac{d^{|m|+2}}{dw^{|m|+2}} - 2(|m|+1)w\frac{d^{|m|+1}}{dw^{|m|+1}} + \{l(l+1)\right.$$
$$\left. -|m|(|m|+1)\}\frac{d^{|m|}}{dw^{|m|}}\right]P_l(w) = 0. \qquad (6.89)$$

Using (6.88) we then find that $P_l^{|m|}(w)$ satisfies the equation

$$\left[(1-w^2)\frac{d^2}{dw^2} - 2w\frac{d}{dw} + l(l+1) - \frac{m^2}{1-w^2}\right]P_l^{|m|}(w) = 0 \qquad (6.90)$$

which is the same as (6.69). Apart from an arbitrary multiplicative constant, the associated Legendre functions $P_l^{|m|}(w)$ are the only physically acceptable solutions of (6.69). We emphasise that the orbital angular momentum quantum number l is restricted to one of the values $l = 0, 1, 2, \ldots$ Regarding the magnetic quantum number m, we note from (6.88) that since $P_l(w)$ is a polynomial of degree, l, its $|m|$th derivative, and hence $P_l^{|m|}(w)$, will vanish if $|m| > l$. Hence for a fixed value of l there are $(2l+1)$ allowed values of m given by

$$m = -l, -l+1, \ldots, l. \qquad (6.91)$$

This result can readily be understood as follows. Since $\mathbf{L}^2 = L_x^2 + L_y^2 + L_z^2$, the expectation value of $\mathbf{L}^2$ in a given state Ψ is

$$\langle \mathbf{L}^2 \rangle = \langle L_x^2 \rangle + \langle L_y^2 \rangle + \langle L_z^2 \rangle. \qquad (6.92)$$

Now, since L_x and L_y are Hermitian, $\langle L_x^2 \rangle \geq 0$ and $\langle L_y^2 \rangle \geq 0$, so that

$$\langle \mathbf{L}^2 \rangle \geq \langle L_z^2 \rangle. \qquad (6.93)$$

Taking Ψ to be such that its angular part is an eigenfunction $Y_{lm}(\theta, \phi)$ common to $\mathbf{L}^2$ and L_z, we have from (6.63), (6.64) and (6.93)

$$l(l+1) \geq m^2 \qquad (6.94)$$

and as l and m can only take the values $l = 0, 1, 2, \ldots$, and $m = 0, \pm 1, \pm 2, \ldots$, we see that m is restricted for a given l to the $(2l+1)$ values given by (6.91).

We note from (6.88) that the associated Legendre function $P_l^{|m|}(w)$ is the product of the quantity $(1-w^2)^{|m|/2}$ and of a polynomial of degree $l - |m|$. Using (6.78) and (6.88) it is straightforward to show that (Problem 6.9)

$$P_l^{|m|}(-w) = (-1)^{l-|m|} P_l^{|m|}(w). \qquad (6.95)$$

A generating function for the $P_l^{|m|}$ can be obtained by differentiating (6.82) $|m|$ times with respect to w and multiplying by $(1-w^2)^{|m|/2}$ as required by (6.88). We obtain in this way

$$T_{|m|}(w,s) = \frac{(2|m|)!(1-w^2)^{|m|/2}s^{|m|}}{2^{|m|}(|m|)!(1-2ws+s^2)^{|m|+1/2}}$$

$$= \sum_{l=|m|}^{\infty} P_l^{|m|}(w)s^l, \quad |s| < 1. \tag{6.96}$$

We also remark that the functions $P_l^{|m|}$ satisfy the recurrence relations (Problem 6.10)

$$(2l+1)wP_l^{|m|}(w) = (l-|m|+1)P_{l+1}^{|m|}(w) + (l+|m|)P_{l-1}^{|m|}(w) \tag{6.97a}$$

$$(2l+1)(1-w^2)^{1/2}P_l^{|m|-1}(w) = P_{l+1}^{|m|}(w) - P_{l-1}^{|m|}(w) \tag{6.97b}$$

$$(1-w^2)\frac{dP_l^{|m|}}{dw} = -lwP_l^{|m|}(w) + (l+|m|)P_{l-1}^{|m|}(w)$$

$$= (l+1)wP_l^{|m|}(w) - (l+1-|m|)P_{l+1}^{|m|}(w) \tag{6.97c}$$

$$(1-w^2)\frac{dP_l^{|m|}}{dw} = (1-w^2)^{1/2}P_l^{|m|+1}(w) - |m|wP_l^{|m|}(w)$$

$$= -(1-w^2)^{1/2}(l+|m|)(l-|m|+1)P_l^{|m|-1}(w) + |m|wP_l^{|m|}(w) \tag{6.97d}$$

and the orthogonality relations

$$\int_{-1}^{+1} P_l^{|m|}(w)P_{l'}^{|m|}(w)dw = \frac{2}{2l+1}\frac{(l+|m|)!}{(l-|m|)!}\delta_{ll'}. \tag{6.98}$$

Finally, it can be shown that the $P_l^{|m|}$ form a complete set.

The first few associated Legendre functions are given explicitly by

$$P_1^1(w) = (1-w^2)^{1/2}$$
$$P_2^1(w) = 3(1-w^2)^{1/2}w$$
$$P_2^2(w) = 3(1-w^2)$$
$$P_3^1(w) = \tfrac{3}{2}(1-w^2)^{1/2}(5w^2-1)$$
$$P_3^2(w) = 15w(1-w^2)$$
$$P_3^3(w) = 15(1-w^2)^{3/2} \tag{6.99}$$

It should be noted from (6.98) that the functions $P_l^{|m|}(w)$ are not normalised to unity. However, by multiplying $P_l^{|m|}(w)$ by an appropriate normalisation factor, physically acceptable solutions $F_{lm}(w)$ of equation (6.69) which are normalised to unity can be determined to within an arbitrary phase factor of modulus one. We shall not quote these, but instead we write down the corresponding physically admissible solutions $\Theta_{lm}(\theta)$ of equation (6.67), normalised so that

$$\int_0^\pi \Theta^*_{l'm}(\theta)\Theta_{lm}(\theta)\sin\theta d\theta = \delta_{ll'}. \qquad (6.100)$$

These functions are given in terms of the associated Legendre functions by

$$\Theta_{lm}(\theta) = (-1)^m \left[\frac{(2l+1)(l-m)!}{2(l+m)!}\right]^{1/2} P_l^m(\cos\theta), \quad m \geq 0$$

$$= (-1)^m \Theta_{l|m|}(\theta), \quad m < 0 \qquad (6.101)$$

where the choice of the phase factor is a conventional one.

Spherical harmonics

The eigenfunctions $Y_{lm}(\theta, \phi)$, common to the operators $\mathbf{L}^2$ and L_z (see (6.63) and (6.64)) and normalised to unity on the unit sphere, are called *spherical harmonics*. Using (6.65), (6.56) and (6.101), we see that they are given for non-negative values of m by

$$Y_{lm}(\theta, \phi) = (-1)^m \left[\frac{(2l+1)(l-m)!}{4\pi(l+m)!}\right]^{1/2} P_l^m(\cos\theta) e^{im\phi}, \quad m \geq 0 \quad (6.102a)$$

and for negative values of m we have

$$Y_{lm}(\theta, \phi) = (-1)^m Y^*_{l,-m}(\theta, \phi) \qquad (6.102b)$$

where the phase convention adopted here is a frequently used one. We recall that the quantum number l can only take the values $l = 0, 1, 2, \ldots$, (see (6.76)) and that for a fixed value of l the allowed values of m are $m = -l, -l+1, \ldots, l$ (see (6.91)).

From (6.65), (6.56) and (6.101) we also note that the spherical harmonics satisfy the *orthonormality relations*

$$\int Y^*_{l'm'}(\theta, \phi) Y_{lm}(\theta, \phi) d\Omega = \int_0^{2\pi} d\phi \int_0^\pi d\theta \sin\theta Y^*_{l'm'}(\theta, \phi) Y_{lm}(\theta, \phi)$$

$$= \delta_{ll'}\delta_{mm'} \qquad (6.103)$$

where we have written $d\Omega \equiv \sin\theta d\theta d\phi$ and the symbol $\int d\Omega$ means that we integrate over the full range of the angular variables (θ, ϕ), namely

$$\int d\Omega \equiv \int_0^{2\pi} d\phi \int_0^\pi d\theta \sin\theta. \qquad (6.104)$$

It can also be proved that the Y_{lm} form a *complete set*, so that a function $f(\theta, \phi)$ can be expanded in terms of them as

$$f(\theta, \phi) = \sum_{l=0}^\infty \sum_{m=-l}^{+l} a_{lm} Y_{lm}(\theta, \phi) \qquad (6.105)$$

By using the orthonormality relation (6.103), the coefficients of the expansion are seen to be given by

$$a_{lm} = \int Y_{lm}^*(\theta, \phi) f(\theta, \phi) d\Omega. \tag{6.106}$$

Substituting (6.106) into (6.105) and using (6.103) we obtain the *closure relation*

$$\sum_{l=0}^{\infty} \sum_{m=-l}^{+l} Y_{lm}^*(\theta', \phi') Y_{lm}(\theta, \phi) = \delta(\Omega - \Omega'),$$

$$\delta(\Omega - \Omega') = \frac{\delta(\theta - \theta')\delta(\phi - \phi')}{\sin \theta} \tag{6.107}$$

It is clear from (6.102) and our study of associated Legendre functions that $Y_{lm}(\theta, \phi)$ is the product of $\exp(im\phi) \sin^{|m|} \theta$ times a polynomial in $\cos \theta$ of degree $(l - |m|)$. In particular, for $m = 0$ and $m = l$, we have, respectively,

$$Y_{l,0}(\theta) = \left(\frac{2l+1}{4\pi}\right)^{1/2} P_l(\cos \theta) \tag{6.108}$$

and

$$Y_{l,l}(\theta, \phi) = (-1)^l \left[\frac{2l+1}{4\pi} \frac{(2l)!}{2^{2l}(l!)^2}\right]^{1/2} \sin^l \theta e^{il\phi}. \tag{6.109}$$

The behaviour of the spherical harmonics under the *parity* operation $\mathbf{r} \to -\mathbf{r}$ is of special interest. Under this operation the spherical polar coordinates (r, θ, ϕ) transform as

$$r \to r, \quad \theta \to \pi - \theta, \quad \phi \to \phi + \pi. \tag{6.110}$$

Thus, if $\mathcal{P}$ is the parity operator defined in (5.336), we have

$$\mathcal{P}[Y_{lm}(\theta, \phi)] = Y_{lm}(\pi - \theta, \phi + \pi). \tag{6.111}$$

Now, from (6.95) we have

$$P_l^{|m|}[\cos(\pi - \theta)] = P_l^{|m|}(-\cos \theta) = (-1)^{l-|m|} P_l^{|m|}(\cos \theta) \tag{6.112}$$

while from (6.56) we deduce that

$$\Phi_m(\phi + \pi) = (-1)^m \Phi_m(\phi). \tag{6.113}$$

Hence

$$Y_{lm}(\pi - \theta, \phi + \pi) = (-1)^l Y_{lm}(\theta, \phi) \tag{6.114}$$

so that Y_{lm} has the parity of l (even for even l, odd for odd l).

Following a notation originally introduced in spectroscopy, it is customary to designate the states corresponding to the values $l = 0, 1, 2, 3, 4, 5, \ldots$, of the orbital angular momentum quantum number l of a particle by the symbols s, p, d, f, g, h, ... When there is more than one particle, capital letters S, P, D, F, G, H, ..., are used for the total orbital angular momentum. The first few spherical harmonics, corresponding

Table 6.1 Low-order spherical harmonics.

l	m	Spherical harmonic $Y_{lm}(\theta, \phi)$
0	0	$Y_{0,0} = \dfrac{1}{(4\pi)^{1/2}}$
1	0	$Y_{1,0} = \left(\dfrac{3}{4\pi}\right)^{1/2} \cos\theta$
	± 1	$Y_{1,\pm 1} = \mp\left(\dfrac{3}{8\pi}\right)^{1/2} \sin\theta e^{\pm i\phi}$
2	0	$Y_{2,0} = \left(\dfrac{5}{16\pi}\right)^{1/2} (3\cos^2\theta - 1)$
	± 1	$Y_{2,\pm 1} = \mp\left(\dfrac{15}{8\pi}\right)^{1/2} \sin\theta \cos\theta e^{\pm i\phi}$
	± 2	$Y_{2,\pm 2} = \left(\dfrac{15}{32\pi}\right)^{1/2} \sin^2\theta e^{\pm 2i\phi}$
3	0	$Y_{3,0} = \left(\dfrac{7}{16\pi}\right)^{1/2} (5\cos^3\theta - 3\cos\theta)$
	± 1	$Y_{3,\pm 1} = \mp\left(\dfrac{21}{64\pi}\right)^{1/2} \sin\theta(5\cos^2\theta - 1)e^{\pm i\phi}$
	± 2	$Y_{3,\pm 2} = \left(\dfrac{105}{32\pi}\right)^{1/2} \sin^2\theta \cos\theta e^{\pm 2i\phi}$
	± 3	$Y_{3,\pm 3} = \mp\left(\dfrac{35}{64\pi}\right)^{1/2} \sin^3\theta e^{\pm 3i\phi}$

to the s, p, d and f states are listed in Table 6.1, and polar plots of the corresponding probability distributions

$$|Y_{lm}(\theta, \phi)|^2 = (2\pi)^{-1}|\Theta_{lm}(\theta)|^2 \tag{6.115}$$

are shown in Fig. 6.2.

It is also interesting to study the effect of the operators L_x and L_y on Y_{lm}. For this purpose, it is convenient to introduce the two operators

$$L_+ = L_x + iL_y, \qquad L_- = L_x - iL_y. \tag{6.116}$$

We note that these operators are not Hermitian, but are mutually adjoint, since $L_+^\dagger = L_x - iL_y = L_-$ and $L_-^\dagger = L_x + iL_y = L_+$. Because L_x and L_y commute with $\mathbf{L}^2$, we have

$$[\mathbf{L}^2, L_\pm] = 0. \tag{6.117}$$

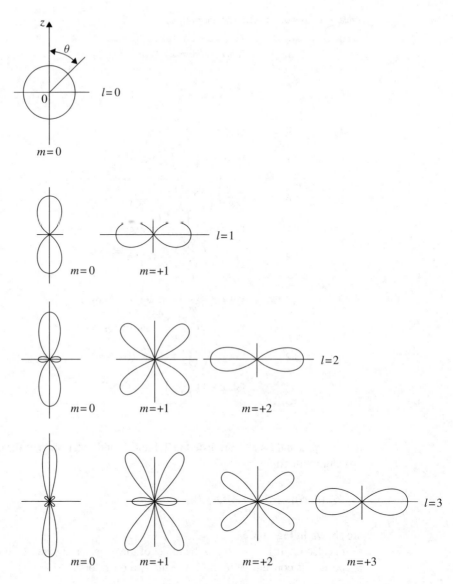

Figure 6.2 Polar plots of the probability distributions $|Y_{lm}(\theta,\phi)|^2 = (2\pi)^{-1}|\Theta_{lm}(\theta)|^2$.

It is also readily verified from the commutation relations (6.9) that (Problem 6.11)

$$L_{\pm}L_{\mp} = \mathbf{L}^2 - L_z^2 \pm \hbar L_z \tag{6.118a}$$

$$[L_+, L_-] = 2\hbar L_z \tag{6.118b}$$

6.3 The eigenvalues and eigenfunctions of $\mathbf{L}^2$ and L_z

and

$$[L_z, L_\pm] = \pm\hbar L_\pm. \tag{6.118c}$$

Acting on both sides of (6.64) with $L_\pm$ and using (6.118c), we find that

$$L_z(L_\pm Y_{lm}) = (m \pm 1)\hbar(L_\pm Y_{lm}). \tag{6.119}$$

Also, if we act on both sides of (6.63) with $L_\pm$ and use the fact that $\mathbf{L}^2$ commutes with $L_\pm$, we have

$$\mathbf{L}^2(L_\pm Y_{lm}) = l(l+1)\hbar^2(L_\pm Y_{lm}). \tag{6.120}$$

Thus the operator L_+ acting on a simultaneous eigenfunction Y_{lm} of $\mathbf{L}^2$ and L_z generates a new common eigenfunction of these two operators, for which the eigenvalue of $\mathbf{L}^2$ remains equal to $l(l+1)\hbar^2$, while the eigenvalue of L_z is increased by one unit of $\hbar$ to become $(m+1)\hbar$. Similarly, (L_-Y_{lm}) is a simultaneous eigenfunction of $\mathbf{L}^2$ and L_z with eigenvalues given, respectively, by $l(l+1)\hbar^2$ and $(m-1)\hbar$. Hence

$$L_\pm Y_{lm}(\theta, \phi) = c_{lm}^\pm Y_{l,m\pm 1}(\theta, \phi) \tag{6.121}$$

where $c_{lm}^\pm$ are constants. With regard to the eigenvalue equation (6.64) the operators L_+ and L_- are therefore respectively *raising* and *lowering* operators; we see that they play a role similar to that of the ladder operators a_+ and a_- introduced in Section 5.6 to study the spectrum of the linear harmonic oscillator.

In order to obtain the constants $c_{lm}^\pm$ of (6.121), we use the expressions (6.16a) and (6.16b) of the operators L_x and L_y to write $L_\pm$ in spherical polar coordinates. The result is

$$L_\pm = \hbar e^{\pm i\phi}\left[\pm\frac{\partial}{\partial\theta} + i\cot\theta\frac{\partial}{\partial\phi}\right]. \tag{6.122}$$

From (6.64) and (6.16c) we know that by acting with $\partial/\partial\phi$ on $Y_{lm}(\theta, \phi)$, one obtains $(im)Y_{lm}(\theta, \phi)$. The action of $\partial/\partial\theta$ on $Y_{lm}(\theta, \phi)$ may be studied by using (6.102) and the recurrence relations (6.97d) satisfied by the associated Legendre functions (in which one sets $w = \cos\theta$). In this way it is found that

$$L_\pm Y_{lm}(\theta, \phi) = \hbar[l(l+1) - m(m \pm 1)]^{1/2} Y_{l,m\pm 1}(\theta, \phi) \tag{6.123a}$$

or

$$L_\pm|lm\rangle = \hbar[l(l+1) - m(m \pm 1)]^{1/2}|lm \pm 1\rangle \tag{6.123b}$$

where in the last line we have used the Dirac ket notation, the ket $|lm\rangle$ being represented in the position representation by the spherical harmonic $Y_{lm}(\theta, \phi)$. From (6.123) and the fact that

$$L_x = \frac{1}{2}(L_+ + L_-), \qquad L_y = \frac{1}{2i}(L_+ - L_-) \tag{6.124}$$

the result of the action of the operators L_x and L_y on Y_{lm} (or $|lm\rangle$) is then immediately obtained.

Let us now consider a particle which is in the orbital angular momentum state $|lm\rangle$, so that the angular part of its wave function Ψ (which we assume to be normalised to unity) is given by $Y_{lm}(\theta, \phi)$. Thus we have $\mathbf{L}^2|lm\rangle = l(l+1)\hbar^2|lm\rangle$ and $L_z|lm\rangle = m\hbar|lm\rangle$. Although we have seen that two components of the orbital angular momentum cannot in general be assigned precise values simultaneously, it is nevertheless possible to say something about the components L_x and L_y. Indeed, from (6.123) and (6.124) the expectation value of L_x in the state $|lm\rangle$ is given by

$$\langle L_x \rangle = \langle lm|L_x|lm\rangle = \tfrac{1}{2}\langle lm|L_+ + L_-|lm\rangle. \tag{6.125}$$

Now, from (6.123b) we have

$$\langle lm|L_\pm|lm\rangle = \hbar[l(l+1) - m(m\pm 1)]^{1/2}\langle lm|lm\pm 1\rangle$$
$$= 0 \tag{6.126}$$

where we have used the fact that the spherical harmonics satisfy the orthonormality relations (6.103) so that $\langle lm|l'm'\rangle = \delta_{ll'}\delta_{mm'}$. Thus

$$\langle L_x \rangle = 0 \tag{6.127a}$$

and similarly

$$\langle L_y \rangle = 0. \tag{6.127b}$$

On the other hand, from (6.124), (6.118a) and (6.126) we obtain

$$\langle L_x^2 \rangle = \langle L_y^2 \rangle = \tfrac{1}{2}\langle \mathbf{L}^2 - L_z^2\rangle$$
$$= \tfrac{1}{2}[l(l+1) - m^2]\hbar^2. \tag{6.128}$$

Note that when $m = +l$ or $m = -l$, so that the orbital angular momentum is respectively 'parallel' or 'anti-parallel' to the z-axis, its x- and y-components are still not zero, although the average values of L_x and L_y always vanish. It is convenient to visualise these results in terms of a *vector model* (see Fig. 6.3). According to this model, the orbital angular momentum vector $\mathbf{L}$, of length $\sqrt{l(l+1)}\hbar$, precesses about the z-axis, the $(2l+1)$ allowed projections of $\mathbf{L}$ on this axis being given by $m\hbar$, with $m = -l, -l+1, \ldots, +l$. Thus the vector $\mathbf{L}$ may be viewed as lying on the surface of a cone with altitude $m\hbar$ which has the z-axis as its symmetry axis, all orientations of $\mathbf{L}$ on the surface of the cone being equally likely.

We conclude this section by mentioning an important property of the Y_{lm} called the *addition theorem* of the spherical harmonics. Let $\mathbf{r}_1$ and $\mathbf{r}_2$ be two vectors having polar angles (θ_1, ϕ_1) and (θ_2, ϕ_2), respectively, and let θ be the angle between them. The addition theorem states that

$$P_l(\cos\theta) = \frac{4\pi}{2l+1} \sum_{m=-l}^{+l} Y_{lm}^*(\theta_1, \phi_1) Y_{lm}(\theta_2, \phi_2). \tag{6.129}$$

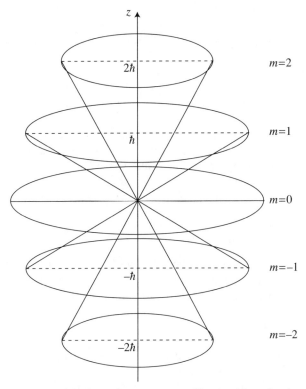

Figure 6.3 The vector model of angular momentum, illustrated here for the orbital angular momentum **L** when $l = 2$. The vector **L**, of length $\sqrt{l(l+1)}\hbar$, precesses about the axis of quantisation, the $(2l+1)$ allowed projections of **L** on this axis being given by $m\hbar$, with $m = -l, l+1, \ldots, +l$.

Note that since the Y_{lm} satisfy (6.102b) and the summation runs over the entire set $m = -l, -l+1, \ldots, +l$, of allowed values of m, the complex-conjugate sign can be put either on $Y_{lm}(\theta_1, \phi_1)$ or on $Y_{lm}(\theta_2, \phi_2)$.

6.4 Particle on a sphere and the rigid rotator

In this section we shall illustrate the results obtained above on a simple example. Let us consider a particle of mass μ and let $\mathbf{p} = -i\hbar \nabla$ be its momentum operator in the position representation of wave mechanics. The kinetic energy operator of the particle is therefore given in this representation by $T = \mathbf{p}^2/2\mu = -(\hbar^2/2\mu)\nabla^2$. Expressing the Laplacian operator ∇^2 in spherical polar coordinates[5], we may also

---

[5] See, for example, Byron and Fuller (1969).

write for $r \neq 0$

$$T = -\frac{\hbar^2}{2\mu}\left[\frac{1}{r^2}\frac{\partial}{\partial r}\left(r^2\frac{\partial}{\partial r}\right) + \frac{1}{r^2\sin\theta}\frac{\partial}{\partial\theta}\left(\sin\theta\frac{\partial}{\partial\theta}\right) + \frac{1}{r^2\sin^2\theta}\frac{\partial^2}{\partial\phi^2}\right] \quad (6.130a)$$

or

$$T = -\frac{\hbar^2}{2\mu}\left[\frac{1}{r^2}\frac{\partial}{\partial r}\left(r^2\frac{\partial}{\partial r}\right) - \frac{\mathbf{L}^2}{\hbar^2 r^2}\right] \quad (6.130b)$$

where we have used the expression (6.17) of the operator $\mathbf{L}^2$.

Let us now assume that the particle is constrained to move on the surface of a sphere of radius a. Taking the centre of the sphere as the origin of coordinates, r has the *fixed* value $r = a$. Hence the first term in the square brackets on the right of (6.130b) does not occur, and the kinetic energy operator of the particle reduces to

$$T = \frac{\mathbf{L}^2}{2\mu a^2} = \frac{\mathbf{L}^2}{2I} \quad (6.131)$$

where $I = \mu a^2$ is the moment of inertia of the particle with respect to the origin. Moreover, its potential energy V is clearly independent of r. If we suppose that V is also time-independent, the Hamiltonian of the particle is given by

$$H = \frac{\mathbf{L}^2}{2I} + V(\theta, \phi) \quad (6.132)$$

and the time-independent Schrödinger equation reads

$$\left[\frac{\mathbf{L}^2}{2I} + V(\theta, \phi)\right]\psi(\theta, \phi) = E\psi(\theta, \phi). \quad (6.133)$$

Rigid rotator

We now focus our attention on the simple case in which the particle is *free* (except for being constrained to move on the sphere) so that $V = 0$ and we are dealing with a rigid rotator. In this case the Hamiltonian (6.132) reduces to

$$H = \frac{\mathbf{L}^2}{2I} \quad (6.134)$$

and the Schrödinger equation (6.133) becomes

$$\frac{\mathbf{L}^2}{2I}\psi(\theta, \phi) = E\psi(\theta, \phi). \quad (6.135)$$

Since the eigenfunctions of $\mathbf{L}^2$ are the spherical harmonics $Y_{lm}(\theta, \phi)$, and its eigenvalues are given by $l(l+1)\hbar^2$ (with $l = 0, 1, 2, \ldots$) we see that the eigenfunctions of the rigid-rotator Hamiltonian (6.134) are also the Y_{lm}, the corresponding energy eigenvalues being

$$E_l = \frac{\hbar^2}{2I}l(l+1), \qquad l = 0, 1, 2, \ldots \quad (6.136)$$

It is worth stressing that these energy eigenvalues are independent of the quantum number m. Thus, the $(2l+1)$ eigenfunctions $Y_{lm}(\theta, \phi)$ with $m = -l, -l+1, \ldots, l$, correspond to the same energy, so that the energy level E_l is $(2l+1)$-*fold degenerate*. This degeneracy arises from the fact that the Hamiltonian (6.134) of the rigid rotator commutes with the orbital angular momentum operator $\mathbf{L}$ (see (6.14)) and is therefore invariant under rotations. Hence for this system all directions of space are physically equivalent. Now we have seen in Section 6.3 that $m\hbar$ measures the projection of the orbital angular momentum $\mathbf{L}$ on the z-axis, and thus is determined by the orientation of $\mathbf{L}$. The $(2l+1)$-fold degeneracy with respect to the quantum number m therefore results from the fact that the Hamiltonian (6.134) is independent of this orientation, so that its energy levels cannot depend on the magnitude of the component of $\mathbf{L}$ in any particular direction.

Rotational energy levels of a diatomic molecule

The result (6.136) giving the eigenvalues of the rigid-rotator Hamiltonian can be used to obtain an approximation for the rotational energy levels of a diatomic molecule. As far as its rotational motion is concerned, a diatomic molecule can be considered (in first approximation) as a rigid 'dumb-bell' with the nuclei of the two atoms A and B held a fixed distance R_0 apart (see Fig. 6.4). Classically the (kinetic) energy of rotation is

$$T = \frac{1}{2}I\omega^2 = \frac{(I\omega)^2}{2I} = \frac{\mathbf{L}^2}{2I} \tag{6.137}$$

where ω is the angular frequency of rotation, I is the moment of inertia and $L = I\omega$ is the magnitude of the angular momentum of the molecule with respect to the axis of rotation, which is perpendicular to the symmetry axis of the molecule (i.e. the line joining the two nuclei) and passes through the centre of mass. Denoting respectively by M_A and M_B the masses of the nuclei A and B, and by R_A and R_B their distances from the centre of mass, we have (see Fig. 6.4)

$$I = M_A R_A^2 + M_B R_B^2 = \mu R_0^2 \tag{6.138}$$

where $\mu = M_A M_B / (M_A + M_B)$ is the reduced mass. In obtaining the above result we have used the relations $M_A R_A = M_B R_B$ and $R_0 = R_A + R_B$. Since there is no potential energy, the Hamiltonian of the system is just $H = \mathbf{L}^2/2I$, so that the energy eigenvalues are given by (6.136), as in the case of the rigid rotator. The only difference is that the quantity μ which appears in the moment of inertia (6.138) is now the reduced mass μ of the two nuclei.

As an example, the constant $\hbar^2/2I$ has the value 1.3×10^{-3} eV in the lowest electronic state of the molecule HCl. Transitions between rotational states which differ in the quantum number l give rise to a series of closely spaced lines, called a *band spectrum* because it appears as a band when the structure due to the lines is not resolved. Pure rotational spectra of this kind lie in the far infrared and microwave

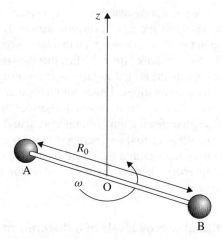

Figure 6.4 Rotation of a diatomic molecule considered as a rigid 'dumb-bell', the nuclei of the two atoms A and B being a distance R_0 apart. The rotation is about an axis Oz perpendicular to the symmetry axis of the molecule and passing through the centre of mass. The classical energy of rotation is $I\omega^2/2$, where I is the moment of inertia (6.138) and ω is the angular frequency of rotation.

regions[6].

Thus far we have assumed that the molecule is completely rigid. In fact, the nuclei can vibrate about their equilibrium position, and we shall see in Chapter 10 that the spacing of the corresponding vibrational energy levels is of the order of 0.1 eV, i.e. is about one hundred times larger than typical spacings of rotational levels. Transitions in which *both* the vibrational and rotational quantum numbers change give rise to the *vibration–rotation* (also called *rovibrational*) spectra of molecules, which appear in the infrared part of the electromagnetic spectrum. Finally, *electronic spectra* of molecules are observed when changes occur in the electronic as well as in the vibrational and rotational states of the molecule. As we shall show in Chapter 10, typical energy separations between low-lying electronic levels of molecules are of the order of a few eV, so that the corresponding electronic spectra appear in the ultraviolet and visible regions.

6.5 General angular momentum. The spectrum of J^2 and J_z

In Chapter 1 we saw that experiments demonstrate that particles can be assigned an intrinsic angular momentum, the *spin* **S**, and a spin quantum number s. For example, the electron has a spin quantum number $s = 1/2$, so that the component of its spin angular momentum **S** in a given direction can only take the values $\pm \hbar/2$.

---

[6] See Bransden and Joachain (1983), Chapter 10.

6.5 General angular momentum. The spectrum of $\mathbf{J}^2$ and J_z

Such values are excluded for the component of the *orbital* angular momentum **L** in a given direction, which can only take the values $m\hbar$, where $m = -l, -l+1, \ldots, l$, is either zero or a positive or negative integer (see (6.76) and (6.91)). This arises from the condition that the wave function should have a unique value at each point in space. Since the spin angular momentum of a particle is not associated with the spatial dependence of the wave function, this restriction need not apply for the spin. The theory developed so far for the orbital angular momentum is therefore not comprehensive enough to include also the spin angular momentum.

Because the commutation relations play a fundamental role in quantum mechanics, the way forward is to adopt the following general definition of angular momentum: *a vector operator* **J** *is an angular momentum if its components are Hermitian operators satisfying the commutation relations*

$$[J_x, J_y] = i\hbar J_z, \qquad [J_y, J_z] = i\hbar J_x, \qquad [J_z, J_x] = i\hbar J_y \qquad (6.139)$$

which are equivalent to the vector commutation relation

$$\mathbf{J} \times \mathbf{J} = i\hbar \mathbf{J}. \qquad (6.140)$$

Moreover, using the commutation relations (6.139), and proceeding as in the cases of the orbital angular momentum (see (6.20) and (6.21)), one finds that if J_u, J_v and J_w, are, respectively, the components of **J** along the three orthogonal unit vectors $\hat{\mathbf{u}}$, $\hat{\mathbf{v}}$ and $\hat{\mathbf{w}} = \hat{\mathbf{u}} \times \hat{\mathbf{v}}$, then

$$[J_u, J_v] = i\hbar J_w, \qquad [J_v, J_w] = i\hbar J_u, \qquad [J_w, J_u] = i\hbar J_v. \qquad (6.141)$$

From the commutation relations (6.139), one also deduces that the square of the angular momentum operator,

$$\mathbf{J}^2 = J_x^2 + J_y^2 + J_z^2 \qquad (6.142)$$

commutes with J_x, J_y and J_z, namely

$$[\mathbf{J}^2, J_x] = [\mathbf{J}^2, J_y] = [\mathbf{J}^2, J_z] = 0. \qquad (6.143)$$

The proof is entirely similar to that given in Section 6.1 for the orbital angular momentum (see (6.13)). We shall summarise the relations (6.143) by writing

$$[\mathbf{J}^2, \mathbf{J}] = 0. \qquad (6.144)$$

Because $\mathbf{J}^2$ commutes with each of the components of **J**, simultaneous eigenfunctions of $\mathbf{J}^2$ and any component of **J** (for example J_z) can be found. Starting with the definition of the angular momentum based on the commutation relations (6.139), and without adopting any specific representation, we shall now obtain the eigenvalues of $\mathbf{J}^2$ and J_z, which will be seen to include both integer and half-odd-integer values of the angular momentum quantum numbers.

The eigenvalues of $\mathbf{J}^2$ and J_z are real, since $\mathbf{J}^2$ and J_z are Hermitian operators. Moreover, the eigenvalues of $\mathbf{J}^2$ must be either positive or zero. For later convenience we shall write the eigenvalues of $\mathbf{J}^2$ in the form $j(j+1)\hbar^2$ and those of J_z in the form[7] $m\hbar$, with $j \geq 0$. Denoting the simultaneous eigenvectors of $\mathbf{J}^2$ and J_z by $|jm\rangle$, we have

$$\mathbf{J}^2|jm\rangle = j(j+1)\hbar^2|jm\rangle \tag{6.145}$$

and

$$J_z|jm\rangle = m\hbar|jm\rangle. \tag{6.146}$$

Because the expectation value of the square of a Hermitian operator must be positive (or zero), we have

$$\langle \mathbf{J}^2 \rangle = \langle J_x^2 \rangle + \langle J_y^2 \rangle + \langle J_z^2 \rangle$$
$$\geq \langle J_z^2 \rangle \tag{6.147}$$

and it follows that

$$j(j+1) \geq m^2. \tag{6.148}$$

We now introduce the raising and lowering operators (compare with (6.116))

$$J_+ = J_x + iJ_y, \qquad J_- = J_x - iJ_y \tag{6.149}$$

and we note that

$$J_+^\dagger = J_-, \qquad J_-^\dagger = J_+. \tag{6.150}$$

As in the case of the orbital angular momentum (see (6.117) and (6.118)) we have the relations

$$[\mathbf{J}^2, J_\pm] = 0 \tag{6.151a}$$
$$J_\pm J_\mp = \mathbf{J}^2 - J_z^2 \pm \hbar J_z \tag{6.151b}$$
$$[J_+, J_-] = 2\hbar J_z \tag{6.151c}$$

and

$$[J_z, J_\pm] = \pm \hbar J_\pm. \tag{6.151d}$$

Using (6.151a) and (6.151d), we see that $J_\pm|jm\rangle$ are simultaneous eigenvectors of $\mathbf{J}^2$ and J_z belonging, respectively, to the eigenvalues $j(j+1)\hbar^2$ and $(m \pm 1)\hbar$. Indeed, since $[\mathbf{J}^2, J_\pm] = 0$ we have

$$\mathbf{J}^2(J_\pm|jm\rangle) = J_\pm \mathbf{J}^2|jm\rangle = j(j+1)\hbar^2(J_\pm|jm\rangle) \tag{6.152}$$

and from (6.151d) we find that

$$J_z(J_\pm|jm\rangle) = J_\pm(J_z|jm\rangle) \pm \hbar J_\pm|jm\rangle$$
$$= (m \pm 1)\hbar(J_\pm|jm\rangle). \tag{6.153}$$

---

[7] For notational simplicity we shall use here the symbol m for the 'magnetic' quantum number associated with the operator J_z. Whenever it will be necessary, we shall add subscripts and write m_l, m_s or m_j to denote the magnetic quantum numbers associated with the operators L_z, S_z or J_z.

6.5 General angular momentum. The spectrum of $\mathbf{J}^2$ and J_z

By operating repeatedly with J_+, a sequence of eigenvectors of J_z can be constructed, with eigenvalues $(m+1)\hbar$, $(m+2)\hbar$, and so on, each of which being an eigenvector of $\mathbf{J}^2$ corresponding to the eigenvalue $j(j+1)\hbar^2$. Similarly, by repeated operation with J_-, a sequence of eigenvectors of J_z can be constructed, with eigenvalues $(m-1)\hbar$, $(m-2)\hbar$, etc., and again each of them is an eigenvector of $\mathbf{J}^2$ with the same eigenvalue $j(j+1)\hbar^2$. Now, in view of (6.148) for each j there must be a maximum (top) eigenvalue, say $m_T \hbar$, and also a minimum (bottom) eigenvalue $m_B \hbar$, so that the ladder of eigenvalues of J_z cannot continue indefinitely in either direction. Therefore, the quantity $m_T - m_B$ must be a positive integer or zero

$$m_T - m_B = n, \qquad n = 0, 1, 2, \ldots \tag{6.154}$$

If J_+ is applied to the state $|jm_T\rangle$, the sequence terminates and

$$J_+|jm_T\rangle = 0. \tag{6.155}$$

From the above result and from (6.151b), we have

$$\begin{aligned} J_-(J_+|jm_T\rangle) &= (\mathbf{J}^2 - J_z^2 - \hbar J_z)|jm_T\rangle \\ &= [j(j+1) - m_T^2 - m_T]\hbar^2 |jm_T\rangle \\ &= 0 \end{aligned} \tag{6.156}$$

where we have used the fact that $J_z|jm_T\rangle = m_T \hbar |jm_T\rangle$. Thus

$$j(j+1) = m_T^2 + m_T. \tag{6.157}$$

Similarly, if J_- is applied to the state $|jm_B\rangle$, the sequence terminates, so that

$$J_-|jm_B\rangle = 0. \tag{6.158}$$

From this relation and from (6.151b), we have in this case

$$\begin{aligned} J_+(J_-|jm_B\rangle) &= (\mathbf{J}^2 - J_z^2 + \hbar J_z)|jm_B\rangle \\ &= [j(j+1) - m_B^2 + m_B]\hbar^2 |jm_B\rangle \\ &= 0 \end{aligned} \tag{6.159}$$

since $J_z|jm_B\rangle = m_B \hbar |jm_B\rangle$. Hence

$$j(j+1) = m_B^2 - m_B. \tag{6.160}$$

The two equations (6.157) and (6.160) require that

$$m_T^2 + m_T = m_B^2 - m_B. \tag{6.161}$$

This equation has two solutions:

$$m_T = -m_B, \qquad m_T = m_B - 1. \tag{6.162}$$

The second solution contradicts (6.154), so that we must have $m_T = -m_B$, and hence also $m_T \geqslant 0$. Going back to (6.157) and using the fact that j is also non-negative, we find that the only acceptable solution of this equation is $m_T = j$. Thus

$$m_T = j, \quad m_B = -j. \tag{6.163}$$

Now, as $m_T - m_B$ is a positive integer or zero (see (6.154)), $2j$ must also be a positive integer or zero, so that the allowed values of the angular momentum quantum number j are

$$j = 0, \tfrac{1}{2}, 1, \tfrac{3}{2}, 2, \tfrac{5}{2}, \ldots \tag{6.164}$$

Moreover, because the maximum value of the magnetic quantum number m is $m_T = j$ and its minimum value is $m_B = -j$, the allowed values of m, for a given j, are the $(2j+1)$ values

$$-j, -j+1, \ldots, j. \tag{6.165}$$

Our general definition of angular momentum, based on the commutation relations (6.139), therefore leads to *integer* (including zero) and *half-odd-integer* values for j (and hence for m). As we have seen above, half-odd-integer values of j are excluded in the case of the orbital angular momentum, where $j = l = 0, 1, 2$, etc. On the other hand, in the case of the spin angular momentum the quantum number j has a fixed value s which is an integer (or zero) for some particles, and a half-odd-integer for others. For example, pions have spin zero ($s = 0$), photons have spin one ($s = 1$), electrons, protons and neutrons have spin 1/2, the Ω^- particle (a particle with a mass roughly 1.8 times that of the proton and a lifetime of about 10^{-10} s) has spin 3/2.

6.6 Matrix representations of angular momentum operators

Let us consider the simultaneous eigenvectors $|jm\rangle$ of $\mathbf{J}^2$ and J_z. These eigenvectors can be chosen to be normalised to unity, $\langle jm|jm\rangle = 1$. Moreover, the eigenvectors of Hermitian operators belonging to different eigenvalues are orthogonal. We can therefore adopt the orthonormality conditions

$$\langle j'm'|jm\rangle = \delta_{jj'}\delta_{mm'}. \tag{6.166}$$

Using as a basis the eigenvectors $|jm\rangle$ of $\mathbf{J}^2$ and J_z, it is easy to find a matrix representation of the angular momentum operators. First, we note that the matrices representing the operators $\mathbf{J}^2$ and J_z are diagonal, with elements

$$\begin{aligned}(\mathbf{J}^2)_{j'm',jm} &= \langle j'm'|\mathbf{J}^2|jm\rangle \\ &= j(j+1)\hbar^2 \delta_{jj'}\delta_{mm'}\end{aligned} \tag{6.167}$$

and

$$(J_z)_{j'm',jm} = \langle j'm'|J_z|jm\rangle$$
$$= m\hbar\delta_{jj'}\delta_{mm'}. \tag{6.168}$$

In order to find the matrix elements of J_x and J_y, it is convenient to obtain first those of the operators J_+ and J_-. To this end, we remark that

$$J_\pm|jm\rangle = N_\pm|jm\pm 1\rangle \tag{6.169}$$

where $N_\pm$ are normalisation factors. Since both $|jm\rangle$ and $|jm\pm 1\rangle$ are normalised to unity

$$|N_\pm|^2 = \langle jm|J_\pm^\dagger J_\pm|jm\rangle$$
$$= \langle jm|J_\mp J_\pm|jm\rangle$$
$$= [j(j+1) - m(m\pm 1)]\hbar^2 \tag{6.170}$$

where we have used (6.150) in the second line and (6.151b) in the last line. Adopting the convention that $N_\pm$ are real and positive, we have

$$N_\pm = [j(j+1) - m(m\pm 1)]^{1/2}\hbar. \tag{6.171}$$

The matrix representation of J_+ follows. It is

$$(J_+)_{j'm',jm} = \langle j'm'|J_+|jm\rangle$$
$$= [j(j+1) - m(m+1)]^{1/2}\hbar\delta_{jj'}\delta_{m'm+1}. \tag{6.172}$$

In the same way, we find that

$$(J_-)_{j'm',jm} = [j(j+1) - m(m-1)]^{1/2}\hbar\delta_{jj'}\delta_{m'm-1}. \tag{6.173}$$

The matrix representations of J_x and J_y can be found at once from those of J_+ and J_- by using the relations (see (6.149))

$$J_x = \frac{1}{2}(J_+ + J_-), \qquad J_y = \frac{1}{2i}(J_+ - J_-). \tag{6.174}$$

It is clear from the above results that the matrices representing the angular momentum operators J_x, J_y, J_z, J_+, J_- and $\mathbf{J}^2$ in the basis $|jm\rangle$ are diagonal in j. We therefore have an infinite number of representations for these matrices, each of them characterised by a given value of j ($j = 0, 1/2, 1, 3/2, \ldots$) and having $2j+1$ columns and rows, labelled respectively by the values of m and m'. One may take all these representations together to form one single representation of infinite rank, or we may consider each of the representations of dimension $2j+1$ separately. The first three finite-dimensional representations, for $j = 0$, $1/2$ and 1, are given below for the operators J_x, J_y, J_z and $\mathbf{J}^2$.

(1) For $j = 0$

$$J_x = (0), \qquad J_y = (0), \qquad J_z = (0), \qquad \mathbf{J}^2 = (0) \tag{6.175a}$$

where (0) is the null matrix of unit rank.

(2) For $j = 1/2$

$$J_x = \frac{\hbar}{2}\begin{pmatrix} 0 & 1 \\ 1 & 0 \end{pmatrix}, \quad J_y = \frac{\hbar}{2}\begin{pmatrix} 0 & -i \\ i & 0 \end{pmatrix}$$

$$J_z = \frac{\hbar}{2}\begin{pmatrix} 1 & 0 \\ 0 & -1 \end{pmatrix}, \quad \mathbf{J}^2 = \frac{3}{4}\hbar^2\begin{pmatrix} 1 & 0 \\ 0 & 1 \end{pmatrix}. \tag{6.175b}$$

(3) For $j = 1$

$$J_x = \frac{\hbar}{\sqrt{2}}\begin{pmatrix} 0 & 1 & 0 \\ 1 & 0 & 1 \\ 0 & 1 & 0 \end{pmatrix}, \quad J_y = \frac{\hbar}{\sqrt{2}}\begin{pmatrix} 0 & -i & 0 \\ i & 0 & -i \\ 0 & i & 0 \end{pmatrix}$$

$$J_z = \hbar\begin{pmatrix} 1 & 0 & 0 \\ 0 & 0 & 0 \\ 0 & 0 & -1 \end{pmatrix}, \quad \mathbf{J}^2 = 2\hbar^2\begin{pmatrix} 1 & 0 & 0 \\ 0 & 1 & 0 \\ 0 & 0 & 1 \end{pmatrix}. \tag{6.175c}$$

Since the orbital angular momentum $\mathbf{L}$ is a particular angular momentum for which the quantum number $j (= l)$ is an integer or zero, one can use the matrix representations with $j = l = 0, 1, 2, \ldots$ for the orbital angular momentum operators. In fact, when $j (= l)$ is an integer or zero, the basic results (6.167), (6.168), (6.172) and (6.173) can be obtained by using the properties of the spherical harmonics $Y_{lm}(\theta, \phi)$, which represent the kets $|lm\rangle$ in the position representation. Indeed, from (6.63), (6.64) and the orthonormality property (6.103) of the spherical harmonics, we have

$$(\mathbf{L}^2)_{l'm',lm} = \int Y^*_{l'm'}(\theta, \phi)\mathbf{L}^2 Y_{lm}(\theta, \phi)d\Omega$$

$$= l(l+1)\hbar^2 \delta_{ll'}\delta_{mm'} \tag{6.176}$$

and

$$(L_z)_{l'm',lm} = \int Y^*_{l'm'}(\theta, \phi)L_z Y_{lm}(\theta, \phi)d\Omega$$

$$= m\hbar \delta_{ll'}\delta_{mm'} \tag{6.177}$$

while from (6.123) we obtain the matrix representations of the operators L_+ and L_-

$$(L_\pm)_{l'm',lm} = \int Y^*_{l'm'}(\theta, \phi)L_\pm Y_{lm}(\theta, \phi)d\Omega$$

$$= [l(l+1) - m(m \pm 1)]^{1/2}\hbar \delta_{ll'}\delta_{m'm\pm1}. \tag{6.178}$$

The vector model introduced in Section 6.3 to visualise the orbital angular momentum can readily be extended to all angular momenta. To see this, we first note from (6.172)–(6.174) that

$$\langle J_x \rangle \equiv \langle jm|J_x|jm\rangle = 0 \tag{6.179a}$$

and similarly

$$\langle J_y \rangle \equiv \langle jm|J_y|jm\rangle = 0. \tag{6.179b}$$

Also, from (6.174), (6.151b) and the fact that $\langle jm|J_\pm|jm\rangle = 0$, we have

$$\langle J_x^2 \rangle = \langle J_y^2 \rangle = \tfrac{1}{2}\langle \mathbf{J}^2 - J_z^2\rangle$$
$$= \tfrac{1}{2}[j(j+1) - m^2]\hbar^2 \tag{6.180}$$

so that an angular momentum $\mathbf{J}$ can be viewed as a vector, of length $\sqrt{j(j+1)}\hbar$, precessing about the z-axis, the $2j+1$ allowed projections of $\mathbf{J}$ on this axis being given by $m\hbar$, with $m = -j, -j+1, \ldots, +j$.

6.7 Spin angular momentum

In Chapter 1, we reviewed some of the evidence showing that 'elementary' particles possess an internal degree of freedom which behaves like an angular momentum and is termed *spin*. In what follows we shall use the symbol $\mathbf{S}$ to denote the spin angular momentum operator of a particle, rather than $\mathbf{J}$ which we used in the two previous sections to denote angular momentum in general.

Since the spin operator $\mathbf{S}$ is an angular momentum, its Cartesian components S_x, S_y and S_z are Hermitian operators satisfying the commutation relations (see (6.139))

$$[S_x, S_y] = i\hbar S_z, \qquad [S_y, S_z] = i\hbar S_x, \qquad [S_z, S_x] = i\hbar S_y \tag{6.181}$$

which may also be rewritten as

$$\mathbf{S} \times \mathbf{S} = i\hbar \mathbf{S}. \tag{6.182}$$

From the general theory of angular momentum discussed in Section 6.5, we know that the square of the spin angular momentum operator,

$$\mathbf{S}^2 = S_x^2 + S_y^2 + S_z^2 \tag{6.183}$$

commutes with S_x, S_y and S_z. Thus simultaneous eigenvectors of $\mathbf{S}^2$ and S_z can be found, corresponding to eigenvalues $s(s+1)\hbar^2$ and $m_s\hbar$, respectively. In the notation of Section 6.5 these eigenvectors, normalised to unity would be written $|sm_s\rangle$. In what follows, however, it will be more convenient to denote them by the symbol χ_{s,m_s}. We therefore have

$$\mathbf{S}^2 \chi_{s,m_s} = s(s+1)\hbar^2 \chi_{s,m_s} \tag{6.184}$$

and

$$S_z \chi_{s,m_s} = m_s \hbar \chi_{s,m_s}. \tag{6.185}$$

It follows from the general discussion of Section 6.5 that the spin quantum number s (usually abbreviated as spin) can be an integer (including zero) or half an odd integer. Particles having integer spin ($s = 0, 1, 2, \ldots$) are called *bosons* while those having half-odd-integer spin ($s = 1/2, 3/2, \ldots$) are called *fermions*. Regarding the quantum number m_s, we know from the general theory developed in Section 6.5 that it has $(2s+1)$ allowed values, given by $-s, -s+1, \ldots, +s$. Furthermore, from the orthonormality condition (6.166), we have

$$\langle \chi_{s,m'_s} | \chi_{s,m_s} \rangle = \delta_{m_s m'_s}. \tag{6.186}$$

If the $(2s+1)$ eigenvectors χ_{s,m_s} are used as a basis, the spin operators S_x, S_y, S_z and $\mathbf{S}^2$ are represented by $(2s+1) \times (2s+1)$ matrices, which are of course the same matrices as those appearing in the $(2j+1)$ dimensional representation of the general angular momentum operators J_x, J_y, J_z and $\mathbf{J}^2$, with $j = s$. The eigenvectors χ_{s,m_s} are then represented by *column vectors* with $(2s+1)$ components, all components being equal to zero except one, which is equal to unity. For example, in the case of spin one ($s=1$) we see from (6.175c) that

$$S_x = \frac{\hbar}{\sqrt{2}} \begin{pmatrix} 0 & 1 & 0 \\ 1 & 0 & 1 \\ 0 & 1 & 0 \end{pmatrix} \quad S_y = \frac{\hbar}{\sqrt{2}} \begin{pmatrix} 0 & -i & 0 \\ i & 0 & -i \\ 0 & i & 0 \end{pmatrix}$$

$$S_z = \hbar \begin{pmatrix} 1 & 0 & 0 \\ 0 & 0 & 0 \\ 0 & 0 & -1 \end{pmatrix} \quad \mathbf{S}^2 = 2\hbar^2 \begin{pmatrix} 1 & 0 & 0 \\ 0 & 1 & 0 \\ 0 & 0 & 1 \end{pmatrix} \tag{6.187}$$

and it is readily verified that the three spin eigenvectors

$$\chi_{1,1} = \begin{pmatrix} 1 \\ 0 \\ 0 \end{pmatrix} \quad \chi_{1,0} = \begin{pmatrix} 0 \\ 1 \\ 0 \end{pmatrix} \quad \chi_{1,-1} = \begin{pmatrix} 0 \\ 0 \\ 1 \end{pmatrix} \tag{6.188}$$

correspond respectively to eigenvalues $+\hbar$, 0 and $-\hbar$ of S_z and to the eigenvalue $2\hbar^2$ of $\mathbf{S}^2$. We remark that in matrix notation the orthonormality relations (6.186) read

$$\chi^\dagger_{s,m'_s} \chi_{s,m_s} = \delta_{m_s m'_s} \tag{6.189}$$

where $\chi^\dagger_{s,m'_s}$, the Hermitian adjoint of χ_{s,m'_s}, is a *row vector*. For example, the Hermitian adjoints of the spin-one eigenvectors (6.188) are

$$\chi^\dagger_{1,1} = \begin{pmatrix} 1 & 0 & 0 \end{pmatrix}, \quad \chi^\dagger_{1,0} = \begin{pmatrix} 0 & 1 & 0 \end{pmatrix}, \quad \chi^\dagger_{1,-1} = \begin{pmatrix} 0 & 0 & 1 \end{pmatrix} \tag{6.190}$$

and the orthonormality relations (6.189) can easily be checked by using the rules of matrix multiplication. For instance

$$\chi^\dagger_{1,1} \chi_{1,1} = \begin{pmatrix} 1 & 0 & 0 \end{pmatrix} \begin{pmatrix} 1 \\ 0 \\ 0 \end{pmatrix} = 1,$$

$$\chi^\dagger_{1,1} \chi_{1,0} = \begin{pmatrix} 1 & 0 & 0 \end{pmatrix} \begin{pmatrix} 0 \\ 1 \\ 0 \end{pmatrix} = 0. \tag{6.191}$$

In the position representation – in which **r** is diagonal – the state vector $|\Psi(t)\rangle$ of a *spinless* particle is represented by a wave function $\Psi(\mathbf{r}, t)$ which is in general a function of the three dynamical variables x, y, z describing its position, and of the time t. From the foregoing discussion it is apparent that in the case of a particle having spin s the wave function must also depend on an additional *spin variable* describing the spin orientation of the particle. In a representation in which **r** and S_z are diagonal, the state vector $|\Psi(t)\rangle$ will therefore be represented by a wave function $\Psi(\mathbf{r}, t, \sigma)$, in which the spin variable σ, denoting the component of the spin along the z-axis, can only take on $2s + 1$ values $-s\hbar, (-s+1)\hbar, \ldots, s\hbar$ corresponding to the $2s + 1$ possible values of the quantum number m_s. We emphasise that in contrast to the position variables x, y and z, which vary in a continuous way, the spin variable σ can only take on *discrete* values; this variable has no classical analogue.

We shall assume that the basic postulates of quantum mechanics discussed in Chapter 5 also apply to the new independent spin variable σ. In particular, we may expand a general wave function $\Psi(\mathbf{r}, t, \sigma)$ for a particle having spin in terms of the basic spin eigenfunctions χ_{s,m_s}. That is

$$\Psi(\mathbf{r}, t, \sigma) = \sum_{m_s=-s}^{+s} \Psi_{m_s}(\mathbf{r}, t) \chi_{s,m_s} \tag{6.192}$$

so that a particle of spin s is described by a wave function Ψ having $2s+1$ components $\Psi_{m_s}(\mathbf{r}, t)$, each of them corresponding to a particular value of the spin variable σ. For example, in the case of spin one we may use the three spin eigenvectors (6.188) to write the wave function Ψ as a column vector:

$$\Psi(\mathbf{r}, t, \sigma) = \Psi_1(\mathbf{r}, t) \begin{pmatrix} 1 \\ 0 \\ 0 \end{pmatrix} + \Psi_0(\mathbf{r}, t) \begin{pmatrix} 0 \\ 1 \\ 0 \end{pmatrix} + \Psi_{-1}(\mathbf{r}, t) \begin{pmatrix} 0 \\ 0 \\ 1 \end{pmatrix}$$

$$= \begin{pmatrix} \Psi_1(\mathbf{r}, t) \\ \Psi_0(\mathbf{r}, t) \\ \Psi_{-1}(\mathbf{r}, t) \end{pmatrix} \tag{6.193}$$

where the three components Ψ_1, Ψ_0 and Ψ_{-1} correspond respectively to the values $+\hbar$, 0 and $-\hbar$ of the spin variable σ. We note that the requirement that the wave function (6.192) be normalised to unity reads

$$\langle \Psi | \Psi \rangle = 1 \tag{6.194a}$$

where

$$\langle \Psi | \Psi \rangle = \int \Psi^\dagger(\mathbf{r}, t, \sigma) \Psi(\mathbf{r}, t, \sigma) d\mathbf{r}$$

$$= \sum_{m_s=-s}^{+s} \int |\Psi_{m_s}(\mathbf{r}, t)|^2 d\mathbf{r} \tag{6.194b}$$

and we have used (6.192) and (6.189). For instance, in the case of the spin-one wave function (6.193) we have

$$\langle \Psi | \Psi \rangle = \int \left(\Psi_1^*(\mathbf{r},t) \; \Psi_0^*(\mathbf{r},t) \; \Psi_{-1}^*(\mathbf{r},t) \right) \begin{pmatrix} \Psi_1(\mathbf{r},t) \\ \Psi_0(\mathbf{r},t) \\ \Psi_{-1}(\mathbf{r},t) \end{pmatrix} d\mathbf{r}$$

$$= \int [|\Psi_1(\mathbf{r},t)|^2 + |\Psi_0(\mathbf{r},t)|^2 + |\Psi_{-1}(\mathbf{r},t)|^2] d\mathbf{r}. \tag{6.195}$$

Having expanded the wave function Ψ according to (6.192), we may interpret the quantity $|\Psi_{m_s}(\mathbf{r},t)|^2 d\mathbf{r}$ as the probability of finding the particle at time t in the volume element $d\mathbf{r}$ *with the component of its spin along the z-axis equal to $m_s \hbar$*, provided the wave function Ψ is normalised to unity. The probability that the particle be found at time t in the volume element $d\mathbf{r}$, independently of its spin orientation, is equal to

$$\sum_{m_s=-s}^{+s} |\Psi_{m_s}(\mathbf{r},t)|^2 d\mathbf{r} \tag{6.196}$$

and the integral

$$\int |\Psi_{m_s}(\mathbf{r},t)|^2 d\mathbf{r} \tag{6.197}$$

gives the probability that the component of the spin along the z-axis is equal to $m_s \hbar$ at the time t. We remark that if the particle is in an eigenstate of S_z corresponding to a definite value $\bar{m}_s$ of m_s, its wave function has the form

$$\Psi(\mathbf{r},t,\sigma) = \Psi_{\bar{m}_s}(\mathbf{r},t) \chi_{s,\bar{m}_s} \tag{6.198}$$

and hence all the components of Ψ are zero except the one corresponding to the value $\bar{\sigma} = \bar{m}_s \hbar$ of the spin variable σ.

The expectation value of an operator A is given by the usual expression (see (5.28))

$$\langle A \rangle = \frac{\langle \Psi | A | \Psi \rangle}{\langle \Psi | \Psi \rangle} \tag{6.199}$$

where $\langle \Psi | \Psi \rangle$ is given by (6.194b). Using (6.192) and (6.189), the matrix element $\langle \Psi | A | \Psi \rangle$ reads

$$\langle \Psi | A | \Psi \rangle = \int \Psi^\dagger(\mathbf{r},t,\sigma) A \Psi(\mathbf{r},t,\sigma) d\mathbf{r}$$

$$= \sum_{m_s=-s}^{+s} \sum_{m_s'=-s}^{+s} \int \Psi_{m_s}^*(\mathbf{r},t) A_{m_s m_s'} \Psi_{m_s'}(\mathbf{r},t) d\mathbf{r} \tag{6.200}$$

where

$$A_{m_s m_s'} = \langle \chi_{s,m_s} | A | \chi_{s,m_s'} \rangle \tag{6.201}$$

is the $(2s+1) \times (2s+1)$ matrix representing the operator A in 'spin space'.

It is important to note that the position and time variables on the one hand, and the spin variable on the other, can be studied separately and can then be assembled by using (6.192). One can therefore focus the attention on the spin properties, and temporarily disregard the dependence of the wave function Ψ on the position and time variable. The (spin) state of a particle of spin s is then determined by a wave function χ_s which is in general an arbitrary superposition of the $(2s+1)$ basic spin eigenstates χ_{s,m_s}, namely

$$\chi_s = \sum_{m_s=-s}^{+s} a_{m_s} \chi_{s,m_s} \qquad (6.202)$$

where the a_{m_s} are complex coefficients. If χ_s is normalised to unity, so that $\chi_s^\dagger \chi_s = 1$, then $|a_{m_s}|^2$ is the probability of finding the particle in the basic spin state χ_{s,m_s}, and from the orthonormality relations (6.189) we have

$$\sum_{m_s=-s}^{+s} |a_{m_s}|^2 = 1. \qquad (6.203)$$

It is also interesting to note from (6.184) and (6.202) that for an *arbitrary* spin function χ_s one has $\mathbf{S}^2 \chi_s = s(s+1)\hbar^2 \chi_s$. More generally, since the operator $\mathbf{S}^2$ does not act on the variables $\mathbf{r}$ and t, we can write the relation

$$\mathbf{S}^2 \Psi = s(s+1)\hbar^2 \Psi \qquad (6.204)$$

for an arbitrary wave function (6.192) corresponding to a particle of spin s. The above relation means that in contrast to the case of the orbital angular momentum, the operator $\mathbf{S}^2$ is a purely numerical one, which can be written as

$$\mathbf{S}^2 = s(s+1)\hbar^2. \qquad (6.205)$$

Finally, we remark that if spin-dependent interactions are negligible, so that the Hamiltonian H of the particle is spin-independent, the wave function $\Psi(\mathbf{r}, t, \sigma)$ of a particle of spin s will simply be the product

$$\Psi(\mathbf{r}, t, \sigma) = \Psi(\mathbf{r}, t) \chi_s \qquad (6.206)$$

where $\Psi(\mathbf{r}, t)$ is a (spinless) solution of the Schrödinger equation $i\hbar \partial \Psi / \partial t = H\Psi$, and χ_s is the spin function of the particle. From (6.192) and (6.202) the $2s+1$ components of the wave function are then given by

$$\Psi_{m_s}(\mathbf{r}, t) = \Psi(\mathbf{r}, t) a_{m_s}. \qquad (6.207)$$

6.8 Spin one-half

In this section we shall now discuss in more detail the simplest non-trivial case of spin quantum mechanics, namely that of particles of spin $s = 1/2$. The non-relativistic theory of spin-1/2 particles was first developed by W. Pauli in 1927. It is a very important case, since *electrons* and *nucleons* (protons and neutrons), the building blocks of atomic and nuclear physics, have spin 1/2. Moreover, it is also believed that all the *hadrons* (the strongly interacting particles), including the proton and the neutron, are made of more elementary constituents, the *quarks*, which also have spin 1/2.

Since for a particle of spin $s = 1/2$ the only possible values of m_s are $m_s = \pm 1/2$, the spin variable σ can only take the *two* values $+\hbar/2$ and $-\hbar/2$, and there are only *two* independent normalised spin eigenfunctions χ_{s,m_s}, namely $\chi_{1/2,1/2}$ and $\chi_{1/2,-1/2}$. We shall denote them by α and β, respectively. Thus

$$\alpha \equiv \chi_{\frac{1}{2},\frac{1}{2}}, \qquad \beta \equiv \chi_{\frac{1}{2},-\frac{1}{2}} \tag{6.208}$$

and we see from (6.184) and (6.185) that

$$\mathbf{S}^2 \alpha = \frac{3}{4}\hbar^2 \alpha, \qquad \mathbf{S}^2 \beta = \frac{3}{4}\hbar^2 \beta \tag{6.209}$$

and

$$S_z \alpha = \frac{\hbar}{2}\alpha, \qquad S_z \beta = -\frac{\hbar}{2}\beta. \tag{6.210}$$

The two basic spin-1/2 eigenfunctions α and β are said to correspond respectively to 'spin-up' ($\uparrow$) and 'spin-down' ($\downarrow$) states. They satisfy the orthonormality relations (see (6.186))

$$\langle \alpha | \alpha \rangle = \langle \beta | \beta \rangle = 1$$
$$\langle \alpha | \beta \rangle = \langle \beta | \alpha \rangle = 0. \tag{6.211}$$

Introducing the raising and lowering operators

$$S_\pm = S_x \pm iS_y \tag{6.212}$$

and using the general relations (6.169) and (6.171) with $j = 1/2$ and $m = \pm 1/2$, we find that

$$S_+ \alpha = 0, \qquad S_+ \beta = \hbar \alpha, \tag{6.213a}$$
$$S_- \alpha = \hbar \beta, \qquad S_- \beta = 0. \tag{6.213b}$$

From (6.210), (6.212) and (6.213) we can construct a table which tells us how the components of $\mathbf{S}$ act on α and β. That is

$$\begin{aligned} S_x \alpha &= \frac{\hbar}{2}\beta, & S_x \beta &= \frac{\hbar}{2}\alpha \\ S_y \alpha &= \frac{i\hbar}{2}\beta, & S_y \beta &= -\frac{i\hbar}{2}\alpha \\ S_z \alpha &= \frac{\hbar}{2}\alpha, & S_z \beta &= -\frac{\hbar}{2}\beta \end{aligned} \tag{6.214}$$

According to (6.192), a general wave function for a particle of spin 1/2 can be expanded in terms of the two basic spin states α and β as

$$\Psi(\mathbf{r}, t, \sigma) = \Psi_{1/2}(\mathbf{r}, t)\alpha + \Psi_{-1/2}(\mathbf{r}, t)\beta. \tag{6.215}$$

If we concentrate our attention on the spin variable only, we can, following (6.202), write a general spin-1/2 function as

$$\chi = a\alpha + b\beta \tag{6.216}$$

where in order to simplify the notation we have set $\chi \equiv \chi_{1/2}, a \equiv a_{1/2}$ and $b \equiv a_{-1/2}$. If χ is normalised to unity, so that $\langle\chi|\chi\rangle = 1$, then using (6.211) we have

$$|a|^2 + |b|^2 = 1. \tag{6.217}$$

The probability of finding the particle in the 'spin-up' state α (i.e. the probability that a measurement of S_z will yield the value $+\hbar/2$) is given by $|a|^2$, while $|b|^2$ is the probability of finding the particle in the 'spin-down' state β (i.e. the probability that the value $-\hbar/2$ will be found upon measuring S_z).

Using (6.204) with $s = 1/2$, we see that for an arbitrary spin-1/2 wave function (6.215) we have $\mathbf{S}^2\Psi = (3/4)\hbar^2\Psi$, so that $\mathbf{S}^2$ is the purely numerical operator

$$\mathbf{S}^2 = \frac{3}{4}\hbar^2. \tag{6.218}$$

In addition, it follows from (6.214) that for any spin-1/2 wave function (6.215) one has

$$S_x^2 \Psi = \frac{\hbar^2}{4}\Psi \tag{6.219}$$

with a similar result for S_y^2 and S_z^2. We can therefore write

$$S_x^2 = S_y^2 = S_z^2 = \frac{\hbar^2}{4}. \tag{6.220}$$

Let us now explore in more detail the algebra of spin-1/2 operators. From (6.213) we deduce at once that $S_\pm^2 = 0$ and hence, using (6.212) and (6.220), we find that

$$0 = (S_x \pm \mathrm{i}S_y)^2 = S_x^2 - S_y^2 \pm \mathrm{i}(S_x S_y + S_y S_x)$$
$$= \pm \mathrm{i}(S_x S_y + S_y S_x) \tag{6.221}$$

so that

$$S_x S_y + S_y S_x = 0. \tag{6.222}$$

It is easy to verify (Problem 6.15) that the same relation holds between any different pair of components. If we introduce the *anticommutator* $[A, B]_+$ of two operators A and B, which is defined by the relation

$$[A, B]_+ = AB + BA \tag{6.223}$$

we have

$$[S_x, S_y]_+ = 0, \qquad [S_y, S_z]_+ = 0, \qquad [S_z, S_x]_+ = 0 \tag{6.224}$$

so that the operators S_x, S_y and S_z *anticommute in pairs*. Combining the above anticommutation relations with the basic commutation relations (6.181), we find that

$$S_i S_j = \frac{i\hbar}{2} S_k \tag{6.225}$$

where $i, j, k = x, y$ or z, in cyclic order. This relation is very useful because when used together with (6.220) it allows one to reduce an arbitrary product of spin-1/2 operators either to a *spin-independent* term or to a term *linear* in S_x, S_y and S_z. For example

$$S_x S_y S_z S_y S_z S_x = S_x S_y S_z S_y \left(\frac{i\hbar}{2}\right) S_y = S_x S_y S_z \frac{\hbar^2}{4}\left(\frac{i\hbar}{2}\right) = S_x\left(\frac{i\hbar}{2}\right) S_x \frac{\hbar^2}{4}\left(\frac{i\hbar}{2}\right)$$

$$= \frac{\hbar^2}{4}\left(\frac{i\hbar}{2}\right)\frac{\hbar^2}{4}\left(\frac{i\hbar}{2}\right) = -\frac{\hbar^6}{64} \tag{6.226a}$$

and

$$S_x S_y S_x S_y S_z S_x = S_x S_y S_x S_y \left(\frac{i\hbar}{2}\right) S_y = S_x S_y S_x \frac{\hbar^2}{4}\left(\frac{i\hbar}{2}\right)$$

$$= S_x\left(-\frac{i\hbar}{2}\right) S_z \frac{\hbar^2}{4}\left(\frac{i\hbar}{2}\right) = \left(-\frac{i\hbar}{2}\right) S_y \left(-\frac{i\hbar}{2}\right) \frac{\hbar^2}{4}\left(\frac{i\hbar}{2}\right)$$

$$= -\frac{i\hbar^5}{32} S_y. \tag{6.226b}$$

As a result, the most general spin-1/2 operator can be written in the form

$$\begin{aligned} A &= A_0 + A_x S_x + A_y S_y + A_z S_z \\ &= A_0 + \mathbf{A}\cdot\mathbf{S} \end{aligned} \tag{6.227}$$

where the operators A_0, A_x, A_y and A_z are spin-independent.

It follows from the general theory of angular momentum which was developed earlier in this chapter that one cannot ascribe any meaning to the statement that the spin vector $\mathbf{S}$ is in a given direction. Indeed, this would imply that its three components S_x, S_y and S_z could be measured simultaneously with arbitrary precision, which is impossible since they do not commute. In particular, we see from (6.220) that when the particle is in the 'spin-up' state α or the 'spin-down' state β the expectation values of S_x^2 and S_y^2 are given by $\langle S_x^2 \rangle = \langle S_y^2 \rangle = \hbar^2/4$, so that the x- and y-components of the spin are not zero. However, one can speak of the *average* spin direction. For example, if the particle is in the 'spin-up' state α, one finds from (6.211) and (6.214) that the average value of S_z is given by $\langle S_z \rangle = \hbar/2$, while $\langle S_x \rangle = \langle S_y \rangle = 0$. Similarly, if the particle is in the 'spin-down' state β, one has $\langle S_z \rangle = -\hbar/2$ while $\langle S_x \rangle = \langle S_y \rangle = 0$. In what follows it is always in this average sense that a spin 1/2 particle will be said to have its spin 'up' or 'down'. These results, which are particular cases of those

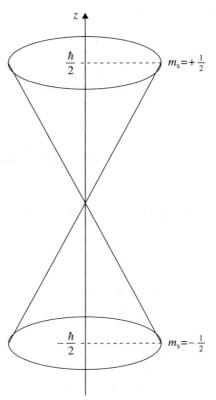

Figure 6.5 The vector model of the spin, for a particle of spin 1/2. The spin vector **S**, of length $\sqrt{3/4}\hbar$, precesses about the z-axis, the only allowed projections of **S** on the z-axis being $m_s\hbar$, with $m_s = \pm 1/2$.

obtained in Section 6.6 for a general angular momentum (see (6.179) and (6.180)) can conveniently be visualised with the help of the vector model, as illustrated in Fig. 6.5.

If the two basic spin-1/2 eigenvectors α and β are used as a basis, the spin operators are represented by 2×2 matrices. Thus, using our previous results (6.175b) with $j = s = 1/2$, we have

$$S_x = \frac{\hbar}{2}\begin{pmatrix} 0 & 1 \\ 1 & 0 \end{pmatrix}, \qquad S_y = \frac{\hbar}{2}\begin{pmatrix} 0 & -i \\ i & 0 \end{pmatrix}, \tag{6.228a}$$

$$S_z = \frac{\hbar}{2}\begin{pmatrix} 1 & 0 \\ 0 & -1 \end{pmatrix}, \qquad \mathbf{S}^2 = \frac{3}{4}\hbar^2\begin{pmatrix} 1 & 0 \\ 0 & 1 \end{pmatrix} \tag{6.228b}$$

and from (6.172) and (6.173) we also deduce that

$$S_+ = \hbar\begin{pmatrix} 0 & 1 \\ 0 & 0 \end{pmatrix}, \qquad S_- = \hbar\begin{pmatrix} 0 & 0 \\ 1 & 0 \end{pmatrix}. \tag{6.229}$$

The normalised spin-1/2 eigenvectors α and β are given by the two component column vectors

$$\alpha = \begin{pmatrix} 1 \\ 0 \end{pmatrix}, \qquad \beta = \begin{pmatrix} 0 \\ 1 \end{pmatrix} \tag{6.230}$$

and may be considered as the basis vectors of a two-dimensional 'spin space'. The orthonormality relations (6.211) can then be written in the form

$$\begin{aligned} \alpha^\dagger \alpha &= \beta^\dagger \beta = 1 \\ \alpha^\dagger \beta &= \beta^\dagger \alpha = 0 \end{aligned} \tag{6.231}$$

where the Hermitian adjoints $\alpha^\dagger$ and $\beta^\dagger$ are the row vectors

$$\alpha^\dagger = \begin{pmatrix} 1 & 0 \end{pmatrix}, \qquad \beta^\dagger = \begin{pmatrix} 0 & 1 \end{pmatrix}. \tag{6.232}$$

From (6.216) and (6.230), we see that a general spin-1/2 function χ is now represented as

$$\chi = a \begin{pmatrix} 1 \\ 0 \end{pmatrix} + b \begin{pmatrix} 0 \\ 1 \end{pmatrix} = \begin{pmatrix} a \\ b \end{pmatrix}. \tag{6.233}$$

If χ is normalised to unity, we have

$$\chi^\dagger \chi = \begin{pmatrix} a^* & b^* \end{pmatrix} \begin{pmatrix} a \\ b \end{pmatrix} = 1 \tag{6.234}$$

so that $|a|^2 + |b|^2 = 1$, in agreement with (6.217). More generally, if we take into account the position coordinates $\mathbf{r}$ of the particle and the time t, we can write a general wave function (6.215) for a particle of spin 1/2 in the form of a two-component wave function (also called a Pauli wave function)

$$\begin{aligned} \Psi(\mathbf{r}, t, \sigma) &= \Psi_{1/2}(\mathbf{r}, t) \begin{pmatrix} 1 \\ 0 \end{pmatrix} + \Psi_{-1/2}(\mathbf{r}, t) \begin{pmatrix} 0 \\ 1 \end{pmatrix} \\ &= \begin{pmatrix} \Psi_{1/2}(\mathbf{r}, t) \\ \Psi_{-1/2}(\mathbf{r}, t) \end{pmatrix}. \end{aligned} \tag{6.235}$$

The Pauli spin matrices

When working with the spin operators for spin 1/2, it is convenient to introduce the Pauli vector operator $\boldsymbol{\sigma}$ by the relation

$$\mathbf{S} = \frac{\hbar}{2} \boldsymbol{\sigma}. \tag{6.236}$$

From the basic commutation relations (6.181) we deduce immediately that

$$[\sigma_x, \sigma_y] = 2i\sigma_z, \quad [\sigma_y, \sigma_z] = 2i\sigma_x, \quad [\sigma_z, \sigma_x] = 2i\sigma_y \tag{6.237}$$

and by using (6.220) we infer that

$$\sigma_x^2 = \sigma_y^2 = \sigma_z^2 = 1. \tag{6.238}$$

We also note from (6.224) that σ_x, σ_y and σ_z anticommute in pairs, so that we may write

$$[\sigma_i, \sigma_j]_+ = 2\delta_{ij} \qquad (i, j = x, y \text{ or } z). \tag{6.239}$$

From (6.225) and (6.236) we also have

$$\sigma_i \sigma_j = i\sigma_k \tag{6.240}$$

where $i, j, k = x, y$ or z, in cyclic order. From (6.239) and (6.240) it also follows (Problem 6.16) that

$$(\boldsymbol{\sigma} \cdot \mathbf{A})(\boldsymbol{\sigma} \cdot \mathbf{B}) = \mathbf{A} \cdot \mathbf{B} + i\boldsymbol{\sigma} \cdot (\mathbf{A} \times \mathbf{B}) \tag{6.241}$$

where $\mathbf{A}$ and $\mathbf{B}$ are two vectors, or two vector operators whose components commute with those of $\boldsymbol{\sigma}$. In the latter case the order of $\mathbf{A}$ and $\mathbf{B}$ on both sides of (6.241) must be respected. For example

$$(\boldsymbol{\sigma} \cdot \mathbf{r})(\boldsymbol{\sigma} \cdot \mathbf{p}) = \mathbf{r} \cdot \mathbf{p} + i\boldsymbol{\sigma} \cdot (\mathbf{r} \times \mathbf{p}). \tag{6.242}$$

Using (6.236) and the matrix representation (6.228) of the spin-1/2 operators S_x, S_y and S_z, we find that σ_x, σ_y and σ_z are represented by the matrices

$$\sigma_x = \begin{pmatrix} 0 & 1 \\ 1 & 0 \end{pmatrix}, \qquad \sigma_y = \begin{pmatrix} 0 & -i \\ i & 0 \end{pmatrix}, \qquad \sigma_z = \begin{pmatrix} 1 & 0 \\ 0 & -1 \end{pmatrix} \tag{6.243}$$

which are known as the *Pauli spin matrices*. We see that the traces of all Pauli spin matrices vanish,

$$\text{Tr}\sigma_x = \text{Tr}\sigma_y = \text{Tr}\sigma_z = 0 \tag{6.244}$$

and that

$$\det \sigma_x = \det \sigma_y = \det \sigma_z = -1 \tag{6.245}$$

where det means the determinant. Moreover, the unit 2×2 matrix I and the three Pauli matrices (6.243) are four linearly independent matrices. They form a complete set of 2×2 matrices, in the sense that an arbitrary 2×2 matrix can always be expressed as a linear combination of I, σ_x, σ_y and σ_z, with coefficients which can be complex.

Eigenvalues and eigenfunctions of an arbitrary spin-1/2 component

Let $\hat{\mathbf{n}}$ be a unit vector with polar angles (θ, ϕ) as shown in Fig. 6.6. The component of the spin vector $\mathbf{S}$ along $\hat{\mathbf{n}}$ is $S_n = \hat{\mathbf{n}} \cdot \mathbf{S}$. In order to find the eigenvalues and eigenfunctions of S_n, we must solve the eigenvalue equation

$$S_n \chi = \nu \hbar \chi \tag{6.246}$$

where we have written the eigenvalue as $\nu\hbar$ for further convenience. Since the Cartesian components of $\hat{\mathbf{n}}$ are $(\sin\theta \cos\phi, \sin\theta \sin\phi, \cos\theta)$, we find with the help

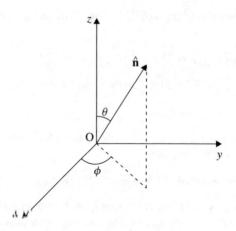

Figure 6.6 The unit vector $\hat{\mathbf{n}}$ of the text, having spherical polar angles (θ, ϕ).

of (6.228) that

$$S_n = S_x \sin\theta \cos\phi + S_y \sin\theta \sin\phi + S_z \cos\theta$$

$$= \frac{\hbar}{2} \begin{pmatrix} \cos\theta & \sin\theta e^{-i\phi} \\ \sin\theta e^{i\phi} & -\cos\theta \end{pmatrix}. \tag{6.247}$$

Writing the spin function as $\chi = \begin{pmatrix} a \\ b \end{pmatrix}$, we have

$$S_n \chi = \frac{\hbar}{2} \begin{pmatrix} a\cos\theta + b\sin\theta e^{-i\phi} \\ a\sin\theta e^{i\phi} - b\cos\theta \end{pmatrix}. \tag{6.248}$$

Hence (6.246) reduces to the set of two linear, homogeneous equations

$$(\cos\theta - 2\nu)a + \sin\theta e^{-i\phi} b = 0 \tag{6.249a}$$

$$\sin\theta e^{i\phi} a - (\cos\theta + 2\nu)b = 0 \tag{6.249b}$$

which have non-trivial solutions only if

$$\begin{vmatrix} \cos\theta - 2\nu & \sin\theta e^{-i\phi} \\ \sin\theta e^{i\phi} & -(\cos\theta + 2\nu) \end{vmatrix} = 0. \tag{6.250}$$

The determinant on the left-hand side being equal to $4\nu^2 - 1$, we must have $\nu = \pm 1/2$. If $\nu = +1/2$, we see from (6.249) that

$$\frac{b}{a} = \tan\frac{\theta}{2} e^{i\phi}. \tag{6.251}$$

Hence, apart from an irrelevant phase factor common to a and b the corresponding spin function $\chi_\uparrow$, normalised according to (6.234), may be written as

$$\chi_\uparrow = \begin{pmatrix} \cos\frac{\theta}{2} \\ \sin\frac{\theta}{2} e^{i\phi} \end{pmatrix}. \tag{6.252}$$

We note that the expectation values of the components of **S** in this state $\chi_\uparrow$ are (Problem 6.17)

$$\langle S_x \rangle = \frac{\hbar}{2} \sin\theta \cos\phi$$

$$\langle S_y \rangle = \frac{\hbar}{2} \sin\theta \sin\phi \tag{6.253}$$

$$\langle S_z \rangle = \frac{\hbar}{2} \cos\theta$$

so that the polar angles θ and ϕ give the direction of $\langle \mathbf{S} \rangle$. In this sense we shall say that the spin is 'up' in the direction of $\hat{\mathbf{n}}$. Using (6.233), we also remark that if the z-component of the spin of a particle in the state $\chi_\uparrow$ is measured, the probability of finding the spin 'up' is $|a|^2 = \cos^2(\theta/2)$. As θ varies between 0 and π, $|a|^2$ varies between 1 and 0.

If $\nu = -1/2$, we have from (6.249),

$$\frac{b}{a} = -\cot\frac{\theta}{2} e^{i\phi} \tag{6.254}$$

and the corresponding normalised spin function $\chi_\downarrow$ may be written (apart from an arbitrary phase factor common to a and b) as

$$\chi_\downarrow = \begin{pmatrix} \sin\frac{\theta}{2} \\ -\cos\frac{\theta}{2} e^{i\phi} \end{pmatrix}. \tag{6.255}$$

Upon calculating the expectation values of the components of **S** in the state $\chi_\downarrow$, it is readily verified (Problem 6.17) that the direction of $\langle \mathbf{S} \rangle$ is now given by the polar angles $(\pi - \theta, \phi + \pi)$, and we shall say that the spin is 'down' in the $\hat{\mathbf{n}}$ direction.

It is a simple matter to check (Problem 6.18) that the two spin functions $\chi_\uparrow$ and $\chi_\downarrow$ form a complete orthonormal set. We also remark that when $\theta = 0$ the spin function $\chi_\uparrow$ reduces to

$$\alpha = \begin{pmatrix} 1 \\ 0 \end{pmatrix},$$

while the spin function $\chi_\downarrow$ reduces (apart from a phase factor) to

$$\beta = \begin{pmatrix} 0 \\ 1 \end{pmatrix},$$

as expected.

6.9 Total angular momentum

Let us conside a particle which possesses a spin angular momentum. If **S** denotes its spin angular momentum operator, and **L** its orbital angular momentum operator, the *total angular momentum operator* of this particle is

$$\mathbf{J} = \mathbf{L} + \mathbf{S}. \tag{6.256}$$

Since the orbital angular momentum $\mathbf{L}$ and the spin angular momentum $\mathbf{S}$ operate on different variables ($\mathbf{L}$ operates on the angular variables and $\mathbf{S}$ on the spin variable), all the components of $\mathbf{L}$ commute with all those of $\mathbf{S}$. As a result, the commutation relations for the components of $\mathbf{J}$ are the same as those for the components of $\mathbf{L}$ and those of $\mathbf{S}$, and are given by (6.139).

More generally, let us consider an isolated system of N particles. Its total angular momentum operator is the sum of the angular momentum operators of the individual particles,

$$\mathbf{J} = \sum_{i=1}^{N} \mathbf{J}_i \tag{6.257}$$

where $\mathbf{J}_i$, the angular momentum operator of the ith particle, is the sum of its orbital angular momentum operator $\mathbf{L}_i$ and of its spin angular momentum operator $\mathbf{S}_i$ (if any)

$$\mathbf{J}_i = \mathbf{L}_i + \mathbf{S}_i. \tag{6.258}$$

Note that (6.257) may also be written in the form of equation (6.256), provided that $\mathbf{L}$ now denotes the *total orbital angular momentum* operator of the system, as defined by (6.23), and $\mathbf{S}$ is now the *total spin* operator

$$\mathbf{S} = \sum_{i=1}^{N} \mathbf{S}_i. \tag{6.259}$$

We observed at the end of Section 6.1 that because operators acting on different, independent variables commute, all the components of $\mathbf{L}_i$ commute with all those of $\mathbf{L}_k$ when $i \neq k$, so that the components of the total orbital angular momentum $\mathbf{L}$ obey the same basic commutation relations (6.25) as those obtained for the case of a single particle. Likewise, when $i \neq k$ all the components of $\mathbf{S}_i$ commute with all those of $\mathbf{S}_k$, and all the components of $\mathbf{J}_i$ commute with all those of $\mathbf{J}_k$. As a consequence, the components of the total spin operator (6.259) and those of the total angular momentum operator (6.257) also obey the basic commutation relations (6.139) characteristic of angular momenta.

Total angular momentum and rotations

In Section 6.2 we considered the effect of spatial rotations on an isolated system of N *spinless* particles, and we found that the component of the total orbital angular momentum $\mathbf{L}$ along an axis is the generator of infinitesimal rotations about that axis (see (6.43)). The *key properties* of the generators L_x, L_y and L_z corresponding respectively to infinitesimal rotations about the x-, y- and z-axes are the basic *commutation relations* (6.25). Similarly, the generators $L_u \equiv \hat{\mathbf{u}}.\mathbf{L}$, $L_v \equiv \hat{\mathbf{v}}.\mathbf{L}$ and $L_w \equiv \hat{\mathbf{w}}.\mathbf{L}$

corresponding respectively to infinitesimal rotations about the three orthogonal unit vectors $\hat{\mathbf{u}}$, $\hat{\mathbf{v}}$ and $\hat{\mathbf{w}} = \hat{\mathbf{u}} \times \hat{\mathbf{v}}$ satisfy the basic commutation relations (6.21).

Likewise, in the case of a system of particles possessing both orbital angular momentum and spin, we can take the generators of infinitesimal rotations about the x-, y- and z-axes to be given respectively by the Cartesian components J_x, J_y and J_z of the *total* angular momentum $\mathbf{J}$ of the system, which satisfy the basic commutation relations (6.139). In the same way, the generators of infinitesimal rotations about the three orthogonal unit vectors $\hat{\mathbf{u}}$, $\hat{\mathbf{v}}$ and $\hat{\mathbf{w}} = \hat{\mathbf{u}} \times \hat{\mathbf{v}}$ can be taken, respectively, to be the components $J_u \equiv \hat{\mathbf{u}}.\mathbf{J}$, $J_v \equiv \hat{\mathbf{v}}.\mathbf{J}$ and $J_w \equiv \hat{\mathbf{w}}.\mathbf{J}$ which also satisfy the basic commutation relations (see (6.141)). Thus, in analogy with (6.43), the unitary operator corresponding to an infinitesimal rotation by an angle $\delta\alpha$ in the positive sense about an axis oriented along an arbitrary vector $\hat{\mathbf{n}}$ is now given by

$$U_{\hat{\mathbf{n}}}(\delta\alpha) = I - \frac{i}{\hbar}\delta\alpha\hat{\mathbf{n}}.\mathbf{J} \tag{6.260}$$

and we shall say that the *total* angular momentum $\mathbf{J}$ of the system is the *generator of infinitesimal rotations* in the combined space of spatial plus spin variables. The unitary operator corresponding to a *finite* rotation of angle α about $\hat{\mathbf{n}}$ is

$$U_{\hat{\mathbf{n}}}(\alpha) = \exp\left(-\frac{i}{\hbar}\alpha\hat{\mathbf{n}}.\mathbf{J}\right) \tag{6.261}$$

which is the direct generalisation of (6.51). From (6.26) and (6.261) the state vector describing the system must therefore transform under a rotation of angle α about $\hat{\mathbf{n}}$ as

$$\Psi' = U_{\hat{\mathbf{n}}}(\alpha)\Psi$$

$$= \exp\left(-\frac{i}{\hbar}\alpha\hat{\mathbf{n}}.\mathbf{J}\right)\Psi. \tag{6.262}$$

Since the Hamiltonian of an isolated system is invariant under rotations, and because $\mathbf{J}$ is the generator of infinitesimal rotations, it follows from (6.31) and (6.260) that for an isolated system

$$[\mathbf{J}, H] = 0 \tag{6.263}$$

so that the *total angular momentum is conserved*. Conversely, if the Hamiltonian H of a system of particles commutes with the total angular momentum operator $\mathbf{J}$, then H is invariant under rotations. Equation (6.263) is the generalisation of equation (6.53) to particles or systems of particles which may possess a spin angular momentum. From (6.263) and (6.144) it follows that the operators H, $\mathbf{J}^2$ and J_z mutually commute, so that the eigenfunctions of H can be searched for among those common to $\mathbf{J}^2$ and J_z, a fact which greatly simplifies the eigenvalue problem of H. If the eigenvalues of $\mathbf{J}^2$ and J_z are denoted respectively by $j(j+1)\hbar^2$ and $m\hbar$, the energy eigenvalues corresponding to the same value of j but different values of m must be identical. This is because there is no preferred direction of space for an isolated system, so that the energy eigenvalues cannot depend on the projection of the total angular momentum on the z-axis, and hence are independent of the quantum number m.

Rotations in 'spin space'

Let us return to the equation (6.262), which tells us how the state vector describing a system transforms under a rotation of angle α about the unit vector $\hat{\mathbf{n}}$. In what follows we shall concentrate on the transformation properties of *spin* functions under rotations. Since $\mathbf{J} = \mathbf{L} + \mathbf{S}$ and the orbital angular momentum $\mathbf{L}$ acts only on the angular variables, it follows from (6.262) that under a rotation of angle α about $\hat{\mathbf{n}}$, a spin function χ_s for a particle of spin s transforms into a new spin function χ_s' such that

$$\chi_s' = \exp\left(-\frac{i}{\hbar}\alpha \hat{\mathbf{n}}.\mathbf{S}\right)\chi_s. \tag{6.264}$$

In particular, for a spin-1/2 particle we have (dropping the subscript 1/2)

$$\chi' = \exp\left(-i\frac{\alpha}{2}\hat{\mathbf{n}}.\boldsymbol{\sigma}\right)\chi$$
$$= \left(I\cos\frac{\alpha}{2} - i\hat{\mathbf{n}}.\boldsymbol{\sigma}\sin\frac{\alpha}{2}\right)\chi \tag{6.265}$$

where I denotes the unit 2×2 matrix and we have used the equations (6.236) and (6.241). For example, in the case of a rotation of angle α about the z-axis we have

$$\chi' = \left(I\cos\frac{\alpha}{2} - i\sigma_z\sin\frac{\alpha}{2}\right)\chi$$
$$= \begin{pmatrix} e^{-i\alpha/2} & 0 \\ 0 & e^{i\alpha/2} \end{pmatrix}\chi. \tag{6.266}$$

An object χ with one column and two rows which transforms under spatial rotations according to (6.265) is called a *spinor*. We remark that if a full rotation of angle $\alpha = 2\pi$ is made, we have $\chi' = -\chi$. This change of sign of spinors under rotations should not deter us from using them, since all the matrix elements giving measurable quantities (such as expectation values) depend bi-linearly on the spinors.

Let us now return to (6.264) giving the transformation property of a general spin function under rotations. For a rotation of angle α about the z-axis, we have

$$\chi_s' = \exp\left(-\frac{i}{\hbar}\alpha S_z\right)\chi_s. \tag{6.267}$$

Using (6.185) and (6.202), we see that for a rotation of angle $\alpha = 2\pi$

$$\chi_s' = \exp(-2\pi i m_s)\chi_s$$
$$= (-1)^{2m_s}\chi_s = (-1)^{2s}\chi_s \tag{6.268}$$

where we have used the fact that $2s$ has the same parity as $2m_s$. Hence, when a full rotation of angle 2π is made, the wave functions corresponding to particles of integer spin return to their original values, while those corresponding to half-odd integer values of s ($s = 1/2, 3/2, \ldots$) change sign.

6.10 The addition of angular momenta

In many situations of physical interest we are dealing with systems whose Hamiltonian H is invariant under rotations, and hence commutes with the total angular momentum operator **J** (see (6.263)). We can then search for the eigenfunctions of H among the simultaneous eigenfunctions of $\mathbf{J}^2$ and J_z. Consequently, it is very important to determine the common eigenfunctions of $\mathbf{J}^2$ and J_z.

Now, the total angular momentum operator **J** of a system is in general a sum of angular momentum operators referring either to distinct sets of dynamical variables of the system (such as the orbital angular momentum **L** and the spin angular momentum **S** of equation (6.256)) or to distinct sub-systems (such as the individual angular momenta $\mathbf{J}_i$ of equation (6.257)). Since these 'separate' angular momenta whose sum gives **J** correspond to distinct, independent quantities, they all mutually commute (i.e. all the components of one of them commute with all the components of the others). In addition, the components of each separate angular momentum satisfy the basic commutation relations (6.139). As observed in Section 6.9, it follows that the components of **J** also satisfy the commutation relations (6.139) characteristic of angular momenta. Thus simultaneous eigenvectors of $\mathbf{J}^2$ and J_z can be found, which correspond to eigenvalues $j(j+1)\hbar^2$ and $m\hbar$, respectively. As usual, for a given value of j the quantum number m can take one of the $2j+1$ values $-j, -j+1, \ldots, j$. Now, in general, we can obtain the eigenvectors and eigenvalues of the 'separate' angular momentum operators. The addition problem consists of determining the eigenvectors and eigenvalues of $\mathbf{J}^2$ and J_z in terms of those of the separate, independent angular momentum operators.

In what follows we shall treat the simplest addition problem, namely that of adding *two* commuting angular momenta. Thus we have

$$\mathbf{J} = \mathbf{J}_1 + \mathbf{J}_2 \tag{6.269}$$

where $\mathbf{J}_1$ and $\mathbf{J}_2$ are any two angular momenta corresponding respectively to the independent sub-systems (or sets of dynamical variables) 1 and 2.

Let $|j_1 m_1\rangle$ be a normalised simultaneous eigenvector of $\mathbf{J}_1^2$ and J_{1z}, so that

$$\mathbf{J}_1^2 |j_1 m_1\rangle = j_1(j_1 + 1)\hbar^2 |j_1 m_1\rangle \tag{6.270}$$

and

$$J_{1z} |j_1 m_1\rangle = m_1 \hbar |j_1 m_1\rangle. \tag{6.271}$$

Similary, let $|j_2 m_2\rangle$ be a normalised simultaneous eigenvector of $\mathbf{J}_2^2$ and J_{2z}, whence

$$\mathbf{J}_2^2 |j_2 m_2\rangle = j_2(j_2 + 1)\hbar^2 |j_2 m_2\rangle \tag{6.272}$$

and

$$J_{2z} |j_2 m_2\rangle = m_2 \hbar |j_2 m_2\rangle. \tag{6.273}$$

A normalised simultaneous eigenvector of $\mathbf{J}_1^2$, $\mathbf{J}_2^2$, J_{1z} and J_{2z} belonging respectively to the eigenvalues $j_1(j_1 + 1)\hbar^2$, $j_2(j_2 + 1)\hbar^2$, $m_1 \hbar$ and $m_2 \hbar$ is therefore given by the 'direct product'

$$|j_1 j_2 m_1 m_2\rangle = |j_1 m_1\rangle |j_2 m_2\rangle. \tag{6.274}$$

For a fixed value of j_1, m_1 can take one of the $2j_1 + 1$ values $-j_1, -j_1 + 1, \ldots, j_1$, and for a fixed value of j_2 the $2j_2 + 1$ allowed values of m_2 are $-j_2, -j_2 + 1, \ldots, j_2$. Hence, for given values of j_1 and j_2 there are $(2j_1 + 1)(2j_2 + 1)$ 'direct products' (6.274) which form a complete orthonormal set, i.e. a basis in the 'product space' of the combined system $(1 + 2)$.

It is instructive to rewrite the direct products (6.274) in a more explicit way. Suppose that instead of using the abstract eigenvector $|j_1 m_1\rangle$ we denote by $\psi_{j_1 m_1}(1)$ a normalised simultaneous eigenfunction of $\mathbf{J}_1^2$ and J_{1z} corresponding respectively to the eigenvalues $j_1(j_1+1)\hbar^2$ and $m_1\hbar$. Likewise, instead of using the abstract eigenvector $|j_2 m_2\rangle$ let us denote by $\psi_{j_2 m_2}(2)$ a normalised simultaneous eigenfunction of $\mathbf{J}_2^2$ and J_{2z} belonging to the eigenvalues $j_2(j_2+1)\hbar^2$ and $m_2\hbar$, respectively. The normalised simultaneous eigenfunctions of $\mathbf{J}_1^2$, $\mathbf{J}_2^2$, J_{1z} and J_{2z} corresponding respectively to the eigenvalues $j_1(j_1 + 1)\hbar^2$, $j_2(j_2 + 1)\hbar^2$, $m_1 \hbar$ and $m_2 \hbar$ are then given by

$$\psi_{j_1 j_2 m_1 m_2}(1, 2) = \psi_{j_1 m_1}(1) \psi_{j_2 m_2}(2). \tag{6.275}$$

We emphasise that when operating on $\psi_{j_1 j_2 m_1 m_2}(1, 2)$ the components of $\mathbf{J}_1$ only act on the variables denoted collectively by 1, while the components of $\mathbf{J}_2$ only act on the variables denoted by 2. In particular, from (6.269) and (6.275) we have

$$\begin{aligned} J_z \psi_{j_1 j_2 m_1 m_2}(1, 2) &= (J_{1z} + J_{2z})\psi_{j_1 m_1}(1)\psi_{j_2 m_2}(2) \\ &= [J_{1z}\psi_{j_1 m_1}(1)]\psi_{j_2 m_2}(2) + \psi_{j_1 m_1}(1)[J_{2z}\psi_{j_2 m_2}(2)] \\ &= m_1 \hbar \psi_{j_1 m_1}(1)\psi_{j_2 m_2}(2) + m_2 \hbar \psi_{j_1 m_1}(1)\psi_{j_2 m_2}(2) \\ &= (m_1 + m_2)\hbar \; \psi_{j_1 j_2 m_1 m_2}(1, 2) \end{aligned} \tag{6.276}$$

which means that $\psi_{j_1 j_2 m_1 m_2}(1, 2)$ is also an eigenfunction of J_z corresponding to the eigenvalue $(m_1 + m_2)\hbar$. In terms of abstract eigenvectors we can write the above result as

$$J_z |j_1 j_2 m_1 m_2\rangle = (m_1 + m_2)\hbar \; |j_1 j_2 m_1 m_2\rangle. \tag{6.277}$$

The fact that simultaneous eigenfunctions of $\mathbf{J}_1^2$, $\mathbf{J}_2^2$, J_{1z} and J_{2z} are also eigenfunctions of J_z could have been anticipated by noting that $J_z = J_{1z} + J_{2z}$ commutes with $\mathbf{J}_1^2$, $\mathbf{J}_2^2$, J_{1z} and J_{2z}.

Let us now consider the operator $\mathbf{J}^2$. From (6.269) we have

$$\mathbf{J}^2 = (\mathbf{J}_1 + \mathbf{J}_2)^2 = \mathbf{J}_1^2 + \mathbf{J}_2^2 + 2\mathbf{J}_1 \cdot \mathbf{J}_2. \tag{6.278a}$$

Because all the components of $\mathbf{J}_1$ commute with all those of $\mathbf{J}_2$, and since $[\mathbf{J}_1^2, \mathbf{J}_1] = [\mathbf{J}_2^2, \mathbf{J}_2] = 0$ it follows that $\mathbf{J}^2$ commutes with $\mathbf{J}_1^2$ and $\mathbf{J}_2^2$. However, as

$$\mathbf{J}_1 \cdot \mathbf{J}_2 = J_{1x} J_{2x} + J_{1y} J_{2y} + J_{1z} J_{2z} \tag{6.278b}$$

and since J_{1z} does not commute with J_{1x} or J_{1y}, we see that $\mathbf{J}^2$ does *not* commute with J_{1z}; likewise $\mathbf{J}^2$ does *not* commute with J_{2z}. Consequently, the simultaneous eigenfunctions of $\mathbf{J}^2$ and J_z are eigenfunctions of $\mathbf{J}_1^2$ and $\mathbf{J}_2^2$, but not (in general) of J_{1z} and J_{2z}. There are therefore two complete, but distinct descriptions of the system: either in terms of the eigenfunctions of $\mathbf{J}_1^2$, $\mathbf{J}_2^2$, J_{1z} and J_{2z}, given by (6.275), or in terms of the eigenfunctions of $\mathbf{J}_1^2$, $\mathbf{J}_2^2$, $\mathbf{J}^2$ and J_z. We shall denote the latter (normalised to unity) by $\Phi^{jm}_{j_1 j_2}(1, 2)$. Thus we have in particular

$$\mathbf{J}^2 \Phi^{jm}_{j_1 j_2}(1, 2) = j(j+1)\hbar^2 \Phi^{jm}_{j_1 j_2}(1, 2) \tag{6.279}$$

and

$$J_z \Phi^{jm}_{j_1 j_2}(1, 2) = m\hbar \Phi^{jm}_{j_1 j_2}(1, 2). \tag{6.280}$$

Like the functions $\psi_{j_1 j_2 m_1 m_2}(1, 2)$, the functions $\Phi^{jm}_{j_1 j_2}(1, 2)$ form a complete orthonormal set, and hence are another basis in the 'product space' of the system $(1 + 2)$. These two basis sets of orthonormal eigenfunctions can therefore be related by a *unitary* transformation

$$\Phi^{jm}_{j_1 j_2}(1, 2) = \sum_{m_1 m_2} \langle j_1 j_2 m_1 m_2 | jm \rangle \psi_{j_1 j_2 m_1 m_2}(1, 2) \tag{6.281}$$

where the summation must only be performed over m_1 and m_2, since j_1 and j_2 are assumed to have fixed values. The coefficients $\langle j_1 j_2 m_1 m_2 | jm \rangle$ of that unitary transformation are called *Clebsch–Gordan* or *vector addition coefficients*. In the Dirac notation we can write the unitary transformation (6.281) as

$$|j_1 j_2 j m\rangle = \sum_{m_1 m_2} \langle j_1 j_2 m_1 m_2 | jm \rangle |j_1 j_2 m_1 m_2\rangle \tag{6.282}$$

where $|j_1 j_2 m_1 m_2\rangle$ is given by (6.274).

In order to find out what are the allowed values of j for given j_1 and j_2 we proceed as follows. We first notice that since $\psi_{j_1 j_2 m_1 m_2}$ is an eigenfunction of J_z belonging to the eigenvalue $(m_1 + m_2)\hbar$ (see (6.276)) we must have

$$m = m_1 + m_2 \tag{6.283}$$

so that the double sum in (6.282) reduces to a single sum. In other words the Clebsch–Gordan coefficients must vanish unless $m = m_1 + m_2$:

$$\langle j_1 j_2 m_1 m_2 | jm \rangle = 0 \quad \text{if} \quad m \neq m_1 + m_2. \tag{6.284}$$

Moreover, because the maximum allowed values of m_1 and m_2 are given respectively by j_1 and j_2, the relation (6.283) tells us that the maximum possible value of m is $j_1 + j_2$. Since m can only take on the $2j + 1$ values $-j, -j + 1, \ldots, j$, it follows that the maximum possible value of j is $j_1 + j_2$. We remark that for $j = j_1 + j_2$ and $m = j_1 + j_2$ there is only one term in the sum on the right of (6.282), namely that corresponding to $m_1 = j_1$ and $m_2 = j_2$. Thus

$$\Phi_{j_1 j_2}^{j_1+j_2, j_1+j_2}(1, 2) = \langle j_1 j_2 j_1 j_2 | j_1 + j_2, j_1 + j_2 \rangle \psi_{j_1 j_2 j_1 j_2}(1, 2) \qquad (6.285)$$

and since the eigenfunctions $\Phi_{j_1 j_2}^{jm}$ and $\psi_{j_1 j_2 m_1 m_2}$ are normalised to unity, we have

$$|\langle j_1 j_2 j_1 j_2 | j_1 + j_2, j_1 + j_2 \rangle| = 1. \qquad (6.286)$$

Next, let us consider a state $\Phi_{j_1 j_2}^{jm}$ for which $m = j_1 + j_2 - 1$. In this case there are two possibilities for the values of m_1 and m_2: we can either have $m_1 = j_1$ and $m_2 = j_2 - 1$ or $m_1 = j_1 - 1$ and $m_2 = j_2$. Thus a state $\Phi_{j_1 j_2}^{j, j_1+j_2-1}$ must be a linear combination of the two linearly independent eigenfunctions $\psi_{j_1 j_2 j_1 j_2-1}$ and $\psi_{j_1 j_2 j_1-1 j_2}$. Moreover, there are two such linear combinations, one of them belonging to the set of eigenfunctions with $j = j_1 + j_2$ while the other (orthogonal) combination must be a member of a set of eigenfunctions for which the maximum value of m is $j_1 + j_2 - 1$; this latter set must therefore be such that $j = j_1 + j_2 - 1$. Proceeding further to states $\Phi_{j_1 j_2}^{jm}$ with $m = j_1 + j_2 - 2$, we see that three linearly independent states of this kind exist, corresponding to the values $j = j_1 + j_2, j_1 + j_2 - 1$ and $j_1 + j_2 - 2$, respectively. Repeating this argument successively, it is found that the minimum value of j is $|j_1 - j_2|$, because when j reaches this smallest value all the combinations have been exhausted. This is readily checked by noting that there must be the same number of eigenfunctions in the basis set $\{\Phi_{j_1 j_2}^{jm}\}$ as in the basis set $\{\psi_{j_1 j_2 m_1 m_2}\}$. As we have shown above, the latter set contains $(2j_1 + 1)(2j_2 + 1)$ eigenfunctions. Now, for each value of j there are $2j + 1$ values of m so that the total number of eigenfunctions $\Phi_{j_1 j_2}^{jm}$ is given by

$$\sum_{j=|j_1-j_2|}^{j_1+j_2} (2j + 1) = (2j_1 + 1)(2j_2 + 1). \qquad (6.287)$$

This is illustrated in Table 6.2 for the case $j_1 = 1$, $j_2 = 3/2$, where the number of $(m_1 m_2)$ and of (jm) combinations are both seen to be equal to 12.

We have therefore obtained the following fundamental result: for given values of j_1 and j_2 the allowed values of j are

$$j = |j_1 - j_2|, |j_1 - j_2| + 1, \ldots, j_1 + j_2 \qquad (6.288)$$

so that the three angular momentum quantum numbers j_1, j_2 and j must satisfy the *triangular condition*

$$|j_1 - j_2| \leqslant j \leqslant j_1 + j_2. \qquad (6.289)$$

Moreover, for each value of j there are $2j + 1$ eigenfunctions $\Phi_{j_1 j_2}^{jm}$, such that $m = -j, -j + 1, \ldots, +j$.

Table 6.2 Allowed values of $(m_1 m_2)$ and (jm) for $j_1 = 1$ and $j_2 = 3/2$.

m_1	m_2	m	j
1	3/2	5/2	5/2
0 1	3/2 1/2	3/2	5/2, 3/2
−1 0 1	3/2 1/2 −1/2	1/2	5/2, 3/2, 1/2
−1 0 1	1/2 −1/2 −3/2	−1/2	5/2, 3/2, 1/2
−1 0	−1/2 −3/2	−3/2	5/2, 3/2
−1	−3/2	−5/2	5/2
Total: $(2j_1+1)(2j_2+1) = 12$		Total: $\sum_{j=|j_1-j_2|}^{j_1+j_2}(2j+1) = 12$	

The Clebsch–Gordan coefficients $\langle j_1 j_2 m_1 m_2 | jm \rangle$ can be determined by applying the raising and lowering operators $J_\pm = J_x \pm iJ_y$ to (6.281) (or (6.282)), so that recursion relations are obtained. First of all, it follows from the foregoing discussion that they must satisfy the *selection rules* (6.283) and (6.289). Furthermore, in order that this determination be unambiguous, the relative phases of the eigenvectors $|j_1 j_2 m_1 m_2\rangle$ and $|j_1 j_2 jm\rangle$ must be specified. A frequently used convention, which we shall adopt here, is that of Condon and Shortely and of Wigner, namely

(1) $J_\pm |j_1 j_2 jm\rangle = [j(j+1) - m(m\pm 1)]^{1/2} \hbar |j_1 j_2 jm \pm 1\rangle$ (6.290a)

with similar relationships for $|j_1 m_1\rangle$ and $|j_2 m_2\rangle$ (see (6.169) and (6.171)); and

(2) $\langle j_1 j_2 j_1 (j-j_1) | jj \rangle$ is real and positive. (6.290b)

With this phase convention, the Clebsch–Gordan coefficients are real. It can also be proved[8] that they satisfy the orthogonality relations

$$\sum_{m_1 m_2} \langle j_1 j_2 m_1 m_2 | jm \rangle \langle j_1 j_2 m_1 m_2 | j'm' \rangle = \delta_{jj'} \delta_{mm'}$$

$$\sum_{jm} \langle j_1 j_2 m_1 m_2 | jm \rangle \langle j_1 j_2 m_1' m_2' | jm \rangle = \delta_{m_1 m_1'} \delta_{m_2 m_2'} \quad (6.291)$$

---

[8] See, for example, Edmonds (1957).

Table 6.3 Clebsch–Gordan coefficients.

	$\langle j_1 \tfrac{1}{2} m_1 m_2 \vert jm \rangle$	
j	$m_2 = \tfrac{1}{2}$	$m_2 = -\tfrac{1}{2}$
$j_1 + \tfrac{1}{2}$	$\left(\dfrac{j_1 + m + \tfrac{1}{2}}{2j_1 + 1}\right)^{1/2}$	$\left(\dfrac{j_1 - m + \tfrac{1}{2}}{2j_1 + 1}\right)^{1/2}$
$j_1 - \tfrac{1}{2}$	$-\left(\dfrac{j_1 - m + \tfrac{1}{2}}{2j_1 + 1}\right)^{1/2}$	$\left(\dfrac{j_1 + m + \tfrac{1}{2}}{2j_1 + 1}\right)^{1/2}$

	$\langle j_1 1 m_1 m_2 \vert jm \rangle$		
j	$m_2 = 1$	$m_2 = 0$	$m_2 = -1$
$j_1 + 1$	$\left[\dfrac{(j_1+m)(j_1+m+1)}{(2j_1+1)(2j_1+2)}\right]^{1/2}$	$\left[\dfrac{(j_1-m+1)(j_1+m+1)}{(2j_1+1)(j_1+1)}\right]^{1/2}$	$\left[\dfrac{(j_1-m)(j_1-m+1)}{(2j_1+1)(2j_1+2)}\right]^{1/2}$
j_1	$-\left[\dfrac{(j_1+m)(j_1-m+1)}{2j_1(j_1+1)}\right]^{1/2}$	$\left[\dfrac{m^2}{j_1(j_1+1)}\right]^{1/2}$	$\left[\dfrac{(j_1-m)(j_1+m+1)}{2j_1(j_1+1)}\right]^{1/2}$
$j_1 - 1$	$\left[\dfrac{(j_1-m)(j_1-m+1)}{2j_1(2j_1+1)}\right]^{1/2}$	$-\left[\dfrac{(j_1-m)(j_1+m)}{j_1(2j_1+1)}\right]^{1/2}$	$\left[\dfrac{(j_1+m+1)(j_1+m)}{2j_1(2j_1+1)}\right]^{1/2}$

and the symmetry relations

$$\begin{aligned}
\langle j_1 j_2 m_1 m_2 \vert jm \rangle &= (-)^{j_1+j_2-j} \langle j_2 j_1 m_2 m_1 \vert jm \rangle \\
&= (-)^{j_1+j_2-j} \langle j_1 j_2 -m_1 -m_2 \vert j -m \rangle \\
&= \langle j_2 j_1 -m_2 -m_1 \vert j -m \rangle \\
&= (-)^{j_1-m_1} \left(\frac{2j+1}{2j_2+1}\right)^{1/2} \langle j_1 j m_1 -m \vert j_2 -m_2 \rangle \\
&= (-)^{j_2+m_2} \left(\frac{2j+1}{2j_1+1}\right)^{1/2} \langle j j_2 -m m_2 \vert j_1 -m_1 \rangle.
\end{aligned} \quad (6.292)$$

Details concerning the explicit computation of the Clebsch–Gordan coefficients, together with tables, can be found for example in Edmonds (1957) or Rose (1957). In Table 6.3 the Clebsch–Gordan coefficients $\langle j_1 j_2 m_1 m_2 \vert jm \rangle$ are given for the cases $j_2 = 1/2$ and $j_2 = 1$. By using the symmetry relations, all the coefficients with any one of j_1, j_2 or j equal to 1/2 or to 1 can be obtained.

Addition of the orbital angular momentum and spin of a particle

As a first example, let us consider a particle of spin s. Let **L** be its orbital angular momentum operator and **S** its spin angular momentum operator. The total angular momentum operator of the particle is therefore $\mathbf{J} = \mathbf{L} + \mathbf{S}$. We shall denote by m_l, m_s and m_j the quantum numbers corresponding to the operators L_z, S_z and J_z, respectively. In the position representation the simultaneous eigenfunctions of the operators $\mathbf{L}^2$ and L_z are the spherical harmonics $Y_{lm_l}(\theta, \phi)$ with $l = 0, 1, 2, \ldots$, and

$m_l = -l, -l+1, \ldots, l$. The simultaneous eigenfunctions of the operators $\mathbf{S}^2$ and S_z are the spin functions χ_{s,m_s} (with $m_s = -s, -s+1, \ldots, s$) which as we have seen in Section 6.7 can be represented by column vectors with $(2s+1)$ positions, having zeros in all positions except one. Hence the simultaneous eigenfunctions of the operators $\mathbf{L}^2, \mathbf{S}^2, L_z$ and S_z are represented in the direct product 'coordinate–spin' space by the spin-angle functions

$$\psi_{lsm_lm_s} = Y_{lm_l}(\theta, \phi)\chi_{s,m_s}. \tag{6.293}$$

According to the fundamental result (6.288), the allowed values of the total angular momentum quantum number j of the particle are

$$j = |l-s|, |l-s|+1, \ldots, l+s. \tag{6.294}$$

The simultaneous normalised eigenfunctions of the operators $\mathbf{L}^2, \mathbf{S}^2, \mathbf{J}^2$ and J_z (which are usually denoted by the symbol $\mathcal{Y}_{ls}^{jm_j}$) are seen from (6.281) to be the spin-angle functions

$$\begin{aligned}\mathcal{Y}_{ls}^{jm_j} &= \sum_{m_l m_s} \langle lsm_l m_s | jm_j\rangle \psi_{lsm_lm_s} \\ &= \sum_{m_l m_s} \langle lsm_l m_s | jm_j\rangle Y_{lm_l}(\theta, \phi)\chi_{s,m_s}.\end{aligned} \tag{6.295}$$

In particular, for a particle of spin $s = 1/2$ we see from (6.294) that for a given value of the orbital angular momentum quantum number l the total angular momentum quantum number j can take the two values

$$j = l - \tfrac{1}{2}, l + \tfrac{1}{2} \tag{6.296}$$

except when $l = 0$ (s-state) in which case the only allowed value of j is $j = 1/2$. By using the Clebsch–Gordan coefficients displayed in Table 6.3 we find that (Problem 6.21)

$$\mathcal{Y}_{l,\frac{1}{2}}^{l\pm\frac{1}{2},m_j} = \begin{pmatrix} \pm\left[\dfrac{l\pm m_j + \frac{1}{2}}{2l+1}\right]^{1/2} Y_{l,m_j-\frac{1}{2}}(\theta,\phi) \\ \left[\dfrac{l\mp m_j + \frac{1}{2}}{2l+1}\right]^{1/2} Y_{l,m_j+\frac{1}{2}}(\theta,\phi) \end{pmatrix} \tag{6.297}$$

Addition of two spins 1/2

As a second example, let us consider two particles whose spin operators are $\mathbf{S}_1$ and $\mathbf{S}_2$, respectively. The total spin angular momentum is thus

$$\mathbf{S} = \mathbf{S}_1 + \mathbf{S}_2. \tag{6.298}$$

If the two particles have spin 1/2, the direct-product spin space of the combined system $(1+2)$ has four dimensions. The simultaneous eigenfunctions of the operators $\mathbf{S}_1^2$

and S_{1z} for the first particle are the two basic spinors $\alpha(1)$ and $\beta(1)$ corresponding respectively to 'spin up' ($m_{s_1} = +1/2$) and 'spin down' ($m_{s_1} = -1/2$) for that particle. Likewise, the simultaneous eigenfunctions of the operators $\mathbf{S}_2^2$ and S_{2z} for particle 2 are the two basic spinors $\alpha(2)$ and $\beta(2)$ corresponding respectively to 'spin up' ($m_{s_2} = +1/2$) and 'spin down' ($m_{s_2} = -1/2$) for the second particle. The direct-product eigenfunctions $\psi_{j_1 j_2 m_1 m_2}(1,2)$ are therefore in the present case the four spin functions

$$\alpha(1)\alpha(2), \qquad \alpha(1)\beta(2), \qquad \beta(1)\alpha(2), \qquad \beta(1)\beta(2) \tag{6.299}$$

which constitute a basis in the four-dimensional spin space of the combined system. If we denote by $M_s \hbar$ the eigenvalues of the operator S_z, we see that $M_s = m_{s_1} + m_{s_2}$, so that the four eigenfunctions (6.299) correspond respectively to the values $M_s = 1, 0, 0$ and -1.

From the basic result (6.288), the allowed values of the total spin quantum number S are given by

$$S = 0, 1. \tag{6.300}$$

The simultaneous normalised eigenfunctions of $\mathbf{S}_1^2$, $\mathbf{S}_2^2$, $\mathbf{S}^2$ and S_z, which we shall denote by χ_{S,M_s}, are readily obtained from (6.281), (6.299) and the Table 6.3 of Clebsch–Gordan coefficients. We find in this way that when $S = 0$ there is only *one* such eigenfunction,

$$\chi_{0,0} = \frac{1}{\sqrt{2}} [\alpha(1)\beta(2) - \beta(1)\alpha(2)] \tag{6.301}$$

which is called a *singlet* spin state. We note that it is an eigenstate of S_z corresponding to the quantum number $M_s = 0$ and that it is *antisymmetric* in the interchange of the spin coordinates of the two particles. When $S = 1$ we obtain *three* simultaneous normalised eigenfunctions of $\mathbf{S}_1^2$, $\mathbf{S}_2^2$, $\mathbf{S}^2$ and S_z, namely

$$\begin{aligned}\chi_{1,1} &= \alpha(1)\alpha(2) \\ \chi_{1,0} &= \frac{1}{\sqrt{2}}[\alpha(1)\beta(2) + \beta(1)\alpha(2)] \\ \chi_{1,-1} &= \beta(1)\beta(2)\end{aligned} \tag{6.302}$$

which are said to form a spin *triplet*. These three states, which are *symmetric* in the interchange of the spin coordinates of the two particles, are respectively eigenstates of S_z corresponding to the values $M_s = +1, 0, -1$ of the quantum number M_s.

As an example, the lowest state of the helium atom (which contains two electrons) is a single state ($S = 0$), while excited states can be either singlet or triplet. Another example is the deuteron, the nucleus composed of one proton and one neutron, whose only bound state is a triplet spin state ($S = 1$). Both the helium atom and the deuteron will be studied in Chapter 10.

Problems

6.1 Prove that the orbital angular momentum operator $\mathbf{L}$ is a Hermitian operator.

6.2 Show that the three commutation relations (6.9) are equivalent to the vector commutation relation (6.10).

6.3 Prove equations (6.16) and (6.17).

6.4 Prove that

(a) $[L_i, x_j] = i\hbar \varepsilon_{ijk} x_k$ $\quad (i, j, k = 1, 2, 3)$
where $L_1 = L_x, L_2 = L_y, L_3 = L_z$ and $x_1 = x, x_2 = y$ and $x_3 = z$. The symbol ε_{ijk}, called the Levi–Civita antisymmetric symbol, is such that

$$\varepsilon_{ijk} = \begin{cases} 1, & \text{if } ijk = 123, 231, 312 \\ -1, & \text{if } ijk = 321, 213, 132 \\ 0, & \text{otherwise.} \end{cases}$$

(b) $[L_i, p_j] = i\hbar \varepsilon_{ijk} p_k,$
where $p_1 = p_x, p_2 = p_y, p_3 = p_z$.

(c) $\mathbf{L} \cdot \mathbf{r} = \mathbf{L} \cdot \mathbf{p} = 0$.

6.5 Prove equations (6.21).

6.6 Show that for a system of N structureless particles the rotation operators U_R are given by (6.43) for infinitesimal rotations, and by (6.51) for finite rotations, where $\mathbf{L}$ is the total orbital angular momentum operator of the system.

6.7 Prove equation (6.80) by using the Rodrigues formula.

6.8 Prove the recurrence relations (6.83) for the Legendre polynomials.

6.9 Prove equation (6.95).

6.10 Establish the recurrence relations (6.97) for the associated Legendre functions.

6.11 Prove the relations (6.118) for L_+ and L_-.

6.12 Let $\hat{\mathbf{n}}$ be a unit vector in a direction specified by the polar angles (θ, ϕ). Show that the component of the angular momentum in the direction $\hat{\mathbf{n}}$ is

$$L_n = \sin\theta \cos\phi L_x + \sin\theta \sin\phi L_y + \cos\theta L_z$$
$$= \tfrac{1}{2} \sin\theta (e^{-i\phi} L_+ + e^{i\phi} L_-) + \cos\theta L_z.$$

If the system is in simultaneous eigenstates of $\mathbf{L}^2$ and L_z belonging to the eigenvalues $l(l+1)\hbar^2$ and $m\hbar$,

(a) what are the possible results of a measurement of L_n?
(b) what are the expectation values of L_n and L_n^2?

6.13 Consider a free particle of mass μ constrained to move on a ring of radius a.

(a) Show that the Hamiltonian of this system is

$$H = L_z^2/2I$$

where the z-axis is through the centre O of the ring and is perpendicular to its plane, and I is the moment of inertia of the particle with respect to the centre O.

(b) Find the energy eigenfunctions for the system and write down a general expression for the solution of the time-dependent Schrödinger equation.

6.14 Using as a basis the eigenvectors $|jm\rangle$ of $\mathbf{J}^2$ and J_z, obtain the matrix representation of the angular momentum operators J_x, J_y, J_z and $\mathbf{J}^2$ for the case $j = \frac{3}{2}$.

6.15 Verify the anticommutation relations (6.224).

6.16 Prove equation (6.241).

6.17 Calculate the expectation values of the components of $\mathbf{S}$ for a spin-$\frac{1}{2}$ particle

(a) in the state $\chi_\uparrow$ given by (6.252); and

(b) in the state $\chi_\downarrow$ given by (6.255).

6.18 (a) Verify that the two eigenvectors $\chi_\uparrow$ and $\chi_\downarrow$ of the operator S_n given respectively by (6.252) and (6.255) are orthogonal.

(b) By using $\alpha = \begin{pmatrix} 1 \\ 0 \end{pmatrix}$ and $\beta = \begin{pmatrix} 0 \\ 1 \end{pmatrix}$ as a basis, prove that $\chi_\uparrow$ and $\chi_\downarrow$ satisfy the closure relation

$$|\chi_\uparrow\rangle\langle\chi_\uparrow| + |\chi_\downarrow\rangle\langle\chi_\downarrow| = I.$$

6.19 Obtain the eigenvalues and the corresponding normalised eigenfunctions of S_x and S_y for a spin-$\frac{1}{2}$ particle and check your results by referring to (6.252) and (6.255).

6.20 Obtain the eigenvalues and corresponding normalised eigenvectors of $S_n = \hat{\mathbf{n}} \cdot \mathbf{S}$ for a particle of spin 1, where $\hat{\mathbf{n}}$ is a unit vector defined by the polar angles (θ, ϕ).

6.21 Using the Clebsch–Gordan coefficients given in Table 6.3, obtain the result (6.297).

6.22 Two particles of spin-$\frac{1}{2}$ have spin operators $\mathbf{S}_1$ and $\mathbf{S}_2$. Find the expectation values of the product $\mathbf{S}_1 \cdot \mathbf{S}_2$ in the singlet and triplet spin states, for which the wave functions are given by (6.301) and (6.302), respectively.

(*Hint*: $\mathbf{S}_1 \cdot \mathbf{S}_2 = (\mathbf{S}^2 - \mathbf{S}_1^2 - \mathbf{S}_2^2)/2$.)

6.23 The spin properties of a system of two spin-$\frac{1}{2}$ particles are described by the singlet and triplet wave functions (6.301) and (6.302). If a third particle of spin-$\frac{1}{2}$ is added to the system, show that the total spin quantum number of the three-particle system can be $S = \frac{3}{2}$ (a quartet state) or $S = \frac{1}{2}$ (a doublet state), and find the corresponding spin functions.

(*Hint*: Treat the original two-particle system as a single entity of either spin 1 or spin 0.)

6.24 Let T be the time-reversal operator introduced in Section 5.9. According to (5.364) this operator is given by $T = U_\tau K$, where U_τ is a unitary operator and K is the operator of complex conjugation.

(a) Using the results (5.365) and (5.367), show that the orbital angular momentum operator $\mathbf{L} = \mathbf{r} \times \mathbf{p}$ changes sign under the operation of time reversal:

$$\mathbf{L}' = T\mathbf{L}T^\dagger = -\mathbf{L}.$$

(b) The natural extension of this result is to require that the spin angular momentum operator $\mathbf{S}$ changes sign under the operation of time reversal:

$$\mathbf{S}' = T\mathbf{S}T^\dagger = -\mathbf{S}.$$

Using this relation, show that for a spin-1/2 particle the operator T is given by

$$T = U_\tau K \quad \text{with} \quad U_\tau = i\sigma_y$$

where σ_y is the second of the Pauli spin matrices (6.243).

7 The Schrödinger equation in three dimensions

7.1 Separation of the Schrödinger equation in Cartesian coordinates 328

7.2 Central potentials. Separation of the Schrödinger equation in spherical polar coordinates 336

7.3 The free particle 341

7.4 The three-dimensional square well potential 347

7.5 The hydrogenic atom 351

7.6 The three-dimensional isotropic oscillator 367

Problems 370

In Chapter 4 we solved the time-independent Schrödinger equation for several one-dimensional problems. We shall now generalise our treatment to study in three dimensions the non-relativistic motion of a particle in a time-independent potential $V(\mathbf{r})$, where $\mathbf{r}$ denotes the position vector of the particle. In fact, we can treat a slightly more general problem at no extra cost. Indeed, we have seen in Section 5.7 that the motion of two particles A and B, of masses m_A and m_B, interacting via a time-independent potential $V(\mathbf{r})$ which depends only on their relative coordinate $\mathbf{r} = \mathbf{r}_A - \mathbf{r}_B$, reduces in the centre-of-mass system to a one-body problem: the motion of a particle of mass $\mu = m_A m_B/(m_A + m_B)$ in the potential $V(\mathbf{r})$. The two problems can therefore be treated on the same footing, the time-independent Schrödinger equation to be solved being

$$\left[-\frac{\hbar^2}{2\mu}\nabla^2 + V(\mathbf{r})\right]\psi(\mathbf{r}) = E\psi(\mathbf{r}). \tag{7.1}$$

This is a second-order partial differential equation, in contrast to the one-dimensional case, where the time-independent Schrödinger equation (4.3) is a second-order ordinary differential equation. As a result, the solution of the three-dimensional Schrödinger equation (7.1) can only be obtained exactly and explicitly in a few simple cases. In particular, the potential may be such that the technique of *separation of variables* may be used. The original three-dimensional problem then reduces to simpler problems of lower dimensionality.

In this chapter we shall examine several cases for which a separation of variables can be made in the three-dimensional Schrödinger equation (7.1), and its solutions can be obtained. We shall first consider some simple examples where the Schrödinger equation is separable in Cartesian coordinates. The most interesting case, however, is that of *central* potentials, that is potentials $V(r)$ which depend only upon the magnitude $r = |\mathbf{r}|$ of the vector $\mathbf{r}$. For such spherically symmetric potentials the Schrödinger equation (7.1) can be separated in spherical polar coordinates. As in classical mechanics, the orbital angular momentum plays an essential role in this problem. In fact, using results obtained in Chapter 6 we shall be able to write down the angular part of the eigenfunctions, so that the solution of the Schrödinger equation (7.1) will reduce to that of a differential equation involving only the radial coordinate r. This radial equation will then be solved for several important problems: the free particle, the spherical square well, the hydrogenic atom and the isotropic three-dimensional harmonic oscillator.

7.1 Separation of the Schrödinger equation in Cartesian coordinates

Let us first consider the simple case where the potential $V(\mathbf{r})$ has the special form

$$V(\mathbf{r}) = V_1(x) + V_2(y) + V_3(z) \tag{7.2}$$

The Hamiltonian $H = -(\hbar^2/2\mu)\nabla^2 + V(\mathbf{r})$ may then be written as

$$H = H_x + H_y + H_z \tag{7.3}$$

where

$$H_x = -\frac{\hbar^2}{2\mu}\frac{\partial^2}{\partial x^2} + V_1(x), \tag{7.4a}$$

$$H_y = -\frac{\hbar^2}{2\mu}\frac{\partial^2}{\partial y^2} + V_2(y), \tag{7.4b}$$

$$H_z = -\frac{\hbar^2}{2\mu}\frac{\partial^2}{\partial z^2} + V_3(z) \tag{7.4c}$$

and the Schrödinger equation (7.1) becomes

$$\left\{\left[-\frac{\hbar^2}{2\mu}\frac{\partial^2}{\partial x^2} + V_1(x)\right] + \left[-\frac{\hbar^2}{2\mu}\frac{\partial^2}{\partial y^2} + V_2(y)\right] \right.$$
$$\left. + \left[-\frac{\hbar^2}{2\mu}\frac{\partial^2}{\partial z^2} + V_3(z)\right]\right\}\psi(x, y, z) = E\psi(x, y, z). \tag{7.5}$$

The form of this equation suggests that we seek solutions which are products of three functions, each of a single variable

$$\psi(x, y, z) = X(x)Y(y)Z(z). \tag{7.6}$$

Substituting (7.6) into (7.5) and dividing through by ψ, we find that

$$\left[-\frac{\hbar^2}{2\mu}\frac{1}{X}\frac{d^2X}{dx^2} + V_1(x)\right] + \left[-\frac{\hbar^2}{2\mu}\frac{1}{Y}\frac{d^2Y}{dy^2} + V_2(y)\right]$$
$$+ \left[-\frac{\hbar^2}{2\mu}\frac{1}{Z}\frac{d^2Z}{dz^2} + V_3(z)\right] = E. \tag{7.7}$$

Each expression in square brackets can only be a function of one of the variables x, y, z. Moreover, the sum of these three expressions must be equal to the constant E. Hence each separate expression must itself be equal to a constant, the sum of these constants being equal to E. We therefore obtain the three ordinary differential equations

$$\left[-\frac{\hbar^2}{2\mu}\frac{d^2}{dx^2} + V_1(x)\right]X(x) = E_x X(x) \tag{7.8a}$$

$$\left[-\frac{\hbar^2}{2\mu}\frac{d^2}{dy^2} + V_2(y)\right]Y(y) = E_y Y(y) \tag{7.8b}$$

$$\left[-\frac{\hbar^2}{2\mu}\frac{d^2}{dz^2} + V_3(z)\right]Z(z) = E_z Z(z) \tag{7.8c}$$

with the condition

$$E = E_x + E_y + E_z. \tag{7.9}$$

Each of the equations (7.8) has the form of the one-dimensional time-independent Schrödinger equation (4.3) discussed in Chapter 4. The three-dimensional states are therefore expressible in terms of familiar one-dimensional states.

The free particle

As a first example, let us consider the motion of a free particle of mass μ. In this case $V = 0$ in all space, and the Schrödinger equation (7.1) becomes

$$-\frac{\hbar^2}{2\mu}\nabla^2\psi(\mathbf{r}) = E\psi(\mathbf{r}). \tag{7.10}$$

This equation is of course separable in Cartesian coordinates with $V_1 = V_2 = V_3 = 0$. In particular, equation (7.8a) reduces to

$$-\frac{\hbar^2}{2\mu}\frac{d^2X(x)}{dx^2} = E_x X(x). \tag{7.11}$$

From our discussion of Chapter 4, we know that acceptable solutions of (7.11) exist for any value $E_x \geq 0$, and are given by

$$X(x) = A\exp(i|k_x|x) + B\exp(-i|k_x|x) \tag{7.12}$$

where A and B are arbitrary complex constants, and $|k_x| = (2\mu E_x/\hbar^2)^{1/2}$.

Proceeding in a similar way with equations (7.8b) and (7.8c), we see that the three-dimensional free-particle solutions $\psi(x, y, z) = X(x)Y(y)Z(z)$ of the Schrödinger equation (7.10) can be written as linear combinations of *plane-wave* states

$$\psi_{\mathbf{k}}(\mathbf{r}) = C \exp[i(k_x x + k_y y + k_z z)]$$
$$= C \exp(i\mathbf{k}.\mathbf{r}) \tag{7.13}$$

where the propagation vector (or wave vector) $\mathbf{k}$ has Cartesian components (k_x, k_y, k_z), and C is a 'normalisation' constant. Moreover, from (7.9) we have

$$E = E_x + E_y + E_z = \frac{\hbar^2}{2\mu}(k_x^2 + k_y^2 + k_z^2) = \frac{\hbar^2 k^2}{2\mu} = \frac{p^2}{2\mu} \tag{7.14}$$

where p is the magnitude of the linear momentum $\mathbf{p} = \hbar \mathbf{k}$.

The fact that the free-particle Schrödinger equation (7.10) has plane-wave solutions of the form (7.13) is in accordance with our discussion in Chapter 3. We also recall that the plane-wave state (7.13) is a *momentum eigenfunction*, that is an eigenstate of the momentum operator $\mathbf{p}_{\text{op}} = -i\hbar \nabla$, with the eigenvalue $\mathbf{p} = \hbar \mathbf{k}$ corresponding to a definite linear momentum. The plane waves (7.13) are thus simultaneous eigenfunctions of energy and linear momentum; more precisely, they are simultaneous eigenfunctions of the free-particle Hamiltonian $H = -(\hbar^2/2\mu)\nabla^2$ and of the three operators $(p_x)_{\text{op}} = -i\hbar \partial/\partial x$, $(p_y)_{\text{op}} = -i\hbar \partial/\partial y$ and $(p_z)_{\text{op}} = -i\hbar \partial/\partial z$.

Because every non-negative value of the energy E is allowed, the spectrum is *continuous*. Moreover, each energy eigenvalue is *infinitely degenerate*, since the condition (7.14) only restricts the magnitude of the vector $\mathbf{k}$, so that there are infinitely many possible orientations of this vector for a given value of E.

The plane wave (7.13) can be 'normalised' by a simple extension of the procedures described in Section 4.2 for the one-dimensional case. Generalising (4.37) to three dimensions, we obtain the 'delta function' normalisation

$$\int \psi_{\mathbf{k}'}^*(\mathbf{r}) \psi_{\mathbf{k}}(\mathbf{r}) d\mathbf{r} = \delta(\mathbf{k} - \mathbf{k}') \tag{7.15}$$

where the three-dimensional Dirac delta function $\delta(\mathbf{k} - \mathbf{k}')$ is given by

$$\delta(\mathbf{k} - \mathbf{k}') = \delta(k_x - k_x')\delta(k_y - k_y')\delta(k_z - k_z')$$
$$= (2\pi)^{-3} \int_{-\infty}^{+\infty} \exp[i(k_x - k_x')x]dx \int_{-\infty}^{+\infty} \exp[i(k_y - k_y')y]dy$$
$$\times \int_{-\infty}^{+\infty} \exp[i(k_z - k_z')z]dz$$
$$= (2\pi)^{-3} \int \exp[i(\mathbf{k} - \mathbf{k}').\mathbf{r}]d\mathbf{r}. \tag{7.16}$$

Taking the arbitrary phase of the normalisation constant C in (7.13) to be zero, we see from (7.13), (7.15) and (7.16) that $C = (2\pi)^{-3/2}$, so that the plane-wave states

$\psi_\mathbf{k}(\mathbf{r})$, normalised in the sense of (7.15), are

$$\psi_\mathbf{k}(\mathbf{r}) = (2\pi)^{-3/2} \exp(i\mathbf{k}\cdot\mathbf{r}). \tag{7.17}$$

Finally, we remark that the plane waves (7.17) form a *complete set*, so that an arbitrary state can be expressed as a superposition of these functions; this is precisely the Fourier expansion which was discussed in Chapter 2. The *closure relation* (5.63a) applied to the plane-wave eigenfunctions (7.17) reads

$$\int \psi_\mathbf{k}^*(\mathbf{r}')\psi_\mathbf{k}(\mathbf{r})d\mathbf{k} = (2\pi)^{-3} \int \exp[i\mathbf{k}\cdot(\mathbf{r}-\mathbf{r}')]d\mathbf{k}$$
$$= \delta(\mathbf{r}-\mathbf{r}'). \tag{7.18}$$

If we elect to work with discrete eigenvalues, we can restrict the domain of the plane waves (7.13) to an arbitrary large but finite cubical box of volume L^3, at the walls of which they obey *periodic boundary conditions*. By a straightforward generalisation of the one-dimensional treatment of Section 4.2 we then find that the components (k_x, k_y, k_z) of the wave vector $\mathbf{k}$ must satisfy the conditions

$$k_x = \frac{2\pi}{L}n_x, \qquad k_y = \frac{2\pi}{L}n_y, \qquad k_z = \frac{2\pi}{L}n_z \tag{7.19}$$

where n_x, n_y, n_z are positive or negative integers, or zero. The energy E, given by (7.14), can therefore take only discrete values in this case. The momentum eigenfunctions (7.13) can now be normalised by requiring that in the basic cube of side L

$$\int_0^L dx \int_0^L dy \int_0^L dz |\psi_\mathbf{k}(\mathbf{r})|^2 = 1. \tag{7.20}$$

Hence, taking the arbitrary phase of the normalisation constant C to be zero, this constant is given by $C = L^{-3/2}$, and the normalised plane waves are

$$\psi_\mathbf{k}(\mathbf{r}) = L^{-3/2} \exp(i\mathbf{k}\cdot\mathbf{r}). \tag{7.21}$$

These eigenfunctions form a complete set of orthonormal functions in terms of which an arbitrary state can be expanded by a Fourier series within the basic cube.

The three-dimensional box

As a second example, let us consider a particle of mass μ which is constrained by *impenetrable walls* to move in a rectangular box of sides L_1, L_2 and L_3. Inside the box the potential energy V is a constant which we choose to be zero, while at the walls V is infinite. This problem is therefore a generalisation of the one-dimensional infinite square well treated in Section 4.5.

The time-independent Schrödinger equation to be solved in the region inside the box is

$$-\frac{\hbar^2}{2\mu}\left(\frac{\partial^2}{\partial x^2} + \frac{\partial^2}{\partial y^2} + \frac{\partial^2}{\partial z^2}\right)\psi(x,y,z) = E\psi(x,y,z). \tag{7.22}$$

In addition, we must have $\psi(x, y, z) = 0$ at each of the walls and beyond. Taking the origin to be at one corner of the box and writing $\psi(x, y, z)$ in the product form (7.6), we find that for $0 \leq x \leq L_1$ the function $X(x)$ satisfies the equation

$$-\frac{\hbar^2}{2\mu} \frac{d^2 X(x)}{dx^2} = E_x X(x) \tag{7.23}$$

and that

$$X(x) = 0 \text{ for } x \leq 0 \text{ and } x \geq L_1. \tag{7.24}$$

From our results of Section 4.5 we see at once that the allowed values of E_x are given by

$$E_{n_x} = \frac{\hbar^2}{2\mu} \frac{\pi^2 n_x^2}{L_1^2}, \qquad n_x = 1, 2, 3, \ldots \tag{7.25}$$

the corresponding normalised eigenfunctions being the standing waves

$$X_{n_x}(x) = \left(\frac{2}{L_1}\right)^{1/2} \sin\left(\frac{n_x \pi}{L_1} x\right). \tag{7.26}$$

Proceeding in a similar way for the functions $Y(y)$ and $Z(z)$, we find that the normalised eigenfunctions of the full three-dimensional problem are the standing waves

$$\psi_{n_x n_y n_z}(x, y, z) = \left(\frac{8}{V}\right)^{1/2} \sin\left(\frac{n_x \pi}{L_1} x\right) \sin\left(\frac{n_y \pi}{L_2} y\right) \sin\left(\frac{n_z \pi}{L_3} z\right) \tag{7.27}$$

where $V = L_1 L_2 L_3$ is the volume of the box and n_x, n_y, n_z are positive integers[1]. The allowed values of the energy $E = E_x + E_y + E_z$ are

$$E_{n_x n_y n_z} = \frac{\hbar^2 \pi^2}{2\mu} \left(\frac{n_x^2}{L_1^2} + \frac{n_y^2}{L_2^2} + \frac{n_z^2}{L_3^2}\right). \tag{7.28}$$

This energy spectrum does not exhibit regularities for general values of L_1, L_2 and L_3. Although it is strictly discrete, we see that as the dimensions of the box increase the spacing of the energy levels decreases, so that for a macroscopic box the spectrum is nearly continuous.

Let us now consider the simpler case of a cubical box of side L. The normalised eigenfunctions are then given by

$$\psi_{n_x n_y n_z}(x, y, z) = \left(\frac{8}{L^3}\right)^{1/2} \sin\left(\frac{n_x \pi}{L} x\right) \sin\left(\frac{n_y \pi}{L} y\right) \sin\left(\frac{n_z \pi}{L} z\right),$$
$$n_x, n_y, n_z = 1, 2, 3, \ldots \tag{7.29}$$

---

[1] Note that if n_x, n_y or n_z is equal to zero one obtains the trivial non-physical solution $\psi = 0$, while changing n_x to $-n_x$ (and similarly for n_y or n_z) changes the sign of the wave function, which is not physically significant.

7.1 Cartesian coordinates

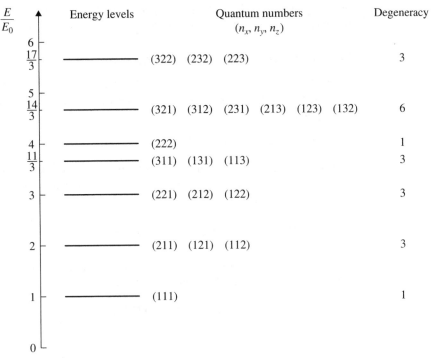

Figure 7.1 The first few energy levels of a particle in a cubical box of side L. Also shown are the quantum numbers (n_x, n_y, n_z) of the corresponding eigenfunctions, and the degeneracy of each energy level. The ground state energy is $E_0 = 3\hbar^2\pi^2/(2\mu L^2)$.

and the allowed energy values are

$$E_n = \frac{\hbar^2\pi^2}{2\mu L^2} n^2 \qquad (7.30)$$

where

$$n^2 = n_x^2 + n_y^2 + n_z^2. \qquad (7.31)$$

The ground state is such that $n_x = n_y = n_z = 1$, so that $n^2 = 3$ and the corresponding energy is $E_0 = 3\hbar^2\pi^2/(2\mu L^2)$; this level is non-degenerate. The next energy level has an energy $2E_0$ (corresponding to $n^2 = 6$), which can be obtained in three different ways by choosing the set of quantum numbers (n_x, n_y, n_z) to be $(2, 1, 1)$ or $(1, 2, 1)$ or $(1, 1, 2)$; this level is therefore three-fold degenerate. The next energy levels can be obtained in a similar way. The first few energy levels, together with their degeneracies, are shown in Fig. 7.1. It is apparent that since the energy eigenvalues (7.30) depend on the quantum numbers n_x, n_y and n_z only through the combination $n^2 = n_x^2 + n_y^2 + n_z^2$, they can generally be obtained from different sets of values of (n_x, n_y, n_z), and hence are usually degenerate. This is in contrast with the energy levels (7.28) corresponding to the rectangular box. Clearly the existence of degeneracies is directly related to the symmetry of the potential.

The three-dimensional harmonic oscillator

As another example of a three-dimensional system for which the Schrödinger equation is separable in Cartesian coordinates, we consider the motion of a particle of mass μ in the potential

$$V(\mathbf{r}) = \tfrac{1}{2}k_1 x^2 + \tfrac{1}{2}k_2 y^2 + \tfrac{1}{2}k_3 z^2 \tag{7.32}$$

corresponding to a three-dimensional harmonic oscillator. The separated Schrödinger equations (7.8) are now of the form of the one-dimensional linear harmonic oscillator equation (4.130), namely

$$-\frac{\hbar^2}{2\mu}\frac{\mathrm{d}^2 X(x)}{\mathrm{d}x^2} + \frac{1}{2}k_1 x^2 X(x) = E_x X(x) \tag{7.33}$$

with similar equations for $Y(y)$ and $Z(z)$. Using the results of Section 4.7 we find that the spectrum is entirely discrete, the energy levels being given by

$$E_{n_x n_y n_z} = \left(n_x + \tfrac{1}{2}\right)\hbar\omega_1 + \left(n_y + \tfrac{1}{2}\right)\hbar\omega_2 + \left(n_z + \tfrac{1}{2}\right)\hbar\omega_3 \tag{7.34a}$$

where

$$\omega_1 = \left(\frac{k_1}{\mu}\right)^{1/2}, \qquad \omega_2 = \left(\frac{k_2}{\mu}\right)^{1/2}, \qquad \omega_3 = \left(\frac{k_3}{\mu}\right)^{1/2} \tag{7.34b}$$

and n_x, n_y, n_z are positive integers, or zero.

The corresponding normalised eigenfunctions can also be readily obtained by using (7.6) and the one-dimensional linear harmonic oscillator wave functions of Section 4.7. They are given by

$$\psi_{n_x n_y n_z}(x,y,z) = N_{n_x} N_{n_y} N_{n_z} \exp\left[-\tfrac{1}{2}(\alpha_1^2 x^2 + \alpha_2^2 y^2 + \alpha_3^2 z^2)\right] \\ \times H_{n_x}(\alpha_1 x) H_{n_y}(\alpha_2 y) H_{n_z}(\alpha_3 z) \tag{7.35a}$$

where $H_n(\xi)$ is the Hermite polynomial defined in (4.154) and

$$\alpha_1 = \left(\frac{\mu k_1}{\hbar^2}\right)^{1/4}, \qquad N_{n_x} = \left(\frac{\alpha_1}{\sqrt{\pi}2^{n_x}n_x!}\right)^{1/2} \tag{7.35b}$$

with similar expressions for α_2, α_3 and N_{n_y}, N_{n_z}. Note that if the three force constants k_1, k_2 and k_3 are all different, so that the harmonic oscillator is *anisotropic*, the energy levels (7.34) are, in general, non-degenerate.

Let us now consider the particular case of an *isotropic* three-dimensional harmonic oscillator, for which $k_1 = k_2 = k_3$. Calling this common value of the force constant k, we see that the potential

$$V(r) = \tfrac{1}{2}k(x^2 + y^2 + z^2) = \tfrac{1}{2}kr^2 \tag{7.36}$$

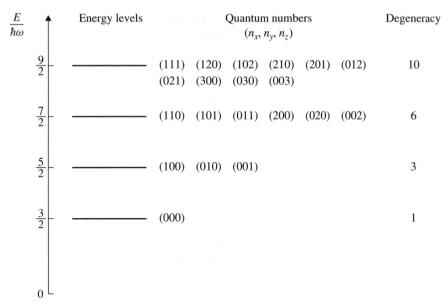

Figure 7.2 The first few energy levels of a three-dimensional isotropic harmonic oscillator. Also shown are the quantum numbers (n_x, n_y, n_z) of the corresponding eigenfunctions, and the degeneracy of each energy level.

is *central*. The motion of a particle in a central potential will be discussed at length in the remaining sections of this chapter. Here we only note that upon setting $k_1 = k_2 = k_3 = k$ in (7.34) the energy levels of the three-dimensional isotropic harmonic oscillator are given by

$$E_n = \left(n + \tfrac{3}{2}\right)\hbar\omega \tag{7.37}$$

where $\omega = (k/\mu)^{1/2}$. The quantum number n is such that

$$n = n_x + n_y + n_z \tag{7.38}$$

and thus can take on the values $n = 0, 1, 2, \ldots$. The eigenfunctions are given by (7.35) with $\alpha_1 = \alpha_2 = \alpha_3 = \alpha = (\mu k/\hbar^2)^{1/4}$. The ground state has the energy $E_{n=0} = 3\hbar\omega/2$ and is not degenerate, while all the other energy levels are degenerate. For example, the first excited level has an energy $E_{n=1} = 5\hbar\omega/2$ and is three-fold degenerate since it can be obtained in three different ways by taking the set of quantum numbers (n_x, n_y, n_z) to be $(1, 0, 0)$ or $(0, 1, 0)$ or $(0, 0, 1)$. The first few energy levels of the three-dimensional isotropic harmonic oscillator and their degeneracies are shown in Fig. 7.2. Note that the energy level $E_n = (n+3/2)\hbar\omega$ is $(n+1)(n+2)/2$-fold degenerate, since this is the number of ways that n can be obtained as the sum of the three non-negative integers n_x, n_y and n_z. Thus the three-dimensional isotropic harmonic oscillator provides another illustration of how the symmetry of the potential results in degeneracies.

7.2 Central potentials. Separation of the Schrödinger equation in spherical polar coordinates

We now turn to the main subject of this chapter: the study of the non-relativistic motion of a spinless particle of mass μ in a *central* potential (that is a potential $V(r)$ which depends only on the magnitude r of the position vector $\mathbf{r}$). We shall see that the properties of the orbital angular momentum $\mathbf{L} = \mathbf{r} \times \mathbf{p}$ obtained in Chapter 6 are of particular importance in this analysis.

Since $V(r)$ is spherically symmetric, it is natural to use the spherical polar coordinates defined in (6.15) and illustrated in Fig. 6.1. Our first task is to express the Hamiltonian operator

$$H = -\frac{\hbar^2}{2\mu}\nabla^2 + V(r) \tag{7.39}$$

in these coordinates. The potential energy $V(r)$ is already given in terms of the polar coordinate r, and the expression of the kinetic-energy operator $T = -(\hbar^2/2\mu)\nabla^2$ in spherical polar coordinates is given by (6.130b) for $r \neq 0$. Hence the Hamiltonian (7.39) may be written in these coordinates (for $r \neq 0$) as

$$H = -\frac{\hbar^2}{2\mu}\left[\frac{1}{r^2}\frac{\partial}{\partial r}\left(r^2\frac{\partial}{\partial r}\right) + \frac{1}{r^2\sin\theta}\frac{\partial}{\partial \theta}\left(\sin\theta\frac{\partial}{\partial \theta}\right) + \frac{1}{r^2\sin^2\theta}\frac{\partial^2}{\partial \phi^2}\right] + V(r) \tag{7.40a}$$

or

$$H = -\frac{\hbar^2}{2\mu}\left[\frac{1}{r^2}\frac{\partial}{\partial r}\left(r^2\frac{\partial}{\partial r}\right) - \frac{\mathbf{L}^2}{\hbar^2 r^2}\right] + V(r) \tag{7.40b}$$

where $\mathbf{L}^2$ is given by (6.17). The corresponding time-independent Schrödinger equation is

$$\left\{-\frac{\hbar^2}{2\mu}\left[\frac{1}{r^2}\frac{\partial}{\partial r}\left(r^2\frac{\partial}{\partial r}\right) - \frac{\mathbf{L}^2}{\hbar^2 r^2}\right] + V(r)\right\}\psi(\mathbf{r}) = E\psi(\mathbf{r}). \tag{7.41}$$

In order to simplify the solution of this equation, we first recall that the operators L_x, L_y, L_z and $\mathbf{L}^2$ do not operate on the radial variable r. Hence, *for a spherically symmetric potential* $V(r)$, we have $[V(r), \mathbf{L}] = [V(r), \mathbf{L}^2] = 0$. Moreover, L_x, L_y, L_z and $\mathbf{L}^2$ also commute with the kinetic energy operator T, as is readily seen by using the expression (6.130b) of T and remembering that $[\mathbf{L}^2, \mathbf{L}] = 0$. Therefore,

$$[H, \mathbf{L}] = [H, \mathbf{L}^2] = 0. \tag{7.42}$$

Remembering that the operators L_x, L_y and L_z do not commute among themselves, we see that a set of commuting operators can be taken to be H, $\mathbf{L}^2$ and any one of L_x, L_y and L_z. Taking this set to be H, $\mathbf{L}^2$ and L_z, it follows that it is possible to find simultaneous eigenfunctions of these three operators, or in other words to obtain solutions of the Schrödinger equation (7.41) which are also eigenfunctions of $\mathbf{L}^2$ and L_z. Since the spherical harmonics $Y_{lm}(\theta, \phi)$ are simultaneous eigenfunctions of $\mathbf{L}^2$

and L_z (see (6.63) and (6.64)) we can look for solutions of the Schrödinger equation having the separable form

$$\psi_{Elm}(\mathbf{r}) = R_{Elm}(r) Y_{lm}(\theta, \phi) \tag{7.43}$$

where $R_{Elm}(r)$ is a *radial function* which remains to be found. It is worth stressing that the angular dependence of the eigenfunction (7.43) is entirely given by the spherical harmonic $Y_{lm}(\theta, \phi)$; it is the proper one associated with eigenvalues of $\mathbf{L}^2$ and L_z characterised by the orbital angular momentum quantum number l and the magnetic quantum number m. These two quantum numbers, together with the energy E, can thus be used to 'label' the eigenfunction ψ_{Elm}. We also remark (see Problem 7.3) that all the solutions of the Schrödinger equation (7.41) can be obtained as linear combinations of the separable solutions (7.43).

Inserting (7.43) into the Schrödinger equation (7.41) and using the fact that $\mathbf{L}^2 Y_{lm}(\theta, \phi) = l(l+1)\hbar^2 Y_{lm}(\theta, \phi)$, we obtain for the radial function the differential equation

$$\left[-\frac{\hbar^2}{2\mu} \left(\frac{d^2}{dr^2} + \frac{2}{r} \frac{d}{dr} \right) + \frac{l(l+1)\hbar^2}{2\mu r^2} + V(r) \right] R_{El}(r) = E R_{El}(r). \tag{7.44}$$

Note that the magnetic quantum number m does not appear in this equation. The radial function is therefore independent of this quantum number. For this reason we have written $R_{El}(r) \equiv R_{Elm}(r)$, and equation (7.43) becomes

$$\psi_{Elm}(\mathbf{r}) = R_{El}(r) Y_{lm}(\theta, \phi). \tag{7.45}$$

Similarly, the eigenvalues E obtained from (7.44) are independent of the quantum number m. Thus, for a given value of l there are $(2l+1)$ eigenfunctions (7.45) corresponding to the $(2l+1)$ possible different values of $m (m = -l, -l+1, \ldots, l)$ which all have the same energy E. The reason for this $(2l+1)$-*fold degeneracy* is that for a spherically symmetric potential no direction of space is physically different from another, or in other words the Hamiltonian (7.39) is *invariant under rotations*. Since $m\hbar$ measures the projection of the orbital angular momentum $\mathbf{L}$ on the z-axis (see Section 6.3) the energy levels cannot depend on the quantum number m.

Using (6.115), we see that the modulus squared of the eigenfunctions (7.45) is given by

$$|\psi_{Elm}(r, \theta, \phi)|^2 = |R_{El}(r)|^2 |Y_{lm}(\theta, \phi)|^2$$
$$= |R_{El}(r)|^2 (2\pi)^{-1} |\Theta_{lm}(\theta)|^2 \tag{7.46}$$

and hence does not depend on the angle ϕ. The behaviour of $|\psi_{Elm}|^2$ is thus completely specified by the *radial* quantity $|R_{El}(r)|^2$, which depends on the central potential $V(r)$ considered (see (7.44)) and on the *angular factor* $(2\pi)^{-1}|\Theta_{lm}(\theta)|^2$ which is independent of $V(r)$ and has been studied in Chapter 6. In particular, we refer the reader to the polar plots of $(2\pi)^{-1}|\Theta_{lm}(\theta)|^2$ shown in Fig. 6.2.

We also note that if an eigenfunction (7.45) corresponds to a *bound state* it is square integrable and thus can be normalised to unity by requiring that

$$\int_0^\infty \mathrm{d}r\, r^2 \int_0^\pi \mathrm{d}\theta \sin\theta \int_0^{2\pi} \mathrm{d}\phi |\psi_{Elm}(r,\theta,\phi)|^2 = 1. \tag{7.47}$$

Since the spherical harmonics are normalised on the unit sphere (see (6.103)), it follows from (7.46) and (7.47) that the radial bound-state eigenfunctions must satisfy the normalisation condition

$$\int_0^\infty |R_{El}(r)|^2 r^2 \mathrm{d}r = 1. \tag{7.48}$$

Another important remark concerns the *parity* of the states (7.45). We recall that under the parity operation $\mathbf{r} \to -\mathbf{r}$ the spherical polar coordinates (r,θ,ϕ) become $(r, \pi - \theta, \phi + \pi)$. The Hamiltonian for a particle in a central potential, given by (7.40), is clearly unaffected by this operation, so that the parity operator $\mathcal{P}$ commutes with the Hamiltonian (7.40). As a result, simultaneous eigenfunctions of the operators H and $\mathcal{P}$ can be found. Applying the parity operator to the wave function (7.45), we have

$$\mathcal{P}\psi_{Elm}(r,\theta,\phi) = \mathcal{P}[R_{El}(r)Y_{lm}(\theta,\phi)]$$
$$= R_{El}(r)Y_{lm}(\pi-\theta,\phi+\pi). \tag{7.49}$$

Now, we have seen in Chapter 6 that $Y_{lm}(\pi-\theta,\phi+\pi) = (-1)^l Y_{lm}(\theta,\phi)$, so that Y_{lm} has the parity of l (see (6.114)). We have therefore

$$\mathcal{P}\psi_{Elm}(r,\theta,\phi) = R_{El}(r)(-1)^l Y_{lm}(\theta,\phi)$$
$$= (-1)^l \psi_{Elm}(r,\theta,\phi). \tag{7.50}$$

As a result, the states (7.45) have a definite parity which, like that of the Y_{lm}, is the parity of l (even for even l and odd for odd l).

It is also interesting to examine the relation between the radial equation (7.44) and the one-dimensional Schrödinger equation (4.3). To this end, we introduce the new radial function

$$u_{El}(r) = rR_{El}(r). \tag{7.51}$$

Using (7.44) and (7.51), we obtain for $u_{El}(r)$ the radial equation

$$-\frac{\hbar^2}{2\mu} \frac{\mathrm{d}^2 u_{El}(r)}{\mathrm{d}r^2} + V_{\mathrm{eff}}(r)u_{El}(r) = Eu_{El}(r) \tag{7.52}$$

where

$$V_{\mathrm{eff}}(r) = V(r) + \frac{l(l+1)\hbar^2}{2\mu r^2} \tag{7.53}$$

is an *effective potential* which, in addition to the interaction potential $V(r)$, also contains the repulsive *centrifugal barrier* term $l(l+1)\hbar^2/2\mu r^2$. The equation (7.52) is then identical in form to the one-dimensional Schrödinger equation (4.3). However, it has significance only for positive values of r, and must be supplemented by a boundary condition at $r = 0$. We shall require that the radial function $R_{El}(r)$ remains finite at the origin[2]. Since $R_{El}(r) = r^{-1} u_{El}(r)$, this implies that we must have

$$u_{El}(0) = 0. \tag{7.54}$$

Let us examine more closely the behaviour of the function $u_{El}(r)$ near the origin. We shall first assume that in the vicinity of $r = 0$ the interaction potential $V(r)$ has the form

$$V(r) = r^p(b_0 + b_1 r + \cdots), \qquad b_0 \neq 0 \tag{7.55}$$

where p is an integer such that $p \geqslant -1$. In other words, the potential cannot be more singular than r^{-1} at the origin, which is the case for nearly all interactions of physical interest. Since $r = 0$ is a regular singular point[3] of the differential equation (7.52), we can expand the solution $u_{El}(r)$ in the vicinity of the origin as

$$u_{El}(r) = r^s \sum_{k=0}^{\infty} c_k r^k, \qquad c_0 \neq 0. \tag{7.56}$$

Substituting this expansion in (7.52), we find by looking at the coefficient of the lowest power of r (i.e. r^{s-2}) that the quantity s must satisfy the *indicial equation*

$$s(s-1) - l(l+1) = 0 \tag{7.57}$$

so that $s = l + 1$ or $s = -l$. The choice $s = -l$ corresponds to *irregular* solutions which do not satisfy the condition (7.54). The other choice $s = l + 1$ corresponds to *regular* solutions which are physically allowed, and are such that

$$u_{El}(r) \underset{r \to 0}{\sim} r^{l+1}. \tag{7.58}$$

Note that the corresponding admissible radial functions $R_{El}(r)$ behave like r^l near $r = 0$. We see that as l increases, the functions $u_{El}(r)$, or $R_{El}(r)$, become smaller and smaller in the neighbourhood of the origin. This is clearly due to the presence of the centrifugal barrier which for $l \neq 0$ 'blocks out' the region near the origin, the effect being more and more pronounced as l increases.

We now consider briefly the case for which the integer p in (7.55) is such that $p < -1$. If the interaction is *repulsive* near $r = 0$ (so that $b_0 > 0$), we may still impose the condition (7.54), since the radial function $R_{El}(r)$ itself must clearly vanish at the infinite wall at $r = 0$. On the other hand, if the potential is *attractive* in the neighbourhood of the origin (so that $b_0 < 0$) the nature of the singularity is important. For example, when $p = -2$ and $b_0 < 0$ in (7.55), it can be shown (Problem 7.4) that physically acceptable solutions of (7.52) exist only when $b_0 > -\hbar^2/8\mu$. In what

---

[2] The requirement of finiteness of $R_{El}(r)$ at $r = 0$ is in fact too stringent and must be relaxed in certain cases. However, it will be fully adequate for all the non-relativistic applications treated in this book.
[3] See, for example, Mathews and Walker (1973).

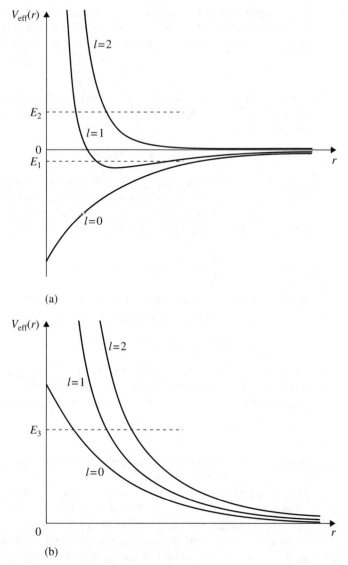

Figure 7.3 Illustration of the effective potential $V_{\text{eff}}(r) = V(r) + l(l+1)\hbar^2/2\mu r^2$ for $l = 0, 1, 2$ (a) in the case of a typical attractive interaction $V(r)$, and (b) in the case of a typical repulsive interaction $V(r)$. The energy $E_1 < 0$ corresponds to a bound state, while the positive energies E_2 and E_3 correspond to states belonging to the continuous spectrum.

follows, the attractive potentials which we shall consider will always be assumed to be less singular than r^{-2} at the origin, so that the results (7.54) and (7.58) hold true.

In the remaining sections of this chapter we shall solve the radial equations (7.44) or (7.52) for several central potentials. Before doing this, however, it is instructive to examine the form of the effective potential (7.53) in some typical cases. In Fig. 7.3(a) we illustrate the effective potential $V_{\text{eff}}(r)$ for the first values of l for an attractive interaction $V(r)$, and in Fig. 7.3(b) this is done for a repulsive interaction. It is seen that in the first case the effect of the centrifugal barrier term $l(l+1)\hbar^2/2\mu r^2$ is to reduce the effective depth of the potential well. This effect obviously increases as l increases. Thus, in the example shown in Fig. 7.3(a), bound states of negative energy (such as E_1) could exist for the values $l = 0$ and 1, but we see that for $l \geqslant 2$ the effective potential is purely repulsive, so that no bound states can exist for $l \geqslant 2$. We also remark that for the example displayed in Fig. 7.3(a), solutions $u_{El}(r)$ of the radial equation (7.52) which exhibit an oscillatory behaviour at infinity exist for every l when $E > 0$; the spectrum is therefore continuous for every l in the case of positive energies (such as E_2). Turning now to the case of a repulsive interaction, it is clear from (7.53) that as l increases the effective potential becomes increasingly repulsive. This is illustrated in Fig. 7.3(b). The energy spectrum will therefore consist only of continuum states of positive energy (such as E_3), for all values of l.

7.3 The free particle

As a first example of motion in a central potential, we shall consider the very simple case of a free particle, for which $V(r) = 0$. Using Cartesian coordinates, we have obtained in Section 7.1 plane-wave solutions (7.13) of the free-particle Schrödinger equation (7.10). These plane-wave states are also momentum eigenfunctions, and hence are simultaneous eigenfunctions of the free-particle Hamiltonian $H = -(\hbar^2/2\mu)\nabla^2$ and of the (linear) momentum operator $\mathbf{p}_{\text{op}} = -i\hbar\nabla$. Using spherical polar coordinates, we shall now look for solutions of the free-particle Schrödinger equation of the form (7.45), that is simultaneous eigenfunctions of H, $\mathbf{L}^2$ and L_z corresponding to definite values of E, l and m. The free-particle radial functions $R_{El}(r)$ are therefore solutions of the radial equation (7.44) with $V(r) = 0$, namely

$$\left[\frac{d^2}{dr^2} + \frac{2}{r}\frac{d}{dr} - \frac{l(l+1)}{r^2} + k^2\right]R_{El}(r) = 0 \tag{7.59}$$

where $k^2 = 2\mu E/\hbar^2$. Using (7.52) and the fact that in the present case the effective potential $V_{\text{eff}}(r)$ is just the centrifugal barrier term $l(l+1)\hbar^2/2\mu r^2$, we see that the free-particle functions $u_{El}(r) = rR_{El}(r)$ are solutions of the differential equation

$$\left[\frac{d^2}{dr^2} - \frac{l(l+1)}{r^2} + k^2\right]u_{El}(r) = 0. \tag{7.60}$$

Let us first consider the case $l = 0$. From (7.60) we see immediately that the solution $u_{E0}(r)$ which vanishes at the origin (in order to satisfy the boundary condition (7.54)) is given, up to a multiplicative constant, by $u_{E0}(r) = \sin kr$, so that $R_{E0}(r) \propto r^{-1} \sin kr$.

The spherical Bessel differential equation and its solutions

For $l \neq 0$ the equation (7.60) is more difficult to solve and it is convenient to return to the radial equation (7.59), which can be solved in terms of known functions for all values of l ($l = 0, 1, 2, \ldots$). Indeed, if we change variables in (7.59) to $\rho = kr$, and write $R_l(\rho) \equiv R_{El}(r)$, we obtain for $R_l(\rho)$ the equation

$$\left[\frac{d^2}{d\rho^2} + \frac{2}{\rho}\frac{d}{d\rho} + \left(1 - \frac{l(l+1)}{\rho^2}\right)\right]R_l(\rho) = 0 \tag{7.61}$$

which is called the *spherical Bessel differential equation*. Particular solutions of this equation are the *spherical Bessel functions*

$$j_l(\rho) = \left(\frac{\pi}{2\rho}\right)^{1/2} J_{l+\frac{1}{2}}(\rho) \tag{7.62a}$$

and the *spherical Neumann functions*

$$n_l(\rho) = (-1)^{l+1}\left(\frac{\pi}{2\rho}\right)^{1/2} J_{-l-\frac{1}{2}}(\rho) \tag{7.62b}$$

where $J_\nu(\rho)$ is an ordinary Bessel function[4] of order ν.

The functions $j_l(\rho)$ and $n_l(\rho)$ are also given by the expressions (Problem 7.5)

$$j_l(\rho) = (-\rho)^l \left(\frac{1}{\rho}\frac{d}{d\rho}\right)^l \frac{\sin\rho}{\rho} \tag{7.63a}$$

and

$$n_l(\rho) = -(-\rho)^l \left(\frac{1}{\rho}\frac{d}{d\rho}\right)^l \frac{\cos\rho}{\rho}. \tag{7.63b}$$

From the above equations, the first few functions $j_l(\rho)$ and $n_l(\rho)$ are readily found to be

$$j_0(\rho) = \frac{\sin\rho}{\rho}, \qquad n_0(\rho) = -\frac{\cos\rho}{\rho} \tag{7.64a}$$

$$j_1(\rho) = \frac{\sin\rho}{\rho^2} - \frac{\cos\rho}{\rho}, \qquad n_1(\rho) = -\frac{\cos\rho}{\rho^2} - \frac{\sin\rho}{\rho} \tag{7.64b}$$

$$j_2(\rho) = \left(\frac{3}{\rho^3} - \frac{1}{\rho}\right)\sin\rho - \frac{3}{\rho^2}\cos\rho, \quad n_2(\rho) = -\left(\frac{3}{\rho^3} - \frac{1}{\rho}\right)\cos\rho - \frac{3}{\rho^2}\sin\rho. \tag{7.64c}$$

These functions are plotted in Fig. 7.4.

---

[4] See, for example, Bell (1968) or Watson (1966).

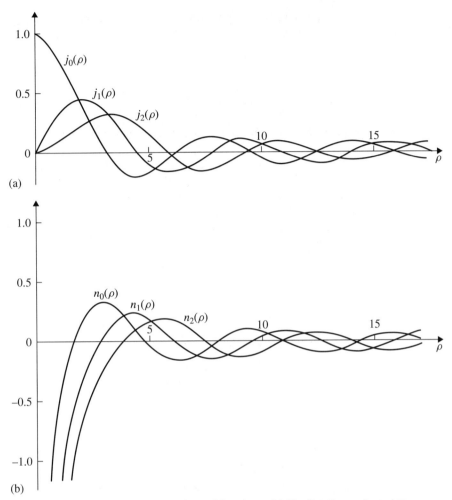

Figure 7.4 (a) The first three spherical Bessel functions. (b) The first three spherical Neumann functions.

We may also use equations (7.63) to obtain the leading term of $j_l(\rho)$ and $n_l(\rho)$ for small and large values of ρ (Problem 7.6). For small ρ, upon expanding $\rho^{-1}\sin\rho$ and $\rho^{-1}\cos\rho$ in a power series in ρ, it is found that

$$j_l(\rho) \underset{\rho \to 0}{\to} \frac{\rho^l}{1.3.5\ldots(2l+1)} \tag{7.65a}$$

$$n_l(\rho) \underset{\rho \to 0}{\to} -\frac{1.1.3.5\ldots(2l-1)}{\rho^{l+1}}. \tag{7.65b}$$

Asymptotically, for large values of ρ, the leading term is that proportional to ρ^{-1}, and one obtains

$$j_l(\rho) \underset{\rho \to \infty}{\to} \frac{1}{\rho} \sin\left(\rho - \frac{l\pi}{2}\right) \tag{7.66a}$$

$$n_l(\rho) \underset{\rho \to \infty}{\to} -\frac{1}{\rho} \cos\left(\rho - \frac{l\pi}{2}\right). \tag{7.66b}$$

Equations (7.66) are in fact useful approximations for ρ somewhat larger than $l(l+1)/2$.

For every l the two functions $\{j_l(\rho), n_l(\rho)\}$ provide a pair of linearly independent solutions of the spherical Bessel differential equation (7.61), so that the general solution of this equation can be written as a linear combination of these two functions. Another pair of linearly independent solutions of equation (7.61) is given by the *spherical Hankel functions* of the first and second kinds, which are defined respectively by

$$h_l^{(1)}(\rho) = j_l(\rho) + in_l(\rho) \tag{7.67a}$$

and

$$h_l^{(2)}(\rho) = j_l(\rho) - in_l(\rho) = [h_l^{(1)}(\rho)]^*. \tag{7.67b}$$

The first few spherical Hankel functions of the first kind are

$$h_0^{(1)}(\rho) = -i\frac{e^{i\rho}}{\rho} \tag{7.68a}$$

$$h_1^{(1)}(\rho) = -\left(\frac{1}{\rho} + \frac{i}{\rho^2}\right)e^{i\rho} \tag{7.68b}$$

$$h_2^{(1)}(\rho) = \left(\frac{i}{\rho} - \frac{3}{\rho^2} - \frac{3i}{\rho^3}\right)e^{i\rho}. \tag{7.68c}$$

Using (7.66) and (7.67) we also find that the asymptotic behaviour of the spherical Hankel functions is given by

$$h_l^{(1)}(\rho) \underset{\rho \to \infty}{\to} -i\frac{\exp[i(\rho - l\pi/2)]}{\rho} \tag{7.69a}$$

and

$$h_l^{(2)}(\rho) \underset{\rho \to \infty}{\to} i\frac{\exp[-i(\rho - l\pi/2)]}{\rho}. \tag{7.69b}$$

The eigenfunctions of the free particle in spherical polar coordinates

As seen from (7.65b), the spherical Neumann function $n_l(\rho)$ has a pole of order $l+1$ at the origin, and is therefore an *irregular* solution of equation (7.61). From their definition (7.67), we note that the spherical Hankel functions, which contain $n_l(\rho)$, are also irregular solutions of (7.61). On the other hand, the spherical Bessel function

$j_l(\rho)$ is finite at the origin and is thus a *regular* solution of (7.61). In fact $j_l(\rho)$ is finite everywhere, so that the radial eigenfunction of the Schrödinger equation (7.59) for a free particle is

$$R_{El}(r) = C j_l(kr) \tag{7.70}$$

where C is a constant. Using (7.65a), we see that near $r = 0$ this free-particle radial wave function behaves like r^l, in agreement with the remark following (7.58).

Having obtained the free-particle radial functions we may now, using (7.45) and (7.70), write down the full eigenfunctions of the free particle in spherical polar coordinates as

$$\psi_{Elm}(\mathbf{r}) = C j_l(kr) Y_{lm}(\theta, \phi). \tag{7.71}$$

These wave functions will be called *spherical waves*.

The eigenvalues k^2 of (7.59) can take on any value in the interval $(0, \infty)$, so that the energy $E = \hbar^2 k^2 / 2\mu$ can assume any value in this interval, and the spectrum is continuous, as we found in Section 7.1. Every free-particle eigenfunction (7.71) can thus be labelled by the two discrete indices l and m and by the continuous index E (or k). It can be shown that the ensemble of the spherical waves forms a complete orthonormal set. Note that each energy eigenvalue is *infinitely degenerate*, since for a fixed value of E the eigenfunctions (7.71) are labelled by the two quantum numbers l and m, such that $l = 0, 1, 2, \ldots$, and $m = -l, -l+1, \ldots, l$.

Expansion of a plane wave in spherical harmonics

We found in Section 7.1 that the stationary states of a free particle can be written as linear combinations of plane waves of the form (7.13). These plane-wave states are simultaneous eigenfunctions of the free-particle Hamiltonian $H = -(\hbar^2/2\mu)\nabla^2$, and of the three Cartesian components $(p_x)_{\text{op}}$, $(p_y)_{\text{op}}$ and $(p_z)_{\text{op}}$ of the linear momentum operator $\mathbf{p}_{\text{op}} = -i\hbar\nabla$. They are characterised by the three well-defined Cartesian components $p_x = \hbar k_x$, $p_y = \hbar k_y$, $p_z = \hbar k_z$ of the momentum $\mathbf{p}$, and by the energy $E = \hbar^2 k^2/2\mu$. Note that since the operators $(p_x)_{\text{op}}$, $(p_y)_{\text{op}}$ and $(p_z)_{\text{op}}$ do *not* simultaneously commute with $\mathbf{L}^2$ and L_z, the plane-wave states (7.13) cannot be labelled by the quantum numbers (l, m), so that the orbital angular momentum is poorly defined in those states. On the other hand, the spherical wave states (7.71) are states of well-defined orbital angular momentum, which are characterised by the quantum numbers (l, m), and for which the linear momentum is poorly defined.

Since both the plane-wave states (7.13) and the spherical-wave states (7.71) form a complete set, an arbitrary state can be expressed as a superposition of either of them. In particular, a plane wave $\exp(i\mathbf{k}\cdot\mathbf{r})$ can be expanded in terms of spherical waves, so that we may write

$$e^{i\mathbf{k}\cdot\mathbf{r}} = \sum_{l=0}^{\infty} \sum_{m=-l}^{+l} c_{lm} j_l(kr) Y_{lm}(\theta, \phi) \tag{7.72}$$

where the coefficients c_{lm} (which are independent of $\mathbf{r}$) must be determined. In order to do this, we first consider the special case in which the vector $\mathbf{k}$ lies along the z-axis. The left-hand side of (7.72) then reads $\exp(i\mathbf{k}\cdot\mathbf{r}) = \exp(ikr\cos\theta)$, which is independent of ϕ, so that the expansion (7.72) reduces to one in terms of the Legendre polynomials $P_l(\cos\theta)$

$$e^{i\mathbf{k}\cdot\mathbf{r}} = \sum_{l=0}^{\infty} a_l j_l(kr) P_l(\cos\theta). \tag{7.73}$$

The coefficients a_l of this expansion can be determined in the following way. Using the relation (6.80) satisfied by the Legendre polynomials, we have

$$\frac{2}{2l+1} a_l j_l(kr) = \int_{-1}^{+1} \exp(ikrw) P_l(w) dw \tag{7.74}$$

where we have set $w = \cos\theta$. Integrating by parts, we find that

$$\frac{2}{2l+1} a_l j_l(kr) = \left[\frac{\exp(ikrw)}{ikr} P_l(w)\right]_{w=-1}^{w=+1} - \int_{-1}^{+1} \frac{\exp(ikrw)}{ikr} \left[\frac{d}{dw} P_l(w)\right] dw. \tag{7.75}$$

Let us now examine this equation in the limit of large r. Using the asymptotic expression (7.66a) for $j_l(kr)$, and noting that the second term on the right of (7.75) is of order r^{-2} (which can be seen by performing a second integration by parts), we find that in the large r limit equation (7.75) reduces to

$$\frac{2}{2l+1} a_l \frac{1}{kr} \sin\left(kr - \frac{l\pi}{2}\right) = \frac{1}{ikr}[e^{ikr} P_l(1) - e^{-ikr} P_l(-1)]$$

$$= \frac{1}{ikr}[e^{ikr} - (-1)^l e^{-ikr}] \tag{7.76}$$

from which we deduce that $a_l = (2l+1)i^l$. The expansion of a plane wave $\exp(i\mathbf{k}\cdot\mathbf{r})$ in Legendre polynomials is therefore given by

$$e^{i\mathbf{k}\cdot\mathbf{r}} = \sum_{l=0}^{\infty} (2l+1) i^l j_l(kr) P_l(\cos\theta). \tag{7.77}$$

Using the addition theorem of the spherical harmonics (see (6.129)) we can also write the above formula in the form of equation (7.72). That is

$$e^{i\mathbf{k}\cdot\mathbf{r}} = 4\pi \sum_{l=0}^{\infty} \sum_{m=-l}^{+l} i^l j_l(kr) Y_{lm}^*(\hat{\mathbf{k}}) Y_{lm}(\hat{\mathbf{r}}) \tag{7.78}$$

where $\hat{\mathbf{v}}$ denotes the polar angles of a vector $\mathbf{v}$. Upon comparison of (7.72) and (7.78) we see that the coefficients c_{lm} of the expansion (7.72) are given by $c_{lm}(\hat{\mathbf{k}}) = 4\pi i^l Y_{lm}^*(\hat{\mathbf{k}})$.

7.4 The three-dimensional square well potential

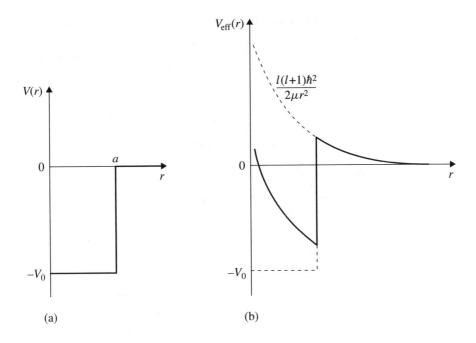

Figure 7.5 (a) The spherically symmetric square well potential of depth V_0 and range a. (b) The corresponding effective potential (solid line) $V_{\text{eff}}(r) = V(r) + l(l+1)\hbar^2/2\mu r^2$, for $l \neq 0$.

7.4 The three-dimensional square well potential

The next example of central potential which we shall analyse is the three-dimensional spherically symmetric square well of depth V_0 and range a

$$
\begin{aligned}
V(r) &= -V_0, \quad r < a \\
     &= 0, \quad\quad\; r > a
\end{aligned}
\tag{7.79}
$$

where V_0 and a are positive constants (see Fig. 7.5(a)). The corresponding effective potential (7.53) is shown in Fig. 7.5(b). Since $V_{\text{eff}}(r)$ vanishes as $r \to \infty$, solutions $u_{El}(r)$ of the radial equation (7.52) for $E \geqslant 0$ will have an oscillatory behaviour at infinity and will be acceptable eigenfunctions for any non-negative value of E. We therefore have a continuous spectrum for $E \geqslant 0$; the corresponding unbound states will be studied in Chapter 13, where we investigate collision phenomena. Here we shall only be interested in the *bound states*, for which the energy is such that $-V_0 \leqslant E < 0$.

Because the potential (7.79) is central we know that there exist solutions of the Schrödinger equation having the form (7.45). The radial functions $R_{El}(r)$ are solutions of the differential equation (7.44), which in the present case becomes

$$\left[-\frac{\hbar^2}{2\mu}\left(\frac{d^2}{dr^2}+\frac{2}{r}\frac{d}{dr}\right)+\frac{l(l+1)\hbar^2}{2\mu r^2}\right]R_{El}(r)=(E+V_0)R_{El}(r), \qquad r<a \tag{7.80a}$$

and

$$\left[-\frac{\hbar^2}{2\mu}\left(\frac{d^2}{dr^2}+\frac{2}{r}\frac{d}{dr}\right)+\frac{l(l+1)\hbar^2}{2\mu r^2}\right]R_{El}(r)=ER_{El}(r), \qquad r>a. \tag{7.80b}$$

Interior solution

Let us first analyse the equation (7.80a), which holds inside the well ($r<a$). If we define the quantity

$$K=\left[\frac{2\mu}{\hbar^2}(E+V_0)\right]^{1/2} \tag{7.81}$$

change variables to $\rho=Kr$, and write $R_l(\rho)\equiv R_{El}(r)$, we find that for $r<a$ the radial function $R_l(\rho)$ satisfies the spherical Bessel differential equation (7.61). Just as in the case of the free particle, the condition that $R_l(\rho)$ must be finite everywhere – including at the origin – restricts us to the spherical Bessel functions j_l, and we have, inside the well,

$$R_{El}(r)=Aj_l(Kr), \qquad r<a \tag{7.82}$$

where A is a constant.

Exterior solution

We now turn to the equation (7.80b), which is valid outside the well ($r>a$). This equation is formally identical to the free-particle equation (7.59), but we must remember that $E<0$ in the present case. It is convenient to write $E=-(\hbar^2/2\mu)\lambda^2$, so that

$$\lambda=\left(-\frac{2\mu}{\hbar^2}E\right)^{1/2}. \tag{7.83}$$

In order to put (7.80b) in the form of the spherical Bessel equation (7.61) we must redefine the variable ρ to be given by $\rho=i\lambda r$, which amounts to replacing k by $i\lambda$ in the free-particle treatment. Note that since $r>a$ in (7.80b), the domain of ρ does *not* extend down to zero, so that there is no reason to limit our choice to the spherical Bessel function, j_l, which is regular at the origin. Instead, a linear combination of the functions j_l and n_l (or $h_l^{(1)}$ and $h_l^{(2)}$) is perfectly admissible. The proper linear combination can be determined by looking at the asymptotic behaviour of the solutions. Using the asymptotic formulae (7.66) and (7.69), with $\rho=i\lambda r$, we see that for large r the functions $j_l(i\lambda r)$, $n_l(i\lambda r)$ and $h_l^{(2)}(i\lambda r)$ all increase exponentially (like $r^{-1}\exp(\lambda r)$) and must therefore be excluded. The only admissible solution is $h_l^{(1)}(\mathrm{i}\lambda r)$, which is proportional to $r^{-1}\exp(-\lambda r)$ for large r. Hence the desired solution of (7.80b), outside the well, is

$$R_{El}(r) = Bh_l^{(1)}(\mathrm{i}\lambda r)$$
$$= B[j_l(\mathrm{i}\lambda r) + \mathrm{i}n_l(\mathrm{i}\lambda r)], \qquad r > a \tag{7.84}$$

where B is a constant. From (7.68) we see that the first three functions $h_l^{(1)}(\mathrm{i}\lambda r)$ are

$$h_0^{(1)}(\mathrm{i}\lambda r) = -\frac{1}{\lambda r}\mathrm{e}^{-\lambda r} \tag{7.85a}$$

$$h_1^{(1)}(\mathrm{i}\lambda r) = \mathrm{i}\left(\frac{1}{\lambda r} + \frac{1}{\lambda^2 r^2}\right)\mathrm{e}^{-\lambda r} \tag{7.85b}$$

$$h_2^{(1)}(\mathrm{i}\lambda r) = \left(\frac{1}{\lambda r} + \frac{3}{\lambda^2 r^2} + \frac{3}{\lambda^3 r^3}\right)\mathrm{e}^{-\lambda r}. \tag{7.85c}$$

Energy levels

As in the analogous one-dimensional problem studied in Section 4.6, the energy levels are obtained by requiring that the eigenfunction and its derivative be continuous at the discontinuity ($r = a$) of the potential. Thus the logarithmic derivative $(1/R_{El})(\mathrm{d}R_{El}/\mathrm{d}r)$ must be continuous at $r = a$. By applying this condition to the interior solution (7.82) and the exterior solution (7.84), we have

$$\left[\frac{\mathrm{d}j_l(Kr)/\mathrm{d}r}{j_l(Kr)}\right]_{r=a} = \left[\frac{\mathrm{d}h_l^{(1)}(\mathrm{i}\lambda r)/\mathrm{d}r}{h_l^{(1)}(\mathrm{i}\lambda r)}\right]_{r=a}. \tag{7.86}$$

This transcendental equation is complicated for arbitrary l, but for $l = 0$ we find with the help of (7.64a) and (7.85a) that it reduces to the equation

$$K \cot Ka = -\lambda \tag{7.87}$$

which we have already analysed in Section 4.6 when we studied the *odd parity* solutions of the one-dimensional square well. Thus, setting $\xi = Ka$ and $\eta = \lambda a$ and using (7.81) and (7.83), we have to solve the equation

$$\xi \cot \xi = -\eta \tag{7.88}$$

with

$$\xi^2 + \eta^2 = \frac{2\mu}{\hbar^2}V_0 a^2 = \gamma^2 \tag{7.89}$$

where $\gamma = (2\mu V_0 a^2/\hbar^2)^{1/2}$ is the 'strength parameter' of the square well. From our work of Section 4.6, we deduce that there is no ($l = 0$) bound state if $\gamma \leqslant \pi/2$, one ($l = 0$) bound state if $\pi/2 < \gamma \leqslant 3\pi/2$, and so on. The radial functions $R_{E0}(r)$ are illustrated in Fig. 7.6 for a spherical square well with $\gamma = 6$ which can support two $l = 0$ bound states.

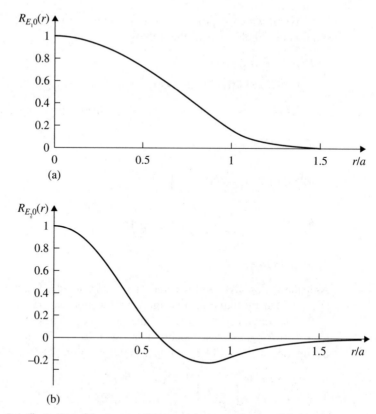

Figure 7.6 Illustration of the two $l = 0$ (s-wave) (unnormalised) radial eigenfunctions $R_{E0}(r)$ for a spherical square well potential such that $\gamma = 6$, corresponding to the bound state energies (a) $E = E_1 = -0.80 V_0$ and (b) $E = E_2 = -0.24 V_0$.

For the case $l = 1$ one readily finds (Problem 7.9) by using (7.64b) and (7.85b) that equation (7.86) reduces to

$$\frac{\cot \xi}{\xi} - \frac{1}{\xi^2} = \frac{1}{\eta} + \frac{1}{\eta^2} \tag{7.90}$$

where again $\xi^2 + \eta^2 = \gamma^2$. This equation must be solved numerically or graphically. It is worth noting that one can easily predict the number of $l = 1$ bound states as a function of the parameter γ without actually solving equation (7.90). Indeed, a new $l = 1$ bound state will appear whenever $\eta = 0$ or $\cot \xi$ is infinite. This will happen when $\xi = \pi, 2\pi, \ldots$, so that there is no $l = 1$ bound state when $\gamma \leqslant \pi$, one $l = 1$ bound state when $\pi < \gamma \leqslant 2\pi$, etc. We also remark that the smallest value of the 'strength parameter' γ for which the spherical square well can support an $l = 1$ bound state is larger than the corresponding value of γ for the $l = 0$ case. In fact, it can be shown (Problem 7.11) that the minimum value of γ

necessary to bind a particle of angular momentum l in the square well increases as l increases. This is physically reasonable in view of the presence of the repulsive centrifugal barrier term in the effective potential $V_{\text{eff}}(r) = V(r) + l(l+1)\hbar^2/2\mu r^2$ (see Fig. 7.5(b)).

Finally, we remark that in general there is no degeneracy between the energy levels obtained from equation (7.86) for different l values. Of course, each energy level of orbital angular momentum quantum number l is $(2l+1)$-fold degenerate with respect to the magnetic quantum number m, a feature which is common to all central potentials, as was shown in Section 7.2.

7.5 The hydrogenic atom

Let us now consider a hydrogenic atom containing an atomic nucleus of charge Ze and an electron of charge $-e$ interacting by means of the Coulomb potential

$$V(r) = -\frac{Ze^2}{(4\pi\varepsilon_0)r} \tag{7.91}$$

where r is the distance between the two particles. We denote by m the mass of the electron and by M the mass of the nucleus. Since the interaction potential (7.91) depends only on the relative coordinate of the two particles, we may use the results of Section 5.7 to separate the motion of the centre of mass. Thus, working in the centre-of-mass system (where the total momentum $\mathbf{P}$ of the atom is equal to zero), the Hamiltonian of the atom reduces to that describing the relative motion of the two particles:

$$H = \frac{\mathbf{p}^2}{2\mu} - \frac{Ze^2}{(4\pi\varepsilon_0)r} \tag{7.92}$$

where $\mathbf{p}$ is the relative momentum and

$$\mu = \frac{mM}{m+M} \tag{7.93}$$

is the reduced mass of the two particles. The corresponding one-body Schrödinger equation describing the relative motion is

$$\left[-\frac{\hbar^2}{2\mu}\nabla^2 - \frac{Ze^2}{(4\pi\varepsilon_0)r}\right]\psi(\mathbf{r}) = E\psi(\mathbf{r}). \tag{7.94}$$

Since the Coulomb potential is central, this equation admits solutions of the form $\psi_{Elm}(\mathbf{r}) = R_{El}(r)Y_{lm}(\theta,\phi)$. Writing $u_{El}(r) = rR_{El}(r)$ and using (7.52) and (7.53), we see that the functions $u_{El}(r)$ must satisfy the equation

$$\frac{d^2 u_{El}(r)}{dr^2} + \frac{2\mu}{\hbar^2}[E - V_{\text{eff}}(r)]u_{El}(r) = 0 \tag{7.95a}$$

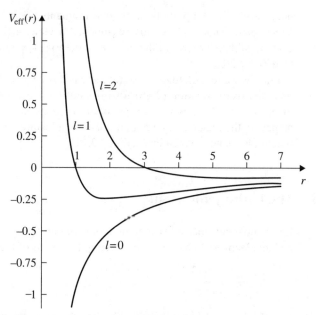

Figure 7.7 The effective potential $V_{\text{eff}}(r)$ given by (7.95b) for the case $Z = 1$ and for the values $l = 0, 1, 2$. The unit of length is $a_\mu = (m/\mu)a_0$ where a_0 is the Bohr radius (1.66). The unit of energy is $e^2/(4\pi\varepsilon_0 a_\mu)$.

where

$$V_{\text{eff}}(r) = -\frac{Ze^2}{(4\pi\varepsilon_0)r} + \frac{l(l+1)\hbar^2}{2\mu r^2} \tag{7.95b}$$

is the effective potential. Fig. 7.7 shows this effective potential for the case $Z = 1$ and for the values $l = 0, 1, 2$ of the orbital angular momentum quantum number. Because $V_{\text{eff}}(r)$ tends to zero for large r, the solution $u_{El}(r)$ for $E > 0$ will have an oscillatory behaviour at infinity and will be an acceptable eigenfunction for any positive value of E. We therefore have a *continuous spectrum* for $E > 0$. The corresponding unbound (scattering) states play an important role in the analysis of collision phenomena between electrons and ions. In what follows, however, we shall focus our attention on the *bound states*, for which $E < 0$.

We now proceed to solve equation (7.95), subject to the condition $u_{El}(0) = 0$ (see (7.54)). It is convenient to introduce the dimensionless quantities

$$\rho = \left(-\frac{8\mu E}{\hbar^2}\right)^{1/2} r \tag{7.96}$$

and

$$\lambda = \frac{Ze^2}{(4\pi\varepsilon_0)\hbar}\left(-\frac{\mu}{2E}\right)^{1/2} = Z\alpha\left(-\frac{\mu c^2}{2E}\right)^{1/2} \tag{7.97}$$

where $\alpha = e^2/(4\pi\varepsilon_0\hbar c) \simeq 1/137$ is the fine-structure constant and we recall that $E < 0$. In terms of the new quantities ρ and λ, (7.95) becomes

$$\left[\frac{d^2}{d\rho^2} - \frac{l(l+1)}{\rho^2} + \frac{\lambda}{\rho} - \frac{1}{4}\right] u_{El}(\rho) = 0. \tag{7.98}$$

Let us first examine the asymptotic behaviour of $u_{El}(\rho)$. To this end, we remark that when $\rho \to \infty$ the terms in ρ^{-1} and ρ^{-2} become negligible with respect to the constant term $(-1/4)$. Hence for large ρ equation (7.98) reduces to the 'asymptotic' equation

$$\left[\frac{d^2}{d\rho^2} - \frac{1}{4}\right] u_{El}(\rho) = 0, \tag{7.99}$$

the solutions of which are proportional to $\exp(\pm\rho/2)$. Since the function $u_{El}(\rho)$ must be bounded everywhere, including at infinity, we must keep only the exponentially decreasing function, so that

$$u_{El}(\rho) \underset{\rho \to \infty}{\sim} \exp(-\rho/2), \tag{7.100}$$

This result suggests that we look for a solution of the radial equation (7.98) having the form

$$u_{El}(\rho) = e^{-\rho/2} f(\rho) \tag{7.101}$$

where we have written $f(\rho) \equiv f_{El}(\rho)$ to simplify the notation. Substituting (7.101) into (7.98), we obtain for $f(\rho)$ the equation

$$\left[\frac{d^2}{d\rho^2} - \frac{d}{d\rho} - \frac{l(l+1)}{\rho^2} + \frac{\lambda}{\rho}\right] f(\rho) = 0. \tag{7.102}$$

We now write a series expansion for $f(\rho)$ in the form

$$f(\rho) = \rho^{l+1} g(\rho) \tag{7.103}$$

where

$$g(\rho) = \sum_{k=0}^{\infty} c_k \rho^k, \qquad c_0 \neq 0 \tag{7.104}$$

and we have used the fact (see (7.58)) that $u_{El}(\rho)$, and thus also $f(\rho)$, behaves like ρ^{l+1} in the vicinity of the origin. Inserting (7.103) into (7.102), we find that the function $g(\rho)$ satisfies the differential equation

$$\left[\rho\frac{d^2}{d\rho^2} + (2l+2-\rho)\frac{d}{d\rho} + (\lambda - l - 1)\right] g(\rho) = 0. \tag{7.105}$$

Using the expansion (7.104) to solve this equation, we have

$$\sum_{k=0}^{\infty} [k(k-1)c_k\rho^{k-1} + (2l+2-\rho)kc_k\rho^{k-1} + (\lambda - l - 1)c_k\rho^k] = 0 \tag{7.106}$$

or

$$\sum_{k=0}^{\infty}\{[k(k+1)+(2l+2)(k+1)]c_{k+1}+(\lambda-l-1-k)c_k\}\rho^k=0 \quad (7.107)$$

so that the coefficients c_k must satisfy the recursion relation

$$c_{k+1} = \frac{k+l+1-\lambda}{(k+1)(k+2l+2)} c_k. \quad (7.108)$$

If the series (7.104) does not terminate, we see from (7.108) that for large k

$$\frac{c_{k+1}}{c_k} \sim \frac{1}{k} \quad (7.109)$$

a ratio which is the same as that of the series for $\rho^p \exp(\rho)$, where p has a finite value. Thus in this case, we deduce by using (7.101) and (7.103) that the function $u_{El}(\rho)$ has an asymptotic behaviour of the type

$$u_{El}(\rho) \underset{\rho \to \infty}{\sim} \rho^{l+1+p} e^{\rho/2} \quad (7.110)$$

which is clearly unacceptable.

The series (7.104) must therefore terminate, which means that $g(\rho)$ must be a *polynomial* in ρ. Let the highest power of ρ appearing in $g(\rho)$ be ρ^{n_r}, where the *radial quantum number* $n_r = 0, 1, 2, \ldots$, is a positive integer or zero. Then the coefficient $c_{n_r+1} = 0$, and from the recursion formula (7.108) we have

$$\lambda = n_r + l + 1. \quad (7.111)$$

Let us introduce the *principal quantum number*

$$n = n_r + l + 1 \quad (7.112)$$

which is a positive integer ($n = 1, 2, \ldots$) since both n_r and l can take on positive integer or zero values. From (7.111) and (7.112) we see that the eigenvalues of equation (7.98) corresponding to the bound-state spectrum ($E < 0$) are given by

$$\lambda = n. \quad (7.113)$$

Energy levels

Replacing in (7.97) the quantity λ by its value (7.113), we obtain the bound-state energy eigenvalues

$$\begin{aligned} E_n &= -\frac{\mu}{2\hbar^2}\left(\frac{Ze^2}{4\pi\varepsilon_0}\right)^2 \frac{1}{n^2} \\ &= -\frac{e^2}{(4\pi\varepsilon_0)a_\mu} \frac{Z^2}{2n^2} \\ &= -\frac{1}{2}\mu c^2 \frac{(Z\alpha)^2}{n^2}, \quad n=1,2,3,\ldots \end{aligned} \quad (7.114)$$

where $\alpha \simeq 1/137$ is the fine-structure constant and a_μ denotes the modified Bohr radius

$$a_\mu = \frac{(4\pi\varepsilon_0)\hbar^2}{\mu e^2} = \frac{m}{\mu}a_0 \tag{7.115}$$

with $a_0 = (4\pi\varepsilon_0)\hbar^2/(me^2)$ being the Bohr radius (1.66).

The energy levels (7.114), which we have obtained here by solving the Schrödinger equation for one-electron atoms, agree exactly with those found in Section 1.4 from the Bohr model. The agreement of this energy spectrum with the main features of the experimental spectrum was pointed out when we analysed the Bohr results. This agreement, however, is not perfect and various corrections (such as the fine structure arising from relativistic effects and the electron spin, the Lamb shift and the hyperfine structure due to nuclear effects) must be taken into account in order to explain the details of the experimental spectrum[5].

We remark from (7.114) that since n may take on all integral values from 1 to $+\infty$, the bound-state energy spectrum corresponding to the Coulomb potential (7.91) contains an *infinite number of discrete energy levels* extending from $-(\mu/2\hbar^2)(Ze^2/4\pi\varepsilon_0)^2$ to zero. This is due to the fact that the magnitude of the Coulomb potential falls off slowly at larger r. On the contrary, short-range potentials – such as the square well studied in the previous section – have a finite (sometimes zero) number of bound states.

Another striking feature of the result (7.114) is that the energy eigenvalues E_n depend only on the principal quantum number n, and are therefore degenerate with respect to the quantum numbers l and m. Indeed, for each value of n the orbital angular momentum quantum number l may take on the values $0, 1, \ldots, n-1$, and for each value of l the magnetic quantum number m may take the $(2l+1)$ possible values $-l, -l+1, \ldots, +l$. The total degeneracy of the bound-state energy level E_n is therefore given by

$$\sum_{l=0}^{n-1}(2l+1) = 2\frac{n(n-1)}{2} + n = n^2. \tag{7.116}$$

As we have shown in Section 7.2, the degeneracy with respect to the quantum number m is present for any central potential $V(r)$. On the other hand, the degeneracy with respect to l is characteristic of the Coulomb potential; it is removed if the dependence of the potential on r is modified. For example, we have seen in Section 7.4 that there is no degeneracy with respect to l in the case of a square well. Another illustration is provided by the spectrum of alkali atoms. Many properties of these atoms can be understood in terms of the motion of a single 'valence' electron in a central potential which is Coulombic (i.e. proportional to r^{-1}) at large distances, but which deviates from the Coulomb behaviour as r decreases because of the presence of the 'inner' electrons. As a result, the energy of the valence electron does depend on l, and the degeneracy with respect to l is removed, leading to n distinct levels

---

[5] See Bransden and Joachain (1983), Chapter 5.

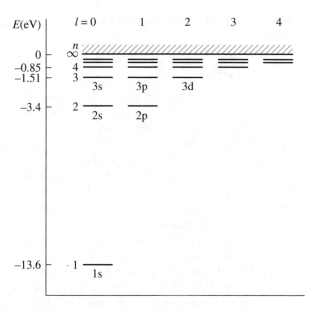

Figure 7.8 The energy-level diagram of atomic hydrogen.

$E_{nl}(l = 0, 1, \ldots, n - 1)$ for a given principal quantum number n. Finally, if an external magnetic field is applied to the atom, we shall see in Chapter 12 that the $(2l + 1)$ degeneracy with respect to the magnetic quantum number m is removed.

Figure 7.8 shows the energy-level diagram of the hydrogen atom; it is similar to that displayed in Fig. 1.11, except that the degenerate levels with the same n but different l are shown separately. Following the usual spectroscopic notation, these levels are labelled by two symbols. The first one gives the value of the principal quantum number n; the second one is a code letter which indicates the value of the orbital angular momentum quantum number l according to the correspondence discussed in Chapter 6, namely

Value of l	0	1	2	3	4	5	...
	↕	↕	↕	↕	↕	↕	
Code letter	s	p	d	f	g	h	....

Looking at the hydrogen atom spectrum illustrated in Fig. 7.8, we see that the ground state ($n = 1$) is a 1s state, the first excited state ($n = 2$) is four-fold degenerate and contains one 2s state and three 2p states (with $m = -1, 0, +1$), the second excited state ($n = 3$) is nine-fold degenerate and contains one 3s state, three 3p states (with $m = -1, 0, +1$) and five 3d states (with $m = -2, -1, 0, +1, +2$), etc.

Having obtained the energy levels of one-electron atoms within the framework of the Schrödinger non-relativistic quantum theory, we may now inquire about the *spectral lines* corresponding to transitions from one energy level to another. This problem has already been discussed in Chapter 1, where the frequencies of the spectral lines were obtained by using the Bohr model. We shall return to this question in detail in Chapter 11, where the interaction of atomic systems with electromagnetic radiation will be studied quantum mechanically. In particular, we shall re-derive in that chapter the Bohr result (1.61) giving the frequencies of the spectral lines, and we shall calculate the transition rates for the most common transitions, the so-called electric dipole transitions.

The eigenfunctions of the bound states

Until now we have seen that the energy levels predicted by the Schrödinger theory for hydrogenic atoms agree with those already obtained in Section 1.4 by using the Bohr model. However, the Schrödinger theory has much more predictive power than the old quantum theory since it also yields the *eigenfunctions* which enable one to calculate probability densities, expectation values of operators, transition rates, etc.

The radial eigenfunctions of the bound states

In order to obtain these eigenfunctions explicitly, let us return to (7.105). This equation can be identified with the *Kummer–Laplace differential equation*

$$z\frac{d^2 w}{dz^2} + (c-z)\frac{dw}{dz} - aw = 0 \tag{7.117}$$

with $z = \rho$, $w = g$, $a = l+1-\lambda$ and $c = 2l+2$. Within a multiplicative constant, the solution of (7.117), regular at the origin, is the *confluent hypergeometric function*

$$\begin{aligned}{}_1F_1(a, c, z) &= 1 + \frac{az}{c1!} + \frac{a(a+1)z^2}{c(c+1)2!} + \cdots \\ &= \sum_{k=0}^{\infty} \frac{(a)_k}{(c)_k} \frac{z^k}{k!}\end{aligned} \tag{7.118a}$$

where

$$\alpha_k = \alpha(\alpha+1)\ldots(\alpha+k-1), \qquad (\alpha)_0 = 1. \tag{7.118b}$$

In general, for large positive values of its argument the confluent hypergeometric series (7.118) behaves asymptotically as

$$_1F_1(a, c, z) \to \frac{\Gamma(c)}{\Gamma(a)} e^z z^{a-c} \tag{7.119}$$

where Γ is Euler's gamma function. Thus, in the present case the series (7.118) for $_1F_1(l+1-\lambda, 2l+2, \rho)$ is in general proportional to $\rho^{-l-1-\lambda} \exp(\rho)$ for large ρ, leading to a function $u_{El}(\rho)$ having the unacceptable asymptotic behaviour $u_{El}(\rho) \sim \rho^{-\lambda} \exp(\rho/2)$ (see (7.101) and (7.103)). The only way to obtain physically acceptable solutions of (7.105) is to require that the hypergeometric series for $_1F_1(l+1-\lambda, 2l+2, \rho)$ terminates, which implies that $l+1-\lambda = -n_r (n_r = 0, 1, 2, \ldots)$, and hence $\lambda = n_r + l + 1 = n$. The confluent hypergeometric function $_1F_1(l+1-\lambda, 2l+2, \rho) \equiv {_1F_1}(-n_r, 2l+2, \rho)$ then reduces to a *polynomial* of degree n_r, namely

$$_1F_1(l+1-n, 2l+2, \rho)$$
$$= \sum_{k=0}^{n-l-1} \frac{(k+l-n)(k-1+l-n)\ldots(1+l-n)}{(k+2l+1)(k-1+2l+1)\ldots(1+2l+1)} \frac{\rho^k}{k!} \tag{7.120}$$

in accordance with our foregoing discussion. We may readily verify the correctness of this result by using the recursion relation (7.108) which we derived above for the coefficients c_k of the function $g(\rho)$. Thus, setting $c_0 = 1$ and using the fact that $\lambda = n$, we find from (7.108) that

$$c_k = \frac{(k+l-n)(k-1+l-n)\ldots(1+l-n)}{(k+2l+1)(k-1+2l+1)\ldots(1+2l+1)} \frac{1}{k!} \tag{7.121}$$

in agreement with (7.120). We also remark from (7.96) and (7.114) that

$$\rho = \frac{2Z}{na_\mu} r = \frac{2Z}{na_0} \frac{\mu}{m} r. \tag{7.122}$$

The physically admissible solutions $g(\rho)$ of (7.105), corresponding to $\lambda = n$, may also be expressed in terms of *associated Laguerre polynomials*. To see how this comes about, we first define the *Laguerre polynomials* $L_q(\rho)$ by the relation

$$L_q(\rho) = e^\rho \frac{d^q}{d\rho^q}(\rho^q e^{-\rho}) \tag{7.123}$$

and we note that these Laguerre polynomials may also be obtained from the generating function

$$U(\rho, s) = \frac{\exp[-\rho s/(1-s)]}{1-s}$$
$$= \sum_{q=0}^{\infty} \frac{L_q(\rho)}{q!} s^q, \quad |s| < 1. \tag{7.124}$$

Differentiation of this generating function with respect to s yields the recurrence formula (Problem 7.12)

$$L_{q+1}(\rho) + (\rho - 1 - 2q)L_q(\rho) + q^2 L_{q-1}(\rho) = 0. \tag{7.125}$$

Similarly, upon differentiation of $U(\rho, s)$ with respect to ρ, we find that

$$\frac{d}{d\rho}L_q(\rho) - q\frac{d}{d\rho}L_{q-1}(\rho) + qL_{q-1}(\rho) = 0. \tag{7.126}$$

Using (7.125) and (7.126), it is readily shown that the lowest order differential equation involving only $L_q(\rho)$ is

$$\left[\rho\frac{d^2}{d\rho^2} + (1-\rho)\frac{d}{d\rho} + q\right]L_q(\rho) = 0. \tag{7.127}$$

7.5 The hydrogenic atom 359

Next, we define the *associated Laguerre polynomials* $L_q^p(\rho)$ by the relation

$$L_q^p(\rho) = \frac{d^p}{d\rho^p} L_q(\rho). \tag{7.128}$$

Differentiating (7.127) p times, we find that $L_q^p(\rho)$ satisfies the differential equation (Problem 7.13)

$$\left[\rho \frac{d^2}{d\rho^2} + (p+1-\rho)\frac{d}{d\rho} + (q-p)\right] L_q^p(\rho) = 0. \tag{7.129}$$

Setting $\lambda = n$ in (7.105) and comparing with (7.129), we see that the physically acceptable solution $g(\rho)$ of (7.105) is given (up to a multiplicative constant) by the associated Laguerre polynomial $L_{n+l}^{2l+1}(\rho)$. Note that this polynomial is of order $(n+l)-(2l+1) = n-l-1 = n_r$, in accordance with the discussion following (7.110).

The generating function for the associated Laguerre polynomials may be obtained by differentiating (7.124) p times with respect to ρ. That is,

$$U_p(\rho, s) = \frac{(-s)^p \exp[-\rho s/(1-s)]}{(1-s)^{p+1}}$$

$$= \sum_{q=p}^{\infty} \frac{L_q^p(\rho)}{q!} s^q, \quad |s| < 1. \tag{7.130}$$

An explicit expression for $L_{n+l}^{2l+1}(\rho)$ is given by

$$L_{n+l}^{2l+1}(\rho) = \sum_{k=0}^{n-l-1} (-1)^{k+1} \frac{[(n+l)!]^2}{(n-l-1-k)!(2l+1+k)!} \frac{\rho^k}{k!} \tag{7.131}$$

and is readily verified by substitution into (7.130), with $q = n+l$ and $p = 2l+1$.

Since the physically admissible solutions $g(\rho)$ of (7.105) are given within a multiplicative constant either by the confluent hypergeometric function $_1F_1(l+1-n, 2l+2, \rho)$ or by the associated Laguerre polynomial $L_{n+l}^{2l+1}(\rho)$, it is clear that these two functions differ only by a constant factor. This factor is readily found by comparing (7.120) and (7.131) at $\rho = 0$ and remembering that the confluent hypergeometric function is equal to unity at the origin (see (7.118)). Thus we have

$$L_{n+l}^{2l+1}(\rho) = -\frac{[(n+l)!]^2}{(n-l-1)!(2l+1)!} {}_1F_1(l+1-n, 2l+2, \rho). \tag{7.132}$$

Using (7.51), (7.101), (7.103) and the foregoing results, we may now write the full hydrogenic radial functions as

$$R_{nl}(r) = N e^{-\rho/2} \rho^l L_{n+l}^{2l+1}(\rho) \tag{7.133a}$$

$$= \tilde{N} e^{-\rho/2} \rho^l {}_1F_1(l+1-n, 2l+2, \rho) \tag{7.133b}$$

where N and $\tilde{N}$ are constants which will be determined below (apart from an arbitrary phase factor) by the normalisation condition. In (7.133) we have used the notation

R_{nl} (which displays explicitly the quantum numbers n and l) instead of the symbol R_{El}, and we recall that $\rho = (2Z/na_\mu)r$ (see (7.122)).

The hydrogenic wave functions of the discrete spectrum

Using (7.45), we see that the full eigenfunctions of the discrete spectrum for hydrogenic atoms may be written as

$$\psi_{nlm}(r, \theta, \phi) = R_{nl}(r) Y_{lm}(\theta, \phi) \tag{7.134}$$

where the radial functions are given by (7.133) and the spherical harmonics provide the angular part of the wave functions. We require that the eigenfunctions (7.134) be normalised to unity, so that

$$\int_0^\infty dr\, r^2 \int_0^\pi d\theta \sin\theta \int_0^{2\pi} d\phi |\psi_{nlm}(r, \theta, \phi)|^2 = 1. \tag{7.135}$$

Because the spherical harmonics are normalised on the unit sphere (see (6.103)), the normalisation condition (7.135) implies that

$$\int_0^\infty |R_{nl}(r)|^2 r^2 dr = 1 \tag{7.136}$$

or

$$|N|^2 \left(\frac{na_\mu}{2Z}\right)^3 \int_0^\infty e^{-\rho} \rho^{2l} [L_{n+l}^{2l+1}(\rho)]^2 \rho^2 d\rho = 1 \tag{7.137}$$

where we have used (7.133a). The integral over ρ can be evaluated by using the generating function (7.130) for the associated Laguerre polynomials (see Problem 7.15). The result is

$$\int_0^\infty e^{-\rho} \rho^{2l} [L_{n+l}^{2l+1}(\rho)]^2 \rho^2 d\rho = \frac{2n[(n+l)!]^3}{(n-l-1)!} \tag{7.138}$$

so that the normalised radial functions for the bound states of hydrogenic atoms may be written as

$$R_{nl}(r) = -\left\{ \left(\frac{2Z}{na_\mu}\right)^3 \frac{(n-l-1)!}{2n[(n+l)!]^3} \right\}^{1/2} e^{-\rho/2} \rho^l L_{n+l}^{2l+1}(\rho) \tag{7.139a}$$

or

$$R_{nl}(r) = \frac{1}{(2l+1)!} \left\{ \left(\frac{2Z}{na_\mu}\right)^3 \frac{(n+l)!}{2n(n-l-1)!} \right\}^{1/2}$$
$$\times e^{-\rho/2} \rho^l {}_1F_1(l+1-n, 2l+2, \rho) \tag{7.139b}$$

with

$$\rho = \frac{2Z}{na_\mu} r, \quad a_\mu = \frac{(4\pi\varepsilon_0)\hbar^2}{\mu e^2} \tag{7.139c}$$

where a constant multiplicative factor of modulus one is still arbitrary. In writing (7.139b) we have used equation (7.132) which relates the associated Laguerre polynomial $L_{n+l}^{2l+1}(\rho)$ to the confluent hypergeometric function $_1F_1(l+1-n, 2l+2, \rho)$.

The first few radial eigenfunctions (7.139) are given by

$$R_{10}(r) = 2(Z/a_\mu)^{3/2} \exp(-Zr/a_\mu)$$

$$R_{20}(r) = 2(Z/2a_\mu)^{3/2}(1 - Zr/2a_\mu) \exp(-Zr/2a_\mu)$$

$$R_{21}(r) = \frac{1}{\sqrt{3}}(Z/2a_\mu)^{3/2}(Zr/a_\mu) \exp(-Zr/2a_\mu)$$

$$R_{30}(r) = 2(Z/3a_\mu)^{3/2}(1 - 2Zr/3a_\mu + 2Z^2r^2/27a_\mu^2) \exp(-Zr/3a_\mu)$$

$$R_{31}(r) = \frac{4\sqrt{2}}{9}(Z/3a_\mu)^{3/2}(1 - Zr/6a_\mu)(Zr/a_\mu) \exp(-Zr/3a_\mu)$$

$$R_{32}(r) = \frac{4}{27\sqrt{10}}(Z/3a_\mu)^{3/2}(Zr/a_\mu)^2 \exp(-Zr/3a_\mu) \quad (7.140)$$

and are illustrated in Fig. 7.9.

Using the radial wave functions (7.140), together with the explicit expressions of the spherical harmonics given in Table 6.1, we display in Table 7.1 the full normalised bound-state hydrogenic eigenfunctions $\psi_{nlm}(r, \theta, \phi)$ for the first three shells (that is, the K, L and M shells corresponding, respectively, to the values $n = 1, 2, 3$ of the principal quantum number). We have also indicated in Table 7.1 the spectroscopic notation, introduced in the discussion of the energy levels, with the subscripts corresponding to the values of the magnetic quantum number m (when $l \neq 0$).

According to the interpretation of the wave function discussed in Chapter 2, the quantity

$$|\psi_{nlm}(r, \theta, \phi)|^2 d\mathbf{r} = \psi_{nlm}^*(r, \theta, \phi)\psi_{nlm}(r, \theta, \phi)r^2 dr \sin\theta d\theta d\phi \quad (7.141)$$

represents the probability of finding the electron in the volume element $d\mathbf{r}$ (given in spherical polar coordinates by $d\mathbf{r} = r^2 dr \sin\theta d\theta d\phi$) when the system is in the stationary state specified by the quantum numbers (n, l, m). In agreement with the discussion following (7.46), the *position probability density*

$$|\psi_{nlm}(r, \theta, \phi)|^2 = |R_{nl}(r)|^2 (2\pi)^{-1} |\Theta_{lm}(\theta)|^2 \quad (7.142)$$

does not depend on the coordinate ϕ. It is the product of the *angular factor* $(2\pi)^{-1}|\Theta_{lm}(\theta)|^2$, studied in Chapter 6 and of the quantity $|R_{nl}(r)|^2$, which gives the *electron density* as a function of r along a given direction. The *radial distribution function*

$$D_{nl}(r) = r^2 |R_{nl}(r)|^2 \quad (7.143)$$

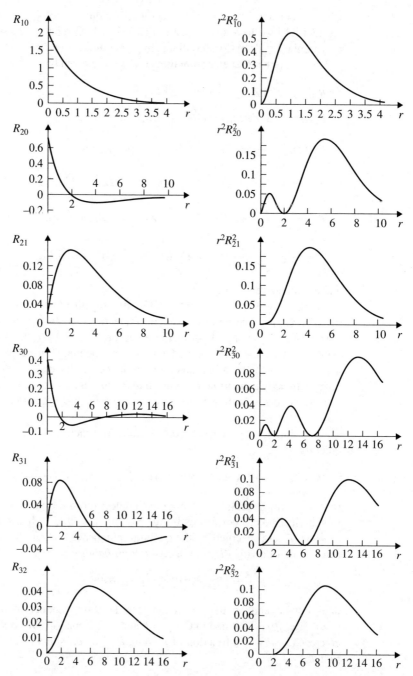

Figure 7.9 Radial functions $R_{nl}(r)$ and radial distribution functions $r^2 R_{nl}^2(r)$ for atomic hydrogen. The unit of length is $a_\mu = (m/\mu)a_0$, where a_0 is the first Bohr radius (1.66).

Table 7.1 The complete normalised hydrogenic wave functions corresponding to the first three shells.

Shell	Quantum numbers			Spectroscopic notation	Wave function $\psi_{nlm}(r, \theta, \phi)$
	n	l	m		
K	1	0	0	1s	$\dfrac{1}{\sqrt{\pi}}(Z/a_\mu)^{3/2}\exp(-Zr/a_\mu)$
L	2	0	0	2s	$\dfrac{1}{2\sqrt{2\pi}}(Z/a_\mu)^{3/2}(1 - Zr/2a_\mu)\exp(-Zr/2a_\mu)$
	2	1	0	$2p_0$	$\dfrac{1}{4\sqrt{2\pi}}(Z/a_\mu)^{3/2}(Zr/a_\mu)\exp(-Zr/2a_\mu)\cos\theta$
	2	1	± 1	$2p_{\pm 1}$	$\mp\dfrac{1}{8\sqrt{\pi}}(Z/a_\mu)^{3/2}(Zr/a_\mu)\exp(-Zr/2a_\mu)\sin\theta\exp(\pm i\phi)$
M	3	0	0	3s	$\dfrac{1}{3\sqrt{3\pi}}(Z/a_\mu)^{3/2}(1 - 2Zr/3a_\mu + 2Z^2r^2/27a_\mu^2)\exp(-Zr/3a_\mu)$
	3	1	0	$3p_0$	$\dfrac{2\sqrt{2}}{27\sqrt{\pi}}(Z/a_\mu)^{3/2}(1 - Zr/6a_\mu)(Zr/a_\mu)\exp(-Zr/3a_\mu)\cos\theta$
	3	1	± 1	$3p_{\pm 1}$	$\mp\dfrac{2}{27\sqrt{\pi}}(Z/a_\mu)^{3/2}(1 - Zr/6a_\mu)(Zr/a_\mu)\exp(-Zr/3a_\mu)\sin\theta\exp(\pm i\phi)$
	3	2	0	$3d_0$	$\dfrac{1}{81\sqrt{6\pi}}(Z/a_\mu)^{3/2}(Z^2r^2/a_\mu^2)\exp(-Zr/3a_\mu)(3\cos^2\theta - 1)$
	3	2	± 1	$3d_{\pm 1}$	$\mp\dfrac{1}{81\sqrt{\pi}}(Z/a_\mu)^{3/2}(Z^2r^2/a_\mu^2)\exp(-Zr/3a_\mu)\sin\theta\cos\theta\exp(\pm i\phi)$
	3	2	± 2	$3d_{\pm 2}$	$\dfrac{1}{162\sqrt{\pi}}(Z/a_\mu)^{3/2}(Z^2r^2/a_\mu^2)\exp(-Zr/3a_\mu)\sin^2\theta\exp(\pm 2i\phi)$

gives the probability per unit length that the electron is to be found at a distance r from the nucleus. Indeed, by integrating (7.141) over the polar angles θ and ϕ and using (7.134) and (6.103), we see that

$$\int_0^\pi d\theta \sin\theta \int_0^{2\pi} d\phi |\psi_{nlm}(r,\theta,\phi)|^2 r^2 dr$$
$$= r^2 |R_{nl}(r)|^2 dr \int_0^\pi d\theta \sin\theta \int_0^{2\pi} d\phi |Y_{lm}(\theta,\phi)|^2$$
$$= r^2 |R_{nl}(r)|^2 dr$$
$$= D_{nl}(r) dr \tag{7.144}$$

represents the probability of finding the electron between the distances r and $r + dr$ from the nucleus, regardless of direction. The radial distribution functions $D_{nl}(r)$ corresponding to the first few radial functions are plotted in Fig. 7.9.

Several interesting features emerge from the examination of the radial eigenfunctions $R_{nl}(r)$ and the radial distribution functions $D_{nl}(r)$.

(1) Only for s states ($l = 0$) are the radial eigenfunctions different from zero at $r = 0$. We also note that since $Y_{00} = (4\pi)^{-1/2}$ is independent of θ and ϕ, one has from (7.134) and (7.139)

$$|\psi_{n00}(0)|^2 = \frac{1}{4\pi} |R_{n0}(0)|^2 = \frac{Z^3}{\pi a_\mu^3 n^3}. \tag{7.145}$$

Moreover, each of the s-state radial eigenfunctions R_{n0} are such that $dR_{n0}/dr \neq 0$ at $r = 0$. This peculiar behaviour is due to the fact that the potential energy (7.91) is infinite at the origin.

(2) We also verify from (7.139) that the radial eigenfunctions $R_{nl}(r)$ are proportional to r^l near the origin, in agreement with our general discussion of Section 7.2. Thus, for $l \neq 0$ the radial wave functions are forced to remain small over distances which increase with l. As we have seen in Section 7.2, this behaviour is due to the centrifugal barrier term $l(l+1)\hbar^2/2\mu r^2$ contained in the effective potential. Among the radial eigenfunctions $R_{nl}(r)$ having the same principal quantum number n, the one with the lowest value of l has the largest amplitude in the vicinity of the nucleus.

(3) The associated Laguerre polynomial $L_{n+l}^{2l+1}(\rho)$ is a polynomial of degree $n_r = n - l - 1$ having n_r radial nodes (zeros). Thus the radial distribution function $D_{nl}(r)$ has $n - l$ maxima. We remark that there is only one maximum when, for a given n, the orbital angular momentum quantum number l has its largest value $l = n - 1$. In that case $n_r = 0$ and we see from (7.131) and (7.139) that

$$R_{n,n-1}(r) \sim r^{n-1} \exp(-Zr/na_\mu). \tag{7.146}$$

Hence $D_{n,n-1}(r) = r^2 R_{n,n-1}^2(r)$ will exhibit a maximum at a value of r obtained by solving the equation

$$\frac{dD_{n,n-1}}{dr} = \left(2nr^{2n-1} - \frac{2Z}{na_\mu}r^{2n}\right)\exp(-2Zr/na_\mu) = 0 \tag{7.147}$$

i.e. at

$$r = \frac{n^2 a_\mu}{Z}. \tag{7.148}$$

This is precisely the value (1.74) given by the Bohr model. However, in contrast to the Bohr model, the diffuseness of the electron cloud implies that the concept of size is less precise in the quantum mechanical theory, so that the value (7.148) should be interpreted as a 'most probable distance'. We see from (7.148) that this most probable distance is proportional to n^2 and is inversely proportional to Z. More generally, the maximum value of $D_{nl}(r)$ recedes from the nucleus with increasing values of n (see Fig. 7.9) and becomes closer to the nucleus (by a factor of Z^{-1}) when Z increases.

To conclude our study of the hydrogenic bound state wave functions we recall that all wave functions of the form (7.45) transform under the *parity* operation according to (7.50). In particular, the action of the parity operator $\mathcal{P}$ on the hydrogenic wave function (7.134) yields

$$\begin{aligned}\mathcal{P}\psi_{nlm}(r,\theta,\phi) &= \mathcal{P}[R_{nl}(r)Y_{lm}(\theta,\phi)] \\ &= R_{nl}(r)(-1)^l Y_{lm}(\theta,\phi) \\ &= (-1)^l \psi_{nlm}(r,\theta,\phi)\end{aligned} \tag{7.149}$$

so that the hydrogenic states ψ_{nlm} have the parity of l, in agreement with the general discussion of Section 7.2.

Hydrogen iso-electronic sequence; hydrogen isotopes; positronium; muonium; antihydrogen

Let us recall some important results we have obtained for hydrogenic systems. The energy eigenvalues are given by (7.114) and the eigenfunctions by (7.134). In particular, the ionisation potential $I_P = |E_{n=1}|$ is

$$I_P = \frac{e^2}{(4\pi\varepsilon_0)a_\mu}\frac{Z^2}{2} \tag{7.150}$$

and the 'extension' a of the wave function describing the relative motion of the system is roughly given in the ground state (see (7.148)) by

$$a = \frac{a_\mu}{Z} = \frac{(4\pi\varepsilon_0)\hbar^2}{Z\mu e^2}. \tag{7.151}$$

The hydrogenic systems we have considered so far correspond to an atomic nucleus of mass M and charge Ze and an electron of mass m and charge $-e$ interacting by means of the Coulomb potential (7.91). The 'normal' hydrogen atom, containing a proton and an electron is the prototype of these hydrogenic systems. The hydrogenic ions He$^+$ ($Z = 2$), Li^{2+} ($Z = 3$), Be^{3+} ($Z = 4$), etc., which belong to the *hydrogen iso-electronic sequence* are also examples of such systems. Note that apart from small reduced mass effects the ionisation potential I_P of these hydrogenic ions is increased by a factor of Z^2, while the value of a for these ions is decreased by a factor of Z with respect to the hydrogen atom.

The (neutral) *isotopes* of atomic hydrogen, deuterium and tritium, also provide examples of hydrogenic systems. Here the proton is replaced by a nucleus having the same charge $+e$, namely a deuteron (containing a proton and a neutron) in the case of deuterium and a triton (containing one proton and two neutrons) in the case of tritium. Since $M_d \simeq 2M_p$ and $M_t \simeq 3M_p$, where M_p is the mass of the proton, M_d the mass of the deuteron and M_t the mass of the triton, we see from (7.93) that the reduced mass μ is slightly different for (normal) hydrogen, deuterium and tritium; the relative differences are of the order of 10^{-3}. Thus the quantities I_P and a are nearly identical for these three atoms, the small differences in the value of μ giving rise to *isotopic shifts* of the spectral lines which we have already discussed in Section 1.4.

In addition to deuterium and tritium, there exist also other 'isotopes' of hydrogen, in which the role of the nucleus is played by another particle. For example, *positronium* (e^+e^-) is a bound hydrogenic system made of a positron e^+ (the antiparticle of the electron, having the same mass m as the electron, but the opposite charge) and an electron e^-. *Muonium* (μ^+e^-) is another 'isotope' of hydrogen, in which the proton has been replaced by a positive muon μ^+, a particle which is very similar to the positron e^+, except that it has a mass $M_\mu \simeq 207m$ and that it is unstable, with a lifetime of about 2.2×10^{-6} s. Positronium and muonium may thus be considered as light 'isotopes' of hydrogen. Positronium was first observed in 1951 and muonium in 1960. Table 7.2 gives the values of the reduced mass μ, the 'radius' a and the ionisation potential I_P for positronium and muonium, compared with those of the hydrogen atom. It should be noted that both positronium and muonium are *unstable*. Indeed, muonium has a lifetime of 2.2×10^{-6} s (which is the lifetime of the muon μ^+ itself) while in positronium the electron and the positron may annihilate, their total energy including their rest mass energy being completely converted into electromagnetic radiation (photons).

Positronium and muonium have attracted a great deal of interest because they contain only *leptons* (i.e. particles which are not affected by the strong interactions) and hence are particularly suitable systems in which the predictions of quantum electrodynamics can be accurately verified.

Finally, we mention that atoms of antihydrogen ($\bar{p}e^+$), a bound system made of an antiproton $\bar{p}$ and a positron e^+, have been observed in an experiment performed at CERN in 1995 by G. Baur *et al.*

Muonic atoms

In all the hydrogenic atoms we have considered until now the negative particle is an electron. In 1947, J. A. Wheeler suggested that other negative particles could form

Table 7.2 The reduced mass μ, 'radius' a and ionisation potential I_P of some special hydrogenic systems, compared with the corresponding quantities for the hydrogen atom (pe^-). The following 'atomic units' are used: unit of mass = electron mass m; unit of length = Bohr radius a_0 (see (1.66)); unit of energy = $e^2/[(4\pi\varepsilon_0)a_0]$, i.e. twice the ionisation potential of atomic hydrogen (with an infinite nuclear mass).

System	Reduced mass μ	'Radius' a	Ionisation potential I_P
$(pe^-), (\bar{p}e^+)$	$\frac{1836}{1837} \simeq 1$	$\simeq a_0 = 1$	$\simeq \frac{e^2}{(4\pi\varepsilon_0)2a_0} = 0.5$
(e^+e^-)	0.5	2	0.25
(μ^+e^-)	$\frac{207}{208} \simeq 1$	$\simeq a_0 = 1$	$\simeq 0.5$
$(p\mu^-)$	$\simeq 186$	$\simeq 5.4 \times 10^{-3}$	$\simeq 93$

a bound system with a nucleus. This, in particular, is the case for a lepton such as the *negative muon* μ^-, which is the antiparticle of the positive muon μ^+, having the same mass $M_\mu \simeq 207m$ and the same lifetime (2.2×10^{-6} s) but a negative charge $-e$. The negative muon μ^- is therefore a kind of 'heavy electron'. As it is slowing down in bulk matter, it can be captured by the Coulomb attraction of a nucleus of charge Z, thus forming a *muonic atom*.

The simplest example of muonic atom is the bound system $(p\mu^-)$ consisting of a proton p and a negative muon μ^-. Since the muon has a mass $M_\mu \simeq 207m$, the reduced mass of the muon with respect to the proton is approximately 186 times the electron mass. As a result, we see from (7.151) that the 'radius' a of muonic hydrogen $(p\mu^-)$ is 186 times smaller than that of the ordinary hydrogen atom (pe^-). On the other hand, using (7.150) we see that the ionisation potential I_P of the muonic atom $(p\mu^-)$ is 186 times larger than the corresponding quantity for ordinary atomic hydrogen (see Table 7.2). The frequencies of the spectral lines corresponding to transitions between the energy levels of $(p\mu^-)$ may thus be obtained from those of the hydrogen atom by multiplying the latter by a factor of 186. For transitions between the lowest energy levels of $(p\mu^-)$ the spectral lines are therefore lying in the X-ray region.

7.6 The three-dimensional isotropic oscillator

As a last example of central field, we shall consider the motion of a particle of mass μ in the potential

$$V(r) = \tfrac{1}{2}kr^2 = \tfrac{1}{2}\mu\omega^2 r^2 \tag{7.152}$$

which corresponds to an isotropic three-dimensional oscillator of classical angular frequency $\omega = (k/\mu)^{1/2}$. In Section 7.1 we solved the corresponding Schrödinger equation in Cartesian coordinates. In particular, we found that the energy levels are given by $E_n = (n + 3/2)\hbar\omega$, where $n = 0, 1, 2, \ldots$, each energy level E_n being $(n + 1)(n + 2)/2$-fold degenerate.

In this section we want to analyse this problem using spherical polar coordinates, so that we look for solutions of the Schrödinger equation having the form (7.45). The radial equation (7.52) for the function $u_{El}(r) = rR_{El}(r)$ becomes in the present case

$$\left[\frac{d^2}{dr^2} - \frac{l(l+1)}{r^2} + \frac{2\mu E}{\hbar^2} - \frac{\mu^2\omega^2 r^2}{\hbar^2}\right]u_{El}(r) = 0. \tag{7.153}$$

It is convenient to introduce the dimensionless variable

$$\rho = \alpha r \tag{7.154}$$

where

$$\alpha = \left(\frac{\mu k}{\hbar^2}\right)^{1/4} = \left(\frac{\mu\omega}{\hbar}\right)^{1/2} \tag{7.155}$$

and the dimensionless eigenvalue

$$\lambda = \frac{2E}{\hbar\omega} \tag{7.156}$$

so that (7.153) now reads

$$\left[\frac{d^2}{d\rho^2} - \frac{l(l+1)}{\rho^2} + \lambda - \rho^2\right]u_{El}(\rho) = 0. \tag{7.157}$$

Let us first study the asymptotic behaviour of $u_{El}(\rho)$. When $\rho \to \infty$ we may neglect the terms $l(l+1)/\rho^2$ and λ compared to ρ^2, so that (7.157) reduces to

$$\left[\frac{d^2}{d\rho^2} - \rho^2\right]u_{El}(\rho) = 0 \tag{7.158}$$

which is the same asymptotic equation as that studied in Section 4.7 in the one-dimensional case (see (4.136)). For sufficiently large ρ the functions $u_{El}(\rho) = \rho^p \exp(\pm\rho^2/2)$ satisfy (7.158) as far as the leading terms, which are of order $\rho^2 u_{El}(\rho)$, are concerned when p has any finite value. Since the function $u_{El}(r)$ must be bounded everywhere, we only keep the minus sign in the exponent. This suggests looking for solutions of (7.157) in the form

$$u_{El}(\rho) = e^{-\rho^2/2}v(\rho) \tag{7.159}$$

where we have written $v(\rho) \equiv v_{El}(\rho)$. Substituting (7.159) in (7.157) we obtain for $v(\rho)$ the equation

$$\left(\frac{d^2}{d\rho^2} - 2\rho\frac{d}{d\rho} - \frac{l(l+1)}{\rho^2} + \lambda - 1\right)v(\rho) = 0. \tag{7.160}$$

From the general discussion of Section 7.2 we know that $u_{El}(\rho)$ – and hence also $v(\rho)$ – behaves like ρ^{l+1} near the origin. We therefore seek a solution of (7.160) of the form

$$v(\rho) = \rho^{l+1} w(\rho). \tag{7.161}$$

Upon substitution of (7.161) in (7.160) we find that the function $w(\rho)$ satisfies the differential equation

$$\frac{d^2 w}{d\rho^2} + 2\left(\frac{l+1}{\rho} - \rho\right)\frac{dw}{d\rho} + \left[\lambda - 2\left(l + \frac{3}{2}\right)\right]w(\rho) = 0. \tag{7.162}$$

By introducing the new independent variable $\zeta = \rho^2$, this equation goes over to the differential equation

$$\zeta \frac{d^2 w}{d\zeta^2} + \left[\left(l + \frac{3}{2}\right) - \zeta\right]\frac{dw}{d\zeta} - \left[\frac{1}{2}\left(l + \frac{3}{2}\right) - \frac{\lambda}{4}\right]w(\zeta) = 0 \tag{7.163}$$

which is just the Kummer–Laplace differential equation (7.117), with $z = \zeta$, $a = (l + 3/2)/2 - \lambda/4$ and $c = l + 3/2$. The solution of (7.163), regular at the origin, is thus given by

$$w(\zeta) = C \, _1F_1\left[\frac{1}{2}\left(l + \frac{3}{2}\right) - \frac{\lambda}{4}, l + \frac{3}{2}, \zeta\right] \tag{7.164}$$

where C is a constant.

Energy levels

Let us examine the solution (7.164) in the limit of large ζ. From (7.119), we remark that in this limit the confluent hypergeometric series (7.118) for $_1F_1[(l + 3/2)/2 - \lambda/4, l + 3/2, \zeta]$ is proportional to $\zeta^{-(l+3/2+\lambda/2)/2} \exp(\zeta)$. Thus, using (7.159) and (7.161), we see that if this series does not terminate, the function $u_{El}(\rho)$ will have an asymptotic behaviour of the type

$$u_{El}(\rho) \underset{\rho \to \infty}{\sim} \rho^{-(\lambda+1)/2} e^{\rho^2/2} \tag{7.165}$$

which is inadmissible. The only way to avoid this divergence is to transform the confluent hypergeometric series into a *polynomial* of degree n_r, by requiring that

$$\frac{1}{2}\left(l + \frac{3}{2}\right) - \frac{\lambda}{4} = -n_r \tag{7.166}$$

where the *radial quantum number* $n_r = 0, 1, 2, \ldots$ is a positive integer or zero. This condition may also be written in the form

$$\lambda = 2\left(n + \frac{3}{2}\right) \tag{7.167}$$

with

$$n = 2n_r + l, \qquad n = 0, 1, 2, \ldots \tag{7.168}$$

and from (7.156) we see that the energy levels are given by

$$E_n = \left(n + \frac{3}{2}\right)\hbar\omega, \qquad n = 0, 1, 2, \ldots \tag{7.169}$$

in agreement with the result (7.37) of Section 7.1.

The eigenfunctions of the three-dimensional isotropic oscillator

Using (7.154), (7.159), (7.161), (7.164) and (7.166), we find that the radial eigenfunctions $R_{El}(r) = r^{-1}u_{El}(r)$ of the three-dimensional isotropic oscillator are given by

$$R_{El}(r) = N e^{-\alpha^2 r^2/2} r^l \, {}_1F_1\left(-n_r, l + \tfrac{3}{2}, \alpha^2 r^2\right) \tag{7.170}$$

where $\alpha = (\mu\omega/\hbar)^{1/2}$ and N is a normalisation constant. Hence, from (7.45) the complete three-dimensional isotropic oscillator eigenfunctions are given in spherical polar coordinates by

$$\psi_{Elm}(r, \theta, \phi) = N e^{-\alpha^2 r^2/2} r^l \, {}_1F_1\left(-n_r, l + \tfrac{3}{2}, \alpha^2 r^2\right) Y_{lm}(\theta, \phi). \tag{7.171}$$

Except for the ground state ($n = 0$), the energy levels (7.169) are *degenerate*. Indeed, for even n there are $n/2 + 1$ partitions of n according to (7.168), and for odd n there are $(n + 1)/2$ partitions. Moreover, for each value of l, there is a $(2l + 1)$ degeneracy with respect to the magnetic quantum number m (which can take on the values $-l, -l + 1, \ldots, l$). As a result, there is only one eigenfunction for $n = 0$, there are three linearly independent eigenfunctions for $n = 1$ (corresponding to $l = 1$ and $m = -1, 0, +1$), six for $n = 2$ (one corresponding to $l = 0$ and five to $l = 2$), etc. In general, there are $(n + 1)(n + 2)/2$ linearly independent eigenfunctions for each value of n, in agreement with the conclusion reached in Section 7.1.

Problems

7.1 Consider a particle of mass μ confined within a box with impenetrable walls of sides L_1, L_2 and L_3. If $L_1 = L_2$, obtain the allowed energies and discuss the degeneracy of the first few energy levels.

7.2 Consider an anisotropic harmonic oscillator described by the Hamiltonian

$$H = \frac{1}{2\mu}(p_x^2 + p_y^2 + p_z^2) + \frac{1}{2}k_1(x^2 + y^2) + \frac{1}{2}k_2 z^2.$$

(a) Find the energy levels and the corresponding energy eigenfunctions using Cartesian coordinates. What are the degeneracies of the levels, assuming that $\omega_1 = (k_1/\mu)^{1/2}$ and $\omega_2 = (k_2/\mu)^{1/2}$ are incommensurable?
(b) Can the stationary states be eigenstates of $\mathbf{L}^2$? of L_z?

7.3 Show that all solutions of the Schrödinger equation (7.41) can be obtained as linear combinations of separable solutions of the form (7.43).

(*Hint*: Use the orthonormality and closure relations satisfied by the spherical harmonics.)

7.4 Show that when $p = -2$ and $b_0 < 0$ in (7.55), physically acceptable solutions of (7.52) exist only if $b_0 > -\hbar^2/8\mu$.

7.5 Prove that the spherical Bessel and Neumann functions defined by (7.63a) and (7.63b) satisfy the differential equation (7.61).

7.6 Using equations (7.63) obtain the leading term of $j_l(\rho)$ and $n_l(\rho)$ for small and large values of ρ.

7.7 Consider a particle of mass μ confined in a spherical box of radius a.

(a) Write down the equation which determines the allowed energy levels.
(b) Solve this equation explicitly for the case $l = 0$.

7.8 Solve equation (7.87) numerically to obtain the $l = 0$ energy levels and corresponding normalised radial functions for a three-dimensional square well such that

$$\gamma = (2\mu V_0 a^2/\hbar^2)^{1/2} = 5.$$

7.9 Prove equation (7.90).

7.10 Consider a particle of mass μ moving in a very deep square well potential for which $Ka \gg 1$. Show that for the bound states with $|E| \ll V_0$, equation (7.86) reduces to

$$Ka - l\pi/2 \simeq (n + \tfrac{1}{2})\pi; \qquad n = 0, 1, 2, \ldots$$

and that the energy levels are given approximately by

$$E \simeq -2V_0\{1 - [n + (l+1)/2]\pi/\gamma\}.$$

7.11 (a) Show that for a square well potential, bound states of zero energy for $l > 0$ occur when

$$j_{l-1}(\gamma) = 0.$$

(b) Deduce from this condition that the value of γ necessary to bind a particle of angular momentum quantum number l increases with increasing l.

7.12 (a) Prove the relations (7.125) and (7.126) satisfied by the Laguerre polynomials.
(b) Show that the lowest order differential equation involving only $L_q(\rho)$ is given by (7.127).

7.13 Prove equation (7.129).

7.14 Any region of space in which the kinetic energy T of a particle would become negative is forbidden for classical motion. For a hydrogen atom in the ground state:

(a) find the classically forbidden region; and
(b) using the ground-state wave function $\psi_{100}(r)$, calculate the probability of finding the electron in this region.

7.15 Using the generating function (7.130) and proceeding as in the case of the linear harmonic oscillator (see Section 4.7) evaluate the integral (7.138).

7.16 Using the generating function (7.130), show that the average values

$$\langle r^k \rangle_{nlm} = \int \psi_{nlm}^*(\mathbf{r}) r^k \psi_{nlm}(\mathbf{r}) d\mathbf{r} = \int_0^\infty |R_{nl}(r)|^2 r^{k+2} dr$$

are given respectively for $k = 1, -1, -2$ and -3 by

$$\langle r \rangle_{nlm} = a_\mu \frac{n^2}{Z} \left\{ 1 + \frac{1}{2} \left[1 - \frac{l(l+1)}{n^2} \right] \right\} \tag{1}$$

$$\langle r^{-1} \rangle_{nlm} = \frac{Z}{a_\mu n^2} \tag{2}$$

$$\langle r^{-2} \rangle_{nlm} = \frac{Z^2}{a_\mu^2 n^3 \left(l + \frac{1}{2} \right)} \tag{3}$$

$$\langle r^{-3} \rangle_{nlm} = \frac{Z^3}{a_\mu^3 n^3 l \left(l + \frac{1}{2} \right)(l+1)}. \tag{4}$$

7.17 Using the result (2) of the preceding problem, together with (7.114) show that for a hydrogenic atom, the average values of the kinetic energy operator T and of the potential energy V in a bound state (nlm) are such that

$$2 \langle T \rangle_{nlm} = -\langle V \rangle_{nlm}.$$

Relate your result to the virial theorem of Section 5.7.

7.18 Consider a hydrogen atom whose wave function at $t = 0$ is the following superposition of energy eigenfunctions $\psi_{nlm}(\mathbf{r})$:

$$\Psi(\mathbf{r}, t=0) = \frac{1}{\sqrt{14}} [2\psi_{100}(\mathbf{r}) - 3\psi_{200}(\mathbf{r}) + \psi_{322}(\mathbf{r})].$$

(a) Is this wave function an eigenfunction of the parity operator?
(b) What is the probability of finding the system in the ground state (100)? In the state (200)? In the state (322)? In another energy eigenstate?
(c) What is the expectation value of the energy; of the operator $\mathbf{L}^2$; of the operator L_z?

7.19 Find the energy levels and the corresponding eigenfunctions of a particle of mass μ in the potential $V(r) = (A/r^2 + Br^2)$, where $A \geq 0$ and $B > 0$. In the case $A = 0$ compare your result with that obtained in Section 7.6 for the three-dimensional isotropic oscillator.

(*Hint*: Set $l'(l'+1) = l(l+1) + 2\mu A/\hbar^2$ and compare the radial equation you obtain with (7.153).)

7.20 A two-dimensional harmonic oscillator has the Hamiltonian

$$H = -\frac{\hbar^2}{2\mu}\left[\frac{\partial^2}{\partial x^2} + \frac{\partial^2}{\partial y^2}\right] + \frac{1}{2}k_1 x^2 + \frac{1}{2}k_2 y^2.$$

(a) Show that the energy levels are given by

$$E_{n_x n_y} = \left(n_x + \frac{1}{2}\right)\hbar\omega_1 + \left(n_y + \frac{1}{2}\right)\hbar\omega_2, \qquad \begin{array}{l} n_x = 0, 1, 2, \ldots \\ n_y = 0, 1, 2, \ldots \end{array}$$

where $\omega_1 = (k_1/\mu)^{1/2}$ and $\omega_2 = (k_2/\mu)^{1/2}$. Obtain the corresponding energy eigenfunctions in terms of the one-dimensional linear harmonic oscillator wave functions of Section 4.7.

(b) Assuming that the harmonic oscillator is isotropic ($k_1 = k_2 = k$), what is the degeneracy of each energy level?

(c) Solve the Schrödinger equation for the two-dimensional isotropic oscillator in plane polar coordinates (r, ϕ). Using the fact that the Laplacian operator is given in these coordinates by

$$\nabla^2 = \frac{\partial^2}{\partial r^2} + \frac{1}{r}\frac{\partial}{\partial r} + \frac{1}{r^2}\frac{\partial^2}{\partial \phi^2}$$

and introducing the dimensionless quantities

$$\rho = \alpha r, \qquad \lambda = 2E/\hbar\omega,$$

where $\alpha = (\mu k/\hbar^2)^{1/4} = (\mu\omega/\hbar)^{1/2}$, show first that the eigenfunctions are of the form

$$\psi = e^{-\rho^2/2}\rho^{|m|}g(\rho)e^{im\phi}, \qquad m = 0, \pm 1, \pm 2, \ldots$$

where the function $g(\rho)$ must satisfy the differential equation

$$\frac{d^2 g}{d\rho^2} + \left[\frac{2|m|+1}{\rho} - 2\rho\right]\frac{dg}{d\rho} - [2(|m|+1) - \lambda]g = 0. \tag{1}$$

Then, taking $v = \rho^2$ as a new variable, prove that equation (1) is transformed into the Kummer–Laplace equation

$$v\frac{d^2 g}{dv^2} + [(|m|+1) - v]\frac{dg}{dv} - \frac{1}{2}\left[(|m|+1) - \frac{\lambda}{2}\right]g = 0 \tag{2}$$

whose solution regular at $v = 0$ is the confluent hypergeometric function

$$g(v) = {}_1F_1(a, c, v) \tag{3}$$

with

$$a = \frac{1}{2}\left(|m| + 1 - \frac{\lambda}{2}\right), \qquad c = |m| + 1. \tag{4}$$

Using these results, show that the eigenfunctions are given by
$$\psi_{n_r,m}(\rho,\phi) = N e^{-\rho^2/2} \rho^{|m|} \,_1F_1(-n_r, |m|+1, \rho^2) e^{im\phi} \tag{5}$$
where $n_r = 0, 1, 2, \ldots$, $m = 0, \pm 1, \pm 2, \ldots$ and N is a normalisation constant. Prove that the energy levels are given by
$$E_n = \hbar\omega(n+1) \tag{6}$$
where
$$n = 2n_r + |m| = 0, 1, 2, \ldots \tag{7}$$

Discuss the degeneracy of the energy levels and compare your results with those obtained by using Cartesian coordinates.

7.21 The parabolic coordinates (ξ, η, ϕ) of a point in three-dimensional space are defined by the relations
$$\begin{aligned} x &= \sqrt{\xi\eta}\cos\phi, & \xi &= r+z \\ y &= \sqrt{\xi\eta}\sin\phi, & \eta &= r-z \\ z &= \tfrac{1}{2}(\xi-\eta), & \phi &= \tan^{-1}(y/x) \\ r &= \tfrac{1}{2}(\xi+\eta) \end{aligned}$$
and the Laplacian operator is given in these coordinates by
$$\nabla^2 = \frac{4}{\xi+\eta}\left[\frac{\partial}{\partial\xi}\left(\xi\frac{\partial}{\partial\xi}\right) + \frac{\partial}{\partial\eta}\left(\eta\frac{\partial}{\partial\eta}\right)\right] + \frac{1}{\xi\eta}\frac{\partial^2}{\partial\phi^2}$$

(a) Prove that the Schrödinger equation (7.94) with a Coulomb potential is separable in these coordinates.
(*Hint*: Write the eigenfunctions in the form
$$\psi = f_1(\xi) f_2(\eta) e^{im\phi}, \quad m = 0, \pm 1, \pm 2, \ldots$$
and show that the Schrödinger equation can be replaced by the two differential equations
$$\frac{d}{d\xi}\left(\xi\frac{df_1}{d\xi}\right) - \left(\frac{1}{4}\lambda^2\xi + \frac{m^2}{4\xi} + \nu_1\right)f_1 = 0$$
$$\frac{d}{d\eta}\left(\eta\frac{df_2}{d\eta}\right) - \left(\frac{1}{4}\lambda^2\eta + \frac{m^2}{4\eta} + \nu_2\right)f_2 = 0$$
where $\lambda^2 = -2\mu E/\hbar^2$ and the constants of separation ν_1 and ν_2 are related by
$$\nu_1 + \nu_2 = -\frac{Ze^2}{4\pi\varepsilon_0}\frac{\mu}{\hbar^2}.)$$

(b) Taking the energy to be negative (discrete spectrum), obtain the functions $f_1(\xi)$ and $f_2(\eta)$, and hence the eigenfunctions and energy levels. Compare your results with those derived in the text in spherical polar coordinates.

8 Approximation methods for stationary problems

8.1 Time-independent perturbation theory for a non-degenerate energy level 375
8.2 Time-independent perturbation theory for a degenerate energy level 386
8.3 The variational method 399
8.4 The WKB approximation 408
Problems 427

As in the case of classical mechanics, there are relatively few physically interesting problems in quantum mechanics which can be solved exactly. Approximation methods are therefore very important in nearly all the applications of the theory. In this chapter and the next one we shall develop several approximation methods to study mainly the bound states of physical systems; approximation methods for collision problems, which deal with the continuous part of the spectrum, will be considered in Chapter 13.

Approximation methods can be conveniently divided into two groups, according to whether the Hamiltonian of the system is time-independent or time-dependent. The latter case will be considered in Chapter 9. In the present chapter we are concerned with the approximate determination of the discrete eigenenergies and corresponding eigenfunctions for the stationary states of a time-independent Hamiltonian. We shall begin by discussing perturbation theory, which studies the changes induced in a system by a small disturbance. In Section 8.1 we shall develop a stationary perturbation method for the case of a non-degenerate energy level. The generalisation to degenerate energy levels will be considered in Section 8.2. The next section is devoted to the variational method. Finally, in Section 8.4 we shall discuss the semi-classical approximation method of Wentzel, Kramers and Brillouin, which is known as the WKB approximation.

8.1 Time-independent perturbation theory for a non-degenerate energy level

In this section and the following one we shall discuss the *Rayleigh–Schrödinger perturbation theory*, which analyses the modifications of discrete energy levels and of the corresponding eigenfunctions of a system when a perturbation is applied.

Let us suppose that the time-independent Hamiltonian H of a system can be expressed as

$$H = H_0 + \lambda H' \tag{8.1}$$

where the 'unperturbed' Hamiltonian H_0 is sufficiently simple so that the corresponding time-independent Schrödinger equation

$$H_0 \psi_n^{(0)} = E_n^{(0)} \psi_n^{(0)} \tag{8.2}$$

can be solved, and $\lambda H'$ is a small perturbation. The quantity λ is a real parameter, which will be used below to distinguish between the various orders of the perturbation calculation. We can let λ tend to zero, in which case the Hamiltonian H reduces to the unperturbed Hamiltonian H_0, or we may let λ reach its full value, which we shall choose to be $\lambda = 1$.

We assume that the known eigenfunctions $\psi_n^{(0)}$ corresponding to the known eigenvalues $E_n^{(0)}$ of H_0 form a complete orthonormal set (which may be partly continuous). Thus, if $\psi_i^{(0)}$ and $\psi_j^{(0)}$ are two members of that set, we have

$$\langle \psi_i^{(0)} | \psi_j^{(0)} \rangle = \delta_{ij} \tag{8.3}$$

where for notational simplicity the meaning of δ_{ij} is implicitly extended to cover the possibility of $\psi_i^{(0)}$ and $\psi_j^{(0)}$ being discrete or continuous states. The eigenvalue problem which we want to solve is

$$H \psi_n = E_n \psi_n. \tag{8.4}$$

Let us consider a *particular* unperturbed, discrete energy level $E_n^{(0)}$, which we assume to be *non-degenerate* (other unperturbed energy levels may be degenerate). We suppose that the perturbation $\lambda H'$ is small enough so that the perturbed energy level E_n is much closer to $E_n^{(0)}$ than to any other unperturbed level. As λ tends to zero, we have

$$\lim_{\lambda \to 0} E_n = E_n^{(0)}. \tag{8.5}$$

Similarly, since the state n is non-degenerate, the perturbed eigenfunction ψ_n must approach the unperturbed eigenfunction $\psi_n^{(0)}$ as λ approaches zero:

$$\lim_{\lambda \to 0} \psi_n = \psi_n^{(0)}. \tag{8.6}$$

The basic idea of perturbation theory is to assume that both the eigenvalues and eigenfunctions of H can be expanded in powers of the perturbation parameter λ. That is

$$E_n = \sum_{j=0}^{\infty} \lambda^j E_n^{(j)} \tag{8.7}$$

and
$$\psi_n = \sum_{j=0}^{\infty} \lambda^j \psi_n^{(j)} \tag{8.8}$$

where j, the power of λ, is the order of the perturbation. Substituting the expansions (8.7) and (8.8) into the Schrödinger equation (8.4), we have

$$(H_0 + \lambda H')(\psi_n^{(0)} + \lambda \psi_n^{(1)} + \lambda^2 \psi_n^{(2)} + \cdots) = (E_n^{(0)} + \lambda E_n^{(1)} + \lambda^2 E_n^{(2)} + \cdots)$$
$$\times (\psi_n^{(0)} + \lambda \psi_n^{(1)} + \lambda^2 \psi_n^{(2)} + \cdots). \tag{8.9}$$

We now equate the coefficients of equal powers of λ on both sides of this equation. Beginning with λ^0, we find, as expected, that $H_0 \psi_n^{(0)} = E_n^{(0)} \psi_n^{(0)}$. Next, the coefficients of λ give

$$H_0 \psi_n^{(1)} + H' \psi_n^{(0)} = E_n^{(0)} \psi_n^{(1)} + E_n^{(1)} \psi_n^{(0)} \tag{8.10}$$

while those of λ^2 yield

$$H_0 \psi_n^{(2)} + H' \psi_n^{(1)} = E_n^{(0)} \psi_n^{(2)} + E_n^{(1)} \psi_n^{(1)} + E_n^{(2)} \psi_n^{(0)}. \tag{8.11}$$

More generally, by equating the coefficients of λ^j in (8.9) we have for $j \geqslant 1$

$$H_0 \psi_n^{(j)} + H' \psi_n^{(j-1)} = E_n^{(0)} \psi_n^{(j)} + E_n^{(1)} \psi_n^{(j-1)} + \cdots + E_n^{(j)} \psi_n^{(0)}. \tag{8.12}$$

In order to obtain the first-order energy correction $E_n^{(1)}$, we premultiply (8.10) by $\psi_n^{(0)*}$ and integrate over all coordinates. This gives

$$\langle \psi_n^{(0)} | H_0 - E_n^{(0)} | \psi_n^{(1)} \rangle + \langle \psi_n^{(0)} | H' - E_n^{(1)} | \psi_n^{(0)} \rangle = 0. \tag{8.13}$$

Now, using (8.2) and the fact that the operator H_0 is Hermitian, we have

$$\langle \psi_n^{(0)} | H_0 | \psi_n^{(1)} \rangle = \langle \psi_n^{(1)} | H_0 | \psi_n^{(0)} \rangle^*$$

$$= E_n^{(0)} \langle \psi_n^{(1)} | \psi_n^{(0)} \rangle^*$$

$$= E_n^{(0)} \langle \psi_n^{(0)} | \psi_n^{(1)} \rangle \tag{8.14}$$

so that the first term on the left of (8.13) vanishes. Moreover, since $\langle \psi_n^{(0)} | \psi_n^{(0)} \rangle = 1$, we see that (8.13) reduces to

$$E_n^{(1)} = \langle \psi_n^{(0)} | H' | \psi_n^{(0)} \rangle. \tag{8.15}$$

This is a *very important result*. It tells us that the first-order correction to the energy for a non-degenerate level is just the perturbation H' averaged over the corresponding unperturbed state of the system.

Proceeding in a similar way with equation (8.11), we have

$$\langle \psi_n^{(0)} | H_0 - E_n^{(0)} | \psi_n^{(2)} \rangle + \langle \psi_n^{(0)} | H' - E_n^{(1)} | \psi_n^{(1)} \rangle - E_n^{(2)} = 0. \tag{8.16}$$

Again the first term vanishes, so that the second-order energy correction is given by

$$E_n^{(2)} = \langle \psi_n^{(0)} | H' - E_n^{(1)} | \psi_n^{(1)} \rangle. \tag{8.17}$$

Expressions for higher-order energy corrections can be obtained in a similar way from (8.12). In particular, one has (Problem 8.2)

$$E_n^{(3)} = \langle \psi_n^{(1)} | H' - E_n^{(1)} | \psi_n^{(1)} \rangle - 2 E_n^{(2)} \langle \psi_n^{(0)} | \psi_n^{(1)} \rangle. \tag{8.18}$$

It is interesting to note that the knowledge of the unperturbed wave function $\psi_n^{(0)}$ yields $E_n^{(0)}$ and $E_n^{(1)}$, while that of the first-order wave function correction $\psi_n^{(1)}$ gives $E_n^{(2)}$ and $E_n^{(3)}$. More generally, if $\psi_n^{(0)}, \psi_n^{(1)}, \ldots, \psi_n^{(s)}$, are known, the energy corrections can be obtained up to and including $E_n^{(2s+1)}$ (see Problem 8.3).

Let us now return to equation (8.10). The Rayleigh–Schrödinger method attempts to obtain the solution $\psi_n^{(1)}$ of this equation in the following way. First, the 'unperturbed' equation (8.2) is solved for all eigenvalues and eigenfunctions (including those belonging to the continuous part of the spectrum, if one exists). The unknown function $\psi_n^{(1)}$ is then expanded in the basis set of the unperturbed eigenfunctions, namely

$$\psi_n^{(1)} = \sum_k a_{nk}^{(1)} \psi_k^{(0)} \tag{8.19}$$

where the sum over k means a summation over the discrete part of the set and an integration over its continuous part. Substituting the expansion (8.19) into (8.10), we obtain

$$(H_0 - E_n^{(0)}) \sum_k a_{nk}^{(1)} \psi_k^{(0)} + (H' - E_n^{(1)}) \psi_n^{(0)} = 0. \tag{8.20}$$

Premultiplying by $\psi_l^{(0)*}$, integrating over all coordinates, and using equations (8.2) and (8.3), we find that

$$a_{nl}^{(1)} (E_l^{(0)} - E_n^{(0)}) + H_{ln}' - E_n^{(1)} \delta_{nl} = 0 \tag{8.21}$$

where we have written $H_{ln}' \equiv \langle \psi_l^{(0)} | H' | \psi_n^{(0)} \rangle$.

For $l = n$, equation (8.21) reduces to $E_n^{(1)} = H_{nn}'$, which is the result (8.15). On the other hand, for $l \neq n$, we have

$$a_{nl}^{(1)} = \frac{H_{ln}'}{E_n^{(0)} - E_l^{(0)}}, \quad l \neq n. \tag{8.22}$$

From this result and equation (8.19), we see that a sufficient condition for the applicability of the Rayleigh–Schrödinger perturbation method is that

$$\left| \frac{H_{ln}'}{E_n^{(0)} - E_l^{(0)}} \right| \ll 1, \quad l \neq n. \tag{8.23}$$

We also remark that the coefficient $a_{nn}^{(1)} = \langle \psi_n^{(0)} | \psi_n^{(1)} \rangle$, which is the 'component' of $\psi_n^{(1)}$ along $\psi_n^{(0)}$, cannot be obtained from equation (8.21). We shall return shortly to the determination of $a_{nn}^{(1)}$.

The Rayleigh–Schrödinger method can also be applied to solve higher-order equations such as (8.11). In this case, expanding $\psi_n^{(2)}$ as

$$\psi_n^{(2)} = \sum_k a_{nk}^{(2)} \psi_k^{(0)} \tag{8.24}$$

substituting in (8.11) and using (8.19), we have

$$(H_0 - E_n^{(0)}) \sum_k a_{nk}^{(2)} \psi_k^{(0)} + (H' - E_n^{(1)}) \sum_k a_{nk}^{(1)} \psi_k^{(0)} - E_n^{(2)} \psi_n^{(0)} = 0. \tag{8.25}$$

Premultiplying by $\psi_l^{(0)*}$, integrating over all coordinates and using (8.2) and (8.3), we find that

$$a_{nl}^{(2)}(E_l^{(0)} - E_n^{(0)}) + \sum_k H'_{lk} a_{nk}^{(1)} - E_n^{(1)} a_{nl}^{(1)} - E_n^{(2)} \delta_{nl} = 0. \tag{8.26}$$

Let us first examine what happens for $l = n$. The above equation then yields

$$E_n^{(2)} = \sum_k H'_{nk} a_{nk}^{(1)} - H'_{nn} a_{nn}^{(1)}$$

$$= \sum_{k \neq n} H'_{nk} a_{nk}^{(1)} \tag{8.27}$$

where we have used the fact that $E_n^{(1)} = H'_{nn}$. With the help of (8.22) we can therefore write

$$E_n^{(2)} = \sum_{k \neq n} \frac{H'_{nk} H'_{kn}}{E_n^{(0)} - E_k^{(0)}} \tag{8.28a}$$

$$= \sum_{k \neq n} \frac{|H'_{kn}|^2}{E_n^{(0)} - E_k^{(0)}}. \tag{8.28b}$$

The second-order energy correction $E_n^{(2)}$ can therefore be obtained by performing a summation over all the states $\psi_k^{(0)}$, with $k \neq n$, as indicated in (8.28). The states $\psi_k^{(0)}$ over which the summation is performed are often called 'intermediate' states. Indeed, we see from (8.28a) that each term in the summation may be viewed as a succession of two first-order transitions, weighted by the energy denominator $E_n^{(0)} - E_k^{(0)}$, in which the system leaves the state $\psi_n^{(0)}$, 'propagates' in the intermediate state $\psi_k^{(0)}$ and then 'falls back' to the state $\psi_n^{(0)}$. We also remark from (8.28b) that if the level n which we are studying corresponds to the *ground state* of the system, then $E_n^{(0)} - E_k^{(0)} < 0$ for $k \neq n$; hence in that case the second-order energy correction $E_n^{(2)}$ is always *negative*, for any perturbation H'.

Summarising the results we have obtained thus far for the perturbed energy level E_n, we see from (8.15) and (8.28) that to second order in the perturbation, we have

$$E_n = E_n^{(0)} + H'_{nn} + \sum_{k \neq n} \frac{|H'_{kn}|^2}{E_n^{(0)} - E_k^{(0)}} \tag{8.29}$$

where we have set $\lambda = 1$.

Let us now return to (8.26). Using (8.22) and the fact that $E_n^{(1)} = H'_{nn}$, we have for $l \neq n$

$$a_{nl}^{(2)} = \frac{1}{E_n^{(0)} - E_l^{(0)}} \sum_{k \neq n} \frac{H'_{lk} H'_{kn}}{E_n^{(0)} - E_k^{(0)}} - \frac{H'_{nn} H'_{ln}}{(E_n^{(0)} - E_l^{(0)})^2} - a_{nn}^{(1)} \frac{H'_{ln}}{E_n^{(0)} - E_l^{(0)}} \quad (8.30)$$

and we note that no information can be obtained from (8.26) – and hence from (8.11) – concerning the coefficient $a_{nn}^{(2)}$. Similarly, we recall that $a_{nn}^{(1)}$ could not be determined from equation (8.10). More generally, if we denote by

$$a_{nn}^{(j)} = \langle \psi_n^{(0)} | \psi_n^{(j)} \rangle, \qquad j \geqslant 1 \quad (8.31)$$

the 'component' of $\psi_n^{(j)}$ along $\psi_n^{(0)}$, we see by pre-multiplying equation (8.12) with $\psi_n^{(0)*}$ and integrating over all coordinates that this equation does not determine $a_{nn}^{(j)}$.

Since our basic perturbation equations (8.10)–(8.12) leave the coefficients $a_{nn}^{(j)}$ undetermined, it is clear that the choice of these quantities can have no physical consequences. We may, for example, require that the $a_{nn}^{(j)}$ be chosen in such a way that the perturbed wave function ψ_n, calculated through order λ^j, be normalised to unity in the sense that

$$\langle \psi_n | \psi_n \rangle \simeq \langle \psi_n^{(0)} + \lambda \psi_n^{(1)} + \cdots + \lambda^j \psi_n^{(j)} | \psi_n^{(0)} + \lambda \psi_n^{(1)} + \cdots + \lambda^j \psi_n^{(j)} \rangle$$
$$= 1 + \mathcal{O}(\lambda^{j+1}) \quad (8.32)$$

where $\mathcal{O}(\lambda^{j+1})$ denotes a correction of order λ^{j+1}. In particular, since $\langle \psi_n^{(0)} | \psi_n^{(0)} \rangle = 1$, we find from (8.32), to first order in λ,

$$\langle \psi_n^{(0)} | \psi_n^{(1)} \rangle + \langle \psi_n^{(1)} | \psi_n^{(0)} \rangle = 0 \quad (8.33\text{a})$$

or, using (8.31),

$$a_{nn}^{(1)} + a_{nn}^{(1)*} = 0 \quad (8.33\text{b})$$

so that the real part of $a_{nn}^{(1)}$ must vanish. Similarly, to second order in λ, the normalisation condition (8.32) gives

$$\langle \psi_n^{(0)} | \psi_n^{(2)} \rangle + \langle \psi_n^{(2)} | \psi_n^{(0)} \rangle + \langle \psi_n^{(1)} | \psi_n^{(1)} \rangle = 0. \quad (8.34\text{a})$$

Using (8.31), (8.19) and (8.3), this equation yields

$$a_{nn}^{(2)} + a_{nn}^{(2)*} + \sum_k |a_{nk}^{(1)}|^2 = 0 \quad (8.34\text{b})$$

which is an equation for the real part of $a_{nn}^{(2)}$. This procedure can be continued to higher orders in λ. Note, however, that the imaginary parts of the coefficients $a_{nn}^{(j)}$ cannot be determined from the normalisation condition (8.32). This remaining arbitrariness corresponds to the fact that without changing its normalisation we can multiply the perturbed wave function ψ_n by an arbitrary phase factor $\exp(i\alpha)$, where α is a real quantity which may depend on λ. We can therefore require without loss of generality that the imaginary parts of the coefficients $a_{nn}^{(j)}$ vanish, in which case we have, through second order in λ,

$$a_{nn}^{(1)} = 0, \qquad a_{nn}^{(2)} = -\frac{1}{2}\sum_{k \neq n}|a_{nk}^{(1)}|^2. \tag{8.35}$$

As a result of this discussion, the normalised perturbed wave function is given to second order by

$$\psi_n = \psi_n^{(0)} + \psi_n^{(1)} + \psi_n^{(2)} \tag{8.36a}$$

where

$$\psi_n^{(1)} = \sum_{l \neq n} \frac{H'_{ln}}{E_n^{(0)} - E_l^{(0)}} \psi_l^{(0)}, \tag{8.36b}$$

$$\psi_n^{(2)} = \sum_{l \neq n}\left[\sum_{k \neq n}\frac{H'_{lk}H'_{kn}}{(E_n^{(0)} - E_l^{(0)})(E_n^{(0)} - E_k^{(0)})} - \frac{H'_{nn}H'_{ln}}{(E_n^{(0)} - E_l^{(0)})^2}\right]\psi_l^{(0)}$$

$$-\frac{1}{2}\sum_{k \neq n}\frac{|H'_{kn}|^2}{(E_n^{(0)} - E_k^{(0)})^2}\psi_n^{(0)} \tag{8.36c}$$

and we have set $\lambda = 1$.

We have discussed above a possible choice of the coefficients $a_{nn}^{(j)}$ which ensures that the perturbed wave function ψ_n be normalised to a given order in λ. Another way of choosing the coefficients $a_{nn}^{(j)}$ which is often used in perturbation calculations is to require that

$$a_{nn}^{(j)} = 0, \qquad j \geqslant 1. \tag{8.37}$$

In this case it is clear that the first correction $\psi_n^{(1)}$ is identical to that found above, equation (8.36b), so that the perturbed wave function ψ_n is still normalised to unity to first order in λ. However, we see from (8.35) that this is not the case beyond first order. Nevertheless, if desired, we can always obtain a normalised perturbed eigenfunction at the end of the calculation (performed to a given order in λ) by multiplying ψ_n by a constant $N(\lambda)$ such that $\langle N(\lambda)\psi_n | N(\lambda)\psi_n \rangle = 1$.

Finally, we remark that if the expression (8.36b) for $\psi_n^{(1)}$ is substituted into (8.17) we retrieve the second-order energy correction $E_n^{(2)}$ given by (8.28). The third-order correction $E_n^{(3)}$ can be obtained in a similar way by using (8.18) and (8.36b). Remembering that $\langle \psi_n^{(0)} | \psi_n^{(1)} \rangle = a_{nn}^{(1)} = 0$, one finds (Problem 8.5) that

$$E_n^{(3)} = \sum_{k \neq n}\sum_{m \neq n}\frac{H'_{nk}H'_{km}H'_{mn}}{(E_n^{(0)} - E_k^{(0)})(E_n^{(0)} - E_m^{(0)})} - H'_{nn}\sum_{k \neq n}\frac{|H'_{kn}|^2}{(E_n^{(0)} - E_k^{(0)})^2}. \tag{8.38}$$

Perturbed harmonic oscillator

As a first example of the application of Rayleigh–Schrödinger perturbation theory to non-degenerate energy levels, we shall consider a problem which can be solved exactly, so that we may check our perturbative results against exact expressions. The unperturbed Hamiltonian is that of a linear harmonic oscillator

$$H_0 = \frac{p_x^2}{2m} + \frac{1}{2}kx^2, \qquad k > 0 \tag{8.39}$$

and the perturbation, proportional to x^2, is written as

$$H' = \frac{1}{2}k'x^2, \qquad k' > 0 \tag{8.40}$$

where we have set $\lambda = 1$ (in fact the parameter λ may be thought of as being absorbed in the quantity k'). The unperturbed energy levels are

$$E_n^{(0)} = \hbar\omega\left(n + \tfrac{1}{2}\right), \qquad n = 0, 1, 2, \ldots \tag{8.41}$$

with $\omega = (k/m)^{1/2}$. The corresponding unperturbed eigenfunctions are the linear harmonic oscillator wave functions $\psi_n(x)$ given by (4.168).

Since the full Hamiltonian

$$H = \frac{p_x^2}{2m} + \frac{1}{2}kx^2 + \frac{1}{2}k'x^2 \tag{8.42}$$

is that of a linear harmonic oscillator with a force constant $K = k + k'$, the perturbed eigenfunctions and eigenvalues can readily be obtained exactly by replacing k by $k + k'$ in the expressions of the unperturbed eigenfunctions and eigenenergies. In particular, the perturbed energy levels are given by

$$E_n = \hbar\omega\left(n + \frac{1}{2}\right)\left(1 + \frac{k'}{k}\right)^{1/2}, \qquad n = 0, 1, 2, \ldots. \tag{8.43}$$

Let us now calculate the perturbed energies through second order in perturbation theory. According to (8.15), the first-order correction to the energy of the nth state is

$$E_n^{(1)} = H'_{nn} = \tfrac{1}{2}k'(x^2)_{nn} \tag{8.44}$$

and from (8.28) we see that the second-order energy correction is

$$E_n^{(2)} = \frac{1}{4}(k')^2 \sum_{k \neq n} \frac{|(x^2)_{kn}|^2}{E_n^{(0)} - E_k^{(0)}} \tag{8.45}$$

where

$$(x^2)_{kn} = \langle \psi_k^{(0)} | x^2 | \psi_n^{(0)} \rangle \tag{8.46}$$

denotes the matrix element of x^2 between a pair of unperturbed linear harmonic oscillator wave functions $\psi_k^{(0)}(x)$ and $\psi_n^{(0)}(x)$.

The matrix elements (8.46) can be evaluated in a variety of ways (see Problem 5.13). The result is

$$(x^2)_{kn} = \frac{1}{2\alpha^2}[(n+1)(n+2)]^{1/2}, \qquad k = n+2$$

$$= \frac{1}{2\alpha^2}[2n+1], \qquad k = n$$

$$= \frac{1}{2\alpha^2}[n(n-1)]^{1/2}, \qquad k = n-2$$

$$= 0 \qquad \text{otherwise} \qquad (8.47)$$

where $\alpha = (mk/\hbar^2)^{1/4} = (m\omega/\hbar)^{1/2}$. Thus

$$E_n^{(1)} = \frac{1}{2}k'\frac{1}{2\alpha^2}(2n+1) = \hbar\omega\left(n+\frac{1}{2}\right)\left(\frac{1}{2}\frac{k'}{k}\right) \qquad (8.48)$$

and

$$E_n^{(2)} = \frac{1}{4}k'^2\left\{\frac{[(x^2)_{n-2,n}]^2}{2\hbar\omega} + \frac{[(x^2)_{n+2,n}]^2}{-2\hbar\omega}\right\}$$

$$= \hbar\omega\left(n+\frac{1}{2}\right)\left(-\frac{1}{8}\frac{k'^2}{k^2}\right) \qquad (8.49)$$

in agreement with the expansion of the exact result (8.43) through second order in k'/k. Note that this expansion converges only if $k'/k < 1$. As the ratio $E_n^{(j+1)}/E_n^{(j)}$ is independent of n, the condition for convergence does not depend on the size of the energy shift, which can be large for large n. This feature is peculiar to this example.

As a second example, let us consider the case of an anharmonic oscillator whose Hamiltonian is

$$H = \frac{p_x^2}{2m} + \frac{1}{2}kx^2 + ax^3 + bx^4. \qquad (8.50)$$

As seen from (4.128), the anharmonic terms ax^3 and bx^4 arise in correcting the approximation to a continuous potential well $W(x)$ given by the linear harmonic oscillator potential $kx^2/2$. In particular, such anharmonic corrections occur in the study of the vibrational spectra of molecules. We shall assume here that $b > 0$ since otherwise the potential energy would tend to $-\infty$ for $x \to \pm\infty$, with the consequence that the energy spectrum would be continuous and unbounded towards negative as well as positive energies.

The unperturbed Hamiltonian H_0 is still that of the linear harmonic oscillator (see (8.39)) and with $\lambda = 1$ we have

$$H' = ax^3 + bx^4. \qquad (8.51)$$

The first-order correction to the energy of the nth state is therefore

$$E_n^{(1)} = a(x^3)_{nn} + b(x^4)_{nn}$$
$$= \int_{-\infty}^{+\infty} (ax^3 + bx^4)|\psi_n^{(0)}(x)|^2 dx. \tag{8.52}$$

Now in Section 4.7 we showed that the linear harmonic oscillator wave functions (4.168) have a *definite* parity (even when n is even, odd when n is odd), so that $|\psi_n^{(0)}(x)|^2$ is always an *even* function of x. On the other hand, x^3 is an *odd* function of x. As a result, the diagonal matrix element $(x^3)_{nn}$ vanishes, and the term ax^3 does not contribute to $E_n^{(1)}$. The matrix element $(x^4)_{nn}$ is given by (Problem 5.13)

$$(x^4)_{nn} = \frac{3}{4\alpha^4}(2n^2 + 2n + 1) \tag{8.53}$$

so that the first-order energy shift is

$$E_n^{(1)} = b(x^4)_{nn} = \frac{3}{4}b\left(\frac{\hbar}{m\omega}\right)^2 (2n^2 + 2n + 1). \tag{8.54}$$

Note that for a fixed value of b the quantity $E_n^{(1)}$ grows rapidly with n. This is not surprising, since the higher excited states of the linear harmonic oscillator extend to larger and larger values of x (see Fig. 4.18), so that the perturbation bx^4 can have an increasingly important effect. Because the validity of the perturbation method requires that the magnitude of the correction term be small compared to the spacing between the unperturbed levels (given in the present case by $\hbar\omega$), we see from (8.54) that the higher the value of n the smaller are the values of the parameter b for which reliable results may be obtained from the perturbative result (8.54).

The second-order correction to the energy of the state n is

$$E_n^{(2)} = \sum_{k \neq n} \frac{|a(x^3)_{kn} + b(x^4)_{kn}|^2}{E_n^{(0)} - E_k^{(0)}}. \tag{8.55}$$

Using the results of Problem 5.13 for the required off-diagonal matrix elements of x^3 and x^4 one finds (Problem 8.8) that

$$E_n^{(2)} = -\frac{15}{4}\frac{a^2}{\hbar\omega}\left(\frac{\hbar}{m\omega}\right)^3 \left(n^2 + n + \frac{11}{30}\right)$$
$$-\frac{1}{8}\frac{b^2}{\hbar\omega}\left(\frac{\hbar}{m\omega}\right)^4 (34n^3 + 51n^2 + 59n + 21) \tag{8.56}$$

and we see that $|E_n^{(2)}|$ grows rapidly with n. Thus, as n increases, reliable results[1] will only be obtained from (8.56) for smaller and smaller values of a^2 and b^2.

---

[1] In fact, the perturbation series for this problem does not converge, but is an *asymptotic*, or *semi-convergent*, series. (See Matthews and Walker, 1973.)

Finally, we remark that the second-order corrections (8.45) and (8.55) could be calculated relatively easily because (a) the spectrum of the unperturbed Hamiltonian (the linear harmonic oscillator) is entirely discrete, and (b) only a small number of the required matrix elements of the perturbation are non-vanishing. When these conditions are not satisfied, the calculation of the second- (and higher) order corrections to the energy can become very difficult and must usually be performed approximately.

Gravitational energy shift in atomic hydrogen

As a final example of the application of perturbation theory for non-degenerate energy levels, let us consider an ordinary hydrogen atom (proton + electron) whose unperturbed Hamiltonian is

$$H_0 = \frac{\mathbf{p}^2}{2\mu} - \frac{e^2}{(4\pi\varepsilon_0)r} \tag{8.57}$$

where $\mu = mM_p/(m+M_p)$ is the reduced mass, m being the mass of the electron and M_p that of the proton. Now the proton and the electron interact not only through the electrostatic potential $-e^2/(4\pi\varepsilon_0 r)$, but also by means of the gravitational interaction. The perturbation H' due to the gravitational force is (with $\lambda = 1$)

$$H' = -G\frac{mM_p}{r} \tag{8.58}$$

where $G = 6.672 \times 10^{-11}$ N m² kg⁻² is the gravitational constant. To first order in perturbation theory, the energy shift of the ground (1s) state of atomic hydrogen (which is non-degenerate) due to this perturbation is

$$\begin{aligned} E_{1s}^{(1)} &= \langle\psi_{1s}|H'|\psi_{1s}\rangle \\ &= \frac{1}{\pi a_\mu^3}\int\left(-G\frac{mM_p}{r}\right)\exp(-2r/a_\mu)d\mathbf{r} \\ &= -\frac{GmM_p}{a_\mu} \end{aligned} \tag{8.59}$$

where $a_\mu = 4\pi\varepsilon_0\hbar^2/\mu e^2$. Since the unperturbed ground state energy of atomic hydrogen is given (see (7.114)) by

$$E_{1s}^{(0)} = -\frac{e^2}{(4\pi\varepsilon_0)2a_\mu} \tag{8.60}$$

the relative energy shift is

$$\frac{E_{1s}^{(1)}}{E_{1s}^{(0)}} = \frac{8\pi\varepsilon_0 GmM_p}{e^2} \simeq 8.8 \times 10^{-40} \tag{8.61}$$

which is a truly small number! Needless to say, it is not necessary to calculate higher-order corrections in the present case.

8.2 Time-independent perturbation theory for a degenerate energy level

Until now we have assumed that the perturbed eigenfunction ψ_n differs slightly from a given function $\psi_n^{(0)}$, a solution of the 'unperturbed' equation (8.2). When the unperturbed energy level $E_n^{(0)}$ is α-fold degenerate, there are *several* 'unperturbed' wave functions $\psi_{nr}^{(0)}$ ($r = 1, 2, \ldots, \alpha$) corresponding to this level, and we do not know *a priori* to which functions the perturbed eigenfunctions tend when $\lambda \to 0$. This implies that the treatment of Section 8.1 – in particular the basic expansion (8.8) – must be modified to deal with the degenerate case.

The α unperturbed wave functions $\psi_{nr}^{(0)}$ corresponding to the level $E_n^{(0)}$ are of course orthogonal to the unperturbed wave functions $\psi_l^{(0)}$ corresponding to *other* energy levels $E_l^{(0)} \neq E_n^{(0)}$. Although they need not be orthogonal among themselves, it is always possible to construct from linear combinations of them a new set of α unperturbed wave functions which are mutually orthogonal and normalised to unity. We may therefore assume without loss of generality that this has already been done, so that

$$\langle \psi_{nr}^{(0)} | \psi_{ns}^{(0)} \rangle = \delta_{rs} \quad (r, s = 1, 2, \ldots, \alpha). \tag{8.62}$$

Let us now introduce the correct zero-order functions $\chi_{nr}^{(0)}$ which yield the first term in the expansion of the exact wave functions ψ_{nr} in powers of λ. That is

$$\psi_{nr} = \chi_{nr}^{(0)} + \lambda \psi_{nr}^{(1)} + \lambda^2 \psi_{nr}^{(2)} + \cdots \tag{8.63}$$

We shall also write the perturbed energy E_{nr} as

$$E_{nr} = E_n^{(0)} + \lambda E_{nr}^{(1)} + \lambda^2 E_{nr}^{(2)} + \cdots \tag{8.64}$$

with $E_n^{(0)} \equiv E_{nr}^{(0)}$ ($r = 1, 2, \ldots, \alpha$) since the unperturbed level $E_n^{(0)}$ is α-fold degenerate. Substituting the expansions (8.63) and (8.64) in equation (8.4) and equating the coefficients of λ, we find that

$$H_0 \psi_{nr}^{(1)} + H' \chi_{nr}^{(0)} = E_n^{(0)} \psi_{nr}^{(1)} + E_{nr}^{(1)} \chi_{nr}^{(0)}. \tag{8.65}$$

Since the functions $\chi_{nr}^{(0)}$ are linear combinations of the unperturbed wave functions $\psi_{nr}^{(0)}$, we may write

$$\chi_{nr}^{(0)} = \sum_{s=1}^{\alpha} c_{rs} \psi_{ns}^{(0)}, \quad (r = 1, 2, \ldots, \alpha) \tag{8.66}$$

where the coefficients c_{rs} are to be determined. Similarly, expanding $\psi_{nr}^{(1)}$ in the basis set of the unperturbed wave functions, we have

$$\psi_{nr}^{(1)} = \sum_{k} \sum_{s} a_{nr,ks}^{(1)} \psi_{ks}^{(0)} \tag{8.67}$$

where the indices r and s refer explicitly to the degeneracy. Substituting the expressions (8.66) and (8.67) in equation (8.65) and using the fact that $H_0\psi_{ks}^{(0)} = E_k^{(0)}\psi_{ks}^{(0)}$, we find that

$$\sum_k \sum_s a_{nr,ks}^{(1)}(E_k^{(0)} - E_n^{(0)})\psi_{ks}^{(0)} + \sum_s c_{rs}(H' - E_{nr}^{(1)})\psi_{ns}^{(0)} = 0. \tag{8.68}$$

Premultiplying by $\psi_{nu}^{(0)*}$ and integrating over all coordinates, we obtain

$$\sum_k \sum_s a_{nr,ks}^{(1)}(E_k^{(0)} - E_n^{(0)})\langle\psi_{nu}^{(0)}|\psi_{ks}^{(0)}\rangle + \sum_s c_{rs}[H'_{nu,ns} - E_{nr}^{(1)}\delta_{us}] = 0,$$

$$(u = 1, 2, \ldots, \alpha) \tag{8.69}$$

where we have used (8.62) and written

$$H'_{nu,ns} \equiv \langle\psi_{nu}^{(0)}|H'|\psi_{ns}^{(0)}\rangle. \tag{8.70}$$

Since $\langle\psi_{nu}^{(0)}|\psi_{ks}^{(0)}\rangle = 0$ when $k \neq n$ and $E_k^{(0)} = E_n^{(0)}$ if $k = n$, we see that (8.69) reduces to

$$\sum_{s=1}^{\alpha} c_{rs}[H'_{nu,ns} - E_{nr}^{(1)}\delta_{us}] = 0, \qquad (u = 1, 2, \ldots, \alpha). \tag{8.71}$$

This is a linear, homogeneous system of equations for the α unknown quantities $c_{r1}, c_{r2}, \ldots, c_{r\alpha}$. A non-trivial solution is obtained if the determinant of the quantity in square brackets vanishes:

$$\det|H'_{nu,ns} - E_{nr}^{(1)}\delta_{us}| = 0, \qquad (s, u = 1, 2, \ldots, \alpha). \tag{8.72}$$

This equation, which is of degree α in $E_{nr}^{(1)}$, is often called a *secular* equation by analogy with similar equations occurring in classical mechanics. It has α real roots $E_{n1}^{(1)}, E_{n2}^{(1)}, \ldots, E_{n\alpha}^{(1)}$. If all these roots are *distinct* the degeneracy is completely removed to first order in the perturbation. On the other hand, if some (or all) roots of (8.72) are identical the degeneracy is only *partially* (or *not at all*) *removed*. The residual degeneracy may then either be removed in higher-order perturbation theory, or it may persist to all orders. The latter case occurs when the operators H_0 and H' share symmetry properties.

For a given value of r, the coefficients c_{rs} ($s = 1, 2, \ldots, \alpha$) which determine the 'correct' unperturbed zero-order wave function $\chi_{nr}^{(0)}$ via (8.66) may be obtained by substituting the value of $E_{nr}^{(1)}$ in the system (8.71) and solving for the coefficients $c_{r1}, c_{r2}, \ldots, c_{r\alpha}$, in terms of one of them. The last coefficient is then obtained (up to a phase) by requiring that the function $\chi_{nr}^{(0)}$ be normalised to unity. Clearly, this procedure does not lead to a unique result when two or more roots $E_{nr}^{(1)}$ of the secular equation (8.72) coincide, since in this case the degeneracy is not fully removed.

It is apparent from the foregoing discussion that the determination of the correct zero-order wave functions $\chi_{nr}^{(0)}(r = 1, 2, \ldots, \alpha)$ amounts to finding the proper orthonormal linear combinations of the original zero-order degenerate states $\psi_{nr}^{(0)}$, such that the matrix

$$\tilde{H}'_{nr,ns} = \langle \chi_{nr}^{(0)} | H' | \chi_{ns}^{(0)} \rangle, \qquad (r, s = 1, 2, \ldots, \alpha) \tag{8.73}$$

be diagonal with respect to the indices r and s. The diagonal elements of this matrix are equal to the corresponding first-order energy corrections

$$E_{nr}^{(1)} = \tilde{H}'_{nr,nr}, \qquad (r = 1, 2, \ldots, \alpha). \tag{8.74}$$

Once the correct zero-order wave functions $\chi_{nr}^{(0)}$ have been determined, the first-order correction $\psi_{nr}^{(1)}$ to the wave function and the second-order energy correction $E_{nr}^{(2)}$ can be obtained in a way similar to that followed in Section 8.1 for the non-degenerate case.

It is also interesting to note that if all the original off-diagonal matrix elements $H'_{nr,ns} = \langle \psi_{nr}^{(0)} | H' | \psi_{ns}^{(0)} \rangle, r \neq s$, vanish (i.e. if the degenerate states $\psi_{nr}^{(0)} (r = 1, 2, \ldots, \alpha)$ belonging to the level $E_n^{(0)}$ are not connected to first order), then the secular equation (8.72) takes the simple diagonal form

$$\begin{vmatrix} H'_{n1,n1} - E_{nr}^{(1)} & 0 & 0 & \ldots & 0 \\ 0 & H'_{n2,n2} - E_{nr}^{(1)} & 0 & \ldots & 0 \\ 0 & 0 & & & \vdots \\ \vdots & \vdots & & & 0 \\ 0 & 0 & \ldots & 0 & H'_{n\alpha,n\alpha} - E_{nr}^{(1)} \end{vmatrix} = 0. \tag{8.75}$$

In this case, we see from (8.66) and (8.71) that our initial unperturbed wave functions $\psi_{nr}^{(0)}$ are already the correct zero-order wave functions for the perturbation H'. Moreover, the roots $E_{nr}^{(1)}$ of equation (8.75) are given immediately by the formula

$$E_{nr}^{(1)} = H'_{nr,nr}, \qquad (r = 1, 2, \ldots, \alpha) \tag{8.76}$$

and upon comparison with (8.15) we see that the degeneracy plays no role in the analysis. This situation occurs whenever the unperturbed states can be uniquely specified in terms of a set of operators which all commute with the perturbation H'.

Doubly degenerate energy level

As a simple illustration of our general discussion, let us consider the case of an unperturbed energy level $E_n^{(0)}$ which is doubly degenerate. Dropping the index n for notational simplicity, we denote by $\psi_1^{(0)}$ and $\psi_2^{(0)}$ two linearly independent orthonormal zero order wave functions corresponding to this level. The system of linear homogeneous equation (8.71) reduces to a set of two equations which we write in matrix form as

$$\begin{pmatrix} H'_{11} - E_r^{(1)} & H'_{12} \\ H'_{21} & H'_{22} - E_r^{(1)} \end{pmatrix} \begin{pmatrix} c_{r1} \\ c_{r2} \end{pmatrix} = 0 \tag{8.77}$$

8.2 Perturbation theory for a degenerate energy level

with $H'_{ij} = \langle \psi_i^{(0)} | H' | \psi_j^{(0)} \rangle$ and $i, j = 1, 2$. The secular equation (8.72) reads

$$\begin{vmatrix} H'_{11} - E_r^{(1)} & H'_{12} \\ H'_{21} & H'_{22} - E_r^{(1)} \end{vmatrix} = 0 \tag{8.78}$$

and yields the two first-order energy corrections

$$E_1^{(1)} = \frac{1}{2}(H'_{11} + H'_{22}) + \frac{1}{2}[(H'_{11} - H'_{22})^2 + 4|H'_{12}|^2]^{1/2} \tag{8.79a}$$

and

$$E_2^{(1)} = \frac{1}{2}(H'_{11} + H'_{22}) - \frac{1}{2}[(H'_{11} - H'_{22})^2 + 4|H'_{12}|^2]^{1/2} \tag{8.79b}$$

where we have used the fact that $H'_{21} = H'^*_{12}$.

Given the above values of $E_r^{(1)}(r = 1, 2)$, we may now return to the system of two homogeneous equations (8.77), which gives

$$\frac{c_{r1}}{c_{r2}} = -\frac{H'_{12}}{H'_{11} - E_r^{(1)}} = -\frac{H'_{22} - E_r^{(1)}}{H'_{21}}, \quad r = 1, 2. \tag{8.80}$$

The normalised correct zero-order wave function for $r = 1$ is therefore given by

$$\chi_1^{(0)} = c_{11} \psi_1^{(0)} + c_{12} \psi_2^{(0)} \tag{8.81a}$$

and that corresponding to $r = 2$ is given by

$$\chi_2^{(0)} = c_{21} \psi_1^{(0)} + c_{22} \psi_2^{(0)} \tag{8.81b}$$

where the coefficients c_{rs} (with $r, s = 1, 2$) are obtained from (8.80) and the normalisation condition

$$|c_{r1}|^2 + |c_{r2}|^2 = 1, \quad r = 1, 2. \tag{8.82}$$

Choosing the phase of c_{r1} so that this coefficient be real and positive, one obtains (Problem 8.9)

$$c_{r1} = \frac{1}{\sqrt{2}} \left[1 - (-1)^r \frac{H'_{11} - H'_{22}}{[(H'_{11} - H'_{22})^2 + 4|H'_{12}|^2]^{1/2}} \right]^{1/2} \tag{8.83a}$$

and

$$c_{r2} = (-1)^{r+1} \frac{|H'_{12}|}{H'_{12}} \frac{1}{\sqrt{2}} \left[1 + (-1)^r \frac{H'_{11} - H'_{22}}{[(H'_{11} - H'_{22})^2 + 4|H'_{12}|^2]^{1/2}} \right]^{1/2}. \tag{8.83b}$$

An interesting special case is that for which

$$H'_{11} = H'_{22} = 0,$$
$$H'_{12} = H'^*_{21} \neq 0. \tag{8.84}$$

We then obtain from (8.79) the simple results

$$E_1^{(1)} = +|H'_{12}|, \quad E_2^{(1)} = -|H'_{12}| \tag{8.85}$$

and we see from (8.80) and (8.85) that

$$\frac{c_{11}}{c_{12}} = -\frac{c_{21}}{c_{22}} = \frac{H'_{12}}{|H'_{12}|}. \tag{8.86}$$

Note that the original degenerate states $\psi_1^{(0)}$ and $\psi_2^{(0)}$ are fully mixed, since $|c_{r1}| = |c_{r2}|$.

Another particular case arises when the off-diagonal matrix elements vanish

$$H'_{12} = H'^{*}_{21} = 0 \tag{8.87}$$

so that the two degenerate states $\psi_1^{(0)}$ and $\psi_2^{(0)}$ are not connected to first order. We then have

$$E_1^{(1)} = H'_{11}, \qquad E_2^{(1)} = H'_{22} \tag{8.88}$$

which is of course a special case of (8.76).

Fine structure of hydrogenic atoms

As a second example of degenerate perturbation theory, we shall calculate the corrections to the Schrödinger (Bohr) energy levels of one-electron atoms due to *relativistic effects*. We assume that the nucleus is 'infinitely heavy' so that the reduced mass μ coincides with m, the mass of the electron. The unperturbed Hamiltonian is then

$$H_0 = \frac{\mathbf{p}^2}{2m} + V(r) \tag{8.89}$$

where $V(r)$ is the Coulomb potential

$$V(r) = -\frac{Ze^2}{(4\pi\varepsilon_0)r}. \tag{8.90}$$

The perturbation H' is given by[2]

$$H' = H'_1 + H'_2 + H'_3 \tag{8.91}$$

where

$$H'_1 = -\frac{\mathbf{p}^4}{8m^3c^2} \tag{8.92}$$

$$H'_2 = \frac{1}{2m^2c^2}\frac{1}{r}\frac{dV}{dr}\mathbf{L}\cdot\mathbf{S} \tag{8.93}$$

and

$$H'_3 = \frac{\pi\hbar^2}{2m^2c^2}\left(\frac{Ze^2}{4\pi\varepsilon_0}\right)\delta(\mathbf{r}). \tag{8.94}$$

---

[2] The Hamiltonian H' can be obtained by starting from the Dirac relativistic wave equation for the electron, and keeping terms up to order v^2/c^2 (where v is the electron velocity and c the velocity of light) in the Dirac Hamiltonian. This will be shown in Chapter 15.

8.2 Perturbation theory for a degenerate energy level

The first term, H'_1, is readily seen to be a *relativistic correction to the kinetic energy* (Problem 8.10). The second term, H'_2, involving the scalar product **L.S** of the orbital angular momentum **L** and the spin operator **S** of the electron, is called the *spin–orbit* term. Its physical origin is the interaction between the intrinsic magnetic dipole moment of the electron, due to its spin, and the internal magnetic field of the atom, related to the electron's orbital angular momentum. The third term, H'_3, is a relativistic correction to the potential energy, called the *Darwin term* which is proportional to the delta function $\delta(\mathbf{r})$ and hence acts only at the origin.

Before we proceed to the evaluation of the energy shifts due to these three terms by using perturbation theory, we remark that the Schrödinger theory of one-electron atoms discussed in Chapter 7 does not include the spin of the electron. In order to calculate corrections involving the spin operator, such as those arising from H'_2, we shall thus start from the 'unperturbed' equation

$$H_0 \psi^{(0)}_{nlm_l m_s} = E_n^{(0)} \psi^{(0)}_{nlm_l m_s} \tag{8.95}$$

where $E_n^{(0)}$ are the Schrödinger energy eigenvalues (7.114) with $\mu = m$. The zero-order wave functions $\psi^{(0)}_{nlm_l m_s}$ are the Pauli two-component hydrogenic spin-orbitals

$$\psi^{(0)}_{nlm_l m_s} = \psi^{(0)}_{nlm_l}(\mathbf{r}) \chi_{\frac{1}{2}, m_s} \tag{8.96}$$

where the functions $\psi^{(0)}_{nlm_l}(\mathbf{r})$ are the hydrogenic Schrödinger eigenfunctions such that

$$H_0 \psi^{(0)}_{nlm_l}(\mathbf{r}) = E_n^{(0)} \psi^{(0)}_{nlm_l}(\mathbf{r}) \tag{8.97}$$

and $\chi_{\frac{1}{2}, m_s}$ are the two-component spin-1/2 eigenfunctions, with $m_s = \pm 1/2$. We remark that the spin-orbitals $\psi^{(0)}_{nlm_l m_s}$ are simultaneous eigenfunctions of the operators H_0, $\mathbf{L}^2$, L_z, $\mathbf{S}^2$ and S_z with eigenvalues $E_n^{(0)}$, $l(l+1)\hbar^2$, $m_l\hbar$, $(3/4)\hbar^2$ and $m_s\hbar$, respectively. To each energy level $E_n^{(0)}$ correspond $2n^2$ spin-orbitals $\psi^{(0)}_{nlm_l m_s}$, the factor of two being due to the two possible values $m_s = \pm 1/2$ of the quantum number m_s.

We shall now calculate the first-order energy corrections to the energy levels $E_n^{(0)}$ due to the three terms (8.92)–(8.94), using the Pauli spin-orbitals $\psi^{(0)}_{nlm_l m_s}$ as our zero-order wave functions.

Energy shift due to the term H'_1 (relativistic correction to the kinetic energy)

Since the unperturbed energy level $E_n^{(0)}$ is $2n^2$-degenerate, we should use degenerate perturbation theory. However, we first note that H'_1 does not act on the spin variable. Furthermore, it commutes with the components of the orbital angular momentum so that the degenerate states belonging to the level $E_n^{(0)}$ are not connected to first order by H'_1. Hence, according to (8.76), the first-order energy correction ΔE_1 due to H'_1

is given by

$$\Delta E_1 = \left\langle \psi_{nlm_lm_s}^{(0)} \left| -\frac{\mathbf{p}^4}{8m^3c^2} \right| \psi_{nlm_lm_s}^{(0)} \right\rangle$$

$$= \left\langle \psi_{nlm_l}^{(0)} \left| -\frac{\mathbf{p}^4}{8m^3c^2} \right| \psi_{nlm_l}^{(0)} \right\rangle$$

$$= -\frac{1}{2mc^2} \langle \psi_{nlm_l}^{(0)} | T^2 | \psi_{nlm_l}^{(0)} \rangle \tag{8.98}$$

where $T = \mathbf{p}^2/2m$ is the kinetic energy operator. From (8.89) and (8.90), we have

$$T = H_0 - V(r) = H_0 + \frac{Ze^2}{(4\pi\varepsilon_0)r} \tag{8.99}$$

so that

$$\Delta E_1 = -\frac{1}{2mc^2} \left\langle \psi_{nlm_l}^{(0)} \left| \left(H_0 + \frac{Ze^2}{(4\pi\varepsilon_0)r}\right)\left(H_0 + \frac{Ze^2}{(4\pi\varepsilon_0)r}\right) \right| \psi_{nlm_l}^{(0)} \right\rangle$$

$$= -\frac{1}{2mc^2}\left[(E_n^{(0)})^2 + 2E_n^{(0)}\left(\frac{Ze^2}{4\pi\varepsilon_0}\right)\left\langle\frac{1}{r}\right\rangle_{nlm_l} + \left(\frac{Ze^2}{4\pi\varepsilon_0}\right)^2\left\langle\frac{1}{r^2}\right\rangle_{nlm_l}\right] \tag{8.100}$$

where we have used (8.97). Now $E_n^{(0)}$ is given by (7.114) with $\mu = m$, and the required average values of r^{-1} and r^{-2} can be obtained from the results of Problem 7.16, in which μ is set equal to m (so that $a_\mu = a_0$, the first Bohr radius). Hence

$$\Delta E_1 = -\frac{1}{2mc^2}\left\{\left[\frac{mc^2(Z\alpha)^2}{2n^2}\right]^2 - 2\left(\frac{Ze^2}{4\pi\varepsilon_0}\right)\frac{mc^2(Z\alpha)^2}{2n^2}\frac{Z}{a_0n^2}\right.$$

$$\left. + \left(\frac{Ze^2}{4\pi\varepsilon_0}\right)^2 \frac{Z^2}{a_0^2 n^3(l+1/2)}\right\}$$

$$= \frac{1}{2}mc^2 \frac{(Z\alpha)^2}{n^2}\frac{(Z\alpha)^2}{n^2}\left[\frac{3}{4} - \frac{n}{l+1/2}\right]$$

$$= -E_n^{(0)}\frac{(Z\alpha)^2}{n^2}\left[\frac{3}{4} - \frac{n}{l+1/2}\right] \tag{8.101}$$

where $\alpha = e^2/(4\pi\varepsilon_0\hbar c) \simeq 1/137$ is the fine-structure constant.

Energy shift due to the term H_2' (spin–orbit term)

Let us first rewrite this term as

$$H_2' = \xi(r)\mathbf{L}\cdot\mathbf{S} \tag{8.102}$$

where the quantity $\xi(r)$ is given from (8.90) and (8.93) by

$$\xi(r) = \frac{1}{2m^2c^2}\frac{1}{r}\frac{dV}{dr}$$
$$= \frac{1}{2m^2c^2}\left(\frac{Ze^2}{4\pi\varepsilon_0}\right)\frac{1}{r^3}. \tag{8.103}$$

Because the operator $\mathbf{L}^2$ does not act on the spin variable, and commutes with the components of $\mathbf{L}$ and with any function of the radial variable r, we see from (8.102) that $\mathbf{L}^2$ commutes with H_2'. Thus the perturbation H_2' does not connect states with different values of the orbital angular momentum quantum number l. Since for a fixed value of n and l there are $2(2l+1)$ degenerate eigenstates of H_0 ($2l+1$ allowed values of m_l given by $m_l = -l, -l+1, \ldots, +l$, and a factor of two due to the two spin states), we see that the calculation of the energy shift arising from H_2' requires the diagonalisation of $2(2l+1) \times 2(2l+1)$ sub-matrices.

This diagonalisation can be greatly simplified by making a judicious choice of the set of zero-order wave functions, the best choice being obviously that for which $\mathbf{L}.\mathbf{S}$ is diagonal. Now, the zero-order wave functions $\psi_{nlm_lm_s}^{(0)}$ given by (8.96) are simultaneous eigenfunctions of $H_0, \mathbf{L}^2, L_z, \mathbf{S}^2$ and S_z and hence are not convenient since $\mathbf{L}.\mathbf{S}$ does not commute with L_z or S_z. However, we shall now show that adequate zero-order wave functions may be obtained by forming appropriate linear combinations of the functions $\psi_{nlm_lm_s}^{(0)}$. To this end, we introduce the total angular momentum of the electron

$$\mathbf{J} = \mathbf{L} + \mathbf{S} \tag{8.104}$$

and since

$$\mathbf{J}^2 = \mathbf{L}^2 + 2\mathbf{L}.\mathbf{S} + \mathbf{S}^2 \tag{8.105}$$

we may write

$$\mathbf{L}.\mathbf{S} = \tfrac{1}{2}(\mathbf{J}^2 - \mathbf{L}^2 - \mathbf{S}^2). \tag{8.106}$$

Consider now wave functions $\psi_{nljm_j}^{(0)}$ which are eigenstates of the operators H_0, $\mathbf{L}^2, \mathbf{S}^2, \mathbf{J}^2$ and J_z, the corresponding eigenvalues being $E_n^{(0)}$, $l(l+1)\hbar^2$, $(3/4)\hbar^2$, $j(j+1)\hbar^2$ and $m_j\hbar$. Using our study of the addition of angular momenta made in Section 6.10, and remembering that $s = 1/2$ in the present case, we see that the allowed values of the total angular momentum quantum number j are

$$\begin{aligned} j &= l \pm 1/2, & l \neq 0 \\ j &= 1/2, & l = 0 \end{aligned} \tag{8.107}$$

and that the corresponding magnetic quantum number m_j can take on the $(2j+1)$ allowed values

$$m_j = -j, -j+1, \ldots, +j. \tag{8.108}$$

Moreover, we can construct the functions $\psi_{nljm_j}^{(0)}$ from linear combinations of the functions $\psi_{nlm_lm_s}^{(0)}$. Using (6.295), we obtain

$$\psi_{nljm_j}^{(0)} = \sum_{m_l m_s} \langle l s m_l m_s | j m_j \rangle \psi_{nlm_l m_s}^{(0)}, \qquad s = 1/2 \tag{8.109}$$

where $\langle l s m_l m_s | j m_j \rangle$ are Clebsch–Gordan coefficients.

Since **L**.**S** commutes with $\mathbf{L}^2, \mathbf{S}^2, \mathbf{J}^2$ and J_z it is clear that the new zero-order wave functions $\psi_{nljm_j}^{(0)}$ form a satisfactory basis set in which the operator **L**.**S** (and hence the perturbation H_2') is diagonal. From (8.102) and (8.106) we see that for $l \neq 0$ the energy shift due to the term H_2' is given by

$$\Delta E_2 = \langle \psi_{nljm_j}^{(0)} | \tfrac{1}{2} \xi(r) [\mathbf{J}^2 - \mathbf{L}^2 - \mathbf{S}^2] | \psi_{nljm_j}^{(0)} \rangle$$

$$= \frac{\hbar^2}{2} \langle \xi(r) \rangle [j(j+1) - l(l+1) - \tfrac{3}{4}] \tag{8.110}$$

where $\langle \xi(r) \rangle$ denotes the average value of $\xi(r)$ in the state $\psi_{nljm_j}^{(0)}$. Using (8.103) and the average value of r^{-3} calculated in Problem 7.16 (with $a_\mu = a_0$), we have

$$\langle \xi(r) \rangle = \frac{1}{2m^2c^2} \left(\frac{Ze^2}{4\pi\varepsilon_0} \right) \left\langle \frac{1}{r^3} \right\rangle$$

$$= \frac{1}{2m^2c^2} \left(\frac{Ze^2}{4\pi\varepsilon_0} \right) \frac{Z^3}{a_0^3 n^3 l(l+1/2)(l+1)}. \tag{8.111}$$

Hence, for $l \neq 0$, we find from (8.110) and (8.111) that

$$\Delta E_2 = \frac{mc^2(Z\alpha)^4}{4n^3 l(l+1/2)(l+1)} \times \begin{cases} l & \text{for } j = l+1/2 \\ -l-1 & \text{for } j = l-1/2 \end{cases}$$

$$= -E_n^{(0)} \frac{(Z\alpha)^2}{2nl(l+1/2)(l+1)} \times \begin{cases} l & \text{for } j = l+1/2 \\ -l-1 & \text{for } j = l-1/2 \end{cases} \tag{8.112}$$

For $l = 0$ the spin–orbit interaction (8.102) vanishes so that $\Delta E_2 = 0$ in that case.

Energy shift due to the term H_3' (Darwin term)

This term does not act on the spin variable and commutes with the operators $\mathbf{L}^2$ and L_z. Moreover, since H_3' acts only at the origin and the hydrogenic wave functions vanish at $r = 0$ when $l \neq 0$, we have only to consider the case $l = 0$. Calling ΔE_3

the corresponding energy correction and using the result (7.145), we find that

$$\Delta E_3 = \frac{\pi \hbar^2}{2m^2 c^2} \left(\frac{Ze^2}{4\pi \varepsilon_0} \right) \langle \psi_{n00} | \delta(\mathbf{r}) | \psi_{n00} \rangle$$

$$= \frac{\pi \hbar^2}{2m^2 c^2} \left(\frac{Ze^2}{4\pi \varepsilon_0} \right) |\psi_{n00}(0)|^2$$

$$= \frac{1}{2} mc^2 \frac{(Z\alpha)^2}{n^2} \frac{(Z\alpha)^2}{n}$$

$$= -E_n^{(0)} \frac{(Z\alpha)^2}{n}, \qquad l = 0. \tag{8.113}$$

We may now combine the effects of the three terms H_1', H_2' and H_3' to obtain the total energy shift $\Delta E = \Delta E_1 + \Delta E_2 + \Delta E_3$ due to relativistic corrections. From (8.101), (8.112) and (8.113), we have for all l

$$\Delta E_{nj} = -\frac{1}{2} mc^2 \frac{(Z\alpha)^2}{n^2} \frac{(Z\alpha)^2}{n^2} \left(\frac{n}{j+1/2} - \frac{3}{4} \right)$$

$$= E_n^{(0)} \frac{(Z\alpha)^2}{n^2} \left(\frac{n}{j+1/2} - \frac{3}{4} \right) \tag{8.114}$$

where the subscripts nj indicate that the correction depends on both the principal quantum number n and the total angular momentum quantum number j, with $j = 1/2, 3/2, \ldots, n - 1/2$. Note that to each value of j correspond two possible values of l given by $l = j \pm 1/2$, except for $j = n - 1/2$ where one can only have $l = j - 1/2 = n - 1$. We also remark that although the three separate contributions ΔE_1, ΔE_2 and ΔE_3 depend on l, the total energy shift ΔE_{nj} does not.

Adding the relativistic correction ΔE_{nj} to the non-relativistic energies $E_n^{(0)}$, we find that the energy levels of one-electron atoms are now given (neglecting reduced mass effects) by

$$E_{nj} = E_n^{(0)} \left[1 + \frac{(Z\alpha)^2}{n^2} \left(\frac{n}{j+1/2} - \frac{3}{4} \right) \right] \tag{8.115}$$

so that the binding energy $|E_{nj}|$ of the electron is slightly increased with respect to the non-relativistic value $|E_n^{(0)}|$, the absolute value $|\Delta E_{nj}|$ of the energy shift becoming smaller as n or j increases, and larger as Z increases. In fact, the formula (8.115) will be shown in Section 15.5 to agree with the result obtained by solving exactly the Dirac equation for the Coulomb potential (8.90) if terms of higher order in $(Z\alpha)^2$ are neglected. The exact Dirac result also yields energy levels which depend only on the quantum numbers n and j.

Remembering that the non-relativistic energy levels $E_n^{(0)}$ are $2n^2$ times degenerate (the factor of two arising from the spin) we see that if relativistic effects are taken into account this degeneracy is partly removed. Indeed, we see from (8.115) that a non-relativistic energy level $E_n^{(0)}$ depending only on the principal quantum number n splits into n different levels E_{nj}, one for each value $j = 1/2, 3/2, \ldots, n - 1/2$ of

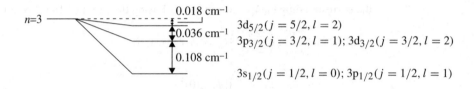

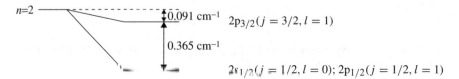

Figure 8.1 Fine structure splitting of the $n = 2$ and $n = 3$ energy levels of the hydrogen atom. The non-relativistic levels are shown on the left and the split levels on the right. For clarity the scales for the two cases $n = 2$ and $n = 3$ are different.

the total angular momentum quantum number j. This splitting is called *fine structure splitting* and the n levels $j = 1/2, 3/2, \ldots, n - 1/2$ are said to form a *fine-structure multiplet*. We also see from (8.115) that the dimensionless constant $\alpha \simeq 1/137$ controls the scale of the splitting; it is for this reason that α has been called the *fine-structure constant*.

The fine-structure splitting of the energy levels corresponding to $n = 2$ and 3 is illustrated in Fig. 8.1 for the case of the hydrogen atom. We have used in that figure the spectroscopic notation nl_j (with the usual convention associating the letters s, p, d, ..., with the values $l = 0, 1, 2, \ldots$, and an additional subscript for the value of j to distinguish the various spectral terms).

It is worth stressing that in Dirac's theory two states having the same value of the pair of quantum numbers (n, j) but with values of l such that $l = j \pm 1/2$ have the same energy. In reality, this degeneracy of the levels with $l = j \pm 1/2$ is removed by small quantum electrodynamics effects, known as *radiative corrections*, which are responsible for an additional shift of the energy levels. The existence of this energy shift, called the *Lamb shift*, was demonstrated experimentally in 1947 by W. E. Lamb and R. C. Retherford. In atomic hydrogen the energy difference between the $2s_{1/2}$ and $2p_{1/2}$ energy levels is about 1058 MHz, which is roughly 10 times less than the fine-structure separation between the $2p_{1/2}$ and $2p_{3/2}$ levels.

Quasi-degenerate states

We now turn to the case for which some unperturbed energy levels are *nearly* (but not completely) *degenerate*. For example, consider a hydrogenic atom to which some external perturbation is applied. We have just seen that when relativistic effects are taken into account a hydrogenic energy level $E_n^{(0)}$ splits into n levels E_{nj} which have slightly different energies. Moreover, even two levels with the same quantum numbers n and j but with values of $l = j \pm 1/2$ are separated by a tiny energy difference (the Lamb shift). It is therefore of interest to examine how the perturbative techniques we have developed can be adapted to this special situation.

We shall assume for simplicity that we are dealing with *two* nearly degenerate states whose unperturbed wave functions are $\psi_1^{(0)}$ and $\psi_2^{(0)}$; the corresponding energies are given respectively by

$$E_1^{(0)} = E^{(0)} + \varepsilon, \qquad E_2^{(0)} = E^{(0)} - \varepsilon, \qquad \varepsilon > 0 \tag{8.116}$$

and hence differ by the small amount 2ε. We want to solve the Schrödinger equation

$$(H_0 + \lambda H')\psi_r = E_r \psi_r, \qquad r = 1, 2. \tag{8.117}$$

Let us expand the unknown wave functions ψ_r in the basis set of the unperturbed eigenfunctions

$$\begin{aligned}\psi_r &= \sum_k a_{rk} \psi_k^{(0)} \\ &= a_{r1}\psi_1^{(0)} + a_{r2}\psi_2^{(0)} + \sum_{k \neq 1,2} a_{rk}\psi_k^{(0)}, \qquad r = 1, 2.\end{aligned} \tag{8.118}$$

From our study of degenerate perturbation theory, we know that if the two states $\psi_1^{(0)}$ and $\psi_2^{(0)}$ were degenerate (so that $\varepsilon = 0$) the correct zero-order wave functions $\chi_1^{(0)}$ and $\chi_2^{(0)}$ would be obtained by forming linear combinations of $\psi_1^{(0)}$ and $\psi_2^{(0)}$ (see (8.81)). We therefore expect that such linear combinations will again play an important role in the present case. Indeed, if we substitute (8.118) into (8.117) and use the fact that $H_0 \psi_k^{(0)} = E_k^{(0)} \psi_k^{(0)}$, we obtain

$$\sum_k a_{rk} E_k^{(0)} \psi_k^{(0)} + \lambda \sum_k a_{rk} H' \psi_k^{(0)} = E_r \sum_k a_{rk} \psi_k^{(0)}, \qquad r = 1, 2. \tag{8.119}$$

Pre-multiplying by $\psi_l^{(0)*}$, integrating over all coordinates and using the fact that $\langle \psi_l^{(0)} | \psi_k^{(0)} \rangle = \delta_{lk}$, we find that

$$a_{rl}(E_l^{(0)} - E_r) = -\lambda \sum_k a_{rk} H'_{lk}, \qquad r = 1, 2 \tag{8.120}$$

where $H'_{lk} = \langle \psi_l^{(0)} | H' | \psi_k^{(0)} \rangle$, as usual. Although the parameter λ could be absorbed into H' (i.e. set equal to one), it will be convenient in what follows to write it explicitly.

Let us rewrite the equations (8.120) in more detail, remembering to treat the two states $\psi_1^{(0)}$ and $\psi_2^{(0)}$ on an equal footing. For $l = 1$ we have

$$a_{r1}(E_1^{(0)} + \lambda H'_{11} - E_r) + \lambda a_{r2} H'_{12} = -\lambda \sum_{k \neq 1,2} a_{rk} H'_{1k} \tag{8.121a}$$

and for $l = 2$ we obtain

$$\lambda a_{r1} H'_{21} + a_{r2}(E_2^{(0)} + \lambda H'_{22} - E_r) = -\lambda \sum_{k \neq 1,2} a_{rk} H'_{2k} \tag{8.121b}$$

while for $l \neq 1, 2$ we see that

$$a_{rl}(E_l^{(0)} - E_r) = -\lambda a_{r1} H'_{l1} - \lambda a_{r2} H'_{l2} - \lambda \sum_{k \neq 1,2} a_{rk} H'_{lk}. \tag{8.121c}$$

Since the only small energy difference is $E_1^{(0)} - E_2^{(0)}$, it is apparent from these equations that the right-hand side of the first two equations is of order λ^2 and hence can be neglected to first order in λ. In this approximation we therefore obtain a system of two homogeneous equations which we write in matrix form as

$$\begin{pmatrix} E_1^{(0)} + \lambda H'_{11} - E_r & \lambda H'_{12} \\ \lambda H'_{21} & E_2^{(0)} + \lambda H'_{22} - E_r \end{pmatrix} \begin{pmatrix} a_{r1} \\ a_{r2} \end{pmatrix} = 0 \tag{8.122}$$

Expressing that the determinant of this pair of equations vanishes, we obtain the two energy values

$$E_r = \tfrac{1}{2}[E_1^{(0)} + E_2^{(0)} + \lambda(H'_{11} + H'_{22})] \pm \tfrac{1}{2}\{[E_1^{(0)} - E_2^{(0)} + \lambda(H'_{11} - H'_{22})]^2$$
$$+ 4\lambda^2 |H'_{12}|^2\}^{1/2} \tag{8.123}$$

with the plus sign corresponding to $r = 1$ and the minus sign to $r = 2$. We also see from (8.122) that

$$\frac{a_{r1}}{a_{r2}} = -\frac{\lambda H'_{12}}{E_1^{(0)} + \lambda H'_{11} - E_r} = -\frac{E_2^{(0)} + \lambda H'_{22} - E_r}{\lambda H'_{21}}, \qquad r = 1, 2. \tag{8.124}$$

Let us examine the result (8.123) more closely. In the limit $\lambda \to 0$ we have of course $E_1 = E_1^{(0)}$ and $E_2 = E_2^{(0)}$. We also remark that if we had applied non-degenerate first-order perturbation theory to each of the two levels, we would have found the energy values

$$E_1 = E_1^{(0)} + \lambda H'_{11}, \qquad E_2 = E_2^{(0)} + \lambda H'_{22}. \tag{8.125}$$

Now, defining the average of these first-order energies,

$$\bar{E} = \tfrac{1}{2}(E_1^{(0)} + \lambda H'_{11} + E_2^{(0)} + \lambda H'_{22}) \tag{8.126a}$$

and the first-order energy difference

$$\Delta E = E_1^{(0)} + \lambda H'_{11} - (E_2^{(0)} + \lambda H'_{22})$$
$$= 2\varepsilon + \lambda(H'_{11} - H'_{22}) \tag{8.126b}$$

we may recast equation (8.123) in the form

$$E_r = \bar{E} \pm \left\{ \left(\frac{\Delta E}{2}\right)^2 + \lambda^2 |H'_{12}|^2 \right\}^{1/2}, \qquad r = 1, 2. \tag{8.127}$$

This result exhibits several interesting features. First of all, we note that the two energy levels E_r are equally spaced about their first-order average $\bar{E}$ (see Fig. 8.2). Second, the effect of the term $\lambda^2 |H'_{12}|^2$ is always to *increase* the relative separation between the two levels E_r. As a result, this term prevents these two energy levels E_r

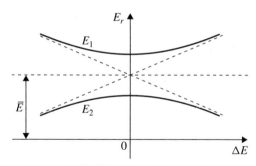

Figure 8.2 Illustration of the result (8.127). The solid curves show the two quasi-degenerate energy levels E_1 and E_2, plotted versus the first-order energy difference ΔE. The dashed curves correspond to the non-degenerate first-order results (8.125), which can be obtained from (8.127) by neglecting the quantity $\lambda^2 |H'_{12}|^2$.

from becoming exactly degenerate, and hence to cross. Third, if the quantity $\lambda^2 |H'_{12}|^2$ can be neglected with respect to $(\Delta E/2)^2$, we retrieve the non-degenerate first-order results (8.125). On the contrary, when $(\Delta E/2)^2 \ll \lambda^2 |H'_{12}|^2$, the two energy levels are given approximately by

$$E_r = \bar{E} \pm \lambda |H'_{12}|, \qquad r = 1, 2. \tag{8.128}$$

Finally, if the two unperturbed energy levels $E_1^{(0)}$ and $E_2^{(0)}$ are exactly degenerate (so that $\varepsilon = 0$) the expression (8.127) (in which λ is set equal to unity) correctly reduces to the first-order result (8.79) which we derived for a doubly degenerate level.

The case for which $H'_{11} = H'_{22} = 0$ is particularly simple. Then $\bar{E} = (E_1^{(0)} + E_2^{(0)})/2 = E^{(0)}$, $\Delta E/2 = \varepsilon$ and equation (8.127) yields

$$E_r = E^{(0)} \pm \{\varepsilon^2 + \lambda^2 |H'_{12}|^2\}^{1/2}, \qquad r = 1, 2. \tag{8.129}$$

Thus for $\varepsilon^2 \gg \lambda^2 |H'_{12}|^2$ we find the 'non-degenerate' results $E_r = E^{(0)} \pm \varepsilon$, while for $\varepsilon^2 \ll \lambda^2 |H'_{12}|^2$ we obtain the energy values $E_r = E^{(0)} \pm \lambda |H'_{12}|$, which upon setting $\lambda = 1$ are seen to be identical with the result derived by using degenerate perturbation theory (see (8.85)).

8.3 The variational method

In this section we shall discuss another method, known as the *variational method*, which is very useful in obtaining approximately the bound-state energies and wave functions of a time-independent Hamiltonian H. We denote by E_n the eigenvalues of this Hamiltonian and by ψ_n the corresponding orthonormal eigenfunctions, and assume that H has at least one bound state. Let ϕ be an arbitrary square-integrable function, and let $E[\phi]$ be the functional

$$E[\phi] = \frac{\langle \phi | H | \phi \rangle}{\langle \phi | \phi \rangle}$$

$$= \frac{\int \phi^* H \phi \, d\tau}{\int \phi^* \phi \, d\tau} \tag{8.130}$$

where the integration extends over the full range of all the coordinates of the system.

It is clear that if the function ϕ is identical to one of the exact eigenfunctions ψ_n of H, so that $H\phi = E_n \phi$, then $E[\phi]$ will be identical to the corresponding exact eigenvalue E_n. Indeed, in this case we have from (8.130)

$$E[\phi] = \frac{\langle \phi | H | \phi \rangle}{\langle \phi | \phi \rangle} = \frac{E_n \langle \phi | \phi \rangle}{\langle \phi | \phi \rangle} = E_n. \tag{8.131}$$

Moreover, we shall now show that any function ϕ for which the functional $E[\phi]$ is stationary is an eigenfunction of the discrete spectrum of H. That is, if ϕ and ψ_n differ by an arbitrary infinitesimal variation $\delta \phi$,

$$\phi = \psi_n + \delta \phi \tag{8.132}$$

then the corresponding first-order variation of $E[\phi]$ vanishes:

$$\delta E = 0 \tag{8.133}$$

and the eigenfunctions of H are solutions of the variational equation (8.133).

To prove this statement, we first note that upon clearing the fractions and varying, we have from (8.130)

$$\delta E \int \phi^* \phi \, d\tau + E \int \delta \phi^* \phi \, d\tau + E \int \phi^* \delta \phi \, d\tau = \int \delta \phi^* H \phi \, d\tau + \int \phi^* H \delta \phi \, d\tau. \tag{8.134}$$

Since $\langle \phi | \phi \rangle$ is finite and non-vanishing, we see that the equation (8.133) is equivalent to

$$\int \delta \phi^* (H - E) \phi \, d\tau + \int \phi^* (H - E) \delta \phi \, d\tau = 0. \tag{8.135}$$

Although the variations $\delta \phi$ and $\delta \phi^*$ are not independent, they may in fact be treated as such, so that the individual terms in (8.135) can be set equal to zero. To see this, we replace the arbitrary variation $\delta \phi$ by $i \delta \phi$ in (8.135), so that we have

$$-i \int \delta \phi^* (H - E) \phi \, d\tau + i \int \phi^* (H - E) \delta \phi \, d\tau = 0. \tag{8.136}$$

Combining (8.135) with (8.136), we obtain the two equations

$$\int \delta \phi^* (H - E) \phi \, d\tau = 0, \qquad \int \phi^* (H - E) \delta \phi \, d\tau = 0 \tag{8.137}$$

which is the desired result. Since H is Hermitian, we see that these two equations are equivalent to the Schrödinger equation

$$(H - E[\phi])\phi = 0. \tag{8.138}$$

Hence any function $\phi = \psi_n$ for which the functional (8.130) is stationary is an eigenfunction of H corresponding to the eigenvalue $E_n \equiv E[\psi_n]$. Conversely, if ψ_n is an eigenfunction of H, and E_n is the corresponding energy, we have $E_n = E[\psi_n]$ and the functional $E[\psi_n]$ is stationary because ψ_n satisfies the equations (8.137). It is worth stressing that if ϕ and ψ_n differ by $\delta\phi$, the variational principle (8.133) implies that the leading term of the difference between $E[\phi]$ and the exact eigenvalue E_n is *quadratic* in $\delta\phi$. Thus errors in the approximate energy values are of *second order* in $\delta\phi$ when the energy is calculated from the functional (8.130).

We also remark that the functional (8.130) is independent of the normalisation and of the phase of ϕ. In particular, it is often convenient to impose the condition that ϕ be normalised to unity, $\langle\phi|\phi\rangle = 1$. The above results may then be retrieved by varying the functional $\langle\phi|H|\phi\rangle$ subject to the condition $\langle\phi|\phi\rangle = 1$, namely

$$\delta \int \phi^* H \phi \, d\tau = 0, \qquad \int \phi^* \phi \, d\tau = 1. \tag{8.139}$$

The constraint $\langle\phi|\phi\rangle = 1$ may be taken care of by introducing a Lagrange multiplier[3] which we denote by E, so that the variational equation reads

$$\delta\left[\int \phi^* H \phi \, d\tau - E \int \phi^* \phi \, d\tau\right] = 0 \tag{8.140}$$

or

$$\int \delta\phi^*(H - E)\phi \, d\tau + \int \phi^*(H - E)\delta\phi \, d\tau = 0. \tag{8.141}$$

This equation is identical to (8.135), and we see that the Lagrange multiplier E has the meaning of an energy eigenvalue.

An important additional property of the functional (8.130) is that it provides an *upper bound* to the exact ground state energy E_0. In order to obtain this result, let us expand the arbitrary, square-integrable function ϕ in the complete set of orthonormal eigenfunctions ψ_n of H. That is,

$$\phi = \sum_n a_n \psi_n. \tag{8.142}$$

Substituting this expression into (8.130), we find that

$$E[\phi] = \frac{\sum_n |a_n|^2 E_n}{\sum_n |a_n|^2} \tag{8.143}$$

---

[3] Lagrange multipliers are discussed, for example, in Byron and Fuller (1969).

where we have used the fact that $H\psi_n = E_n\psi_n$ and $\langle\phi|\phi\rangle = \sum_n |a_n|^2$. If we now substract E_0 (the lowest energy eigenvalue) from both sides of (8.143), we have

$$E[\phi] - E_0 = \frac{\sum_n |a_n|^2 (E_n - E_0)}{\sum_n |a_n|^2}. \tag{8.144}$$

Since $E_n \geq E_0$, the right-hand side of (8.144) is non-negative, whence

$$E_0 \leq E[\phi] \tag{8.145}$$

and the functional $E[\phi]$ gives an upper bound (or in other words a *minimum principle*) for the ground-state energy.

The property (8.145) constitutes the basis of the *Rayleigh–Ritz variational method* for the approximate calculation of E_0. This method consists of evaluating the quantity $E[\phi]$ by using *trial functions* ϕ which depend on a certain number of variational parameters. The functional $E[\phi]$ then becomes a function of these variational parameters, which is minimised in order to obtain the best approximation of E_0 allowed by the form chosen for ϕ.

How should one choose the form of the trial function ϕ? The answer to this equestion depends on the specific problem which is investigated. Clearly, a good trial function is one which contains as many features of the exact wave function as possible. Some of these features can often be readily deduced, for example by using symmetry arguments. One should also try to construct trial wave functions which lead to reasonably simple calculations. Of course, if high accuracy is required, one will almost inevitably have to construct trial functions possessing a great amount of flexibility, i.e. containing many variational parameters.

Particle in a one-dimensional infinite square well

As an example, let us consider a particle of mass m moving in the one-dimensional infinite square well defined by (4.86). This problem was solved exactly in Section 4.5. Denoting by E_0 the exact ground-state energy, we have by setting $n = 1$ in (4.95),

$$E_0 = E_{n=1} = \frac{\hbar^2 \pi^2}{8ma^2} = 1.23370 \frac{\hbar^2}{ma^2}, \tag{8.146}$$

the corresponding exact ground-state wave function $\psi_0(x)$ being given by (4.93) with $n = 1$. We shall now apply the Rayleigh–Ritz variational method to estimate the value of E_0, pretending that we do not know the exact wave function $\psi_0(x)$, but only some of its general features. In particular, since $\psi_0(x)$ must be an even function of x which vanishes for $|x| \geq a$, we choose a trial function of the form

$$\phi(c, x) = \begin{cases} (a^2 - x^2)(1 + cx^2), & -a \leq x \leq a \\ 0, & |x| > a \end{cases} \tag{8.147}$$

where c is a variational parameter. Note that there is no loss of generality in assuming the wave function ψ_0 – and hence ϕ – to be real.

The first step of the calculation consists of evaluating the expression $E[\phi]$. Substituting the trial function (8.147) on the right of equation (8.130), we see that the functional $E[\phi]$ becomes a function of the variational parameter c, namely

$$E[\phi] \equiv E(c) = \frac{-(\hbar^2/2m)\int_{-a}^{+a}(a^2-x^2)(1+cx^2)\frac{d^2}{dx^2}[(a^2-x^2)(1+cx^2)]dx}{\int_{-a}^{+a}(a^2-x^2)^2(1+cx^2)^2 dx} \tag{8.148}$$

where we have used the fact that the Hamiltonian of the particle is given by $H = -(\hbar^2/2m)d^2/dx^2$ for $-a < x < a$. Performing the integrals, we find that

$$E(c) = \frac{3\hbar^2}{4ma^2} \frac{11a^4c^2 + 14a^2c + 35}{a^4c^2 + 6a^2c + 21}. \tag{8.149}$$

Let us now minimise $E(c)$ with respect to the variational parameter c. Thus, writing

$$\frac{dE(c)}{dc} = 0 \tag{8.150}$$

we obtain the quadratic equation

$$26a^4c^2 + 196a^2c + 42 = 0 \tag{8.151}$$

which admits the two roots

$$c^{(1)} = -0.22075a^{-2}, \qquad c^{(2)} = -7.31771a^{-2}. \tag{8.152}$$

Substituting these values in the expression (8.149) for $E(c)$, we have

$$E(c^{(1)}) = 1.23372\frac{\hbar^2}{ma^2}, \qquad E(c^{(2)}) = 12.7663\frac{\hbar^2}{ma^2} \tag{8.153}$$

so that the minimum of $E(c)$ is reached for the value $c = c^{(1)}$ of the variational parameter c. Because we have a minimum principle for the ground-state energy E_0 (see (8.145)), we are able to conclude that $E(c^{(1)})$ is the best approximation to E_0 which can be obtained with trial functions of the type given by (8.147). In fact, upon comparison with the exact result (8.146), we see that $E(c^{(1)})$ is only slightly larger than E_0 and hence is an excellent approximation to the ground-state energy. The corresponding 'optimum' trial function $\phi(c^{(1)}, x)$, obtained by substituting the value $c = c^{(1)}$ in (8.147), is readily seen to agree closely with the exact wave function $\psi_0(x)$. In particular, we note that $\phi(c^{(1)}, x)$, like $\psi_0(x)$, has no nodes inside the well ($|x| < a$). By contrast, the function $\phi(c^{(2)}, x)$, obtained by setting $c = c^{(2)}$ in (8.147) has nodes at $x = \pm[-1/c^{(2)}]^{1/2} \simeq \pm 0.37a$, and we see from (8.153) that it gives an energy $E(c^{(2)})$ much higher than $E(c^{(1)})$. We shall return shortly to the physical meaning of the function $\phi(c^{(2)}, x)$.

Excited states

The Rayleigh–Ritz variational method can also be used to obtain an upper bound for the energy of an excited state, provided that the trial function ϕ is made orthogonal to all the energy eigenfunctions corresponding to states having a lower energy than the energy level considered. To prove this property, let us arrange the energy levels in an ascending sequence: $E_0, E_1, E_2, \ldots$, and let the trial function ϕ be orthogonal to the energy eigenfunctions $\psi_n(n = 0, 1, \ldots, i)$, namely

$$\langle \psi_n | \phi \rangle = 0, \qquad n = 0, 1, \ldots, i. \tag{8.154}$$

Then, if we expand ϕ in the orthonormal set $\{\psi_n\}$ as in (8.142), we have $a_n = \langle \psi_n | \phi \rangle = 0$ $(n = 0, 1, \ldots, i)$ and the functional $E[\phi]$ becomes

$$E[\phi] = \frac{\sum_{n=i+1} |a_n|^2 E_n}{\sum_{n=i+1} |a_n|^2} \tag{8.155}$$

so that

$$E_{i+1} \leqslant E[\phi]. \tag{8.156}$$

As an example, suppose that the lowest energy eigenfunction ψ_0 is known and let ϕ be a trial function. The function

$$\tilde{\phi} = \phi - \psi_0 \langle \psi_0 | \phi \rangle \tag{8.157}$$

is orthogonal to ψ_0 since

$$\langle \psi_0 | \tilde{\phi} \rangle = \langle \psi_0 | \phi \rangle - \langle \psi_0 | \psi_0 \rangle \langle \psi_0 | \phi \rangle = 0 \tag{8.158}$$

where we have used the fact that $\langle \psi_0 | \psi_0 \rangle = 1$. Thus $\tilde{\phi}$ can be used to obtain an upper limit of E_1, the exact energy of the first excited state. More generally, the Schmidt orthogonalisation procedure (see Section 3.7) can be used to construct trial functions satisfying the $(i + 1)$ orthogonality requirements (8.154).

Unfortunately, in many practical applications the lower eigenfunctions $\psi_n(n = 0, 1, \ldots, i)$ are not known exactly and one only has approximations (obtained for example from a variational calculation) of these functions. In this case the orthogonality conditions (8.154) cannot be achieved exactly, so that the relation (8.156) could be violated. For example, suppose that the function ϕ_0 (which we assume to be normalised to unity) is an approximation to the true ground state eigenfunction ψ_0. If ϕ_1 is a trial function orthogonal to ϕ_0 (so that $\langle \phi_0 | \phi_1 \rangle = 0$), and if ε_0 is the positive quantity

$$\varepsilon_0 = 1 - |\langle \psi_0 | \phi_0 \rangle|^2 \tag{8.159}$$

it may be shown (Problem 8.16) that

$$E_1 - \varepsilon_0(E_1 - E_0) \leqslant E[\phi_1]. \tag{8.160}$$

As a result, $E[\phi_1]$ does not necessarily provide an upper bound to E_1. However, if ϕ_0 is a good approximation to ψ_0, we see from (8.159) that ε_0 will be small, so that any violation of the relation $E_1 \leqslant E[\phi_1]$ will also be small.

The application of the variational method to excited states is greatly facilitated if the Hamiltonian H of the system has certain *symmetry properties*, since in this case the orthogonality relations (8.154) can be satisfied exactly for certain states. Suppose for example that there exists a Hermitian operator A which commutes with H:

$$[A, H] = 0 \tag{8.161}$$

so that the operators A and H can be diagonalised simultaneously and have common eigenfunctions. On the other hand, we know that eigenfunctions of the operator A corresponding to different eigenvalues are orthogonal. Hence, if we construct a trial function ϕ entirely from eigenfunctions that correspond to a given eigenvalue α of A, we are certain that this trial function will be orthogonal to all the other eigenfunctions corresponding to different eigenvalues of A; it will therefore give an upper bound for the lowest energy level associated with the eigenvalue α of A. These considerations are particularly useful when the eigenfunctions of the operator A have simple symmetry properties, for instance when the operator A is the parity or the angular momentum. Then, if the excited state under study is of different parity or angular momentum than the states of lower energy, the orthogonality conditions (8.154) will automatically be satisfied.

As an illustration of the above discussion, let us consider again a particle in the one-dimensional infinite square well (4.86). Since this potential is symmetric about $x = 0$, $V(-x) = V(x)$, we know from the analysis of Section 4.5 that the energy eigenfunctions are either even or odd. Thus, choosing the operator A to be the parity operator $\mathcal{P}$ (such that $\mathcal{P} f(x) = f(-x)$), we may divide the energy eigenfunctions into two classes: the even functions belonging to the eigenvalue $+1$ and the odd functions belonging to the eigenvalue -1 of $\mathcal{P}$. Any even function is of course orthogonal to any odd function. Moreover, if the energy levels are ordered by increasing energy values, the corresponding eigenfunctions are alternately even or odd, with the ground state being always even. Thus a trial function of even parity (such as (8.147)) will provide an upper bound for the lowest energy level whose eigenfunction is even, namely for the ground-state energy, as we have already seen. On the other hand, a trial function of odd parity will give an upper bound for the lowest energy level whose eigenfunction is odd, namely the first excited level E_1. For example, by substituting the trial function

$$\phi(c, x) = \begin{cases} (a^2 - x^2)(x + cx^3), & -a \leqslant x \leqslant a \\ 0, & |x| > a \end{cases} \tag{8.162}$$

into the functional (8.130) and minimising the resulting expression with respect to the variational parameter c, an approximate value of E_1 can be calculated (see Problem 8.17). Upon comparison with the exact result (obtained by setting $n = 2$ in (4.95)), it is readily checked that this variational estimate gives an upper bound for E_1.

Trial functions with linear variational parameters and the Hylleraas–Undheim theorem

Particularly useful trial functions ϕ can be constructed by choosing a certain number N of linearly independent functions $\chi_1, \chi_2, \ldots, \chi_N$, and forming the linear combination

$$\phi = \sum_{n=1}^{N} c_n \chi_n \tag{8.163}$$

where the coefficients $c_1, c_2, \ldots, c_N$ are linear variational parameters. For the moment we make no particular assumptions about possible symmetries of the Hamiltonian H or the trial function ϕ. Substituting (8.163) into the functional (8.130), we find that

$$E[\phi] = \frac{\sum_{n=1}^{N} \sum_{n'=1}^{N} c_{n'} c_n H_{n'n}}{\sum_{n=1}^{N} \sum_{n'=1}^{N} c_{n'} c_n \Delta_{n'n}} \tag{8.164}$$

where we have assumed the variational parameters to be real and we have set

$$H_{n'n} = \langle \chi_{n'} | H | \chi_n \rangle$$
$$\Delta_{n'n} = \langle \chi_{n'} | \chi_n \rangle. \tag{8.165}$$

Note that if the functions χ_n are orthonormal, then $\Delta_{n'n} = \delta_{n'n}$.

In order to find the values of the variational parameters $c_1, c_2, \ldots, c_N$ which minimise $E[\phi]$, let us rewrite (8.164) as

$$E[\phi] \sum_{n=1}^{N} \sum_{n'=1}^{N} c_{n'} c_n \Delta_{n'n} = \sum_{n=1}^{N} \sum_{n'=1}^{N} c_{n'} c_n H_{n'n}. \tag{8.166}$$

Differentiating with respect to each c_n and expressing that $\partial E / \partial c_n = 0$ (with $n = 1, 2, \ldots, N$) we obtain a system of N linear and homogeneous equations in the variables $c_1, c_2, \ldots, c_N$, namely

$$\sum_{n=1}^{N} c_n (H_{n'n} - \Delta_{n'n} E) = 0, \qquad n' = 1, 2, \ldots, N. \tag{8.167}$$

The necessary and sufficient condition for this system to have a non-trivial solution is that the determinant of the coefficients vanishes. That is

$$\det |H_{n'n} - \Delta_{n'n} E| = 0. \tag{8.168}$$

Let $E_0^{(N)}, E_1^{(N)}, \ldots, E_{N-1}^{(N)}$ be the N roots of this equation arranged in an ascending sequence, the superscript (N) indicating that we are dealing with a $N \times N$ determinant. By virtue of the minimum principle (8.145), the lowest root $E_0^{(N)}$ will clearly be an upper bound to the ground state energy E_0. Substituting the value of $E_0^{(N)}$ in the system of equations (8.167) and solving for the coefficients c_n in terms of one of them (for example c_1, which may be chosen arbitrarily since $E[\phi]$ does not depend on the normalisation of ϕ), we then obtain the corresponding 'optimum' variational approximation ϕ_0 to the ground-state wave function ψ_0. It can be shown by induction

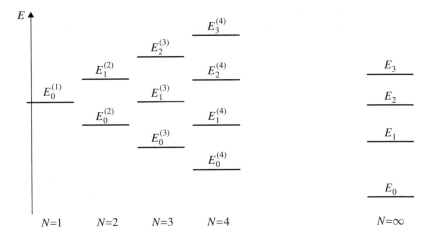

Figure 8.3 Illustration of the Hylleraas–Undheim theorem. The approximate energy eigenvalues given by the Rayleigh–Ritz variational method with linear trial functions are shown for various values of N. Each root $E_i^{(N)}$ of the determinantal equation (8.168) is an upper bound to the corresponding exact eigenvalue E_i. If N is increased by one unit, the 'new' $(N+1)$ roots $E_0^{(N+1)}, E_1^{(N+1)}, \ldots, E_N^{(N+1)}$, of the determinantal equation are separated by the 'old' roots $E_0^{(N)}, E_1^{(N)}, \ldots, E_{N-1}^{(N)}$.

that the other roots $E_i^{(N)}$ (with $i = 1, 2, \ldots, N-1$) of equation (8.168) are upper bounds to the successive excited state energies of H. The corresponding 'optimum' variational wave functions $\phi_i (i = 1, 2, \ldots, N-1)$ are obtained in the same manner as ϕ_0, i.e. by substituting $E_i^{(N)}$ in (8.167) and solving for the coefficients c_n in terms of one of them. From a general property of linear, homogeneous equations, the functions $\phi_0, \phi_1, \ldots, \phi_{N-1}$, are mutually orthogonal. Moreover, if we construct a new trial function $\hat{\phi}$ containing an additional basis function χ_{N+1}, namely

$$\hat{\phi} = \sum_{n=1}^{N+1} c_n \chi_n \qquad (8.169)$$

it can be proved (see Problem 8.19) that the 'new' $(N+1)$ roots $E_0^{(N+1)}$, $E_1^{(N+1)}, \ldots, E_N^{(N+1)}$, of the determinantal equation (8.168) are separated by the 'old' roots $E_0^{(N)}, E_1^{(N)}, \ldots, E_{N-1}^{(N)}$. This property, which is illustrated in Fig. 8.3, is known as the *Hylleraas–Undheim theorem*.

Let us assume that the Hamiltonian H commutes with a Hermitian operator A. In that case we have shown above that the original variational problem can be divided into separate variational calculations, one for each eigenvalue of the operator A. The foregoing discussion can thus be applied to each of these separate variational calculations. In particular, if the trial function (8.163) has been constructed from eigenfunctions corresponding to a given eigenvalue α of A, the N roots $E_0^{(N)}, E_1^{(N)}, \ldots, E_{N-1}^{(N)}$ of the determinantal equation (8.168) will give upper bounds to the first N energy levels associated with this eigenvalue α, arranged in an ascending sequence. For example, returning once again to the problem of a particle in the one-dimensional infinite square well (4.86), we see that for $-a \leqslant x \leqslant a$ the trial function (8.147) may be rewritten in the form

$$\phi = c_1 \chi_1 + c_2 \chi_2 \tag{8.170}$$

where χ_1 and χ_2 are *even* functions given respectively by $\chi_1 = a^2 - x^2$ and $\chi_2 = x^2(a^2 - x^2)$, $c_1 = 1$ (we recall that one of the coefficients can always be chosen arbitrarily) and c_2 is the variational parameter c of equation (8.147). The 2×2 determinantal equation (8.168) corresponding to the trial function (8.170) is then readily solved, giving the two energy values denoted respectively by $E(c^{(1)})$ and $E(c^{(2)})$ in (8.153). The lowest one, $E(c^{(1)})$, is of course an upper bound to the ground state energy E_0, as we have seen. According to the foregoing discussion, the second energy value, $E(c^{(2)})$ must give an upper bound for the next energy level corresponding to an eigenfunction of *even* parity, namely the second excited level of the system. Comparison of $E(c^{(2)})$ with the exact value of this level (obtained by setting $n = 3$ in (4.95)) shows that this is indeed the case. We also remark that the variational function $\phi(c^{(2)}, x)$, like the corresponding exact wave function $\psi_{n=3}(x)$, has two nodes in the domain $-a < x < a$ located inside the infinite square well.

8.4 The WKB approximation

We shall conclude this chapter with a discussion of an approximation method which can be used when the potential energy is a slowly varying function of position. The method we shall study is called the *WKB approximation*, after G. Wentzel, H. A. Kramers and L. Brillouin who independently introduced it in quantum mechanics in 1926[4]. It can only be readily applied to one-dimensional problems (and of course to problems of higher dimensionality that can be reduced to the solution of a one-dimensional Schrödinger equation). We shall therefore focus our attention on the one-dimensional motion of a particle of mass m in a potential $V(x)$, the corresponding time-independent Schrödinger equation being (see (4.3))

$$-\frac{\hbar^2}{2m} \frac{d^2 \psi(x)}{dx^2} + V(x)\psi(x) = E\psi(x). \tag{8.171}$$

If the potential were constant, $V(x) = V_0$, solutions of (8.171) would be written as linear combinations of the basic plane waves

$$\psi(x) = A \exp\left(\pm \frac{i}{\hbar} p_0 x\right) \tag{8.172}$$

---

[4] The required mathematical techniques, concerned with asymptotic solutions of differential equations, had been developed much earlier, in particular by J. Liouville, G. Stokes, Lord Rayleigh and H. Jeffreys.

where

$$p_0 = [2m(E - V_0)]^{1/2} \tag{8.173}$$

and A is a constant. The fact that the potential is 'slowly varying' means that $V(x)$ changes only slightly over the de Broglie wavelength

$$\lambda(x) = \frac{h}{p(x)} \tag{8.174}$$

associated with the particle (having an energy $E > V$) in this potential, where

$$p(x) = \{2m[E - V(x)]\}^{1/2} \tag{8.175}$$

is the classical momentum at the point x. Now in the classical limit the de Broglie wavelength λ tends to zero; the slow variation of the potential therefore implies that we are dealing with a quasi-classical situation, so that the WKB method is also called a *semi-classical approximation*.

The result (8.172) obtained for constant potentials suggests that in the case of slowly varying potentials, solutions of the Schrödinger equation (8.171) should be sought which have the form

$$\psi(x) = A \exp\left[\frac{i}{\hbar} S(x)\right]. \tag{8.176}$$

Substituting (8.176) into (8.171), we obtain for $S(x)$ the equation

$$-\frac{i\hbar}{2m} \frac{d^2 S(x)}{dx^2} + \frac{1}{2m} \left[\frac{dS(x)}{dx}\right]^2 + V(x) - E = 0. \tag{8.177}$$

So far, no approximation has been made, this equation being strictly equivalent to the original Schrödinger equation (8.171). Unfortunately, equation (8.177) is a non-linear equation which is in fact more complicated than (8.171) itself! We must therefore try to solve (8.177) approximately. To this end, we first remark that if the potential is constant then $S(x) = \pm p_0 x$ (see (8.172)) and the first term on the left of (8.177) vanishes. Moreover, this term is proportional to $\hbar$, and hence vanishes in the classical limit ($\hbar \to 0$). This suggests that we treat $\hbar$ as a parameter of smallness and expand the function $S(x)$ in the power series

$$S(x) = S_0(x) + \hbar S_1(x) + \frac{\hbar^2}{2} S_2(x) + \cdots \tag{8.178}$$

Inserting the expansion (8.178) into (8.177) and equating to zero the coefficients of each power of $\hbar$ separately, we find the set of equations

$$\frac{1}{2m}\left[\frac{dS_0(x)}{dx}\right]^2 + V(x) - E = 0 \tag{8.179a}$$

$$\frac{dS_0(x)}{dx}\frac{dS_1(x)}{dx} - \frac{i}{2}\frac{d^2 S_0(x)}{dx^2} = 0 \tag{8.179b}$$

$$\frac{dS_0(x)}{dx}\frac{dS_2(x)}{dx} + \left[\frac{dS_1(x)}{dx}\right]^2 - i\frac{d^2 S_1(x)}{dx^2} = 0 \tag{8.179c}$$

$$\vdots$$

which must be solved successively to find $S_0(x)$, $S_1(x)$, $S_2(x)$, and so on. As we shall shortly see, the WKB method yields the first *two* terms[5] of the expansion (8.178).

Let us begin by solving the equation (8.179a). Assuming first that $E > V(x)$, so that we are in a classically allowed region of positive kinetic energy, we find that

$$S_0(x) = \pm \int^x p(x')dx' \tag{8.180}$$

where $p(x)$ is the classical momentum given by (8.175) and an arbitrary integration constant (which can be absorbed in the coefficient A on the right of (8.176)) has been omitted. The above expression for $S_0(x)$ clearly provides the classical limit of $S(x)$. Using this result in equation (8.179b), the function $S_1(x)$ is found to be

$$S_1(x) = \frac{i}{2}\log p(x) \tag{8.181}$$

where an arbitrary constant of integration has again been omitted. Next, by using the foregoing expressions for $S_0(x)$ and $S_1(x)$ in (8.179c), we have

$$S_2(x) = \frac{1}{2}m[p(x)]^{-3}\frac{dV(x)}{dx} - \frac{1}{4}m^2\int^x [p(x')]^{-5}\left[\frac{dV(x')}{dx'}\right]^2 dx'. \tag{8.182}$$

From this expression and that of $p(x)$ (see (8.175)), it is clear that S_2 will be small whenever $dV(x)/dx$ is small, and provided that $E - V$ is not too close to zero. If, in addition, all the higher derivatives of $V(x)$ are small, it can be verified that $S_3, S_4, \ldots$, will also be small. Note that in the case of a constant potential $S_0 = \pm p_0 x$ (apart from an irrelevant additive constant) and $S_1, S_2, \ldots$, are all zero.

These considerations show that if $V(x)$ is a slowly varying function of position, and provided $E - V$ is not too small, one can retain only the first two terms on the

---

[5] It can be shown that the series (8.178) does not converge, but is an *asymptotic series* for the function $S(x)$. As a result, the best approximation to $S(x)$ is obtained by keeping a finite number of terms on the right of (8.178).

right of the expansion (8.178). Using (8.180) and (8.181) we then obtain the *WKB approximation* to the wave function (8.176) in a classically allowed region, namely

$$\psi(x) = A[p(x)]^{-1/2} \exp\left[\pm \frac{i}{\hbar} \int^x p(x') dx'\right], \qquad E > V. \tag{8.183}$$

Of course, the general WKB solution in this region is a linear combination of such solutions. That is,

$$\psi(x) = [p(x)]^{-1/2} \left\{ A \exp\left[\frac{i}{\hbar} \int^x p(x') dx'\right] \right.$$
$$\left. + B \exp\left[-\frac{i}{\hbar} \int^x p(x') dx'\right] \right\}, \qquad E > V \tag{8.184}$$

where A and B are arbitrary constants. The first exponential corresponds to a wave moving in the positive direction, the second exponential to a wave moving in the opposite direction. We also remark that if the potential is constant ($V = V_0$) these exponentials reduce to the plane waves $\exp(i p_0 x/\hbar)$ and $\exp(-i p_0 x/\hbar)$, respectively, where p_0 is given by (8.173).

A criterion of validity for the WKB approximation may be obtained by requiring that the contribution to the wave function (8.176) arising from the third term on the right of (8.178) be negligible. This will be the case if

$$|(\hbar/2) S_2(x)| \ll 1. \tag{8.185}$$

Using the expression (8.182) of $S_2(x)$ and taking into account that both terms on the right of (8.182) are of the same order of magnitude, we may rewrite the condition (8.185) more explicitly as

$$\left| \frac{\hbar m \, dV(x)/dx}{[2m(E - V(x))]^{3/2}} \right| \ll 1 \tag{8.186}$$

where we have used the expression (8.175) of $p(x)$. This criterion is seen to be satisfied if the potential energy varies slowly enough, provided that the kinetic energy $E - V$ is sufficiently large. An alternative way of writing the condition (8.186) is

$$\left| \frac{\hbar}{[p(x)]^2} \frac{dp(x)}{dx} \right| = \left| \frac{1}{p(x)} \frac{dp(x)}{dx} \bar\lambda(x) \right| \ll 1 \tag{8.187}$$

where $\bar\lambda(x) = \hbar/p(x) = \lambda(x)/2\pi$ is the reduced de Broglie wavelength of the particle. The inequality (8.187) means that the fractional change in the momentum must be small in a wavelength. In other words, the potential must change so slowly that the momentum of the particle is nearly constant over many wavelengths. We remark that the condition (8.187) may equally be written in the form

$$\left| \frac{d\bar\lambda(x)}{dx} \right| \ll 1. \tag{8.188}$$

Now $\delta \lambdabar = (d\lambdabar/dx)\delta x$ is the change occurring in the reduced wavelength $\lambdabar$ in the distance δx. Upon setting $\delta x = \lambdabar$, we see that the condition (8.188) is equivalent to

$$|\delta \lambdabar (x)| = \left| \frac{d\lambdabar (x)}{dx} \lambdabar (x) \right| \ll \lambdabar (x) \tag{8.189}$$

showing that $\lambdabar$ must only change by a small fraction of itself over a distance of the order of $\lambdabar$. This condition is well known in wave optics as the one to be satisfied if a wave is to propagate in a medium of varying index of refraction without giving rise to significant reflection.

Until now we have only considered classically allowed regions for which $E > V$. However, we can equally well obtain the WKB approximation to the wave function (8.176) in classically forbidden regions of negative kinetic energy ($E < V$), where $p(x)$ becomes purely imaginary (see (8.175)). Solving equations (8.179a) and (8.179b) in this case, we obtain the WKB solutions

$$\psi(x) = A|p(x)|^{-1/2} \exp\left[\pm \frac{1}{\hbar} \int^x |p(x')|dx'\right], \quad E < V \tag{8.190}$$

where A is a constant. The general WKB solution for $E < V$ is a linear combination of such solutions, namely

$$\psi(x) = |p(x)|^{-1/2} \left\{ C \exp\left[-\frac{1}{\hbar} \int^x |p(x')|dx'\right] \right.$$
$$\left. + D \exp\left[\frac{1}{\hbar} \int^x |p(x')|dx'\right] \right\}, \quad E < V \tag{8.191}$$

where C and D are arbitrary constants. These WKB solutions are accurate provided the criterion (8.186) is satisfied, which requires that $V(x)$ varies slowly enough and that $V - E$ be sufficiently large.

The connection formulae

It is apparent from (8.186) that the WKB approximation breaks down in the vicinity of a *classical turning point*, for which $E = V$. The two kinds of WKB wave functions (8.184) and (8.191) we have obtained are therefore 'asymptotically' valid, because they may only be used sufficiently far away from the nearest turning point. Since the exact wave function is continuous and smooth for all x, it should be possible to obtain *connection formulae* which allow one to join the two types of WKB solutions across a turning point. This interpolation is based on the fact that the original Schrödinger equation (8.171) can be modified slightly so that an 'exact' solution at and near a turning point can be written down. Since this is a rather technical point, we shall quote directly the main results here, leaving a detailed discussion for Appendix B.

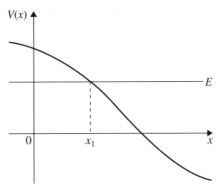

Figure 8.4 A left-hand barrier. The point x_1 is a classical turning point, such that $V(x_1) = E$. For $x < x_1$ one has $V(x) > E$ and for $x > x_1$ one has $V(x) < E$.

Left-hand barrier

Let us start by considering the case in which $V(x)$ is a left-hand barrier (see Fig. 8.4), with $V(x_1) = E$, $V(x) > E$ for $x < x_1$ and $V(x) < E$ for $x > x_1$. The region $x > x_1$ (to the right of the classical turning point $x = x_1$) is the classically accessible region, where $p(x)$ is real. In this region, the general WKB solution is given by (8.184). Choosing x_1 as the lower limit of the integrals in (8.184), we may write this general WKB solution as

$$\psi(x) = [p(x)]^{-1/2} \left\{ A \exp\left[\frac{i}{\hbar} \int_{x_1}^{x} p(x')dx'\right] \right.$$
$$\left. + B \exp\left[-\frac{i}{\hbar} \int_{x_1}^{x} p(x')dx'\right] \right\}, \quad x > x_1. \tag{8.192}$$

On the other hand, in the classically forbidden region $x < x_1$, where $p(x)$ is purely imaginary, the general WKB solution is given from (8.191) by

$$\psi(x) = |p(x)|^{-1/2} \left\{ C \exp\left[-\frac{1}{\hbar} \int_{x_1}^{x} |p(x')|dx'\right] \right.$$
$$\left. + D \exp\left[\frac{1}{\hbar} \int_{x_1}^{x} |p(x')|dx'\right] \right\}, \quad x < x_1. \tag{8.193}$$

where x_1 has been chosen again as the lower limit of integration, so that the integrals in (8.193) are negative. We recall that both (8.192) and (8.193) are 'asymptotic' solutions which are good approximations to the true solution only if x is not too close to the turning point x_1. We shall now use the results of Appendix B to 'join' the WKB solutions (8.192) and (8.193) across the turning point.

Let us first consider the case for which the approximate solution (8.193) is a decaying exponential as $x \to -\infty$, so that $C = 0$. We then find from Appendix B that $iA = B = D \exp(i\pi/4)$. Setting the overall normalisation constant, which is arbitrary, equal to unity, the connection formula can be expressed as

$$|p(x)|^{-1/2}\exp\left[\frac{1}{\hbar}\int_{x_1}^{x}|p(x')|\mathrm{d}x'\right] \to 2[p(x)]^{-1/2}\cos\left[\frac{1}{\hbar}\int_{x_1}^{x}p(x')\mathrm{d}x' - \frac{\pi}{4}\right] \quad (8.194)$$

where the left-hand side holds for $x < x_1$ and the right-hand side for $x > x_1$, provided x is not too close to x_1.

Although the wave function (8.193) cannot increase indefinitely as $x \to -\infty$, it is nevertheless useful to consider the case for which it increases exponentially as x recedes to the left of the turning point. The appropriate connection formula is then

$$-|p(x)|^{-1/2}\exp\left[-\frac{1}{\hbar}\int_{r_1}^{x}|p(x')|\mathrm{d}x'\right] \leftarrow [p(x)]^{-1/2}\sin\left[\frac{1}{\hbar}\int_{r_1}^{x}p(x')\mathrm{d}x' - \frac{\pi}{4}\right] \quad (8.195)$$

where again the left-hand side is valid in the region $x < x_1$ and the right-hand side in the region $x > x_1$, sufficiently far away from the turning point.

It is worth stressing that connection formulae such as (8.194) and (8.195) can only be employed in the direction indicated by the arrows (see Appendix B). For example, if we know that the solution is exponentially decreasing far to the left of the turning point, we are entitled to infer from (8.194) that the cosine solution should be used sufficiently far to the right of x_1. However, knowing that the exact solution is represented approximately by the increasing exponential as x recedes to the left of the turning point, we cannot use (8.195) to deduce that the sine solution should be used for $x > x_1$. This is because a small admixture of the exponentially decreasing solution in the region $x < x_1$ might be negligible there but would lead to an appreciable admixture of the cosine solution in the region $x > x_1$. Similarly, if the exact solution is approximated by the cosine function for $x > x_1$, we cannot infer that for $x < x_1$ the solution is exponentially decreasing because a small admixture of the sine solution would provide an exponentially increasing component for $x < x_1$, which would dominate the solution when x is sufficiently far to the left of x_1. As a consequence, if we know that the solution sufficiently far to the right of the turning point x_1 is oscillatory and of the general form

$$\psi(x) = [p(x)]^{-1/2}\sin\left[\frac{1}{\hbar}\int_{x_1}^{x}p(x')\mathrm{d}x' - \frac{\pi}{4} + \delta\right], \qquad x > x_1 \quad (8.196)$$

where δ is a given phase, then the solution sufficiently far to the left of x_1 must be the increasing exponential

$$\psi(x) = -|p(x)|^{-1/2}\cos\delta\exp\left[-\frac{1}{\hbar}\int_{x_1}^{x}|p(x')|\mathrm{d}x'\right], \qquad x < x_1. \quad (8.197)$$

Exceptional cases occur when

$$\delta = (2n+1)\frac{\pi}{2}, \qquad n = 0, 1, 2, \ldots \quad (8.198)$$

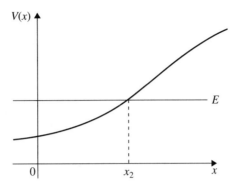

Figure 8.5 A right-hand barrier. The point x_2 is a classical turning point, such that $V(x_2) = E$. For $x < x_2$ one has $V(x) < E$ and for $x > x_2$ one has $V(x) > E$.

for which the asymptotic form to the left of the turning point is a decreasing exponential. It should be noted once again that the connection is from (8.196) to (8.197) and that the knowledge of the exponentially increasing asymptotic solution (8.197) for $x < x_1$ does not allow us to infer the asymptotic solution (8.196) for $x > x_1$.

Right-hand barrier

Consider now a right-hand barrier, as in Fig. 8.5, with a classical turning point at $x = x_2$. In this case $p(x)$ is real for $x < x_2$ and purely imaginary for $x > x_2$. The appropriate connection formulae are given by

$$2[p(x)]^{-1/2} \cos\left[-\frac{1}{\hbar}\int_{x_2}^{x} p(x')dx' - \frac{\pi}{4}\right] \leftarrow |p(x)|^{-1/2} \exp\left[-\frac{1}{\hbar}\int_{x_2}^{x} |p(x')|dx'\right] \tag{8.199}$$

and

$$[p(x)]^{-1/2} \sin\left[-\frac{1}{\hbar}\int_{x_2}^{x} p(x')dx' - \frac{\pi}{4} + \delta\right] \rightarrow$$
$$-|p(x)|^{-1/2} \cos\delta \exp\left[\frac{1}{\hbar}\int_{x_2}^{x} |p(x')|dx'\right] \tag{8.200}$$

where the WKB solution for $x < x_2$ is placed on the left-hand side and that for $x > x_2$ on the right-hand side. Once again the arrows indicate in what direction the connections can be made safely. It should also be noted that (8.200) leads to a decreasing exponential for $x > x_2$ if δ is given by (8.198).

Energy levels in a potential well

As a first application of the WKB approximation, we shall estimate the energies of the bound states in a simple one-dimensional potential $V(x)$ such as that illustrated in Fig. 8.6. For each assumed energy level E it is supposed that there are just two

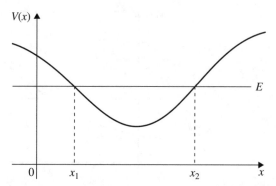

Figure 8.6 A simple one-dimensional potential well. For each assumed energy level E it is supposed that there are just two classical turning points x_1 and x_2, such that $V(x_1) = V(x_2) = E$.

classical turning points at x_1 and x_2. Provided these are well separated, the connection formulae for a left-hand barrier can be used at x_1 and for a right-hand barrier at x_2. Now, since bound state wave functions are square integrable, we must have $\psi(x) \to 0$ as $x \to \pm\infty$. As a result, we only have the exponentially decreasing WKB solutions for $x < x_1$ and $x > x_2$, namely

$$\psi(x) = C_1 |p(x)|^{-1/2} \exp\left[\frac{1}{\hbar}\int_{x_1}^{x} |p(x')|dx'\right], \qquad x < x_1 \qquad (8.201)$$

and

$$\psi(x) = C_2 |p(x)|^{-1/2} \exp\left[-\frac{1}{\hbar}\int_{x_2}^{x} |p(x')|dx'\right], \qquad x > x_2 \qquad (8.202)$$

where C_1 and C_2 are constants. Hence across the turning point at $x = x_1$ we must use the connection formula (8.194), and across the turning point at $x = x_2$ we must use the formula (8.199). Inside the well, the WKB solution must be given equally by a constant times the right-hand side of (8.194) or by a constant times the left-hand side of (8.199), so that we have

$$N_1 \cos\left[\frac{1}{\hbar}\int_{x_1}^{x} p(x')dx' - \frac{\pi}{4}\right] = N_2 \cos\left[-\frac{1}{\hbar}\int_{x_2}^{x} p(x')dx' - \frac{\pi}{4}\right] \qquad (8.203)$$

where N_1 and N_2 are constants. This condition is fulfilled provided that

$$\frac{1}{\hbar}\int_{x_1}^{x_2} p(x)dx = \left(n + \frac{1}{2}\right)\pi, \qquad n = 0, 1, 2, \ldots \qquad (8.204)$$

and $N_2 = (-1)^n N_1$. Equation (8.204) can only be satisfied at discrete values of the energy $E = E_0, E_1, E_2, \ldots$, as $n = 0, 1, 2, \ldots$, and these are the WKB estimates of the bound-state energies. It is also readily seen from the form of the WKB solution inside the well that n is the number of nodes of the WKB wave function between the two classical turning points.

It is interesting to note that the condition (8.204) can be rewritten in the form

$$\oint p\mathrm{d}x = \left(n + \tfrac{1}{2}\right)h, \qquad n = 0, 1, 2, \ldots \tag{8.205}$$

where

$$\oint p\mathrm{d}x = 2\int_{x_1}^{x_2} p(x)\mathrm{d}x \tag{8.206}$$

denotes the integral of the momentum $p(x)$ around a complete cycle of the classical motion (i.e. from x_1 to x_2 and back to x_1). Except for the fact that $(n+1/2)$ and not n appears on the right of (8.205), this equation is identical to the quantisation condition postulated in 1916 for a periodic system by W. Wilson and A. Sommerfeld within the framework of the old quantum theory.

As we pointed out above, the two turning points should, in principle, be well separated, i.e. be at least several wavelengths apart. The application of the condition (8.204) then requires n to be large, in agreement with the fact that the WKB method is a semi-classical approximation, valid in the quasi-classical limit of large quantum numbers. However, in many instances the WKB method also gives quite good results even for small n. An extreme case is that of the linear harmonic oscillator (see Problem 8.20) for which the WKB approximation actually provides the exact energy spectrum.

Penetration of a potential barrier

A second interesting application of the WKB approximation is provided by the penetration of a potential barrier. This problem was solved in Section 4.4 for the special case of a rectangular barrier. However, when the barrier has a more complicated shape the Schrödinger equation cannot in general be solved exactly, and approximation methods must be used. We shall now see that the WKB method is often well suited for this problem.

Let us consider a potential barrier $V(x)$ such as that represented in Fig. 8.7. It is assumed that a beam of particles is incident from the left, with an energy E smaller than V_0 (the top of the barrier), and such that there are two classical turning points at $x = x_1$ and $x = x_2$. As indicated in Fig. 8.7 we shall denote by 1, 2 and 3 the regions corresponding respectively to $x < x_2$, $x_2 < x < x_1$ and $x > x_1$. Some of the particles are reflected back to region 1, others are transmitted and emerge to the right, in region 3. In that region the transmitted wave function is given in the WKB approximation by

$$\psi_3(x) = A[p(x)]^{-1/2} \exp\left[\frac{i}{\hbar}\int_{x_1}^{x} p(x')\mathrm{d}x' - i\frac{\pi}{4}\right], \qquad x > x_1 \tag{8.207}$$

where the factor $\exp(-i\pi/4)$, which could, of course, be included in the constant A, has been written explicitly in order to facilitate the application of the connection formulae. For the same purpose, we rewrite (8.207) in terms of cosine and sine

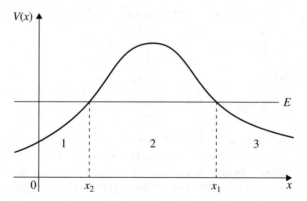

Figure 8.7 A one-dimensional potential barrier. The energy E of the incident particles is assumed to be less than the top of the barrier. There are supposed to be just two classical turning points x_1 and x_2 such that $V(x_1) = V(x_2) = E$. The regions corresponding to $x < x_2$, $x_2 < x < x_1$ and $x > x_1$ are denoted respectively by 1, 2 and 3.

functions, i.e.

$$\psi_3(x) = A[p(x)]^{-1/2} \left\{ \cos\left[\frac{1}{\hbar}\int_{x_1}^{x} p(x')dx' - \frac{\pi}{4}\right] \right.$$
$$\left. + i\sin\left[\frac{1}{\hbar}\int_{x_1}^{x} p(x')dx' - \frac{\pi}{4}\right] \right\}, \qquad x > x_1. \tag{8.208}$$

Using the connection formulae for the case of a barrier to the left, we find that the connecting wave function of exponential type in region 2 is given by

$$\psi_2(x) = -iA|p(x)|^{-1/2} \exp\left[\frac{1}{\hbar}\int_{x}^{x_1} |p(x')|dx'\right], \qquad x_2 < x < x_1. \tag{8.209}$$

In obtaining this formula we have used the fact that as x recedes to the left of the turning point x_1 only the exponentially increasing function is significant. It is convenient to rewrite (8.209) in the form of a decreasing exponential for $x > x_2$, namely

$$\psi_2(x) = -iA|p(x)|^{-1/2} e^{\Lambda} \exp\left[-\frac{1}{\hbar}\int_{x_2}^{x} |p(x')|dx'\right], \qquad x_2 < x < x_1 \tag{8.210}$$

where

$$\Lambda = \frac{1}{\hbar}\int_{x_2}^{x_1} |p(x)|dx = \frac{1}{\hbar}\int_{x_2}^{x_1} \{2m[V(x) - E]\}^{1/2} dx. \tag{8.211}$$

In order to find the approximate wave function in region 1, we apply to (8.210) the connection formula (8.199) for the case of a barrier to the right. This yields the connecting oscillatory wave function

$$\psi_1(x) = -2iA[p(x)]^{-1/2}e^\Lambda \cos\left[-\frac{1}{\hbar}\int_{x_2}^{x} p(x')dx' - \frac{\pi}{4}\right]$$

$$= -iA[p(x)]^{-1/2}e^\Lambda \left\{\exp\left[i\left(\frac{1}{\hbar}\int_{x_2}^{x} p(x')dx' + \frac{\pi}{4}\right)\right]\right.$$

$$\left.+ \exp\left[-i\left(\frac{1}{\hbar}\int_{x_2}^{x} p(x')dx' + \frac{\pi}{4}\right)\right]\right\}, \qquad x < x_2. \qquad (8.212)$$

The first term in the curly brackets represents a wave moving to the right, and hence corresponds to the incident wave, while the second term corresponds to the reflected wave, moving to the left.

As we have seen in Chapter 4, the *transmission coefficient T* is the ratio of the intensity of the transmitted probability current density to that of the incident probability current density. The former is the product of the transmitted intensity times the velocity after transmission, while the latter is the incident intensity times the velocity of the incident particles. Since the ratio of the velocities is equal to the ratio of the corresponding momenta, we find from (8.207) and (8.212) that the WKB approximation yields the transmission coefficient

$$T = e^{-2\Lambda} = \exp\left[-\frac{2}{\hbar}\int_{x_2}^{x_1} \{2m[V(x) - E]\}^{1/2} dx\right]. \qquad (8.213)$$

We recall that this method of calculation is only justified if $V(x)$ is slowly varying. Moreover, the turning points must be sufficiently well separated, so that the barrier has a thickness $(x_1 - x_2)$ of at least several wavelengths. As a result, the quantity Λ must be large, which in turn means that the transmission coefficient T is very small. The *reflection coefficient* $R = 1 - T$, calculated from (4.69) (which expresses the conservation of the probability flux) is therefore very close to unity. This is consistent with the WKB value $R = 1$ obtained directly from (8.212) by noting that the incident and reflected waves have the same intensity.

It is also interesting to compare the approximate WKB transmission coefficient (8.213) obtained for a thick, smooth barrier of variable height with the result (4.83) derived in Section 4.4 for a thick rectangular barrier of height V_0 and thickness a. From (8.213) we have

$$\log T = -\frac{2}{\hbar}\int_{x_2}^{x_1} \{2m[V(x) - E]\}^{1/2} dx \qquad (8.214)$$

while equation (4.83) yields

$$\log T = \log\frac{16E(V_0 - E)}{V_0^2} - 2\kappa a \qquad (8.215)$$

where $\kappa = [2m(V_0 - E)]^{1/2}/\hbar$. Since the second term on the right of (8.215) dominates the first one for a thick barrier, equation (8.215) reduces approximately to

$$\log T \simeq -\frac{2}{\hbar}[2m(V_0 - E)]^{1/2} a \tag{8.216}$$

and we see that the WKB expression (8.214) is the natural generalisation of the result (8.216) to a barrier whose height is variable.

Cold emission of electrons from a metal

As a first example of the application of our WKB result (8.213) for the transmission coefficient, we shall consider the cold emission of electrons from a metal. Within a metal the electrons are bound by a potential which, as shown in Fig. 8.8(a) may in first approximation be described by a box of finite depth. Because this box is large the energy levels are very dense. Moreover, as we shall see in Chapter 10, when the metal is in its lowest energy state (at an absolute temperature $T = 0$) all the energy levels in the well are filled up to a certain energy called the *Fermi energy* E_F, with no more than two electrons occupying any given energy level[6]. When the temperature rises, a fraction of the electrons are excited to higher energy levels, but even at room temperature this fraction is small and will be neglected here. The difference between the top of the well (V_0) and the Fermi energy E_F is the minimum energy required to remove an electron from the metal; it is the *work function W* introduced in Section 1.2 in our discussion of the photoelectric effect.

Let us now assume that a constant electric field of strength $\mathcal{E}$ is applied to the metal. The potential function will then have the form displayed in Fig. 8.8(b), where the electric potential $-e\mathcal{E}x$ has been added to the original potential, x being the distance from the surface of the metal. As seen from Fig. 8.8(b), the potential barrier now has a finite width, so that electrons have the possibility of tunnelling through the barrier and escaping from the metal. This phenomenon is known as *cold emission of electrons*, in contrast with thermal emission which occurs when electrons acquire enough energy from thermal motion to overcome the barrier.

A crude estimate of the variation of cold emission with the work function W and the applied electric field $\mathcal{E}$ can readily be obtained from equation (8.213). For an electron at the top of the sea of levels, having an energy $E = E_F = V_0 - W$, the turning point x_1 is found by solving the equation

$$V_0 - e\mathcal{E}x_1 = V_0 - W \tag{8.217}$$

which gives $x_1 = W/e\mathcal{E}$. The second turning point is of course at the surface of the metal, so that $x_2 = 0$. Since the potential varies sharply in the vicinity of this turning point, the WKB approximation may well break down in this region. This, however, should not introduce important errors because the distance x_1 is usually quite large compared to interatomic distances, and the fraction of the transmission coefficient

---

[6] This property is a consequence of the *Pauli exclusion principle*, which will be discussed in Section 10.2.

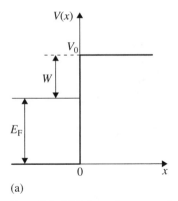
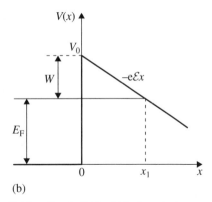

Figure 8.8 (a) Schematic representation of the potential function (solid line) binding electrons in a metal. The energy E_F is the Fermi energy and $W = V_0 - E_F$ is the work function. (b) The potential function modified by the addition of an external electric field $\mathcal{E}$. The classical turning point x_1 is such that $x_1 = W/e\mathcal{E}$.

coming from the region of rapid variation of the potential near $x = 0$ is quite small. Thus, using (8.213) with $x_1 = W/e\mathcal{E}$, $x_2 = 0$ and

$$V(x) - E = V_0 - e\mathcal{E}x - E = W - e\mathcal{E}x \tag{8.218}$$

we find that the transmission coefficient is given approximately by

$$T = \exp\left[-\frac{2}{\hbar}\int_0^{W/e\mathcal{E}}\{2m(W - e\mathcal{E}x)\}^{1/2}dx\right]$$
$$= \exp\left[-\frac{4}{3}\frac{(2m)^{1/2}W^{3/2}}{\hbar e\mathcal{E}}\right]. \tag{8.219}$$

This expression, known as the *Fowler–Nordheim formula*, is in qualitative agreement with experiment.

Alpha-particle decay of nuclei

As a final application of the WKB method, we shall now discuss the decay of a nucleus into an alpha particle (a He nucleus with charge $2e$) plus a daughter nucleus, using a simple, successful theory proposed in 1928 by G. Gamow and independently by R. W. Gurney and E. U. Condon. Since the parent nuclei emit alpha particles, the latter may be supposed to exist as entities within these nuclei, at least for a short time before emission. The potential energy $V(r)$ of an alpha particle inside and outside a nucleus of radius R may then be represented schematically by the curve shown in Fig. 8.9. Inside the nucleus (i.e. for $r < R$), the alpha particle is acted upon by the nuclear forces, which are strongly attractive. In this region the function $V(r)$ is not well known and its precise shape will not be needed in what follows. For simplicity the potential $V(r)$ shown in Fig. 8.9 has been taken to be a square well of (constant)

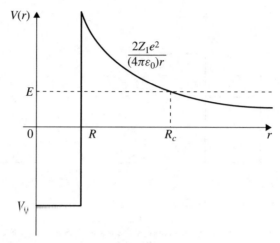

Figure 8.9 Schematic representation of the potential energy (solid curve) of an alpha particle inside and outside a nucleus of radius R. The total energy E of the alpha particle (dashed curve) is also the kinetic energy with which it emerges at large distances from the nucleus. The classical turning point R_c is given by (8.221).

depth V_0 for $r < R$. Outside the nucleus (i.e. for $r > R$) we shall assume that the nuclear forces (which are of short range) may be neglected, so that the alpha particle experiences only the Coulomb repulsion with the daughter nucleus. Let Ze be the charge carried by the parent nucleus and $Z_1 e$ that on the daughter nucleus, with $Z = Z_1 + 2$. The potential energy $V(r)$ of an alpha particle outside the nuclear surface is thus given approximately by

$$V(r) = \frac{2Z_1 e^2}{(4\pi\varepsilon_0)r}, \qquad r > R. \tag{8.220}$$

We note that in the vicinity of the nuclear radius R the potential $V(r)$ must dip sharply because of the strongly attractive nuclear forces. In the simple model illustrated in Fig. 8.9 this sharp dip of $V(r)$ has been idealised as a potential step.

Also shown in Fig. 8.9 is the total energy E of the alpha particle. Since the Coulomb potential energy (8.220) effectively vanishes at large r, the energy E is also the kinetic energy with which the alpha particle emerges at large distances from the nucleus. Note that because E is positive, the alpha particle is not considered as being initially in a bound state. Indeed, if this were the case the nucleus could not decay.

It is apparent from Fig. 8.9 that in order to escape from the nucleus the alpha particle has to penetrate a *potential barrier* in the region from R to R_c, where R_c is a classical turning point such that $E = V(R_c)$. Using (8.220) we have

$$R_c = \frac{2Z_1 e^2}{(4\pi\varepsilon_0)E}. \tag{8.221}$$

As a typical example, considered the α-decay of the ^{204}Po (polonium 204) nucleus, for which $Z = 84$. The energy of the emitted alpha particle is measured to be 5.4 MeV. Using (8.221) with $Z_1 = 82$ we then find that

$$R_c = \frac{2 \times 82 \times (1.6)^2 \times 10^{-38}}{4\pi \times 8.9 \times 10^{-12} \times 5.4 \times 1.6 \times 10^{-13}} \text{m}$$

$$\simeq 4.3 \times 10^{-14} \text{ m} = 43 \text{ fm} \tag{8.222}$$

where 1 fm (1 femtometer or fermi) $= 10^{-15}$ m. On the other hand, the radius R of a nucleus of mass number A is given approximately by

$$R = r_0 A^{1/3} \tag{8.223}$$

where $r_0 = 1.1$ fm. In the case of the ^{204}Po nucleus, for which $A = 204$, we have $R \simeq 6.5$ fm. The potential barrier is therefore very thick and the ratio R/R_c is small with respect to unity. Since the energies E of the alpha particles emitted in α-decays typically lie in the range 4–10 MeV, and the values of Z_1 and R do not vary much (all alpha emitters being heavy nuclei) the quantity R/R_c is generally quite small. It is also clear that since the maximum of the potential barrier is larger than the energy E of the alpha particle, the explanation of the α-decay of a nucleus is completely beyond the scope of classical mechanics; it is a *tunnelling* phenomenon which can only be understood by using quantum mechanics.

Let us suppose that the alpha particle is in a state of zero orbital angular momentum. Using spherical polar coordinates, we may then write its wave function as $r^{-1} u(r)$. The function $u(r) [\equiv u_{l=0}(r)]$ satisfies the radial equation (see (7.52))

$$-\frac{\hbar^2}{2\mu} \frac{d^2 u(r)}{dr^2} + V(r) u(r) = E u(r) \tag{8.224}$$

where the reduced mass μ may be taken to be the mass M of the alpha particle, since we are dealing with heavy nuclei. The equation (8.224) being identical in form to the one-dimensional Schrödinger equation (4.3), we can readily obtain the *transmission coefficient* T for the penetration of the potential barrier by using the one-dimensional WKB result (8.213), suitably modified. Thus we have

$$T = e^{-2\Lambda} \tag{8.225a}$$

where

$$\Lambda = \left(\frac{2M}{\hbar^2}\right)^{1/2} \int_R^{R_c} \left[\frac{2Z_1 e^2}{(4\pi\varepsilon_0)r} - E\right]^{1/2} dr. \tag{8.225b}$$

Using (8.221) we can also rewrite (8.225b) in the form

$$\Lambda = \left(\frac{4M Z_1 e^2}{\hbar^2 (4\pi\varepsilon_0)}\right)^{1/2} \int_R^{R_c} \left(\frac{1}{r} - \frac{1}{R_c}\right)^{1/2} dr. \tag{8.226}$$

The integral on the right of this equation can be done exactly, and one finds that

$$\Lambda = \left(\frac{4MZ_1e^2 R_c}{\hbar^2(4\pi\varepsilon_0)}\right)^{1/2}\left[\cos^{-1}\left(\frac{R}{R_c}\right)^{1/2} - \left(\frac{R}{R_c} - \frac{R^2}{R_c^2}\right)^{1/2}\right]. \tag{8.227}$$

We have seen above that the quantity (R/R_c) is generally fairly small compared to unity. Setting $x = (R/R_c)^{1/2}$, using the fact that

$$\cos^{-1} x = \frac{\pi}{2} - x + \mathcal{O}(x^3) \tag{8.228}$$

and keeping terms through order x in the square bracket on the right of (8.227), we have

$$\Lambda \simeq \left(\frac{4MZ_1e^2 R_c}{\hbar^2(4\pi\varepsilon_0)}\right)^{1/2}\left[\frac{\pi}{2} - 2\left(\frac{R}{R_c}\right)^{1/2}\right]. \tag{8.229}$$

Hence, from (8.221), (8.225a) and (8.229) we see that the transmission coefficient becomes

$$T \simeq \exp\left[-\frac{2\pi Z_1 e^2}{\hbar(4\pi\varepsilon_0)}\left(\frac{2M}{E}\right)^{1/2} + \frac{8}{\hbar}\left(\frac{Z_1 e^2}{4\pi\varepsilon_0} MR\right)^{1/2}\right]. \tag{8.230}$$

Let us denote by $Z_2 e$ (with $Z_2 = 2$) the charge carried by the alpha particle and write its energy as $E = Mv^2/2$, where v is the velocity with which it emerges at large distances from the nucleus. The result (8.230) may then be written as

$$T \simeq T_G \exp\left[\frac{32 Z_1 Z_2 e^2}{(4\pi\varepsilon_0)\hbar^2} MR\right]^{1/2} \tag{8.231}$$

where the quantity

$$T_G = \exp\left[-\frac{2\pi Z_1 Z_2 e^2}{(4\pi\varepsilon_0)\hbar v}\right] \tag{8.232}$$

is called the *Gamow factor* and is seen to inhibit the transmission of the alpha particle through the barrier.

With the help of the expression (8.230) for the transmission coefficient, we shall now estimate the lifetime of the alpha emitter. In order to do this, we consider a very simple model, in which the surface of the nucleus is considered as a 'wall', and the alpha particle bounces back and forth inside the nucleus before the emission occurs. The probability of getting through the barrier on a single collision with the 'wall' is equal to the transmission coefficient T, so that the number of collisions required to get out of the nucleus is of the order of T^{-1}. The time interval between successive collisions is roughly equal to the diameter of the nucleus $(2R)$ divided by the velocity of the alpha particle inside the nucleus, which we shall assume to be the same as its velocity $v = (2E/M)^{1/2}$ after emission. Hence the lifetime is given approximately by

$$\tau \simeq \frac{\Delta t}{T} \tag{8.233}$$

where

$$\Delta t = \frac{2R}{v} = 2R\left(\frac{M}{2E}\right)^{1/2}. \quad (8.234)$$

Using the expression (8.230) of the transmission coefficient, and taking the common logarithm (in base 10) of both members of (8.233) we obtain

$$\text{Log}_{10}\tau \simeq \text{Log}_{10}\Delta t + \frac{C}{\sqrt{E}} - D \quad (8.235)$$

where C and D are positive constants.

Let us evaluate the quantities Δt, C and D for the typical case of the α-decay of ^{204}Po. The energy E of the emitted alpha particle being $E = 5.4$ MeV, we have $v \simeq 1.6 \times 10^7$ m s^{-1}, and since $R \simeq 6.5$ fm we find that $\Delta t \simeq 8 \times 10^{-22}$ s. From (8.230) we obtain $C \simeq 141$ if we agree to express E in MeV, and we also have $D \simeq 30$. Thus we may write in that case

$$\text{Log}_{10}[\tau(\text{s})] \simeq \frac{141}{\sqrt{E(\text{MeV})}} - 51. \quad (8.236)$$

Although the quantity $\text{Log}_{10}\Delta t$ (and hence the second term on the right of the above equation) also depends on the energy E, this dependence is logarithmic and hence is very weak compared to that contained in the first term. In other words, the energy dependence of the lifetime τ is almost completely controlled by the term $CE^{-1/2}$ which depends entirely on the tunnelling through the barrier. This is fortunate since the model we have used to obtain the quantity Δt is a very naive one.

The foregoing discussion shows that in first approximation the lifetime τ and the alpha-particle energy E of an alpha-particle emitter satisfy a relation of the form

$$\text{Log}_{10}[\tau(\text{s})] = \frac{C_1}{\sqrt{E(\text{MeV})}} - C_2 \quad (8.237)$$

where the constants C_1 and C_2 should be close to those appearing in (8.236). That this is indeed the case is clear from Fig. 8.10, where the formula (8.237) (represented by a straight line) is seen to give a good fit to the lifetime data of a large number of alpha emitters. Note that these lifetimes vary over an enormous range, from 3×10^{-7} s for ^{212}Po to 4.5×10^9 years for ^{238}U.

The Gamow factor and thermonuclear fusion

We have seen above that the Gamow factor (8.232) inhibits the transmission of alpha particles through the Coulomb barrier. Let us now consider a collision experiment in which a nucleus is bombarded by alpha particles (and more generally by positively charged particles of charge $Z_2 e$) having an energy lower than the maximum of the Coulomb barrier at $r \simeq R$. Using a simple one-dimensional model of this process it is clear from the above discussion that the Gamow factor (8.232) will also inhibit the approach of this charged particle to the nucleus. As a result, the probability of reactions between nuclei is small at low relative velocities and (or) when Z_1 and

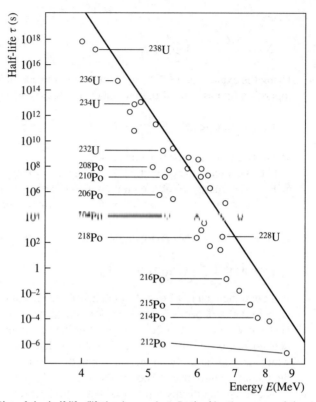

Figure 8.10 Plot of the half-life (lifetime) τ against E, the kinetic energy of the emitted alpha particle. The formula (8.237) yields the solid straight line. The small circles correspond to the experimental data for various alpha-radioactive nuclei.

(or) Z_2 is large. This is one of the reasons for selecting reactions between hydrogen isotopes (for which $Z = 1$) in controlled thermonuclear fusion research. In addition, nuclei having small values of the mass number A release the most energy per nucleon. Possible reactions for controlled thermonuclear fusion are

$$D + D \to T + p + 4.04 \text{ MeV} \tag{8.238a}$$

$$D + D \to {}^3\text{He} + n + 3.27 \text{ MeV} \tag{8.238b}$$

$$D + T \to {}^4\text{He} + n + 17.58 \text{ MeV} \tag{8.238c}$$

where p denotes a proton, n a neutron, D a deuteron (the nucleus of deuterium, the isotope of hydrogen with $A = 2$), and T a triton (the nucleus of tritium, the isotope of hydrogen with $A = 3$). The D–T reaction is considered at the present time as the best candidate, because it has a higher probability than the D–D reactions at the energies of interest for fusion reactors, and also because the D–T process produces more energy per reaction.

Problems

8.1 Show that the second-order energy correction $E_n^{(2)}$ given by (8.17) can also be expressed as

$$E_n^{(2)} = -\langle \psi_n^{(1)} | H_0 - E_n^{(0)} | \psi_n^{(1)} \rangle.$$

8.2 Derive equation (8.18) for the third-order correction $E_n^{(3)}$.

8.3 Obtain expressions for $E_n^{(4)}$ and $E_n^{(5)}$ in terms $\psi_n^{(0)}$, $\psi_n^{(1)}$ and $\psi_n^{(2)}$.

8.4 Show that the second-order correction to the energy in non-degenerate perturbation theory can be written as

$$E_n^{(2)} = \frac{(H'^2)_{nn}}{E_n^{(0)}} - \frac{(H'_{nn})^2}{E_n^{(0)}} + \sum_{k \neq n} \frac{E_k^{(0)} |H'_{nk}|^2}{E_n^{(0)}(E_n^{(0)} - E_k^{(0)})}.$$

8.5 Obtain the expression (8.38) for $E_n^{(3)}$.

8.6 Consider a one-dimensional linear harmonic oscillator perturbed by a Gaussian perturbation $H' = \lambda \exp(-ax^2)$. Calculate the first-order correction to the ground-state energy and to the energy of the first excited state.

8.7 Consider an electron in the electrostatic field of a nucleus of charge Ze. Let the nuclear charge be distributed uniformly within a sphere of radius R, so that the electrostatic potential due to the nucleus is

$$V(r) = \frac{Ze^2}{(4\pi\varepsilon_0)2R}\left(\frac{r^2}{R^2} - 3\right), \quad r \leqslant R$$

$$= -\frac{Ze^2}{(4\pi\varepsilon_0)r}, \quad r > R$$

Taking the unperturbed Hamiltonian to be the hydrogenic Hamiltonian $H_0 = -(\hbar^2/2\mu)\nabla^2 - Ze^2/(4\pi\varepsilon_0 r)$ and the perturbation to be the difference between $V(r)$ and the Coulomb interaction $-Ze^2/[(4\pi\varepsilon_0)r]$:

(a) Show that the first-order energy shift due to this perturbation is

$$\Delta E = \frac{Ze^2}{(4\pi\varepsilon_0)2R} \int_0^R [R_{nl}(r)]^2 \left(\frac{r^2}{R^2} + \frac{2R}{r} - 3\right) r^2 dr$$

where $R_{nl}(r)$ is a radial hydrogenic wave function.

(b) Taking $R_{nl}(r) \simeq R_{nl}(0)$ inside the nucleus show that

$$\Delta E = \frac{e^2}{(4\pi\varepsilon_0)} \frac{2}{5} R^2 \frac{Z^4}{a_\mu^3 n^3} \delta_{l0}.$$

8.8 Derive equation (8.56).

8.9 Derive equations (8.83) for c_{r1} and c_{r2}.

8.10 Starting from the relativistic expression for the energy of a particle $E = (m^2c^4 + \mathbf{p}^2c^2)^{1/2}$, and expanding in powers of $\mathbf{p}^2$, obtain the result (8.92) for the leading relativistic correction to the kinetic energy.

8.11 Consider a two-dimensional isotropic harmonic oscillator having the Hamiltonian

$$H_0 = -\frac{\hbar^2}{2m}\left(\frac{\partial^2}{\partial x^2} + \frac{\partial^2}{\partial y^2}\right) + \frac{1}{2}k(x^2 + y^2)$$

so that its energy levels are $E_n = \hbar\omega(n+1)$, with $\omega = (k/m)^{1/2}$, $n = n_x + n_y$ and $n_x, n_y = 0, 1, \ldots$. Assume that a perturbation $H' = \lambda xy$ is added, where λ is constant. Find the first-order modifications of the energy of the ground state and of the first excited state.

8.12 Consider a 'slightly anisotropic' two-dimensional harmonic oscillator, the Hamiltonian of which is

$$H_0 = -\frac{\hbar^2}{2m}\left(\frac{\partial^2}{\partial x^2} + \frac{\partial^2}{\partial y^2}\right) + \frac{1}{2}(k_1 x^2 + k_2 y^2)$$

where $|k_1 - k_2|$ is small. Given the energy levels $E_{n_x n_y} = \hbar\omega_1\left(n_x + \frac{1}{2}\right) + \hbar\omega_2\left(n_y + \frac{1}{2}\right)$, where $\omega_1 = (k_1/m)^{1/2}$ and $\omega_2 = (k_2/m)^{1/2}$, find the change in energy of the levels with $n_x = 1$, $n_y = 0$ and $n_x = 0$, $n_y = 1$ when a perturbation $H' = \lambda xy$ is added, where λ is a constant.

8.13 (a) By varying the parameter c in the trial function

$$\phi_0(x) = (c^2 - x^2)^2, \quad |x| < c$$
$$= 0, \quad |x| \geq c$$

obtain an upper bound for the ground-state energy of a linear harmonic oscillator having the Hamiltonian

$$H = -\frac{\hbar^2}{2m}\frac{d^2}{dx^2} + \frac{1}{2}m\omega^2 x^2.$$

(b) Show that the function $\phi_1(x) = x\phi_0(x)$ is a suitable trial function for the first excited state, and obtain a variational estimate of the energy of this level.

8.14 Using the trial functions of Problem 8.13 find estimates of the ground and first excited state energies of the cut-off harmonic oscillator

$$V(x) = \frac{1}{2}m\omega^2(x^2 - b^2) \quad |x| \leq b$$
$$= 0, \quad |x| \geq b.$$

Consider the two cases $c < b$ and $b < c$ for a fixed value of b, and explain which results are more accurate.

8.15 Find an estimate for the energy of the ground state of atomic hydrogen by taking as a trial function $\phi_0(r) = \exp(-cr^2)$, where c is a variational parameter.

8.16 Prove the inequality (8.160).
(*Hint*: Prove first that $|\langle \psi_0 | \phi_1 \rangle|^2 \leqslant \varepsilon_0$.)

8.17 By substituting the trial function (8.162) into the functional (8.130) and minimising the resulting expression with respect to the parameter c, obtain a variational estimate of the first excited level E_1 of the one-dimensional infinite square well (4.86). Compare this variational estimate with the exact result, obtained by setting $n = 2$ in (4.95), and verify that the variational estimate gives an upper bound for E_1.

8.18 Let $E_{n,l}$ denote the discrete energy levels of a particle of mass m in a central potential $V(r)$, corresponding to a given orbital angular momentum quantum number l, and let $E_{n,l}^{\min}$ be their minimum value. Prove that $E_{n,l}^{\min} \leqslant E_{n,l+1}^{\min}$.
(*Hint*: Write the Hamiltonian of the particle as

$$H = H_r + \frac{\mathbf{L}^2}{2mr^2}, \qquad H_r = -\frac{\hbar^2}{2m} \frac{1}{r^2} \frac{\partial}{\partial r}\left(r^2 \frac{\partial}{\partial r}\right) + V(r)$$

and note that H_r is a purely radial operator.)

8.19 Prove the Hylleraas–Undheim theorem for the case $N = 2$, namely

$$E_0^{(3)} \leqslant E_0^{(2)} \leqslant E_1^{(3)} \leqslant E_1^{(2)} \leqslant E_2^{(3)}.$$

8.20 Using equation (8.204) show that the WKB approximation gives the correct energy levels for all states of the linear harmonic oscillator.

8.21 Consider the motion of a particle of mass m in a potential well $V(x)$ such that

$$V(x) = 0, \qquad |x| > a$$
$$= -V_0\left(1 - \frac{|x|}{a}\right) \qquad |x| < a$$

where $V_0 > 0$. Use the WKB approximation to find all the bound energy levels if $mV_0 a^2/\hbar^2 = 50$.

9 Approximation methods for time-dependent problems

9.1 Time-dependent perturbation theory. General features 431
9.2 Time-independent perturbation 435
9.3 Periodic perturbation 443
9.4 The adiabatic approximation 447
9.5 The sudden approximation 458
Problems 466

In this chapter we shall continue our study of approximation methods by considering systems for which the Hamiltonian depends on the time. In this case it is generally impossible to obtain exact solutions of the Schrödinger equation. We shall examine three approximation schemes, which are all based on the assumption that the actual Hamiltonian of the system can in some way be approximated by a time-independent Hamiltonian. The first case to be considered will be that for which the time-dependent part of the Hamiltonian is small enough compared to the stationary part, so that *perturbation theory* can be used. After discussing the general features of the time-dependent perturbation method in Section 9.1, we apply it successively in the next two sections to perturbations which are constant and periodic in time. In Section 9.4 we discuss the *adiabatic approximation*, which applies to Hamiltonians changing very slowly on a time scale set by the periodic times associated with the approximate stationary solutions. Finally, we develop in Section 9.5 the *sudden* approximation in order to deal with the opposite situation of a very rapidly varying Hamiltonian.

9.1 Time-dependent perturbation theory. General features

Let us consider a system for which the total, time-dependent Hamiltonian H can be split as

$$H = H_0 + \lambda H'(t) \tag{9.1}$$

where the 'unperturbed' Hamiltonian H_0 is time-independent and $\lambda H'(t)$ is a *time-dependent* perturbation. As in the case of time-independent perturbations, the parameter λ, which may be taken to have the value unity for the actual physical problem, has been introduced for convenience. It will allow us again to identify the different orders of the perturbation calculation and to pass smoothly from the physical problem to the unperturbed one by letting λ tend to zero.

We assume that the eigenvalues $E_k^{(0)}$ of the unperturbed Hamiltonian H_0 are known, together with the corresponding eigenfunctions $\psi_k^{(0)}$, which we assume to be orthonormal and to form a complete set. Thus, since

$$H_0 \psi_k^{(0)} = E_k^{(0)} \psi_k^{(0)} \tag{9.2}$$

the general solution of the time-dependent unperturbed Schrödinger equation

$$i\hbar \frac{\partial \Psi_0}{\partial t} = H_0 \Psi_0 \tag{9.3}$$

is given by

$$\Psi_0 = \sum_k c_k^{(0)} \psi_k^{(0)} \exp(-iE_k^{(0)} t/\hbar) \tag{9.4}$$

where the coefficients $c_k^{(0)}$ are *constants* and the summation symbol implies a sum over the entire set (discrete plus continuous) of eigenfunctions $\psi_k^{(0)}$, as usual. Assuming that Ψ_0 is normalised to unity and remembering the discussion of Section 5.3 (see (5.68)), we can interpret $|c_k^{(0)}|^2$ as the probability of finding the system in the unperturbed energy state $\psi_k^{(0)}$ in the absence of perturbation ($\lambda = 0$).

Since the Hamiltonian H is time-dependent, there are no stationary solutions of the Schrödinger equation

$$i\hbar \frac{\partial \Psi}{\partial t} = H\Psi \tag{9.5}$$

so that energy is not conserved. It is therefore meaningless to seek corrections to the energy eigenvalues. The problem, instead, is to calculate approximately the wave function Ψ, given the unperturbed wave functions $\psi_k^{(0)}$ corresponding to the unperturbed energies $E_k^{(0)}$. We shall do this by using an approach known as *Dirac's method of variation of constants*.

We first note that because the functions $\psi_k^{(0)}$ form a complete set, the general solution Ψ of the time-dependent Schrödinger equation (9.5) can be expanded as

$$\Psi = \sum_k c_k(t) \psi_k^{(0)} \exp(-iE_k^{(0)} t/\hbar) \tag{9.6}$$

where the unknown coefficients $c_k(t)$ must now be regarded as *functions of time*. Since the functions $\psi_k^{(0)}$ are orthonormal, and provided Ψ is normalised to unity, we see that $|c_k(t)|^2 = |\langle \psi_k^{(0)} | \Psi(t) \rangle|^2$ is the probability of finding the system in the unperturbed state $\psi_k^{(0)}$ at time t, and $c_k(t)$ is the corresponding probability amplitude. Moreover,

$$\sum_k |c_k(t)|^2 = 1. \tag{9.7}$$

Upon comparison of (9.4) and (9.6) we also see that if there is no perturbation ($\lambda = 0$) the coefficients c_k reduce to the constants $c_k^{(0)}$, which are therefore the *initial values* of the c_k, specifying the state of the system before the perturbation is applied.

In order to find equations for the unknown coefficients $c_k(t)$, we substitute the expansion (9.6) into the Schrödinger equation (9.5) and use (9.1). We then have

$$i\hbar \sum_k \dot{c}_k(t)\psi_k^{(0)} \exp(-iE_k^{(0)}t/\hbar) + \sum_k c_k(t) E_k^{(0)} \psi_k^{(0)} \exp(-iE_k^{(0)}t/\hbar)$$
$$= \sum_k [H_0 + \lambda H'(t)] c_k(t) \psi_k^{(0)} \exp(-iE_k^{(0)}t/\hbar) \qquad (9.8)$$

where the dot denotes a derivative with respect to time. Now, according to (9.2) $H_0 \psi_k^{(0)} = E_k^{(0)} \psi_k^{(0)}$ so that equation (9.8) simplifies to yield

$$i\hbar \sum_k \dot{c}_k(t) \psi_k^{(0)} \exp(-iE_k^{(0)}t/\hbar) = \lambda \sum_k H'(t) c_k(t) \psi_k^{(0)} \exp(-iE_k^{(0)}t/\hbar). \qquad (9.9)$$

Taking the scalar product with a function $\psi_b^{(0)}$ belonging to the set $\{\psi_k^{(0)}\}$ of unperturbed energy eigenfunctions, and using the fact that $\langle \psi_b^{(0)} | \psi_k^{(0)} \rangle = \delta_{bk}$, we find that

$$\dot{c}_b(t) = (i\hbar)^{-1} \lambda \sum_k H'_{bk}(t) \exp(i\omega_{bk} t) c_k(t) \qquad (9.10)$$

where $H'_{bk}(t)$ denotes the matrix element

$$H'_{bk}(t) = \langle \psi_b^{(0)} | H'(t) | \psi_k^{(0)} \rangle \qquad (9.11)$$

and ω_{bk} is the *Bohr angular frequency*

$$\omega_{bk} = \frac{E_b^{(0)} - E_k^{(0)}}{\hbar}. \qquad (9.12)$$

The set of equations (9.10) for all b constitutes a system of first-order coupled differential equations which is strictly equivalent to the Schrödinger equation (9.5), since no approximation has been made thus far. We remark in particular that for any b the time rate $\dot{c}_b(t)$ depends upon *all* states $\psi_k^{(0)}$ connected with $\psi_b^{(0)}$ via the matrix element (9.11) of the perturbation. This can be understood in terms of the equation (9.7), which expresses probability conservation. Indeed, a change in one of the coefficients, for example c_b, implies that other coefficients must also change in order to satisfy the constraint imposed by (9.7).

Let us now assume that the perturbation $\lambda H'$ is weak. We can then expand the coefficients c_k in powers of the parameter λ as

$$c_k = c_k^{(0)} + \lambda c_k^{(1)} + \lambda^2 c_k^{(2)} + \cdots \qquad (9.13)$$

Substituting this expansion into the system (9.10) and equating the coefficients of equal powers of λ, we find that

$$\dot{c}_b^{(0)} = 0 \tag{9.14a}$$

$$\dot{c}_b^{(1)} = (i\hbar)^{-1} \sum_k H'_{bk}(t) \exp(i\omega_{bk}t) c_k^{(0)} \tag{9.14b}$$

$$\vdots \qquad \vdots$$

$$\dot{c}_b^{(s+1)} = (i\hbar)^{-1} \sum_k H'_{bk}(t) \exp(i\omega_{bk}t) c_k^{(s)}, \qquad s = 0, 1, \ldots \tag{9.14c}$$

These equations can now in principle be integrated successively to any given order in the perturbation.

The first equation (9.14a) simply confirms that the coefficients $c_k^{(0)}$ are time-independent. As we have seen above, the constants $c_k^{(0)}$ define the initial conditions of the problem. In what follows, we shall assume that the system is initially (that is, for $t \leqslant t_0$) in a particular unperturbed state $\psi_a^{(0)}$ of energy $E_a^{(0)}$. Thus we write

$$c_k^{(0)} = \delta_{ka} \qquad \text{or} \qquad \delta(k-a) \tag{9.15}$$

according to whether the state $\psi_a^{(0)}$ is discrete or continuous.

Substituting (9.15) into (9.14b), we have

$$\dot{c}_b^{(1)}(t) = (i\hbar)^{-1} H'_{ba}(t) \exp(i\omega_{ba}t) \tag{9.16}$$

where $\omega_{ba} = (E_b^{(0)} - E_a^{(0)})/\hbar$. The equation is readily solved to give

$$c_a^{(1)}(t) = (i\hbar)^{-1} \int_{t_0}^t H'_{aa}(t') dt' \tag{9.17a}$$

for $b = a$, while for $b \neq a$ one has

$$c_b^{(1)}(t) = (i\hbar)^{-1} \int_{t_0}^t H'_{ba}(t') \exp(i\omega_{ba}t') dt', \qquad b \neq a \tag{9.17b}$$

where the integration constant has been chosen in such a way that $c_a^{(1)}(t)$ and $c_b^{(1)}(t)$ vanish at $t = t_0$, before the perturbation is applied. To first order in the perturbation, the *transition probability* corresponding to the transition $a \to b$ (i.e. the probability that the system, initially in the state a, be found at time t in the state $b \neq a$) is therefore

$$P_{ba}^{(1)}(t) = |c_b^{(1)}(t)|^2 = \hbar^{-2} \left| \int_{t_0}^t H'_{ba}(t') \exp(i\omega_{ba}t') dt' \right|^2, \qquad b \neq a. \tag{9.18}$$

It is also worth noting that for $t > t_0$ the coefficient c_a of the state a is given to first order in the perturbation by

$$\begin{aligned} c_a(t) &\simeq c_a^{(0)} + c_a^{(1)}(t) \\ &\simeq 1 + (i\hbar)^{-1} \int_{t_0}^{t} H'_{aa}(t') dt' \\ &\simeq \exp\left[-\frac{i}{\hbar} \int_{t_0}^{t} H'_{aa}(t') dt'\right] \end{aligned} \quad (9.19)$$

so that $|c_a(t)|^2 \simeq 1$ and the principal effect of the perturbation on the initial state is to change its phase.

9.2 Time-independent perturbation

The results (9.17) take a particularly simple form if the perturbation H' is independent of time, except for being switched on suddenly at a given time (say $t_0 = 0$). We then obtain

$$c_a^{(1)}(t) = (i\hbar)^{-1} H'_{aa} t \quad (9.20a)$$

and

$$c_b^{(1)}(t) = \frac{H'_{ba}}{\hbar \omega_{ba}} [1 - \exp(i\omega_{ba} t)], \quad b \neq a. \quad (9.20b)$$

Note that if the perturbation H' is 'switched off' at time t, the above amplitudes are also those at any subsequent time $t_1 > t$.

From (9.19) and (9.20a) we find that the coefficient $c_a(t)$ is given to first order in the perturbation by

$$c_a(t) \simeq \exp\left(-\frac{i}{\hbar} H'_{aa} t\right). \quad (9.21)$$

Hence

$$c_a(t) \psi_a^{(0)} \exp(-iE_a^{(0)} t/\hbar) \simeq \psi_a^{(0)} \exp\left[-\frac{i}{\hbar}(E_a^{(0)} + H'_{aa}) t\right] \quad (9.22)$$

and we see that in first approximation, the energy of the perturbed state is

$$E \simeq E_a^{(0)} + H'_{aa} \quad (9.23)$$

where H'_{aa} is the average value of the perturbation in the state a. This is in agreement with the result (8.15) which we obtained in Chapter 8 using stationary-state perturbation theory.

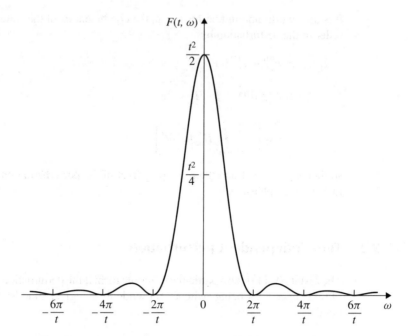

Figure 9.1 The function $F(t, \omega)$ of equation (9.25), for fixed t.

Using (9.18) and (9.20b), the first-order transition probability from state a to state $b \neq a$ is given by

$$P_{ba}^{(1)}(t) = \frac{2}{\hbar^2}|H'_{ba}|^2 F(t, \omega_{ba}) \tag{9.24}$$

where we have introduced the function

$$F(t, \omega) = \frac{1 - \cos \omega t}{\omega^2} = \frac{2\sin^2(\omega t/2)}{\omega^2} \tag{9.25}$$

and we remark that $F(t, \omega = 0) = t^2/2$.

It is useful for further purposes to study in detail the function $F(t, \omega)$. This function is plotted in Fig. 9.1 for fixed t. We see that it exhibits a sharp peak about the value $\omega = 0$. The height of this peak is proportional to t^2, while its width is approximately $2\pi/t$. Setting $x = \omega t/2$, we note that

$$\int_{-\infty}^{+\infty} F(t, \omega) d\omega = t \int_{-\infty}^{+\infty} \frac{\sin^2 x}{x^2} dx = \pi t \tag{9.26}$$

where we have used a standard integral. From equation (A.24) of Appendix A we also deduce that in the limit $t \to \infty$

$$F(t, \omega) \underset{t \to \infty}{\sim} \pi t \delta(\omega). \tag{9.27}$$

Let us now return to the result (9.24) and analyse it first for a *fixed* value of t. Since the function $F(t, \omega_{ba})$ is sharply peaked about the value $\omega_{ba} = 0$, with a width approximately given by $2\pi/t$, it is clear that transitions to those final states b for which ω_{ba} does not deviate from zero by more than $\delta\omega_{ba} \simeq 2\pi/t$ will be strongly favoured. Hence the transitions $a \to b$ will occur mainly towards those final states whose energy $E_b^{(0)}$ is located in a band of width

$$\delta E \simeq 2\pi\hbar/t \tag{9.28}$$

about the initial energy $E_a^{(0)}$, so that the unperturbed energy is conserved within $2\pi\hbar/t$. This result can readily be related to the time–energy uncertainty relation $\Delta E \Delta t \gtrsim \hbar$. Indeed, since the perturbation gives a way of measuring the energy of the system by inducing transitions $a \to b$, and because this perturbation acts during a time t, the uncertainty related to this energy measurement should be of order $\hbar/t$, in agreement with (9.28).

We now examine the transition probability (9.24) *as a function of t*. We begin by considering transitions to a *given state $b(\neq a)$*, and distinguish two cases.

Non-degenerate case

If $\omega_{ba} \neq 0$ (i.e. if the state b is not degenerate with the initial state a) we have from (9.24) and (9.25)

$$P_{ba}^{(1)}(t) = \frac{4|H_{ba}'|^2}{\hbar^2 \omega_{ba}^2} \sin^2(\omega_{ba} t/2) \tag{9.29}$$

and we see that $P_{ba}^{(1)}(t)$ oscillates with a period $T = 2\pi/|\omega_{ba}|$ about the average value $2|H_{ba}'|^2/\hbar^2\omega_{ba}^2$ and is such that

$$P_{ba}^{(1)}(t) \leqslant \frac{4|H_{ba}'|^2}{\hbar^2 \omega_{ba}^2}. \tag{9.30}$$

We note that for times t small with respect to the period of oscillation we have $\sin(\omega_{ba}t/2) \simeq \omega_{ba}t/2$ so that $P_{ba}^{(1)}(t) \simeq |H_{ba}'|^2 t^2/\hbar^2$ increases quadratically with time.

The total (first-order) probability $P^{(1)}(t)$ that the system has made at time t a transition away from the initial state a is given by summing the transition probabilities (9.29) over all final states $b \neq a$ (assuming that none of the states b is degenerate with the initial state a). Thus

$$P^{(1)}(t) = \sum_{b \neq a} P_{ba}^{(1)}(t) = \sum_{b \neq a} \frac{4|H_{ba}'|^2}{(E_b^{(0)} - E_a^{(0)})^2} \sin^2[(E_b^{(0)} - E_a^{(0)})t/2\hbar]. \tag{9.31}$$

Clearly, in order that our perturbation treatment be valid, it is necessary that

$$P^{(1)}(t) \ll 1. \tag{9.32}$$

If this condition is fulfilled, the individual transition probabilities will all be small

$$P_{ba}^{(1)}(t) \ll 1 \tag{9.33}$$

so that there will be little change in the initial state for times t of physical interest. From (9.30), (9.31) and the fact that $\sin^2 x \leqslant 1$ we can readily obtain a sufficient condition for the validity of our perturbative treatment, namely

$$\sum_{b \neq a} \frac{4|H'_{ba}|^2}{(E_b^{(0)} - E_a^{(0)})^2} \ll 1. \tag{9.34}$$

Remembering that all the states b have been assumed to be non-degenerate with the initial state a (so that $E_b^{(0)} \neq E_a^{(0)}$), we see that the condition (9.34) can always be satisfied if the perturbation H' is sufficiently weak.

Degenerate case

We now turn to the case for which $\omega_{ba} = 0$ so that $E_b^{(0)} = E_a^{(0)}$ and the states a and b are degenerate. From (9.20b) we then have

$$c_b^{(1)}(t) = -\frac{\mathrm{i}}{\hbar} H'_{ba} t. \tag{9.35}$$

The first-order transition probability is given in this case by

$$P_{ba}^{(1)}(t) = \frac{|H'_{ba}|^2}{\hbar^2} t^2 \tag{9.36}$$

and hence increases indefinitely with the time. Thus, after a sufficient length of time the quantity $P_{ba}^{(1)}(t)$ will no longer satisfy the inequality (9.33) required by our perturbative approach, and will eventually exceed unity, which is clearly absurd. We therefore conclude that the present perturbative treatment cannot be applied to degenerate systems which are perturbed over long periods of time.

Two-level system with time-independent perturbation

As a simple illustration of the foregoing discussion, let us consider a quantum system having only two stationary states a and b, with unperturbed energies $E_a^{(0)}$ and $E_b^{(0)}$ and corresponding eigenfunctions $\psi_a^{(0)}$ and $\psi_b^{(0)}$, respectively. Such a two-level system is obviously an idealised model, but in a number of interesting situations it is reasonable to 'single out' two levels of a quantum system, as we shall see later in various applications.

Non-degenerate case

We begin by considering the non-degenerate case (see Fig. 9.2), so that $\omega_{ba} = (E_b^{(0)} - E_a^{(0)})/\hbar \neq 0$. We assume that $E_b^{(0)} > E_a^{(0)}$ and that the two-level system is initially in the ground state a. We also suppose that a constant perturbation H' is switched on at the time $t = 0$. The system of equations (9.10) reduces in this case to the set of two coupled first-order equations

$$\mathrm{i}\hbar \dot{c}_a(t) = H'_{aa} c_a(t) + H'_{ab} \exp(-\mathrm{i}\omega_{ba} t) c_b(t) \tag{9.37a}$$

$$\mathrm{i}\hbar \dot{c}_b(t) = H'_{ba} \exp(\mathrm{i}\omega_{ba} t) c_a(t) + H'_{bb} c_b(t) \tag{9.37b}$$

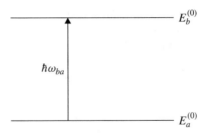

Figure 9.2 A two-level system with unperturbed energies $E_a^{(0)} \neq E_b^{(0)}$. The Bohr angular frequency ω_{ba} is such that $\omega_{ba} = (E_b^{(0)} - E_a^{(0)})/\hbar$.

where we have set $\lambda = 1$. The initial condition that the quantum system be in the state a before the perturbation is applied implies that

$$c_a(t \leqslant 0) = 1, \qquad c_b(t \leqslant 0) = 0. \tag{9.38}$$

The equations (9.37) are sufficiently simple so that they can be solved exactly in closed form. After some algebra (see Problem 9.4), it is found that the solutions $c_a(t)$ and $c_b(t)$, satisfying the initial conditions (9.38), are given by

$$c_a(t) = \exp(-i\alpha t)\left[\cos \beta t + i\frac{\gamma}{\beta}\sin \beta t\right] \tag{9.39a}$$

and

$$c_b(t) = -i\frac{H'_{ba}}{\hbar \beta}\exp[-i(\alpha - \omega_{ba})t]\sin \beta t \tag{9.39b}$$

where

$$\alpha = \frac{1}{2\hbar}(H'_{aa} + H'_{bb} + \hbar\omega_{ba}) \tag{9.40a}$$

$$\beta = \frac{1}{\hbar}\left[\frac{1}{4}(H'_{bb} - H'_{aa} + \hbar\omega_{ba})^2 + |H'_{ba}|^2\right]^{1/2} \tag{9.40b}$$

and

$$\gamma = \frac{1}{2\hbar}(H'_{bb} - H'_{aa} + \hbar\omega_{ba}) = \alpha - H'_{aa}/\hbar. \tag{9.40c}$$

The probability of finding the system in the ground state a at time $t > 0$ is therefore given by

$$|c_a(t)|^2 = \cos^2 \beta t + \left(\frac{\gamma}{\beta}\right)^2 \sin^2 \beta t$$

$$= 1 - \frac{|H'_{ba}|^2}{\hbar^2 \gamma^2 + |H'_{ba}|^2}\sin^2 \beta t \tag{9.41}$$

where we have used the fact that $\beta^2 - \gamma^2 = |H'_{ba}|^2/\hbar^2$. The probability of finding the system in the excited state b at time t (the probability that the transition $a \to b$ has occurred) is

$$P_{ba}(t) = |c_b(t)|^2 = \frac{|H'_{ba}|^2}{\hbar^2 \beta^2} \sin^2 \beta t$$

$$= \frac{|H'_{ba}|^2}{\hbar^2 \gamma^2 + |H'_{ba}|^2} \sin^2 \beta t \tag{9.42}$$

and we verify immediately from (9.41) and (9.42) that $|c_a(t)|^2 + |c_b(t)|^2 = 1$. We also note that the system oscillates between the two levels with a period $T = \pi/\beta$.

Having obtained exact results for this model problem, we can now check the accuracy of a first-order perturbative approach. From the initial condition we have $c_a^{(0)} = 1$ and $c_b^{(0)} = 0$. Hence, using (9.20) we find that to first order in the perturbation

$$c_a(t) \simeq c_a^{(0)} + c_a^{(1)}(t) = 1 + (i\hbar)^{-1} H'_{aa} t \simeq \exp\left(-\frac{i}{\hbar} H'_{aa} t\right) \tag{9.43a}$$

and

$$c_b(t) \simeq c_b^{(1)}(t) = \frac{H'_{ba}}{\hbar \omega_{ba}} [1 - \exp(i\omega_{ba} t)]. \tag{9.43b}$$

These results are readily seen to agree with the exact results (9.39) to first order in H' and for small times t (Problem 9.5).

Degenerate case

We now turn to the situation for which the two energy levels are degenerate, so that $E_a^{(0)} = E_b^{(0)} = E^{(0)}$. As a simple example, we shall consider the case such that $H'_{aa} = H'_{bb} = 0$ and H'_{ba} is real and positive. Using the results of our study of doubly degenerate energy levels made in Section 8.2, we find that the correct normalised zero-order wave functions are

$$\chi_a^{(0)} = \frac{1}{\sqrt{2}} (\psi_a^{(0)} + \psi_b^{(0)}) \tag{9.44a}$$

and

$$\chi_b^{(0)} = \frac{1}{\sqrt{2}} (\psi_a^{(0)} - \psi_b^{(0)}) \tag{9.44b}$$

the corresponding energies being given to first order in the perturbation by $E^{(0)} + H'_{ba}$ and $E^{(0)} - H'_{ba}$, respectively.

Let $\Psi(t)$ be the time-dependent wave function of the system[1], and suppose that at the initial time $t = 0$ the system is in the eigenstate $\psi_a^{(0)}$, so that

$$\Psi(t = 0) = \psi_a^{(0)}. \tag{9.45}$$

---

[1] In order to simplify the notation, only the time-dependence of Ψ is indicated explicitly.

9.2 Time-independent perturbation

Since $\chi_a^{(0)} \exp[-i(E^{(0)}+H'_{ba})t/\hbar]$ and $\chi_b^{(0)} \exp[-i(E^{(0)}-H'_{ba})t/\hbar]$ are (approximate) stationary states, we can write the wave function $\Psi(t)$ for $t > 0$ as the linear superposition

$$\Psi(t) = c_1 \chi_a^{(0)} \exp[-i(E^{(0)} + H'_{ba})t/\hbar] + c_2 \chi_b^{(0)} \exp[-i(E^{(0)} - H'_{ba})t/\hbar], \tag{9.46}$$

The coefficients c_1 and c_2 are readily obtained from the initial condition (9.45) and the relations (9.44). Indeed,

$$\Psi(t=0) = c_1 \chi_a^{(0)} + c_2 \chi_b^{(0)}$$
$$= c_1 \frac{1}{\sqrt{2}} (\psi_a^{(0)} + \psi_b^{(0)}) + c_2 \frac{1}{\sqrt{2}} (\psi_a^{(0)} - \psi_b^{(0)}) = \psi_a^{(0)} \tag{9.47}$$

giving $c_1 = c_2 = 2^{-1/2}$. Hence

$$\Psi(t) = \frac{1}{\sqrt{2}} \chi_a^{(0)} \exp[-i(E^{(0)} + H'_{ba})t/\hbar] + \frac{1}{\sqrt{2}} \chi_b^{(0)} \exp[-i(E^{(0)} - H'_{ba})t/\hbar]$$
$$= \exp(-iE^{(0)}t/\hbar) \frac{1}{\sqrt{2}} [\chi_a^{(0)} \exp(-iH'_{ba}t/\hbar) + \chi_b^{(0)} \exp(iH'_{ba}t/\hbar)]$$
$$= \exp(-iE^{(0)}t/\hbar)[\psi_a^{(0)} \cos(H'_{ba}t/\hbar) - i\psi_b^{(0)} \sin(H'_{ba}t/\hbar)]. \tag{9.48}$$

At $t = 0$ this wave function is clearly equal to $\psi_a^{(0)}$. As time goes on, the function $\psi_b^{(0)}$ 'comes in', which means that transitions from $\psi_a^{(0)}$ to $\psi_b^{(0)}$ occur. When $t = \pi\hbar/(2H'_{ba})$, the system is entirely in the state $\psi_b^{(0)}$; subsequently it reverts to $\psi_a^{(0)}$ and is completely in that state after a time $t = \pi\hbar/H'_{ba}$. Thus the system *oscillates* between the two states $\psi_a^{(0)}$ and $\psi_b^{(0)}$ with a frequency

$$\nu = \frac{H'_{ba}}{\pi\hbar} \tag{9.49}$$

proportional to the coupling term H'_{ba}. This oscillation is also referred to as a '*resonance*' between the two degenerate levels.

Let us now compare the above results with the predictions of first-order time-dependent perturbation theory. Following this approach, the coefficient c_b of the stationary state $\psi_b^{(0)} \exp(-iE^{(0)}t/\hbar)$ grows linearly with time according to (9.35). Looking at (9.48), we see that

$$c_b(t) = -i \sin(H'_{ba}t/\hbar) \tag{9.50}$$

which indeed reduces to (9.35) for small times. However, as time passes the result (9.35) starts to deviate from (9.50), indicating that perturbation theory is breaking down.

Transitions to a group of final states. The Golden Rule

Instead of considering transitions to a particular state b, as we have done so far, it is often necessary to deal with transitions to a *group* of final states n whose energy E_n lies within a given interval $(E_b^{(0)} - \eta, E_b^{(0)} + \eta)$ centred about the value $E_b^{(0)}$. This is the case, for example, when one studies transitions to states belonging to the *continuous spectrum*.

Let us denote by $\rho_n(E_n)$ the density of levels on the energy scale, so that $\rho_n(E_n)\mathrm{d}E_n$ is the number of final states n in an interval $\mathrm{d}E_n$ containing the energy E_n. We shall continue to assume that the perturbation H' is constant in time, except that it is switched on at $t = 0$. The first-order transition probability $P_{ba}^{(1)}(t)$ from the state a to the group of final states n having energies in the interval $(E_b^{(0)} - \eta, E_b^{(0)} + \eta)$ centred about $E_b^{(0)}$ is obtained by multiplying $P_{na}^{(1)}(t)$, as given by (9.24), by $\rho_n(E_n)\mathrm{d}E_n$ and integrating with respect to E_n

$$P_{ba}^{(1)}(t) = \frac{2}{\hbar^2} \int_{E_b^{(0)}-\eta}^{E_b^{(0)}+\eta} |H'_{na}|^2 F(t, \omega_{na}) \rho_n(E_n) \mathrm{d}E_n \tag{9.51}$$

where $\omega_{na} = (E_n - E_a^{(0)})/\hbar$. Assuming that η is small enough so that H'_{na} and $\rho_n(E_n)$ are nearly constant within the integration range, we have

$$P_{ba}^{(1)}(t) = \frac{2}{\hbar^2} |H'_{ba}|^2 \rho_b(E_b^{(0)}) \int_{E_b^{(0)}-\eta}^{E_b^{(0)}+\eta} F(t, \omega_{na}) \mathrm{d}E_n. \tag{9.52}$$

We shall also assume that t is large enough so that the quantity η satisfies the condition

$$\eta \gg 2\pi\hbar/t. \tag{9.53}$$

Now, it is clear from the form of the function $F(t, \omega)$ (see (9.25) and Fig. 9.1) that the integral on the right of (9.52) will be small except for transitions which conserve energy (within $\delta E = 2\pi\hbar/t$). Because of (9.53) we can then write

$$\int_{E_b^{(0)}-\eta}^{E_b^{(0)}+\eta} F(t, \omega_{na}) \mathrm{d}E_n \simeq \hbar \int_{-\infty}^{+\infty} F(t, \omega_{na}) = \pi\hbar t \tag{9.54}$$

where we have used the result (9.26). Hence (9.52) reduces to

$$P_{ba}^{(1)}(t) = \frac{2\pi}{\hbar} |H'_{ba}|^2 \rho_b(E)\, t \tag{9.55}$$

with $E = E_a^{(0)} = E_b^{(0)}$. Thus the transition probability increases linearly with time for energy-conserving transitions to a group of states. It is worth recalling that the above result is only soundly based if the condition (9.33) required by our perturbative approach is obeyed.

Introducing the *transition probability per unit time* or *transition rate*

$$W_{ba} = \frac{\mathrm{d}P_{ba}}{\mathrm{d}t} \tag{9.56}$$

we see from (9.55) that to first order in perturbation theory we have

$$W_{ba} = \frac{2\pi}{\hbar} |H'_{ba}|^2 \rho_b(E). \tag{9.57}$$

This formula, first obtained by P. A. M. Dirac, was later dubbed by E. Fermi the 'Golden Rule' of perturbation theory. We have derived it here for a perturbation H' which is constant in time (except for being switched on at $t = 0$). However, as we shall see below, it can be generalised to other cases and has wide applications in quantum physics.

9.3 Periodic perturbation

Another case for which the equations (9.17) take a simple form occurs when the perturbation $H'(t)$ is a *periodic* function of time, except for being turned on at $t = 0$. We shall assume first that the perturbation changes *sinusoidally* in time with an angular frequency ω. That is

$$\begin{aligned} H'(t) &= \hat{H}' \sin \omega t \\ &= A \exp(i\omega t) + A^\dagger \exp(-i\omega t) \end{aligned} \tag{9.58}$$

where $\hat{H}'$ is a time-independent Hermitian operator and $A = (1/2i)\hat{H}'$. It is assumed that the system is initially (say for $t \leqslant 0$) in the unperturbed bound state $\psi_a^{(0)}$, of energy $E_a^{(0)}$, so that the initial conditions are $c_a(t \leqslant 0) = 1$ and $c_b(t \leqslant 0) = 0$ for $b \neq a$.

According to (9.19) we have $|c_a(t)|^2 \simeq 1$ for $t > 0$. In order to find the coefficient $c_b^{(1)}(t)$ for $t > 0$ and $b \neq a$, we substitute (9.58) into (9.17b) and use the fact that $t_0 = 0$. This gives

$$c_b^{(1)}(t) = (i\hbar)^{-1} \left\{ A_{ba} \int_0^t \exp[i(\omega_{ba} + \omega)t']dt' + A_{ba}^\dagger \int_0^t \exp[i(\omega_{ba} - \omega)t']dt' \right\} \tag{9.59}$$

where $A_{ba} = \langle \psi_b^{(0)} | A | \psi_a^{(0)} \rangle = (2i)^{-1} \hat{H}'_{ba}$ and $A_{ba}^\dagger = A_{ab}^*$. Performing the integrals, we find that

$$\begin{aligned} c_b^{(1)}(t) = A_{ba} &\frac{1 - \exp[i(E_b^{(0)} - E_a^{(0)} + \hbar\omega)t/\hbar]}{E_b^{(0)} - E_a^{(0)} + \hbar\omega} \\ &+ A_{ba}^\dagger \frac{1 - \exp[i(E_b^{(0)} - E_a^{(0)} - \hbar\omega)t/\hbar]}{E_b^{(0)} - E_a^{(0)} - \hbar\omega} \end{aligned} \tag{9.60}$$

where we have used the fact that $\hbar\omega_{ba} = E_b^{(0)} - E_a^{(0)}$. From (9.18), the corresponding first-order transition probability is

$$P_{ba}^{(1)}(t) = \left| A_{ba} \frac{1 - \exp[i(E_b^{(0)} - E_a^{(0)} + \hbar\omega)t/\hbar]}{E_b^{(0)} - E_a^{(0)} + \hbar\omega} \right.$$
$$\left. + A_{ba}^\dagger \frac{1 - \exp[i(E_b^{(0)} - E_a^{(0)} - \hbar\omega)t/\hbar]}{E_b^{(0)} - E_a^{(0)} - \hbar\omega} \right|^2. \tag{9.61}$$

It is clear from the above equations that if t is large enough the probability of finding the system in the state b will only be appreciable if the denominator of one or the other of the two terms on the right of (9.60) is close to zero. Moreover, assuming that $E_b^{(0)} \neq E_a^{(0)}$ (so that the levels $E_a^{(0)}$ and $E_b^{(0)}$ are not degenerate), both denominators cannot simultaneously be close to zero. A good approximation is therefore to neglect the interference between the two terms in calculating the transition probability (we shall come back to this point in more detail below). Thus, if the energy $E_b^{(0)}$ lies in a small band about the value

$$E = E_a^{(0)} + \hbar\omega \tag{9.62}$$

only the second term in (9.60) will have an appreciable magnitude, the corresponding transition probability being given by

$$P_{ba}^{(1)}(t) = \frac{2}{\hbar^2} |A_{ba}^\dagger|^2 F(t, \omega_{ba} - \omega). \tag{9.63}$$

The main difference with the expression (9.24), obtained for a time-independent perturbation, is that the angular frequency ω_{ba} has now been replaced by $\omega_{ba} - \omega$. From the properties of the function $F(t, \omega)$ discussed after equation (9.25), it is apparent that the transition probability (9.63) will only be significant if $E_b^{(0)}$ is located in an interval of width $2\pi\hbar/t$ about the value $E_a^{(0)} + \hbar\omega$. Hence the first-order transition probability (9.63) will be appreciable if the system has *absorbed* an amount of energy given (to within $2\pi\hbar/t$) by $\hbar\omega = E_b^{(0)} - E_a^{(0)}$. This, of course, is nothing but the Bohr frequency rule (1.52). When it is exactly satisfied, *resonance* is said to occur and we see from (9.63) and (9.25) that the first-order transition probability increases quadratically with time according to the formula

$$P_{ba}^{(1)}(t) = \frac{|A_{ba}^\dagger|^2}{\hbar^2} t^2 = \frac{|\hat{H}_{ba}'|^2}{4\hbar^2} t^2. \tag{9.64}$$

In the same way, if the energy $E_b^{(0)}$ lies in a small interval in the neighbourhood of the value

$$E = E_a^{(0)} - \hbar\omega \tag{9.65}$$

only the first term on the right of (9.60) will be significant. The corresponding first-order transition probability will be given by

$$P_{ba}^{(1)}(t) = \frac{2}{\hbar^2} |A_{ba}|^2 F(t, \omega_{ba} + \omega) \tag{9.66}$$

and will only be significant if the system has *emitted* an amount of energy given (to within $2\pi\hbar/t$) by $\hbar\omega = E_a^{(0)} - E_b^{(0)}$. Again resonance occurs if this condition is exactly satisfied, in which case the transition probability (9.66) increases quadratically with time. It should be noted that in practice t is large enough ($t \gg 2\pi/\omega$) so that the two bands of width $2\pi\hbar/t$ about the values (9.62) and (9.65) do not overlap. Thus our neglect of the interference between the two terms on the right of (9.61) in calculating $P_{ba}^{(1)}$ is indeed justified.

As in the case of a time-independent perturbation, one can consider transitions to a *group* of final states n whose energy E_n lies within an interval $(E_b^{(0)} - \eta, E_b^{(0)} + \eta)$ about the value $E_b^{(0)} = E_a^{(0)} + \hbar\omega$ (for absorption) or $E_b^{(0)} = E_a^{(0)} - \hbar\omega$ (for emission) with $\eta \gg 2\pi\hbar/t$. Let $\rho_n(E_n)$ be the density of levels E_n on the energy scale. Under conditions similar to those discussed above for a time-independent perturbation, a transition probability per unit time (transition rate) can be defined. For transitions in which the system *absorbs* an energy $\hbar\omega \simeq E_b^{(0)} - E_a^{(0)}$, this transition rate is given to first order by (Problem 9.6)

$$W_{ba} = \frac{2\pi}{\hbar}|A_{ba}^\dagger|^2 \rho_b(E) \qquad (9.67)$$

where $E = E_a^{(0)} + \hbar\omega$. The above expression is clearly the direct generalisation of the Golden Rule (9.57). Similarly, for transitions in which the system *emits* an energy $\hbar\omega \simeq E_a^{(0)} - E_b^{(0)}$ by making transitions to a group of final states, the corresponding transition rate is given (under the same conditions) by

$$W_{ba} = \frac{2\pi}{\hbar}|A_{ba}|^2 \rho_b(E) \qquad (9.68)$$

with $E = E_a^{(0)} - \hbar\omega$.

Until now we have considered a perturbation $H'(t)$ which is a sinusoidal function of time. However, the generalisation of the above results to a perturbation $H'(t)$ which is a general periodic function of time is straightforward. Indeed, we can then develop $H'(t)$ in the Fourier series (see Appendix A)

$$H'(t) = \sum_{n=1}^{\infty}[A_n \exp(in\omega t) + A_n^\dagger \exp(-in\omega t)] \qquad (9.69)$$

where the operators A_n are time-independent. For large enough times ($t \gg 2\pi/\omega$), there is no interference between the contributions of the various terms of this development to the transition probability, because each term corresponds to a different energy transfer. A term of the type $A_n \exp(in\omega t)$ will therefore lead to the *emission* by the system of an energy given (to within $2\pi\hbar/t$) by $n\hbar\omega = E_a^{(0)} - E_b^{(0)}$ while a term of the type $A_n^\dagger \exp(-in\omega t)$ corresponds to the *absorption* by the system of the energy $n\hbar\omega = E_b^{(0)} - E_a^{(0)}$ (to within $2\pi\hbar/t$). The corresponding first-order transition probabilities are readily obtained, either for transitions to a given state b or to a group of final states.

Two-level system with harmonic perturbation

As a simple application of some of the concepts developed above, let us consider a two-level system, with unperturbed energies $E_a^{(0)} < E_b^{(0)}$ and corresponding eigenfunctions $\psi_a^{(0)}$ and $\psi_b^{(0)}$, respectively. The system being initially in the state a, a perturbation of the form (9.58) is switched on at time $t = 0$. Setting $\lambda = 1$ in (9.10), we obtain the two coupled equations

$$i\hbar \dot{c}_a(t) = \{A_{aa}\exp(i\omega t) + A_{aa}^\dagger \exp(-i\omega t)\}c_a$$
$$+ \{A_{ab}\exp(i(\Delta\omega)t) + A_{ab}^\dagger \exp[-i(\omega+\omega_{ba})t]\}c_b \quad (9.70a)$$

and

$$i\hbar \dot{c}_b(t) = \{A_{ba}\exp[i(\omega_{ba}+\omega)t] + A_{ba}^\dagger \exp(-i(\Delta\omega)t)\}c_a$$
$$+ \{A_{bb}\exp(i\omega t) + A_{bb}^\dagger \exp(-i\omega t)\}c_b \quad (9.70b)$$

where $\omega_{ba} = (E_b^{(0)} - E_a^{(0)})/\hbar$ and we have introduced the 'detuning' angular frequency

$$\Delta\omega = \omega - \omega_{ba}. \quad (9.71)$$

The system (9.70) must be solved subject to the initial conditions

$$c_a(t \leqslant 0) = 1 \qquad c_b(t \leqslant 0) = 0. \quad (9.72)$$

The equations (9.70) cannot be solved exactly, but if it is assumed that $|\Delta\omega| \ll \omega$ (so that the angular frequency ω is always close to its resonant value $\omega = \omega_{ba}$) then the terms in $\exp(\pm i(\Delta\omega)t)$ will be much more important than those in $\exp[\pm i(\omega+\omega_{ba})t]$ and $\exp(\pm i\omega t)$. This is because the latter terms oscillate much more rapidly and on the average make little contribution to $\dot{c}_a$ or $\dot{c}_b$. It is therefore reasonable to neglect the higher frequency terms. This is known as the *rotating wave approximation* because the only terms which are kept are those in which the time dependence of the system and of the perturbation are in phase. In this approximation, the system of equations (9.70) reduces to

$$i\hbar \dot{c}_a(t) = A_{ab}\exp(i(\Delta\omega)t)c_b \quad (9.73a)$$
$$i\hbar \dot{c}_b(t) = A_{ba}^\dagger \exp(-i(\Delta\omega)t)c_a \quad (9.73b)$$

where $A_{ab} = (2i)^{-1}\hat{H}'_{ab} = (2i)^{-1}\hat{H}'^*_{ba}$ and $A_{ba}^\dagger = A_{ab}^*$. This system, which is much simpler than (9.70), can be solved exactly (see Problem 9.7). The solutions $c_a(t)$ and $c_b(t)$ satisfying the initial conditions (9.72) are

$$c_a(t) = \exp(i(\Delta\omega)t/2)\left[\cos(\omega_R t/2) - i\left(\frac{\Delta\omega}{\omega_R}\right)\sin(\omega_R t/2)\right] \quad (9.74a)$$

and

$$c_b(t) = \frac{\hat{H}'_{ba}}{\hbar\omega_R}\exp(-i(\Delta\omega)t/2)\sin(\omega_R t/2) \quad (9.74b)$$

where

$$\omega_R = \left[(\Delta\omega)^2 + \frac{|\hat{H}'_{ba}|^2}{\hbar^2}\right]^{1/2}$$

$$= \left[(\Delta\omega)^2 + \frac{4|A^\dagger_{ba}|^2}{\hbar^2}\right]^{1/2} \quad (9.75)$$

is called the Rabi *'flopping frequency'*.

The probability of finding the system at time $t > 0$ in the state a is therefore given by

$$|c_a(t)|^2 = \cos^2(\omega_R t/2) + \frac{(\Delta\omega)^2}{\omega_R^2} \sin^2(\omega_R t/2) \quad (9.76)$$

while that of finding it in the state b (i.e. the probability for the transition $a \to b$ to take place) is

$$P_{ba}(t) = |c_b(t)|^2 = \frac{|\hat{H}'_{ba}|^2}{\hbar^2 \omega_R^2} \sin^2(\omega_R t/2). \quad (9.77)$$

As expected, the excitation is a typical *resonance* process, since the probability (9.77) rapidly decreases when the absolute value $|\Delta\omega|$ of the detuning angular frequency increases. It is also readily verified from (9.74)–(9.77) that $|c_a(t)|^2 + |c_b(t)|^2 = 1$, and that the system oscillates between the two levels with a period $T = 2\pi/\omega_R$.

Having obtained 'exact' results for this problem (within the framework of the rotating wave approximation), we can compare them with those arising from first-order perturbation theory. Using (9.19) we find that to first order in the perturbation $|c_a(t)|^2 \simeq 1$, in agreement with (9.76) when $\omega_R t \ll 1$. From (9.63) and (9.25) we also note that the first-order transition probability is given by

$$P^{(1)}_{ba}(t) = \frac{2}{\hbar^2}|A^\dagger_{ba}|^2 F(t, \Delta\omega)$$

$$= \frac{|\hat{H}'_{ba}|^2}{\hbar^2(\Delta\omega)^2} \sin^2((\Delta\omega)t/2). \quad (9.78)$$

If $\Delta\omega \neq 0$ this result agrees with (9.77) provided that the perturbation is weak enough so that one can write $\omega_R \simeq \Delta\omega$ (see (9.75)). On resonance ($\Delta\omega = 0$), we see that $P^{(1)}_{ba}(t)$ increases quadratically with time according to (9.64). As expected, this result is only in accord with the 'exact' expression (9.77) for small enough times.

9.4 The adiabatic approximation

The time-dependent perturbation method we have studied above is based on the assumption that the magnitude of the time-dependent part of the Hamiltonian, $\lambda H'(t)$, is small. In this section and the next one we shall study approximation methods in which the key parameter is the *rate of change* of the Hamiltonian H of the system.

We begin here by assuming that H varies *very slowly* with time. We then expect that approximate solutions of the time-dependent Schrödinger equation

$$i\hbar \frac{\partial \Psi}{\partial t} = H(t)\Psi \tag{9.79}$$

can be obtained in terms of the eigenfunctions $\psi_k(t)$ of the 'instantaneous' Hamiltonian, such that[2]

$$H(t)\psi_k(t) = E_k(t)\psi_k(t) \tag{9.80}$$

at any particular time t. We then expect on physical grounds that if the Hamiltonian $H(t)$ changes very slowly with time, a system which is initially (at $t = t_0$) in a discrete non-degenerate state $\psi_a(t_0)$ with energy $E_a(t_0)$ is very likely to go over to the corresponding state $\psi_a(t)$ with energy $E_a(t)$ at time t, without making any transition.

In order to prove this statement, which is often referred to as the *adiabatic theorem*, we shall use the *adiabatic approximation* developed by M. Born and V. Fock in 1928. Assuming that the wave function Ψ is known at $t = t_0$, we expand Ψ for $t \geq t_0$ in terms of the 'instantaneous' eigenfunctions $\psi_k(t)$ as

$$\Psi = \sum_k C_k(t)\psi_k(t) \exp\left[-\frac{i}{\hbar}\int_{t_0}^{t} E_k(t')dt'\right] \tag{9.81}$$

where the ψ_k are assumed to form an orthonormal and complete set. The instantaneous energy levels $E_k(t)$ are supposed to be non-degenerate and to form a purely discrete spectrum. We emphasise that the terms 'energy levels' have only a formal meaning since the energy is not conserved for a time-dependent Hamiltonian.

Substituting the expansion (9.81) into the Schrödinger equation (9.79), we have

$$i\hbar \sum_k \left(\dot{C}_k \psi_k + C_k \frac{\partial \psi_k}{\partial t} - \frac{i}{\hbar} C_k \psi_k E_k\right) \exp\left[-\frac{i}{\hbar}\int_{t_0}^{t} E_k(t')dt'\right]$$
$$= H(t) \sum_k C_k \psi_k \exp\left[-\frac{i}{\hbar}\int_{t_0}^{t} E_k(t')dt'\right] \tag{9.82}$$

where the dot denotes a derivative with respect to time, as usual. Using (9.80) we see that the right-hand side of this equation cancels the last term on the left-hand side. Pre-multiplying both sides of (9.82) by $\psi_b^*(t)$, where the function $\psi_b(t)$ is a member of the set of 'instantaneous' eigenfunctions $\{\psi_k(t)\}$, integrating over the coordinates of the system and using the fact that $\langle \psi_b|\psi_k\rangle = \delta_{bk}$, we find that

$$\dot{C}_b(t) = -\sum_k C_k(t) \exp\left\{\frac{i}{\hbar}\int_{t_0}^{t} [E_b(t') - E_k(t')]dt'\right\}\left\langle\psi_b\left|\frac{\partial \psi_k}{\partial t}\right.\right\rangle. \tag{9.83}$$

---

[2] In writing the equation (9.80) the dependence of H and ψ_k on other variables than t has not been indicated for notational simplicity.

The set of equations (9.83) for all values of b constitutes a system of coupled first-order differential equations for the coefficients $C_k(t)$. We shall now show that the diagonal terms ($k = b$) in this system can be removed. To prove this, let us first consider the quantity

$$\alpha_k(t) = \left\langle \psi_k \left| \frac{\partial \psi_k}{\partial t} \right. \right\rangle. \tag{9.84}$$

Using the normalisation condition

$$\langle \psi_k(t) | \psi_k(t) \rangle = 1 \tag{9.85}$$

and differentiating this equation with respect to the time, we have

$$\left\langle \frac{\partial \psi_k}{\partial t} \bigg| \psi_k \right\rangle + \left\langle \psi_k \bigg| \frac{\partial \psi_k}{\partial t} \right\rangle = \alpha_k^*(t) + \alpha_k(t) = 0 \tag{9.86}$$

so that $\alpha_k(t)$ is purely imaginary and hence may be written in the form $\alpha_k(t) = \mathrm{i}\beta_k(t)$, where $\beta_k(t)$ is real. Next, let us perform on the coefficients $C_k(t)$ the phase transformation

$$C'_k(t) = C_k(t) \exp\left[\mathrm{i} \int_{t_0}^{t} \beta_k(t')\mathrm{d}t'\right]. \tag{9.87}$$

Using (9.83), (9.87) and the fact that $\alpha_b(t) = \langle \psi_b | \partial \psi_b / \partial t \rangle = \mathrm{i}\beta_b(t)$, we find that the new coefficients $C'_k(t)$ satisfy the set of equations

$$\dot{C}'_b = -\sum_{k \neq b} C'_k(t) \exp\left\{ \frac{\mathrm{i}}{\hbar} \int_{t_0}^{t} [E'_b(t') - E'_k(t')]\mathrm{d}t' \right\} \left\langle \psi_b \bigg| \frac{\partial \psi_k}{\partial t} \right\rangle \tag{9.88}$$

where the new 'instantaneous' energies $E'_k(t)$ are given by

$$E'_k(t) = E_k(t) + \hbar \beta_k(t). \tag{9.89}$$

It should be noted that the phase transformation (9.87) on the coefficients $C_k(t)$ amounts to the change of phase

$$\psi'_k = \psi_k \exp\left[-\mathrm{i} \int_{t_0}^{t} \beta_k(t')\mathrm{d}t'\right] \tag{9.90}$$

on the eigenfunctions ψ_k. Assuming that the phases of the eigenfunctions ψ_k are arbitrary at each instant of time[3], this change of phase can be performed on all the ψ_k. From here on we shall assume that it has already been made and shall omit the primes.

We now proceed to examine the quantities $\langle \psi_b | \partial \psi_k / \partial t \rangle$ when $k \neq b$. First of all, by differentiating (9.80) with respect to the time, we have

$$\frac{\partial H}{\partial t} \psi_k + H \frac{\partial \psi_k}{\partial t} = \frac{\partial E_k}{\partial t} \psi_k + E_k \frac{\partial \psi_k}{\partial t}. \tag{9.91}$$

---

[3] This assumption is not correct for the case of cyclic systems, as will be discussed below in connection with the Berry phase.

Taking the scalar product with ψ_b, we obtain for $b \neq k$

$$\left\langle \psi_b \left| \frac{\partial H}{\partial t} \right| \psi_k \right\rangle + \left\langle \psi_b | H | \frac{\partial \psi_k}{\partial t} \right\rangle = E_k \left\langle \psi_b \Big| \frac{\partial \psi_k}{\partial t} \right\rangle. \tag{9.92}$$

Remembering that H is Hermitian and using again (9.80) we can rewrite the second term on the left of (9.91) as

$$\left\langle \psi_b | H | \frac{\partial \psi_k}{\partial t} \right\rangle = \left\langle H \psi_b \Big| \frac{\partial \psi_k}{\partial t} \right\rangle = E_b \left\langle \psi_b \Big| \frac{\partial \psi_k}{\partial t} \right\rangle. \tag{9.93}$$

Substitution of this result into (9.92) then yields

$$\left\langle \psi_b \Big| \frac{\partial \psi_k}{\partial t} \right\rangle = -\frac{(\partial H/\partial t)_{bk}}{E_b(t) - E_k(t)} = -\frac{(\partial H/\partial t)_{bk}}{\hbar \omega_{bk}(t)}, \quad b \neq k \tag{9.94}$$

where we have introduced the notation

$$\left(\frac{\partial H}{\partial t} \right)_{bk} = \left\langle \psi_b \left| \frac{\partial H}{\partial t} \right| \psi_k \right\rangle \tag{9.95}$$

and

$$\omega_{bk}(t) = \frac{E_b(t) - E_k(t)}{\hbar}, \quad b \neq k. \tag{9.96}$$

Note that $\omega_{bk}(t)$ is always different from zero since the energy levels are non-degenerate.

Making use of (9.94) and remembering that we omit the primes, we can therefore rewrite (9.88) as

$$\dot{C}_b(t) = \sum_{k \neq b} \frac{C_k(t)}{\hbar \omega_{bk}(t)} \left(\frac{\partial H}{\partial t} \right)_{bk} \exp\left[i \int_{t_0}^{t} \omega_{bk}(t') dt' \right]. \tag{9.97}$$

This system of equations (for all b), like the system (9.10) obtained in Section 9.1, is strictly equivalent to the original time-dependent Schrödinger equation. We recall that the system (9.10) was obtained by assuming that the Hamiltonian H could be split as $H(t) = H_0 + \lambda H'(t)$, where H_0 is time-independent, and by expanding the full wave function Ψ in terms of the eigenfunctions $\psi_k^{(0)}$ of H_0 according to (9.6). In the present case the expansion (9.81) is used, where $\psi_k(t)$ and $E_k(t)$ are time-dependent. As a result, the quantities $\omega_{bk}(t)$ appearing in (9.97) depend upon the time, in contrast to the Bohr angular frequencies ω_{bk} occurring in (9.10).

Both systems of equations (9.10) and (9.97) are equally difficult to solve exactly, but provide convenient starting points for approximate solutions if there are 'small factors' on their right-hand sides. As we have seen above, the system (9.10) is an adequate starting point for the perturbative approach, based on the fact that $\lambda H'(t)$ is small. In contrast, the system (9.97) is useful when $\partial H/\partial t$ is small, that is when the Hamiltonian $H(t)$ changes slowly with time. If the Hamiltonian of the system were independent of the time ($\partial H/\partial t = 0$), then clearly the equations (9.97) would yield the solution $C_b =$ constant for all b. Hence, if $\partial H/\partial t$ is different from zero but is sufficiently small, we can attempt to solve the equations (9.97) approximately by setting the C_k equal to constants on the right-hand side. Assuming that the system is initially (at $t = t_0$) in the state a, we therefore substitute the values $C_k = \delta_{ka}$ in (9.97) and obtain

$$\dot{C}_b(t) = \hbar^{-1} \omega_{ba}^{-1}(t) \left(\frac{\partial H}{\partial t}\right)_{ba} \exp\left[i \int_{t_0}^{t} \omega_{ba}(t') dt'\right] \tag{9.98}$$

for all $b \neq a$. For $b = a$ we have $\dot{C}_a = 0$ to the same approximation. Integrating the equation (9.98) subject to the initial conditions

$$C_b(t \leqslant t_0) = 0, \qquad b \neq a \tag{9.99}$$

we find that

$$C_b(t) = \hbar^{-1} \int_{t_0}^{t} dt' \omega_{ba}^{-1}(t') \left(\frac{\partial H(t')}{\partial t'}\right)_{ba} \exp\left[i \int_{t_0}^{t'} \omega_{ba}(t'') dt''\right], \qquad b \neq a. \tag{9.100}$$

We shall refer to this result as the *adiabatic approximation* for the probability amplitude $C_b(t)$.

It is apparent from the above derivation that the adiabatic result (9.100) will only be valid if $|C_b(t)|$ remains small. Thus, if $P_{ba}(t) = |C_b(t)|^2$ denotes the transition probability from the initial state a to a state $b \neq a$ we must have

$$P_{ba}(t) \ll 1. \tag{9.101}$$

A crude estimate of $C_b(t)$ may be readily obtained by assuming that both ω_{ba} and $\partial H/\partial t$ are time-independent. From (9.100) we then have

$$C_b(t) \simeq (i\hbar)^{-1} \omega_{ba}^{-2} \left(\frac{\partial H}{\partial t}\right)_{ba} \{\exp[i\omega_{ba}(t - t_0)] - 1\} \tag{9.102}$$

and hence

$$P_{ba}(t) \simeq 4\hbar^{-2} \omega_{ba}^{-4} \left|\left(\frac{\partial H}{\partial t}\right)_{ba}\right|^2 \sin^2[\omega_{ba}(t - t_0)/2]. \tag{9.103}$$

We note that this transition probability does not exhibit any steady increase over a long time interval. Moreover, since $\sin^2 x \leqslant 1$, an upper bound for $P_{ba}(t)$ is given by

$$P_{ba}(t) \leqslant \frac{4|(\partial H/\partial t)_{ba}|^2}{\hbar^2 \omega_{ba}^4}. \tag{9.104}$$

A criterion of validity of the adiabatic approximation may now be formulated as follows. Since $T = 2\pi/|\omega_{ba}|$ is the period corresponding to the transition $a \to b$, we see that

$$|\omega_{ba}|^{-1} \left(\frac{\partial H}{\partial t}\right)_{ba} = \frac{T}{2\pi} \left(\frac{\partial H}{\partial t}\right)_{ba} \tag{9.105}$$

is the matrix element of the change of the Hamiltonian occurring during the time $T/2\pi$. From (9.101), (9.103) and (9.105) we then conclude that the adiabatic approximation will be valid if

$$\left|\hbar^{-1}\omega_{ba}^{-2}\left(\frac{\partial H}{\partial t}\right)_{ba}\right| = \left|\frac{(T/2\pi)(\partial H/\partial t)_{ba}}{E_b - E_a}\right| \ll 1 \tag{9.106}$$

which means that the matrix element of the change of H during the time $T/2\pi$ must be small compared with the energy difference $|E_b - E_a|$.

We remark that in the case for which the Hamiltonian varies extremely slowly, one has $(\partial H/\partial t)_{ba} \simeq 0$. In that case $P_{ba} \simeq 0$, so that the system, initially in the state $\psi_a(t_0)$, will essentially remain in that state, which at time t has evolved into $\psi_a(t)$. As we mentioned above, this property is known as the *adiabatic theorem*.

Comparison with time-dependent perturbation theory

Let us now investigate the link between the adiabatic approximation and the first-order time-dependent perturbation theory developed at the beginning of this chapter. In order to do this we shall assume that the Hamiltonian of the system may be split as in (9.1) into a time-independent unperturbed part H_0 and a small time-dependent perturbation which we shall write as $H'(t)$, thus setting $\lambda = 1$ in (9.1). We suppose that $H'(t)$, and hence the full Hamiltonian

$$H(t) = H_0 + H'(t) \tag{9.107}$$

varies slowly with time. Furthermore, we assume that the perturbation H' is 'switched on' at the time t_0 and off at the time t in a smooth way. According to first-order time-dependent perturbation theory, the probability amplitude that the system, initially (at $t = t_0$) in the (unperturbed) state a, be found at time t (or any subsequent time) in the (unperturbed) state $b \neq a$ is given by equation (9.17b). Performing the integral on the right of that equation by parts, we have

$$c_b^{(1)}(t) = (i\hbar)^{-1}\left\{-i\omega_{ba}^{-1}[H'_{ba}(t')\exp(i\omega_{ba}t')]_{t_0}^t\right.$$

$$\left.+i\omega_{ba}^{-1}\int_{t_0}^t\left(\frac{\partial H'(t')}{\partial t'}\right)_{ba}\exp(i\omega_{ba}t')dt'\right\}. \tag{9.108}$$

Using the fact that $\partial H'/\partial t = \partial H/\partial t$ (see (9.107)) and remembering that the perturbation $H'(t')$ goes smoothly to zero at $t' = t_0$ and $t' = t$, we find that[4]

$$c_b^{(1)}(t) = \hbar^{-1}\omega_{ba}^{-1}\int_{t_0}^t\left(\frac{\partial H(t')}{\partial t'}\right)_{ba}\exp(i\omega_{ba}t')dt'. \tag{9.109}$$

---

[4] Note that if the perturbation H' is constant in time, except for being 'switched on' suddenly at time t_0 and off at time t, then the second term on the right of (9.108) vanishes, and upon choosing $t_0 = 0$ we retrieve the result (9.20b) of Section 9.2.

This expression bears a close resemblance to the adiabatic result (9.100). We recall, however, that in the perturbation formulae the Bohr angular frequency $\omega_{ba} = (E_b^{(0)} - E_a^{(0)})/\hbar$ corresponds to the unperturbed energies $E_a^{(0)}$ and $E_b^{(0)}$ and is therefore independent of the time, while in the adiabatic result (9.100) the angular frequency $\omega_{ba}(t) = [E_b(t) - E_a(t)]/\hbar$ corresponds to the 'instantaneous' energies $E_a(t)$ and $E_b(t)$. If in (9.100) the angular frequency $\omega_{ba}(t)$ can be replaced by its unperturbed value, then both formulae (9.100) and (9.109) lead to the same transition probability

$$P_{ba}(t) = \hbar^{-2}\omega_{ba}^{-2}\left|\int_{t_0}^{t}\left(\frac{\partial H(t')}{\partial t'}\right)_{ba}\exp(i\omega_{ba}t')dt'\right|^2. \quad (9.110)$$

If in addition $\partial H/\partial t$ is also assumed to be time-independent, then (9.110) reduces to the estimate (9.103) obtained above for the transition probability.

An interesting special case is that for which the perturbation $H'(t)$ oscillates in time with an angular frequency ω according to (9.58). Then

$$\frac{\partial H}{\partial t} = i\omega[A\exp(i\omega t) - A^{\dagger}\exp(-i\omega t)]. \quad (9.111)$$

In order for the adiabatic approximation to be valid, we assume that the oscillation of H' is slow and hence $\partial H/\partial t$ is small. As in Section 9.3, we suppose that the perturbation is switched on at $t_0 = 0$. Inserting (9.111) into the adiabatic formula (9.100) and neglecting the time dependence of ω_{ba}, we find that

$$C_b(t) \simeq \frac{\omega}{\omega_{ba}}\left[A_{ba}\frac{\exp[i(E_b^{(0)} - E_a^{(0)} + \hbar\omega)t/\hbar] - 1}{E_b^{(0)} - E_a^{(0)} + \hbar\omega}\right.$$
$$\left. - A_{ba}^{\dagger}\frac{\exp[i(E_b^{(0)} - E_a^{(0)} - \hbar\omega)t/\hbar] - 1}{E_b^{(0)} - E_a^{(0)} - \hbar\omega}\right]. \quad (9.112)$$

It is apparent from this result that the transition probability $P_{ba}(t) = |C_b(t)|^2$ will not increase steadily over long periods of time, except in the two *resonant* cases $E_b^{(0)} = E_a^{(0)} - \hbar\omega$ and $E_b^{(0)} = E_a^{(0)} + \hbar\omega$, for which $P_{ba}(t)$ increases quadratically with time. Thus *when resonance occurs the adiabatic theorem will be violated*. If $E_b^{(0)} = E_a^{(0)} - \hbar\omega$ (resonant emission), the second term in the bracket on the right of (9.112) can be neglected, and we have $\omega/\omega_{ba} = -1$. On the other hand, if $E_b^{(0)} = E_a^{(0)} + \hbar\omega$ (resonant absorption), the first bracket term in (9.112) can be neglected, and $\omega/\omega_{ba} = +1$. Hence in both resonant cases we see that (9.112) agrees with the result (9.60) which we obtained from first-order time-dependent perturbation theory.

It should be stressed that in the above discussion the perturbation $H'(t)$ in (9.107) has been assumed to be weak *and* slowly varying in time, in which case we found that first-order perturbation theory and the adiabatic approximation lead to very similar results. Of course, if $H'(t)$ is weak but does not vary slowly with time, then we are only entitled to use perturbation theory. On the other hand, if the full Hamiltonian $H(t)$ cannot be split as in (9.107) into an unperturbed part H_0 and a weak perturbation $H'(t)$, but does vary slowly with time, then only the adiabatic approach should be used.

Charged harmonic oscillator in a time-dependent electric field

As an application of the adiabatic approximation, let us consider a charged-particle linear harmonic oscillator, acted upon by a spatially uniform, time-dependent electric field $\mathcal{E}(t)$. If m is the mass and q the charge of the particle, the Hamiltonian of the system is

$$H(t) = -\frac{\hbar^2}{2m}\frac{\partial^2}{\partial x^2} + \frac{1}{2}kx^2 - q\mathcal{E}(t)x \tag{9.113}$$

or

$$H(t) = -\frac{\hbar^2}{2m}\frac{\partial^2}{\partial x^2} + \frac{1}{2}k[x - a(t)]^2 - \frac{1}{2}ka^2(t) \tag{9.114}$$

where

$$a(t) = \frac{q}{k}\mathcal{E}(t). \tag{9.115}$$

Except for the term $-ka^2(t)/2$ which is constant at a given time, the Hamiltonian (9.114) is that of a linear harmonic oscillator of the same angular frequency $\omega = (k/m)^{1/2}$ as the undisturbed oscillator (without the electric field), but whose equilibrium position is displaced at $x = a(t)$. The instantaneous energy eigenfunctions of (9.114) can thus be obtained by making the substitution $x \to x - a$ in the linear harmonic oscillator wave functions (4.168), and hence are given by

$$\psi_n(x - a) = \left(\frac{\alpha}{\sqrt{\pi}2^n n!}\right)^{1/2} \exp[-\alpha^2(x - a)^2/2] H_n[\alpha(x - a)]. \tag{9.116}$$

The corresponding instantaneous energies are

$$E_n(t) = \left(n + \frac{1}{2}\right)\hbar\omega - \frac{1}{2}ka^2(t), \quad n = 0, 1, 2, \ldots, \tag{9.117}$$

and we note that the angular frequencies

$$\omega_{n'n} = \frac{E_{n'}(t) - E_n(t)}{\hbar} = (n' - n)\omega \tag{9.118}$$

are independent of the time and equal to their 'unperturbed' values $(n' - n)\omega$.

Suppose that the electric field $\mathcal{E}(t)$ is applied at the time t_0 and that it varies slowly with time. Assuming that the harmonic oscillator is initially in its ground state ($n = 0$), we shall use (9.100) to evaluate the probability that it will be left in an excited state at time t_1. From (9.114) the time derivative of the Hamiltonian is

$$\frac{\partial H}{\partial t} = -k\dot{a}x \tag{9.119a}$$

where

$$\dot{a} = \frac{da(t)}{dt} = \frac{q}{k}\frac{d\mathcal{E}(t)}{dt}. \tag{9.119b}$$

9.4 The adiabatic approximation

Now, using (4.174) the matrix element $x_{n0} \equiv \langle \psi_n | x | \psi_0 \rangle$ is seen to vanish when $n \neq 1$, while for $n = 1$ we have

$$x_{10} = 2^{-1/2}\alpha^{-1} = \left(\frac{\hbar}{2m\omega}\right)^{1/2} \tag{9.120}$$

where we have used the fact that $\alpha = (m\omega/\hbar)^{1/2}$ (see (4.134)). The only non-vanishing transition probability is therefore that corresponding to the transition $0 \to 1$. Making use of (9.100) and (9.118)–(9.120), this transition probability at the time t_1 is given by

$$P_{10}(t_1) = |C_b(t_1)|^2 = \hbar^{-2}\omega^{-2}\left|\exp(-i\omega t_0)\int_{t_0}^{t_1}\left(\frac{\partial H}{\partial t}\right)_{10}\exp(i\omega t)\mathrm{d}t\right|^2$$

$$= \frac{q^2}{2m\hbar\omega^3}\left|\int_{t_0}^{t_1}\frac{\mathrm{d}\mathcal{E}(t)}{\mathrm{d}t}\exp(i\omega t)\mathrm{d}t\right|^2. \tag{9.121}$$

Looking back at (9.110), we note that the same result may be obtained by using first-order perturbation theory, where

$$H_0 = -\frac{\hbar^2}{2m}\frac{\mathrm{d}^2}{\mathrm{d}x^2} + \frac{1}{2}kx^2 \tag{9.122}$$

is the Hamiltonian of the linear harmonic oscillator in the absence of the electric field, provided that the perturbation

$$H'(t) = -q\mathcal{E}(t)x = -ka(t)x \tag{9.123}$$

is assumed to be 'switched on' at $t = t_0$ and 'switched off' at time $t = t_1$ in a smooth way.

A quick estimate of the probability (9.121) may be obtained by assuming that $\mathrm{d}\mathcal{E}(t)/\mathrm{d}t$ is constant. Then

$$P_{10}(t_1) \simeq \frac{q^2}{2m\hbar\omega^5}\left|\frac{\mathrm{d}\mathcal{E}}{\mathrm{d}t}\right|^2|\exp(i\omega t_1) - \exp(i\omega t_0)|^2 \tag{9.124}$$

Now

$$\frac{q^2}{2m\hbar\omega^5}\left|\frac{\mathrm{d}\mathcal{E}}{\mathrm{d}t}\right|^2 = \frac{k^2\dot{a}^2}{2m\hbar\omega^5} = \left(\frac{\dot{a}}{(2\hbar\omega/m)^{1/2}}\right)^2 \tag{9.125}$$

and we note that the quantity $v_c = (\hbar\omega/m)^{1/2}$ is the maximum speed of a 'classical' oscillator having the zero-point energy $\hbar\omega/2$. Hence, the condition of validity (9.101) of the adiabatic approximation will be satisfied if the equilibrium point moves slowly with respect to the speed v_c. In the limit of a very slow motion of the equilibrium point (corresponding to an electric field $\mathcal{E}(t)$ which changes very slowly) we see that $P_{10} \simeq 0$ so that the harmonic oscillator remains in the ground state ($n = 0$), in agreement with the adiabatic theorem.

The Berry phase

In discussing the adiabatic approximation, it has been assumed that the phases of the eigenfunctions $\psi_k(t)$ are arbitrary at each instant of time. This was the generally accepted position until 1984, when M. V. Berry showed that in a cyclic system in which the Hamiltonian at time t_f is the same as at time t_0 there is a relative change in phase between $\psi_k(t_0)$ and $\psi_k(t_f)$ which cannot be removed by a phase transformation and which has observable consequences. In fact, before Berry demonstrated this result generally, a particular case had been discovered by C. A. Mead and D. Truhlar which explained some anomalies observed in molecular spectra.

Let us consider the case in which $H(t)$ is varying so slowly that the adiabatic theorem applies and the system always remains in the initial non-degenerate state with energy $E_a(t)$ and eigenfunction $\psi_a(t)$. The (approximate) solution of the Schrödinger equation (9.79) is then a single term of the expansion (9.81), and

$$\Psi(t) = C_a(t)\psi_a(t) \exp\left[-\frac{i}{\hbar}\int_{t_0}^{t} E_a(t')dt'\right]. \tag{9.126}$$

The wave function $\Psi(t)$ can be normalised to unity and as the system is in the state $\psi_a(t)$ at $t = t_0$ we can set $C_a(t_0) = 1$. Since $\psi_a(t)$ is also normalised to unity, $C_a(t)$ for $t \neq t_0$ is a phase factor, so that

$$C_a(t) = \exp[i\gamma_a(t)] \tag{9.127}$$

where $\gamma_a(t)$ is real and $\gamma_a(t_0) = 0$. By inserting $\Psi(t)$ given by (9.126) into (9.79) and using (9.80), $\gamma_a(t)$ is seen to satisfy

$$i\dot{\gamma}_a(t)\psi_a(t) = -\frac{\partial}{\partial t}\psi_a(t) \tag{9.128}$$

with the solution

$$\gamma_a(t) = i\int_{t_0}^{t} \langle\psi_a(t')|\frac{\partial\psi_a(t')}{\partial t'}\rangle dt'. \tag{9.129}$$

If the system is cyclic so that $H(t)$ returns to its value at $t = t_0$ at a later time t_f,

$$H(t_f) = H(t_0) \tag{9.130}$$

with the consequence that

$$E_a(t_f) = E_a(t_0), \qquad \psi_a(t_f) = \psi_a(t_0). \tag{9.131}$$

The accumulated phase change over the period from t_0 to t_f is the *Berry phase* $\bar{\gamma}_a$, where

$$\bar{\gamma}_a = i\int_{t_0}^{t_f} \langle\psi_a(t')|\frac{\partial\psi_a(t')}{\partial t'}\rangle dt'. \tag{9.132}$$

Let us try to eliminate the phase $\bar{\gamma}_a$ by making a transformation in which $\psi_a(t)$ is replaced by $\psi'_a(t)$, where

$$\psi'_a(t) = \psi_a(t)\exp[i\eta(t)]. \tag{9.133}$$

Under this transformation the Berry phase becomes $\bar{\gamma}_a'$, where

$$\begin{aligned}\bar{\gamma}_a' &= i\int_{t_0}^{t_f}\langle\psi_a'(t')|\frac{\partial\psi_a'(t')}{\partial t'}\rangle dt'\\ &= i\int_{t_0}^{t_f}\langle\psi_a(t')|\frac{\partial\psi_a(t')}{\partial t'}\rangle dt' - \int_{t_0}^{t_f}\frac{d\eta(t')}{dt'}dt'\\ &= \bar{\gamma}_a - \eta(t_f) + \eta(t_0).\end{aligned} \qquad (9.134)$$

Since $\psi_a(t_f) = \psi_a(t_0)$ and $\psi_a'(t_f) = \psi_a'(t_0)$, it follows that $\eta(t_f) - \eta(t_0) = 2\pi n$, where n is zero or an integer. Consequently the factor $\exp(i\bar{\gamma}_a) = \exp(i\bar{\gamma}_a')$ cannot be removed by a phase transformation of the eigenfunction ψ_a. The phase $\bar{\gamma}_a$ is an observable and its existence gives rise to experimentally verifiable consequences, an example of which will be discussed in Chapter 12.

The Hamiltonian $H(t)$ may derive its time dependence through a number of parameters, each of which is slowly varying with time. A frequently encountered case is when the parameters are the components of external electric or magnetic fields which interact with the system under consideration. Another example arises in developing the theory of diatomic molecules. In this case the Hamiltonian describing the motion of the electrons depends on the relative position vector **R** of the two nuclei. Because of the large mass of the nuclei compared with that of the electrons, **R** can be treated as a slowly varying parameter. This will be studied in Chapter 10.

Let us consider the case in which $H(t)$ depends on t through three parameters $c_1(t), c_2(t), c_3(t)$, so that

$$H(t) = H(c_i(t)), \quad i = 1, 2, 3. \qquad (9.135)$$

Since $H(t_f) = H(t_0)$, we have $c_i(t_f) = c_i(t_0)$ and using a vector notation (9.132) can be written as

$$\bar{\gamma}_a = i\oint\langle\psi_a(\mathbf{c})|\boldsymbol{\nabla}_{\mathbf{c}}\psi_a(\mathbf{c})\rangle\cdot d\mathbf{c} \qquad (9.136)$$

where $\boldsymbol{\nabla}_{\mathbf{c}}$ is the gradient in parameter space. The integral in (9.136) is a line integral taken around the closed curve $\mathbf{c} \equiv \mathbf{c}(t)$ in parameter space. The quantity $\mathbf{A}(\mathbf{c})$ defined as

$$\mathbf{A}(\mathbf{c}) = i\langle\psi_a(\mathbf{c})|\boldsymbol{\nabla}_{\mathbf{c}}\psi_a(\mathbf{c})\rangle \qquad (9.137)$$

is known as the *Berry connection* or as the *Berry gauge potential*. Since the Berry phase depends on the closed curve $\mathbf{c}$, it is often called a *geometrical phase* and such phases have been shown to arise in many non-adiabatic situations in addition to the strictly adiabatic context discussed here. By applying Stokes's theorem $\bar{\gamma}_a$ can be expressed as

$$\bar{\gamma}_a = \oint\mathbf{A}(\mathbf{c})\cdot d\mathbf{c} = \int\mathcal{B}\cdot d\mathbf{S} \qquad (9.138)$$

where the integral on the right is over a surface S bounded by the closed curve $\mathbf{c}$ and

$$\mathcal{B} = \nabla_{\mathbf{c}} \times \mathbf{A}(\mathbf{c}). \tag{9.139}$$

The Berry connection $\mathbf{A}$ acts as a kind of vector potential and $\mathcal{B}$ is the corresponding field (in parameter space) known as the *Berry curvature*. The significance of these considerations will become apparent when we discuss the Aharonov–Bohm effect in Chapter 12, where $\mathcal{B}$ will be taken to be a magnetic field, but it should be emphasized that the formalism is quite general and can be extended to systems depending on any number of parameters $c_i(t)$.

9.5 The sudden approximation

In this section we shall consider the case for which the Hamiltonian H of the system changes with time *very rapidly*. We have already treated in Section 9.2 the case of a small perturbation which is 'switched on' suddenly at a given time; we shall now generalise this treatment so that we may handle large disturbances as well. In order to do this, let us first consider the situation in which the Hamiltonian H changes *instantaneously* at $t = 0$ from H_0 to H_1, where H_0 and H_1 are both constants in time. Thus for $t < 0$ we have $H = H_0$, with

$$H_0 \psi_k^{(0)} = E_k^{(0)} \psi_k^{(0)} \tag{9.140}$$

where the superscript zero emphasises that we are dealing with eigenfunctions and eigenvalues of H_0. The eigenfunctions $\psi_k^{(0)}$ are assumed to be orthonormal and to form a complete set (which is not necessarily discrete). For $t > 0$, we have $H = H_1$, with

$$H_1 \phi_n^{(1)} = E_n^{(1)} \phi_n^{(1)} \tag{9.141}$$

and the eigenfunctions $\phi_n^{(1)}$ are also assumed to form a complete, orthonormal set. We can therefore write the general solution of the time-dependent Schrödinger equation for this problem as

$$\Psi(t) = \sum_k c_k^{(0)} \psi_k^{(0)} \exp(-\mathrm{i} E_k^{(0)} t/\hbar), \qquad t < 0 \tag{9.142a}$$

and

$$\Psi(t) = \sum_n d_n^{(1)} \phi_n^{(1)} \exp(-\mathrm{i} E_n^{(1)} t/\hbar), \qquad t > 0 \tag{9.142b}$$

where the summation symbol implies a sum over the entire set (discrete plus continuous) of eigenfunctions. Assuming that Ψ is normalised to unity, the time-independent coefficients $c_k^{(0)}$ and $d_n^{(1)}$ are respectively the probability amplitudes of finding the system in the state $\psi_k^{(0)}$ at $t < 0$ and in the state $\phi_n^{(1)}$ at $t > 0$.

We now recall that since the time-dependent Schrödinger equation, $\mathrm{i}\hbar \partial \Psi / \partial t = H\Psi$, is of first order in the time, the wave function $\Psi(t)$ must be a continuous function of t. This is true in particular at $t = 0$. We may therefore equate the two solutions (9.142a) and (9.142b) at $t = 0$, and we find in this way that

$$\sum_k c_k^{(0)} \psi_k^{(0)} = \sum_n d_n^{(1)} \phi_n^{(1)}. \tag{9.143}$$

This relation allows one to express the d coefficients in terms of the c coefficients. Indeed, taking the scalar product of both members of (9.143) with a particular eigenfunction $\phi_n^{(1)}$ of H_1, we find that

$$d_n^{(1)} = \sum_k c_k^{(0)} \langle \phi_n^{(1)} | \psi_k^{(0)} \rangle. \tag{9.144}$$

The above relation is an exact one for the ideal case of an Hamiltonian H which changes instantaneously at $t = 0$. In practice, the change in H happens during a finite interval of time τ. If this interval is very short (we shall give a more precise meaning to this statement below), we may in first approximation set $\tau = 0$ and continue to use equation (9.144) to obtain the probability amplitudes $d_n^{(1)}$; this procedure is known as the *sudden approximation*.

A simple criterion of validity of the sudden approximation may be obtained as follows. Let us assume that $H = H_0$ for $t < 0$, $H = H_1$ for $t > \tau$ and that during the 'intermediate' period $0 < t < \tau$ we have $H = H_i$, where H_i is also time-independent. If $\{\chi_l^{(i)}\}$ denotes a complete, orthonormal set of eigenfunctions of H_i, such that

$$H_i \chi_l^{(i)} = E_l^{(i)} \chi_l^{(i)} \tag{9.145}$$

we can expand the general solution of the time-dependent Schrödinger equation as

$$\Psi(t) = \sum_k c_k^{(0)} \psi_k^{(0)} \exp(-iE_k^{(0)} t/\hbar), \qquad t < 0 \tag{9.146a}$$

$$= \sum_l a_l^{(i)} \chi_l^{(i)} \exp(-iE_l^{(i)} t/\hbar), \qquad 0 < t < \tau \tag{9.146b}$$

$$= \sum_n d_n^{(1)} \phi_n^{(1)} \exp(-iE_n^{(1)} t/\hbar), \qquad t > \tau \tag{9.146c}$$

where the coefficients $a_l^{(i)}$ are also time-independent. Expressing that the wave function $\Psi(t)$ is continuous at $t = 0$ we obtain from (9.146a) and (9.146b)

$$\sum_k c_k^{(0)} \psi_k^{(0)} = \sum_l a_l^{(i)} \chi_l^{(i)} \tag{9.147}$$

so that

$$a_l^{(i)} = \sum_k c_k^{(0)} \langle \chi_l^{(i)} | \psi_k^{(0)} \rangle. \tag{9.148}$$

Similarly, the continuity of $\Psi(t)$ at $t = \tau$ yields from (9.146b) and (9.146c)

$$\sum_l a_l^{(i)} \chi_l^{(i)} \exp(-iE_l^{(i)} \tau/\hbar) = \sum_n d_n^{(1)} \phi_n^{(1)} \exp(-iE_n^{(1)} \tau/\hbar) \tag{9.149}$$

and by taking the scalar product of both members of this equation with a particular eigenfunction $\phi_n^{(1)}$ of H_1, we find that

$$d_n^{(1)} = \sum_l a_l^{(i)} \langle \phi_n^{(1)} | \chi_l^{(i)} \rangle \exp[i(E_n^{(1)} - E_l^{(i)})\tau/\hbar]. \tag{9.150}$$

Using (9.148), we can also write the above relation as

$$d_n^{(1)} = \sum_k \sum_l c_k^{(0)} \langle \phi_n^{(1)} | \chi_l^{(i)} \rangle \langle \chi_l^{(i)} | \psi_k^{(0)} \rangle \exp[i(E_n^{(1)} - E_l^{(i)})\tau/\hbar]. \tag{9.151}$$

Let us compare the above result with the sudden approximation expression for $d_n^{(1)}$ given by (9.144). First of all, we note that if $\tau = 0$ we can use the closure relation satisfied by the eigenfunctions $\chi_l^{(i)}$ to perform the summation on l in (9.151). The resulting expression for $d_n^{(1)}$ is then given by (9.144), as expected. When $\tau \neq 0$ it is apparent that the difference between the exact expression (9.151) and the sudden approximation result (9.144) arises from the fact that $\exp[i(E_n^{(1)} - E_l^{(i)})\tau/\hbar]$ differs from unity. In order for the sudden approximation to be valid, it is therefore necessary that τ be small in comparison with all the quantities $\hbar/|E_n^{(1)} - E_l^{(i)}|$. This rather complicated condition can be simplified by making use of the fact that highly energetic new states of motion (having short periodic times) usually have a relatively small probability of being excited. A useful criterion of validity of the sudden approximation is then obtained by requiring that the time τ be small with respect to a typical period associated with the initial motion.

Let us now return to the sudden approximation result (9.144). An interesting special case is that for which the system is initially (for $t < 0$) in a particular stationary state $\psi_a^{(0)} \exp(-iE_a^{(0)}t/\hbar)$, where $\psi_a^{(0)}$ is an eigenstate of H_0. Then $c_k^{(0)} = \delta_{ka}$, so that the probability amplitude of finding the system in the eigenstate $\phi_n^{(1)}$ of H_1 after the sudden change in the Hamiltonian has occurred is

$$d_n^{(1)} = \langle \phi_n^{(1)} | \psi_a^{(0)} \rangle. \tag{9.152}$$

Comparison with time-dependent perturbation theory

It is interesting to compare the method used above to obtain the sudden approximation results (9.144) and (9.152) with that employed in Section 9.2 to treat the case of a perturbation $H'(t)$ which is 'switched on' suddenly at $t = 0$ and is then constant in time. In both cases the Hamiltonian of the system is equal to H_0 at $t < 0$, and the solution of the time-dependent Schrödinger equation is expanded in the complete set of eigenfunctions $\psi_k^{(0)}$ of H_0 as (see (9.4) and (9.142a))

$$\Psi(t) = \sum_k c_k^{(0)} \psi_k^{(0)} \exp(-iE_k^{(0)}t/\hbar), \qquad t < 0. \tag{9.153}$$

The two methods, however, differ in the way of handling the situation for $t > 0$. In the case of the sudden approximation, the wave function $\Psi(t)$ is expanded for $t > 0$ in terms of the eigenfunctions $\phi_n^{(1)}$ of H_1, namely (see (9.142b))

$$\Psi(t) = \sum_n d_n^{(1)} \phi_n^{(1)} \exp(-iE_n^{(1)}t/\hbar), \qquad t > 0 \tag{9.154}$$

and one obtains the (time-independent) probability amplitudes $d_n^{(1)}$ of finding the system in the eigenstate $\phi_n^{(1)}$ of H_1. It should be stressed that the sudden approximation results (9.144) and (9.152) are valid whatever the magnitude of the disturbance, provided that the time interval τ is sufficiently short. On the other hand, in the perturbative approach one starts from the exact expansion of $\Psi(t)$ in terms of the eigenfunctions $\psi_k^{(0)}$ of H_0 (see (9.6))

$$\Psi(t) = \sum_k c_k(t) \psi_k^{(0)} \exp(-iE_k^{(0)}t/\hbar), \qquad t > 0 \tag{9.155}$$

in which the coefficients $c_k(t)$ depend on the time because the perturbation $H'(t)$ can cause transitions between the eigenstates of the unperturbed Hamiltonian H_0. In Section 9.2 the coefficients $c_k(t)$ were determined to first order in the perturbation, assuming that a weak enough perturbation $H'(t)$ is turned on at $t = 0$ and is subsequently constant in time. If the system is initially in the eigenstate $\psi_a^{(0)}$ of H_0 (so that $c_k^{(0)} = \delta_{ka}$), the first-order probability amplitude $c_b^{(1)}$ of finding it at $t > 0$ in a different eigenstate $\psi_b^{(0)}$ of H_0 is then given by (9.17b).

It is also interesting to note that since (9.154) and (9.155) are two expansions of the same wave function $\Psi(t)$ for $t > 0$, we may write

$$\sum_k c_k(t) \psi_k^{(0)} \exp(-iE_k^{(0)}t/\hbar) = \sum_n d_n^{(1)} \phi_n^{(1)} \exp(-iE_n^{(1)}t/\hbar), \qquad t > 0. \tag{9.156}$$

Multiplying both members of this equation on the left by $\psi_b^{(0)*}$, integrating over the coordinates of the system, and using the orthonormality of the functions $\psi_k^{(0)}$, we find that

$$c_b(t) = \sum_n d_n^{(1)} \langle \psi_b^{(0)} | \phi_n^{(1)} \rangle \exp[i(E_b^{(0)} - E_n^{(1)})t/\hbar], \qquad t > 0. \tag{9.157}$$

In particular, if the system is initially in the eigenstate $\psi_a^{(0)}$ of H_0 and the time interval τ is short enough so that the sudden approximation is valid, the probability amplitudes $d_n^{(1)}$ are given by $d_n^{(1)} = \langle \phi_n^{(1)} | \psi_a^{(0)} \rangle$ (see (9.152)), and we have

$$c_b(t) = \sum_n \langle \psi_b^{(0)} | \phi_n^{(1)} \rangle \langle \phi_n^{(1)} | \psi_a^{(0)} \rangle \exp[i(E_b^{(0)} - E_n^{(1)})t/\hbar], \qquad t > 0. \tag{9.158}$$

Charged harmonic oscillator in a time-dependent electric field

As a first application of the sudden approximation, we shall consider once again a charged-particle linear harmonic oscillator acted upon by a spatially uniform, time-dependent electric field $\mathcal{E}(t)$. The Hamiltonian of this system is therefore given by (9.114). At $t = 0$ the electric field is switched on suddenly (i.e. in a time τ much shorter than ω^{-1}, where ω is the angular frequency of the oscillator), and afterwards it is assumed that it has the constant value $\mathcal{E}_0$. Thus we have

$$H = H_0 = -\frac{\hbar^2}{2m}\frac{d^2}{dx^2} + \frac{1}{2}kx^2, \qquad t < 0 \tag{9.159}$$

and

$$H = H_1 = -\frac{\hbar^2}{2m}\frac{d^2}{dx^2} + \frac{1}{2}k(x-a)^2 - \frac{1}{2}ka^2, \qquad t > \tau \tag{9.160}$$

with

$$a = \frac{q\mathcal{E}_0}{k} = \frac{q\mathcal{E}_0}{m\omega^2}. \tag{9.161}$$

The Hamiltonian H_0 being that of a linear-harmonic oscillator, its eigenfunctions are the harmonic oscillator wave functions $\psi_n(x)$ given by (4.168). We shall assume that the oscillator is initially (at $t \leqslant 0$) in its ground state $\psi_0(x)$. The Hamiltonian H_1 is, except for the constant term $-ka^2/2$, that of a linear-harmonic oscillator of the same angular frequency ω as H_0, but oscillating about the point $x = a$. The eigenfunctions $\phi_n^{(1)}(x)$ of H_1 are therefore given by

$$\phi_n^{(1)}(x) = \psi_n(x-a) \tag{9.162}$$

where $\psi_n(x-a)$ are the 'displaced' linear-harmonic oscillator wave functions (9.116), with the constant a given by (9.161). The corresponding eigenenergies of H_1 are

$$\begin{aligned}E_n^{(1)} &= \left(n+\frac{1}{2}\right)\hbar\omega - \frac{1}{2}ka^2\\ &= \left(n+\frac{1}{2}\right)\hbar\omega - \frac{q^2\mathcal{E}_0^2}{2m\omega^2}, \qquad n = 0, 1, 2, \ldots.\end{aligned} \tag{9.163}$$

Let us now calculate the probability amplitude that the oscillator, initially (at $t \leqslant 0$) in the ground state $\psi_0(x)$ of H_0, be found in the eigenstate $\phi_n^{(1)}$ of H_1 after the sudden change in the Hamiltonian has occurred. Using the sudden approximation expression (9.152) with $\psi_a^{(0)} \equiv \psi_0(x)$ and $\phi_n^{(1)} \equiv \psi_n(x-a)$, we have

$$d_n^{(1)} = \int_{-\infty}^{+\infty} \psi_n^*(x-a)\psi_0(x)dx. \tag{9.164}$$

This integral can be evaluated with the help of the generating function (4.156) for the Hermite polynomials. The result is (Problem 9.9)

$$d_n^{(1)} = \frac{(-1)^n(\alpha a)^n \exp(-\alpha^2 a^2/4)}{(2^n n!)^{1/2}} \tag{9.165}$$

where $\alpha = (m\omega/\hbar)^{1/2}$. The corresponding probability is

$$P_n^{(1)} = |d_n^{(1)}|^2 = \frac{(\alpha a)^{2n} \exp(-\alpha^2 a^2/2)}{2^n n!} \tag{9.166}$$

and we note that it may be written in the form of a Poisson distribution[5]

$$P_n^{(1)} = \frac{(\bar{n})^n \exp(-\bar{n})}{n!} \tag{9.167}$$

---

[5] See, for example, Feller (1970).

9.5 The sudden approximation

where the mean $\bar{n}$ has the value

$$\bar{n} = \frac{1}{2}(\alpha a)^2 = \frac{q^2 \mathcal{E}_0^2}{2m\hbar\omega^3}. \tag{9.168}$$

The above calculation of $P_n^{(1)}$ is valid if the time τ during which the Hamiltonian changes is very short ($\tau \ll \omega^{-1}$), whatever the strength $\mathcal{E}_0$ of the electric field. In the limiting case for which $|q\mathcal{E}_0| \gg (m\hbar\omega^3)^{1/2}$, corresponding to *strong coupling*, we have $\bar{n} \gg 1$ and we see from (9.167) that

$$P_0^{(1)} = \exp(-\bar{n}) \ll 1. \tag{9.169}$$

The probability of finding the system in the ground state of H_1 after the electric field has been applied is therefore very small, in complete contrast to the result obtained in Section 9.4 under adiabatic conditions (see the discussion following (9.125)). In fact, for large $\bar{n}$ the maximum of $P_n^{(1)}$ occurs for the value $n \simeq \bar{n}$, so that the energy level of H_1 which will be excited with the highest probability is that having the energy $E_{\bar{n}}^{(1)}$.

In the opposite limiting case, $|q\mathcal{E}_0| \ll (m\hbar\omega^3)^{1/2}$, corresponding to *weak coupling*, we expect the results of the sudden approximation to agree with those of first-order time-dependent perturbation theory. To verify that this is indeed the case, we first note that the weak coupling condition $|q\mathcal{E}_0| \ll (m\hbar\omega^3)^{1/2}$ implies that $\bar{n} \ll 1$. Hence, from (9.167) the probability of finding the system in the eigenstate $\phi_n^{(1)} \equiv \psi_n(x-a)$ of H_1 after the electric field has been turned on is given by

$$P_n^{(1)} \simeq \frac{(\bar{n})^n}{n!}. \tag{9.170}$$

We see from this equation that $P_0^{(1)} \simeq 1$. We also have

$$P_1^{(1)} \simeq \bar{n} = \frac{q^2\mathcal{E}_0^2}{2m\hbar\omega^3} \tag{9.171}$$

so that $P_1^{(1)} \ll 1$. For higher values of n the probabilities $P_n^{(1)}$ become utterly negligible. Now, from (9.157) and (9.165), we can also obtain the probability of finding the system at time $t > 0$ in a particular eigenstate of the *unperturbed* harmonic oscillator Hamiltonian H_0. Using the fact that $|\alpha a| = |q\mathcal{E}_0|/(m\hbar\omega^3)^{1/2}$ is small with respect to unity in the weak coupling limit, it is straightforward to show (Problem 9.10) that the only eigenstate of H_0 which will be excited with a sizeable probability is the first excited state $\psi_1(x)$, the corresponding transition probability being

$$P_{10}(t) = |c_1(t)|^2 \simeq \frac{2q^2\mathcal{E}_0^2}{m\hbar\omega^3} \sin^2(\omega t/2). \tag{9.172}$$

This result agrees exactly with that obtained from the first-order perturbation theory formula (9.24), as is readily checked by noting that (a) the perturbation is $H'(t) = -q\mathcal{E}_0 x$ and (b) the only non-vanishing matrix element of $H'(t)$ connecting the ground state to another state of H_0 is

$$H'_{10} = -q\mathcal{E}_0 x_{10}$$
$$= -q\mathcal{E}_0 \left(\frac{\hbar}{2m\omega}\right)^{1/2} \tag{9.173}$$

where we have used (9.120).

Beta decay of the tritium nucleus

A second interesting illustration of the sudden approximation is provided by the following problem. Consider a tritium atom containing a nucleus ^{3}H (the triton, also denoted by the symbol T) and an electron. The triton nucleus, which has a mass number $A = 3$, contains one proton and two neutrons. It is an unstable nucleus, since by beta emission[6] it decays into the nucleus ^{3}He (which contains two protons and one neutron), an electron e^- and an electron antineutrino $\bar{\nu}_e$. That is,

$$^3\text{H} \rightarrow {}^3\text{He} + e^- + \bar{\nu}_e. \tag{9.174}$$

Assuming that the tritium atom is in its ground state before the β-decay of ^{3}H takes place, what is the influence of this decay on the atomic electron?

In order to answer that question, we first note that in the β-decay process (9.174) the electron is emitted from the triton nucleus with an energy which, in most cases[7], is of the order of several keV. The velocity v of the β-electron is therefore usually much larger than the velocity $v_0 = \alpha c \simeq c/137$ of the atomic electron in the ground state of tritium. Thus, if a_0 denotes the first Bohr radius (1.66), the β-electron will leave the atom in a time $\tau \simeq a_0/v$, which is much shorter than the periodic time $T = 2\pi a_0/v_0$ associated with the motion of the atomic electron. As a result, one can say for all practical purposes that when the tritium nucleus ^{3}H decays by beta emission into ^{3}He, the nuclear charge 'seen' by the atomic electron changes instantaneously from Ze to $Z'e$, where $Z = 1$ and $Z' = 2$. The Hamiltonian of the system is therefore given for $t < 0$ by that of the original tritium atom (^{3}H + e^-):

$$H = H_0 = -\frac{\hbar^2}{2m}\nabla^2 - \frac{Ze^2}{(4\pi\varepsilon_0)r}, \quad Z = 1 \tag{9.175}$$

and for $t > 0$ by that of the (^{3}He + e^-) ion:

$$H = H_1 = -\frac{\hbar^2}{2m}\nabla^2 - \frac{Z'e^2}{(4\pi\varepsilon_0)r}, \quad Z' = 2 \tag{9.176}$$

---

[6] Beta decay is discussed, for example, in Burcham and Jobes (1995).

[7] More precisely, the energy of the β-decay is shared between the three particles in the final state, so that there is a whole spectrum of electron energies, extending from zero to about 17 keV. However, only a very small fraction of the electrons emitted in the process (9.174) have a velocity less than or comparable to that of the atomic electron; the effect of these low-velocity β electrons may therefore be neglected in first approximation.

where m is the mass of the electron. In writing the above equations we have made the infinite nuclear mass approximation, so that small recoil effects are neglected.

The eigenfunctions of the Hamiltonians H_0 and H_1 are clearly hydrogenic wave functions corresponding respectively to the atomic number $Z = 1$ and $Z' = 2$. Since the tritium atom is assumed to be initially in its ground state (with quantum numbers $n = 1, l = 0, m = 0$) the probability amplitude $d^{(1)}_{n'l'm'}$ of finding the atomic electron in a discrete eigenstate $(n'l'm')$ of the $(^3\text{He} + e^-)$ hydrogenic Hamiltonian H_1 at $t > 0$ is given from (9.152) by

$$\begin{aligned} d^{(1)}_{n'l'm'} &= \langle \psi^{(Z'=2)}_{n'l'm'} | \psi^{(Z=1)}_{100} \rangle \\ &= \int \psi^{(Z'=2)*}_{n'l'm'}(\mathbf{r}) \psi^{(Z=1)}_{100}(\mathbf{r}) d\mathbf{r} \end{aligned} \qquad (9.177)$$

where $\psi^{(Z)}_{nlm}(\mathbf{r})$ denotes a discrete hydrogenic wave function, the atomic number Z being indicated explicitly as a superscript.

The overlap integral (9.177) can be simplified by remembering that the hydrogenic wave functions $\psi^{(Z)}_{nlm}(\mathbf{r})$ are given by (see (7.134))

$$\psi^{(Z)}_{nlm}(\mathbf{r}) = R^{(Z)}_{nl}(r) Y_{lm}(\theta, \phi) \qquad (9.178)$$

where $R^{(Z)}_{nl}(r)$ are normalised radial hydrogenic wave functions (see (7.139)) and the functions $Y_{lm}(\theta, \phi)$ are the spherical harmonics. From the orthonormality property (6.103) of the Y_{lm}, we find that the only non-vanishing probability amplitudes $d^{(1)}_{n'l'm'}$ are those corresponding to s-states ($l' = m' = 0$), which are given by

$$d^{(1)}_{n'00} = \int_0^\infty R^{(Z'=2)}_{n'0}(r) R^{(Z=1)}_{10}(r) r^2 dr \qquad (9.179)$$

Let us consider first the particular case for which $n' = 1$. Using the first radial function (7.140) with $a_\mu = a_0$, we find that

$$\begin{aligned} d^{(1)}_{100} &= \int_0^\infty R^{(Z'=2)}_{10}(r) R^{(Z=1)}_{10}(r) r^2 dr \\ &= 2^{7/2} a_0^{-3} \int_0^\infty dr\, r^2 \exp(-3r/a_0) \\ &= \frac{16\sqrt{2}}{27}. \end{aligned} \qquad (9.180)$$

Hence the probability that the $(^3\text{He} + e^-)$ ion be found in its ground state is

$$P^{(1)}_{100} = |d^{(1)}_{100}|^2 = \frac{512}{729} \simeq 0.702. \qquad (9.181)$$

The total probability for excitation and ionisation of the $(^3\text{He} + e^-)$ ion is therefore equal to $[1 - P^{(1)}_{100}] \simeq 0.298$.

More generally, the integral (9.179) can be performed for any value of the quantum number n'. The resulting probability of finding the (^{3}He + e$^-$) ion in an s-state with principal quantum number n' is found to be

$$P^{(1)}_{n'00} = |d^{(1)}_{n'00}|^2 = \frac{2^9 n'^5 (n'-2)^{2n'-4}}{(n'+2)^{2n'+4}}. \tag{9.182}$$

In particular, the probabilities of excitation of the first few s-states are given by

$$P^{(1)}_{200} = 0.250, \qquad P^{(1)}_{300} = \frac{2^9 3^5}{5^{10}} \simeq 0.013, \qquad P^{(1)}_{400} = \frac{2^{23}}{6^{12}} \simeq 0.004 \tag{9.183}$$

and the ionisation probability is

$$P^{(1)}_{ion} = 1 - \sum_{n'=1}^{\infty} P^{(1)}_{n'00} \simeq 0.026. \tag{9.184}$$

Problems

9.1 Consider a particle of charge q and mass m, which is in simple harmonic motion along the x-axis so that its Hamiltonian is given by

$$H_0 = -\frac{\hbar^2}{2m}\frac{d^2}{dx^2} + \frac{1}{2}kx^2$$

A homogeneous electric field $\mathcal{E}(t)$ directed along the x-axis is switched on at time $t = 0$, so that the system is perturbed by the interaction

$$H'(t) = -qx\mathcal{E}(t).$$

If $\mathcal{E}(t)$ has the form

$$\mathcal{E}(t) = \mathcal{E}_0 \exp(-t/\tau)$$

where $\mathcal{E}_0$ and τ are constants, and if the oscillator is in the ground state for $t \leq 0$, find the probability that it will be found in an excited state as $t \to \infty$, using (9.17).

9.2 A particle of charge q, in simple harmonic motion along the x-axis, is acted on by a time-dependent homogeneous electric field

$$\mathcal{E}(t) = \mathcal{E}_0 \exp[-(t/\tau)^2]$$

where $\mathcal{E}_0$ and τ are constants. If the oscillator is in its ground state at $t = -\infty$, find the probability that it will be found in an excited state as $t \to \infty$.
(*Hint*: Use the integral (2.48).)

9.3 A hydrogen atom is placed in a time-dependent homogeneous electric field given by

$$\mathcal{E}(t) = \mathcal{E}_0(t^2 + \tau^2)^{-1}$$

where $\mathcal{E}_0$ and τ are constants. If the atom is in the ground state at $t = -\infty$, obtain the probability that it will be found in a 2p state at $t = +\infty$.

(*Hint*: Use the integral $\int_{-\infty}^{\infty} e^{i\omega t}(t^2 + \tau^2)^{-1} dt = (\pi/\tau)e^{-\omega\tau}$.)

9.4 Verify that $c_a(t)$ and $c_b(t)$ given by (9.39) satisfy the equations (9.37).

9.5 Show that the expressions (9.43) agree with the exact results given by (9.39) to first order in H' and for small times t.

9.6 Prove equation (9.67).

9.7 Verify that $c_a(t)$ and $c_b(t)$ given by (9.74) satisfy the equations (9.73).

9.8 Consider a charged-particle linear-harmonic oscillator acted upon by a homogeneous time-dependent electric field

$$\mathcal{E}(t) = \mathcal{E}_0 \exp[-(t/\tau)^2]$$

where $\mathcal{E}_0$ and τ are constants. Assuming that $d\mathcal{E}(t)/dt$ is small, and that at $t = -\infty$ the oscillator is in the ground state, use the adiabatic approximation to obtain the probability that it will be found in an excited state as $t \to \infty$. Compare your result with that obtained in Problem 9.2 by using perturbation theory.

9.9 Using the generating function (4.156) for the Hermite polynomials, perform the integral on the right of (9.164) to obtain the result (9.165).

9.10 Consider a charged-particle linear-harmonic oscillator, the Hamiltonian H_0 of which is given by (9.159). At $t = 0$ a constant electric field $\mathcal{E}_0$ is suddenly switched on, so that for $t > 0$ the Hamiltonian H_1 of the system is given by (9.160). Show that in the weak coupling limit the only eigenstate of H_0 which will be excited with a sizeable probability is the first excited state $\psi_1(x)$, and that the corresponding transition probability $P_{10}(t)$ is given by (9.172).

9.11 A particle is in the ground state of a one-dimensional infinite square well with walls at $x = 0$ and $x = L$. At time $t = 0$, the width of the well is suddenly increased to $2L$. Find the probability that the particle will be found in the nth stationary state of the expanded well.

9.12 A particle is in the ground state of a one-dimensional infinite square well with walls at $x = 0$ and $x = L$. At time $t = 0$ the walls are suddenly removed so that the particle becomes free.

(a) Find the probability $\Pi(p_x)dp_x = |\phi(p_x)|^2 dp_x$ that a measurement of the momentum of the particle will produce a result between p_x and $p_x + dp_x$.

(b) Explain why a measurement of the energy of the particle after the walls are removed need not give the same result as the initial energy.

(c) Calculate the corresponding probability $\Pi(p_x)dp_x = |\phi(p_x)|^2 dp_x$ for the case in which the particle is initially in the nth energy eigenstate. Show that your result is in agreement with the uncertainty principle and that for large n it is in accord with the correspondence principle.

10 Several- and many-particle systems

10.1 Introduction 469
10.2 Systems of identical particles 472
10.3 Spin-1/2 particles in a box. The Fermi gas 478
10.4 Two-electron atoms 485
10.5 Many-electron atoms 492
10.6 Molecules 498
10.7 Nuclear systems 506
     Problems 511

In this chapter some illustrative examples of applications of quantum mechanics to systems containing two or more particles will be described. Following a review of the general formalism, we discuss in Section 10.2 the special features which arise for systems of identical particles. In Section 10.3 we study the ground state of a gas of spin-1/2 fermions. In the next two sections we show how the theory can be applied to analyse the structure of atoms, and the following section deals with the elements of molecular structure. Finally, in Section 10.7 we consider some examples drawn from nuclear physics.

10.1 Introduction

Let us consider a quantum mechanical system of N particles. The basic dynamical variables of the ith particle are its position vector $\mathbf{r}_i$, momentum $\mathbf{p}_i$ and – if this particle possesses spin – its spin angular momentum $\mathbf{S}_i$. Let q_i be a complete set of commuting observables for particle i, for example its position $\mathbf{r}_i$ and the third component of its spin S_{iz}. In the Schrödinger picture and the position representation, the system is described by a wave function $\Psi(q_1, q_2, \ldots, q_N, t)$ which satisfies the time-dependent Schrödinger equation

$$i\hbar \frac{\partial}{\partial t} \Psi(q_1, q_2, \ldots, q_N, t) = H \Psi(q_1, q_2, \ldots, q_N, t) \tag{10.1}$$

where the Hamiltonian H is the sum of the total kinetic-energy operator T and the potential-energy operator V

$$H = T + V. \tag{10.2}$$

In terms of the momentum operators $\mathbf{p}_i$ and the masses m_i of the particles, the kinetic energy operator is given by

$$T = \sum_{i=1}^{N} \frac{\mathbf{p}_i^2}{2m_i}. \tag{10.3}$$

The potential energy V can in principle depend on all the basic dynamical variables of the N particles (coordinates $\mathbf{r}_i$, spins $\mathbf{S}_i \ldots$). In the presence of an external field V can also be a function of the time variable t, but in this chapter we shall restrict our attention to systems for which V (and hence the Hamiltonian H) is time-independent. As a result, there exist *stationary-state* solutions of the Schrödinger equation (10.1) having the form

$$\Psi(q_1, \ldots, q_N, t) = \psi(q_1, \ldots, q_N) e^{-iEt/\hbar} \tag{10.4}$$

where E is the total energy of the system and $\psi(q_1, \ldots, q_N)$ is a solution of the time-independent Schrödinger equation $H\psi = E\psi$.

The observables $\mathbf{r}_i$ and $\mathbf{p}_i$ satisfy the commutation rules discussed in Section 5.4: all operators referring to one particle commute with all those referring to another, while for an individual particle i the Cartesian components (x_i, y_i, z_i) of $\mathbf{r}_i$ and (p_{ix}, p_{iy}, p_{iz}) of $\mathbf{p}_i$ satisfy the basic commutation relations (5.89). Moreover, we recall that in the position representation the operators $\mathbf{r}_i$ and $\mathbf{p}_i$ are given respectively by $(\mathbf{r}_i)_{\text{op}} = \mathbf{r}_i$ and $(\mathbf{p}_i)_{\text{op}} = -i\hbar \nabla_i$.

The total orbital angular momentum of the system is the vector sum of the individual orbital angular momenta $\mathbf{L}_i = \mathbf{r}_i \times \mathbf{p}_i$:

$$\mathbf{L} = \sum_{i=1}^{N} \mathbf{L}_i. \tag{10.5}$$

Similarly, if the particles possess spins $\mathbf{S}_i$, the total spin angular momentum is

$$\mathbf{S} = \sum_{i=1}^{N} \mathbf{S}_i. \tag{10.6}$$

The total angular momentum $\mathbf{J}$ is the vector sum of the total orbital- and total spin-angular momentum vectors:

$$\mathbf{J} = \mathbf{L} + \mathbf{S}. \tag{10.7}$$

In what follows we shall as a rule use *capital* letters to denote the values of the quantum numbers associated with the total orbital angular momentum, total spin and total angular momentum of systems containing more than one particle.

In an isolated system (that is a system not acted on by external fields), the total angular momentum must be conserved. This implies that $\mathbf{J}^2$ and J_z commute with

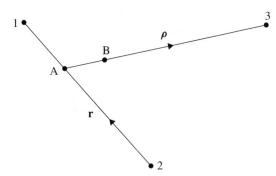

Figure 10.1 A set of Jacobi centre-of-mass coordinates for a three-particle system. The centre of mass of particles 1 and 2 is located at A and the centre of mass of the whole three-particle system is at B.

H and simultaneous eigenfunctions of H, $\mathbf{J}^2$ and J_z can be found. The energy eigenvalues depend, in general, on the value of the total angular momentum quantum number J, but not on the quantum number M_J associated with J_z. This is because in the absence of external fields there is no preferred direction in space, the energy cannot depend on the arbitrary choice of z-axis and hence cannot depend on the value of M_J. The system is said to be rotationally invariant.

In general, the motion of the centre of mass of the system is not of interest, since in the absence of external fields the centre of mass is either at rest or moves with a constant velocity. We have already seen how, in the case of a two-particle system, the motion relative to the centre of mass is described by the one-body Schrödinger equation

$$i\hbar \frac{\partial}{\partial t} \Psi(\mathbf{r}, t) = \left[-\frac{\hbar^2}{2\mu} \nabla^2 + V(\mathbf{r}) \right] \Psi(\mathbf{r}, t) \tag{10.8}$$

where $\mu = m_1 m_2/(m_1 + m_2)$ is the reduced mass and $\mathbf{r} = \mathbf{r}_1 - \mathbf{r}_2$ is the relative position vector of the particles 1 and 2. Suitable sets of centre-of-mass coordinates (which are not unique) can be constructed for systems containing any number of particles. For example, for a three-particle system with masses m_1, m_2, m_3, we can start by introducing the relative position vector $\mathbf{r}$ and reduced mass μ of a two-particle sub-system, for instance particles 1 and 2. The motion of this sub-system relative to the third particle, in the centre-of-mass system, is then described in terms of ρ, the relative position vector of particle 3 and the centre of mass of particles 1 and 2 (see Fig. 10.1), and a corresponding reduced mass $\mu' = (m_1 + m_2)m_3/[(m_1 + m_2) + m_3]$. In this case, the Schrödinger equation is

$$i\hbar \frac{\partial}{\partial t} \Psi(\mathbf{r}, \boldsymbol{\rho}, t) = \left[-\frac{\hbar^2}{2\mu} \nabla_\mathbf{r}^2 - \frac{\hbar^2}{2\mu'} \nabla_\rho^2 + V(\mathbf{r}, \boldsymbol{\rho}) \right] \Psi(\mathbf{r}, \boldsymbol{\rho}, t). \tag{10.9}$$

A four-particle system can be treated either by considering the motion of particle 4 relative to the centre of mass of the three-particle sub-system, or by considering the relative motion of two sub-systems each containing two particles. In coordinate systems of this kind, called Jacobi coordinates, the kinetic energy is separable, no cross terms like $\nabla_r \cdot \nabla_\rho$ appearing. This approach, while straightforward for systems containing a few particles, is not usually convenient for many-particle systems. A more suitable general procedure for N-electron atoms can be found in Appendix 8 of Bransden and Joachain (1983).

An interesting special case is one in which there is no interaction between any pair of particles but each particle may interact with an external potential. This occurs, for instance, when the particles are so far from each other than their mutual interaction is negligible. In this case the Hamiltonian of a system of N particles is the sum of N single-particle Hamiltonians h_i:

$$H = \sum_{i=1}^{N} h_i \tag{10.10}$$

Let us denote by $u_\lambda(q_i)$ the normalised eigenfunction of h_i corresponding to the eigenenergy E_λ:

$$h_i u_\lambda(q_i) = E_\lambda u_\lambda(q_i) \tag{10.11}$$

A solution of the Schrödinger eigenvalue equation

$$H\psi(q_1, q_2, \ldots, q_N) = E\psi(q_1, q_2, \ldots, q_N) \tag{10.12}$$

is then given by the product

$$\psi(q_1, q_2, \ldots, q_N) = u_\alpha(q_1) u_\beta(q_2) \ldots u_\lambda(q_i) \ldots u_\nu(q_N) \tag{10.13}$$

where each of the symbols $\alpha, \beta, \ldots \lambda, \ldots \nu$ represents a set of quantum numbers characterising the individual (one-particle) states $u_\alpha, u_\beta, \ldots u_\lambda, \ldots u_\nu$. In addition, the total energy E is just the sum of the individual energies

$$E = E_\alpha + E_\beta + \cdots + E_\lambda + \cdots + E_\nu. \tag{10.14}$$

The product wave function (10.13) describes an *uncorrelated* system such that measurements of the properties of one particle can be made independently of the others. Wave functions for systems of mutually interacting particles can never be expressed, except approximately, as a single product of one-particle wave functions. They are said to be *entangled*; the corresponding states are also said to be entangled and measurements cannot be made on one particle without affecting the others.

10.2 Systems of identical particles

Two particles are said to be identical when they cannot be distinguished by means of any intrinsic property. While in classical physics the existence of sharp trajectories makes it possible, in principle, to distinguish 'classical' particles by their paths, in quantum mechanics there is no way of keeping track of individual particles when the wave functions of identical particles overlap. As a result, it is impossible in quantum mechanics to distinguish between identical particles in regions of space where they may be found simultaneously, such as in their interaction region. As we shall now see, this quantum mechanical indistinguishability of identical particles has profound consequences.

Let us consider a quantum-mechanical system of N identical particles, and let q_i denote the ensemble of the (continuous) spatial coordinates $\mathbf{r}_i$ and (discrete) spin coordinate of particle i. Because the N particles are identical, all observables corresponding to this system must be symmetric functions of the basic dynamical variables. In particular, the Hamiltonian H of the system must be symmetric in the interchange of any pair of particles i and j. Specifically, if we denote by P_{ij} an operator which interchanges the variables q_i and q_j of particles i and j, then this operator commutes with the Hamiltonian:

$$[P_{ij}, H] = 0. \tag{10.15}$$

In general, an exact eigenfunction $\psi(q_1, \ldots, q_N)$ of H has no particular symmetry property under the interchanges of the variables q_i. However, it is important to recognise that if $\psi(q_1, \ldots, q_i, \ldots, q_j, \ldots, q_N)$ is an eigenfunction of H corresponding to the eigenvalue E, then so is $P_{ij}\psi$, where

$$P_{ij}\psi(q_1, \ldots, q_i, \ldots, q_j, \ldots, q_N) = \psi(q_1, \ldots, q_j, \ldots, q_i, \ldots, q_N). \tag{10.16}$$

Since two successive interchanges of q_i and q_j bring back the original configuration,

$$P_{ij}^2 = I \tag{10.17}$$

so that the eigenvalues of the operator P_{ij} are $\varepsilon = \pm 1$. Wave functions corresponding to the eigenvalue $\varepsilon = 1$ are such that

$$\begin{aligned} P_{ij}\psi(q_1, \ldots, q_i, \ldots, q_j, \ldots, q_N) &= \psi(q_1, \ldots, q_j, \ldots, q_i, \ldots, q_N) \\ &= \psi(q_1, \ldots, q_i, \ldots, q_j, \ldots, q_N) \end{aligned} \tag{10.18}$$

and are said to be *symmetric* under the interchange P_{ij}. On the other hand, wave functions which correspond to the eigenvalue $\varepsilon = -1$ are such that

$$\begin{aligned} P_{ij}\psi(q_1, \ldots, q_i, \ldots, q_j, \ldots, q_N) &= \psi(q_1, \ldots, q_j, \ldots, q_i, \ldots, q_N) \\ &= -\psi(q_1, \ldots, q_i, \ldots, q_j, \ldots, q_N) \end{aligned} \tag{10.19}$$

and are said to be *antisymmetric under the interchange* P_{ij}.

More generally, there are $N!$ different permutations of the variables $q_1, \ldots, q_N$. Defining P as the permutation operator that replaces q_1 by q_{P1}, q_2 by $q_{P2}, \ldots, q_N$ by q_{PN} and noting that P can be obtained as a succession of interchanges, we have

$$[P, H] = 0. \tag{10.20}$$

A permutation P is said to be *even* or *odd* depending on whether the number of interchanges leading to it is even or odd. If we let the operator P act on a wave function $\psi(q_1, \ldots, q_N)$ we have

$$P\psi(q_1, \ldots, q_N) = \psi(q_{P1}, \ldots, q_{PN}). \tag{10.21}$$

It is important to note that except for the case $N = 2$ the $N!$ permutations P do *not* commute among themselves. This is due to the fact that the interchange operators P_{ij} and $P_{ik} (k \neq j)$ do not mutually commute. Therefore the eigenfunctions $\psi(q_1, \ldots, q_N)$ of H are not *in general* eigenfunctions of all the $N!$ permutation operators P. However, there are *two* exceptional states which *are* eigenstates of H and of the $N!$ permutation operators P. The first one is the *totally symmetric state* $\psi_S(q_1, \ldots, q_N)$ which satisfies (10.18) for any interchange P_{ij}, so that for all P

$$P\psi_S(q_1, \ldots, q_N) = \psi_S(q_1, \ldots, q_N). \tag{10.22}$$

The second one is the *totally antisymmetric state* $\psi_A(q_1, \ldots, q_N)$ which satisfies (10.19) for any interchange P_{ij}. Thus, for all P

$$P\psi_A(q_1, \ldots, q_N) = \begin{cases} \psi_A(q_1, \ldots, q_N) & \text{for an even permutation} \\ -\psi_A(q_1, \ldots, q_N) & \text{for an odd permutation} \end{cases}. \tag{10.23}$$

We also remark that the equation (10.20) implies that P is a constant of the motion, so that a system of identical particles represented by a wave function of a given symmetry (S or A) will keep that symmetry at all times.

Bosons and fermions

According to our present knowledge of particles occurring in nature, the two types of states ψ_S and ψ_A are thought to be sufficient to describe all systems of identical particles. This is called the *symmetrisation postulate*. Particles having states described by completely symmetric wave functions are called *bosons*; they are said to obey Bose–Einstein statistics. Experiment shows that particles of *zero* or *integer* spin ($s = 0, 1, 2, \ldots$) are bosons. For example, all the mesons (such as the π and K mesons, which have spin $s = 0$, the ρ meson, which has spin $s = 1$, etc.) are bosons. The photon (which has spin $s = 1$) is also a boson, as are the newly discovered W and Z particles, which also have spin $s = 1$, and are called 'intermediate vector bosons'. On the other hand, particles having states described by totally antisymmetric wave functions are called *fermions*; they are said to obey Fermi–Dirac statistics. Experiment shows that particles having half-odd-integer spin ($s = 1/2, 3/2, \ldots$) are fermions. For example, all the leptons (such as the electron, the muon and the neutrinos, which have spin $1/2$) are fermions, as are the baryons (such as the proton, the neutron and the hyperons).

If a system is composed of different kinds of bosons (fermions) then the corresponding wave function must be separately totally symmetric (antisymmetric) with respect to interchanges of each kind of identical particle. For example, in the case of the hydrogen molecule H_2, the total wave function must be antisymmetric under the interchange of the (space plus spin) coordinates of the two electrons and also antisymmetric under the interchange of the (space plus spin) coordinates of the two protons.

Construction of totally symmetric and antisymmetric wave functions

We now turn to the problem of constructing totally symmetric and antisymmetric wave functions, starting with wave functions which do not possess symmetry properties under the interchange of the variables q_i.

Let us begin by considering a system of two identical particles. We denote by $\psi(q_1, q_2)$ an unsymmetrised eigenfunction of the Hamiltonian of this system corresponding to the eigenvalue E. The symmetric and antisymmetric eigenfunctions corresponding to the same eigenvalue E are given respectively by

$$\psi_S(q_1, q_2) = \frac{1}{\sqrt{2}}[\psi(q_1, q_2) + \psi(q_2, q_1)] \tag{10.24a}$$

and

$$\psi_A(q_1, q_2) = \frac{1}{\sqrt{2}}[\psi(q_1, q_2) - \psi(q_2, q_1)] \tag{10.24b}$$

where the factor $2^{-1/2}$ has been introduced for further convenience. According to the foregoing discussion, the symmetric wave function (10.24a) must be used in the case of a system of two identical bosons while the antisymmetric wave function (10.24b) must be used when the system contains two identical fermions.

The generalisation to systems of N identical particles is straightforward. For example, in the case $N = 3$, we can construct from the unsymmetrised wave function $\psi(q_1, q_2, q_3)$ the completely symmetric wave function

$$\psi_S(q_1, q_2, q_3) = \frac{1}{\sqrt{6}}[\psi(q_1, q_2, q_3) + \psi(q_2, q_1, q_3) + \psi(q_2, q_3, q_1) \\ + \psi(q_3, q_2, q_1) + \psi(q_3, q_1, q_2) + \psi(q_1, q_3, q_2)] \tag{10.25a}$$

which is appropriate for a system of three identical bosons. For a system of three identical fermions, we must use the completely antisymmetric wave function

$$\psi_A(q_1, q_2, q_3) = \frac{1}{\sqrt{6}}[\psi(q_1, q_2, q_3) - \psi(q_2, q_1, q_3) + \psi(q_2, q_3, q_1) \\ - \psi(q_3, q_2, q_1) + \psi(q_3, q_1, q_2) - \psi(q_1, q_3, q_2)]. \tag{10.25b}$$

Let us consider again the special case in which the Hamiltonian H is the sum of N single-particle Hamiltonians h_i, as in (10.10), but where the N particles are now all identical. In general, the wave function (10.13) has no particular symmetry with respect to interchanges of the coordinates q_i. However, the total symmetric (antisymmetric) wave functions required to describe systems of identical bosons (fermions) can be readily constructed from the individual wave functions $u_\alpha(q_1), u_\beta(q_2), \ldots, u_\nu(q_N)$.

Consider first the simplest case, $N = 2$. The symmetric and antisymmetric wave functions corresponding to the energy $E = E_\alpha + E_\beta$ are given by

$$\psi_S(q_1, q_2) = \frac{1}{\sqrt{2}}[u_\alpha(q_1)u_\beta(q_2) + u_\alpha(q_2)u_\beta(q_1)] \qquad (10.26a)$$

and

$$\psi_A(q_1, q_2) = \frac{1}{\sqrt{2}}[u_\alpha(q_1)u_\beta(q_2) - u_\alpha(q_2)u_\beta(q_1)]. \qquad (10.26b)$$

The symmetric and antisymmetric wave functions ψ_S and ψ_A are not in the form of a single product and are thus *entangled*. This has the consequence that whereas the two-particle states are eigenstates of energy corresponding to a definite energy $E = E_\alpha + E_\beta$, neither of the particles 1 and 2 is in an energy eigenstate, being partly in the state α and partly in the state β. A measurement of the energy of one of the particles in the ensemble of identically prepared two-particle states represented by the wave functions ψ_S or ψ_A will either produce the energy E_α or the energy E_β on a fifty–fifty basis.

For the general case of N identical particles, the totally antisymmetric wave function built from the one-particle states $u_\alpha(q_1), u_\beta(q_2), \ldots, u_\nu(q_N)$ is given by the *Slater determinant*

$$\psi_A(q_1, q_2, \ldots, q_N) = \frac{1}{\sqrt{N!}} \begin{vmatrix} u_\alpha(q_1) & u_\beta(q_1) & \cdots & u_\nu(q_1) \\ u_\alpha(q_2) & u_\beta(q_2) & \cdots & u_\nu(q_2) \\ \vdots & \vdots & & \vdots \\ u_\alpha(q_N) & u_\beta(q_N) & \cdots & u_\nu(q_N) \end{vmatrix}. \qquad (10.27)$$

This wave function is indeed completely antisymmetric, because if we interchange the (spatial and spin) coordinates of two particles (say q_1 and q_2) this is equivalent to interchanging two rows, so that the determinant changes sign. The corresponding totally symmetric wave function $\psi_S(q_1, \ldots, q_N)$ can be obtained by expanding the determinant (10.27), and making all the signs positive. The factor $(N!)^{-1/2}$ appearing in (10.27) is a normalisation factor, arising from the fact that there are $N!$ permutations of the coordinates $q_1, \ldots, q_N$. Recalling that P denotes a permutation of these coordinates, we may rewrite the Slater determinant (10.27) in the compact form

$$\psi_A(q_1, q_2, \ldots, q_N) = \frac{1}{\sqrt{N!}} \sum_P (-1)^P P u_\alpha(q_1) u_\beta(q_2) \cdots u_\nu(q_N) \qquad (10.28a)$$

where the symbol $(-1)^P$ is equal to $+1$ when P is an even permutation and to -1 when P is an odd permutation, and the sum is over all permutations P. Likewise, we can write the totally symmetric wave function constructed from the one-particle states $u_\alpha(q_1), u_\beta(q_2) \ldots u_\nu(q_N)$ as

$$\psi_S(q_1, q_2, \ldots, q_N) = \frac{1}{\sqrt{N!}} \sum_P P u_\alpha(q_1) u_\beta(q_2) \cdots u_\nu(q_N). \qquad (10.28b)$$

The Pauli exclusion principle

Let us return to the Slater determinant (10.27) giving the totally antisymmetric wave function, built from one-particle states, which describes a system of N identical fermions. It is apparent that if two or more sets of individual quantum numbers $\alpha, \beta, \ldots$, are identical, the wave function (10.27) vanishes. As a result, *only one fermion can occupy a given individual quantum state*. This statement expresses the *Pauli exclusion principle*, formulated in 1925 by W. Pauli in order to explain the structure of complex atoms.

Quarks and colour

As an example, let us consider the discovery of '*colour*' in the quark[1] model of elementary particles. This model was introduced in 1964 by M. Gell-Mann and independently by G. Zweig in order to explain the diversity of the strongly interacting particles called *hadrons*. The hadrons are classified into the baryons (which are fermions) and the mesons (which are bosons). According to the quark model, quarks can combine in two different ways to form hadrons: three quarks bound together constitute a baryon while a bound system of a quark and an antiquark is a meson. The quarks have spin 1/2 and are therefore fermions. Originally, Gell-Mann and Zweig introduced three kinds (now called 'flavours') of quarks, denoted by u ('up'), d ('down') and s ('strange'). The u quark has charge $\frac{2}{3}e$, the d quark $-\frac{1}{3}e$ and the s quark also $-\frac{1}{3}e$. All the hadrons known at that time could be 'explained' as combinations of these three quarks and the corresponding antiquarks $\bar{u}$, $\bar{d}$ and $\bar{s}$. For example, the proton can be considered as a (uud) bound state, the neutron as a (udd) bound state, the π^+ meson as a (u$\bar{d}$) bound state and the K^+ meson as a (u$\bar{s}$) bound state.

Among the hadrons there is a particle called the Δ^{++} baryon, which has charge $2e$ and a total angular momentum quantum number $J = 3/2$. This particle is expected to be a (uuu) bound state. Indeed, this system has charge $\frac{2}{3}e + \frac{2}{3}e + \frac{2}{3}e = 2e$; in addition, since quarks have spin 1/2, the total angular momentum $J = 3/2$ could be accounted for if each of the three u quarks were in a completely symmetric spatial state of zero orbital angular momentum. Now, if the three u quarks were identical in all respects, then by the Pauli exclusion principle no more than two of them could occupy such a state (one with spin 'up' and the other with spin 'down'). A way out of this difficulty is the suggestion that the quarks possess an additional internal quantum number called 'colour' and that each of the three u quarks in the Δ^{++} particle has a different colour. Since the quantum numbers of the three quarks are no longer the same, the Pauli exclusion principle is not violated. This suggestion, which looks rather *ad hoc*, has been confirmed by its success when applied to many other situations and is now well established.

---

[1] The term 'quark' was taken by M. Gell-Mann from a passage in James Joyce's *Finnegans Wake*:

"Three quarks for Muster Mark
Sure he hasn't got much of a bark"

10.3 Spin-1/2 particles in a box. The Fermi gas

There are a number of physical systems which can be modelled approximately by a large number of identical non-interacting spin-1/2 particles contained in a box with impenetrable walls. If the box is large enough, the properties of the system are independent of the shape of the box, and if the number of particles is very large such a system is called a *Fermi gas*. After considering the basic properties of the state of lowest energy of a Fermi gas, we shall discuss briefly applications to the thermal properties of metals, and to the structure of white dwarfs and neutron stars.

Since the particles in the Fermi gas are non-interacting and are contained in a box having impenetrable walls, the spatial part of the wave function of each particle (having mass m) satisfies inside the box the free-particle Schrödinger equation

$$-\frac{\hbar^2}{2m}\left(\frac{\partial^2}{\partial x^2} + \frac{\partial^2}{\partial y^2} + \frac{\partial^2}{\partial z^2}\right)\psi(x, y, z) = E\psi(x, y, z) \tag{10.29}$$

while $\psi = 0$ at the boundary. Let us take the box to be a large cube of side L and choose the origin of coordinates to be at one corner of the cube. Using the result (7.29), the spatial part of the wave function of a particle is given by the standing wave

$$\psi_{n_x n_y n_z}(x, y, z) = C \sin\left(\frac{n_x \pi}{L}x\right) \sin\left(\frac{n_y \pi}{L}y\right) \sin\left(\frac{n_z \pi}{L}z\right) \tag{10.30}$$

where $C = (8/L^3)^{1/2}$ is a normalisation constant and n_x, n_y, n_z are positive integers. The corresponding allowed values of the energy of a particle are given by (see (7.30) and (7.31))

$$E_n = \frac{\hbar^2 \pi^2}{2mL^2}n^2 \tag{10.31}$$

with

$$n^2 = n_x^2 + n_y^2 + n_z^2 \tag{10.32}$$

and we recall that since each energy level (10.31) can in general be obtained from a number of different sets of values of (n_x, n_y, n_z), it is usually degenerate.

Because the particles have spin 1/2, we must multiply the spatial part (10.30) of their wave function by the spin functions $\chi_{\frac{1}{2}, m_s}$, with $m_s = \pm 1/2$. The individual particle wave functions are therefore the spin-orbitals

$$\psi_{n_x n_y n_z m_s}(q) = \psi_{n_x n_y n_z}(x, y, z)\chi_{\frac{1}{2}, m_s} \tag{10.33}$$

where q denotes the ensemble of the spatial and spin coordinates, as usual. Clearly, for every spatial wave function $\psi_{n_x n_y n_z}$ there are two spin-orbitals $\psi_{n_x n_y n_z m_s}$, one corresponding to spin 'up' ($m_s = 1/2$) and the other to spin 'down' ($m_s = -1/2$). As a result, the degeneracy of the individual energy levels (10.31) is multiplied by two.

Since the energy spacings are very small for a macroscopic box of side L, it is a good approximation to consider that the energy levels are distributed nearly continuously.

We may then introduce the *density of states* or *density of orbitals* $D(E)$, which is defined as the number of particle quantum states (i.e. the number of spin-orbitals) per unit energy range. Thus $D(E)dE$ is the number of particle states for which the energy of a particle lies in the interval $(E, E + dE)$.

In order to obtain the quantity $D(E)$, we consider the space formed by the axes n_x, n_y and n_z (see Fig. 10.2). Since n_x, n_y and n_z are positive integers, we are interested only in the octant for which $n_x > 0$, $n_y > 0$ and $n_z > 0$. As seen from Fig. 10.2, each set of spatial quantum numbers (n_x, n_y, n_z) corresponds to a point of a cubical lattice, and every elementary cube of the lattice has unit volume. Hence, apart from small values of (n_x, n_y, n_z), the total number of spatial orbitals for all energies up to a certain value E is closely equal to the volume of the octant of a sphere of radius $n = (n_x^2 + n_y^2 + n_z^2)^{1/2}$. The total number N_s of individual particle states for energies up to E is therefore given approximately by

$$N_s = 2 \frac{1}{8} \frac{4}{3} \pi n^3 = \frac{1}{3} \pi n^3 \tag{10.34}$$

where the factor 2 is due to the two spin states per spatial orbital. Using (10.31) and setting $V = L^3$, we may rewrite this result as

$$N_s = \frac{1}{3\pi^2} \left(\frac{2m}{\hbar^2} \right)^{3/2} V E^{3/2}. \tag{10.35}$$

The number $D(E)dE$ of particle states within the energy range $(E, E + dE)$ is thus given by dN_s, so that

$$D(E) = \frac{dN_s}{dE} = \frac{1}{2\pi^2} \left(\frac{2m}{\hbar^2} \right)^{3/2} V E^{1/2}. \tag{10.36}$$

The total wave function describing the system of N identical spin-1/2 particles must be totally antisymmetric in the (space and spin) coordinates q_i of the particles. It is therefore a Slater determinant constructed from the individual spin-orbitals (10.33). In this way only one particle can occupy each state (specified by the quantum numbers n_x, n_y, n_z and m_s), in agreement with the Pauli exclusion principle. The total energy is the sum of the individual energies of the particles. Assuming that the system is in the *ground state* (i.e. that the Fermi gas is at an absolute temperature $T = 0$), the lowest total energy is obtained when the N particles fill all the spin-orbitals up to an energy E_F, called the *Fermi energy*, the remaining orbitals (with energies $E > E_F$) being vacant. This is illustrated in Fig. 10.3, which shows the density of states $D(E)$ as a function of E, the occupied spin-orbitals corresponding to the ground state of the Fermi gas being represented by the shaded area. It is worth stressing that if the particles were identical bosons instead of fermions, or if they were all distinguishable, the Pauli exclusion principle would not apply, so that all the N particles would be in the level of lowest energy at zero absolute temperature.

The Fermi energy E_F can be evaluated by requiring that the total number N of

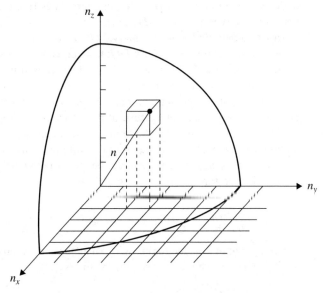

Figure 10.2 Three-dimensional n-space used in the calculation of $D(E)$. With each set of integers (n_x, n_y, n_z) is associated a cube of unit volume. For fairly large values of (n_x, n_y, n_z) the total number of spatial orbitals within $n = (n_x^2 + n_y^2 + n_z^2)^{1/2}$ equals the volume of one octant of a sphere of radius n in n-space.

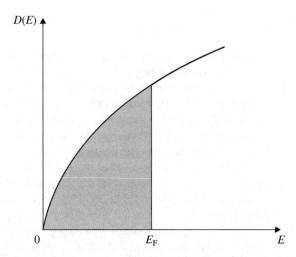

Figure 10.3 The density of states $D(E) = dN_s/dE$ as a function of the energy E. The occupied levels corresponding to the ground state of a Fermi gas are represented by the shaded area.

particles in the system should be equal to

$$N = \int_0^{E_F} D(E) dE. \tag{10.37}$$

In writing this equation we have used the fact that the system contains many particles, so that the integral (10.37) is a good approximation to the corresponding sum over discrete states. Moreover, since N is large, it does not matter whether the highest occupied energy level contains one or more electrons. Using the result (10.36), we have

$$N = \frac{1}{2\pi^2} \left(\frac{2m}{\hbar^2} \right)^{3/2} V \int_0^{E_F} E^{1/2} dE$$

$$= \frac{1}{3\pi^2} \left(\frac{2m}{\hbar^2} \right)^{3/2} V E_F^{3/2} \tag{10.38}$$

so that

$$E_F = \frac{\hbar^2}{2m} (3\pi^2 \rho)^{2/3} \tag{10.39}$$

where

$$\rho = \frac{N}{V} \tag{10.40}$$

is the number of particles per unit volume, i.e. the density of particles. The total energy of a Fermi gas in the ground state (at absolute temperature $T = 0$) is

$$E_{\text{tot}} = \int_0^{E_F} E D(E) dE$$

$$= \frac{1}{2\pi^2} \left(\frac{2m}{\hbar^2} \right)^{3/2} V \int_0^{E_F} E^{3/2} dE$$

$$= \frac{1}{5\pi^2} \left(\frac{2m}{\hbar^2} \right)^{3/2} V E_F^{5/2}$$

$$= \frac{3}{5} N E_F \tag{10.41}$$

where we have used (10.36) and (10.38). The average particle energy at $T = 0$ is therefore

$$\bar{E} = \frac{E_{\text{tot}}}{N} = \frac{3}{5} E_F. \tag{10.42}$$

It is also instructive to study the problem of the Fermi gas by imposing *periodic boundary conditions* on the spatial part of the wave functions of the particles, that is by requiring these wave functions to be periodic in x, y and z with period L. Instead of the standing waves (10.30), we then have travelling, plane-wave solutions of the Schrödinger equation (10.29), having the form

$$\psi_\mathbf{k}(\mathbf{r}) = \exp[i(k_x x + k_y y + k_z z)] \tag{10.43}$$

where the allowed components of the wave vector $\mathbf{k} \equiv (k_x, k_y, k_z)$ are given by (7.19), namely

$$k_x = \frac{2\pi}{L} n_x, \qquad k_y = \frac{2\pi}{L} n_y, \qquad k_z = \frac{2\pi}{L} n_z \tag{10.44}$$

where n_x, n_y and n_z are positive or negative integers, or zero. The number of spatial wave functions in the range $d\mathbf{k} = dk_x dk_y dk_z$ is $(L/2\pi)^3 dk_x dk_y dk_z$, and this number must be multiplied by 2 to take into account the two possible spin states. A unit volume of $\mathbf{k}$-space will therefore accommodate $V/4\pi^3$ fermions (with $V = L^3$). Thus, the individual particle states having energies up to $E = \hbar^2 k^2/2m$ are contained within a sphere in $\mathbf{k}$ space, of radius k. The number of these states is given by

$$N_s = \frac{V}{4\pi^3} \frac{4}{3}\pi k^3 = \frac{1}{3\pi^2} V k^3$$

$$= \frac{1}{3\pi^2} \left(\frac{2m}{\hbar^2}\right)^{3/2} V E^{3/2} \tag{10.45}$$

in agreement with (10.35).

We have seen above that in the ground state of the Fermi gas the N fermions fill all the levels up to the Fermi energy E_F. Thus in $\mathbf{k}$-space all states up to a maximum value of k equal to k_F are then filled, while the states for which $k > k_F$ are empty. In other words, all occupied single-particle states of a Fermi gas at zero absolute temperature fill a sphere in $\mathbf{k}$-space having a radius k_F. This sphere, called the *Fermi sphere*, therefore contains

$$\frac{1}{3\pi^2} V k_F^3 = N \tag{10.46}$$

single-particle states, so that

$$k_F = (3\pi^2 \rho)^{1/3}. \tag{10.47}$$

At the surface of the Fermi sphere, known as the *Fermi surface*, the energy is the Fermi energy

$$E_F = \frac{\hbar^2}{2m} k_F^2 \tag{10.48}$$

and we see that the result (10.39) follows upon substitution of (10.47) in (10.48). It is also convenient to introduce the Fermi momentum p_F and Fermi velocity v_F such that

$$E_F = \frac{p_F^2}{2m} = \frac{1}{2} m v_F^2. \tag{10.49}$$

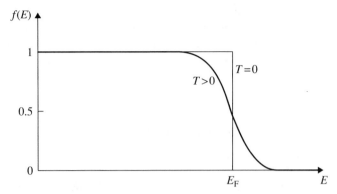

Figure 10.4 The probability $f(E)$ that a level of energy E in a Fermi gas is occupied at $T = 0$ and at $T > 0$, where $T \ll E_F/k$.

Thermal properties of metals

As an example, consider the conduction electrons in a metal. In metals, some of the electrons are free to move and are responsible for the electrical conductivity. These are called *conduction electrons*, in contrast to those electrons bound to the nuclei situated at fixed centres in the metallic crystals. Classical statistical mechanics predicts that a free particle should contribute to the heat capacity of a system an amount $3k/2$, where k is Boltzmann's constant. If the conduction electrons move freely through the metal, then the electronic contribution to the heat capacity should be $(3/2)kN$, where N is the number of such electrons. The observed electronic contribution is less than 1% of this value and this proved impossible to explain using classical statistical mechanics. This difficulty was overcome in the free-electron model developed in 1928 by A. Sommerfeld, in which the conduction electrons are considered to form a Fermi gas confined by the boundaries of the metal. In the common conductors, silver, gold and copper there is one conduction electron per atom so that the number density ρ is equal to the number density of the atoms. For silver $\rho \simeq 5.8 \times 10^{28}$ atoms m^{-3} and using (10.39) with m set equal to the electron mass, one finds that the Fermi energy is given by $E_F \simeq 8.8 \times 10^{-19}$ J $\simeq 5.5$ eV. From (10.42), the average energy per electron at $T = 0$ is $\bar{E} \simeq 3.3$ eV. Now, in order for a classical gas to have an average energy per particle equal to that of a Fermi gas at zero absolute temperature, $\bar{E}$, this classical gas would have to be raised to a temperature T_c such that

$$\tfrac{3}{2}kT_c = \bar{E}. \tag{10.50}$$

Thus, in the present case, with $\bar{E} \simeq 3.3$ eV, we see that $T_c \simeq 2.6 \times 10^4$ K.

When thermal energy is supplied to a Fermi gas of electrons, some electrons make transitions to occupy energy levels with $E > E_F$. At room temperature $T_r \simeq 300$ K the thermal energy available is $(3/2)kT_r \simeq 0.039$ eV. It is so small compared with E_F that only very few electrons with energies close to E_F can be excited. Thus the energy distribution remains close to that at absolute zero (see Fig. 10.4), the fraction of electrons excited being of the order $T_r/T_F \simeq 0.005$, where $T_F = E_F/k$ is the *Fermi temperature*. This in turn means that the conduction electrons only contribute a very small proportion of the specific heat of a metal, which is almost entirely determined by the thermal motion of the ions (the nuclei + bound electrons). This discovery of Sommerfeld cleared up what had been a long-standing puzzle.

Degenerate matter in stars

As a second example of the use of the Fermi-gas model near zero temperature, we consider briefly stars near the end of the evolutionary process – white dwarfs and neutron stars. A star of mass M and radius R can be considered to be in hydrostatic equilibrium, the gravitational pressure being balanced by the pressures due to the thermal agitation of its constituents and to radiation. As the nuclear fuel of the stars is used up, the star contracts and the pressure increases. Ultimately, the constituent fermions in the core of the star are found in the lowest possible energy state of the system, and are said to form a *degenerate* Fermi gas. Stars of mass smaller than, or comparable to, that of the Sun collapse to white dwarfs, of the order of 10 000 km in diameter. Under these conditions the main source of the pressure balancing the gravitational pressure is due to the Fermi gas of *electrons*. Stars with much larger densities can be formed in supernova explosions. At a typical density of 10^{18} kg m^{-3}, most of the electrons have been absorbed by protons (inverse beta decay), so forming neutrons. These stars are called *neutron stars* and are some tens of kilometres in diameter. In this case, the pressure balancing the gravitational potential is due to the Fermi gas of *neutrons*.

Consider first the case of white dwarf stars, for which the mass densities d are typically in the range 10^7 kg m$^{-3} \leqslant d \leqslant 10^{12}$ kg m^{-3} and the temperatures T are less than 10^7 K. Since these stars are towards the end of their lives, nuclear fusion reactions will have converted most of the hydrogen nuclei to ^{12}C or ^{16}O nuclei in the core of the star surrounded by outer layers of ^{4}He nuclei. For the sake of an order-of-magnitude model, we shall suppose that the nuclei are all of the same kind with atomic number Z and mass number $A = 2Z$. For densities $d > 10^7$ kg m^{-3}, and $A \leqslant 16$, the internuclear spacing is less than the radius of a one-electron atom with nuclear charge Ze. For this reason the atoms are totally ionised (pressure ionisation) and the electrons form a free Fermi gas, the Fermi energy being given by (10.39), where ρ is the number density of the electrons and m is the electron mass. Since there are Z electrons per nucleus and each nucleus is approximately of mass AM_p, where M_p is the proton mass, we have that $\rho \simeq Zd/(AM_p)$. For typical densities and temperatures one has $kT < E_F$, so that the electron gas can be taken to be degenerate and contributes a total kinetic energy $E_K = (3/5)NE_F$, where N is the total number of electrons. This is a non-relativistic expression which must be modified if the energy becomes large. The total energy contributed by the thermal motion of the ions is small compared with the kinetic energy of the electrons and can be neglected. In this approximation we find by using (10.39) that the total (kinetic plus gravitational) energy of the star is

$$E_T = E_K - \frac{3}{5}\frac{GM^2}{R}$$

$$= N\frac{3\hbar^2}{10m}\left(\frac{3\pi^2 Zd}{AM_p}\right)^{2/3} - \frac{3}{5}\frac{GM^2}{R} \tag{10.51}$$

where R and M are the radius and mass of the star, respectively, and G is the gravitational constant, $G = 6.67 \times 10^{-11}$ N m^2 kg^{-2}. Since $N \simeq ZM/(AM_p)$, and $d = M/(4\pi R^3/3)$ we can write E_T in terms of R and M as

$$E_T = a\frac{M^{5/3}}{R^2} - \frac{3}{5}\frac{GM^2}{R} \tag{10.52}$$

where

$$a = \frac{1}{5}\left(\frac{3}{2}\right)^{7/3}\pi^{2/3}\frac{\hbar^2}{m}\left(\frac{Z}{AM_p}\right)^{5/3}. \tag{10.53}$$

For equilibrium at a given value of the mass M, E_T must be a minimum as a function of R, which is treated as a parameter, so that $dE_T/dR = 0$. This condition provides the relationship between the mass of a star and its radius. Thus using (10.52), we find for equilibrium the relation

$$R = \frac{10a}{3G}M^{-1/3}. \tag{10.54}$$

It is seen that an equilibrium configuration always exists for any star mass M in this non-relativistic case. This is not correct for higher densities, for which relativistic expressions must be used. It is then found (see Problem 10.3) that there is an upper limit to the star mass, called the *Chandrasekhar limit*, above which the star becomes unstable.

For neutron stars, in a relativistic treatment, one finds (as in the case of white dwarfs) that there is a critical mass – the Chandrasekhar limit – above which equilibrium is not possible (see Problem 10.3). If no allowance is made for the interaction between neutrons the critical mass is found to be about twice the mass of the Sun, which is a little larger than the Chandrasekhar mass limit for white dwarfs.

10.4 Two-electron atoms

We now turn to systems composed of a small number of particles, starting with atoms (or ions) consisting of a nucleus of mass M and charge Ze, and two electrons each of mass m and charge $-e$. Thus if $Z = 1$ the system is a negative hydrogen ion (H$^-$), if $Z = 2$ it is a helium atom, if $Z = 3$ a positively charged lithium ion (Li$^+$), and so on. To a good approximation the nuclear mass M can be taken to be infinitely large compared with the electronic mass m. Let us choose the origin of our coordinate system to be the nucleus, and denote by $\mathbf{r}_1$ and $\mathbf{r}_2$ the position vectors of the two electrons. The Hamiltonian of the system is then

$$H = -\frac{\hbar^2}{2m}\nabla_1^2 - \frac{\hbar^2}{2m}\nabla_2^2 - \frac{Ze^2}{(4\pi\varepsilon_0)r_1} - \frac{Ze^2}{(4\pi\varepsilon_0)r_2} + \frac{e^2}{(4\pi\varepsilon_0)r_{12}} \quad (10.55)$$

where $r_{12} = |\mathbf{r}_1 - \mathbf{r}_2|$. In writing (10.55), we have taken into account the Coulomb interactions between the particles, but have neglected small corrections arising from spin–orbit and spin–spin interactions. The Schrödinger eigenvalue equation to be solved reads

$$H\psi(q_1, q_2) = E\psi(q_1, q_2) \quad (10.56)$$

where q_i denotes collectively the position and spin variables of the ith electron. Since the Hamiltonian (10.55) is spin independent, the operators $\mathbf{S}^2$ and S_z commute with H, where $\mathbf{S} = \mathbf{S}_1 + \mathbf{S}_2$ denotes the total spin operator and $\mathbf{S}_i$ is the spin operator of electron i. The operators $\mathbf{L}^2$ and L_z also commute with H, where $\mathbf{L} = \mathbf{L}_1 + \mathbf{L}_2$ is the total orbital angular momentum operator, $\mathbf{L}_i$ being the orbital angular momentum of electron i. The simultaneous eigenfunctions of H, $\mathbf{L}^2$, L_z, $\mathbf{S}^2$ and S_z can be written as products of spatial eigenfunctions $\phi(\mathbf{r}_1, \mathbf{r}_2)$ satisfying the Schrödinger eigenvalue equation

$$H\phi(\mathbf{r}_1, \mathbf{r}_2) = E\phi(\mathbf{r}_1, \mathbf{r}_2) \quad (10.57)$$

and of two-electron spin functions $\chi(1, 2)$, which are eigenfunctions of $\mathbf{S}^2$ and S_z. Thus we have explicitly

$$\psi(q_1, q_2) = \phi_{L,M_L}(\mathbf{r}_1, \mathbf{r}_2)\chi_{S,M_S}(1, 2) \quad (10.58)$$

where the eigenvalues of $\mathbf{L}^2$, L_z, $\mathbf{S}^2$ and S_z are $L(L+1)\hbar^2$, $M_L\hbar$, $S(S+1)\hbar^2$ and $M_S\hbar$, respectively.

The spin functions χ_{S,M_S} have already been obtained in Section 6.10. The two possible values of the total spin quantum number are $S = 0$ and $S = 1$. When $S = 0$, there is only one spin function, the *singlet* spin state χ_{00} given by (6.301), which has $M_S = 0$ and is *antisymmetric* in the interchange of the spin coordinates of the two electrons. When $S = 1$, there are three spin states χ_{1,M_S}, with $M_S = 1, 0, -1$, given by (6.302), which form a spin *triplet*. These three spin states are *symmetric* in the interchange of the spin coordinates of the two electrons.

Let us now introduce the *space-symmetric* wave function $\phi^+_{L,M_L}(\mathbf{r}_1, \mathbf{r}_2)$ such that

$$\phi^+_{L,M_L}(\mathbf{r}_1, \mathbf{r}_2) = \phi^+_{L,M_L}(\mathbf{r}_2, \mathbf{r}_1) \quad (10.59)$$

and the *space-antisymmetric* wave function $\phi^-_{L,M_L}(\mathbf{r}_1, \mathbf{r}_2)$ satisfying the relation

$$\phi^-_{L,M_L}(\mathbf{r}_1, \mathbf{r}_2) = -\phi^-_{L,M_L}(\mathbf{r}_2, \mathbf{r}_1). \quad (10.60)$$

Since electrons are fermions, the complete wave function $\psi(q_1, q_2)$ of a system of two electrons must be *antisymmetric* under the interchange of the (space plus spin) coordinates q_1 and q_2. From (10.58) and the foregoing discussion, it follows that in order to obtain such antisymmetric wave functions $\psi(q_1, q_2)$ we must either multiply space-symmetric wave functions ϕ^+_{L,M_L} by the antisymmetric (singlet) $S = 0$ spin state (6.301)

$$\psi^{S=0}(q_1, q_2) = \phi^+_{L,M_L}(\mathbf{r}_1, \mathbf{r}_2) \frac{1}{\sqrt{2}} [\alpha(1)\beta(2) - \beta(1)\alpha(2)] \tag{10.61}$$

or multiply space-antisymmetric wave functions ϕ^-_{L,M_L} by one of the three symmetric spin functions (6.302) belonging to the $S = 1$ spin triplet

$$\psi^{S=1}(q_1, q_2) = \phi^-_{L,M_L}(\mathbf{r}_1, \mathbf{r}_2) \times \begin{cases} \alpha(1)\alpha(2) \\ \frac{1}{\sqrt{2}}[\alpha(1)\beta(2) + \beta(1)\alpha(2)] \\ \beta(1)\beta(2) \end{cases} \tag{10.62}$$

Perturbation theory

Our next task is to determine the spatial solutions $\phi^\pm_{L,M_L}(\mathbf{r}_1, \mathbf{r}_2)$ of the Schrödinger equation (10.57), and the corresponding eigenenergies. Because of the presence of the electron–electron interaction term $e^2/(4\pi\varepsilon_0 r_{12})$ in the Hamiltonian (10.55), this equation cannot be solved exactly. Approximation methods must therefore be used. We shall begin by investigating the problem with the help of time-independent perturbation theory.

Let us rewrite the Hamiltonian (10.55) as

$$H = H_0 + H' \tag{10.63}$$

where H_0, the 'unperturbed' Hamiltonian, is chosen to be

$$H_0 = -\frac{\hbar^2}{2m}\nabla_1^2 - \frac{\hbar^2}{2m}\nabla_2^2 - \frac{Ze^2}{(4\pi\varepsilon_0)r_1} - \frac{Ze^2}{(4\pi\varepsilon_0)r_2} \tag{10.64}$$

and the 'perturbation' is the electron–electron interaction

$$H' = \frac{e^2}{(4\pi\varepsilon_0)r_{12}}. \tag{10.65}$$

We see that H_0 is the sum of two Hamiltonians for one-electron hydrogenic atoms:

$$H_0 = h_1 + h_2 \tag{10.66}$$

where

$$h_i = -\frac{\hbar^2}{2m}\nabla_i^2 - \frac{Ze^2}{(4\pi\varepsilon_0)r_i}, \quad i = 1, 2. \tag{10.67}$$

The normalised wave functions $\psi_{n_i l_i m_i}(\mathbf{r}_i)$ and the corresponding eigenenergies E_{n_i} of the one-electron Schrödinger equation

$$h_i \psi_{n_i l_i m_i}(\mathbf{r}_i) = E_{n_i} \psi_{n_i l_i m_i}(\mathbf{r}_i) \tag{10.68}$$

have been obtained in Section 7.5. Using these functions $\psi_{n_i l_i m_i}$, one can construct the space symmetric (+) and space-antisymmetric (−) zero-order wave functions

$$\phi_\pm^{(0)}(\mathbf{r}_1, \mathbf{r}_2) = \psi_{n_1 l_1 m_1}(\mathbf{r}_1)\psi_{n_2 l_2 m_2}(\mathbf{r}_2) \pm \psi_{n_1 l_1 m_1}(\mathbf{r}_2)\psi_{n_2 l_2 m_2}(\mathbf{r}_1) \tag{10.69}$$

which are eigenfunctions of H_0 corresponding to the unperturbed total energy

$$E^{(0)}_{n_1 n_2} = E_{n_1} + E_{n_2} = -\frac{Z^2 e^2}{(4\pi\varepsilon_0)2a_0}\left(\frac{1}{n_1^2} + \frac{1}{n_2^2}\right) \tag{10.70}$$

where $a_0 = 4\pi\varepsilon_0\hbar^2/(me^2)$ is the Bohr radius. The total orbital angular momentum quantum number L can take the values $L = |l_1 - l_2|, |l_1 - l_2| + 1, \ldots, l_1 + l_2$, and for a given L the possible values of the quantum number M_L are $M_L = -L, -L+1, \ldots, L$. It is customary to associate to the value of the total orbital angular momentum quantum number L a code letter according to the correspondence (see Section 6.3)

$$L = 0\ 1\ 2\ 3\ 4\ 5$$
$$\updownarrow \updownarrow \updownarrow \updownarrow \updownarrow \updownarrow$$
$$S\ P\ D\ F\ G\ H,\ \ldots.$$

The ground state

Let us focus our attention on the ground state. It is apparent from (10.69) and (10.70) that the unperturbed state of lowest energy is obtained when each of the two electrons is in the ground state of the one-electron system. That is, each electron is in a (1s) level so that $n_1 = n_2 = 1$ and $l_1 = l_2 = m_1 = m_2 = 0$ (which implies that $L = M_L = 0$). Inspecting (10.69), we see that only the space-symmetric wave function can be of that form, since the space-antisymmetric wave function vanishes when both electrons are in the same state. Within the framework of this independent-particle model the ground state of two-electron atoms (ions) can thus be designated by $(1s)^2$, which means that both electrons are in the 1s level. Using (10.70), the unperturbed ground-state energy is seen to be

$$E^{(0)} = -\frac{Z^2 e^2}{(4\pi\varepsilon_0)a_0}. \tag{10.71}$$

Since $e^2/[(4\pi\varepsilon_0)2a_0] \simeq 13.6$ eV is the ionisation potential of a hydrogen atom (with infinite nuclear mass), we see that the value of $E^{(0)}$ is $E^{(0)} \simeq -2Z^2 \times 13.6$ eV. The corresponding unperturbed wave function, normalised to unity is

$$\phi^{(0)}(r_1, r_2) = \psi_{1s}(r_1)\psi_{1s}(r_2) = \frac{1}{\pi}\left(\frac{Z}{a_0}\right)^3 \exp[-Z(r_1 + r_2)/a_0] \tag{10.72}$$

where the $L = 0, M_L = 0$ and $(+)$ indices have been omitted for notational simplicity. To first order in the perturbation, the ground-state energy is

$$E = E^{(0)} + E^{(1)} \tag{10.73}$$

where $E^{(1)}$, the first-order energy correction, is given from (8.15) and (10.65) by

$$E^{(1)} = \int |\phi^{(0)}(r_1, r_2)|^2 \frac{e^2}{(4\pi\varepsilon_0)r_{12}} d\mathbf{r}_1 d\mathbf{r}_2. \tag{10.74}$$

We shall calculate this integral by using a general procedure which is very useful in many atomic physics calculations. With the help of the generating function (6.82) of the Legendre polynomials, we first expand the quantity $1/r_{12}$ as

$$\frac{1}{r_{12}} = \frac{1}{(r_1^2 - 2r_1 r_2 \cos\theta + r_2^2)^{1/2}} = \frac{1}{r_1} \sum_{l=0}^{\infty} \left(\frac{r_2}{r_1}\right)^l P_l(\cos\theta), \qquad r_1 > r_2$$

$$= \frac{1}{r_2} \sum_{l=0}^{\infty} \left(\frac{r_1}{r_2}\right)^l P_l(\cos\theta), \qquad r_1 < r_2$$

(10.75)

where θ is the angle between the vectors $\mathbf{r}_1$ and $\mathbf{r}_2$ so that

$$\cos\theta = \cos\theta_1 \cos\theta_2 + \sin\theta_1 \sin\theta_2 \cos(\phi_1 - \phi_2).$$

(10.76)

Here (θ_1, ϕ_1) and (θ_2, ϕ_2) are the polar angles of the vectors $\mathbf{r}_1$ and $\mathbf{r}_2$, respectively. The formula (10.75) can be rewritten in the more compact form

$$\frac{1}{r_{12}} = \sum_{l=0}^{\infty} \frac{(r_<)^l}{(r_>)^{l+1}} P_l(\cos\theta)$$

(10.77)

where $r_<$ is the smaller and $r_>$ the larger of r_1 and r_2. Using the addition theorem (6.129) of the spherical harmonics, we also have

$$\frac{1}{r_{12}} = \sum_{l=0}^{\infty} \sum_{m=-l}^{+l} \frac{4\pi}{2l+1} \frac{(r_<)^l}{(r_>)^{l+1}} Y_{lm}^*(\theta_1, \phi_1) Y_{lm}(\theta_2, \phi_2).$$

(10.78)

Let us now substitute this expansion in (10.74) and use the fact that the spherical harmonics are orthonormal on the unit sphere (see (6.103)). Since the function $\phi^{(0)}(r_1, r_2)$, given by (10.72), is spherically symmetric, and because $Y_{00} = (4\pi)^{-1/2}$, we obtain at once by integrating over the polar angles (θ_1, ϕ_1) and (θ_2, ϕ_2),

$$E^{(1)} = \frac{e^2}{4\pi\varepsilon_0} \sum_{l=0}^{\infty} \sum_{m=-l}^{+l} \frac{(4\pi)^2}{2l+1} \int_0^{\infty} dr_1 r_1^2 \int_0^{\infty} dr_2 r_2^2 |\phi^{(0)}(r_1, r_2)|^2 \frac{(r_<)^l}{(r_>)^{l+1}}$$

$$\times \int d\Omega_1 Y_{lm}^*(\theta_1, \phi_1) Y_{00} \int d\Omega_2 Y_{00} Y_{lm}(\theta_2, \phi_2)$$

$$= \frac{e^2}{4\pi\varepsilon_0} \sum_{l=0}^{\infty} \sum_{m=-l}^{+l} \frac{(4\pi)^2}{2l+1} \int_0^{\infty} dr_1 r_1^2 \int_0^{\infty} dr_2 r_2^2 |\phi^{(0)}(r_1, r_2)|^2 \frac{(r_<)^l}{(r_>)^{l+1}} \delta_{l,0} \delta_{m,0}.$$

(10.79)

Table 10.1 Values of the ground-state energy of various two-electron atoms and ions (in eV).

Z	System	A $E^{(0)}$	B $E^{(0)} + E^{(1)}$	C Variational	D Accurate values
2	He	−108.85	−74.83	−77.50	−79.02
3	Li$^+$	−244.90	−193.88	−196.52	−198.10
4	Be^{2+}	−435.39	−367.36	−370.08	−371.71
5	B^{3+}	−680.29	−595.39	−597.84	−599.47
6	C^{4+}	−979.62	−877.57	−880.30	−881.93

A = Unperturbed values from (10.71).
B = First-order perturbation theory using the expression (10.81) of $E^{(1)}$.
C = Rayleigh–Ritz method using (10.87).
D = Rayleigh–Ritz method using elaborate trial functions.

Thus all the terms in this double sum vanish, except the first one, for which $l = m = 0$. Using the expression (10.72) for $\phi^{(0)}(r_1, r_2)$, we then have

$$E^{(1)} = 16 \frac{e^2}{4\pi\varepsilon_0} \left(\frac{Z}{a_0}\right)^6 \int_0^\infty dr_1 r_1^2 \int_0^\infty dr_2 r_2^2 \exp[-2Z(r_1+r_2)/a_0] \frac{1}{r_>}$$

$$= 16 \frac{e^2}{4\pi\varepsilon_0} \left(\frac{Z}{a_0}\right)^6 \int_0^\infty dr_1 r_1^2 \exp(-2Zr_1/a_0) \left[\frac{1}{r_1} \int_0^{r_1} dr_2 r_2^2 \exp(-2Zr_2/a_0)\right.$$

$$\left. + \int_{r_1}^\infty dr_2 r_2 \exp(-2Zr_2/a_0)\right]. \qquad (10.80)$$

The integrals are now straightforward, and yield the result

$$E^{(1)} = \frac{5}{8} \frac{e^2}{(4\pi\varepsilon_0)a_0} Z. \qquad (10.81)$$

In Table 10.1, the unperturbed energies $E^{(0)}$ and the first-order perturbed values $E^{(0)} + E^{(1)}$ are given for several values of Z and are compared with accurate values obtained using the Rayleigh–Ritz variational method with elaborate trial functions. As Z increases the perturbation $1/r_{12}$ becomes relatively less important compared with the interactions between the electrons and the nucleus, and the accuracy of the simple first-order result improves.

The variational method

The variational method discussed in Section 8.3 can also be used to determine the energy levels of atoms. We recall that if H denotes the Hamiltonian of a quantum system, and ϕ a physically admissible trial function, the functional

$$E[\phi] = \frac{\langle \phi | H | \phi \rangle}{\langle \phi | \phi \rangle} \qquad (10.82)$$

provides a variational principle for the discrete eigenvalues of the Hamiltonian. Moreover, it also yields a minimum principle for the ground state energy (see (8.145)). Thus, if we follow the Rayleigh–Ritz variational method and use trial functions ϕ depending on variational parameters $\lambda_1, \lambda_2, \ldots$, the expression (10.82) becomes a function $E(\lambda_1, \lambda_2, \ldots)$ of the parameters $\lambda_1, \lambda_2, \ldots$, and the optimum ground-state energy is obtained by minimising $E(\lambda_1, \lambda_2, \ldots)$ with respect to the parameters λ_i.

In the case of two-electron atoms (ions) which we are studying here, the Hamiltonian H is given by (10.55). The basic defect of the zero-order (unperturbed) wave function (10.72) – from which the first-order energy correction (10.81) was obtained – is that each electron moves in the fully unscreened field of the nucleus. In order to take into account approximately the screening effect of each electron on the other one, we choose a simple one-parameter trial function of the form

$$\phi(\lambda, r_1, r_2) = \frac{1}{\pi}\left(\frac{\lambda}{a_0}\right)^3 \exp[-\lambda(r_1 + r_2)/a_0] \tag{10.83}$$

where λ is a variational parameter. Upon comparison with (10.72) we see that λe may be interpreted as an 'effective charge'. Since the trial function (10.83) is normalised to unity, the variational expression (10.82) becomes

$$E(\lambda) = \int \phi^*(\lambda, r_1, r_2) H \phi(\lambda, r_1, r_2) \mathrm{d}\mathbf{r}_1 \mathrm{d}\mathbf{r}_2. \tag{10.84}$$

This is readily evaluated (see Problem 10.4), giving

$$E(\lambda) = \left(\lambda^2 - 2Z\lambda + \frac{5}{8}\lambda\right)\frac{e^2}{(4\pi\varepsilon_0)a_0}. \tag{10.85}$$

The minimum of E as a function of λ is obtained when $\mathrm{d}E/\mathrm{d}\lambda = 0$, from which condition we obtain

$$\lambda = Z - \frac{5}{16}. \tag{10.86}$$

With this value of λ, the minimum of the energy is given by

$$E = -\left(Z - \frac{5}{16}\right)^2 \frac{e^2}{(4\pi\varepsilon_0)a_0}. \tag{10.87}$$

The values of E for various two-electron atoms are included in Table 10.1. By recollecting that the variational estimate of a ground-state energy is an upper bound, it is seen that the variational values in Table 10.1 are more accurate than those provided by first-order perturbation theory. To obtain the accurate values shown in the last column of Table 10.1 much more elaborate trial functions are required depending on many variational parameters. It is worth noting that the accurate value -79.02 eV of the non-relativistic ground-state energy of helium quoted in Table 10.1 corresponds to an ionisation potential of 198 345 cm^{-1} which is in very good agreement with the experimental value $(198\,310.82 \pm 0.15)$ cm^{-1}. When allowance is made for the finite mass of the nucleus and for further relativistic and radiative corrections, the theoretical helium ionisation potential is found to be 198 310.699 cm^{-1}, which is in phenomenal agreement with the experimental value. As Z increases, the relativistic corrections become more important, and for highly charged helium-like ions the non-relativistic approximation used in this section breaks down.

Details of further calculations of ground-state and excited-state energies of two-electron atoms and ions may be consulted in Bransden and Joachain (1983).

10.5 Many-electron atoms

Let us consider an atom or ion containing a nucleus of charge Ze and N electrons. Neglecting all but the Coulomb interactions, and taking the mass of the nucleus to be infinitely large compared with the electron mass, we write the Hamiltonian of this system as

$$H = \sum_{i=1}^{N}\left(-\frac{\hbar^2}{2m}\nabla_i^2 - \frac{Ze^2}{(4\pi\varepsilon_0)r_i}\right) + \sum_{i<j=1}^{N}\frac{e^2}{(4\pi\varepsilon_0)r_{ij}} \tag{10.88}$$

where $\mathbf{r}_i$ denotes the position vector of the electron i with respect to the nucleus, $r_{ij} = |\mathbf{r}_i - \mathbf{r}_j|$ and the last summation is over all pairs of electrons. The corresponding time-independent Schrödinger equation for the N-electron atom wave functions $\psi(q_1, \ldots, q_N)$ reads

$$H\psi(q_1, \ldots, q_N) = E\psi(q_1, \ldots, q_N) \tag{10.89}$$

and since electrons are fermions we want totally antisymmetric solutions of (10.89).

Because of the presence of the electron–electron interaction terms $e^2/(4\pi\varepsilon_0 r_{ij})$ in (10.88) the Schrödinger equation (10.89) is not separable, and approximation methods must be used. In the previous section we saw that very accurate results can be obtained for the energy levels of two-electron atoms (ions) by performing Rayleigh–Ritz variational calculations in which elaborate trial wave functions containing a large number of variational parameters are used. Extended variational calculations of this kind have also been carried out for other light atoms (such as lithium), but this approach becomes increasingly tedious when the number of electrons increases.

The starting point of calculations on many-electron atoms is the *central-field approximation*. This approximation is based on an *independent-particle model*, in which each electron moves in an effective central potential $V(r)$ which represents the attraction of the nucleus and the average effect of the repulsive interactions between this electron and the other ones.

We may readily obtain the form of the central potential $V(r)$ at small and large distances. Indeed, when an electron is close to the nucleus it experiences the full Coulomb potential due to the nuclear charge Ze, and we have

$$V(r) \underset{r\to 0}{\to} -\frac{Ze^2}{(4\pi\varepsilon_0)r}. \tag{10.90}$$

On the other hand, at large distances, an electron experiences a Coulomb potential due to the nucleus screened by the $(N-1)$ other electrons. Thus

$$V(r) \underset{r\to\infty}{\to} -\frac{[Z-(N-1)]e^2}{(4\pi\varepsilon_0)r} \qquad (10.91)$$

The determination of the effective potential at intermediate distances is a much more difficult problem, to which we shall return shortly. At this point, however, it is important to stress that for intermediate values of r the potential $V(r)$ – which represents the attraction of the nucleus plus the average repulsion of the other electrons – must depend on the details of the charge distribution of the electrons, or in other words on the dynamical state of the electrons. As a result, the same effective potential $V(r)$ cannot account for the full spectrum of a complex atom (ion). However, if we limit our attention to the ground state and the first excited states, it is reasonable to assume that a given central potential $V(r)$ can be used as a starting point.

Let us now return to the Hamiltonian (10.88). From the foregoing discussion it is clear that a meaningful separation of H into an unperturbed part and a perturbation can be achieved by writing

$$H = H_c + H_1 \qquad (10.92)$$

where

$$H_c = \sum_{i=1}^{N}\left(-\frac{\hbar^2}{2m}\nabla_i^2 + V(r_i)\right) \qquad (10.93a)$$

$$= \sum_{i=1}^{N} h_i, \qquad h_i = -\frac{\hbar^2}{2m}\nabla_i^2 + V(r_i) \qquad (10.93b)$$

is the Hamiltonian corresponding to the central-field approximation and

$$H_1 = \sum_{i<j=1}^{N}\frac{e^2}{(4\pi\varepsilon_0)r_{ij}} - \sum_{i=1}^{N}\left(\frac{Ze^2}{(4\pi\varepsilon_0)r_i} + V(r_i)\right) \qquad (10.94)$$

is the remaining part of the Hamiltonian.

Let us focus our attention on the central-field Hamiltonian H_c. We see from (10.93b) that it is the sum of N identical, individual Hamiltonians h_i. The (normalised) eigenfunctions of h_i are one-electron, central-field orbitals $u_{n_i l_i m_i}(\mathbf{r})$ satisfying the equation

$$\left(-\frac{\hbar^2}{2m}\nabla^2 + V(r)\right)u_{nlm_l}(\mathbf{r}) = E_{nl}u_{nlm_l}(\mathbf{r}) \qquad (10.95)$$

where the subscript i has been omitted. From our study of central-field problems in Chapter 7, it follows that the energy eigenvalues E_{nl} are independent of the magnetic quantum number m_l and that the orbitals $u_{nlm_l}(\mathbf{r})$ are products of radial functions times spherical harmonics

$$u_{nlm_l}(\mathbf{r}) = R_{nl}(r)Y_{lm_l}(\theta,\phi). \qquad (10.96)$$

The principal quantum number n can take the values $n = 1, 2, \ldots$, while the allowed values of the orbital angular momentum quantum number l are $l = 0, 1, \ldots, n-1$ and those of the magnetic quantum number m_l are $m_l = -l, -l+1, \ldots, l$. In order to take into account the electron spin, we multiply each one-electron spatial orbital $u_{nlm_l}(\mathbf{r})$ by a spin-1/2 function $\chi_{\frac{1}{2}, m_s}$, thus forming the (normalised) spin-orbitals

$$u_{nlm_lm_s}(q) = u_{nlm_l}(\mathbf{r})\chi_{\frac{1}{2}, m_s}, \qquad m_s = \pm 1/2 \tag{10.97}$$

characterised by the four quantum numbers n, l, m_l and m_s. Since the energy E_{nl} does not depend on the quantum numbers m_l and m_s, we see that each individual electron energy level is $2(2l+1)$ degenerate.

Because the central-field Hamiltonian H_c is just the sum of the individual Hamiltonians h_i, the total energy E_c of the atom in the central-field approximation is the sum of the individual electron energies

$$E_c = \sum_{i=1}^{N} E_{n_i l_i}. \tag{10.98}$$

The corresponding total central-field wave function $\psi_c(q_1, \ldots, q_N)$ is a totally antisymmetric wave function built from the single electron spin-orbitals (10.97). It is therefore a Slater determinant of the form (10.27), in which each of the indices $\alpha, \beta, \ldots, \nu$, designates a different set of four quantum numbers (n, l, m_l, m_s), so that the Pauli exclusion principle is satisfied.

The order of the individual energy levels E_{nl} does not depend in a crucial way on the form of the potential $V(r)$. Starting with the most tightly bound, this order is given by

$$\text{1s, 2s, 2p, 3s, 3p, [4s, 3d], 4p, [5s, 4d], 5p, [6s, 4f, 5d], \ldots} \tag{10.99}$$

where as usual the figure represents the principal quantum number and the letter the value of l, with $l = 0, 1, 2, 3, \ldots$, corresponding to s, p, d, f, ... The brackets in (10.99) enclose levels which have so nearly the same energy that their order can vary from one atom to another. Electrons in orbitals having the same value of the quantum number n are said to belong to the same *shell*. The shells are labelled K, L, M, N, ..., according to whether $n = 1, 2, 3, 4, \ldots$. Electrons having the same value of n and l are said to belong to the same *subshell*. According to our foregoing discussion there are $2(2l+1)$ electron states having the same value of n and l but different values of m_l and m_s, so that the maximum number of electrons in a subshell is $2(2l+1)$. As seen from (10.98), in the central-field approximation the total energy E_c of the atom depends only on the number of electrons occupying each of the individual energy levels E_{nl}. Therefore, this total energy is determined by the *electron configuration*, that is the distribution of the electrons with respect to the quantum numbers n and l.

The periodic system of the elements

We are now equipped with the necessary information to discuss the electronic structure of atoms or ions. In what follows we shall restrict our attention to the *ground state* of *neutral* atoms. The Z electrons of an atom of atomic number Z then occupy the lowest individual energy levels in accordance with the requirements of the Pauli exclusion principle. The *ground-state configuration* of an atom is therefore obtained by distributing the Z electrons among a certain number f of subshells, the first $(f-1)$ subshells being filled and the last one (corresponding to the highest energy) generally not, except for particular values of Z (2, 4, 10, 12, 18, ...). The least tightly bound electrons, which are in the subshell of highest energy, and in insufficient numbers to form another closed subshell, are called *valence* electrons. In going from one atom having atomic number Z to the next one, with atomic number $Z + 1$, the number of electrons increases by one, the $Z + 1$ electrons occupying the lowest energy levels allowed by the exclusion principle.

The first element $(Z = 1)$ is atomic hydrogen, which has the ground-state configuration 1s. For $Z = 2$ (helium), both electrons occupy the 1s level and the configurations is $(1s)^2$, where the superscript denotes the number of electrons occupying the (nl) orbital. Since no more than two electrons can occupy 1s orbitals, the two electrons of helium fill the K shell $(n = 1)$, which is said to be 'closed'. For $Z = 3$ (lithium), the ground-state configuration is $(1s)^2 2s$ because the configuration $(1s)^3$ is forbidden by the Pauli exclusion principle. The next element is beryllium with $Z = 4$. The fourth electron can also occupy the 2s level, so that the ground state configuration of beryllium is $(1s)^2(2s)^2$. For $Z = 5$ (boron) the K shell and the 2s subshell are full so that the fifth electron must occupy a 2p level. As Z increases from 5 to 10 (neon), the 2p subshell progressively fills up. At $Z = 10$ (neon) the L shell $(n = 2)$ is full and for $Z = 11$, (sodium), the eleventh electron must go into a 3s level, belonging to the M shell.

From $Z = 11$ to $Z = 18$ (argon), the 3p levels fill progressively. At $Z = 19$ (potassium), it might be expected that the nineteenth electron would go in the 3d level, but in fact the 4s level has a lower energy than the 3d level, so the configuration of potassium is $(1s)^2(2s)^2(2p)^6(3s)^2(3p)^6 4s$.

In the same way, the electronic configuration of all the elements with higher values of Z can be studied. Additional information is provided by the *term value* which indicates the total orbital and spin angular momentum quantum numbers (L and S, respectively), and also the total angular momentum quantum number J. The term is written in the so-called *Russell–Saunders notation* as

$$^{2S+1}L_J$$

where we recall that the code letters S, P, D, F, ..., correspond to $L = 0, 1, 2, 3, \ldots$.

The *noble gases* (He, Ne, Ar, Kr, Xe) have a full K shell or p subshell and are chemically inert because it takes an appreciable amount of energy to perturb these atoms. If there is one valence electron outside a closed shell as in the *alkalis* (Li, Na, K, Rb, Cs), it is easily detached and these elements are very reactive. Similarly, the *halogens* (F, Cl, Br, I) which lack one electron to make up a closed shell, also are highly reactive because of their tendency to capture an electron from another atom.

The noble gases (He, Ne, Ar, Kr, ...), the alkalis (Li, Na, K, ...) and the halogens (F, Cl, Br, I) are all examples of the recurrence as Z increases of similar chemical properties. These recurrences were classified by Mendeleev in 1869 in the *periodic table* of the elements, and are basically consequences of the Pauli exclusion principle.

The determination of the central potential

Several methods have been proposed to obtain the central potential $V(r)$, the most important of them being the *Thomas–Fermi* and the *Hartree–Fock* models. We shall only describe here the main features of these two approaches[2].

The theory developed independently by Thomas and Fermi for the ground state of complex atoms (ions) having a *large* number of electrons is based on *statistical* and *semi-classical* considerations. The N electrons of the system are treated as a Fermi electron gas in the ground state, confined to a region of space by a central potential $V(r)$ which vanishes at infinity. It is assumed that this potential is slowly varying over a distance comparable with the de Broglie wavelengths of the electrons, so that many electrons are present in a volume where $V(r)$ is nearly constant, and the statistical approach used in studying the Fermi electron gas can be applied. In addition, since the number of electrons is large, many of them have high principal quantum numbers, and semi-classical methods can be employed. The Thomas–Fermi method has the advantage of simplicity: it leads to a 'universal' equation, which can readily be solved numerically, and from which the central potential $V(r)$ and the electron density $\rho(r)$ can be obtained for any multi-electron atom. The method is particularly useful in calculating quantities which depend on the 'average electron' (such as the total energy of the atom). On the other hand, quantities which rely on the properties of the 'outer' electrons (such as the ionisation potential) are poorly given in the Thomas–Fermi model.

A much more elaborate approximation for complex atoms is the *Hartree–Fock* or *self-consistent field* method. In this approach, it is assumed, in accordance with the independent particle approximation and the Pauli exclusion principle, that the N-electron wave function is a *Slater determinant* (10.27), in which $u_\alpha(q_1), u_\beta(q_2), \ldots, u_\nu(q_N)$, are individual electron spin-orbitals, where each of the subscripts $\alpha, \beta, \ldots, \nu$, denotes a different set of four quantum numbers (n, l, m_l, m_s). The *optimum* Slater determinant is then obtained by using the *variational method* to determine the best individual electron spin-orbitals. The Hartree–Fock method is therefore a particular case of the variational method, in which the trial function ϕ for the N-electron atom is a Slater determinant whose individual spin-orbitals are optimised. It should be noted that the N-electron-atom wave function $\psi(q_1, q_2, \ldots, q_N)$, solution of the Schrödinger equation (10.89), can only be represented by an infinite sum of Slater determinants, so that the Hartree–Fock method can be considered as a 'first step' in the determination of atomic wave functions and energies. We also remark that the application of the Hartree–Flock method is not confined to atoms or ions, but can also be made to other systems such as the electrons of a molecule or a solid, or the nucleons of a nucleus.

---

[2] A more detailed account can be found in Bransden and Joachain (1983), Chapter 7.

Let us consider the functional $E[\phi] = \langle \phi | H | \phi \rangle$ of the variational method, in which H is the Hamiltonian (10.88), and the trial function ϕ is a Slater determinant (normalised to unity) containing the one-electron orbitals $u_\alpha(q_1), u_\beta(q_2), \ldots, u_\nu(q_N)$, to be determined. Upon variation of $E[\phi]$ with respect to the spin-orbitals, it is found that the spin-orbitals must satisfy a system of coupled equations, known as the *Hartree–Fock equations*, which have the form

$$\left[-\frac{\hbar^2}{2m} \nabla_i^2 + \mathcal{V}(q_i) \right] u_\lambda(q_i) = E_\lambda u_\lambda(q_i), \qquad \lambda = \alpha, \beta, \ldots, \nu \qquad \textbf{(10.100)}$$

where the Hartree–Fock potential $\mathcal{V}(q_i)$ represents the field in which the electron i moves when the atom is in a given state (for example the ground state).

The Hartree–Fock equations (10.100) look similar to individual Schrödinger eigenvalue equations for each of the spin-orbitals u_λ. They are *not* genuine eigenvalue equations, however, since the Hartree–Fock potential $\mathcal{V}(q_i)$ depends on the $(N-1)$ other spin-orbitals u_μ, with $\mu \neq \lambda$. In fact, to solve the system of Hartree–Fock equations (10.100) one proceeds by *iteration*. Starting from approximate individual spin-orbitals $u_\alpha^{(1)}, u_\beta^{(1)}, \ldots, u_\nu^{(1)}$, one first calculates the corresponding approximate expression $\mathcal{V}^{(1)}$ of the Hartree–Fock potential. The Hartree–Fock equations are then solved with this potential $\mathcal{V}^{(1)}$ to obtain new spin-orbitals $u_\alpha^{(2)}, u_\beta^{(2)}, \ldots, u_\nu^{(2)}$, which in turn yield a new potential $\mathcal{V}^{(2)}$. This procedure is then repeated until the final spin-orbitals give a potential $\mathcal{V}^{(n)}$ which is identical (within the desired approximation) to the potential $\mathcal{V}^{(n-1)}$ obtained from the preceding cycle. The Hartree–Fock potential determined in this way is known as the *self-consistent field* of the atom (ion). In general, the Hartree–Fock potential $\mathcal{V}$ is spin-dependent and depends also on the angular coordinates. A central potential $V(r)$ of the type discussed above can then be obtained by averaging over the angular variables and the spin states.

Corrections to the central-field approximation. L–S coupling and j–j coupling

Let us now consider the main corrections to the central-field approximation. The first important correction to the central-field Hamiltonian H_c is the term H_1, given by (10.94), which represents the difference between the actual Coulomb repulsion of the electrons and the average electron repulsion contained in the central field $V(r)$. In particular, if $V(r)$ is a (central) Hartree–Fock potential, the term H_1 leads to the so-called *correlation effects*, the *correlation energy* E_{corr} being defined as the difference between the exact (non-relativistic) energy E of a given atomic state and the corresponding Hartree–Fock energy E_{HF}:

$$E_{\text{corr}} = E - E_{\text{HF}}. \qquad \textbf{(10.101)}$$

For light atoms (ions) the relative error E_{corr}/E in the Hartree–Fock energy of the ground state is about 1%. Correlation effects can be studied by using *perturbation theory*, starting with the Hartree–Fock energies and wave functions as the 'zero-order' approximation. Alternatively, the variational method can be used with a trial function ϕ which is a linear combination of Slater determinants

$$\phi = \sum_i c_i \phi_i \tag{10.102}$$

where the coefficients c_i are variational parameters. The various Slater determinants ϕ_i differ in the choice of the spin-orbitals occupied by the electrons, and correspond therefore to different electronic configurations. This approach is known as the *configuration-interaction method*.

A second important correction to the central-field approximation is due to the *spin–orbit interactions* of the electrons, which we have ignored so far. Working within the framework of the independent particle model, we write the spin–orbit correction term in a form which is an extension of (8.102), namely

$$H_2 = \sum_i \xi(r_i) \mathbf{L}_i \cdot \mathbf{S}_i \tag{10.103}$$

where $\mathbf{L}_i$ is the orbital angular momentum operator of electron i, $\mathbf{S}_i$ is its spin angular momentum operator, and

$$\xi(r_i) = \frac{1}{2m^2 c^2} \frac{1}{r_i} \frac{dV(r_i)}{dr_i}. \tag{10.104}$$

Adding the corrections H_1 and H_2 to the central-field Hamiltonian H_c, we obtain the total Hamiltonian

$$\mathcal{H} = H_c + H_1 + H_2. \tag{10.105}$$

In order to study the effects of the terms H_1 and H_2, one can use perturbation theory, starting from the eigenfunctions and eigenenergies of H_c as our zero-order approximation. The manner in which the perturbation calculation is to be carried out depends on the relative magnitude of the two perturbing terms H_1 and H_2. The case for which $|H_1| \gg |H_2|$ occurs for atoms with small and intermediate values of Z and is called the *L–S coupling* case. The situation for which $|H_2| \gg |H_1|$ arises for atoms with large Z and is known as the *j–j coupling* case. The situation for which both perturbations H_1 and H_2 are comparable is known as *intermediate coupling*.

10.6 Molecules

The study of molecular structure is considerably more difficult than that of isolated atoms, but fortunately the problem is simplified because the mass of the electrons is much smaller than that of the nuclei, while the forces to which the electrons and the nuclei are submitted are of comparable intensity. As a result, the motion of the nuclei is much slower than that of the electrons, and the nuclei occupy nearly fixed positions within the molecule.

Evidence from X-ray diffraction and molecular spectra shows that when atoms associate to form molecules, the tightly bound inner shells of electrons are nearly undisturbed, and remain localised about each nucleus. The outer electrons, on the other hand, are distributed throughout the molecule, and it is the charge distribution of these *valence* electrons that provides the binding force. The order of magnitude of the separation of energy levels for the electronic motion of the valence electrons can be estimated by using the following argument. Let a be a typical average distance of the nuclei in a molecule. From the uncertainty principle, the magnitude of the momentum of the valence electrons is of the order of $\hbar/a$, so that a rough estimate of their kinetic energy, and hence of the magnitude E_e of the electronic energies, is given by

$$E_e \simeq \frac{\hbar^2}{ma^2} \tag{10.106}$$

where m is the electron mass. Since neutron diffraction experiments indicate that $a \simeq 1$ Å, we see that E_e is of the order of a few eV. The result (10.106) clearly gives also an estimate of the energy separation between low-lying electronic energy levels of the molecule. The corresponding line spectra are observed in the ultraviolet and visible regions.

Let us now consider the nuclear motions. These can be classified into translations and rotations of the entire (quasi-rigid) equilibrium arrangement, and internal vibrations of the nuclei about their equilibrium positions. The translational motion can be separated by introducing the centre of mass, which moves as a free particle in the absence of external fields. In what follows we shall assume that the separation of the centre of mass motion has been performed, and we shall only consider the vibrational and rotational motions of the nuclei. To estimate the vibrational energy, we can use the following classical argument. If the electrons are bound to the molecule by a force of magnitude F, the nuclei must be bound by an equal and opposite force. Taking this force to be simple harmonic, with a force constant k, the angular frequency of the electronic motion will be $\omega_e = (k/m)^{1/2}$ and that of the vibrational nuclear motion will be $\omega_N = (k/M)^{1/2}$, where M is of the order of a typical nuclear mass. The ratio of the energy of the vibrational motion to that of the electronic motion is thus $\hbar\omega_N/\hbar\omega_e \simeq (m/M)^{1/2}$ and the energy E_v associated with a low mode of vibration is given approximately by

$$E_v \simeq \left(\frac{m}{M}\right)^{1/2} E_e. \tag{10.107}$$

The ratio m/M being in the range 10^{-3} to 10^{-5}, we see from (10.107) that E_v is roughly a hundred times smaller than E_e, so that typical vibrational transitions between adjacent vibrational levels, without change of the electronic state, lie in the infrared part of the spectrum.

To estimate the rotational energy E_r, let us consider the simple case of a diatomic molecule, with each of the nuclei having the same mass M and being a distance a apart. The moment of inertia of the molecule is then $I = Ma^2/2$. Using the result (6.136) which was obtained in Chapter 6 for the rigid rotator, we see that the order of magnitude of the energy associated with a fairly low mode of rotation is

$$E_r \simeq \frac{\hbar^2}{Ma^2} \simeq \frac{m}{M} E_e. \tag{10.108}$$

Thus the rotational molecular energies are smaller than electronic energies by a factor of the order of m/M, and smaller than vibrational energies by a factor of the order of $(m/M)^{1/2}$. The rotational motion leads to a 'second-order' splitting of the electronic spectrum, the spacing being of the order of 0.001 eV, which is very small compared with the 'first-order' splitting of about 0.1 eV produced by the vibrational motion. The rotational motion has also the effect of splitting the vibrational spectra for which the electronic state does not change, thus giving rise to the rovibrational band spectra referred to in Section 6.4. Transitions between rotational levels, belonging to the same electronic and vibrational level give rise to the rotational band spectra which are observed in the far infrared and microwave regions at wave numbers of 1 to 10^7 cm^{-1}.

Because of the small ratio of the electronic mass to the nuclear mass ($m/M \simeq 10^{-3} - 10^{-5}$), and since the period of a motion is of the order of h divided by its energy, we see from (10.106)–(10.108) that the nuclear periods are much longer than the electronic periods. Thus the electronic and nuclear motions can essentially be treated independently, and it is a good approximation to determine the electronic states at each value of the internuclear separation by treating the nuclei as fixed. The charge distribution of the electrons is then a function of the nuclear positions and determines the nuclear motions. This decoupling of the electronic and nuclear motions is known as the *Born–Oppenheimer approximation*. In what follows we shall apply these ideas to the study of the simplest molecule: the hydrogen molecular ion.

The hydrogen molecular ion

The hydrogen molecular ion H_2^+ is a system composed of two protons and one electron. A centre-of-mass system of coordinates can be constructed by introducing the relative position vector $\mathbf{R}$ of the two protons A and B, and the position vector $\mathbf{r}$ of the electron with respect to the centre of mass of A and B (see Fig. 10.5). The Schrödinger equation is

$$\left[-\frac{\hbar^2}{2\mu_{AB}}\nabla_\mathbf{R}^2 - \frac{\hbar^2}{2\mu_e}\nabla_\mathbf{r}^2 - \frac{e^2}{(4\pi\varepsilon_0)r_A} - \frac{e^2}{(4\pi\varepsilon_0)r_B} + \frac{e^2}{(4\pi\varepsilon_0)R} - E\right]\psi(\mathbf{r},\mathbf{R}) = 0 \tag{10.109}$$

where r_A and r_B are the distances of the electron from the protons A and B. The reduced mass of the two-proton system is $\mu_{AB} = M_p/2$, where M_p is the proton mass and μ_e is the reduced mass of the electron with respect to the two-proton system

$$\mu_e = \frac{m(2M_p)}{m + 2M_p} \simeq m \tag{10.110}$$

where m is the mass of the electron.

Since the nuclear motion is very slow compared with that of the electrons, an electronic Hamiltonian $H_e(\mathbf{r}, \mathbf{R})$ can be introduced in which $\mathbf{R}$ can be considered to

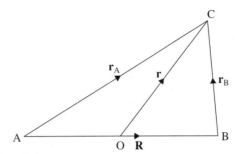

Figure 10.5 A coordinate system for the hydrogen molecular ion H_2^+.

be a slowly varying parameter, and

$$H_e(\mathbf{r}, \mathbf{R}) = -\frac{\hbar^2}{2m}\nabla_\mathbf{r}^2 - \frac{e^2}{(4\pi\epsilon_0)r_A} - \frac{e^2}{(4\pi\epsilon_0)r_B} + \frac{e^2}{(4\pi\epsilon_0)R}. \tag{10.111}$$

For each (fixed) value of $\mathbf{R}$ a set of eigenenergies $E_k(\mathbf{R})$ with corresponding orthonormal eigenfunctions $\chi_k(\mathbf{r}, \mathbf{R})$ can be found which satisfy the *electronic* Schrödinger equation

$$H_e(\mathbf{r}, \mathbf{R})\chi_k(\mathbf{r}, \mathbf{R}) = E_k(\mathbf{R})\chi_k(\mathbf{r}, \mathbf{R}). \tag{10.112}$$

The functions χ_k are known as *molecular orbitals*, in analogy with the atomic orbitals introduced in the previous section.

The adiabatic theorem (Section 9.4) tells us that as $\mathbf{R}$ is slowly varying, if the system is prepared in the level k with wave function χ_k it will remain in that level and not make a transition to any other. It follows that the full time-independent wave functions $\psi(\mathbf{r}, \mathbf{R})$ can be written for a particular level k in the Born–Oppenheimer form:

$$\psi(\mathbf{r}, \mathbf{R}) = F_k(\mathbf{R})\chi_k(\mathbf{r}, \mathbf{R}), \tag{10.113}$$

where $F_k(\mathbf{R})$ is a wave function describing the nuclear motion. Inserting (10.113) into (10.109) and using (10.112) we find that

$$\chi_k(\mathbf{r}, \mathbf{R})\left(-\frac{\hbar^2}{2\mu_{AB}}\nabla_\mathbf{R}^2 + E_k(\mathbf{R}) - E\right)F_k(\mathbf{R}) - \frac{\hbar^2}{2\mu_{AB}}[F_k(\mathbf{R})\nabla_\mathbf{R}^2\chi_k(\mathbf{r}, \mathbf{R})$$

$$+ 2\nabla_\mathbf{R} F_k(\mathbf{R}) \cdot \nabla_\mathbf{R}\chi_k(\mathbf{r}, \mathbf{R})] = 0. \tag{10.114}$$

Multiplying on the left by $\chi_k^*(\mathbf{r}, \mathbf{R})$ and integrating over $\mathbf{r}$, the equation satisfied by $F_k(\mathbf{R})$ is found to be

$$\left[-\frac{\hbar^2}{2\mu_{AB}}\nabla_\mathbf{R}^2 + \frac{\hbar^2}{\mu_{AB}}i\,\mathbf{A}(\mathbf{R})\cdot\nabla_\mathbf{R} + U(\mathbf{R}) - E\right]F_k(\mathbf{R}) = 0 \tag{10.115}$$

where

$$\mathbf{A}(\mathbf{R}) = i \langle \chi_k(\mathbf{r}, \mathbf{R}) | \nabla_\mathbf{R} \chi_k(\mathbf{r}, \mathbf{R}) \rangle \tag{10.116}$$

and

$$U(\mathbf{R}) = E_k(\mathbf{R}) - \frac{\hbar^2}{2\mu_{AB}} \langle \chi_k(\mathbf{r}, \mathbf{R}) | \nabla_\mathbf{R}^2 \chi_k(\mathbf{r}, \mathbf{R}) \rangle \tag{10.117}$$

On comparing (10.117) with (9.137), taking $\mathbf{c} = \mathbf{R}$ and $\psi_a = \chi_k$, it is seen that $\mathbf{A}(\mathbf{R})$ is a Berry connection which in (10.115) is a vector potential coupling to the heavy particle momentum $\mathbf{P} = -i\hbar \nabla_\mathbf{R}$. In addition a scalar potential $U(\mathbf{R})$ is also generated. If χ_k is non-degenerate and is chosen to be real then by differentiating the normalisation condition $\langle \chi_k(\mathbf{r}, \mathbf{R}) | \chi_k(\mathbf{r}, \mathbf{R}) \rangle = 1$, it is seen that $A(\mathbf{R}) = 0$. The second term in (10.117) is small since χ_k is slowly varying with $\mathbf{R}$, and if it is neglected the wave function $F_k(\mathbf{R})$ satisfies the *nuclear* Schrödinger equation

$$\left[-\frac{\hbar^2}{2\mu_{AB}} \nabla_\mathbf{R}^2 + E_k(\mathbf{R}) - E \right] F_k(\mathbf{R}) = 0 \tag{10.118}$$

in which $E_k(\mathbf{R})$ serves as a potential.

Although the Berry connection $\mathbf{A}(\mathbf{R})$ does not appear in (10.118) it is not always the case that it can be removed. For example the component of the angular momentum along the internuclear axis of a diatomic molecule can take the values $\pm \Lambda \hbar$ where $\Lambda = 0, 1, 2 \ldots$, so that except for $\Lambda = 0$ the energy levels $E_k(\mathbf{R})$ are doubly degenerate. If the adiabatic equation (10.115) is modified to take this into account, it is found that $\mathbf{A}$ becomes a 2×2 matrix, and cannot be removed by a phase transformation. Very often the Hamiltonian for a complex system can be expressed as the sum of a fast varying part H_f and a slowly varying part H_s. Under these circumstances the wave equation for the slowly varying motion always takes a similar form to (10.115) with both vector and scalar potentials generated by the fast motion.

The electronic ground state

We shall investigate the lowest electronic levels of H_2^+ using the Rayleigh–Ritz variational method of Section 8.3. For this purpose, we first note that since $\mathbf{r}_A = \mathbf{r} + \mathbf{R}/2$ and $\mathbf{r}_B = \mathbf{r} - \mathbf{R}/2$ the electronic Hamiltonian

$$H_e = -\frac{\hbar^2}{2m} \nabla_\mathbf{r}^2 - \frac{e^2}{(4\pi \varepsilon_0)} \left(\frac{1}{r_A} + \frac{1}{r_B} - \frac{1}{R} \right) \tag{10.119}$$

is invariant under the reflection $\mathbf{r} \to -\mathbf{r}$. If $\mathcal{P}$ is the parity operator which changes $\mathbf{r}$ to $-\mathbf{r}$, $\mathcal{P}$ commutes with H, and hence we can find simultaneous eigenfunctions of $\mathcal{P}$ and H. In the present context the eigenfunctions of $\mathcal{P}$ are called *gerade* if the parity is even and *ungerade* if the parity is odd

$$\mathcal{P} \chi_{k,g}(\mathbf{r}, \mathbf{R}) = \chi_{k,g}(\mathbf{r}, \mathbf{R}), \qquad \mathcal{P} \chi_{k,u}(\mathbf{r}, \mathbf{R}) = -\chi_{k,u}(\mathbf{r}, \mathbf{R}). \tag{10.120}$$

If R is large, the system separates into a hydrogen atom and a proton. As we are interested in the states of lowest energy, we shall look at those functions $\chi_k(\mathbf{r}, \mathbf{R})$

which separate as $R \to \infty$ to $\psi_{1s}(r_A)$ or $\psi_{1s}(r_B)$, where ψ_{1s} is the ground-state wave function of a hydrogen atom. This suggests that approximations to χ_g and χ_u (we drop the subscript k as we are only interested in the lowest states of each symmetry), can be taken as

$$\chi_g = \psi_{1s}(r_A) + \psi_{1s}(r_B), \qquad \chi_u = \psi_{1s}(r_A) - \psi_{1s}(r_B). \tag{10.121}$$

In (10.121) the particular combinations of $\psi_{1s}(r_A)$ and $\psi_{1s}(r_B)$ are chosen to provide the gerade or ungerade symmetry. This approximation in which atomic orbitals are linearly combined to represent molecular wave functions is known as the *linear combination of atomic orbitals* (LCAO) method and can be extended to more complicated molecules. Substituting the functions (10.121) into the Rayleigh–Ritz functional (8.130), we see that

$$E_{g,u}(\mathbf{R}) = \frac{\int \chi^*_{g,u}(\mathbf{r},\mathbf{R}) H \chi_{g,u}(\mathbf{r},\mathbf{R}) d\mathbf{r}}{\int |\chi_{g,u}(\mathbf{r},\mathbf{R})|^2 d\mathbf{r}} \tag{10.122}$$

provides an upper bound on the energies of $E_g(\mathbf{R})$ and $E_u(\mathbf{R})$ for each value of $\mathbf{R}$.

We shall not go into the details of the evaluation of the integrals of (10.122). The results are

$$E_g(R) = E_{1s} + \frac{e^2}{(4\pi\varepsilon_0)R}$$
$$\times \frac{(1+R/a_0)\exp(-2R/a_0) + [1-(2/3)(R/a_0)^2]\exp(-R/a_0)}{1 + [1+(R/a_0)+(1/3)(R/a_0)^2]\exp(-R/a_0)} \tag{10.123a}$$

and

$$E_u(R) = E_{1s} + \frac{e^2}{(4\pi\varepsilon_0)R}$$
$$\times \frac{(1+R/a_0)\exp(-2R/a_0) - [1-(2/3)(R/a_0)^2]\exp(-R/a_0)}{1 - [1+(R/a_0)+(1/3)(R/a_0)^2]\exp(-R/a_0)} \tag{10.123b}$$

where a_0 is the Bohr radius and E_{1s} is the ground-state energy of atomic hydrogen.

The two curves $[E_g(R) - E_{1s}]$ and $[E_u(R) - E_{1s}]$ are plotted in Fig. 10.6 as a function of R. The curve $E_g(R) - E_{1s}$ arising from the symmetrical (gerade) orbital exhibits a minimum at $\bar{R}_0 = 2.49$ a.u. (1.32 Å) and it is found that $E_g(\bar{R}_0) - E_{1s} = -0.065$ a.u. $= -1.77$ eV. As a result, this curve represents an effective attraction between the two nuclei, leading to the formation of a stable molecular ion. Since the nuclear vibration about the stable position is small, the value 1.77 eV is an approximation to the ground-state binding energy of H_2^+. In contrast, the function $E_u(R) - E_{1s}$ is always repulsive and has no minimum, so that a molecular ion H_2^+ in this state will dissociate into a proton and a hydrogen atom. A more accurate calculation leads to an equilibrium distance of $R_0 = 2$ a.u. (1.06 Å) and a value of $E_g(R_0) - E_{1s} = -0.103$ a.u. $= -2.79$ eV.

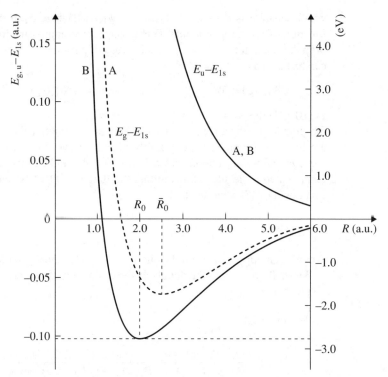

Figure 10.6 The lowest electronic potential energy curves of H_2^+. The lines labelled A show $E_g(R) - E_{1s}$ and $E_u(R) - E_{1s}$ calculated from the approximate expressions (10.123). The lines labelled B show the same quantities calculated by a more accurate method. Energies are given in atomic units (1 a.u. of energy = 27.2 eV) and electronvolts, while the internuclear distance R is given in atomic units (1 a.u. of length = $a_0 = 5.29 \times 10^{-11}$ m).

Rotational and vibrational motion of H_2^+

To examine the motion of the nuclei in H_2^+ about the equilibrium position $R = R_0$, we return to the Schrödinger nuclear equation (10.118) setting $E_k(R) = E_g(R)$ for the ground state. The energy $E_g(R)$, being a function of the radial coordinate only, acts as a central potential. We can therefore obtain solutions of the form

$$F_g(\mathbf{R}) = R^{-1}\mathcal{F}_{v,L}(R)Y_{L,M_L}(\theta, \phi) \qquad (10.124)$$

where $\mathcal{F}_{v,L}(R)$ is a radial function and the spherical harmonic Y_{L,M_L} which depends on the polar angles (θ, ϕ) of $\mathbf{R}$, is a simultaneous eigenfunction of $\mathbf{L}^2$ and L_z. Inserting (10.124) into (10.118), we find the radial equation

$$\left[-\frac{\hbar^2}{2\mu_{AB}}\left(\frac{d^2}{dR^2} - \frac{L(L+1)}{R^2}\right) + E_g(R) - E\right]\mathcal{F}_{v,L}(R) = 0. \qquad (10.125)$$

Since the nuclear motion is confined to a small distance about $R = R_0$, it is useful to approximate $E_g(R)$ by expanding it about $R = R_0$

$$E_g(R) = E_g(R_0) + (R - R_0)\left[\frac{dE_g(R)}{dR}\right]_{R=R_0} + \frac{1}{2}(R - R_0)^2 \left[\frac{d^2 E_g(R)}{dR^2}\right]_{R=R_0}$$
$$+ \cdots$$
$$\simeq E_g(R_0) + \frac{1}{2}k(R - R_0)^2 \tag{10.126}$$

where in the second line we have used the fact that $dE_g(R)/dR$ is equal to zero at the equilibrium position $R = R_0$ (a minimum of $E_g(R)$) and we have set $k = [d^2 E_g(R)/dR^2]_{R=R_0}$. At the same time the rotational energy can be approximated by evaluating the centrifugal term in (10.125) at $R = R_0$,

$$E_r = \frac{\hbar^2}{2\mu_{AB} R_0^2} L(L+1). \tag{10.127}$$

In this approximation, the molecule behaves like a rigid rotator, as we have already discussed in Section 6.4. The rotational energies are of the order of 10^{-3} eV, and are very small compared with the spacing of the electronic levels. With these approximations the eigenvalue equation (10.125) for the vibrational motion becomes

$$\left[-\frac{\hbar^2}{2\mu_{AB}}\frac{d^2}{dR^2} + \frac{1}{2}k(R - R_0)^2 - E_v\right]\mathcal{F}_{v,L}(R) = 0 \tag{10.128}$$

where $E_v = E - E_g(R_0) - E_r$. This is the equation for simple harmonic motion, and it follows that in this approximation the nuclei vibrate with energies

$$E_v = \hbar\omega_0\left(v + \frac{1}{2}\right), \qquad v = 0, 1, 2, \ldots \tag{10.129}$$

where $\omega_0 = (k/\mu_{AB})^{1/2}$. The vibrational energies are of the order of 10^{-1} eV, confirming that the vibrational and rotational motions can be ignored (to a good approximation) when calculating the energies of the electronic levels. The treatment we have outlined can be extended to other diatomic molecules (Bransden and Joachain, 1983).

Transitions between the energy levels of a molecular system can take place with the emission or absorption of radiation, giving rise to *molecular spectra*. The simplest of these are the *pure rotational* spectra involving only transitions between rotational states of the molecule which belong to a given electronic and vibrational level. The pure rotational spectrum of a molecule consists of a set of closely spaced lines in the infrared or the microwave region. Transitions involving changes in the vibrational as well as in the rotational state of the molecule give rise to *vibrational–rotational band spectra* which are observed in the infrared part of the spectrum. When spectrographs of sufficient resolving power are used, one finds that the bands exhibit a fine structure due to the rotational transitions. Finally, molecular spectra for which changes in the electronic as well as in the vibrational and rotational states of the molecule occur are called *electronic spectra*. As the energy differences between electronic levels are much larger than those corresponding to transitions without a change in the electronic state, the lines associated with electronic spectra lie in the visible or in the ultraviolet part of the spectrum. When observed with small dispersion, electronic spectra appear to consist of bands. However, when a better resolving power is used, it is found as in the case of vibrational–rotational spectra that the bands exhibit a fine structure in the sense that they consist of many individual lines, due to the rovibrational transitions.

10.7 Nuclear systems

We conclude this chapter by discussing briefly two applications to nuclear physics. The first concerns the calculation of the binding energy of the deuteron (the nucleus composed of a single proton and a single neutron), and the second is a model of heavier nuclei based on a Fermi-gas picture. Later, in Chapter 11, the theory of the photodisintegration of the deuteron is outlined, and in Chapter 13 neutron–proton scattering is treated.

In this book, we are interested in nuclear physics as providing examples of the applicability of quantum theory. For a systematic development of the subject the reader is referred to the text by Burcham and Jobes (1995). We shall note here only those facts which are strictly necessary for an understanding of our examples.

Neutrons and protons (collectively called nucleons) are fermions of spin one-half, neutrons being electrically neutral and protons possessing a positive charge e which is equal in magnitude to the electronic charge. Both particles have roughly equal mass. In fact, the mass of the neutron is $M_n = 1.67492 \times 10^{-27}$ kg and that of the proton is $M_p = 1.67265 \times 10^{-27}$ kg, which are ~ 1840 times the mass of the electron. The masses of all nuclei are very nearly integral multiples of the mass of the proton, from which it can be inferred that nuclei are composed of nucleons interacting with one another through attractive forces. As discussed earlier in Section 10.2, modern theories show that the nucleons themselves are built from fundamental particles called quarks, but this inner structure can be ignored in the first instance in building a picture of the nucleus and its properties.

Each nucleus can be characterised by its charge Ze, and a *mass number* A, which is defined as the closest integer to the ratio M/M_p, where M is the mass of the nucleus and M_p is the mass of the proton. Thus a nucleus characterised by A and Z contains $(A - Z)$ neutrons and Z protons. In naturally occurring nuclei, Z ranges from 1 (hydrogen) to 92 (uranium), and A from 1 to 238. On average, there are about three stable nuclei called *isotopes* with differing A for each value of Z. As A increases the ratio of the number of protons to the number of neutrons in stable nuclei decreases and this can be attributed to the repulsive Coulomb forces acting because of the electric charge on the protons.

Scattering experiments, between neutrons and protons, and between protons and protons, show that the internuclear force is of short range and is negligible for internuclear distances $r > a$, where a, the *range*, is of the order 2×10^{-15} m.

Confirmation of the short-range nature of the nuclear force comes from the study of binding energies. The binding energy, E_B, is defined as the energy required to dissociate a nucleus completely into Z free protons and $(A - Z)$ free neutrons. Using the Einstein mass–energy relationship, E_B is equal to $[ZM_p + (A - Z)M_n - M]c^2$. E. Wigner showed that it was impossible to explain the increase in binding energy[3] *per particle* in the sequence deuteron ($A = 2, Z = 1, E_B/A = 1.11$ MeV), triton ($A = 3, Z = 1, E_B/A = 2.83$ MeV) and alpha particle ($A = 4, Z = 2, E_B/A = 7$ MeV), if the internuclear forces were of long range and decreased as an inverse power of the internuclear separation. For nuclei heavier than the alpha particle, the binding energy per particle does not continue to increase and becomes nearly constant, so that the total binding energy becomes proportional to A. This *saturation* of the nuclear force is quite different from the expected behaviour if the nuclear force were uniformly attractive, in which case the binding energy would increase with the number of pairs of interactions between nucleons, and be proportional to A^2. There is evidence that at very short distances the internuclear potential changes to a repulsion which is large in magnitude and which prevents nucleons becoming much closer than $\sim 0.5 \times 10^{-15}$ m. This is sufficient to explain the saturation of nuclear forces. However, since this distance is smaller than the average spacing of nucleons in a deuteron, or that probed by low energy nucleon–nucleon scattering experiments, the two-nucleon system at low energies can be discussed approximately in terms of an effective short-range central potential $V(r)$ which is everywhere attractive.

The deuteron

The deuteron, the bound state of a neutron and a proton, is found to have a total angular momentum quantum number $J = 1$. Since the magnetic moment of the deuteron is approximately the sum of the magnetic moments of the neutron and of the proton, it can be inferred (see Chapter 11), that the orbital angular momentum quantum number is $L = 0$, and hence the total spin quantum number is $S = 1$. Using the Russell–Saunders notation, the corresponding state is designated as $^3S_1$. From measurements of the non-zero electric quadrupole moment, it can be shown that there is a small admixture of an $L = 2$ ($^3D_1$) term in the wave function, which arises from a non-central component in the internuclear potential. However, as this only amounts to about 5%, in a discussion of the binding energy it can be neglected to a first approximation.

From this discussion, we see that the effective potential between a neutron and a proton in the triplet state, ^3V, can be taken to be central, attractive and of short range. Various model potential shapes can be adopted, for example a square well

$$\begin{aligned} ^3V(r) &= -\,^3V_0, \quad r < a \\ &= 0, \quad r > a, \end{aligned} \qquad (10.130)$$

---

[3] It is usual to express nuclear energies in MeV (million electron volts), where 1 MeV $\equiv$ 1.60219×10^{-13} J.

a Yukawa potential

$$^3V(r) = -\,^3V_0 r^{-1} \exp(-r/a) \tag{10.131}$$

or an exponential potential

$$^3V(r) = -\,^3V_0 \exp(-r/a). \tag{10.132}$$

In each case the two parameters, the well depth $^3V_0$ and the range a can be adjusted to obtain the observed binding energy of 2.22 MeV.

Since no singlet bound state exists in the zero-spin state, it is clear that the potential in this state must be weaker and that the corresponding depth and range parameters are different (see Chapter 13).

In the centre-of-mass system, the Schrödinger equation for a neutron–proton system in the triplet state is

$$\left[-\frac{\hbar^2}{2\mu}\nabla_\mathbf{r}^2 + {}^3V(r) - E\right]\psi(\mathbf{r}) = 0 \tag{10.133}$$

where μ is the reduced mass

$$\mu = \frac{M_\mathrm{p} M_\mathrm{n}}{M_\mathrm{p} + M_\mathrm{n}} \simeq \frac{1}{2} M_\mathrm{p} \tag{10.134}$$

and $\mathbf{r}$ is the internuclear coordinate. Note that we do not need to display the spin part of the wave function because we are dealing only with the $S = 1$ state. As the wave function for the bound state corresponds to $L = 0$, it has no angular dependence and we can write

$$\psi(\mathbf{r}) = r^{-1} Y(r). \tag{10.135}$$

Introducing (10.135) into (10.133), we find that the function $Y(r)$ satisfies the equation

$$\left[\frac{d^2}{dr^2} - {}^3U(r) - \lambda^2\right] Y(r) = 0 \tag{10.136}$$

where

$$^3U(r) = \frac{2\mu}{\hbar^2}\,{}^3V(r) \tag{10.137}$$

and

$$\lambda^2 = -\frac{2\mu}{\hbar^2} E = \frac{2\mu}{\hbar^2} E_\mathrm{B}, \tag{10.138}$$

E_B being the binding energy. Since $E < 0$ for a bound system, the parameter λ^2 is positive. The boundary condition that $\psi(\mathbf{r})$ should be finite everywhere implies that

$$Y(0) = 0 \quad \text{and} \quad Y(r) \to 0 \text{ as } r \to \infty. \tag{10.139}$$

This problem has been discussed in Section 7.4 for a square well potential such as (10.130). The solution satisfying the boundary conditions (10.139) is

$$Y(r) = C_1 \sin Kr, \qquad r < a$$
$$ = C_2 \exp(-\lambda r), \qquad r > a \tag{10.140}$$

where

$$K^2 = {}^3U_0 - \lambda^2, \qquad {}^3U_0 = \frac{2\mu}{\hbar^2}\,{}^3V_0. \tag{10.141}$$

Joining the solution inside the well smoothly to that outside gives the condition (see (7.87))

$$K \cot Ka = -\lambda. \tag{10.142}$$

To obtain a binding energy $E_B = 2.22$ MeV, with a range of the order of 2×10^{-15} m, the depth of the well has to be of the order 30–40 MeV. It follows that E_B is small compared with ${}^3V_0$, that is, λ is small compared with K. Thus $\cot Ka = (-\lambda/K)$ is small and negative, so that Ka is a little larger than $\pi/2$. (The value $3\pi/2$ would not do as this would give rise to a node in the wave function.) Setting $Ka \simeq \pi/2$ and neglecting λ^2 compared with K^2 in (10.141) we see that the condition for a bound state reduces to

$$K^2 a^2 \simeq {}^3U_0 a^2 \simeq \pi^2/4. \tag{10.143}$$

A similar condition for the product ${}^3U_0 a^2$ can be found for any other potential shape, provided that it is of short range.

The extent of the wave function is measured by λ^{-1}. This quantity is determined by the value of the binding energy $E_B = 2.22$ MeV, and from (10.138) $\lambda^{-1} = 4.32 \times 10^{-15}$ m. We see that $\lambda^{-1} > a$, which is consistent with the inequality $K > \lambda$. This implies that most of the wave function of the deuteron is outside the potential well. For many purposes, the wave function can thus be approximated by its form outside the well which is

$$\psi(\mathbf{r}) = C_2 \exp(-\lambda r)/r. \tag{10.144}$$

This function does not obey the proper boundary condition at $r = 0$. Nevertheless it can be normalised by requiring that

$$\int |\psi(\mathbf{r})|^2 d\mathbf{r} = 1. \tag{10.145}$$

Integrating over angles we have

$$4\pi (C_2)^2 \int_0^\infty \exp(-2\lambda r) dr = 1 \tag{10.146}$$

from which

$$C_2 = (\lambda/2\pi)^{1/2}. \tag{10.147}$$

Defining the radius of the deuteron to be the expectation value of the radial coordinate r, and using the approximation (10.144) for $\psi(\mathbf{r})$, with the normalisation constant C_2 determined from (10.147) we have

$$\langle r \rangle = \int |\psi(\mathbf{r})|^2 \, r \, d\mathbf{r}$$

$$= \frac{1}{2\lambda} = 2.15 \times 10^{-15} \text{ m}. \tag{10.148}$$

The Fermi model of the nucleus

The size of nuclei can be determined through scattering experiments with electrons. Since electrons are charged these experiments measure the charge distribution, that is the distribution of the protons in the nucleus. It is found that the nuclear radius can be expressed as

$$R = r_0 A^{1/3} \tag{10.149}$$

where A is the mass number and r_0 is a constant. A similar result is obtained from nucleon scattering (which depends on the neutron as well as the proton distributions). The average value of r_0 is

$$r_0 = 1.1 \times 10^{-15} \text{ m}. \tag{10.150}$$

A model of a heavy nucleus can be obtained by treating the protons and neutrons as independent Fermi gases, each confined within a sphere of radius R, by a constant potential (within the sphere) $V = -V_0$. In a better approximation, the neutron and proton gases can be considered to move in slightly different potentials, allowing for the Coulomb repulsion which affects only the protons.

From (10.150), the nucleon number density is constant, being

$$\rho = \frac{3A}{4\pi R^3} = \frac{3}{4\pi r_0^3} = 0.18 \times 10^{45} \text{ nucleons m}^{-3}. \tag{10.151}$$

If we take the case of nuclei with equal numbers of neutrons and protons, the number density both of the neutron gas and the proton gas is

$$\rho_p = \rho_n = \rho/2. \tag{10.152}$$

In the ground state the nucleons fill successively the lowest levels in the well. The energy of the last occupied level, the Fermi energy, is given by (10.39), where m must be set equal to M, the nucleon mass, and ρ must be set equal to the number density ρ_p or ρ_n. We obtain in this way

$$E_F = \frac{\hbar^2}{2M}(3\pi^2 \rho_p)^{2/3}$$

$$= \frac{\hbar^2}{2M}\frac{(9\pi)^{2/3}}{(2r_0)^2} \tag{10.153}$$

and by using (10.42) the average kinetic energy per nucleon is found to be

$$\bar{E} = \frac{3}{5}E_F = \frac{3}{5}\left(\frac{\hbar^2}{2M}\right)\frac{(9\pi)^{2/3}}{(2r_0)^2}. \tag{10.154}$$

On evaluating (10.154) using $r_0 = 1.1 \times 10^{-15}$ m, $\bar{E}$ is found to be 23 MeV. Empirically, the energy required to remove one nucleon from a heavy nucleus is roughly constant, and is about 16 MeV. Using this estimate together with the 23 MeV for the average kinetic energy per particle, one finds the depth of the attractive well in which the nucleons move to be $(23 + 16)$ MeV $= 39$ MeV. In principle, this could be related to the parameters of the internucleon potential which we discussed in connection with the binding energy of the deuteron. This, however, is a very difficult task.

Problems

10.1 Write down the Schrödinger equation for a four-particle system using three different sets of centre-of-mass (Jacobi) coordinates.

10.2 Consider a Fermi gas in which N particles confined within a large cube with sides of length L move in a constant potential $V = V_0$. Show that the properties of the gas are the same as in the cases discussed in Section 10.3, where we took $V_0 = 0$, with the exception that a particle now has energy

$$E'_n = V_0 + E_n$$

where E_n is given by (10.31), and the total energy is now

$$E'_{\text{tot}} = E_{\text{tot}} + NV_0$$

where E_{tot} is given by (10.41).

10.3 (a) By using the ultra-relativistic expression $E_F = \hbar c k_F = \hbar c(3\pi^2 \rho)^{1/3}$ for the Fermi energy of an electron gas, and approximating the total kinetic energy of a white dwarf star by NE_F, where N is the total number of electrons, show that the total energy E_T is

$$E_T = \frac{bM^{4/3}}{R} - \frac{3}{5}\frac{GM^2}{R}; \qquad b = \left(\frac{3}{2}\right)^{2/3} \hbar c \frac{\pi^{1/3}}{M_p^{4/3}}\left(\frac{Z}{A}\right)^{4/3}.$$

For the star to be in equilibrium E_T must be such that $E_T \leqslant 0$. Using this condition, show that the Chandrasekhar limit (i.e. the critical mass M_c above which the star becomes unstable) is given by

$$M_c = \left(\frac{5b}{3G}\right)^{3/2} = \frac{5^{3/2}\pi^{1/2}}{2\times 3^{1/2}}\left(\frac{\hbar c}{G}\right)^{3/2}\left(\frac{Z}{AM_p}\right)^2.$$

(b) Obtain the corresponding expression for the Chandrasekhar limit of a neutron star.

10.4 Obtain the result (10.85) by substituting the trial function (10.83) in the expression (10.84).

(*Hint*: To evaluate the quantity $\langle\phi|r_{12}^{-1}|\phi\rangle$, follow the method used to evaluate the first-order energy correction (10.74)).

10.5 Using the rules of addition of angular momenta find the possible values of the total angular momentum J of an atom in the following states: $^1S, ^3S, ^1P, ^3P, ^2D, ^4D$.

10.6 List the possible spectral terms $^{2S+1}L_J$ resulting from the following two-electron configurations:

$(nsn's), (nsn'p), (nsn'd), (np, n'p)$

10.7 Show that if in the hydrogen molecular ion H_2^+ the electron is replaced by a muon μ^-, the equilibrium separation of the nuclei in the ground state is $R_0 = 6.5\times 10^{-13}$ m.

(*Note*: The mass of the muon μ^- is 207 times the mass of the electron, and the charge on the muon μ^- is the same as that on the electron).

This system is of interest, since if the protons are replaced by a deuteron and a triton, the overlap of the nuclear wave functions may be sufficient for the fusion reaction

$$D + T \rightarrow {}^4He + n$$

to take place at a useful rate.

10.8 The moment of inertia $\mu_{AB}R_0^2$ of the molecule H ^{79}Br is 3.30×10^{-47} kg m^2. Using (10.127) calculate the first five rotational levels of this molecule in eV. Find the internuclear distance R_0 in ångstroms.

10.9 The wave number of the fundamental vibrational motion of H ^{79}Br is 2650 cm^{-1}. Calculate (a) the energy of the lowest state and first excited state in eV, (b) the force constant k in SI units.

10.10 In the text we considered a model in which the nucleon–nucleon interaction was taken to be central. Consider a more general case in which the nucleon–nucleon potential depends on spin and orbital angular momentum, as well as on the internucleon distance. Show that the possible states for low values of the total angular momentum are:

$J = 0$ $^1S_0$ $^3P_0$

$J = 1$ $^1P_1$ $^3P_1$ $^3S_1 + {}^3D_1$

$J = 2$ $^1D_2$ $^3D_2$ $^3P_2 + {}^3F_2$.

11 The interaction of quantum systems with radiation

11.1 The electromagnetic field and its interaction with one-electron atoms   516
11.2 Perturbation theory for harmonic perturbations and transition rates   522
11.3 Spontaneous emission   527
11.4 Selection rules for electric dipole transitions   533
11.5 Lifetimes, line intensities, widths and shapes   538
11.6 The spin of the photon and helicity   544
11.7 Photoionisation   545
11.8 Photodisintegration   550
Problems   555

One of the most striking achievements of quantum theory is the explanation of the observed emission and absorption spectra of atoms and molecules. We have already seen that the frequencies of spectral lines are determined by atomic energy levels which can be calculated accurately by quantum mechanics. However, not all pairs of energies correspond to observed lines, and there are marked differences in intensity between one line and another. To explain these phenomena the interaction of electromagnetic radiation with systems of charged particles must be discussed. In this chapter we will examine the interaction of atoms with electromagnetic radiation, concentrating on atomic hydrogen and one-electron ions. Transitions between discrete levels will be discussed first and then photoionisation. Finally, as an example of the interaction of radiation with a nuclear system the photodisintegration of the deuteron will be considered.

An atomic or nuclear system can interact with electromagnetic radiation in three ways. First, these systems can make spontaneous transitions from an excited state to a state of lower energy with the emission of photons. This process is called *spontaneous emission* and takes place in the absence of any external field. It can be looked upon as the quantum analogue of the effect predicted by classical electromagnetism that any accelerated particle of charge q emits radiation, the power emitted being determined by the Larmor formula[1]

$$P = \frac{q^2|\mathbf{a}|^2}{6\pi c^3 \varepsilon_0} \tag{11.1}$$

where $\mathbf{a}$ is the acceleration of the particle and c is the velocity of light *in vacuo*. Second, a system can *emit* radiation under the influence of an applied radiation field. This process is called *stimulated emission* and has an important practical application in *lasers* which produce intense beams of coherent radiation. Finally, a system can *absorb* photons from radiation making a transition from a state of lower to a state of higher energy.

In an exact treatment, we would have to start by developing the quantum theory of the electromagnetic field, in which the field is expressed in terms of its quanta – the photons. However, even in comparatively weak fields the photon density is very large (see Problem 11.1). If this is the case, the number of photons can be treated as a continuous variable and the field behaves classically, being described by Maxwell's equations. The interaction of the field with an atom can then be accounted for by a semi-classical theory in which the radiation field is described classically, but the atomic system is quantised. Since usually only one photon at a time is emitted or absorbed, the influence of the atom on the applied radiation, which contains many photons, can be neglected. Clearly none of these assumptions hold for spontaneous emission, which takes place in the absence of an applied field and for which a rigorous treatment requires the development of quantum electrodynamics. Fortunately in this case the transition rate can be found by using a statistical argument due to Einstein.

11.1 The electromagnetic field and its interaction with one-electron atoms

The classical electromagnetic field is described by electric and magnetic field vectors $\mathcal{E}$ and $\mathcal{B}$, which satisfy Maxwell's equations, and which can be generated from scalar and vector potentials ϕ and $\mathbf{A}$ by

$$\mathcal{E}(\mathbf{r}, t) = -\nabla \phi(\mathbf{r}, t) - \frac{\partial}{\partial t} \mathbf{A}(\mathbf{r}, t), \tag{11.2a}$$
$$\mathcal{B}(\mathbf{r}, t) = \nabla \times \mathbf{A}(\mathbf{r}, t) \tag{11.2b}$$

The potentials are not completely defined by these equations. In particular, $\mathcal{E}$ and $\mathcal{B}$ are unaltered by the substitutions $\mathbf{A} \to \mathbf{A} + \nabla \chi, \phi \to \phi - \partial \chi/\partial t$, where χ is an arbitrary real differentiable function of $\mathbf{r}$ and t. This property of *gauge invariance* allows us to impose a further condition on $\mathbf{A}$, which we shall choose to be

$$\nabla \cdot \mathbf{A} = 0 \tag{11.3}$$

---

[1] Useful texts on electromagnetism are those by Duffin (1968) and Jackson (1998).

When **A** satisfies this condition, we are said to be using the *Coulomb gauge*. This choice of gauge is convenient when no sources are present, which is the case considered here. One may then take $\phi = 0$, and **A** satisfies the wave equation

$$\nabla^2 \mathbf{A} - \frac{1}{c^2}\frac{\partial^2 \mathbf{A}}{\partial t^2} = 0, \tag{11.4}$$

where c is the velocity of light *in vacuo*.

A monochromatic plane wave solution of equation (11.4) corresponding to the angular frequency ω (i.e. to the frequency $\nu = \omega/2\pi$) is

$$\mathbf{A}(\mathbf{r}, t) = A_0(\omega)\hat{\boldsymbol{\epsilon}} \cos(\mathbf{k}\cdot\mathbf{r} - \omega t + \delta_\omega), \tag{11.5}$$

where **k** is the propagation vector, δ_ω is a real constant phase and

$$\omega = kc. \tag{11.6}$$

The vector potential **A** has an amplitude $|A_0(\omega)|$ and is in the direction specified by the unit vector $\hat{\boldsymbol{\epsilon}}$, called the *polarisation vector*. In addition, equation (11.3) is satisfied if

$$\mathbf{k}\cdot\hat{\boldsymbol{\epsilon}} = 0 \tag{11.7}$$

so that $\hat{\boldsymbol{\epsilon}}$ is perpendicular to **k** and the wave is *transverse*.

Using the Coulomb gauge, with $\phi = 0$, the electric and magnetic fields are given from (11.2) and (11.5) by

$$\boldsymbol{\mathcal{E}}(\mathbf{r}, t) = \mathcal{E}_0(\omega)\hat{\boldsymbol{\epsilon}} \sin(\mathbf{k}\cdot\mathbf{r} - \omega t + \delta_\omega), \tag{11.8a}$$

$$\boldsymbol{\mathcal{B}}(\mathbf{r}, t) = \mathcal{E}_0(\omega)\omega^{-1}(\mathbf{k} \times \hat{\boldsymbol{\epsilon}}) \sin(\mathbf{k}\cdot\mathbf{r} - \omega t + \delta_\omega) \tag{11.8b}$$

where $\mathcal{E}_0(\omega) = -\omega A_0(\omega)$. The electric field vector $\boldsymbol{\mathcal{E}}$ has an amplitude $|\mathcal{E}_0(\omega)|$ and is in the direction of the polarisation vector $\hat{\boldsymbol{\epsilon}}$. We also note that the vectors $\boldsymbol{\mathcal{E}}$, $\boldsymbol{\mathcal{B}}$ and **k** are mutually perpendicular. An electromagnetic plane wave such as (11.8), for which the electric field vector points in a fixed direction $\hat{\boldsymbol{\epsilon}}$, is said to be *linearly polarised*. A general state of polarisation for a plane wave propagating in the direction $\hat{\mathbf{k}}$ can be described by combining two independent linearly polarised plane waves with polarisation vectors $\hat{\boldsymbol{\epsilon}}_\lambda (\lambda = 1, 2)$ perpendicular to $\hat{\mathbf{k}}$, where the phases of the two component waves are, in general, different. Any radiation field can be expressed as a superposition of monochromatic fields with different polarisations, so that it is only necessary to study the interaction of the field defined by (11.8) with quantum systems.

It is interesting to relate the energy density of the field to the photon density, keeping in mind that each photon at a frequency ν carries a quantum of energy of magnitude $h\nu = \hbar\omega$. The energy density of the field is given by

$$\frac{1}{2}(\varepsilon_0|\boldsymbol{\mathcal{E}}|^2 + |\boldsymbol{\mathcal{B}}|^2/\mu_0) = \varepsilon_0 \mathcal{E}_0^2(\omega) \sin^2(\mathbf{k}\cdot\mathbf{r} - \omega t + \delta_\omega) \tag{11.9}$$

where ε_0 and μ_0 are the permittivity and permeability of free space, and $\varepsilon_0\mu_0 = c^{-2}$. The average of $\sin^2(\mathbf{k}\cdot\mathbf{r} - \omega t + \delta_\omega)$ over a period $T = 2\pi/\omega$ is given by

$$\frac{1}{T}\int_0^T \sin^2(\mathbf{k}\cdot\mathbf{r} - \omega t + \delta_\omega)\mathrm{d}t = \frac{1}{2}. \tag{11.10}$$

Using this result the average energy density $\rho(\omega)$ is

$$\rho(\omega) = \tfrac{1}{2}\varepsilon_0 \mathcal{E}_0^2(\omega). \tag{11.11}$$

If the number of photons of angular frequency ω within a volume V is $N(\omega)$, the energy density is $\hbar\omega N(\omega)/V$, and equating this with (11.11) the amplitude of the electric field is found to be

$$|\mathcal{E}_0(\omega)| = (2\rho(\omega)/\varepsilon_0)^{1/2} = [2\hbar\omega N(\omega)/(\varepsilon_0 V)]^{1/2}. \tag{11.12}$$

The average rate of energy flow through a unit cross-sectional area, normal to the direction of propagation of the radiation, defines the intensity $I(\omega)$. Since the velocity of electromagnetic waves in free space is c, we have

$$I(\omega) = \rho(\omega)c = \hbar\omega N(\omega)c/V. \tag{11.13}$$

No radiation is perfectly monochromatic, although it may be nearly so, but $\mathcal{E}$ and $\mathcal{B}$ can always be represented by a superposition of monochromatic plane waves. If each term in the superposition has the same direction of propagation and is polarised in the same direction, $\hat{\epsilon}$, the most general expression for $\mathcal{E}$ is

$$\mathcal{E}(\mathbf{r},t) = \hat{\epsilon}\int_0^\infty \mathcal{E}_0(\omega)\sin(\mathbf{k}\cdot\mathbf{r} - \omega t + \delta_\omega)\mathrm{d}\omega \tag{11.14}$$

and $\mathcal{B}$ can be expressed correspondingly. When the radiation is nearly monochromatic, the amplitude $\mathcal{E}_0(\omega)$ is sharply peaked about some value ω_0 of ω. The radiation from a hot gas or glowing filament arises from many atoms each emitting photons independently. As a result, within the integral over ω the phases δ_ω are distributed completely at random, and the radiation is said to be *incoherent*. This is characteristic of light from all sources with the exception of lasers. Because of the random phase distribution, when the average energy density is calculated from the squares of $\mathcal{E}$ and $\mathcal{B}$, the contributions to the energy density from each frequency can be added together, the cross terms averaging to zero[2]. The average energy density ρ and intensity I for radiation composed of a range of frequencies can then be expressed as

$$\rho = \int_0^\infty \rho(\omega)\mathrm{d}\omega, \quad I = \int_0^\infty I(\omega)\mathrm{d}\omega \tag{11.15}$$

where $\rho(\omega)$ and $I(\omega)$ are the *energy density* and *intensity per unit angular frequency range*, given by (11.11) and (11.13), respectively.

---

[2] While this is intuitively clear, a detailed proof is too long to be given here. This point is discussed by Marion and Heals (1980).

Charged particles in an electromagnetic field

In Chapter 5, equation (5.231), we wrote down the Hamiltonian operator for a spinless particle of mass m and charge q moving in an electromagnetic field described by a vector potential $\mathbf{A}(\mathbf{r}, t)$ and a scalar potential $\phi(\mathbf{r}, t)$. It is

$$H = \frac{\mathbf{p}^2}{2m} - \frac{q}{2m}(\mathbf{A}\cdot\mathbf{p} + \mathbf{p}\cdot\mathbf{A}) + \frac{q^2}{2m}\mathbf{A}^2 + q\phi. \tag{11.16}$$

In the position representation, $\mathbf{p}$ is the operator $-i\hbar\nabla$ and the time-dependent Schrödinger equation is

$$i\hbar\frac{\partial}{\partial t}\Psi(\mathbf{r}, t) = \left[-\frac{\hbar^2}{2m}\nabla^2 + i\hbar\frac{q}{2m}(\mathbf{A}\cdot\nabla + \nabla\cdot\mathbf{A}) + \frac{q^2}{2m}\mathbf{A}^2 + q\phi\right]\Psi(\mathbf{r}, t). \tag{11.17}$$

We note that this equation can be obtained by starting from the classical field free relation $E = \mathbf{p}^2/2m$ between the energy and momentum of the particle, making in it the substitutions $E \to E - q\phi$, $\mathbf{p} \to \mathbf{p} - q\mathbf{A}$ (see (5.230)) and using the correspondence rule (see (3.25))

$$E \to E_{\text{op}} = i\hbar\frac{\partial}{\partial t}, \quad \mathbf{p} \to \mathbf{p}_{\text{op}} = -i\hbar\nabla. \tag{11.18}$$

An important property of equation (11.17) is that its form is unchanged under the *gauge transformation*

$$\mathbf{A}(\mathbf{r}, t) = \mathbf{A}'(\mathbf{r}, t) + \nabla\chi(\mathbf{r}, t), \tag{11.19a}$$

$$\phi(\mathbf{r}, t) = \phi'(\mathbf{r}, t) - \frac{\partial}{\partial t}\chi(\mathbf{r}, t), \tag{11.19b}$$

$$\Psi(\mathbf{r}, t) = \Psi'(\mathbf{r}, t)\exp[iq\chi(\mathbf{r}, t)/\hbar] \tag{11.19c}$$

where χ is an arbitrary real, differentiable function of $\mathbf{r}$ and t. That is, the wave function $\Psi'(\mathbf{r}, t)$ satisfies the equation (see Problem 11.2)

$$i\hbar\frac{\partial}{\partial t}\Psi'(\mathbf{r}, t) = \left[-\frac{\hbar^2}{2m}\nabla^2 + i\hbar\frac{q}{2m}(\mathbf{A}'\cdot\nabla + \nabla\cdot\mathbf{A}') + \frac{q^2}{2m}\mathbf{A}'^2 + q\phi'\right]\Psi'(\mathbf{r}, t). \tag{11.20}$$

Since, as seen from (11.19c), a gauge transformation is a particular case of a unitary transformation, measurable quantities (such as expectation values or transition probabilities) calculated in different gauges must be the same. The property of gauge invariance allows us to adopt the Coulomb gauge defined by (11.3) and to take $\phi = 0$, as we have seen above. The Schrödinger equation (11.17) then reduces to

$$i\hbar\frac{\partial}{\partial t}\Psi(\mathbf{r}, t) = \left[-\frac{\hbar^2}{2m}\nabla^2 + i\hbar\frac{q}{m}\mathbf{A}\cdot\nabla + \frac{q^2}{2m}\mathbf{A}^2\right]\Psi(\mathbf{r}, t) \tag{11.21}$$

where we have used the fact that

$$\nabla.(A\Psi) = A.(\nabla\Psi) + (\nabla.A)\Psi$$
$$= A.(\nabla\Psi). \tag{11.22}$$

Interaction of one-electron atoms with an electromagnetic field

Let us now consider the interaction of the electromagnetic field (11.8) with a one-electron atom, containing a nucleus of charge Ze and mass M and an electron of charge $q = -e$ and mass m. Since M is very large compared with m, the interaction between the nucleus and the electromagnetic field can be ignored to a high degree of accuracy. However, we must include in the Hamiltonian the electrostatic Coulomb potential $-Ze^2/(4\pi\varepsilon_0 r)$ between the electron and the nucleus. It is convenient to regard this electrostatic interaction as an additional potential energy term, while the radiation field is described in terms of a vector potential alone, as discussed above. The time-dependent Schrödinger equation for a one-electron atom in an electromagnetic field then reads

$$i\hbar\frac{\partial}{\partial t}\Psi(\mathbf{r}, t) = \left[H_0 - \frac{i\hbar e}{\mu}\mathbf{A}.\nabla + \frac{e^2}{2\mu}\mathbf{A}^2\right]\Psi(\mathbf{r}, t) \tag{11.23}$$

where

$$H_0 = -\frac{\hbar^2}{2\mu}\nabla^2 - \frac{Ze^2}{(4\pi\varepsilon_0)r} \tag{11.24}$$

is the Hamiltonian of the unperturbed atom and $\mu = mM/(m + M)$ is the reduced mass (7.93).

We shall treat the weak field case in which the term in $\mathbf{A}^2$ is small compared with the term linear in $\mathbf{A}$. Accordingly we shall drop the term in $\mathbf{A}^2$ and treat the linear term as a small perturbation. In terms of photons, this means that we shall only treat the emission or absorption of one photon at a time. The simultaneous emission or absorption of two (or more) photons is indeed negligible for weak fields. This is not the case, however, for strong fields which can be generated by lasers.

The dipole approximation

A simplification arises because the radiation field across an atom can usually be taken to be uniform. For example, the wavelength of the transition $n = 2 \to n = 1$ in a one-electron atom is $\lambda = (1.2/Z^2) \times 10^{-7}$ m. This value can be compared with the size of the charge distribution of the atom, typified by the expectation value $\langle r \rangle$ of the radial coordinate r, which depends mainly on n (see Problem 7.16). In particular, for $l = 0$, $\langle r \rangle = (0.8n^2/Z) \times 10^{-10}$ m. It follows that the wavelength of the radiation is much larger than the size of the atom and the electromagnetic field can be taken to be uniform over the whole atom, and equal to its value at the nucleus. This is called the *dipole approximation*, a terminology which will be explained a little later. Making this approximation, we find that the vector potential (11.5) of a monochromatic plane wave reduces to

$$\mathbf{A}(t) = A_0(\omega)\hat{\epsilon}\cos(\omega t - \delta_\omega). \tag{11.25}$$

It is interesting to note that within the dipole approximation the term in $\mathbf{A}^2$ in the time-dependent Schrödinger equation (11.23) can be eliminated by performing the gauge transformation

$$\Psi(\mathbf{r}, t) = \Psi'(\mathbf{r}, t)\exp\left[-\frac{\mathrm{i}}{\hbar}\frac{e^2}{2\mu}\int^t \mathbf{A}^2(t')\mathrm{d}t'\right]. \tag{11.26}$$

This gives for $\Psi'(\mathbf{r}, t)$ the new Schrödinger equation (Problem 11.3)

$$\mathrm{i}\hbar\frac{\partial}{\partial t}\Psi'(\mathbf{r}, t) = \left[H_0 - \frac{\mathrm{i}\hbar e}{\mu}\mathbf{A}\cdot\nabla\right]\Psi'(\mathbf{r}, t) \tag{11.27}$$

which is said to be in the *velocity gauge* since the interaction term $-\mathrm{i}\hbar e\mathbf{A}\cdot\nabla/\mu$ couples the vector potential $\mathbf{A}(t)$ to the velocity operator $-\mathrm{i}\hbar\nabla/\mu$.

Another form of the time-dependent Schrödinger equation in the dipole approximation can be obtained by returning to equation (11.23) and performing a gauge transformation specified by taking $\chi(\mathbf{r}, t) = \mathbf{A}(t)\cdot\mathbf{r}$. We then see from equations (11.19) that

$$\mathbf{A}' = 0 \tag{11.28a}$$

$$\phi' = \frac{\partial}{\partial t}\mathbf{A}(t)\cdot\mathbf{r} = -\boldsymbol{\mathcal{E}}(t)\cdot\mathbf{r} \tag{11.28b}$$

$$\Psi'(\mathbf{r}, t) = \Psi(\mathbf{r}, t)\exp\left[\frac{\mathrm{i}e}{\hbar}\mathbf{A}(t)\cdot\mathbf{r}\right]. \tag{11.28c}$$

The new time-dependent Schrödinger equation for $\Psi'(\mathbf{r}, t)$ is

$$\mathrm{i}\hbar\frac{\partial}{\partial t}\Psi'(\mathbf{r}, t) = [H_0 + e\boldsymbol{\mathcal{E}}\cdot\mathbf{r}]\Psi'(\mathbf{r}, t) \tag{11.29}$$

which is said to be in the *length gauge* because the interaction term $e\boldsymbol{\mathcal{E}}\cdot\mathbf{r}$ now couples the electric field $\boldsymbol{\mathcal{E}}(t)$ to the position operator $\mathbf{r}$. Denoting this interaction by W, we see that it can also be written as

$$W = -\boldsymbol{\mathcal{E}}\cdot\mathbf{D} \tag{11.30}$$

where $\mathbf{D} = -e\mathbf{r}$ is the *electric dipole operator* of a one-electron atom. This is the reason why the approximation in which the electromagnetic field is taken to be uniform over the atom is called the *dipole approximation*.

An alternative way of deriving equation (11.29) is as follows. In classical electromagnetism an electron of velocity $\mathbf{v}$ moving in electric and magnetic fields experiences the Lorentz force

$$\mathbf{F} = -e(\boldsymbol{\mathcal{E}} + \mathbf{v}\times\boldsymbol{\mathcal{B}}). \tag{11.31}$$

For an electron in a light atom the ratio v/c is small. This can be inferred from the Bohr model of a one-electron atom. In this model, for an orbit with principal quantum number n, the electron velocity is $v = Ze^2/(4\pi\varepsilon_0\hbar n)$ (see (1.56)). The largest value of v is for $n = 1$, for which $v = Z\alpha c \simeq Zc/137$ where α is the fine structure constant. From (11.6) and (11.8) we see that for a radiation field of a given frequency, $|\boldsymbol{\mathcal{B}}|/|\boldsymbol{\mathcal{E}}| = 1/c$. It follows that when v/c is small, the magnetic term in the Lorentz force for a radiation field is always small in comparison with the electric term. Apart from special circumstances in which the effect of the electric field vanishes, it is a good approximation to neglect the magnetic term in (11.31) and this will be done in what follows. Moreover, taking the electric field to be uniform over the atom (the dipole approximation), we have

$$\boldsymbol{\mathcal{E}}(t) = \mathcal{E}_0(\omega)\hat{\boldsymbol{\epsilon}} \sin(-\omega t + \delta_\omega) = -\mathcal{E}_0(\omega)\hat{\boldsymbol{\epsilon}} \sin(\omega t - \delta_\omega) \tag{11.32}$$

and

$$\mathbf{F}(t) = e\mathcal{E}_0(\omega)\hat{\boldsymbol{\epsilon}} \sin(\omega t - \delta_\omega). \tag{11.33}$$

The electric force given by (11.33) is conservative and can be obtained from a scalar potential W, with

$$\mathbf{F} = -\nabla W \tag{11.34}$$

so that

$$W = e\boldsymbol{\mathcal{E}}\cdot\mathbf{r} = -\boldsymbol{\mathcal{E}}\cdot\mathbf{D}. \tag{11.35}$$

In order to describe a one-electron atom in a radiation field we must add the time-dependent potential $W(\mathbf{r}, t)$ to the Coulomb potential between the electron and the nucleus. The corresponding time-dependent Schrödinger equation becomes

$$i\hbar\frac{\partial \Psi(\mathbf{r}, t)}{\partial t} = [H_0 + W(\mathbf{r}, t)]\Psi(\mathbf{r}, t) \tag{11.36}$$

where H_0 is the Hamiltonian (11.24) for the atom in the absence of the perturbing radiation field. This equation is identical to the equation (11.29) which we previously derived in the length gauge.

We shall now use time-dependent perturbation theory to obtain approximate solutions of the Schrödinger equation (11.36), following the method developed in Chapter 9.

11.2 Perturbation theory for harmonic perturbations and transition rates

Referring to equation (9.2), and dropping the superscript (0) for notational simplicity, we denote by E_k the eigenvalues and by ψ_k the corresponding eigenfunctions of the hydrogenic Hamiltonian (11.24):

$$H_0\psi_k = E_k\psi_k. \tag{11.37}$$

Since the set of eigenfunctions ψ_k is complete, the general solution Ψ of the Schrödinger equation (11.36) can be expressed as (see (9.6))

$$\Psi(\mathbf{r}, t) = \sum_k c_k(t) \psi_k(\mathbf{r}) \exp(-iE_k t/\hbar) \tag{11.38}$$

where the sum is over both the discrete and continuous hydrogenic eigenfunctions. Assuming that the function $\Psi(\mathbf{r}, t)$ is normalised to unity, we have

$$\sum_k |c_k(t)|^2 = 1 \tag{11.39}$$

and $|c_k(t)|^2$ can be interpreted as the probability that the system is found in the state k at time t. We have already shown that the coefficients $c_k(t)$ satisfy the coupled equations (9.10). Setting $\lambda = 1$ and identifying the perturbation H' with W, given by (11.35), we have

$$\dot{c}_b(t) = (i\hbar)^{-1} \sum_k H'_{bk}(t) \exp(i\omega_{bk} t) c_k(t) \tag{11.40}$$

where

$$H'_{bk}(t) = \langle \psi_b | H'(t) | \psi_k \rangle = \int \psi_b^*(\mathbf{r}) W(\mathbf{r}, t) \psi_k(\mathbf{r}) d\mathbf{r} \tag{11.41}$$

and

$$\omega_{bk} = (E_b - E_k)/\hbar. \tag{11.42}$$

Let us suppose that the system is initially in a well-defined stationary bound state of energy E_a described by the wave function ψ_a and that the applied radiation field is switched on at time $t = 0$. These initial conditions are expressed by

$$c_k(t \leq 0) = \delta_{ka}. \tag{11.43}$$

Then, to first order in the perturbation $H' \equiv W$, we have (see (9.17b))

$$c_b^{(1)}(t) = (i\hbar)^{-1} \int_0^t H'_{ba}(t') \exp(i\omega_{ba} t') dt', \qquad b \neq a. \tag{11.44}$$

From (11.41), (11.35) and (11.32), it follows that

$$c_b^{(1)}(t) = (i\hbar)^{-1} \mathcal{E}_0(\omega) \langle \psi_b | \hat{\boldsymbol{\epsilon}} \cdot \mathbf{D} | \psi_a \rangle \int_0^t \sin(\omega t' - \delta_\omega) \exp(i\omega_{ba} t') dt' \tag{11.45}$$

where

$$\langle \psi_b | \hat{\boldsymbol{\epsilon}} \cdot \mathbf{D} | \psi_a \rangle = -e \int \psi_b^*(\mathbf{r}) (\hat{\boldsymbol{\epsilon}} \cdot \mathbf{r}) \psi_a(\mathbf{r}) d\mathbf{r}. \tag{11.46}$$

The integral over t' in (11.45) can be written as

$$\int_0^t \sin(\omega t' - \delta_\omega)\exp(i\omega_{ba}t')dt'$$

$$= \frac{1}{2i}\left[\exp(-i\delta_\omega)\int_0^t \exp[i(\omega_{ba}+\omega)t']dt' - \exp(i\delta_\omega)\int_0^t \exp[i(\omega_{ba}-\omega)t']dt'\right]$$

$$= \frac{1}{2}\exp(-i\delta_\omega)\left[\frac{1-\exp[i(\omega_{ba}+\omega)t]}{\omega_{ba}+\omega}\right] - \frac{1}{2}\exp(i\delta_\omega)\left[\frac{1-\exp[i(\omega_{ba}-\omega)t]}{\omega_{ba}-\omega}\right]. \tag{11.47}$$

For transitions in the infrared $|\omega_{ba}|$ is of the order 10^{12} s^{-1} to 10^{14} s^{-1}, and $|\omega_{ba}|$ is even larger in the visible and ultraviolet regions. As the duration of the wave train of incident radiation is, in general, such that the product $|\omega_{ba}|t$ is much greater than unity, it follows that the first term in square brackets on the right-hand side of (11.47) is negligible unless $\omega_{ba} \simeq -\omega$, and the second term in square brackets is negligible unless $\omega_{ba} \simeq +\omega$. Since only one of these two conditions can be satisfied for a given pair of states a and b (with $E_a \neq E_b$), we can deal with the two terms separately. The first condition $\omega_{ba} \simeq -\omega$ implies that $E_b \simeq E_a - \hbar\omega$ so that a photon of energy $\hbar\omega \simeq E_a - E_b$ has been *emitted* by the atom under the influence of the field. This first condition, therefore, corresponds to *stimulated emission*. On the other hand, the second condition $\omega_{ba} \simeq \omega$ implies that $E_b \simeq E_a + \hbar\omega$ so that a photon of energy $\hbar\omega \simeq E_b - E_a$ has been *absorbed* by the atom under the influence of the field.

We shall consider each case separately, starting with absorption, and confining our attention at present to the case that both a and b represent bound atomic states. The photoelectric effect, which corresponds to an initial bound state and a final state in the continuum will be studied in Section 11.7.

Absorption

We start with the second term in (11.47), describing absorption. Using (11.45), the probability of finding an atom, initially in the state a, in a state $b(E_b > E_a)$ when exposed for a time t to radiation of angular frequency ω, is

$$|c_b^{(1)}(t)|^2 = \tfrac{1}{2}(\mathcal{E}_0(\omega)/\hbar)^2|\langle\psi_b|\hat{\epsilon}\cdot\mathbf{D}|\psi_a\rangle|^2 F(t,\omega_{ba}-\omega) \tag{11.48}$$

where (see (9.25))

$$F(t,\bar{\omega}) = \frac{1-\cos\bar{\omega}t}{\bar{\omega}^2}, \qquad \bar{\omega} = \omega_{ba}-\omega. \tag{11.49}$$

Introducing the notation

$$\mathbf{D}_{ba} = \langle\psi_b|\mathbf{D}|\psi_a\rangle = -e\int \psi_b^*(\mathbf{r})\,\mathbf{r}\,\psi_a(\mathbf{r})d\mathbf{r} \tag{11.50}$$

11.2 Harmonic perturbations and transition rates

and noting from (11.11) and (11.13) that $\mathcal{E}_0^2(\omega)/2 = I(\omega)/c\varepsilon_0$, we can rewrite (11.48) in the form

$$|c_b^{(1)}(t)|^2 = \frac{I(\omega)}{\hbar^2 c\varepsilon_0} \cos^2\theta |\mathbf{D}_{ba}|^2 F(t, \omega_{ba} - \omega) \tag{11.51}$$

where θ is the angle between the direction of polarisation $\hat{\epsilon}$ and the electric dipole vector $\mathbf{D} = -e\mathbf{r}$. We note that in terms of the Cartesian components of $\mathbf{r}$

$$|\mathbf{D}_{ba}|^2 = e^2 \{|\langle\psi_b|x|\psi_a\rangle|^2 + |\langle\psi_b|y|\psi_a\rangle|^2 + |\langle\psi_b|z|\psi_a\rangle|^2\}. \tag{11.52}$$

We must now take into account the fact that the incident radiation is never absolutely monochromatic, but is distributed over a range of frequencies. Assuming that the radiation is incoherent, we can add the probabilities for different frequencies, rather than adding amplitudes, which would be the case for coherent radiation. Thus the first-order probability $P_{ba}^{(1)}(t)$ for exciting the atom from the state a to the state b, when subjected to radiation with the intensity distribution $I(\omega)$ per unit angular frequency range, is

$$P_{ba}^{(1)}(t) = \int_0^\infty |c_b^{(1)}(t)|^2 d\omega = \frac{1}{\hbar^2 c\varepsilon_0} \cos^2\theta |\mathbf{D}_{ba}|^2 \int_0^\infty I(\omega) F(t, \omega_{ba} - \omega) d\omega. \tag{11.53}$$

For t large compared with $(2\pi/\omega_{ba})$, we have already seen that $F(t, \bar{\omega})$ is sharply peaked at $\bar{\omega} = 0$, that is at $\omega = \omega_{ba}$. In general the intensity distribution $I(\omega)$ varies slowly over the peak of $F(t, \bar{\omega})$ so that $I(\omega)$ can be replaced by $I(\omega_{ba})$ in the integrand of (11.53) and removed from the integral. We then have

$$P_{ba}^{(1)}(t) = \frac{I(\omega_{ba})}{\hbar^2 c\varepsilon_0} \cos^2\theta |\mathbf{D}_{ba}|^2 \int_{-\infty}^\infty F(t, \omega_{ba} - \omega) d\omega \tag{11.54}$$

where the limits of the integration have been extended to $\pm\infty$, which is permissible because the integrand is very small except in the region $\omega \simeq \omega_{ba}$. Using the result (9.26), we obtain

$$P_{ba}^{(1)}(t) = \frac{\pi I(\omega_{ba})}{\hbar^2 c\varepsilon_0} \cos^2\theta |\mathbf{D}_{ba}|^2 t \tag{11.55}$$

and we see that the first-order probability of absorption increases linearly with t. Clearly this is only valid for times t that are sufficiently short so that $P_{ba}^{(1)}(t) \ll 1$, which is the condition for the first-order perturbation theory we have used to be accurate. When the probability P_{ba} for the transition $a \to b$ is not small, the depletion of the initial state must be allowed for and, instead of using perturbation theory, the coupled equations (9.10) must be solved. In the case of light from incoherent sources, the duration of each pulse of radiation with a constant frequency and phase is sufficiently short for the formula (11.55) to be accurate; it is also sufficiently long compared with the period $(2\pi/\omega_{ba})$ for our approximations to be valid.

Since the first-order probability of absorption is given in the dipole approximation by (11.55), we can define a corresponding *transition rate* for absorption in the dipole approximation as $W_{ba} = dP_{ba}^{(1)}(t)/dt$, so that

$$W_{ba} = \frac{\pi I(\omega_{ba})}{\hbar^2 c \varepsilon_0} |\hat{\epsilon} \cdot \mathbf{D}_{ba}|^2 = \frac{\pi I(\omega_{ba})}{\hbar^2 c \varepsilon_0} \cos^2 \theta |\mathbf{D}_{ba}|^2. \tag{11.56}$$

If the radiation is unpolarised and isotropic, the orientation of the polarisation vector $\hat{\epsilon}$ is at random, in which case $\cos^2 \theta$ can be replaced by its average of $1/3$ (see Problem 11.5) giving

$$W_{ba} = \frac{\pi I(\omega_{ba})}{3\hbar^2 c \varepsilon_0} |\mathbf{D}_{ba}|^2. \tag{11.57}$$

The rate of absorption of energy from the beam, per atom, is $(\hbar \omega_{ba}) W_{ba}$ for the transition $a \to b$. It is convenient to define an absorption cross-section σ_{ba} (integrated over ω) which is the rate of absorption of energy, per atom, divided by $I(\omega_{ba})$

$$\sigma_{ba} = \frac{(\hbar \omega_{ba}) W_{ba}}{I(\omega_{ba})}. \tag{11.58}$$

Since the incident flux of photons of angular frequency ω_{ba} is obtained by dividing the intensity $I(\omega_{ba})$ by $\hbar \omega_{ba}$, we see that the cross-section σ_{ba} may also be defined as the transition probability per unit time and per unit atom, W_{ba}, divided by the incident photon flux.

Stimulated emission

To calculate the transition rate for emission, we return to (11.47) and repeat the calculation this time for the first term on the right, which corresponds to a transition in which the atom loses energy so that $E_b < E_a$ and $\omega \simeq -\omega_{ba} \simeq (E_a - E_b)/\hbar$. It is convenient to relabel the states a and b, so that b is again the state with higher energy. In this case b becomes the initial, and a the final state. The transition rate for stimulated emission $\bar{W}_{ab}$ is given by

$$\bar{W}_{ab} = \frac{\pi I(\omega_{ba})}{\hbar^2 c \varepsilon_0} |\hat{\epsilon} \cdot \mathbf{D}_{ab}|^2 = \frac{\pi I(\omega_{ba})}{\hbar^2 c \varepsilon_0} \cos^2 \theta |\mathbf{D}_{ab}|^2 \tag{11.59}$$

where θ is the angle between the vectors $\hat{\epsilon}$ and $\mathbf{D}$, and

$$\mathbf{D}_{ab} = \langle \psi_a | \mathbf{D} | \psi_b \rangle = -e \int \psi_a^*(\mathbf{r}) \, \mathbf{r} \, \psi_b(\mathbf{r}) d\mathbf{r}. \tag{11.60}$$

It is clear, by inspection, that

$$\mathbf{D}_{ab} = \mathbf{D}_{ba}^* \quad \text{and} \quad |\mathbf{D}_{ab}|^2 = |\mathbf{D}_{ba}|^2 \tag{11.61}$$

and comparing (11.59) with (11.57), we see that

$$\bar{W}_{ab} = W_{ba}. \tag{11.62}$$

Thus under the same radiation field, the number of transitions per second exciting an atom from the state a to the state b is the same as the number de-exciting the atom from the state b to the state a. Notice that we have assumed that the states a and b are non-degenerate. The case of degenerate states is left for a problem (Problem 11.6). The result (11.62) is consistent with the thermodynamical *principle of detailed balancing*, which states that in an enclosure containing atoms and radiation in equilibrium, the transition rate from a to b is the same as that from b to a, where a and b are any pair of states.

A *stimulated emission cross-section* $\bar{\sigma}_{ab}$ can be defined in analogy with the absorption cross section (11.58) by dividing the rate at which energy is radiated by the atom by stimulated emission ($\hbar \omega_{ba} \bar{W}_{ab}$) by the intensity $I(\omega_{ba})$. From (11.62) it follows that

$$\bar{\sigma}_{ab} = \sigma_{ba}. \tag{11.63}$$

11.3 Spontaneous emission

As already noted, the development of the quantum theory of spontaneous emission requires a description of the electromagnetic field in terms of its quanta, the photons. Since this is beyond the scope of this book, we shall use an indirect statistical method, developed by Einstein in 1916 before the advent of quantum mechanics, which relates the transition rate for spontaneous emission to those for absorption and stimulated emission.

The Einstein A and B coefficients

Consider an enclosure, containing atoms (of a single kind) and radiation, in thermal equilibrium at an absolute temperature T. Let a and b denote two atomic states of energies E_a and E_b, with $E_a < E_b$. The states a and b will be assumed to be non-degenerate. The number of atoms per unit time, $\dot{N}_{ba}$, making a transition from a to b by absorbing radiation of angular frequency $\omega = \omega_{ba} = (E_b - E_a)/\hbar$, is proportional to the total number, N_a, of atoms in the state a and also to the energy density of the radiation per unit angular frequency range $\rho(\omega_{ba})$. Thus we have

$$\dot{N}_{ba} = B_{ba} N_a \rho(\omega_{ba}) \tag{11.64}$$

where B_{ba} is called the *Einstein coefficient for absorption*. In terms of the transition rate for absorption per atom, W_{ba}, we have

$$\dot{N}_{ba} = W_{ba} N_a \tag{11.65}$$

and using (11.13) we see that

$$B_{ba} = W_{ba}/\rho(\omega_{ba}) = c W_{ba}/I(\omega_{ba}). \tag{11.66}$$

In the dipole approximation B_{ba} becomes, upon using (11.57)

$$B_{ba} = \frac{\pi}{3\hbar^2 \varepsilon_0} |\mathbf{D}_{ba}|^2. \tag{11.67}$$

On the other hand, the number of atoms making the transition $b \to a$ per unit time, $\dot{N}_{ab}$, is the sum of the number of spontaneous transitions per unit time, which is independent of ρ, and the number of stimulated transitions per unit time, which is proportional to ρ. Thus

$$\dot{N}_{ab} = A_{ab} N_b + B_{ab} N_b \rho(\omega_{ba}) \tag{11.68}$$

where A_{ab} is the *Einstein coefficient for spontaneous emission*, and B_{ab} is the *Einstein coefficient for stimulated emission*.

Since at equilibrium $\dot{N}_{ba} = \dot{N}_{ab}$, we find from (11.64) and (11.68) that

$$\frac{N_a}{N_b} = \frac{A_{ab} + B_{ab}\rho(\omega_{ba})}{B_{ba}\rho(\omega_{ba})}. \tag{11.69}$$

From a standard result in thermodynamics[3], it is known that in thermal equilibrium, the ratio N_a/N_b is given by

$$\frac{N_a}{N_b} = \exp[-(E_a - E_b)/kT] = \exp(\hbar\omega_{ba}/kT) \tag{11.70}$$

where k is Boltzmann's constant. Combining (11.69) and (11.70), we find that

$$\rho(\omega_{ba}) = \frac{A_{ab}}{B_{ba}\exp(\hbar\omega_{ba}/kT) - B_{ab}}. \tag{11.71}$$

An alternative expression for the energy density per unit angular frequency at temperature T is given by Planck's law, discussed in Chapter 1. Therein we obtained the energy density per unit wavelength interval $\rho(\lambda, T)$. Writing for simplicity $\rho(\lambda) \equiv \rho(\lambda, T)$, we have the relation

$$\rho(\omega) = \rho(\lambda)\left|\frac{d\lambda}{d\omega}\right| \tag{11.72a}$$

and since $\lambda = 2\pi c/\omega$

$$\rho(\omega) = \rho(\lambda)\frac{2\pi c}{\omega^2}. \tag{11.72b}$$

Using (1.13), $\rho(\omega_{ba})$ can be expressed as

$$\rho(\omega_{ba}) = \frac{\hbar\omega_{ba}^3}{\pi^2 c^3} \frac{1}{\exp(\hbar\omega_{ba}/kT) - 1}. \tag{11.73}$$

Comparing (11.71) and (11.73), we see that

$$B_{ab} = B_{ba} \tag{11.74a}$$

and

$$A_{ab} = \frac{\hbar\omega_{ba}^3}{\pi^2 c^3} B_{ab}. \tag{11.74b}$$

---

[3] See, for example, Kittel and Kroemer (1980).

The first relationship we have already seen to be correct in the dipole approximation using first-order perturbation theory. It expresses the principle of detailed balancing. It is easy to generalise these results to the case in which the energy levels E_a and (or) E_b are degenerate. Denoting by g_a and g_b the degeneracy of these levels, one finds (see Problem 11.6) that (11.74a) becomes

$$g_a B_{ba} = g_b B_{ab} \tag{11.75}$$

while the relation (11.74b) remains unchanged. Using (11.74), (11.61) and (11.67) we find that in the dipole approximation the transition rate for spontaneous emission $W_{ab}^s \equiv A_{ab}$ is given by

$$W_{ab}^s = \frac{\omega_{ba}^3}{3\pi c^3 \hbar \varepsilon_0} |\mathbf{D}_{ba}|^2. \tag{11.76}$$

We note that, had we not averaged over the directions of polarisation (see (11.56) and (11.57)), W_{ab}^s would have been proportional to $|\hat{\boldsymbol{\epsilon}} \cdot \mathbf{D}_{ba}|^2$.

From (11.70) it is seen that under equilibrium conditions a population inversion with $N_b > N_a$ can never be obtained because of the Boltzmann factor $\exp(\hbar \omega_{ba} / kT)$. On the other hand, to obtain *laser* or *maser* action which yields strong coherent radiation through stimulated emission from a highly populated excited state, such a population inversion must be produced. How this can be done will be described in Chapter 16.

Spontaneous emission and Larmor's formula

It is interesting to compare the result (11.76), with what might be inferred from classical ideas. The energy emitted per unit time by an accelerated charge is given in classical electrodynamics by Larmor's formula (11.1). Setting the charge q equal to the charge of the electron, $-e$, and assuming that the radiation of angular frequency $\omega = \omega_{ba}$ arises from an electron oscillating harmonically with the angular frequency ω_{ba} so that

$$\mathbf{r} = \mathbf{r}_0 \cos(\omega_{ba} t) \tag{11.77}$$

we find from (11.1) that the average power radiated is given by

$$\bar{P} = \frac{e^2 \omega_{ba}^4 |\mathbf{r}_0|^2}{12 \pi c^3 \varepsilon_0} \tag{11.78}$$

where we have replaced $\cos^2(\omega_{ba} t)$ by its average over one period, which is equal to $1/2$. Since in each transition a single photon of energy $\hbar \omega_{ba}$ is emitted, the transition rate is

$$W_{ab}^s = \frac{\bar{P}}{\hbar \omega_{ba}} = \frac{e^2 \omega_{ba}^3}{12 \pi c^3 \hbar \varepsilon_0} |\mathbf{r}_0|^2. \tag{11.79}$$

Comparing (11.79) with the quantum result (11.76), we see that the two results agree provided that $e^2 |\mathbf{r}_0|^2$ is equated with $|2\mathbf{D}_{ba}|^2$, or in other words if $|\mathbf{r}_0|^2$ is replaced by $|2\mathbf{r}_{ba}|^2$. In matrix mechanics, the observable quantities are matrix elements of operators and one might expect that the classical variable $\mathbf{r}_0$ would be replaced by a corresponding matrix element of the position operator **r**. However, whereas in quantum theory we have a precise method by which $\mathbf{r}_{ba}$ can be calculated, there was no way in classical physics by which the value of $\mathbf{r}_0$ could be predicted, although arguments could be given, based on atomic sizes, for its order of magnitude. Another point of interest lies in the factor of 2 which occurs in making the identification of $|\mathbf{r}_0|$ with $|2\mathbf{r}_{ba}|$. The origin of this factor lies in the fact that we used $\cos(\omega_{ba}t)$ to describe the classical oscillations of an electron. However, in the quantum theory of spontaneous emission, just as for stimulated emission, only one component, say $\exp(i\omega_{ba}t)$, is effective. Although the analogy between (11.79) and (11.76) is very close it must be emphasised that in no way can the correct quantum result be proved by classical methods, although by the correspondence principle quantum results must approach classical results when classical conditions apply, for example in the limit of large quantum numbers.

Spontaneous emission from the 2p level of hydrogenic atoms

As an example of the calculation of a transition rate for spontaneous emission, we shall consider the transition from the 2p level of a hydrogenic atom with nuclear charge Ze to the ground state, starting from equation (11.76), which we write for later convenience in the form

$$W_{ab}^s = \frac{4}{3}\frac{\alpha}{e^2 c^2} \omega_{ba}^3 |\mathbf{D}_{ba}|^2, \tag{11.80}$$

where $\alpha = e^2/(4\pi\varepsilon_0\hbar c)$ is the fine structure constant. In (11.80), b represents the 2p state of the atom with magnetic quantum number m and a is the 1s ground state. The angular frequency ω_{ba} can be found from (11.42) and (7.114). It is

$$\begin{aligned}\omega_{ba} &= \frac{1}{\hbar}(E_b - E_a) \\ &= \frac{3}{8}\frac{Z^2 e^2}{4\pi\varepsilon_0 \hbar a_0} = \frac{3}{8}\frac{\mu c^2}{\hbar}(Z\alpha)^2.\end{aligned} \tag{11.81}$$

Using (11.50) and (7.134) the dipole matrix element $\mathbf{D}_{ba}$ is seen to be

$$\mathbf{D}_{ba} = -e \int_0^\infty R_{21}(r) R_{10}(r) r^3 dr \int Y_{1m}^*(\theta,\phi)\,\hat{\mathbf{r}}\, Y_{00}(\theta,\phi) d\Omega \tag{11.82}$$

where $\hat{\mathbf{r}}$ is a unit vector in the direction of **r**.

The radial integral in (11.82) can be evaluated with the help of (7.140),

$$\begin{aligned}\int_0^\infty R_{21}(r) R_{10}(r) r^3 dr &= \left(\frac{Z}{a_\mu}\right)^4 \frac{1}{\sqrt{6}} \int_0^\infty r^4 \exp(-3Zr/2a_\mu) dr \\ &= \left(\frac{a_\mu}{Z}\right) \frac{24}{\sqrt{6}}\left(\frac{2}{3}\right)^5.\end{aligned} \tag{11.83}$$

To evaluate the angular integral in (11.82), we note that the x, y and z components of $\hat{\mathbf{r}}$ can be expressed as

$$(\hat{\mathbf{r}})_x = \sin\theta\cos\phi = \sqrt{\frac{2\pi}{3}}[-Y_{11}(\theta,\phi) + Y_{1-1}(\theta,\phi)]$$

$$(\hat{\mathbf{r}})_y = \sin\theta\sin\phi = \sqrt{\frac{2\pi}{3}}i\,[Y_{11}(\theta,\phi) + Y_{1-1}(\theta,\phi)]$$

$$(\hat{\mathbf{r}})_z = \cos\theta = \sqrt{\frac{4\pi}{3}}Y_{10}(\theta,\phi). \tag{11.84}$$

Since $Y_{00}(\theta,\phi) = (4\pi)^{-1/2}$, and using (6.103), it is seen that

$$\int Y_{1m}^*(\theta,\phi)(\hat{\mathbf{r}})_x Y_{00}(\theta,\phi)d\Omega = \frac{1}{\sqrt{6}}(-\delta_{m,1} + \delta_{m,-1})$$

$$\int Y_{1m}^*(\theta,\phi)(\hat{\mathbf{r}})_y Y_{00}(\theta,\phi)d\Omega = \frac{1}{\sqrt{6}}i\,(\delta_{m,1} + \delta_{m,-1})$$

$$\int Y_{1m}^*(\theta,\phi)(\hat{\mathbf{r}})_z Y_{00}(\theta,\phi)d\Omega = \frac{1}{\sqrt{3}}\delta_{m,0}. \tag{11.85}$$

From (11.82), (11.83) and (11.85), $|\mathbf{D}_{ba}|^2$ becomes

$$|\mathbf{D}_{ba}|^2 = e^2\left(\frac{a_\mu}{Z}\right)^2 \frac{2^{15}}{3^{10}}[\delta_{m,1} + \delta_{m,-1} + \delta_{m,0}]$$

$$= \frac{e^2}{Z^2}\left(\frac{\hbar}{\mu c \alpha}\right)^2 \frac{2^{15}}{3^{10}}[\delta_{m,1} + \delta_{m,-1} + \delta_{m,0}]. \tag{11.86}$$

The transition rate from each magnetic substate is the same, and if each state is equally populated the full transition rate is from (11.80), (11.81) and (11.86)

$$W_{ab}^s = \frac{1}{3}\sum_{m=-1}^{1} W_{1s,2pm}^s$$

$$\simeq \left(\frac{2}{3}\right)^8 \frac{\mu\alpha^5 Z^4 c^2}{\hbar}$$

$$\simeq 6.27 \times 10^8 Z^4 \text{s}^{-1} \quad \text{for } \mu = m. \tag{11.87}$$

The velocity and acceleration forms of the dipole matrix element

The matrix elements $\mathbf{D}_{ba}$ of the electric dipole operator have been obtained in terms of what is often called the *length* matrix elements $\mathbf{r}_{ba}$, where

$$\mathbf{r}_{ba} = \langle\psi_b|\mathbf{r}|\psi_a\rangle. \tag{11.88}$$

They can also be expressed in terms of matrix elements of the momentum $\mathbf{p}$ or the gradient of the potential energy V, in which case the matrix elements are said to be expressed in *velocity* or *acceleration* forms, respectively. To see this, we first remark that by using (11.88) and (11.37) we can write

$$\mathbf{r}_{ba} = \frac{1}{E_b - E_a} \langle \psi_b | H_0 \mathbf{r} - \mathbf{r} H_0 | \psi_a \rangle. \tag{11.89}$$

Taking $H_0 = -(\hbar^2/2\mu)\nabla^2 + V(\mathbf{r})$, and using the fact that $V(\mathbf{r})$ commutes with $\mathbf{r}$, we have

$$\mathbf{r}_{ba} = -\frac{\hbar^2}{2\mu} \frac{1}{E_b - E_a} \langle \psi_b | \nabla^2 \mathbf{r} - \mathbf{r} \nabla^2 | \psi_a \rangle. \tag{11.90}$$

Noting that

$$\nabla^2 \{\mathbf{r}\psi_a(\mathbf{r})\} = 2\nabla \psi_a(\mathbf{r}) + \mathbf{r}\nabla^2 \psi_a(\mathbf{r}) \tag{11.91}$$

and that $\mathbf{p} = -i\hbar\nabla$, we obtain the identity

$$\mathbf{r}_{ba} = -\frac{i\hbar}{\mu} \frac{1}{E_b - E_a} \langle \psi_b | \mathbf{p} | \psi_a \rangle = -\frac{i}{\mu\omega_{ba}} \mathbf{p}_{ba}. \tag{11.92}$$

In a similar way, we can write $\mathbf{p}_{ba}$ in the form

$$\begin{aligned}
\mathbf{p}_{ba} &= -\frac{i\hbar}{E_b - E_a} \langle \psi_b | H_0 \nabla - \nabla H_0 | \psi_a \rangle \\
&= -\frac{i\hbar}{E_b - E_a} \langle \psi_b | V\nabla - \nabla V | \psi_a \rangle \\
&= \frac{i\hbar}{E_b - E_a} \langle \psi_b | (\nabla V) | \psi_a \rangle = \frac{i}{\omega_{ba}} (\nabla V)_{ba}.
\end{aligned} \tag{11.93}$$

Using these results the length, velocity and acceleration forms of $\mathbf{D}_{ba}$ are, respectively,

$$\mathbf{D}_{ba}^L = -e\mathbf{r}_{ba}, \tag{11.94a}$$

$$\mathbf{D}_{ba}^V = \frac{ie}{\mu\omega_{ba}} \mathbf{p}_{ba} \tag{11.94b}$$

and

$$\mathbf{D}_{ba}^A = -\frac{e}{\mu\omega_{ba}^2} (\nabla V)_{ba}. \tag{11.94c}$$

Provided that the wave functions ψ_a and ψ_b are the exact eigenfunctions of H_0, for example in the case of hydrogenic atoms, the three forms of $\mathbf{D}_{ba}$ are equal. However, for other systems such that approximate wave functions are employed, the three forms of $\mathbf{D}_{ba}$ will have different numerical values.

11.4 Selection rules for electric dipole transitions

In the last paragraph, we found that, in the electric dipole approximation, the transition rates for absorption, stimulated emission and spontaneous emission between two states a and b depend on the quantity $|\hat{\epsilon}\cdot\mathbf{D}_{ba}|^2$ where $\hat{\epsilon}$ is the polarisation vector, $\mathbf{D}_{ba}$ is the matrix element of the electric dipole moment, and we recollect that $|\mathbf{D}_{ab}|^2 = |\mathbf{D}_{ba}|^2$.

The wave functions for the discrete states of one-electron atoms were discussed in Chapter 7. If the states a and b are defined by quantum numbers (nlm) and $(n'l'm')$, respectively,

$$\psi_a(\mathbf{r}) = R_{nl}(r)Y_{lm}(\theta, \phi), \qquad \psi_b(\mathbf{r}) = R_{n'l'}(r)Y_{l'm'}(\theta, \phi) \tag{11.95}$$

where $R_{nl}(r)$ is a radial hydrogenic function and $Y_{lm}(\theta, \phi)$ is a spherical harmonic. The Cartesian components of the matrix element of the electric dipole moment $\mathbf{D}_{ba}^L$ are given by

$$(D_x^L)_{ba} = -e \int \psi_b^*(\mathbf{r}) x \psi_a(\mathbf{r}) d\mathbf{r} \tag{11.96}$$

with similar expressions for $(D_y^L)_{ba}$, $(D_z^L)_{ba}$ in terms of matrix elements of y and z. Using spherical polar coordinates to perform the integrals, the dipole matrix elements can be expressed as products of a radial integral $J(n'l', nl)$ and angular integrals $I_j(l'm', lm)$ with $j = x, y, z$. That is

$$(D_j^L)_{ba} = -eJ(n'l', nl)I_j(l'm', lm) \tag{11.97}$$

where

$$J(n'l', nl) = \int_0^\infty R_{n'l'}(r) r^3 R_{nl}(r) dr \tag{11.98}$$

and

$$I_x(l'm', lm) = \int_0^\pi d\theta \sin\theta \int_0^{2\pi} d\phi Y_{l'm'}^*(\theta, \phi) \sin\theta \cos\phi Y_{lm}(\theta, \phi) \tag{11.99a}$$

$$I_y(l'm', lm) = \int_0^\pi d\theta \sin\theta \int_0^{2\pi} d\phi Y_{l'm'}^*(\theta, \phi) \sin\theta \sin\phi Y_{lm}(\theta, \phi) \tag{11.99b}$$

$$I_z(l'm', lm) = \int_0^\pi d\theta \sin\theta \int_0^{2\pi} d\phi Y_{l'm'}^*(\theta, \phi) \cos\theta Y_{lm}(\theta, \phi). \tag{11.99c}$$

The radial integrals (11.98) are always non-zero, but the angular integrals I_j are only non-zero for certain values of (lm) and $(l'm')$, giving rise to *selection rules* which we shall now investigate.

Parity

Let us first consider how the dipole matrix element behaves under the *reflection* or *parity operation* in which **r** is replaced by $-\mathbf{r}$, which is the same as replacing x, y and z and by $-x$, $-y$ and $-z$. In terms of polar coordinates this corresponds to replacing r, θ, ϕ by r, $\pi - \theta$, $\phi + \pi$. We have already seen in Chapter 6 (see (6.114)) that under reflections the spherical harmonics $Y_{l,m}$ have the parity $(-1)^l$,

$$Y_{lm}(\pi - \theta, \phi + \pi) = (-1)^l Y_{lm}(\theta, \phi). \tag{11.100}$$

Using this result we find that

$$I_j(l'm', lm) = -(-1)^{l+l'} I_j(l'm', lm) \tag{11.101}$$

from which it follows that the integrals I_x, I_y and I_z all vanish unless $(l + l')$ is odd. In other words, *the electric dipole operator only connects states of different parity.*

Magnetic quantum numbers

Using the expression (6.102) for the spherical harmonics we see that the integral over ϕ contained in I_x is proportional to the expression

$$K(m', m) = \int_0^{2\pi} \exp[i(m - m')\phi] \cos\phi \, d\phi \tag{11.102}$$

which vanishes unless $m - m' = \pm 1$. In the same way we find that I_y vanishes unless $m - m' = \pm 1$ and I_z vanishes unless $m' = m$. For a given transition between a and b, only one of these conditions can be satisfied so that either $(D_z^L)_{ba}$ is zero or $(D_x^L)_{ba}$ and $(D_y^L)_{ba}$ are zero.

Orbital angular momentum

Using again the expression (6.102) of the spherical harmonics, we see from (11.99) that both I_x and I_y contain the integrals

$$L^{\pm}(l', l, m) = \int_0^{\infty} d\theta \sin\theta P_{l'}^{m \pm 1}(\cos\theta) \sin\theta P_l^m(\cos\theta) \tag{11.103}$$

where, in view of (6.102), we need only deal with magnetic quantum numbers which are positive or zero. Employing the recurrence relation (6.97b) together with the orthogonality relation (6.98) we find that either $l' = l + 1$ or $l' = l - 1$.

The remaining expression, I_z, contains the integral

$$M(l', l, m) = \int_0^{\pi} d\theta \sin\theta P_{l'}^m(\cos\theta) \cos\theta P_l^m(\cos\theta) \tag{11.104}$$

which, upon using the recurrence relation (6.97a), is found to vanish unless $l' = l \pm 1$.

Selection rules

In summary we have found that an electric dipole transition can only take place if

$$\Delta l = \pm 1 \tag{11.105}$$

and either

$$\Delta m = 0 \quad \text{or} \quad \Delta m = \pm 1 \tag{11.106}$$

in which case the transition is said to be *allowed*. It should be noted that the condition (11.105) ensures that the states a and b differ in parity.

The dipole operator has no effect on the spin of the electron, so the component of the spin along the direction of quantisation (which we usually take to be the z-axis) is unaltered in an electric dipole transition.

If the selection rules (11.105) and (11.106) are not obeyed, the transition is said to be *forbidden*. However, in actual spectra very faint lines can be seen corresponding to some forbidden transitions. These arise from the magnetic interactions, and from the variation of the electric field (which we took to be uniform) across the atom. These additional interactions can be classified as *magnetic dipole, electric quadrupole* and so on. Furthermore, transitions involving more than one photon can also occur, which are not taken into account in the present first-order perturbation treatment. Each higher-order interaction gives rise to its own set of selection rules. For weak fields, the corresponding transition rates are very small compared with those arising from the electric dipole terms we have studied.

The spectrum of atomic hydrogen

In the non-relativistic approximation, the bound states of one-electron atoms are degenerate in l and m, and the energy of each level only depends on the principal quantum number n. Since there is no selection rule limiting n, the spectrum consists of all frequencies $v_{n'n}$ such that

$$h v_{n'n} = E_{n'} - E_n, \quad n' > n \tag{11.107}$$

where the energies E_n and $E_{n'}$ are given by (7.114). However, the selection rules limit the other quantum numbers of the levels concerned. For example, a transition from the $n = 1$ to the $n = 2$ level is from the 1s ground state to a $2p_m (m = 0, \pm 1)$ final state. The allowed transitions between the lower lying levels are illustrated in Fig. 11.1. Under very high resolution the spectral lines can be split into a number of components forming a multiplet. This *fine structure* is due to relativistic effects (see Section 8.2 and Chapter 15).

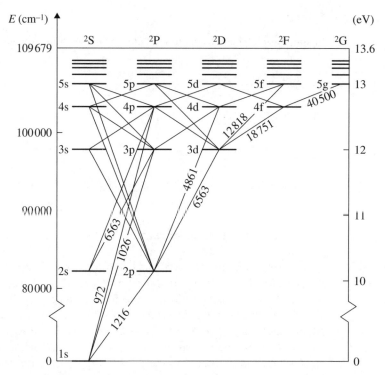

Figure 11.1 The allowed transitions among the lower levels of atomic hydrogen. The ordinate shows the energy above the 1s ground state in cm^{-1} (8065 cm^{-1} = 1 eV) on the left and in eV on the right. The numbers against the lines indicate the wavelength in ångström units (1 Å = 10^{-8} cm). For clarity the wavelengths are shown only for a selection of lines. The splitting due to fine structure is too small to be shown on a diagram of this scale.

N-electron atoms

The electric dipole operator for an atom containing N electrons each at position $\mathbf{r}_i (i = 1, 2, \ldots, N)$ is

$$\mathbf{D} = -e \sum_{i=1}^{N} \mathbf{r}_i \qquad (11.108)$$

and the electric dipole transitions between states a and b are determined by the quantity $|\hat{\boldsymbol{\epsilon}} \cdot \mathbf{D}_{ba}|^2$, where $\hat{\boldsymbol{\epsilon}}$ is the polarisation vector and

$$\mathbf{D}_{ba} = \int \psi_b^*(\mathbf{r}_1, \mathbf{r}_2, \ldots, \mathbf{r}_N) \mathbf{D} \psi_a(\mathbf{r}_1, \mathbf{r}_2, \ldots, \mathbf{r}_N) d\mathbf{r}_1 d\mathbf{r}_2 \ldots d\mathbf{r}_N. \qquad (11.109)$$

11.4 Selection rules for electric dipole transitions

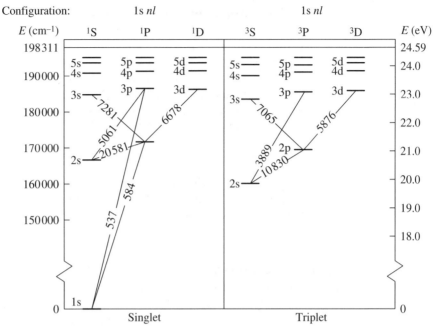

Figure 11.2 The allowed transitions among the lower levels of helium. Wavelengths in ångström units (1 Å = 10^{-8} cm) are shown against the lines representing the transitions. Energies in cm^{-1} (8065 cm^{-1} = 1 eV) and eV are relative to the ground state.

The alternative forms (11.94) of the dipole matrix element remain valid for an N-electron atom provided the operators **r** and **p** are replaced by

$$\mathbf{R} = \sum_{i=1}^{N} \mathbf{r}_i, \qquad \mathbf{P} = \sum_{i=1}^{N} \mathbf{p}_i \tag{11.110}$$

and the potential V now denotes the total potential appearing in H_0.

Those transitions in which the quantum numbers of only one electron change are by far the most important, and if the initial and final one-electron orbitals are specified by quantum numbers (nlm) and $(n'l'm')$, the selection rules are $\Delta l = \pm 1$ and $\Delta m = 0, \pm 1$ as before. Since the energy levels of many-electron atoms are not degenerate in l, not all frequencies corresponding to pairs of energy levels are observed. For example, in the spectrum due to transitions of the 2s valence electron of lithium, the 2s ↔ 2p transition is *allowed* and gives rise to a strong line, but the transition 2s ↔ 3s is *forbidden*. Because in the electric dipole approximation the electron spin is unaltered by a transition, separate systems of strong spectral lines are seen for helium in the singlet ($S = 0$) and in the triplet ($S = 1$) states (see Fig. 11.2). In this case, some *very* weak lines (not illustrated) are observed connecting triplet states with singlet states; these *intercombination* lines arise from higher-order effects.

11.5 Lifetimes, line intensities, widths and shapes

As we saw in Section 11.3, the transition rate between a pair of states a and b is proportional, in the electric dipole approximation, to the quantity $|\mathbf{D}_{ba}|^2$. Thus the relative intensities of a series of transitions from a given initial state a to various final states k are determined by the quantities $|\mathbf{D}_{ka}|^2$. In discussions of intensities it is customary to introduce a related dimensionless quantity f_{ka}, called the *oscillator strength*. For a one-electron atom it is defined as

$$f_{ka} = \frac{2\mu\omega_{ka}}{3\hbar e^2}|\mathbf{D}_{ka}|^2 \tag{11.111}$$

with $\omega_{ka} = (E_k - E_a)/\hbar$. This definition implies that $f_{ka} > 0$ for absorption and $f_{ka} < 0$ for emission. It is not difficult to show (Problem 11.8) that the oscillator strengths (11.111) satisfy the Thomas–Reiche–Kuhn sum rule

$$\sum_k f_{ka} = 1 \tag{11.112}$$

where the sum is over all levels, including the continuum.

In obtaining the transition rates for absorption or stimulated emission, we found that the probability for a transition between an atomic state a and an atomic state b depended on the factor $F(t, \omega_{ba} - \omega)$, which was sharply peaked about the angular frequency $\omega = \omega_{ba}$. Although the peaking approximation we made for obtaining the transition rate is very accurate, there is in fact a finite probability that a photon of angular frequency $\omega \neq \omega_{ba}$ will be emitted or absorbed, although the distribution in ω about ω_{ba} is very narrow. The same is true for spontaneous emission. If an atom in an excited state b emits a photon, making a transition to a state of lower energy a, the angular frequency of the photon will not necessarily be exactly ω_{ba} but will be in a narrow band centred about ω_{ba}. We shall now obtain the distribution in frequency, using an indirect argument.

The decay law

In general a time-dependent system can be described by the set of coupled equations (9.10) which connect the amplitudes $c_k(t)$, and $|c_k(t)|^2$ is the probability of finding the system in a particular state k at time t. Consider a system of two equations connecting the amplitudes $c_a(t)$ and $c_b(t)$, where the system is definitely in the state a at $t = 0$ so that $c_a(0) = 1$ and $c_b(0) = 0$. As time goes on, $|c_a(t)|$ will decrease because of the coupling to the state b, and correspondingly $|c_b(t)|$ must increase because the total probability of finding the system in either the state a or the state b must be unity: $|c_a(t)|^2 + |c_b(t)|^2 = 1$. As $|c_b(t)|$ increases, the back coupling between the state b and the state a becomes important, and ultimately the reverse transition becomes dominant: $|c_b(t)|$ decreases and $|c_a(t)|$ increases. Thus, as we saw in Chapter 9, a typical behaviour of a two-state system is an oscillation between the probabilities $|c_a(t)|^2$ and $|c_b(t)|^2$. When an atom makes a spontaneous transition from a state b to a state a, this is *not* a two-level system, because the final state is composed of the atom in state a together with a photon which may be of any angular frequency ω, although the probability of finding a value of ω outside a very small region about ω_{ba} is very small. For this reason, the initial state b is coupled to a *continuum* of final states. These final states are incoherent and cannot act cooperatively to build up the reverse transitions, so that the probability of finding the atom in the state b decreases steadily with time. Under these circumstances, and only under these circumstances, can we make the following argument, showing that the rate of decay is exponential. Suppose the transition rate for spontaneous emission from state b to state a is W_{ab}^s, which is independent of time. Then the probability $P_b(t + \mathrm{d}t)$ of finding the atom in state b at time $t + \mathrm{d}t$ is equal to the probability $P_b(t)$ of finding it in state b at time t multiplied by the probability that no transition from b has taken place in time $\mathrm{d}t$. The probability that no transition has taken place is $(1 - W_{ab}^s \mathrm{d}t)$, so that

$$P_b(t + \mathrm{d}t) = P_b(t)(1 - W_{ab}^s \mathrm{d}t) \tag{11.113a}$$

or

$$\frac{\mathrm{d}P_b(t)}{\mathrm{d}t} = -W_{ab}^s P_b(t). \tag{11.113b}$$

Since $P_b(0) = 1$, this equation has the solution

$$P_b(t) = \exp(-W_{ab}^s t) = \exp(-t/\tau) \tag{11.114}$$

where

$$\tau = \frac{1}{W_{ab}^s} \tag{11.115}$$

is called the *lifetime* (or *half-life*) of the level b. For example, using the result (11.87) it is seen that the lifetime of the 2p level of a hydrogenic atom is $\tau = (0.16 \times 10^{-8}/Z^4)$ s.

Taking $c_b(t)$ as a real quantity, we can write

$$c_b(t) = \exp(-t/2\tau) \tag{11.116}$$

so that for $t \geqslant 0$ the component of the total wave function which describes the initial state b can be expressed as

$$\Psi_b(\mathbf{r}, t) = c_b(t)\psi_b(\mathbf{r})\exp(-\mathrm{i}E_b t/\hbar) = \psi_b(\mathbf{r})\exp[-\mathrm{i}(E_b - \mathrm{i}\hbar/2\tau)t/\hbar] \tag{11.117}$$

where ψ_b is a time-independent atomic wave function satisfying (11.37). For later convenience, since we are not interested in $\Psi_b(\mathbf{r}, t)$ for $t < 0$, we set $\Psi_b = 0$ for $t < 0$.

In the absence of any coupling to the radiation field, an excited atomic state b would be stable and the wave function would be $\Phi_b(\mathbf{r}, t) = \psi_b(\mathbf{r})\exp(-\mathrm{i}E_b t/\hbar)$. This is a stationary state which is an eigenstate of the energy operator since

$$\mathrm{i}\hbar \frac{\partial}{\partial t}\Phi_b(\mathbf{r}, t) = E_b \Phi_b(\mathbf{r}, t) \tag{11.118}$$

and the system possesses a well defined real energy E_b. In contrast, when the coupling to the radiation field is taken into account, the time variation of the initial wave function is given by (11.117) and

$$i\hbar \frac{\partial}{\partial t} \Psi_b(\mathbf{r}, t) = \left(E_b - i\frac{\hbar}{2\tau} \right) \Psi_b(\mathbf{r}, t) \tag{11.119}$$

showing that $\Psi_b(\mathbf{r}, t)$ does not describe a state with a well defined *real* energy. This is a general result: a decaying state is never a state with a definite real energy. However, *any* wave function can be expressed as a superposition of energy eigenstates, and we can write

$$\exp[-i(E_b - i\hbar/2\tau)t/\hbar] = (2\pi\hbar)^{-1/2} \int_{-\infty}^{\infty} a(E') \exp(-iE't/\hbar) dE'. \tag{11.120}$$

Using the relation (see equation (A.18) of Appendix A)

$$(2\pi\hbar)^{-1} \int_{-\infty}^{\infty} \exp[i(E - E')t/\hbar] dt = \delta(E - E') \tag{11.121}$$

we find from (11.120) that

$$a(E) = (2\pi\hbar)^{-1/2} \int_0^{\infty} \exp[-i(E_b - i\hbar/2\tau)t/\hbar] \exp(iEt/\hbar) dt \tag{11.122}$$

where the lower limit of the integral at $t = 0$ arises because we have set $\Psi_b = 0$ for $t < 0$. Evaluating the integral we find

$$a(E) = (2\pi\hbar)^{-1/2} \frac{-i\hbar}{E_b - E - i\hbar/2\tau}. \tag{11.123}$$

Thus the probability of finding the initial state b in a state of definite energy E is proportional to $|a(E)|^2$, where

$$|a(E)|^2 = \frac{\hbar}{2\pi} \frac{1}{(E_b - E)^2 + \hbar^2/4\tau^2}. \tag{11.124}$$

Conservation of energy requires that an eigenstate of energy E should make a transition to a final state a of energy E_a and a photon of energy $\hbar\omega$, so that

$$E = E_a + \hbar\omega \tag{11.125}$$

where we have assumed that the final state a is stable and does not decay. Thus the quantity $|a(E)|^2$ determines the intensity distribution as a function of angular frequency of the spectral line emitted in the transition $b \to a$:

$$|a(E)|^2 = \frac{\hbar}{2\pi} \frac{1}{(E_b - E_a - \hbar\omega)^2 + \hbar^2/4\tau^2}. \tag{11.126}$$

This distribution in the intensity is called a *Lorentzian distribution*. It is proportional to the function $f(\omega)$ defined by

$$f(\omega) = \frac{\Gamma^2/(4\hbar^2)}{(\omega - \omega_{ba})^2 + \Gamma^2/(4\hbar^2)} \tag{11.127}$$

11.5 Lifetimes, line intensities, widths and shapes 541

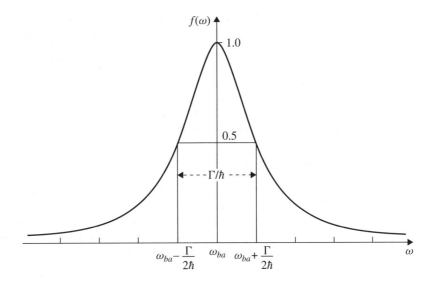

Figure 11.3 A plot of the Lorentzian intensity distribution

$$f(\omega) = \frac{\Gamma^2/(4\hbar^2)}{(\omega - \omega_{ba})^2 + \Gamma^2/(4\hbar^2)}$$

Note that $f(\omega_{ba}) = 1$ and $f(\omega) = \frac{1}{2}$ when $\omega = \omega_{ba} \pm 1/(2\tau) = \omega_{ba} \pm \Gamma/(2\hbar)$.

where $\omega_{ba} = (E_b - E_a)/\hbar$ and $\Gamma = \hbar/\tau$. This function is plotted in Fig. 11.3. The distribution $f(\omega)$ is equal to half its maximum value of unity at $\omega = \omega_{ba} \pm \Gamma/(2\hbar)$. The width of the distribution at half maximum is $\Gamma/\hbar$, where Γ (which has the dimensions of an energy) is called the *natural width* of the line.

Quite generally, any state which decays with a finite lifetime τ has a distribution in energy measured by a natural width Γ, where

$$\Gamma = \frac{\hbar}{\tau}. \tag{11.128}$$

This relationship is consistent with the uncertainty relation

$$\Delta E \Delta t \gtrsim \hbar \tag{11.129}$$

since one may assume that $\Delta E \simeq \Gamma$ and $\Delta t \simeq \tau$. If the final state a is not stable then it can be shown that

$$\Gamma = \hbar(1/\tau_a + 1/\tau_b), \tag{11.130}$$

where τ_a and τ_b are the lifetimes of the two states a and b, respectively. If the state b can decay to more than one state the quantity W^s_{ab} in (11.113)–(11.115) must be replaced by $\sum_k W^s_{kb}$ where the sum is over all the possible final states k.

The natural width of atomic energy levels is very small. For example, the width of the 2p level of atomic hydrogen, which has an unperturbed energy of $E = -3.40$ eV and a lifetime of 0.16×10^{-8} s, is $\Gamma = 4 \times 10^{-7}$ eV. The profile of a spectral line can be measured, either by recording the spectrum on a photographic plate and subsequently measuring the density of the image as a function of wavelength, or by scanning the spectrum with a photoelectric detector. It is found that observed spectral lines usually have much greater widths than the natural width (after allowing for the finite resolving power of the spectrograph employed). The reasons for this will now be explored.

Pressure broadening

In deriving the exponential law (11.114), we assumed that the transition rate between the atomic state b of a higher energy E_b and the state a of lower energy E_a was entirely due to spontaneous emission. However, the population of the state b must also decrease if there are any other mechanisms which lead to transitions out of b. Thus in (11.114) W^s_{ab} should be replaced by the sum of the transition rates for all processes depleting this particular level. If W_{tot} is this sum, then the lifetime of the state b is given by

$$\tau_b = \frac{1}{W_{\text{tot}}}. \tag{11.131}$$

The observed width Γ of a spectral line from b to a is again determined by τ_b and τ_a through equation (11.130) and the line intensity is proportional to the Lorentzian function $f(\omega)$ given by (11.127), with this new, larger, value of Γ. The principal mechanism of this type, broadening lines of radiation from atoms in a gas, arises from *collisions* between the atoms. In each collision there is a certain probability that an atom initially in a state b will make a radiationless transition to some other state. The corresponding transition rate W_c is proportional to the number density of the atoms concerned, n, and to the relative velocity between pairs of atoms, v, so that

$$W_c = nv\sigma \tag{11.132}$$

where σ is a quantity with dimensions of area called the collision cross-section. The cross-section depends on the species of atom and on v. Since n depends on the pressure, the broadening of a spectral line due to this cause is called *pressure broadening* or, alternatively, *collisional broadening*. In addition, both n and v depend on the temperature, so that information about the pressure and temperature of a gas can be obtained by measuring the profiles of spectral lines. This is a major source of information about physical conditions in stellar atmospheres, which in turn provides most of our knowledge about stellar structure.

Doppler broadening

The wavelength of the light emitted by a moving atom is shifted by the Doppler effect. If the emitting atoms are moving at non-relativistic speeds, and v is the component of the velocity of a particular atom along the line of sight, the wavelength of the emitted light is

$$\lambda = \lambda_0 \left(1 \pm \frac{v}{c}\right) \tag{11.133}$$

where λ_0 is the wavelength emitted by a stationary atom. The plus sign corresponds to an atom receding from the observer, and the minus sign to an approaching atom. The angular frequency $\omega = 2\pi c/\lambda$ of the light emitted by the moving atom is thus related to $\omega_0 = 2\pi c/\lambda_0$ by

$$\omega = \omega_0 \left(1 \pm \frac{v}{c}\right)^{-1} \simeq \omega_0 \left(1 \mp \frac{v}{c}\right). \tag{11.134}$$

If the light is emitted from a gas at absolute temperature T, the number of atoms, dN, with velocities between v and $v + dv$ is given by Maxwell's distribution

$$dN = N_0 \exp(-Mv^2/2kT) dv \tag{11.135}$$

where k is Boltzmann's constant, M is the atomic mass and N_0 is a constant. The intensity $\mathcal{I}(\omega)$ of light emitted in an angular frequency interval ω to $\omega + d\omega$ is proportional to the number of atoms with velocities between v and $v + dv$. Hence, using (11.134) and (11.135), we obtain the *Gaussian distribution*

$$\mathcal{I}(\omega) = \mathcal{I}(\omega_0) \exp\left[-\frac{Mc^2}{2kT}\left(\frac{\omega - \omega_0}{\omega_0}\right)^2\right]. \tag{11.136}$$

If ω_1 is the angular frequency at half-maximum, then

$$(\omega_1 - \omega_0)^2 = \frac{2kT}{Mc^2}\omega_0^2 \log 2. \tag{11.137}$$

The total Doppler width at half maximum $\Delta\omega^D$ is $2|\omega_1 - \omega_0|$ and hence

$$\Delta\omega^D = \frac{2\omega_0}{c}\left[\frac{2kT}{M}\log 2\right]^{1/2}. \tag{11.138}$$

This width increases with temperature and with the frequency of the line, and decreases as the atomic mass increases. While pressure broadening increases the width of a spectral line, but preserves the Lorentzian profile, the Gaussian profile produced by Doppler broadening is quite different. The two shapes are compared in Fig. 11.4. In general, both pressure and Doppler broadening are present, and the observed profile, called a *Voigt profile*, is due to a combination of both effects. The decrease of the Gaussian distribution away from $\omega_0 (= \omega_{ba})$ is so rapid that the 'wings' of spectral lines are determined by the residual Lorentzian distribution. Thus if both Doppler and pressure broadening are present, the characteristics of each effect can be distinguished provided sufficiently accurate experimental profiles can be obtained.

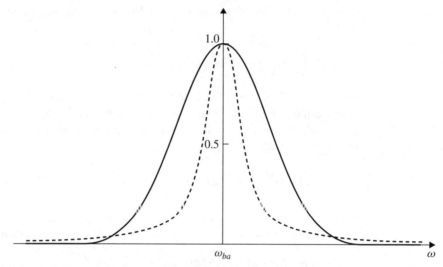

Figure 11.4 A comparison of a Gaussian distribution (solid line) of the form

$$F(\omega) = \exp[-\alpha(\omega - \omega_{ba})^2],$$

with a Lorentzian distribution (dashed line) of the form

$$f(\omega) = \frac{\Gamma^2/(4\hbar^2)}{(\omega - \omega_{ba})^2 + \Gamma^2/(4\hbar^2)}$$

11.6 The spin of the photon and helicity

The spin of a particle with non-zero mass can be defined as the angular momentum it possesses in its rest frame. The spin of the photon cannot be defined in this way, since a photon always moves with the velocity of light, which is the same in all references frames. A photon can possess both intrinsic and orbital angular momentum, the orbital angular momentum being at right angles to the direction of motion, defined by the direction of the propagation vector **k** in (11.5). Any component of angular momentum along the direction of propagation can only be due to an intrinsic angular momentum and this can be used to define the spin of a massless particle travelling with the speed of light. The component of the spin in the direction of motion of a particle is called the *helicity* of that particle.

The selection rules for electric dipole transitions tell us that the angular momentum quantum number of an atom changes by one unit when a photon is emitted or absorbed. In the electric dipole approximation the variation of the electric field over the atom is ignored, and in this approximation the electric field is spherically symmetric. Since the electric field must be connected with the wave function of the photon, this in turn can be taken to be spherically symmetric, and spherically symmetric wave functions describe systems with zero orbital angular momentum. For this reason the change in angular momentum of the atom can be ascribed to the photon spin, which must therefore be equal to unity. Relativistic quantum electrodynamics confirms that indeed the photon is a particle of spin one. From this one might expect that the possible values of the photon helicity, the component of its spin along **k** would be $\pm\hbar$ and 0. In fact it can be shown that the value 0 is excluded, because of the transverse nature of the electromagnetic field.

11.7 Photoionisation

In the previous sections, we have seen how time-dependent perturbation theory can be applied to obtain the transition rates for the absorption or emission of radiation from a system of charged particles. So far we have concentrated on the excitation or de-excitation of atoms in which both the initial and final states of the system are discrete. The same method can be applied to the interaction of radiation with molecules or with atomic nuclei. In this section and the following one, we shall see how to develop the theory of transitions for which one state is discrete but the other lies in the continuum. Two special cases will be taken as examples. The first is the ionisation of a one-electron atom when radiation is absorbed; this photoionisation process is responsible for the photoelectric effect discussed in Section 1.2. The second example is the photodisintegration of the deuteron, to be discussed in Section 11.8.

Let us consider the ejection of an electron from the ground state of a one-electron atom, by the absorption of radiation of angular frequency ω, or in other words by the absorption of a photon of energy $\hbar\omega$. The electric dipole approximation will be made and this is expected to be accurate if the wavelength of the radiation is greater than atomic dimensions. This condition is satisfied for radiation of longer wavelength than X-rays. From Table 7.1 with $a_\mu = a_0$ (the infinite nuclear mass approximation), the wave function of the initial (ground) state of the atom is

$$\psi_a(\mathbf{r}) = \psi_{1s}(r) = \left(\frac{Z^3}{\pi a_0^3}\right)^{1/2} e^{-Zr/a_0} \tag{11.139}$$

where Ze is the nuclear charge. The energy of this state is $E_a \equiv E_{1s} = -Z^2 e^2/[(4\pi\varepsilon_0)2a_0]$ (see (7.114)). The final atomic state b represents a free electron of momentum $\hbar\mathbf{k}_b$ moving in the Coulomb field of the nucleus. The wave function of this continuum state $\psi_b(\mathbf{k}_b, \mathbf{r})$ satisfies the Schrödinger equation

$$\left(-\frac{\hbar^2}{2m}\nabla^2 - \frac{Ze^2}{(4\pi\varepsilon_0)r} - E_b\right)\psi_b(\mathbf{k}_b, \mathbf{r}) = 0 \tag{11.140}$$

where E_b is the electron energy. In the non-relativistic regime which we are studying (i.e. for $\hbar\omega$ or $E_b \ll mc^2$) we have

$$E_b = \frac{\hbar^2 k_b^2}{2m}. \tag{11.141}$$

We shall assume that E_b is sufficiently large with respect to the binding energy $|E_a|$ of the electron. In first approximation the Coulomb interaction with the nucleus can then be neglected and $\psi_b(\mathbf{k}_b, \mathbf{r})$ approximated by a plane wave

$$\psi_b(\mathbf{k}_b, \mathbf{r}) \simeq (2\pi)^{-3/2} e^{i\mathbf{k}_b \cdot \mathbf{r}} \tag{11.142}$$

where we have chosen the Dirac delta function normalisation

$$\int \psi_b^*(\mathbf{k}_b', \mathbf{r}) \psi_b(\mathbf{k}_b, \mathbf{r}) d\mathbf{r} = \delta(\mathbf{k}_b - \mathbf{k}_b'). \tag{11.143}$$

In order to obtain the transition probability, we first recall that if P_{ba} is the probability for excitation of a discrete state b from an initial discrete state a, it can be expressed in the form of the modulus squared of the matrix element

$$P_{ba} = \left| \int \psi_b^*(\mathbf{r}) \mathcal{O} \psi_a(\mathbf{r}) d\mathbf{r} \right|^2, \tag{11.144}$$

where $\mathcal{O}$ is the appropriate operator and $\psi_a(\mathbf{r})$, $\psi_b(\mathbf{r})$ are normalised to unity. If now b represents a state which is an eigenstate of some quantity, having a continuous range of eigenvalues q, then the probability that a transition occurs between the discrete state a and a state for which q lies in the range $(q, q + dq)$ can be expressed as

$$P_{ba} \, dq = \left| \int \psi_b^*(q, \mathbf{r}) \mathcal{O} \psi_a(\mathbf{r}) d\mathbf{r} \right|^2 dq \tag{11.145}$$

and this is consistent with the normalisation

$$\int \psi_b^*(q, \mathbf{r}) \psi_b(q', \mathbf{r}) d\mathbf{r} = \delta(q - q'). \tag{11.146}$$

In the present example, the final state is labelled by the three continuous components $(k_{b,x}, k_{b,y}, k_{b,z})$ of $\mathbf{k}_b$. Using (11.51) and taking into account the normalisation (11.143) we can write down the first-order probability that at time t a photon has been absorbed and a transition has occurred to a final state b for which $\mathbf{k}_b$ lies in the range $(\mathbf{k}_b, \mathbf{k}_b + d\mathbf{k}_b)$. That is

$$P_{ba}^{(1)}(t) \, d\mathbf{k}_b = \frac{I(\omega)}{\hbar^2 c \varepsilon_0} \cos^2 \theta |\mathbf{D}_{ba}|^2 F(t, \omega_{ba} - \omega) d\mathbf{k}_b \tag{11.147}$$

where as before $\omega_{ba} = (E_b - E_a)/\hbar$ and θ is the angle between $\mathbf{D}_{ba}$ and the direction of the polarisation vector $\hat{\epsilon}$. Using (11.141) we can express the volume element $d\mathbf{k}_b$ as

$$d\mathbf{k}_b = k_b^2 dk_b d\Omega = \frac{mk_b}{\hbar} d\omega_{ba} d\Omega \tag{11.148}$$

where $d\Omega \equiv \sin\Theta d\Theta d\Phi$ and (Θ, Φ) are the polar angles of $\mathbf{k}_b$.

11.7 Photoionisation

In the limit of large t, the function $F(t, \omega_{ba} - \omega)$ is sharply peaked about $\omega_{ba} = \omega$. Hence, for $t \to \infty$ transitions occur to final states for which $k_b = k_f$, so that the energy-conservation condition

$$E_f = \frac{\hbar^2 k_f^2}{2m} = \hbar\omega + E_a \tag{11.149}$$

is satisfied. Using this result, the total first-order transition probability is given by

$$\int P_{ba}^{(1)}(t) d\mathbf{k}_b = \frac{m k_f I(\omega)}{\hbar^3 c \varepsilon_0} \int \cos^2\theta |\mathbf{D}_{ba}|^2 d\Omega \int_0^\infty F(t, \omega_{ba} - \omega) d\omega_{ba} \tag{11.150}$$

where $\mathbf{D}_{ba}$ is to be evaluated with $k_b = k_f$. Extending the limits of integration over $d\omega_{ba}$ from $-\infty$ to $+\infty$ (see (11.53) and (11.54)) and using the result (9.26) we obtain

$$\int P_{ba}^{(1)}(t) d\mathbf{k}_b = \frac{m k_f I(\omega)}{\hbar^3 c \varepsilon_0} \pi t \int \cos^2\theta |\mathbf{D}_{ba}|^2 d\Omega. \tag{11.151}$$

Accordingly, the total transition rate W_{ba} is, to first order,

$$W_{ba} = \frac{d}{dt} \int P_{ba}^{(1)}(t) d\mathbf{k}_b = \frac{\pi m k_f I(\omega)}{\hbar^3 c \varepsilon_0} \int \cos^2\theta |\mathbf{D}_{ba}|^2 d\Omega. \tag{11.152}$$

This expression can equally well be obtained by using the Golden Rule formula (9.67). Since in the present case $H' = W = -\boldsymbol{\mathcal{E}} \cdot \mathbf{D} = \hat{\boldsymbol{\epsilon}} \cdot \mathbf{D} \mathcal{E}_0(\omega) \sin(\omega t - \delta_\omega)$ we have from (9.58)

$$A^\dagger = \frac{i}{2} \hat{\boldsymbol{\epsilon}} \cdot \mathbf{D} \mathcal{E}_0(\omega) \exp(i\delta_\omega) \tag{11.153}$$

and therefore the first-order total transition rate is

$$W_{ba} = \frac{2\pi}{\hbar} \frac{\mathcal{E}_0^2(\omega)}{4} \rho_b(E_f) \int \cos^2\theta |\mathbf{D}_{ba}|^2 d\Omega \tag{11.154}$$

where we recall that θ is the angle between $\hat{\boldsymbol{\epsilon}}$ and $\mathbf{D}_{ba}$, and $\rho_b(E_f)$ is the density of final states, with $E_f = \hbar\omega + E_a$ (see (11.149)). In the present case the final states are plane waves normalised according to (11.143), so that if $\rho_b(E_f) d\Omega dE_f$ denotes the number of states whose wave vector $\mathbf{k}_f$ lies within the solid angle $d\Omega$ and the energy of which is in the interval $(E_f, E_f + dE_f)$, we have

$$\rho_b(E_f) d\Omega dE_f = d\mathbf{k}_f = k_f^2 dk_f d\Omega \tag{11.155}$$

so that

$$\rho_b(E_f) = k_f^2 \frac{dk_f}{dE_f} = \frac{m k_f}{\hbar^2}. \tag{11.156}$$

Inserting (11.156) into (11.154) and using the fact that $\mathcal{E}_0^2(\omega) = 2I(\omega)/c\varepsilon_0$, we recover the result (11.152).

Cross-sections

Using (11.58) and (11.152), the photoionisation cross-section $\sigma(\omega)$ is given by

$$\sigma(\omega) = \frac{\pi m k_f \omega}{\hbar^2 c \varepsilon_0} \int \cos^2\theta |\mathbf{D}_{ba}|^2 d\Omega. \tag{11.157}$$

It is convenient to use the velocity form (11.94b) of the dipole matrix element $\mathbf{D}_{ba}$. From (11.139), (11.142) and recalling that $k_b = k_f$ and $\omega = \omega_{ba}$, we have

$$\mathbf{D}_{ba}^V = (\mathbf{D}_{ab}^V)^* = \frac{ie}{m\omega} \mathbf{p}_{ab}^*$$

$$= \frac{ie}{m\omega} (2\pi)^{-3/2} \left(\frac{Z^3}{\pi a_0^3}\right)^{1/2} \left[-i\hbar \int e^{-Zr/a_0} \nabla e^{i\mathbf{k}_f \cdot \mathbf{r}} d\mathbf{r}\right]^*$$

$$= i\frac{e\hbar}{m\omega} (2\pi)^{-3/2} \left(\frac{Z^3}{\pi a_0^3}\right)^{1/2} \mathbf{k}_f \int e^{-Zr/a_0} e^{-i\mathbf{k}_f \cdot \mathbf{r}} d\mathbf{r}. \tag{11.158}$$

The integral on the right of this equation is proportional to the Fourier transform of $\exp(-Zr/a_0)$ and is given by

$$\int e^{-Zr/a_0} e^{-i\mathbf{k}_f \cdot \mathbf{r}} d\mathbf{r} = \frac{8\pi (Z/a_0)}{[(Z/a_0)^2 + k_f^2]^2}. \tag{11.159}$$

Substituting this result in (11.158), we obtain

$$\mathbf{D}_{ba}^V = i\frac{e\hbar}{m\omega} \frac{2\sqrt{2}}{\pi} \left(\frac{Z}{a_0}\right)^{5/2} \frac{\mathbf{k}_f}{[(Z/a_0)^2 + k_f^2]^2} \tag{11.160}$$

and it follows that

$$\sigma(\omega) = 32\alpha \frac{\hbar}{m\omega} \left(\frac{Z}{a_0}\right)^5 \frac{k_f^3}{[(Z/a_0)^2 + k_f^2]^4} \int \cos^2\theta d\Omega \tag{11.161}$$

where $\alpha = e^2/(4\pi\varepsilon_0 \hbar c)$ is the fine-structure constant.

In obtaining the above results we have assumed that the ejected electron energy E_f (and hence the photon energy $\hbar\omega$) is well in excess of the binding energy $|E_a|$ of the electron. Thus, since E_f (or $\hbar\omega$) $\gg |E_a|$, the conservation of energy equation (11.149) can be written to good approximation as

$$\frac{\hbar^2 k_f^2}{2m} \simeq \hbar\omega \tag{11.162}$$

and hence $k_f^2 \simeq 2m\omega/\hbar$. Moreover, since $E_a = -Z^2 e^2/[(4\pi\varepsilon_0)2a_0]$ and $a_0 = (4\pi\varepsilon_0)\hbar^2/(me^2)$, the condition $\hbar\omega \gg |E_a|$ implies that we can neglect $(Z/a_0)^2$ in comparison with k_f^2. As a result, the total cross-section (11.161) becomes, in the (non-relativistic) high-energy limit,

$$\sigma(\omega) = 32\alpha \frac{\hbar}{m\omega} \left(\frac{Z}{k_f a_0}\right)^5 \int \cos^2\theta d\Omega. \tag{11.163}$$

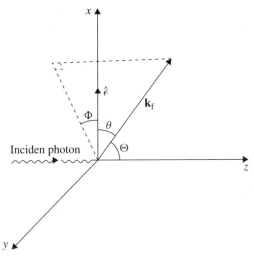

Figure 11.5 The angles employed in the discussion of the photoelectric effect. The incident radiation is in the z direction, with polarisation vector $\hat{\epsilon}$ in the x-direction. The momentum of the ejected electron is $\hbar \mathbf{k}_f$.

In order to study the angular distribution of the ejected electrons, a differential cross-section is defined as $d\sigma/d\Omega$, and we see from (11.163) that

$$\frac{d\sigma}{d\Omega} = 32\alpha \frac{\hbar}{m\omega} \left(\frac{Z}{k_f a_0}\right)^5 \cos^2 \theta. \tag{11.164}$$

Let us take the direction of the propagation of the radiation as the z-axis and the polarisation vector $\hat{\epsilon}$ to be along the x-direction (see Fig. 11.5). We see from (11.160) that $\mathbf{D}_{ba}^V$ is in the direction of $\mathbf{k}_f$, defined by the polar angles (Θ, Φ), so that

$$\cos^2 \theta = \sin^2 \Theta \cos^2 \Phi \tag{11.165}$$

and

$$\frac{d\sigma}{d\Omega} = 32\alpha \frac{\hbar}{m\omega} \left(\frac{Z}{k_f a_0}\right)^5 \sin^2 \Theta \cos^2 \Phi. \tag{11.166}$$

We remark that the ejected electrons have a cosine-squared distribution with respect to the polarisation vector $\hat{\epsilon}$ of the incident radiation. For an *unpolarised* photon beam an average must be made over all directions of the polarisation vector $\hat{\epsilon}$ in the xy plane, which is the same as averaging over the $\cos^2 \Phi$ factor and gives a factor of $1/2$ (Problem 11.10). Thus, in that case,

$$\frac{d\sigma}{d\Omega} = 16\alpha \frac{\hbar}{m\omega} \left(\frac{Z}{k_f a_0}\right)^5 \sin^2 \Theta. \tag{11.167}$$

We see that both differential cross-sections (11.166) and (11.167) exhibit a sine-squared distribution with respect to the angle Θ, which favours the ejection of electrons at right angles to the incident photon beam.

Upon integration of the differential cross-section (11.167) over the polar angles (Θ, Φ) of the ejected electron, we find that the total photoionisation cross-section for an unpolarised photon beam incident on an hydrogenic atom target is given at high (non-relativistic) energies by

$$\sigma(\omega) = \frac{128\pi}{3}\alpha\frac{\hbar}{m\omega}\left(\frac{Z}{k_f a_0}\right)^5$$
$$= \frac{16\pi\sqrt{2}}{3}\alpha\left(\frac{Z}{a_0}\right)^5\left(\frac{\hbar}{m\omega}\right)^{7/2}. \tag{11.168}$$

We note from this result that $\sigma(\omega)$ decreases like $\omega^{-7/2}$ when ω increases, and increases like Z^5 with increasing nuclear charge. The above formulae are applicable not only to one-electron atoms (provided that $\hbar\omega \gg |E_a|$) but also approximately to the ejection of electrons from the inner shells of atoms by X-rays, although the dipole approximation worsens with increasing photon energy.

11.8 Photodisintegration

The theory of Section 11.7 can be applied with little modification to find the cross-section for the photodisintegration of a deuteron. As we saw in Chapter 10, the deuteron is a bound state of a proton and a neutron. The nuclear force acting between the proton and the neutron is spin-dependent, producing a single bound triplet state with binding energy $E_B = 2.22$ MeV. This state is predominantly an S state with zero orbital angular momentum, although there is a small admixture of D state with orbital angular momentum $L = 2$. We shall ignore the D-state component in calculating the photodisintegration cross-section. As we explained in Chapter 10, the range of the nuclear force is short and is given approximately by $a \simeq 2 \times 10^{-15}$ m. Since the deuteron wave function $\psi(\mathbf{r})$ extends over a considerable distance outside the potential well we shall approximate it by its form outside the well given by (10.144)

$$\psi(\mathbf{r}) = C_2 \exp(-\lambda r)/r \tag{11.169}$$

where from (10.147) $C_2 = (\lambda/2\pi)^{1/2}$ with $\lambda^{-1} = 4.30 \times 10^{-15}$ m. After disintegration the final-state wave function can be approximated by a plane wave as in (11.142). From the discussion in the last section, we find, after averaging over the directions of the polarisation vector $\hat{\epsilon}$ and integrating over all angles of ejection that the electric dipole contribution to the photodisintegration cross-section is given for an unpolarised incident beam by

$$\sigma^E(\omega) = \frac{4\pi^2 \mu k_f \omega}{3\hbar^2 c \varepsilon_0}|\mathbf{D}_{ba}|^2 \tag{11.170}$$

where the electron mass m has been replaced by the neutron-proton reduced mass μ given by (10.134). Conservation of energy requires that

$$\frac{\hbar^2 k_f^2}{2\mu} = \hbar\omega - E_B. \tag{11.171}$$

The neutron is uncharged while the proton, which is at a position $\mathbf{r}/2$ relative to the centre of mass, possesses charge $+e$, so that the electric dipole moment of the deuteron is equal to $e\mathbf{r}/2$. Using the length form (11.94a) of the electric dipole matrix element $\mathbf{D}_{ba}$, we have

$$\mathbf{D}_{ba}^L = e(2\pi)^{-3/2}(\lambda/2\pi)^{1/2} \int e^{-i\mathbf{k}_f\cdot\mathbf{r}} \left(\frac{\mathbf{r}}{2}\right) r^{-1} e^{-\lambda r} d\mathbf{r}. \tag{11.172}$$

The x-component of $\mathbf{D}_{ba}^L$ contains the integral

$$J_x = \int e^{-i\mathbf{k}_f\cdot\mathbf{r}} x r^{-1} e^{-\lambda r} d\mathbf{r} \tag{11.173a}$$

$$= i\frac{\partial}{\partial k_{f,x}} \int e^{-i\mathbf{k}_f\cdot\mathbf{r}} r^{-1} e^{-\lambda r} d\mathbf{r}. \tag{11.173b}$$

where $k_{f,x}$ is the x-component of $\mathbf{k}_f$. The integral on the right of (11.173b) is proportional to the Fourier transform of the function $r^{-1}\exp(-\lambda r)$, and is given by

$$\int e^{-i\mathbf{k}_f\cdot\mathbf{r}} r^{-1} e^{-\lambda r} d\mathbf{r} = \frac{4\pi}{\lambda^2 + k_f^2}. \tag{11.174}$$

Using this result we find that

$$J_x = -i\frac{8\pi k_{f,x}}{(\lambda^2 + k_f^2)^2} \tag{11.175}$$

Hence

$$(\mathbf{D}_x^L)_{ba} = -\frac{ie}{\pi}\lambda^{1/2}\frac{k_{f,x}}{(\lambda^2 + k_f^2)^2}. \tag{11.176}$$

Treating the y- and z-components in the same way, we find that

$$\mathbf{D}_{ba}^L = -\frac{ie}{\pi}\lambda^{1/2}\frac{\mathbf{k}_f}{(\lambda^2 + k_f^2)^2}. \tag{11.177}$$

Finally, from (11.170), the electric dipole photodisintegration cross-section is

$$\sigma^E(\omega) = \frac{4\mu\omega e^2}{3\hbar^2 c\varepsilon_0}\frac{\lambda k_f^3}{(\lambda^2 + k_f^2)^4}. \tag{11.178}$$

The computed cross-section is illustrated in Fig. 11.6 as a function of the energy $E_\gamma = \hbar\omega$ of the incident photon. The cross-section increases at first like k_f^3 above the threshold at $E_\gamma = E_B$, rising to a maximum at an energy near $E_\gamma = 2E_B$. Although the electric dipole contribution dominates the cross-section near the cross-section

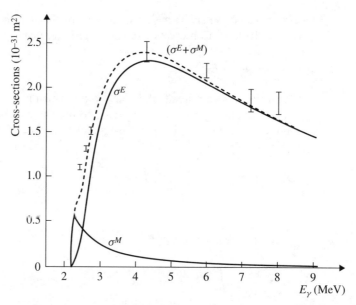

Figure 11.6 Cross-sections for the photodisintegration of the deuteron: σ^E is the electric dipole contribution, σ^M the magnetic dipole contribution, and the total cross-section is $(\sigma^E + \sigma^M)$. The experimental data is due to Bishop and Wilson.

maximum, the magnetic dipole contribution is more important near the threshold, where it is proportional to k_f. We shall discuss this contribution briefly.

Magnetic dipole transitions

We have seen in Chapter 1 that the interaction energy W between a magnetic moment $\mathcal{M}$ and a magnetic field $\mathcal{B}$ is given by $W = -\mathcal{M}.\mathcal{B}$ (see (1.86)). If the magnetic moment is placed at the origin, we have

$$W = -\mathcal{M}.\mathcal{B}(0) \tag{11.179}$$

where, using (11.8b),

$$\mathcal{B}(0) = -\mathcal{E}_0(\omega)c^{-1}\hat{\eta}\sin(\omega t - \delta_\omega). \tag{11.180}$$

In writing (11.180) we have introduced the unit vector $\hat{\eta} = \hat{\mathbf{k}} \times \hat{\boldsymbol{\epsilon}}$, which defines the direction of the magnetic field. The theory now follows, step by step, that developed above for the electric dipole interaction. In the final formulae for the cross-section, we need only replace the matrix element of the electric dipole moment $\mathbf{D}_{bu}$ by $(1/c)$ times the matrix element of the magnetic dipole moment $\mathcal{M}_{ba}$, and the direction of polarisation $\hat{\boldsymbol{\epsilon}}$ by the vector $\hat{\eta}$, which is at right angles to $\hat{\boldsymbol{\epsilon}}$.

The magnetic moment operator for a nuclear system is

$$\mathcal{M} = \mu_N \sum_i (\mathbf{L}_l + g_l \mathbf{S}_l)/\hbar \tag{11.181}$$

where $\mathbf{L}_i$ is the orbital angular momentum and $\mathbf{S}_i$ the spin angular momentum of the ith nucleon, and the sum is over all the nucleons. The constant μ_N is the nuclear magneton

$$\mu_N = e\hbar/2M_p = 5.05 \times 10^{-27} \text{ J T}^{-1} \tag{11.182}$$

and the quantities g_i are the spin gyromagnetic ratios. For a neutron (n) and a proton (p) they are given, respectively, by

$$g_n = -3.8263, \qquad g_p = 5.5883. \tag{11.183}$$

In order to obtain the magnetic contribution to the deuteron photodisintegration cross-section, the matrix element of the magnetic moment is required between the initial state a, which is the ground state of the deuteron, and the final (unbound) state b. Since, in our approximation, the deuteron ground state has zero orbital angular momentum, the terms in $\mathbf{L}_i$ in (11.181) do not contribute. The required matrix element is then

$$\mathcal{M}_{ba} = \mathbf{A} \int \psi_b^*(\mathbf{r}) \psi_a(\mathbf{r}) d\mathbf{r} \tag{11.184}$$

where $\mathbf{A}$ is defined as

$$\mathbf{A} = \mu_N \tfrac{1}{2} \langle \chi_b | g_n \boldsymbol{\sigma}_n + g_p \boldsymbol{\sigma}_p | \chi_a \rangle \tag{11.185}$$

Here χ_a and χ_b are the spin wave functions of the initial and final states and we have introduced the Pauli spin matrices $\boldsymbol{\sigma}_n$ (for the neutron) and $\boldsymbol{\sigma}_p$ (for the proton), with

$$\mathbf{S}_n = \frac{\hbar}{2} \boldsymbol{\sigma}_n, \qquad \mathbf{S}_p = \frac{\hbar}{2} \boldsymbol{\sigma}_p. \tag{11.186}$$

If χ_b were to be a triplet state ($S = 1$) the spatial wave function $\psi_b(\mathbf{r})$ would be orthogonal to the initial wave function $\psi_a(\mathbf{r})$ and the matrix element $\mathcal{M}_{ba}$ would vanish. It follows that the final state b must be a singlet state with $S = 0$.

The spin wave functions for the initial triplet states $\chi_a \equiv \chi_{1,M_s}$ are

$$\chi_{1,1} = \alpha(n)\alpha(p), \qquad\qquad M_s = 1 \tag{11.187a}$$

$$\chi_{1,0} = \frac{1}{\sqrt{2}}[\alpha(n)\beta(p) + \beta(n)\alpha(p)], \qquad M_s = 0 \tag{11.187b}$$

$$\chi_{1,-1} = \beta(n)\beta(p), \qquad\qquad M_s = -1. \tag{11.187c}$$

For the final singlet state the spin wave function is $\chi_b \equiv \chi_{0,0}$, where

$$\chi_{0,0} = \frac{1}{\sqrt{2}}[\alpha(n)\beta(p) - \beta(n)\alpha(p)]. \tag{11.188}$$

To evaluate the quantity $\mathbf{A}$ it is convenient to express the spin operators as

$$\tfrac{1}{2}\mu_N[g_n \boldsymbol{\sigma}_n + g_p \boldsymbol{\sigma}_p] = \tfrac{1}{4}\mu_N[(g_n + g_p)(\boldsymbol{\sigma}_n + \boldsymbol{\sigma}_p) + (g_n - g_p)(\boldsymbol{\sigma}_n - \boldsymbol{\sigma}_p)]. \tag{11.189}$$

Since the final state spin wave function is antisymmetric in n and p, while the initial-state spin wave function is symmetric, the first term in (11.189), which is also symmetric, must vanish and only the term $\mu_N(g_n - g_p)(\sigma_n - \sigma_p)/4$ contributes. We find that if $M_s = 1$.

$$A_x = -\frac{1}{2\sqrt{2}}\mu_N(g_n - g_p), \qquad A_y = -\frac{i}{2\sqrt{2}}\mu_N(g_n - g_p), \qquad A_z = 0. \tag{11.190a}$$

On the other hand, if $M_s = 0$, we have

$$A_x = A_y = 0, \qquad A_z = \tfrac{1}{2}\mu_N(g_n - g_p) \tag{11.190b}$$

and if $M_s = -1$, we obtain

$$A_x = -\frac{1}{2\sqrt{2}}\mu_N(g_n - g_p), \qquad A_y = \frac{i}{2\sqrt{2}}\mu_N(g_n - g_p), \qquad A_z = 0. \tag{11.190c}$$

The spatial part of the matrix element (11.184) is given in the plane wave approximation by

$$\int \psi_b^*(\mathbf{r})\psi_a(\mathbf{r})d\mathbf{r} = (2\pi)^{-3/2}(\lambda/2\pi)^{1/2}\int e^{-i\mathbf{k}_f \cdot \mathbf{r}} r^{-1} e^{-\lambda r} d\mathbf{r}$$

$$= \frac{1}{\pi}\frac{\lambda^{1/2}}{\lambda^2 + k_f^2}. \tag{11.191}$$

Thus, using (11.184), (11.190) and (11.191), we find that

$$|\mathcal{M}_{ba}|^2 = \frac{1}{4}\mu_N^2(g_n - g_p)^2 \frac{1}{\pi^2}\frac{\lambda}{(\lambda^2 + k_f^2)^2} \tag{11.192}$$

for each value of M_s.

The total cross-section for the magnetic dipole contribution to photodisintegration, $\sigma^M(\omega)$, can be obtained from (11.157) by replacing m by the reduced mass μ and $\mathbf{D}_{ba}$ by $\mathcal{M}_{ba}/c$:

$$\sigma^M(\omega) = \frac{\pi\mu k_f \omega}{\hbar^2 c^3 \varepsilon_0}\int \cos^2\theta |\mathcal{M}_{ba}|^2 d\Omega. \tag{11.193}$$

The angle θ is the angle between the vector $\hat{\boldsymbol{\eta}}$ and the direction of $\mathcal{M}_{ba}$ and the integration is over the polar angles of the vector $\mathbf{k}_f$. Since both $\cos\theta$ and $\mathcal{M}_{ba}$ are independent of the direction of $\mathbf{k}_f$ this integration gives a factor 4π. If the radiation is incident along the z-axis, $\hat{\boldsymbol{\eta}}$ lies in the xy plane. We see that for $M_s = \pm 1$, $\mathcal{M}_{ba}$ also lies in the xy plane and averaging over $\cos^2\theta$, for an unpolarised beam, we obtain a factor of $1/2$. For $M_s = 0$, $\mathcal{M}_{ba}$ is parallel to the z-axis and $\cos\theta = 0$. Thus

$$\sigma^M(\omega) = \frac{4\pi^2\mu k_f |\mathcal{M}_{ba}|^2 \omega}{\hbar^2 c^3 \varepsilon_0} \times \begin{cases} 1/2, & M_s = 1 \\ 0, & M_s = 0 \\ 1/2, & M_s = -1. \end{cases} \tag{11.194}$$

If the deuteron in the initial state is unpolarised, each of its sub-levels with $M_s = 1$, 0, and -1 will be equally populated. Averaging over the three initial states with $M_s = 1, 0, -1$, we obtain

$$\sigma^M(\omega) = \frac{4\pi^2 \mu k_f \omega}{3\hbar^2 c^3 \varepsilon_0} |\mathcal{M}_{ba}|^2$$

$$= \frac{\mu \mu_N^2 (g_n - g_p)^2 \omega}{3\hbar^2 c^3 \varepsilon_0} \frac{k_f \lambda}{[\lambda^2 + k_f^2]^2}. \tag{11.195}$$

We see that this cross-section is proportional to k_f near the threshold. The computed cross-section is shown in Fig. 11.6 as a function of $E_\gamma = \hbar\omega$. The total cross-section $(\sigma^E + \sigma^M)$ is also shown in Fig. 11.6 over an energy range extending from threshold to $E_\gamma = 9$ MeV, together with the experimental data. It is seen that the agreement between experiment and theory is very good.

The inverse process of radiative capture

$$n + p \rightarrow d + \gamma \tag{11.196}$$

is only important near the threshold and is therefore controlled by the magnetic dipole interaction. The cross-section can be obtained by applying the principle of detailed balancing to the results obtained above (see Problem 11.11).

Problems

11.1 Find how many photons are radiated per second from a monochromatic source, 1 watt in power, for the wavelengths: (a) 10 m (radio wave), (b) 10 cm (microwave), (c) 5890 Å (yellow sodium light) and (d) 1 Å (soft X-ray). At a distance of 10 m from the source, calculate the number of photons passing through unit area, normal to the direction of propagation, per unit time.

11.2 Prove that the form of the Schrödinger equation (11.17) is unchanged under the gauge transformation (11.19).

11.3 Prove that, within the dipole apprroximation, the term in $\mathbf{A}^2$ in the time-dependent Schrödinger equation (11.23) can be eliminated by performing the gauge transformation (11.26).

11.4 Prove that, in the dipole approximation, the time-dependent Schrödinger equation (11.23) can be written in the form (11.29) by performing the gauge transformation (11.28).

11.5 Show that if the radiation is unpolarised and isotropic, $\cos^2\theta$ in (11.56) can be replaced by its average value of $1/3$.

11.6 In Section 11.3 the Einstein A and B coefficients were obtained for transitions between two levels a and b which were assumed to be non-degenerate. If the level a is g_a times degenerate and the level b is g_b times degenerate, show that

$$g_a B_{ba} = g_b B_{ab}, \qquad A_{ab} = \frac{\hbar \omega_{ba}^3}{\pi^2 c^3} B_{ab}.$$

11.7 Calculate the rates of spontaneous and stimulated emission of radiation arising from the 2p→1s transition in atomic hydrogen (the Lyman α line) contained in a large cavity at 2000 K. Repeat the calculation for a one-electron ion with nuclear charge Ze, and study the Z behaviour of the result.

11.8 Prove that the oscillator strengths f_{ka} defined by (11.111) obey the sum rule

$$\sum_k f_{ka} = 1$$

where the sum is over all levels including the continuum.

11.9 For a given initial level, a, of a hydrogen atom prove that

$$\sum_b \sigma_{ba} - \sum_b{}' \bar\sigma_{ba} = 2\pi^2 r_0 c$$

where σ_{ba} is the absorption cross-section and the sum $\sum_b$ is over all states with $E_b > E_a$, while $\bar\sigma_{ba}$ is the cross-section for stimulated emission, and the sum $\sum_b{}'$ is over all states with $E_b < E_a$. On the right-hand side r_0 is the classical electron radius $r_0 = e^2/(4\pi \varepsilon_0 mc^2)$.

11.10 Starting from (11.166) carry out the average over all directions of polarisation to obtain the result (11.167).

11.11 Use the principle of detailed balancing to relate the cross-section for photodisintegration of the deuteron by the magnetic dipole interaction to that for radiative capture of neutrons by protons.

12 The interaction of quantum systems with external electric and magnetic fields

12.1 The Stark effect 557
12.2 Interaction of particles with magnetic fields 563
12.3 One-electron atoms in external magnetic fields 574
12.4 Magnetic resonance 576
　　　 Problems 585

In the previous chapter, we showed how quantum mechanics can be used to calculate transition rates for atomic and nuclear systems acted on by external electromagnetic radiation. A related set of problems arises when the external fields are electric fields or magnetic fields and these are the topics discussed in this chapter.

12.1 The Stark effect

We shall start by considering how static electric fields affect the energy levels of atoms. The corresponding change in frequency of the associated spectral lines is called the *Stark effect*. We shall concentrate on the case of one-electron atoms.

Let us consider a one-electron atom in a time-independent electric field which is uniform over a region of atomic dimensions. The atomic nucleus will be taken to have a charge Ze and to be of infinite mass compared with the mass of the electron. If the field is of strength $\mathcal{E}$ and is directed parallel to the z-axis, the potential energy of an electron of charge $-e$ situated in the field is

$$W = e\mathcal{E}z = -\mathcal{E}D_z \qquad (12.1)$$

where D_z is the z-component of the electric dipole moment $\mathbf{D} = -e\mathbf{r}$. The only difference between (12.1) and the interaction between a radiation field and a one-electron atom given approximately by (11.35) is that we are now considering the special case in which the electric field vector is directed along the z-axis and is independent of time. The identity of (11.35) and (12.1) is not surprising because in obtaining the interaction with a radiation field we neglected the magnetic part of the interaction arising from the second term in (11.31).

The interaction energy, W, given by (12.1), must be added to the potential energy due to the Coulomb attraction between the electron and the nucleus contained in the Hamiltonian of the one-electron atom. The effect of this additional interaction energy on the atomic energy levels can be investigated using time-independent perturbation theory, setting $H' = W = e\mathcal{E}z$. Thus the zero-order wave functions are the hydrogenic wave functions ψ_{nlm} (see (7.134) and Table 7.1) and the corresponding zero-order energies are the hydrogenic energy levels E_n given by (7.114) with $\mu = m$. We shall first assume that $\mathcal{E}$ is large enough for fine-structure effects (see Section 8.2) to be unimportant, which is correct for field strengths usually encountered, which are of the order 10^7 V m^{-1}.

The ground state

Since the ground state $\psi_{1s}(r) \equiv \psi_{100}(r)$ is non-degenerate, it follows from (8.15) that the first-order correction to the energy is

$$E_{100}^{(1)} = e\mathcal{E}\langle\psi_{100}|z|\psi_{100}\rangle$$

$$= e\mathcal{E}\int |\psi_{100}(r)|^2 z \, d\mathbf{r} \tag{12.2}$$

where the superscripts (0) have been omitted. Consider now the integrals

$$I_{nlm} = \int |\psi_{nlm}(\mathbf{r})|^2 z \, d\mathbf{r}. \tag{12.3}$$

Under the interchange $\mathbf{r} \to -\mathbf{r}$, we have

$$I_{nlm} = \int |\psi_{nlm}(-\mathbf{r})|^2 (-z) d\mathbf{r}. \tag{12.4}$$

However, the hydrogenic wave functions ψ_{nlm} have the parity $(-1)^l$

$$\psi_{nlm}(-\mathbf{r}) = (-1)^l \psi_{nlm}(\mathbf{r}) \tag{12.5}$$

from which we see that $|\psi_{nlm}(-\mathbf{r})|^2 = |\psi_{nlm}(\mathbf{r})|^2$. Hence, comparing (12.3) and (12.4), we see that $I_{nlm} = -I_{nlm}$ so that $I_{nlm} = 0$. In particular, the first-order energy shift given by (12.2) vanishes.

The second-order term of the perturbation series is (see (8.28))

$$E_{100}^{(2)} = e^2\mathcal{E}^2 \sum_{n \neq 1} \sum_{l,m} \frac{|\langle\psi_{nlm}|z|\psi_{100}\rangle|^2}{E_1 - E_n} \tag{12.6}$$

where the sum implies a summation over all the discrete set of hydrogenic eigenstates together with an integration over the continuum set. Since the energy differences $(E_1 - E_n)$ with $n \geq 2$ are always negative, and $E_{100}^{(1)} = 0$, the ground-state energy is always lowered by the interaction with the electric field. The *quadratic* Stark effect (12.6) is very small, being of the order of -2.5×10^{-6} eV for atomic hydrogen in a field of strength $\mathcal{E} = 10^8$ V m^{-1} (see Problem 12.1) and is usually unimportant.

The excited states

Let us now examine the first-order Stark effect for the first excited level of a hydrogenic atom. Since we are neglecting fine-structure effects, the $n = 2$ level is four-fold degenerate because each of the four eigenfunctions

$$\psi_{200}, \psi_{210}, \psi_{211}, \psi_{21-1} \tag{12.7}$$

corresponds to the same energy $E_{n=2}$. It follows that we now have to use degenerate perturbation theory, and in principle we have to solve a system of four homogeneous linear equations (see (8.71)). However $(-ez)$ is the z-component of the electric dipole moment operator and we found in Section 11.4 that matrix elements of the form $\langle n'l'm'|z|nlm \rangle$ vanish unless $l' = l \pm 1$ and $m' = m$. The only non-vanishing matrix elements of the perturbation $H' = e\mathcal{E}z$ between the four states (12.7) are therefore those connecting the 2s (200) and 2p$_0$ (210) states. It follows that the linear homogeneous equations (8.71) reduce to a set of two equations, which we can write as (see (8.77))

$$-E_r^{(1)} c_{r1} + H'_{12} c_{r2} = 0$$
$$H'_{21} c_{r1} - E_r^{(1)} c_{r2} = 0 \tag{12.8}$$

where $r = 1, 2$ and the (real) matrix element H'_{12} is given by:

$$H'_{12} = H'_{21} = e\mathcal{E} \int \psi_{210}(\mathbf{r}) \, z \, \psi_{200}(r) \mathrm{d}\mathbf{r}. \tag{12.9}$$

The condition that a non-zero solution of equations (12.8) exists is that the determinantal equation

$$\begin{vmatrix} -E_r^{(1)} & H'_{12} \\ H'_{21} & -E_r^{(1)} \end{vmatrix} = 0 \tag{12.10}$$

should be satisfied. The two roots of this equation are

$$E_r^{(1)} = \pm |H'_{12}|. \tag{12.11}$$

To evaluate the energy shifts $E_r^{(1)} = \pm|H'_{12}|$, we use the hydrogenic 2s and 2p$_0$ wave functions given in Chapter 7. Since $z = r \cos \theta$, we have

$$H'_{12} = e\mathcal{E} \frac{Z^3}{16\pi a_0^3} \int_0^\infty \mathrm{d}r \, r^3 \left(\frac{Zr}{a_0}\right) \left(1 - \frac{Zr}{2a_0}\right) e^{-Zr/a_0} \int_0^\pi \mathrm{d}\theta \sin\theta \cos^2\theta \int_0^{2\pi} \mathrm{d}\phi$$

$$= e\mathcal{E} \frac{Z^3}{8a_0^3} \frac{2}{3} \int_0^\infty \mathrm{d}r \, r^3 \left(\frac{Zr}{a_0}\right) \left(1 - \frac{Zr}{2a_0}\right) e^{-Zr/a_0}$$

$$= -3e\mathcal{E}a_0/Z. \tag{12.12}$$

The change in energy is proportional to $\mathcal{E}$ so that the effect is called the *linear Stark effect*. The linear Stark effect is non-zero only for degenerate states and is exhibited by the excited states of hydrogenic atoms. In other atoms, both the ground and excited states are usually non-degenerate with respect to the orbital angular

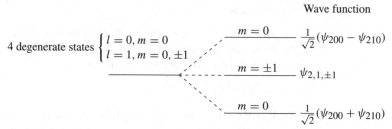

Figure 12.1 Splitting of the degenerate $n=2$ level of atomic hydrogen due to the linear Stark effect.

momentum quantum number so that only the much smaller quadratic Stark effect occurs.

Upon returning to (12.8), we see that for the upper root $E_1^{(1)} = |H'_{12}|$ one has $c_{11} = -c_{12}$. The corresponding normalised wave function is given by (see (8.81) and (8.86))

$$\chi_1 = \frac{1}{\sqrt{2}}(\psi_{200} - \psi_{210}). \tag{12.13}$$

The second root $E_2^{(1)} = -|H'_{12}|$ yields $c_{21} = c_{22}$ and the normalised wave function is

$$\chi_2 = \frac{1}{\sqrt{2}}(\psi_{200} + \psi_{210}). \tag{12.14}$$

It should be emphasised that the wave functions (12.13) and (12.14) are neither eigenfunctions of parity nor of $\mathbf{L}^2$, and neither parity nor l is a 'good' quantum number in this case. On the other hand, m is a good quantum number because H' (and hence H) commutes with L_z and the system is invariant under rotations about the z-axis.

The splitting of the $n=2$ level of a hydrogenic atom due to the linear Stark effect is illustrated in Fig. 12.1. The degeneracy is partly removed by the perturbation which picks out a particular direction of space so that the rotational invariance of the system is destroyed. However, the energies of the $2p_{\pm 1}$ (i.e. 211 and 21 − 1) states remain unaltered. Thus the level splits symmetrically into three sub-levels, one of which (corresponding to $m = \pm 1$) is two-fold degenerate.

The splitting of higher levels can be treated in a similar way. For example one can show that the $n=3$ level is split into five equally spaced levels as shown in Fig. 12.2. Also shown in this figure are the various possible electric dipole transitions between the levels $n=2$ and $n=3$ in the presence of an electric field. The selection rules for the magnetic quantum number m are $\Delta m = 0, \pm 1$. The $\Delta m = 0$ transitions are said to correspond to π components and the $\Delta m = \pm 1$ to σ components.

Up to this point we have assumed that the external field $\mathcal{E}$ is large enough for fine-structure effects to be neglected. We now consider the opposite case where $\mathcal{E}$ is weak. The unperturbed states can no longer be treated as degenerate, but some simplification occurs because the splitting due to the fine-structure effects is very

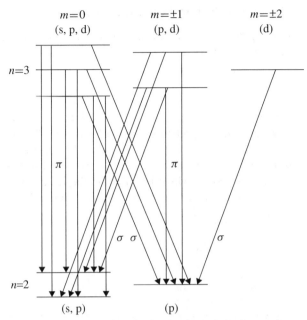

Figure 12.2 Splitting of the $n = 2$ and $n = 3$ levels of atomic hydrogen due to the linear Stark effect. The various possible electric dipole transitions are shown; those with $\Delta m = 0$ are called π lines and those with $\Delta m = \pm 1$, σ lines.

small and the levels are quasi-degenerate. To take a simple example, consider the Stark perturbation of the $2s_{1/2}$ and $2p_{3/2}$ levels of atomic hydrogen which are split by $\Delta E \simeq 4.5 \times 10^{-5}$ eV by the fine-structure interactions. Let us denote these two levels by the labels 1 and 2. Following the discussion of quasi-degenerate states at the end of Section 8.2 and noting that $H'_{11} = H'_{22} = 0$ in the present case, we see from (8.129) that

$$E_r = E^{(0)} \pm \left\{ \frac{(\Delta E)^2}{4} + |H'_{12}|^2 \right\}^{1/2} \qquad r = 1, 2 \qquad (12.15)$$

where $E^{(0)} = E_{n=2}$, $\lambda = 1$ and

$$H'_{12} = e\mathcal{E}\langle 2s_{1/2}|z|2p_{3/2}\rangle. \qquad (12.16)$$

The splitting between the $2s_{1/2}$ and $2p_{3/2}$ levels before the external field is applied is ΔE. When the fine structure is ignored and $\Delta E = 0$ the energy shift $(E_r - E^{(0)})$ depends *linearly* on $\mathcal{E}$. However if $(\Delta E)^2$ is not very small compared with $|H'_{12}|^2$, $(E_r - E^{(0)})$ is a function of $\mathcal{E}^2$ given by $\pm[(\Delta E)^2/4 + e^2|\langle 2s_{1/2}|z|2p_{3/2}\rangle|^2\mathcal{E}^2]^{1/2}$.

The Stark effect and the rigid rotator

In Chapter 6, we saw that the rotational motion of a diatomic molecule could be described by the Hamiltonian

$$H = \frac{\mathbf{L}^2}{2I} \tag{12.17}$$

where $\mathbf{L}$ is the angular momentum operator and I is the moment of inertia of the molecule with respect to the axis of rotation. The orientation of the molecule is described by the angles (θ, ϕ) which are the polar angles of the internuclear line AB of Fig. 6.4. Many diatomic molecules possess a permanent electric dipole moment, for example NaCl, which is mainly in the ionic configuration $Na^+ - Cl^-$. If such a molecule possesses an electric dipole moment $\mathbf{D}$, then in an electric field $\mathcal{E}$, the Hamiltonian (12.17) becomes

$$\begin{aligned} H &= \frac{\mathbf{L}^2}{2I} - \mathbf{D}.\mathcal{E} \\ &= \frac{\mathbf{L}^2}{2I} - D\mathcal{E}\cos\theta \end{aligned} \tag{12.18}$$

where in the second line, we have taken $\mathcal{E}$ parallel to the z-axis. To examine the effect of the additional interaction

$$H' = -D\mathcal{E}\cos\theta \tag{12.19}$$

on the energy levels, time-independent perturbation theory can be used.

The unperturbed energy levels are

$$E_l^{(0)} = \frac{l(l+1)\hbar^2}{2I}; \qquad l = 0, 1, 2\ldots \tag{12.20}$$

corresponding to the zero-order wave functions

$$\psi_{lm}^{(0)}(\theta, \phi) = Y_{lm}(\theta, \phi). \tag{12.21}$$

It is seen that the energy levels (12.20) are degenerate with respect to m. However, since both H_0 and H commute with L_z, non-degenerate perturbation theory is applicable.

The first-order correction to $E_l^{(0)}$ is

$$E_{lm}^{(1)} = \int_0^\pi \sin\theta \, d\theta \int_0^{2\pi} d\phi \, Y_{lm}^*(\theta, \phi) H' Y_{lm}(\theta, \phi) \equiv \langle l, m|H'|l, m\rangle \tag{12.22}$$

and this vanishes because H' is of odd parity. The second-order correction is

$$E_{lm}^{(2)} = \sum_{l' \neq l} \sum_{m'=-l'}^{+l'} \frac{|\langle l, m|H'|l', m'\rangle|^2}{E_l - E_{l'}}. \tag{12.23}$$

The dipole matrix elements $\langle l, m|H'|l', m'\rangle$ vanish except for $m' = m$ and $l' = l \pm 1$ (see Section 11.4) giving

$$E_{lm}^{(2)} = \frac{D^2 \mathcal{E}^2 I}{\hbar^2} \left\{ -\frac{|\langle l, m| \cos\theta |l+1, m\rangle|^2}{l+1} + \frac{|\langle l, m| \cos\theta |l-1, m\rangle|^2}{l} \right\}. \quad (12.24)$$

Note that for $l = 0$, we must have $l' = 1$, and the second term in brackets does not appear. On evaluating the matrix elements, the general expression for $E_l^{(2)}$ is (see Problem 12.3)

$$E_{lm}^{(2)} = \frac{D^2 \mathcal{E}^2 I}{\hbar^2} \frac{l(l+1) - 3m^2}{l(l+1)(2l-1)(2l+3)}, \quad l \neq 0 \quad (12.25a)$$

$$= -\frac{D^2 \mathcal{E}^2 I}{3\hbar^2}, \quad l = 0. \quad (12.25b)$$

As seen from (12.25a), the degeneracy is only partly removed. Indeed, for $l \neq 0$ the corrected energy levels $E_l^{(0)} + E_{lm}^{(2)}$ depend only on l and m^2, so that states differing by the sign of m still correspond to the same energy.

12.2 Interaction of particles with magnetic fields

In Chapter 1, we saw that the Stern–Gerlach experiment demonstrates that electrons possess an intrinsic magnetic moment $\mathcal{M}_s$ which is proportional to the spin angular momentum **S**. In terms of the Bohr magneton μ_B where $\mu_B = e\hbar/2m$, m being the mass of the electron, we have (see (1.93))

$$\mathcal{M}_s = -g_s \mu_B \mathbf{S}/\hbar = -g_s \mu_B \boldsymbol{\sigma}/2 \quad (12.26)$$

where g_s is the electron spin *gyromagnetic ratio* which has the approximate value[1] $g_s = 2$ and $(\sigma_x, \sigma_y, \sigma_z)$ are the Pauli spin matrices. Similarly, other particles possessing non-zero spin have an intrinsic magnetic moment $\mathcal{M}_s$ proportional to their spin **S**. For example, in the case of a proton

$$\mathcal{M}_s = g_p \mu_N \boldsymbol{\sigma}/2 \quad (12.27)$$

where $g_p = 5.5883$ is the gyromagnetic ratio (or g-factor) of the proton p and μ_N is the nuclear magneton defined by (see (11.182))

$$\mu_N = \frac{e\hbar}{2M_p}, \quad (12.28)$$

---

[1] The value $g_s = 2$ is that predicted by Dirac's relativistic theory of the electron (see Section 15.6), and is in good agreement with experiment. The corrections to the Dirac result $g_s = 2$ come from quantum electrodynamics.

M_p being the mass of the proton. The magnetic moment of the proton is parallel to its spin. In contrast, for a neutron (as for an electron) the magnetic moment is anti-parallel to the spin and

$$\mathcal{M}_\mathrm{s} = g_\mathrm{n}\mu_\mathrm{N}\sigma/2 \tag{12.29}$$

where $g_\mathrm{n} = -3.8263$ is the neutron g-factor.

Weak fields

In Section 1.5, the energy of a system having a magnetic moment $\mathcal{M}$ interacting with a magnetic field $\mathcal{B}$ was taken as

$$W = -\mathcal{M}.\mathcal{B}. \tag{12.30}$$

For weak magnetic fields this is a good approximation, but it is not accurate for strong fields and we shall later obtain the exact (but still non-relativistic) form of the Schrödinger equation in that case.

Particle of spin one-half in a uniform magnetic field

Let us now consider the interaction of a particle of spin one-half with a magnetic field, taking the case for which there is no other interaction and the particle is at rest. Then the interaction energy with the magnetic field (and hence the Hamiltonian) is given by

$$H = -\mathcal{M}_\mathrm{s}.\mathcal{B} = g\mu_\mathrm{B}\mathcal{B}.\sigma/2 = \gamma\mathcal{B}.\sigma \tag{12.31}$$

where $g = g_\mathrm{s} \simeq 2$ for an electron, $g = -g_\mathrm{p}$ for a proton, $g = -g_\mathrm{n}$ for a neutron, and $\gamma = g\mu_\mathrm{B}/2$. Assuming that the field is constant and orientated in the z-direction, $\mathcal{B} = \mathcal{B}\hat{z}$, the Hamiltonian is

$$H = g\mu_\mathrm{B}\mathcal{B}\sigma_z/2 = \gamma\mathcal{B}\sigma_z \tag{12.32}$$

and the time-independent Schrödinger equation reads

$$\gamma\mathcal{B}\sigma_z\chi = E\chi. \tag{12.33}$$

Since the eigenvalues of $S_z = (\hbar/2)\sigma_z$ are $m_s\hbar$, with $m_s = \pm 1/2$, the energy eigenvalues are

$$E_{m_s} = 2\gamma\mathcal{B}m_s, \qquad m_s = \pm 1/2 \tag{12.34}$$

the corresponding eigenvectors being the basic spin-1/2 functions α and β, given by (6.230).

The time-dependent Schrödinger equation

$$i\hbar\frac{\partial\Psi}{\partial t} = H\Psi \tag{12.35}$$

therefore admits the two stationary state solutions $\Psi_{m_s}(t)$, namely

$$\Psi_{1/2}(t) = \alpha e^{-i\omega t/2}, \qquad \Psi_{-1/2}(t) = \beta e^{i\omega t/2} \tag{12.36}$$

where $\omega = 2\gamma \mathcal{B}/\hbar$. The general solution of (12.35) is therefore

$$\Psi(t) = c_1 \alpha e^{-i\omega t/2} + c_2 \beta e^{i\omega t/2}$$

$$= \begin{pmatrix} c_1 e^{-i\omega t/2} \\ c_2 e^{i\omega t/2} \end{pmatrix} \tag{12.37}$$

where c_1 and c_2 are constants. If $\Psi(t)$ is normalised to unity,

$$|c_1|^2 + |c_2|^2 = 1. \tag{12.38}$$

Let us consider some particular cases corresponding to different initial conditions at $t = 0$.

(1) If the system is in an eigenstate of S_z at $t = 0$, corresponding to the eigenvalue $+\hbar/2$, then $c_1 = 1$ and $c_2 = 0$, and the system remains in that eigenstate for all t. In the same way, if at $t = 0$ the system is in an eigenstate of S_z corresponding to the eigenvalue $-\hbar/2$, then $c_1 = 0$, $c_2 = 1$ and the system remains in that eigenstate for all t.

(2) A different situation arises if the system is in an eigenstate of S_x at $t = 0$. Using the representation (6.228), it is a simple matter to verify that the normalised eigenvectors of S_x, corresponding respectively to the eigenvalues $+\hbar/2$ and $-\hbar/2$, are given by

$$\frac{1}{\sqrt{2}}(\alpha + \beta) = \frac{1}{\sqrt{2}}\begin{pmatrix} 1 \\ 1 \end{pmatrix} \quad \text{and} \quad \frac{1}{\sqrt{2}}(\alpha - \beta) = \frac{1}{\sqrt{2}}\begin{pmatrix} 1 \\ -1 \end{pmatrix}. \tag{12.39}$$

As a result, the wave function (12.37) is an eigenvector of S_x at $t = 0$ corresponding to the eigenvalue $\hbar/2$ provided $c_1 = c_2$. To satisfy the normalisation condition (12.38) we can set $c_1 = c_2 = 2^{-1/2}$, so that the wave function (12.37) becomes

$$\Psi(t) = \frac{1}{\sqrt{2}}(\alpha e^{-i\omega t/2} + \beta e^{i\omega t/2}). \tag{12.40}$$

We see that the system will again be in an eigenstate of S_x belonging to the eigenvalue $\hbar/2$ at subsequent times $t = 2n\pi/\omega$, where $n = 1, 2, 3, \ldots$, and it will be in an eigenstate of S_x corresponding to the eigenvalue $-\hbar/2$ at times $t = (2n - 1)\pi/\omega$. The expectation value of S_x in the state represented by the wave function (12.40) is

$$\langle S_x \rangle = \frac{1}{2}(\alpha e^{-i\omega t/2} + \beta e^{i\omega t/2})^\dagger S_x (\alpha e^{-i\omega t/2} + \beta e^{i\omega t/2})$$

$$= \frac{\hbar}{4}(\alpha^\dagger e^{i\omega t/2} + \beta^\dagger e^{-i\omega t/2})(\beta e^{-i\omega t/2} + \alpha e^{i\omega t/2})$$

$$= \frac{\hbar}{2}\cos \omega t \tag{12.41}$$

where we have used (6.214) and (6.231).

Since the normalised eigenvectors of S_y, corresponding to the eigenvalues $+\hbar/2$ and $-\hbar/2$, are given respectively by

$$\frac{1}{\sqrt{2}}(\alpha + i\beta) = \frac{1}{\sqrt{2}}\begin{pmatrix}1\\i\end{pmatrix} \quad \text{and} \quad \frac{1}{\sqrt{2}}(\alpha - i\beta) = \frac{1}{\sqrt{2}}\begin{pmatrix}1\\-i\end{pmatrix} \quad (12.42)$$

the wave function (12.40) will be an eigenfunction of S_y with an eigenvalue $+\hbar/2$ at times $t = (4n+1)\pi/2\omega$ and with an eigenvalue $-\hbar/2$ at times $(4n+3)\pi/2\omega$, with $n = 0, 1, 2, \ldots$. The expectation value of S_y in the state (12.40) is, using (6.214) and (6.231),

$$\langle S_y \rangle = \frac{1}{2}(\alpha e^{-i\omega t/2} + \beta e^{i\omega t/2})^{\dagger} S_y (\alpha e^{-i\omega t/2} + \beta e^{i\omega t/2})$$

$$= \frac{\hbar}{4}(\alpha^{\dagger} e^{i\omega t/2} + \beta^{\dagger} e^{-i\omega t/2})(i\beta e^{-i\omega t/2} - i\alpha e^{i\omega t/2})$$

$$= \frac{\hbar}{2}\sin \omega t. \quad (12.43)$$

It is also readily verified that $\langle S_z \rangle = 0$, so that the expectation value of the spin $\mathbf{S}$ in the state (12.40) precesses about the field direction (the z-axis) in the xy plane with an angular frequency $\omega = 2\gamma \mathcal{B}/\hbar$. In that plane, the component of the spin in the direction which makes an angle ϕ with the x-axis is

$$S_\phi = S_x \cos\phi + S_y \sin\phi. \quad (12.44)$$

Using the explicit representation (6.228) of S_x and S_y, it is easy to verify that the wave function (12.40) is an eigenfunction of S_ϕ belonging to the eigenvalue $\hbar/2$ when $\phi = \omega t$. We can say loosely (in the average sense discussed in Section 6.8) that the spin 'points' in the direction ϕ when $t = \phi/\omega$.

Particle of spin one-half in a rotating magnetic field and the Berry phase

If the magnetic field entering into the Hamiltonian (12.31) is pointing in a direction $\hat{\mathbf{n}}$ specified by polar coordinates (θ, ϕ) there are two stationary solutions to the Schrödinger equation (12.35),

$$\Psi_\uparrow(t) = \chi_\uparrow \exp(-i\omega t/2), \qquad \Psi_\downarrow(t) = \chi_\downarrow \exp(i\omega t/2). \quad (12.45)$$

The spinors $\chi_\uparrow$ and $\chi_\downarrow$ are the eigenfunctions of the component of the spin S_n in the direction (θ, ϕ) corresponding to eigenvalues $\hbar/2$ and $-\hbar/2$ respectively, which have been given in (6.252) and (6.255). The angular frequency ω is, as before, given by $\omega = 2\gamma \mathcal{B}/\hbar$. Now let us suppose that the magnitude of $\mathcal{B}$ is constant, but the direction of $\mathcal{B}$ rotates slowly making a complete rotation in an interval T so that $\hat{\mathbf{n}} = (\theta, \phi = 0)$ at time $t_0 = 0$ and $\hat{\mathbf{n}} = (\theta, \phi = 2\pi)$ at time $t_f = T$. If $T \gg 2\pi/\omega$ the adiabatic theorem applies (see Section 9.4) and if the system is in the state $\Psi_\uparrow(t)$ at $t = 0$ it will remain in this state as the magnetic field changes direction and not make a transition to $\Psi_\downarrow(t)$. The Berry phase acquired by the wave function during the complete rotation of the field can be calculated from (9.132). The eigenfunction

$\chi_{\uparrow}(\theta, \phi)$ depends on t through ϕ which can be treated as a slowly varying parameter so that the Berry phase $\bar{\gamma}_{\uparrow} \equiv \bar{\gamma}_a$ is given by

$$\bar{\gamma}_{\uparrow} = i \int_0^T \left\langle \chi_{\uparrow}(\theta, \phi(t')) \middle| \frac{\partial}{\partial t'} \chi_{\uparrow}(\theta, \phi(t')) \right\rangle dt'$$

$$= i \int_0^{2\pi} \left\langle \chi_{\uparrow}(\theta, \phi) \middle| \frac{\partial}{\partial \phi} \chi_{\uparrow}(\theta, \phi) \right\rangle d\phi$$

$$= -\pi(1 - \cos\theta) \tag{12.46}$$

where in the last line (6.252) has been employed.

The phase $\bar{\gamma}_{\uparrow}$ is an observable and can be measured. A good example is provided by an experiment carried out in 1987 by J. Bitter and D. Dibbers based on the result (12.46). In their experiment a spin-polarised beam of neutrons of intensity I_0 is split into two components. One component is passed through a constant magnetic field which interacts with the magnetic moments of the neutrons so that the neutron spins precess about the direction of the field with angular frequency ω. The second component is subject to a magnetic field of the same magnitude, but the direction of which is rotated very slowly compared with the precession rate. After the interval T the magnetic field acting on the second component returns to its original direction. The two components are then recombined. Because of the phase difference $\bar{\gamma}_{\uparrow}$ they interfere and the observed intensity I as a function of the angle θ is

$$I(\theta) = \frac{I_0}{4} \left| 1 + \exp(i\bar{\gamma}_{\uparrow}) \right|^2$$

$$= I_0 \cos^2 \left[\frac{\pi}{2}(1 - \cos\theta) \right] \tag{12.47}$$

The predicted result was obtained, confirming that the Berry phase of the wave function can have observable effects for cyclic systems.

The Aharonov–Bohm effect

In classical physics only the field vectors $\mathcal{E}$ and $\mathcal{B}$ have a direct physical significance; neither the fields nor the equations of motion are altered by a gauge transformation. In contrast the potentials $\mathbf{A}$ and ϕ can be changed at will by a gauge transformation and are not directly measurable. In quantum mechanics the situation is different: the vector potential $\mathbf{A}$ has a different status and is of direct physical relevance. Let us first consider a particle of mass m and charge q confined to a box by some scalar potential $V(\mathbf{r})$. We shall further suppose that no magnetic or electric fields intersect the box. Since in the region in which the particle moves $\mathcal{B} = 0$ we have from (11.2 b) that

$$\nabla \times \mathbf{A}(\mathbf{r}, t) = 0 \tag{12.48}$$

which is satisfied either by taking $\mathbf{A} = 0$ or by setting

$$\mathbf{A}(\mathbf{r}, t) = \nabla \chi(\mathbf{r}, t) \tag{12.49}$$

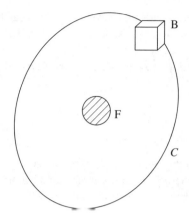

Figure 12.3 A box B containing a particle is moved along a path C enclosing the shaded area A which contains the magnetic flux Φ.

where χ is an arbitrary scalar function. If $\mathbf{A}$ and V are time-independent the Schrödinger energy eigenvalue equation in the first case ($\mathbf{A} = 0$) is

$$\left[-\frac{\hbar^2}{2m}\nabla^2 + V(\mathbf{r})\right]\psi(\mathbf{r}) = E\psi(\mathbf{r}) \tag{12.50}$$

and in the second case it is

$$\left[-\frac{\hbar^2}{2m}\nabla^2 + i\hbar\frac{q}{2m}(\mathbf{A}\cdot\nabla + \nabla\cdot\mathbf{A}) + \frac{q^2}{2m}\mathbf{A}^2 + V(\mathbf{r})\right]\psi'(\mathbf{r}) = E\psi'(\mathbf{r}), \tag{12.51}$$

where $\mathbf{A}$ is given by (12.49).

The functions ψ and ψ' are connected by the gauge (phase) transformation

$$\psi'(\mathbf{r}) = \psi(\mathbf{r})\exp[iq\chi(\mathbf{r})/\hbar] \tag{12.52}$$

where from (12.49)

$$\chi(\mathbf{r}) = \int_{\mathbf{r}_0}^{\mathbf{r}} \mathbf{A}(\mathbf{r}')\cdot d\mathbf{r}'. \tag{12.53}$$

The integral in (12.53) is a line integral from some fixed point $\mathbf{r}_0$ to $\mathbf{r}$. For $\chi(\mathbf{r})$ to exist and to depend only on $\mathbf{r}$ the integral (12.53) must be independent of the path of integration from $\mathbf{r}_0$ to $\mathbf{r}$. This is correct (so that $\mathbf{A}(\mathbf{r})$ can be removed from the problem by a gauge transformation) only if the region in which the particle moves is not multiply connected. To show what can happen in the case of a multiply connected region let us suppose the box containing the particle is moved along a path C (see Fig. 12.3) which encloses an area in which the magnetic flux is not zero in such a way that the magnetic field $\mathcal{B}$ does not intersect the box. The phase change in the wave function on making a complete circuit back to the original position is

$$\delta = \frac{i}{\hbar}q\oint \mathbf{A}(\mathbf{r})\cdot d\mathbf{r} \tag{12.54}$$

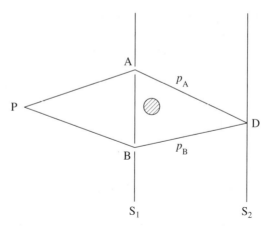

Figure 12.4 An Aharonov–Bohm experiment. Charged particles form a source P impinge on a screen S_1 with two slits A and B and are detected at D on the second screen S_2. The shaded area represents a thin solenoid which is perpendicular to the plane of the paper.

and by Stokes's theorem

$$\delta = \frac{i}{\hbar} q \int_S [\boldsymbol{\nabla} \times \mathbf{A}(\mathbf{r})] \cdot d\mathbf{S}$$

$$= \frac{i}{\hbar} q \int_S \boldsymbol{\mathcal{B}} \cdot d\mathbf{S} = \frac{i}{\hbar} q \Phi \qquad (12.55)$$

where the integral is over the surface S enclosed by the path C and where Φ is the magnetic flux through S. This is an example of a geometrical or Berry phase (see Section 9.4). As a consequence, interference will occur between the wave functions of the particles which remain in the original position and those which have been moved along the curve back to that position. This is the Aharonov–Bohm effect. It should be emphasised that within the box the magnetic field $\boldsymbol{\mathcal{B}}$ always vanishes, but the vector potential $\mathbf{A}$ is not zero and cannot be transformed away by a gauge transformation for a multiply connected region.

The existence of the Aharonov–Bohm effect has been verified experimentally in double slit diffraction experiments represented schematically in Fig. 12.4. A thin solenoid is placed between the slits of a diffraction experiment. Charged particles from a source at P fall on a screen S_1 containing two slits A and B and are detected on the second screen S_2 by the particle detector D. The solenoid contains a magnetic flux Φ and is impenetrable to the charged particles, while no magnetic field extends to the region outside the solenoid. The interference pattern on the screen S_2 is due to the superposition of Ψ_A and Ψ_B, where Ψ_A is the part of the wave function associated with the upper path p_A and Ψ_B is the part associated with the lower path p_B. When $\Phi = 0$, we have

$$\Psi = \Psi_A + \Psi_B. \qquad (12.56)$$

When $\Phi \neq 0$, Ψ_A acquires a phase factor $\exp[(iq/\hbar)\int_{p_A} \mathbf{A}(\mathbf{r}).d\mathbf{r}]$, where the integral is from P to D along the path p_A and similarly Ψ_B acquires the phase factor $\exp[(iq/\hbar)\int_{p_B} \mathbf{A}(\mathbf{r}).d\mathbf{r}]$ where the integral is along the path p_B, so that now

$$\Psi = \Psi_A \exp\left(\frac{iq}{\hbar}\int_{p_A} \mathbf{A}(\mathbf{r}).d\mathbf{r}\right) + \Psi_B \exp\left(\frac{iq}{\hbar}\int_{p_B} \mathbf{A}(\mathbf{r}).d\mathbf{r}\right)$$

$$= \left[\Psi_A \exp\left(\frac{iq}{\hbar}\oint \mathbf{A}(\mathbf{r}).d\mathbf{r}\right) + \Psi_B\right] \exp\left(\frac{iq}{\hbar}\int_{p_B} \mathbf{A}(\mathbf{r}).d\mathbf{r}\right) \qquad (12.57)$$

where $\oint d\mathbf{r}$ is from P to D along p_A and then from D to P along p_B. Since this integral completely encircles the solenoid we may use (12.55) and write

$$\Psi = \left[\Psi_A \exp\left(\frac{iq}{\hbar}\Phi\right) + \Psi_B\right] \exp\left[\frac{iq}{\hbar}\int_{p_B} \mathbf{A}(\mathbf{r}).d\mathbf{r}\right]. \qquad (12.58)$$

The interference pattern now depends on the relative phase difference between Ψ_A and Ψ_B, which is proportional to Φ. This was observed in 1960 by R. G. Chambers.

It is also of interest to examine the Aharonov–Bohm effect in a bound state problem. It is sufficient to consider a simple system in which a particle of mass m and charge q is constrained to move in the xy plane on a circle of radius a, the centre of the circle being the origin of the coordinate system. A thin solenoid containing a magnetic flux Φ is placed along the z-axis. If cylindrical coordinates are employed (ρ, ϕ, z) with $\rho = (x^2 + y^2)^{1/2}$ and $\phi = \tan^{-1}(y/x)$, the components of the vector potential can be taken to be

$$A_\rho = 0, \quad A_\phi = \frac{\Phi}{2\pi\rho}, \quad A_z = 0. \qquad (12.59)$$

When $\Phi = 0$ the Hamiltonian of the system is (see (6.131))

$$H = \frac{\mathbf{L}^2}{2ma^2} = \frac{L_z^2}{2ma^2} = -\frac{\hbar^2}{2ma^2}\frac{\partial^2}{\partial \phi^2}. \qquad (12.60)$$

The solution of the energy eigenvalue equation $H\psi = E\psi$ is $\psi(\phi) = N\exp(in\phi)$, where N is a normalisation constant. In order for $\psi(\phi)$ to be single-valued n must be zero or an integer. The energy eigenvalues are

$$E_n = \frac{\hbar^2 n^2}{2ma^2}. \qquad (12.61)$$

The coupling to the vector potential modifies the Hamiltonian. From (5.231b), it is found, with $\mathbf{A}.\nabla = A_\phi a^{-1}\partial/\partial\phi$ and $A_\phi = \Phi/(2\pi a)$, that

$$H = -\frac{\hbar^2}{2ma^2}\frac{\partial^2}{\partial \phi^2} + \frac{i\hbar q\Phi}{2\pi ma^2}\frac{\partial}{\partial \phi} + \frac{q^2\Phi^2}{8\pi^2 ma^2}. \qquad (12.62)$$

The solution of the energy eigenvalue equation is still $\psi(\phi) = N\exp(in\phi)$ with n being zero or an integer, but the energy eigenvalues are now

$$E_n = \frac{1}{2ma^2}\left(\hbar n - \frac{q\Phi}{2\pi}\right)^2 \tag{12.63}$$

Again we see that the physical parameters of a system in a region where $\mathcal{B} = 0$ can depend on the vector potential if the region is multiply connected.

Interaction of charged particles with constant and uniform magnetic fields

If $\mathcal{E} = 0$ and the magnetic field $\mathcal{B}$ is constant and uniform, the vector potential $\mathbf{A}$ can be written as

$$\mathbf{A}(\mathbf{r}) = \frac{1}{2}(\mathcal{B} \times \mathbf{r}) \tag{12.64a}$$

and the scalar potential ϕ vanishes:

$$\phi = 0. \tag{12.64b}$$

These expressions satisfy the Coulomb gauge condition (11.3) in addition to equations (11.2). Using (11.22), the term linear in $\mathbf{A}$ in the Schrödinger equation (11.17) is

$$i\hbar \frac{q}{m}\mathbf{A}\cdot\nabla = \frac{i\hbar q}{2m}(\mathcal{B} \times \mathbf{r})\cdot\nabla$$

$$= \frac{i\hbar q}{2m}\mathcal{B}\cdot(\mathbf{r} \times \nabla)$$

$$= -\frac{q}{2m}\mathcal{B}\cdot\mathbf{L} \tag{12.65}$$

where $\mathbf{L} = \mathbf{r} \times \mathbf{p} = -i\hbar(\mathbf{r} \times \nabla)$ is the orbital angular momentum operator of the particle. The quadratic term in $\mathbf{A}$ appearing in (11.17) can be reduced as follows:

$$\frac{q^2}{2m}\mathbf{A}^2 = \frac{q^2}{8m}(\mathcal{B} \times \mathbf{r})^2$$

$$= \frac{q^2}{8m}[\mathcal{B}^2 r^2 - (\mathcal{B}\cdot\mathbf{r})^2]. \tag{12.66}$$

The relative magnitude of the two terms (12.65) and (12.66) can now be estimated. Assuming the dimensions of the atomic system to be of the order d, the quadratic term is of the order $q^2 d^2 \mathcal{B}^2/8m$, while if we are dealing with states of low angular momentum ($\simeq \hbar$) the linear term is approximately $-q\hbar\mathcal{B}/2m$. The magnitude of the ratio of the quadratic to the linear term is then $|qd^2\mathcal{B}/4\hbar|$. If the particle is an electron $q = -e$. Moreover, if the region concerned is of the order of atomic dimensions, so that $d \simeq a_0$ (the first Bohr radius), then this ratio is

$$\frac{ea_0^2\mathcal{B}}{4\hbar} \simeq \mathcal{B} \times 10^{-6} \tag{12.67}$$

where $\mathcal{B}$ is expressed in tesla (T). In the laboratory, the largest constant fields encountered do not usually exceed a few tens of tesla although fields of the order of 10^3 T can be obtained for short times. Even for these fields the quadratic term is negligible. However, in some astrophysical situations very large fields are believed to exist. For example, in neutron stars field strengths at the surface may exceed 10^8 T and in such a case the quadratic term cannot be neglected.

The linear term (12.65) is of the form (12.30) since, as we saw in Chapter 1, the magnetic moment of a particle of charge q due to its orbital motion is $\mathcal{M}_L = q\mathbf{L}/2m$. As described earlier in this chapter, for a particle with non-zero spin an additional contribution to the magnetic moment arises, which is proportional to the spin operator $\mathbf{S}$. For an electron in motion, the total magnetic moment operator $\mathcal{M}$ is therefore

$$\mathcal{M} = \mathcal{M}_L + \mathcal{M}_s = -\frac{\mu_B}{\hbar}(\mathbf{L} + 2\mathbf{S}) \tag{12.68}$$

where we have taken $g_s = 2$. As a result, the linear term in (11.17) becomes $-\mathcal{M}\cdot\mathcal{B}$. Using this expression together with (12.66), the time-independent Schrödinger equation for a 'free' electron in a strong uniform magnetic field is

$$H\psi(\mathbf{r},\sigma) \equiv \left[-\frac{\hbar^2}{2m}\nabla^2 - \mathcal{M}\cdot\mathcal{B} + \frac{e^2}{8m}(\mathcal{B}^2 r^2 - (\mathcal{B}\cdot\mathbf{r})^2)\right]\psi(\mathbf{r},\sigma) = E\psi(\mathbf{r},\sigma) \tag{12.69}$$

where the wave function $\psi(\mathbf{r},\sigma)$ depends on both the space and spin variables. Taking $\mathcal{B}$ to be in a direction parallel to the z-axis, we can write the Hamiltonian H appearing in (12.69) as

$$H = -\frac{\hbar^2}{2m}\nabla^2 + \frac{\mu_B}{\hbar}(L_z + 2S_z)\mathcal{B} + \frac{e^2}{8m}(x^2 + y^2)\mathcal{B}^2. \tag{12.70}$$

To obtain the energy eigenvalues E, we first observe that the Hamiltonian H commutes with the operators L_z, S_z and $p_z = -i\hbar\partial/\partial z$. Each of these observables is thus a constant of the motion and simultaneous eigenfunctions of H, L_z, S_z, and p_z can be found. Let us denote these by $\psi_q(\mathbf{r},\sigma)$, where the subscript q labels the corresponding quantum numbers. Thus we have

$$H\psi_q(\mathbf{r},\sigma) = E\psi_q(\mathbf{r},\sigma), \tag{12.71}$$

$$L_z\psi_q(\mathbf{r},\sigma) = m_l\hbar\psi_q(\mathbf{r},\sigma), \qquad m_l = 0, \pm 1, \pm 2, \ldots, \tag{12.72}$$

$$S_z\psi_q(\mathbf{r},\sigma) = m_s\hbar\psi_q(\mathbf{r},\sigma), \qquad m_s = \pm 1/2, \tag{12.73}$$

and

$$p_z\psi_q(\mathbf{r},\sigma) = k\hbar\psi_q(\mathbf{r},\sigma), \qquad -\infty < k < \infty \tag{12.74}$$

so that the index q stands collectively for $\{E, m_l, m_s, k\}$. Using (12.70)–(12.74) we can write the Schrödinger equation as

$$\left[-\frac{\hbar^2}{2m}\left(\frac{\partial^2}{\partial x^2}+\frac{\partial^2}{\partial y^2}\right)+\frac{\hbar^2 k^2}{2m}+\mu_B(m_l+2m_s)\mathcal{B}+\frac{e^2}{8m}(x^2+y^2)\mathcal{B}^2\right]\psi_q(\mathbf{r},\sigma)$$
$$=E\psi_q(\mathbf{r},\sigma). \tag{12.75}$$

Since $\psi_q(\mathbf{r},\sigma)$ satisfies (12.73) and (12.74), it must be of the form

$$\psi_q(\mathbf{r},\sigma)=\phi_q(x,y)\exp(ikz)\times\begin{cases}\alpha & \text{for } m_s=1/2\\ \beta & \text{for } m_s=-1/2\end{cases} \tag{12.76}$$

which shows that the electron is free in the z-direction, the spin wave function α corresponding to $m_s=1/2$ ('spin up') and β to $m_s=-1/2$ ('spin down'). Combining (12.75) and (12.76) we find that the wave function $\phi_q(x,y)$ satisfies the equation

$$H_{xy}\phi_q(x,y)\equiv\left[-\frac{\hbar^2}{2m}\left(\frac{\partial^2}{\partial x^2}+\frac{\partial^2}{\partial y^2}\right)+\frac{1}{2}m\omega_L^2(x^2+y^2)\right]\phi_q(x,y)$$
$$=E'\phi_q(x,y) \tag{12.77}$$

where $\omega_L=\mu_B\mathcal{B}/\hbar$ is the Larmor angular frequency (1.90) and

$$E=E'+\frac{\hbar^2 k^2}{2m}+\hbar\omega_L(m_l+2m_s). \tag{12.78}$$

The equation (12.77) is the Schrödinger equation for a two-dimensional harmonic oscillator. The energy eigenvalues E' are therefore given by (see Problem 7.20)

$$\begin{aligned}E'&=(n_x+n_y+1)\hbar\omega_L, & n_x&=0,1,2,\ldots, & n_y&=0,1,2,\ldots\\ &=(n+1)\hbar\omega_L, & n&=n_x+n_y=0,1,2,\ldots\end{aligned} \tag{12.79}$$

and it is clear from (12.77) that the eigenfunctions $\phi_q(x,y)$ do not depend on the quantum numbers m_s and k. They can, in fact, be labelled by the quantum numbers n_x and n_y or alternatively by n and m_l. Using (12.79), the energy eigenvalues (12.78) for an electron in a strong uniform magnetic field become

$$E=\frac{\hbar^2 k^2}{2m}+\hbar\omega_L(n+m_l+2m_s+1). \tag{12.80}$$

As in this case the scalar potential ϕ is zero (see (12.64b)), the Hamiltonian (5.231) can be written as a square, $H=(\mathbf{p}-q\mathbf{A})^2/2m$. It follows that the eigenvalues E must be positive or zero, and this condition must hold in particular for $k=0$ and for $m_s=-1/2$. Hence

$$n+m_l\geqslant 0. \tag{12.81}$$

The Hamiltonian H_{xy} for the two-dimensional oscillator is invariant under the reflection $x \to -x$ and $y \to -y$ (see (12.77)), from which we infer that we can take the functions $\phi_q(x, y)$ to be eigenstates of the reflection operator:

$$\phi_q(-x, -y) = \pm \phi_q(x, y). \tag{12.82}$$

Inspection of the two-dimensional oscillator wave functions (see Problem 12.7) shows that ϕ_q is even if the quantum number n is even and ϕ_q is odd if n is odd. Now, $\phi_q(x, y)$ can also be expressed in terms of plane polar coordinates ρ and ϕ, where $x = \rho \cos \phi$, $y = \rho \sin \phi$. Since ϕ_q is an eigenfunction of L_z belonging to the eigenvalue $m_l \hbar$ it must be of the form $\phi_q = f(\rho) \exp(i m_l \phi)$. Under the reflection operation $x \to -x$, $y \to -y$, we see that $\rho \to \rho$ and $\phi \to \phi + \pi$. Thus ϕ_q is even when m_l is even and odd when m_l is odd, so that either both n and m_l are even, or both n and m_l are odd. Combining this fact with (12.81), we see that $n + m_l$ is even and

$$n + m_l = 2r, \qquad r = 0, 1, 2, \ldots \tag{12.83}$$

Finally the energy eigenvalues are

$$E = \frac{\hbar^2 k^2}{2m} + \hbar \omega_L (2r + 2m_s + 1), \qquad r = 0, 1, 2, \ldots, \qquad m_s = \pm 1/2. \tag{12.84}$$

The first term on the right-hand side represents the kinetic energy of uniform motion parallel to the z-axis. For given values of k and m_s, the discrete energy levels labelled by the quantum number r are called the *Landau levels*.

12.3 One-electron atoms in external magnetic fields

The splitting of the spectral lines of atoms in an external magnetic field was first observed by P. Zeeman in 1896. The fields encountered in the laboratory are not strong enough for the quadratic term in $\mathcal{B}$ to be of importance, and indeed are usually so weak that the magnetic interaction is smaller than or comparable with the spin–orbit interaction, which we have discussed in Section 8.2. Taking the spin–orbit interaction into account, and neglecting quadratic terms in $\mathcal{B}$, we obtain the following Hamiltonian for a one-electron atom in a constant and uniform magnetic field directed parallel to the z-axis

$$H = -\frac{\hbar^2}{2m}\nabla^2 - \frac{Ze^2}{(4\pi\varepsilon_0)r} + \frac{\mu_B}{\hbar}(L_z + 2S_z)\mathcal{B} + \xi(r)\mathbf{L}\cdot\mathbf{S} \tag{12.85}$$

where the quantity $\xi(r)$ is given by (8.103).

We can distinguish three cases: (a) the strong-field Zeeman effect in which the magnetic term is large enough for the spin–orbit term to be neglected; (b) the Paschen–Back effect in which the magnetic field is again strong, but the spin–orbit interaction is significant and can be treated as a perturbation; and (c) the weak-field Zeeman effect (the anomalous Zeeman effect) in which the spin–orbit interaction is much greater than the magnetic term which is treated as a perturbation.

The strong-field Zeeman effect

In this case the **L.S** term is omitted from the Hamiltonian (12.85). In this approximation H commutes with $\mathbf{L}^2, \mathbf{S}^2, L_z$ and S_z so that these are constants of the motion and l, s, m_l and m_s are all 'good quantum numbers'. If E_n are the hydrogenic eigenenergies (7.114) in the absence of the field, the eigenenergies with $\mathcal{B} \neq 0$ are (with $m_s = \pm 1/2, m_l = -l, -l+1, \ldots, +l$)

$$E_{nm_lm_s} = E_n + \mu_B(m_l + 2m_s)\mathcal{B}. \tag{12.86}$$

The introduction of the magnetic field does not remove the degeneracy of the hydrogenic energy levels in l, but it does remove partly the degeneracy in m_l and m_s, splitting each level with a given n and l into equally spaced terms. Using the selection rules $\Delta m_l = 0, \pm 1$ and $\Delta m_s = 0$, a spectral line for the transition $n' \to n$ is split into three components. That for $\Delta m_l = 0$ with frequency $\nu_{nn'} = (E_{n'} - E_n)/h$ is called the π line, while those for $\Delta m_l = \pm 1$ are called σ lines with frequencies

$$\nu_{nn'}^{\pm} = \nu_{nn'} \pm \nu_L \tag{12.87}$$

where $\nu_L = \omega_L/(2\pi) = \mu_B \mathcal{B}/h$ is the Larmor frequency. The three lines are said to form a *Lorentz triplet*.

The Paschen–Back effect

Taking the wave functions corresponding to the strong-field case as the unperturbed functions and the term in $\xi(r)\mathbf{L.S}$ as the perturbation, and noting that this perturbation does not connect the remaining degenerate states, first-order *non-degenerate* perturbation theory can be applied. The energies $E_{nlm_lm_s}$ of the perturbed levels are

$$E_{nlm_lm_s} = E_{nm_lm_s} + \langle lsm_lm_s|\xi(r)\mathbf{L.S}|lsm_lm_s\rangle \tag{12.88}$$

where $E_{nm_lm_s}$ is given by (12.86), $s = 1/2$ and we have used the Dirac notation to indicate that the unperturbed functions are eigenfunctions of $\mathbf{L}^2, \mathbf{S}^2, L_z$ and S_z. Because the expectation values of L_x and L_y for an eigenstate of L_z are zero, we may replace $\mathbf{L.S}$ by $L_z S_z$ to obtain

$$E_{nlm_lm_s} = E_{nm_lm_s} + \lambda_{nl} m_l m_s \tag{12.89}$$

where

$$\lambda_{nl} = \hbar^2 \int_0^\infty r^2 [R_{nl}(r)]^2 \xi(r) dr \tag{12.90}$$

and $R_{nl}(r)$ is a radial hydrogenic wave function. Evaluating the integral (12.90) it is found that for $l \neq 0$

$$\lambda_{nl} = -\frac{\alpha^2 Z^2}{n} E_n [l(l+1/2)(l+1)]^{-1} \tag{12.91}$$

where $\alpha = e^2/(4\pi\varepsilon_0\hbar c)$ is the fine-structure constant. Looking at (12.89) we see that the degeneracy in l is now removed. Note that for $l = 0$ the energy shift vanishes.

The weak-field Zeeman effect

If the spin–orbit term is treated as part of the unperturbed Hamiltonian, the zero-order wave functions are eigenfunctions of $\mathbf{J}^2, \mathbf{L}^2, \mathbf{S}^2$ and J_z but not of L_z and S_z (see Section 8.2). A detailed treatment (see Bransden and Joachain 1983) shows that the additional energy due to the magnetic interaction $-\mathcal{M}\cdot\mathcal{B}$ can be expressed as

$$\Delta E_{m_j} = g\mu_B \mathcal{B} m_j, \qquad m_j = -j, -j+1, \ldots, j \tag{12.92}$$

where $m_j\hbar$ are the eigenvalues of J_z and g is the *Landé g-factor* given by

$$g = 1 + \frac{j(j+1) + s(s+1) - l(l+1)}{2j(j+1)} \tag{12.93}$$

where in this case $s = 1/2$. Thus the total energy of a level with quantum numbers n, j, m_j of a hydrogenic atom in a constant magnetic field is

$$E_{njm_j} = E_n + \Delta E_{nj} + \Delta E_{m_j} \tag{12.94}$$

where E_n is the non-relativistic energy (7.114), ΔE_{nj} is the fine-structure correction (8.114) and ΔE_{m_j} is the correction (12.92) due to the (weak) magnetic field. We note from (12.92) that the degeneracy in m_j is completely removed in the presence of a weak magnetic field, so that a level with a given value of the quantum numbers n and j splits into $2j + 1$ distinct levels.

As the magnitude of the magnetic field $\mathcal{B}$ increases from the weak-field to the strong-field limit, the energy varies smoothly. This is illustrated in Fig. 12.5 for the case of the 2p levels.

12.4 Magnetic resonance

The overall angular momentum of an atom, arising from both the orbital and the spin angular momenta of the electrons, can either be zero, as for closed shell atoms, or non-zero, as for the ground state of atomic hydrogen ($j = s = 1/2, l = 0$). When the total angular momentum $\mathbf{J}$ is non-zero, an atom possesses a permanent magnetic dipole moment $\mathcal{M}$ which can be expressed for weak fields in $L - S$ coupling, consistently with (12.92), as

$$\mathcal{M} = -g\mu_B \mathbf{J}/\hbar \tag{12.95}$$

where g is the Landé g-factor (12.93). In principle, the magnetic moment can be measured in a Stern–Gerlach experiment as described in Chapter 1. An important consequence of the existence of permanent magnetic dipole moments is the *paramagnetism* observed when the magnetic moments of the atoms in a material are partially aligned by a magnetic field.

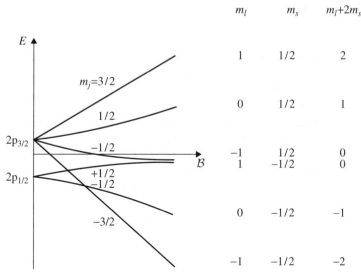

Figure 12.5 The energy levels of a hydrogen atom in a magnetic field are a smooth function of $\mathcal{B}$. A schematic diagram is shown for the 2p levels. For small $\mathcal{B}$ (the anomalous Zeeman effect) the degeneracy of the 2p levels is completely removed, but as $\mathcal{B}$ becomes large (the Paschen–Back effect) the $j = 3/2$, $m_j = -1/2$ level coincides with the $j = 1/2$, $m_j = 1/2$ level so that this degeneracy is only partly removed. In the strong field limit the 2p level is split into five equally spaced levels.

If a single atom is placed in a uniform, constant, weak magnetic field directed along the z-axis, of magnitude $\mathcal{B}_z$, the system will possess stationary states, which are eigenstates of $\mathbf{J}^2$ and J_z. As we saw above (see (12.92)), the energy due to the magnetic interaction $-\mathcal{M}\cdot\mathcal{B}$ is

$$E_{m_j} = g\mu_B \mathcal{B}_z m_j. \tag{12.96}$$

Let us denote by $j(j+1)\hbar^2$ the eigenvalues $\mathbf{J}^2$ and by $m_j\hbar$ those of J_z, where m_j can take any one of the $(2j+1)$ values $-j, -j+1, \ldots, j$. The nuclear magnetic moments also contribute, in principle, to the total magnetic energy. However, since the nuclear magnetic moments are several orders of magnitude smaller than the atomic ones, this contribution is entirely unimportant in the present context.

The time dependence of the wave function corresponding to a stationary state of energy E_{m_j} is

$$\Psi_{m_j}(t) = A\exp(-iE_{m_j}t/\hbar) \tag{12.97}$$

which can be written as

$$\Psi_{m_j}(t) = A\exp(-i\omega_0 m_j t) \tag{12.98}$$

where ω_0 is given by

$$\omega_0 = g\mu_B B_z/\hbar = g\omega_L, \tag{12.99}$$

ω_L being the Larmor angular frequency. Classically, a magnetic dipole placed in a constant magnetic field precesses about the direction of the field with the angular frequency ω_0.

We now consider the response of the system to an additional weak oscillating magnetic field, $\mathcal{B}_x \cos \omega t$, directed along the x-axis. It will be shown that when the angular frequency ω of this second field is close to the angular frequency ω_0, the system is strongly disturbed and there is a large probability of a transition from the initial state. This is called *paramagnetic resonance* (or electron spin resonance, ESR) and detection of the resonant frequency affords an accurate method of measuring gyromagnetic ratios.

In the presence of both components of the magnetic field, the Schrödinger equation is

$$i\hbar \frac{\partial}{\partial t} \Psi(t) = H\Psi(t) \tag{12.100}$$

where, using (12.95)

$$H = -\mathcal{M}.\mathcal{B} = \frac{g\mu_B}{\hbar}(J_z \mathcal{B}_z + J_x \mathcal{B}_x \cos \omega t). \tag{12.101}$$

To simplify the discussion, consider a case for which $j = 1/2$, so that $\mathbf{J}$ can be written in terms of the Pauli spin matrices $\sigma_x, \sigma_y, \sigma_z$ as

$$\mathbf{J} = \mathbf{S} = \frac{\hbar}{2}\boldsymbol{\sigma}. \tag{12.102}$$

The wave function is now a two-component spinor:

$$\Psi(t) = \begin{pmatrix} a_+(t) \\ a_-(t) \end{pmatrix} \tag{12.103}$$

Assuming that this wave function is normalised to unity, the quantity $|a_+(t)|^2$ gives the probability of finding the system in the state with $m_j = +1/2$, and $|a_-(t)|^2$ yields the probability that the system is to be found in the state with $m_j = -1/2$. Hence, we have, for all t

$$|a_+(t)|^2 + |a_-(t)|^2 = 1. \tag{12.104}$$

When the perturbation $\mathcal{B}_x \cos \omega t$ is absent, the unperturbed eigenfunctions are

$$\Psi_{1/2}(t) = \begin{pmatrix} \exp(-i\omega_0 t/2) \\ 0 \end{pmatrix}; \quad \Psi_{-1/2}(t) = \begin{pmatrix} 0 \\ \exp(i\omega_0 t/2) \end{pmatrix}. \tag{12.105}$$

If the system was originally in the $m_j = 1/2$ state, and the perturbation was switched on at time $t = 0$, we can calculate the probability of finding the system in the $m_j = -1/2$ state at some later time t. Since we shall be interested in large values of t, we cannot use the perturbation theory described in Chapter 9. Instead, we shall proceed by using the explicit form of the Pauli matrices to write the Schrödinger equation as a pair of coupled equations for $a_\pm(t)$. That is,

$$i\hbar\frac{\partial}{\partial t}\begin{pmatrix}a_+\\a_-\end{pmatrix} = \frac{g\mu_B}{2}\left[\begin{pmatrix}1 & 0\\0 & -1\end{pmatrix}\mathcal{B}_z + \begin{pmatrix}0 & 1\\1 & 0\end{pmatrix}\mathcal{B}_x\cos\omega t\right]\begin{pmatrix}a_+\\a_-\end{pmatrix}. \quad (12.106)$$

In terms of $\omega_0 = g\mu_B\mathcal{B}_z/\hbar$ and $\bar{\omega}_0 = g\mu_B\mathcal{B}_x/\hbar$, we have,

$$i\dot{a}_+ = \frac{1}{2}\omega_0 a_+ + \frac{1}{2}\bar{\omega}_0\cos(\omega t)a_-$$

$$i\dot{a}_- = -\frac{1}{2}\omega_0 a_- + \frac{1}{2}\bar{\omega}_0\cos(\omega t)a_+. \quad (12.107)$$

A phase transformation

$$A_+ = a_+e^{i\omega_0 t/2}, \qquad A_- = a_-e^{-i\omega_0 t/2} \quad (12.108)$$

removes the secular terms in $\pm\frac{1}{2}\omega_0$, giving

$$i\dot{A}_+ = \frac{1}{2}\bar{\omega}_0\cos(\omega t)e^{+i\omega_0 t}A_-$$

$$i\dot{A}_- = \frac{1}{2}\bar{\omega}_0\cos(\omega t)e^{-i\omega_0 t}A_+. \quad (12.109)$$

These equations cannot be solved exactly, but if ω is close to ω_0 an accurate approximation can be obtained by recognising that in the products $\cos(\omega t)\exp(\pm i\omega_0 t)$, terms in $\exp[\pm i(\omega-\omega_0)t]$ are much more important than those in $\exp[\pm i(\omega+\omega_0)t]$. This is because the latter terms oscillate extremely rapidly and on the average make little contribution to $\dot{A}_+$ or $\dot{A}_-$. Dropping these terms[2], we find the approximate equations:

$$i\dot{A}_+ = \frac{1}{4}\bar{\omega}_0\exp[i(\omega_0-\omega)t]A_-$$

$$i\dot{A}_- = \frac{1}{4}\bar{\omega}_0\exp[-i(\omega_0-\omega)t]A_+. \quad (12.110)$$

Exact resonance

It is easy to verify that in the case of exact resonance $\omega = \omega_0$, the equations (12.110) have the general solution,

$$A_+ = \lambda\cos(\bar{\omega}_0 t/4) + \mu\sin(\bar{\omega}_0 t/4)$$

$$A_- = i\mu\cos(\bar{\omega}_0 t/4) - i\lambda\sin(\bar{\omega}_0 t/4) \quad (12.111)$$

where λ and μ are constants, which are determined by the initial conditions.

Suppose that at $t = 0$ the system is in the state with $m_j = 1/2$; then $A_+(0) = a_+(0) = 1$ and $A_-(0) = a_-(0) = 0$ so that $\lambda = 1$ and $\mu = 0$. Thus the probabilities

---

[2] This is the *rotating wave approximation* discussed in Section 9.3.

$P(+ \rightarrow +)$ and $P(+ \rightarrow -)$ for finding the system in levels with $m_j = 1/2$ or $m_j = -1/2$ at time t, are

$$P(+ \rightarrow +) = |A_+(t)|^2 = \cos^2(\bar{\omega}_0 t/4)$$
$$P(+ \rightarrow -) = |A_-(t)|^2 = \sin^2(\bar{\omega}_0 t/4). \tag{12.112}$$

Both $P(+ \rightarrow +)$ and $P(+ \rightarrow -)$ range between 0 and 1. Clearly one always has $P(+ \rightarrow +) + P(+ \rightarrow -) = 1$.

In the same way, if at $t = 0$, the system is the state with $m_j = -1/2$ so that $A_+(0) = a_+(0) = 0$ and $A_-(0) = a_-(0) = 1$, the probabilities $P(- \rightarrow +)$ and $P(- \rightarrow -)$ for finding the system in the levels with $m_j = 1/2$ and $m_j = -1/2$ after time t are

$$P(- \rightarrow +) = \sin^2(\bar{\omega}_0 t/4),$$
$$P(- \rightarrow -) = \cos^2(\bar{\omega}_0 t/4). \tag{12.113}$$

It should be noticed that, in conformity with time-reversal invariance

$$P(- \rightarrow +) = P(+ \rightarrow -). \tag{12.114}$$

General solution

It is possible to solve equations (12.110) exactly, even when $\omega \neq \omega_0$. The general solution is (see Problem 12.8)

$$A_+ = p e^{i\eta_+ t} + q e^{i\eta_- t},$$
$$A_- = -\frac{4}{\bar{\omega}_0}[p\eta_+ e^{i\eta_+ t} + q\eta_- e^{i\eta_- t}] e^{i(\omega - \omega_0)t} \tag{12.115}$$

where p and q are constants of integration (different from λ and μ!) and $\eta_\pm$ are given by

$$\eta_\pm = \frac{1}{2}\{(\omega_0 - \omega) \pm [(\omega_0 - \omega)^2 + \bar{\omega}_0^2/4]^{1/2}\}. \tag{12.116}$$

With the initial conditions at $t = 0$, $A_+(0) = 1$, $A_-(0) = 0$, we find that

$$p = \frac{\eta_-}{\eta_- - \eta_+}, \qquad q = \frac{-\eta_+}{\eta_- - \eta_+}. \tag{12.117}$$

The probabilities $P(+ \rightarrow +)$ and $P(+ \rightarrow -)$ become

$$P(+ \rightarrow +) = \cos^2(\omega_R t/2) + \frac{(\omega_0 - \omega)^2}{(\omega_0 - \omega)^2 + (\bar{\omega}_0)^2/4}\sin^2(\omega_R t/2),$$

$$P(+ \rightarrow -) = \frac{\bar{\omega}_0^2/4}{(\omega_0 - \omega)^2 + \bar{\omega}_0^2/4}\sin^2(\omega_R t/2). \tag{12.118}$$

The frequency ω_R is the Rabi 'flopping frequency' (see (9.75)) and is given by

$$\omega_R = (\eta_+ - \eta_-) = \sqrt{(\omega_0 - \omega)^2 + \bar{\omega}_0^2/4}. \tag{12.119}$$

Under the conditions in which $B_x \ll B_z$ and $\bar{\omega}_0 \ll \omega_0$, the probability that the system will be found in the second state with $m_j = -1/2$ remains small unless ω is close to ω_0. If the oscillating magnetic field $B_x \cos \omega t$ is applied for a time T which is short ($\bar{\omega}_0 T \ll 1$), equations (12.110) can be solved by using first-order perturbation theory and it is left as an exercise (Problem 12.9) to obtain this solution and compare it with (12.115).

Resonance occurs when the frequency of the oscillating field ω is such that $\hbar \omega$ is equal to the difference in energy of the two Zeeman levels of the system, ΔE. In our spin-1/2 case

$$\Delta E = \hbar \omega_0 = g \mu_B B_z. \tag{12.120}$$

In general, for atoms with non-zero angular momentum, resonance can be produced by matching the frequency of the applied field to the frequency of a transition between particular Zeeman sub-levels, and the theory can be generalised to treat a system of $(2j+1)$ equations rather than the pair of equations treated here.

The Rabi molecular beam apparatus

A molecular beam experiment, which is much more accurate than the Stern–Gerlach experiment discussed in Chapter 1, has been devised by I. I. Rabi to measure magnetic moments of atoms based on paramagnetic resonance. A schematic diagram is shown in Fig. 12.6. A beam of atoms from an oven is passed through a system of three magnets M1, M2 and M3. We shall suppose that the atoms have total angular momentum one-half (examples of which are silver, the alkalis, or copper), although the principle of the experiment is the same for other non-zero values of the angular momentum. The magnets M1 and M3 produce inhomogeneous fields as in the Stern–Gerlach experiment, identical in magnitude but opposite in sign. If the field gradient in M1 is positive upwards, those atoms with $m_j = 1/2$ will be deflected downwards and those with $m_j = -1/2$ will be deflected upwards. If a slit S2 is placed, as shown, only two trajectories are possible from the source (slit S1) into the region to the right of S2. The trajectory B1 will contain the atoms with $m_j = 1/2$ and B2 will contain those with $m_j = -1/2$. Since the magnet M3 has an equal and opposite effect on the two trajectories, the atoms in B1 and B2 will be brought together at the slit S3 and detected at D. Now let us see what happens if the magnet M2 is switched on, which produces a large uniform static field in the z-direction, B_z, and a small oscillating field, $B_x \cos \omega t$, in the x-direction. When ω is close to the resonance angular frequency ω_0, some of the atoms in the beam B1 will have the z-component of their spin changed from $\hbar/2$ to $-\hbar/2$. These atoms will now be deflected downwards in the magnet M3 and will miss the slit S3. Similarly, atoms in the beam B2 for which the z-component of the spin changes from $-\hbar/2$ to $\hbar/2$ will also miss the slit. The net effect is that as ω approaches the resonant angular frequency, the intensity of the beam entering the detector drops sharply. Under the condition $\omega_0 \gg \bar{\omega}_0$, the resonance region is very narrow and well defined, and since frequencies can be measured very accurately, this method can provide correspondingly accurate values for atomic magnetic moments and gyromagnetic ratios.

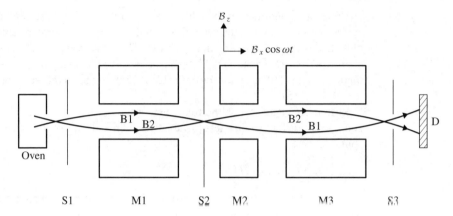

Figure 12.6 Schematic diagram of a Rabi molecular beam resonance apparatus.

Paramagnetic resonance in bulk samples

It is also possible to detect resonance phenomena in bulk samples of materials. Let us consider the case of a material composed of atoms of total angular momentum one-half. In the absence of a magnetic field, the two states of each atom with $m_j = \pm 1/2$ have the same energy, and in a bulk sample as many atoms have $m_j = 1/2$ as have $m_j = -1/2$. If the sample is placed in a uniform magnetic field $\mathcal{B}_z$, directed along the z-axis, then those atoms with $m_j = 1/2$ possess the energy (see (12.96))

$$E_+ = g\mu_B \mathcal{B}_z/2 \tag{12.121}$$

and those with $m_j = -1/2$ have the lower energy

$$E_- = -g\mu_B \mathcal{B}_z/2. \tag{12.122}$$

In thermal equilibrium, the ratio of the number N_+ of atoms per unit volume with $m_j = 1/2$ to the number N_- with $m_j = -1/2$ is

$$\frac{N_+}{N_-} = \frac{\exp[-E_+/kT]}{\exp[-E_-/kT]} \tag{12.123}$$

where T is the absolute temperature and k is Boltzmann's constant. Let $N = N_+ + N_-$ be the number of atoms per unit volume. Then

$$N_+ = N\frac{\exp[-g\mu_B \mathcal{B}_z/2kT]}{\exp[g\mu_B \mathcal{B}_z/2kT] + \exp[-g\mu_B \mathcal{B}_z/2kT]},$$

$$N_- = N\frac{\exp[g\mu_B \mathcal{B}_z/2kT]}{\exp[g\mu_B \mathcal{B}_z/2kT] + \exp[-g\mu_B \mathcal{B}_z/2kT]}. \tag{12.124}$$

The atoms with $m_j = 1/2$ contribute a magnetic moment in the z-direction of magnitude $(-g\mu_B/2)$ and those with $m_j = -1/2$ contribute a magnetic moment

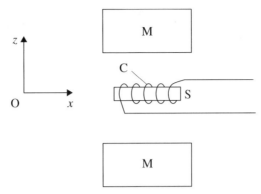

Figure 12.7 Paramagnetic resonance. The specimen S is placed between the poles M of a magnet producing a uniform static field in the z direction. A small oscillatory field in the x direction is produced by the coil C.

$(g\mu_B/2)$, so that the magnitude of the net magnetic moment in the z-direction per unit volume is[3]

$$\mathcal{M}_z = N_+(-g\mu_B/2) + N_-(g\mu_B/2)$$

$$= \left(\frac{g\mu_B}{2}\right) N \frac{\exp[g\mu_B \mathcal{B}_z/2kT] - \exp[-g\mu_B \mathcal{B}_z/2kT]}{\exp[g\mu_B \mathcal{B}_z/2kT] + \exp[-g\mu_B \mathcal{B}_z/2kT]}. \quad (12.125)$$

Now if a small field $\mathcal{B}_x \cos \omega t$ is applied in the x-direction, when ω is close to the resonance frequency transitions are induced between the two states with $m_j = 1/2$ and $m_j = -1/2$. If these two levels were equally populated, these transitions could not be detected, but since the number of atoms in the state of lower energy is greater than the number in the state of higher energy, more transitions occur which absorb energy from the external field than transitions which feed energy into the external field. This net loss of energy at resonance is small, but can be detected in various ways. For example the oscillating magnetic field may be produced by a coil which is placed in a bridge circuit (see Fig. 12.7). At resonance, the energy loss to the medium gives rise to an apparent change in the self-induction of the coil, which can be detected by the bridge.

The technique of paramagnetic resonance is important in industry and research, to give information about the constituents of a sample. On the other hand, if these are known, the apparatus can serve as a magnetometer to measure small fields $\mathcal{B}_x$.

As an example, an actual electron spin resonance spectrum is shown in Fig. 12.8. The material under investigation is Fe^{3+}/MgO containing 310 parts per million of Fe^{3+} ions, and it is these ions which give rise to the spectrum. The field $\mathcal{B}_x$ oscillates at a fixed frequency of 9.10 GHz and the energy absorption is measured as a function of the field $\mathcal{B}_z$. The ordinate on the trace is proportional to the derivative $(dI/d\mathcal{B}_z)$ where

---

[3] This result will be obtained again using an alternative method in Section 14.5.

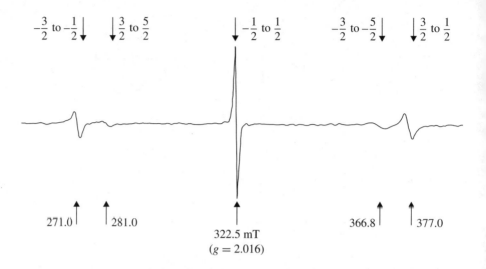

Figure 12.8 An electron spin resonance spectrum of Fe^{3+}/MgO. (*By courtesy of J. S. Thorp.*)

I is the energy loss, and resonance occurs at the points for which I is a maximum, that is for which $(dI/d\mathcal{B}_z) = 0$. The ion Fe^{3+} has angular momentum $j = 5/2$ so that various transitions can be observed corresponding to $m_j = 1/2 \leftrightarrow -1/2$, $\pm 3/2 \leftrightarrow \pm 5/2$ and $\pm 3/2 \leftrightarrow \pm 1/2$. It is easy to calculate from the transition $1/2 \to -1/2$ that $g = 2.016$, using equation (12.99).

Nuclear magnetic resonance

Although the magnetic moments of nuclei are smaller than those of atoms by a factor of the order of m/M_p (where m is the mass of the electron and M_p that of the proton) and nuclear paramagnetism is too small to be observed directly, nuclear magnetic resonance experiments are perfectly possible and are, in fact, of great importance. A substance is chosen for which the total angular momentum due to the electrons is zero, for otherwise the nuclear effect would be completely masked. The experiment can then be performed in exactly the same way as for paramagnetic resonance, with the resonant frequency being determined by the strength of the static field $\mathcal{B}_z$, where now (see (12.99))

$$\omega_0 = g\mu_N \mathcal{B}_z/\hbar \tag{12.126}$$

and μ_N is the nuclear magneton (12.28).

For a field of strength 0.5 T, the resonant frequency of a proton (having a gyromagnetic ratio of 5.5883) is 21.3 MHz, which is in the radio-frequency region. The frequencies associated with paramagnetic resonance are about 2000 times higher and are in the microwave region.

Problems

12.1 According to the result of Problem 8.4, the second-order correction to the energy of a non-degenerate level is given by

$$E_n^{(2)} = \frac{1}{E_n^{(0)}}[(H'^2)_{nn} - (H'_{nn})^2] + \sum_{k \neq n} \frac{E_k^{(0)}|H'_{nk}|^2}{E_n^{(0)}(E_n^{(0)} - E_k^{(0)})}.$$

By neglecting the sum on the right-hand side, obtain an estimate of the quadratic Stark effect on the ground-state energy of atomic hydrogen.

12.2 Suppose that at $t = 0$ a hydrogen atom is in an arbitrary superposition of the 2s and $2p_0$ states. A constant electric field of strength 10^7 V m^{-1} is then applied parallel to the z-axis. Show that during the lifetime of the $2p_0$ state due to radiative decay to the ground state ($\tau = 1.6 \times 10^{-9}$ s), the average populations of the 2s and $2p_0$ states are nearly the same.

12.3 Starting from the expression (12.24) for the second-order Stark effect for the rigid rotator, obtain the results (12.25).

(*Hint*: Use the recurrence relations (6.97) to evaluate the required matrix elements.)

12.4 Show that the Hamiltonian (12.70) commutes with the operators L_z, S_z, $\mathbf{S}^2$ and p_z but not with $\mathbf{L}^2$.

12.5 Apply the gauge transformation generated by taking

$$\chi(\mathbf{r}, t) = -\frac{1}{2}\mathcal{B}xy$$

to the potentials (12.64), where $\mathcal{B}$ is taken to be parallel to the z-axis, and show that the transformed time-independent Schrödinger equation, for a spinless particle of charge $q = -e$ and mass m, is

$$\left(-\frac{\hbar^2}{2m}\nabla^2 + \frac{i\hbar e\mathcal{B}y}{m}\frac{\partial}{\partial x} + \frac{e^2\mathcal{B}^2 y^2}{2m}\right)\Phi(\mathbf{r}) = E\Phi(\mathbf{r}).$$

12.6 By substituting

$$\Phi(\mathbf{r}) = \exp i(Kx + kz) f(y)$$

into the Schrödinger equation of the previous problem, show that the energy eigenvalues are given by

$$E = \frac{\hbar^2 k^2}{2m} + (2r + 1)\hbar\omega_L, \quad r = 0, 1, 2, \ldots$$

where $\omega_L = (\mu_B/\hbar)\mathcal{B}$ is the Larmor angular frequency. This is the same result as (12.84) if the term in m_s is omitted.

12.7 Using the results of Problem 7.20 for the two-dimensional isotropic harmonic oscillator, show that the corresponding eigenfunctions are even under the interchange $x \to -x$, $y \to -y$ if $n = n_x + n_y$ is even and are odd if n is odd.

12.8 Verify by substitution that the expressions (12.115) solve the equations (12.110).

12.9 Solve the equations (12.110) approximately by using first-order time-dependent perturbation theory and compare your results with the exact solution (12.115).

13 Quantum collision theory

13.1 Scattering experiments and cross-sections 588
13.2 Potential scattering. General features 592
13.3 The method of partial waves 595
13.4 Applications of the partial-wave method 599
13.5 The integral equation of potential scattering 608
13.6 The Born approximation 615
13.7 Collisions between identical particles 620
13.8 Collisions involving composite systems 627
Problems 635

Our knowledge of microscopic physics – the physics of molecules, atoms, nuclei and elementary particles – is obtained in two ways. The first one consists in analysing the electromagnetic radiation absorbed or emitted in transitions between bound states of quantum systems; this is the subject of spectroscopy. Alternatively, the interactions between atomic or sub-atomic particles can be probed by performing experiments in which one particle is scattered by another. A famous example is the deduction of the existence of atomic nuclei by Rutherford, using the data from experiments in which alpha particles were scattered by atoms within metallic foils. Many of the properties of the nuclear force have been inferred from the study of nuclear collisions and, while the first mesons were discovered in the cosmic radiation, the majority of the wide spectrum of new particles, hadrons and leptons, have been found among the products of collision experiments. The systematic study of collision processes is the main source of information about the strong, electromagnetic and weak interactions and on this experimental evidence rests the modern picture of matter as being ultimately composed of quarks and leptons. In this chapter we shall lay the foundation of non-relativistic quantum collision theory. We begin by considering the simple case in which structureless particles are scattered by a fixed centre of force. We then discuss the scattering of two identical particles and conclude the chapter with an introduction to the study of collisions involving composite systems.

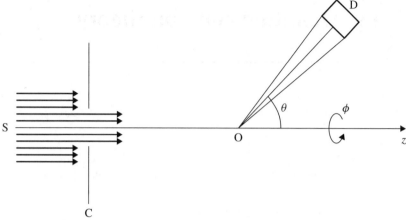

Figure 13.1 The scattering of a beam of particles by a target situated at the origin O. Particles scattered into an element of solid angle centred about a direction specified by the angles θ and ϕ are detected at D.

13.1 Scattering experiments and cross-sections

A particular form of scattering experiment is shown schematically in Fig. 13.1. An incident beam of monoenergetic particles from a source S is collimated by slits C so that all the particles in the beam move in the same direction. The beam is scattered by the target situated at O, and the number of particles scattered per unit time into an element of solid angle centred about a direction specified by polar angles (θ, ϕ) is measured. To take an example, in an experiment to study the interaction between protons and the nuclei of atoms of gold, a beam of protons is allowed to interact with a target consisting of a thin film of gold. In order to ensure that the scattering characterises single collisions between protons and gold nuclei, the target is made so thin that the probability that any proton in the beam makes more than one collision with a gold nucleus is negligible. These single-scattering conditions are in sharp contrast, for example, to those of electron diffraction from crystals, in which the atoms of the target crystal act cooperatively.

In classical mechanics the motion of each particle in the beam is along a well defined trajectory and each particle is scattered into a particular direction (θ, ϕ) which can be calculated from the initial conditions. Quantum theory does not allow such a detailed picture of a collision to be given. The angle of scattering in a collision between two particles cannot be calculated, and only the probability of scattering into a certain direction can be predicted. This can be seen as follows. Suppose the interaction between the two particles is of limited range and vanishes when the two particles are separated by a distance greater than a. For an incident particle to be scattered, it must approach the target particle within the distance a. Since in this way the trajectory of the particle is located with an uncertainty of order a, the momentum transverse to the direction of incidence, p_T, must be uncertain after the collision by an amount $\Delta p_T \simeq \hbar/a$. If p is the momentum of a particle in the incident beam, the angle of scattering is determined by p_T/p and the uncertainty in the scattering angle is of the order $\hbar/(ap)$. In general (see Problem 13.1), this quantity is not small so that the angle of scattering and the trajectory cannot be computed from the initial conditions. Of course, in macroscopic situations, such as the scattering of billiard balls, $\hbar/(ap)$ is very small, the scattering angle is well defined, and classical mechanics can be used, in accordance with the correspondence principle. Since the beam in a collision experiment contains a very large number of particles and is switched on for a time that is very long compared with the time during which a beam particle interacts with the target, the intensity of the scattered beam in a given direction is constant and predictable, even though the results of an individual microscopic collision cannot be predicted.

To count the scattered particles in a collision experiment a detector D (see Fig. 13.1) is placed outside the path of the incident beam. If the detector subtends a solid angle $d\Omega$ at the scattering centre O in the direction (θ, ϕ), the number of particles $N d\Omega$ entering the detector per unit time can be measured. This number is proportional to the flux of particles F in the incident beam, defined as the number of particles per unit time crossing a unit area placed normal to the direction of incidence. It is assumed that the beam is uniform and that the density of particles in the beam is sufficiently small so that the possibility of interactions between the beam particles themselves can be ignored. Under these conditions the collision can be characterised by the *differential scattering cross-section* $d\sigma/d\Omega$ which is defined as the ratio of the number of particles scattered into the direction (θ, ϕ) per unit time, per unit solid angle, divided by the incident flux:

$$\frac{d\sigma}{d\Omega} = \frac{N}{F}. \tag{13.1}$$

The *total scattering cross-section* σ_{tot} is defined as the integral of the differential scattering cross-section over all solid angles:

$$\sigma_{\text{tot}} = \int \left(\frac{d\sigma}{d\Omega}\right) d\Omega = \int_0^{2\pi} d\phi \int_0^{\pi} d\theta \sin\theta \frac{d\sigma}{d\Omega}. \tag{13.2}$$

Since N is a number of particles per unit time, while F is a number of particles per unit time per unit area, the dimensions of $d\sigma/d\Omega$ or of σ_{tot} are those of an area, which accounts for the term 'scattering cross-section'.

We shall be mainly concerned with *elastic scattering* in which the internal energies of the incident and target particles do not change and in which no further particles are created. An example is the scattering of neutrons by protons at low energies:

$$\text{n} + \text{p} \rightarrow \text{n} + \text{p}. \tag{13.3}$$

Many collisions are not elastic, involving for example the excitation of the projectile and (or) the target, or the creation or destruction of particles. Particle creation and destruction in scattering experiments are beyond the scope of this book, but non-relativistic scattering theory can be extended to treat other types of inelastic collisions, for example the excitation of atoms by electron impact:

$$e^- + A \to e^- + A^*. \tag{13.4}$$

The laboratory and centre-of-mass systems

Let us consider a non-relativistic collision between a 'beam' particle A of mass m_A and a target particle B of mass m_B. The *laboratory system* (L) is the frame in which the target particle B is at rest *before* the collision. In what follows, we shall use the subscript L to denote quantities in the laboratory system. The *centre-of-mass* (CM) *system* is the coordinate system in which the centre of mass of the composite system (A + B) is (always) at rest. If the target particles are very heavy compared with the incident particles, as in the scattering of electrons by atoms or molecules, the laboratory and CM systems nearly coincide, and are identical in the limit $m_A/m_B \to 0$. If, on the other hand, the mass m_B of the target particle is not very large compared with the mass m_A of the incident particle, the theoretical description of the collision is best carried out in the centre-of-mass system of coordinates, just as for bound-state problems. In that system the projectile A and target particle B move initially with respect to the centre of mass C with equal and opposite momenta, $\mathbf{p}_A = -\mathbf{p}_B = \mathbf{p}$, where $\mathbf{p}$ is the relative momentum introduced in (5.270). The initial kinetic energy available in the CM system is therefore

$$T = \frac{\mathbf{p}_A^2}{2m_A} + \frac{\mathbf{p}_B^2}{2m_B} = \frac{\mathbf{p}^2}{2\mu} \tag{13.5}$$

where μ is the reduced mass

$$\mu = \frac{m_A m_B}{m_A + m_B}. \tag{13.6}$$

After the collision the two particles A and B emerge with equal and opposite momenta $\mathbf{p}'_A = -\mathbf{p}'_B = \mathbf{p}'$.

The angle of scattering θ_L measured in the laboratory frame is not the same as the corresponding angle of scattering θ in the centre-of-mass frame. Since the apparatus used to detect and measure the flux of scattered particles is macroscopic, the relationship between the two angles can be found using ordinary classical mechanics. We shall now determine this relationship for the case of elastic scattering. In the centre-of-mass frame the collision appears as in Fig. 13.2(a). Since the collision is elastic, the energy of the particles is the same before and after the collision, and the magnitude of the momentum of each particle also remains the same: $p' = p$. If v is the magnitude of the initial relative velocity of A and B, then the magnitude of the momentum p is related to v by

$$p = \mu v. \tag{13.7}$$

In the laboratory system, shown in Fig. 13.2(b), the target particle B is at rest before the collision, and the projectile A moves with a momentum of magnitude $q_A = m_A v$.

13.1 Scattering experiments and cross-sections

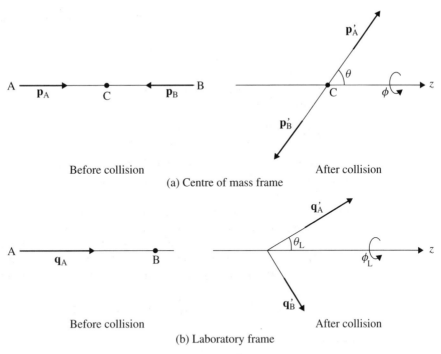

Figure 13.2 Elastic scattering of a projectile A by a target B: (a) in the centre-of-mass frame; and (b) in the laboratory frame.

After the collision A and B move with momenta $\mathbf{q}'_A$ and $\mathbf{q}'_B$, as shown. Throughout the collision the centre of mass of the two particles moves with a constant velocity $\mathbf{V}_c$ along the direction of incidence, and we have

$$(m_A + m_B)\mathbf{V}_c = \mathbf{q}_A. \tag{13.8}$$

It follows that, before the collision, the momentum of the projectile in the centre-of-mass frame, $\mathbf{p}_A$, is related to its momentum in the laboratory frame, $\mathbf{q}_A$, by

$$\mathbf{p}_A = \mathbf{q}_A - m_A \mathbf{V}_c = \frac{m_B}{m_A + m_B}\mathbf{q}_A. \tag{13.9}$$

After the collision, the components of the momentum of the projectile A along the direction of incidence in the two coordinate systems are related by

$$p'_A \cos\theta = q'_A \cos\theta_L - m_A V_c. \tag{13.10}$$

The components of the momenta normal to the direction of incidence must be the same in both coordinate systems so that

$$p'_A \sin\theta = q'_A \sin\theta_L. \tag{13.11}$$

From (13.10) and (13.11), and the fact that $p_A = p'_A (= p)$ for elastic scattering, we obtain

$$\tan\theta_L = \frac{p\sin\theta}{p\cos\theta + m_A V_c} \tag{13.12}$$

and since from (13.7) and (13.8), $m_A V_c = m_A p/m_B$, we arrive at the desired relationship between θ_L and θ:

$$\tan\theta_L = \frac{\sin\theta}{\cos\theta + \tau}, \qquad \tau = \frac{m_A}{m_B}. \tag{13.13}$$

As the plane of scattering is the same in both coordinate systems, the azimuthal angle ϕ is unaltered in transforming from the centre of mass to the laboratory system ($\phi_L = \phi$).

The total cross-section σ_{tot} is a measure of the probability that scattering has occurred irrespective of the angle of scattering. It follows that while the differential cross-sections in the two coordinate systems differ, the total cross-sections are the same. This fact can be used to relate $(d\sigma/d\Omega_L)$, the laboratory differential cross-section, to $d\sigma/d\Omega$, the cross-section in the centre-of-mass system. Indeed, since

$$\sigma_{tot} = \int_0^{2\pi} d\phi_L \int_{-1}^{+1} d(\cos\theta_L) \left(\frac{d\sigma}{d\Omega}\right)_L = \int_0^{2\pi} d\phi \int_{-1}^{+1} d(\cos\theta) \frac{d\sigma}{d\Omega} \tag{13.14}$$

we have

$$\left(\frac{d\sigma}{d\Omega}\right)_L = \left|\frac{d(\cos\theta)}{d(\cos\theta_L)}\right| \frac{d\sigma}{d\Omega}. \tag{13.15}$$

Evaluating $|d(\cos\theta)/d(\cos\theta_L)|$ with the help of (13.13), we find that

$$\left(\frac{d\sigma}{d\Omega}\right)_L = \frac{(1+\tau^2+2\tau\cos\theta)^{3/2}}{|1+\tau\cos\theta|}\frac{d\sigma}{d\Omega}, \qquad \tau = \frac{m_A}{m_B}. \tag{13.16}$$

Although many scattering experiments are carried out in which incident particles are scattered from a stationary target, this is not always the case. Frequently the two beams of particles are allowed to intersect and scatter from each other. The results of such experiments can also be transformed to the centre-of-mass system using kinematical arguments similar to those used above.

13.2 Potential scattering. General features

In this section the theory of the scattering of a beam of particles by a fixed centre of force will be developed. The system is described by the Schrödinger equation

$$i\hbar\frac{\partial}{\partial t}\Psi(\mathbf{r},t) = \left[-\frac{\hbar^2}{2m}\nabla^2 + V(\mathbf{r})\right]\Psi(\mathbf{r},t) \tag{13.17}$$

where m is the mass of the projectile. To apply the theory to the scattering of particles A of mass m_A by target particles B of mass m_B interacting through a potential

13.2 Potential scattering. General features

$V(\mathbf{r})$, it is only necessary to replace m by the reduced mass μ given by (13.6) (see Section 5.7). In the type of experiment schematically shown in Fig. 13.1, the incident beam of particles is switched on for times very long compared with the time a particle would take to cross the interaction region, so that steady-state conditions apply. Since, in addition, the potential $V(\mathbf{r})$ does not depend on the time, we can look for stationary solutions of the form

$$\Psi(\mathbf{r}, t) = \psi(\mathbf{r}) \exp(-iEt/\hbar). \tag{13.18}$$

The wave function $\psi(\mathbf{r})$ satisfies the time-independent Schrödinger equation

$$\left[-\frac{\hbar^2}{2m}\nabla^2 + V(\mathbf{r})\right]\psi(\mathbf{r}) = E\psi(\mathbf{r}) \tag{13.19}$$

where E, the energy of the particle, has the definite value

$$E = \frac{\mathbf{p}^2}{2m} = \frac{\hbar^2 \mathbf{k}^2}{2m} = \frac{1}{2}mv^2. \tag{13.20}$$

Here $\mathbf{p} = \hbar \mathbf{k} = m\mathbf{v}$ is the incident momentum of the particle, $\mathbf{k}$ is its incident wave vector and $\mathbf{v}$ is its incident velocity. It is convenient to rewrite (13.19) as

$$[\nabla^2 + k^2 - U(\mathbf{r})]\psi(\mathbf{r}) = 0 \tag{13.21}$$

where $U(\mathbf{r})$ is the reduced potential given by

$$U(\mathbf{r}) = \frac{2m}{\hbar^2} V(\mathbf{r}). \tag{13.22}$$

Let us consider potentials which decrease faster than r^{-1} as $r \to \infty$. In that case, for large r, $U(\mathbf{r})$ can be neglected and (13.21) reduces to the free-particle Schrödinger equation

$$[\nabla^2 + k^2]\psi(\mathbf{r}) = 0. \tag{13.23}$$

In the large r region, $\psi(\mathbf{r})$ must describe both the incident beam of particles and the particles which have been scattered, so that we can write

$$\psi(\mathbf{r}) \underset{r \to \infty}{\to} \psi_{\text{inc}}(\mathbf{r}) + \psi_{\text{sc}}(\mathbf{r}). \tag{13.24}$$

The particles in the incident beam all have the same momentum of magnitude $p = \hbar k$ and all travel in the same direction, which we take to be the z-axis. The eigenfunctions of momentum being plane waves, the incident wave function can be written as

$$\psi_{\text{inc}}(\mathbf{r}) = e^{i\mathbf{k} \cdot \mathbf{r}} = e^{ikz}. \tag{13.25}$$

With this normalisation $|\psi_{\text{inc}}(\mathbf{r})|^2 = 1$, so that ψ_{inc} represents a beam with one particle per unit volume. Since each particle travels with a velocity $v = p/m$, the incident flux F is

$$F = v. \tag{13.26}$$

The plane wave (13.25) is infinite in extent, but in any real experiment the incident beam is collimated (see Fig. 13.1) and its transverse width is finite. Nevertheless, the beam width is very large compared with atomic or nuclear dimensions, so that the beam can be described accurately by a plane wave. Far from the scattering centre, the scattered wave function ψ_{sc} must represent an *outward* radial flow of particles. It has the form

$$\psi_{\text{sc}}(\mathbf{r}) = f(k,\theta,\phi)\frac{e^{ikr}}{r} \tag{13.27}$$

where (r,θ,ϕ) are the polar coordinates of the position vector $\mathbf{r}$ of the scattered particle. The amplitude f of the outgoing spherical wave $r^{-1}\exp(ikr)$, called the *scattering amplitude*, depends on the direction (θ,ϕ), and determines the flux of particles scattered in that direction. It is left as a problem (Problem 13.5) to verify that in general both incoming and outgoing waves of the form $r^{-1}\exp(\pm ikr)f(k,\theta,\phi)$ satisfy the free-particle equation (13.23) in the limit of large r. The physical conditions dictate the use of outgoing waves for $\psi_{\text{sc}}(\mathbf{r})$. Thus, using (13.24), (13.25) and (13.27), we see that the stationary scattering wave function, which we shall denote by $\psi_{\mathbf{k}}(\mathbf{r})$, is a particular solution of the Schrödinger equation (13.21) satisfying the asymptotic boundary condition

$$\psi_{\mathbf{k}}(\mathbf{r}) \underset{r\to\infty}{\to} e^{i\mathbf{k}\cdot\mathbf{r}} + f(k,\theta,\phi)\frac{e^{ikr}}{r}, \tag{13.28}$$

where the subscript $\mathbf{k}$ indicates that the wave function $\psi_{\mathbf{k}}$ corresponds to the incident plane wave $\exp(i\mathbf{k}\cdot\mathbf{r})$.

The probability current density $\mathbf{j}$ has been obtained in Section 3.2. For the stationary state (13.18) it is given by (see (3.34))

$$\mathbf{j}(\mathbf{r}) = \frac{\hbar}{2mi}[\psi^*(\nabla\psi) - (\nabla\psi^*)\psi]. \tag{13.29}$$

The gradient operator can be expressed in polar coordinates (r,θ,ϕ) as

$$\nabla = \frac{\partial}{\partial r}\hat{\mathbf{r}} + \frac{1}{r}\frac{\partial}{\partial\theta}\hat{\boldsymbol{\theta}} + \frac{1}{r\sin\theta}\frac{\partial}{\partial\phi}\hat{\boldsymbol{\phi}}. \tag{13.30}$$

Since the second and third terms in (13.30) are small when r is large, the current, for large r, is in the radial direction, and upon using (13.27) the radial current of scattered particles in the direction (θ,ϕ) is found to be

$$j_r = \frac{\hbar k}{m}|f(k,\theta,\phi)|^2/r^2. \tag{13.31}$$

Since j_r represents the number of particles crossing unit area per unit time, and the detector presents a cross-sectional area $r^2 d\Omega$ to the scattered beam, the number of particles entering the detector per unit time, $N d\Omega$, is

$$N d\Omega = \frac{\hbar k}{m}|f(k,\theta,\phi)|^2 d\Omega. \tag{13.32}$$

Since $v = \hbar k/m$ is the particle velocity, we find by using (13.1), (13.26) and (13.32) that the differential cross-section is given by

$$\frac{d\sigma}{d\Omega} = |f(k, \theta, \phi)|^2. \tag{13.33}$$

13.3 The method of partial waves

In the particular case of scattering by a central potential $V(r)$, the system is completely symmetrical about the direction of incidence (the z-axis), so that the wave function, and hence the scattering amplitude depend on θ, but not on the azimuthal angle ϕ. For this reason both $\psi_{\mathbf{k}}(r, \theta)$ and $f(k, \theta)$ can be expanded in series of Legendre polynomials, which form a complete set in the interval $-1 \leqslant \cos\theta \leqslant +1$. That is,

$$\psi_{\mathbf{k}}(r, \theta) = \sum_{l=0}^{\infty} R_l(k, r) P_l(\cos\theta), \tag{13.34}$$

and

$$f(k, \theta) = \sum_{l=0}^{\infty} f_l(k) P_l(\cos\theta). \tag{13.35}$$

Each term in the series (13.34) is known as a *partial wave* and is a simultaneous eigenfunction of the operators $\mathbf{L}^2$, and L_z belonging to eigenvalues $l(l+1)\hbar^2$ and 0, respectively. Following the spectroscopic notation the $l = 0, 1, 2, 3, \ldots$, partial waves are usually known as s, p, d, f, ..., waves. In the series (13.35), the *partial wave amplitudes* $f_l(k)$ are determined by the radial function $R_l(k, r)$, as we shall now see. The equations satisfied by the radial functions can be found as in Section 7.2. Remembering that $E = \hbar^2 k^2/2m$, we have

$$\left[\frac{d^2}{dr^2} + \frac{2}{r}\frac{d}{dr} - \frac{l(l+1)}{r^2} - U(r) + k^2\right] R_l(k, r) = 0 \tag{13.36}$$

where $U(r) = (2m/\hbar^2)V(r)$ is the reduced potential. For potentials less singular than r^{-2} at the origin, the behaviour of $R_l(k, r)$ near $r = 0$ can be determined by expanding R_l in a power series:

$$R_l(k, r) = r^s \sum_{n=0}^{\infty} a_n r^n. \tag{13.37}$$

Examination of the indicial equation shows that there are two solutions, one with $s = l$ and one with $s = -(l+1)$. Since the wave function $\psi(r, \theta)$ must be finite everywhere including the origin at $r = 0$, the solution with $s = l$ is the one which must be adopted. For sufficiently large r, say for $r > d$, the potential $U(r)$ can be neglected and the equation satisfied by R_l becomes

$$\left[\frac{d^2}{dr^2} + \frac{2}{r}\frac{d}{dr} - \frac{l(l+1)}{r^2} + k^2\right] R_l(k, r) = 0 \tag{13.38}$$

which is the same as equation (7.59).

As we saw in Section 7.3, for each l the linearly independent solutions of (13.38) are the spherical Bessel and Neumann functions $j_l(kr)$ and $n_l(kr)$. The general solution is a linear combination of these functions so that the radial function R_l is given in the region $r > d$ in which the potential can be neglected by

$$R_l(k, r) = B_l(k) j_l(kr) + C_l(k) n_l(kr), \qquad r > d \tag{13.39}$$

where $B(k)$ and $C(k)$ are independent of r. Using the asymptotic expressions (7.66) for $j_l(kr)$ and $n_l(kr)$, we also have

$$R_l(k, r) \underset{r\to\infty}{\to} \frac{1}{kr}\left[B_l(k)\sin\left(kr - \frac{l\pi}{2}\right) - C_l(k)\cos\left(kr - \frac{l\pi}{2}\right)\right]. \tag{13.40}$$

It is convenient to rewrite this equation as

$$R_l(k, r) \underset{r\to\infty}{\to} A_l(k)\frac{1}{kr}\sin\left(kr - \frac{l\pi}{2} + \delta_l(k)\right) \tag{13.41}$$

where

$$A_l(k) = [B_l^2(k) + C_l^2(k)]^{1/2} \tag{13.42}$$

and

$$\delta_l(k) = -\tan^{-1}[C_l(k)/B_l(k)]. \tag{13.43}$$

The real quantities $\delta_l(k)$ are called the *phase shifts*, and characterise the strength of the scattering in the lth partial wave by the potential $U(r)$, at the energy $E = \hbar^2 k^2/2m$. This follows because if $U(r)$ is zero, the physical solution of equation (13.36), that is the solution which behaves like r^l at the origin, is the function $j_l(kr)$, and this function has the asymptotic form (13.41) with $\delta_l(k) = 0$. The phase shifts can be found either from an analytical or, if necessary, a numerical solution to (13.36), subject to the proper boundary condition at $r = 0$.

We now have to relate the phase shifts to the partial wave amplitudes $f_l(k)$ and hence to the scattering amplitude through (13.35). Returning to (13.28), and remembering that in the present case the scattering amplitude $f(k, \theta)$ is independent of ϕ, we see that the asymptotic form of $\psi_\mathbf{k}(r, \theta)$ for large r is

$$\psi_\mathbf{k}(r, \theta) \underset{r\to\infty}{\to} e^{i\mathbf{k}\cdot\mathbf{r}} + f(k, \theta)\frac{e^{ikr}}{r}. \tag{13.44}$$

In Section 7.3, equation (7.77), we saw that the term $\exp(i\mathbf{k}\cdot\mathbf{r})$ can be expanded as

$$e^{i\mathbf{k}\cdot\mathbf{r}} = \sum_{l=0}^{\infty}(2l+1)i^l j_l(kr) P_l(\cos\theta) \tag{13.45}$$

which upon using (7.66a) becomes for large r

$$e^{i\mathbf{k}\cdot\mathbf{r}} \underset{r\to\infty}{\to} \sum_{l=0}^{\infty}(2l+1)i^l(kr)^{-1}\sin\left(kr - \frac{l\pi}{2}\right)P_l(\cos\theta). \tag{13.46}$$

From (13.34), (13.35), (13.44) and (13.46) it follows that the asymptotic form of the radial function $R_l(k, r)$ is

$$R_l(k, r) \underset{r \to \infty}{\to} (2l + 1)i^l(kr)^{-1} \sin\left(kr - \frac{l\pi}{2}\right) + r^{-1} \exp(ikr) f_l(k). \tag{13.47}$$

By equating this expression with (13.41), we find immediately that

$$A_l(k) = (2l + 1)i^l \exp[i\delta_l(k)] \tag{13.48}$$

and

$$f_l(k) = \frac{2l + 1}{2ik} \{\exp[2i\delta_l(k)] - 1\}. \tag{13.49}$$

It should be noted that the scattering amplitude $f(k, \theta)$ given by

$$f(k, \theta) = \sum_{l=0}^{\infty} f_l(k) P_l(\cos\theta) = \frac{1}{2ik} \sum_{l=0}^{\infty} (2l + 1)\{\exp[2i\delta_l(k)] - 1\} P_l(\cos\theta) \tag{13.50}$$

depends only on the phase shifts $\delta_l(k)$ and not on the particular normalisation of the radial function expressed by (13.48).

The significance of the phase shifts $\delta_l(k)$ can be seen in a different way by rewriting the asymptotic form of $R_l(k, r)$ given by (13.41) as

$$R_l(k, r) \underset{r \to \infty}{\to} -\frac{1}{2ik} A_l(k) \exp[-i\delta_l(k)] \left[\frac{e^{-i(kr - l\pi/2)}}{r} - S_l(k) \frac{e^{i(kr - l\pi/2)}}{r}\right] \tag{13.51}$$

where

$$S_l(k) = \exp[2i\delta_l(k)]. \tag{13.52}$$

In the square brackets on the right of (13.51) the first term represents an incoming, and the second term an outgoing, spherical wave. Since in elastic scattering the number of particles per second entering the scattering region must equal the number of particles per second leaving that region, the moduli of the amplitudes of the incoming and outgoing wave must be the same. As a result, the effect of the potential can only be to produce a phase difference between the ingoing and outgoing spherical waves. We see from (13.52) that the quantity $S_l(k)$, which is called an *S-matrix element*, is of unit modulus. Its interest lies in the fact that the S-matrix can be generalised to describe non-elastic scattering processes. Conservation of probability is then expressed by requiring the S-matrix to be a unitary matrix.

Total cross-section

The total cross-section can be calculated in terms of the phase shifts δ_l. From (13.2), (13.33) and the fact that the scattering amplitude is independent of the azimuthal angle ϕ for scattering by a central potential, we first have

$$\sigma_{\text{tot}} = \int |f(k, \theta)|^2 d\Omega = 2\pi \int_{-1}^{+1} d(\cos\theta) f^*(k, \theta) f(k, \theta). \tag{13.53}$$

Using the partial-wave expansion (13.50) and the orthogonality property of the Legendre polynomials (see (6.80))

$$\int_{-1}^{+1} d(\cos\theta) P_l(\cos\theta) P_{l'}(\cos\theta) = \frac{2}{2l+1}\delta_{ll'} \tag{13.54}$$

we obtain

$$\sigma_{\text{tot}} = \sum_{l=0}^{\infty} \sigma_l \tag{13.55}$$

where

$$\sigma_l = \frac{4\pi}{2l+1}|f_l(k)|^2 = \frac{4\pi}{k^2}(2l+1)\sin^2\delta_l. \tag{13.56}$$

We also note that by setting $\theta = 0$ in (13.50) and using the fact that $P_l(1) = 1$ we find from (13.55) and (13.56) that

$$\sigma_{\text{tot}} = \frac{4\pi}{k}\text{Im}\,f(k,\theta=0). \tag{13.57}$$

This relation is known as the *optical theorem*; it can be shown to follow from the *conservation of the probability flux*[1].

Convergence of the partial-wave series

Neither the expansion (13.55) of σ_{tot} in terms of the *partial cross-sections* σ_l, nor the expansion (13.50) of the scattering amplitude are useful unless the series in l converges reasonably rapidly. From the radial equation (13.36), we see that as l increases the centrifugal barrier term $l(l+1)/r^2$ becomes more important than the reduced potential $U(r)$, so that for sufficiently large l, $U(r)$ can be neglected and the corresponding phase shift δ_l or partial-wave amplitude f_l is negligible. Since at higher energies the incident particles will more easily penetrate the repulsive centrifugal barrier, the number of important partial waves increases as the energy increases.

A simple (non-rigorous) semi-classical argument can be used to estimate the number of important partial waves. If a potential vanishes beyond some distance a, then according to classical mechanics an incident particle having an impact parameter b will be deflected or not according to whether $b < a$ or $b > a$ (see Fig. 13.3). Now the magnitude of the classical angular momentum L of an incident particle of momentum p is given by $L = bp$, so that if $L > pa$ the particle will be undeflected. In the limit of large l, the classical angular momentum L can be identified with $l\hbar$. Using this relationship as a guide and setting $p = \hbar k$, we see that scattering in the lth partial wave is expected to be small if

$$l > ka. \tag{13.58}$$

In fact, although this relation is derived in the classical limit of large l, it is also valid[1] for smaller values of l. It can also be shown (Problem 13.8) that usually, in the limit

---

[1] For a detailed discussion, see Bransden (1983) or Joachain (1983).

13.4 Applications of the partial-wave method

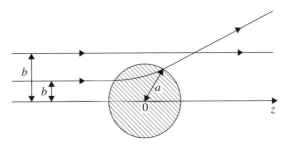

Figure 13.3 Classical particles with impact parameters b greater than the range a of the potential are unscattered.

$k \to 0$, only the $l = 0$ partial wave is of importance. The scattering amplitude is then independent of the scattering angle θ, and the differential cross-section is isotropic. Defining the *scattering length* as

$$\alpha = -\lim_{k \to 0} \frac{\tan \delta_0(k)}{k} \tag{13.59}$$

one finds (Problem 13.8) that the scattering amplitude then reduces to

$$f \underset{k \to 0}{\to} -\alpha \tag{13.60}$$

while the differential cross-section is given by

$$\frac{d\sigma}{d\Omega} \underset{k \to 0}{\to} \alpha^2 \tag{13.61}$$

and the total cross-section becomes

$$\sigma_{\text{tot}} \underset{k \to 0}{\to} 4\pi \alpha^2. \tag{13.62}$$

These conclusions apply generally not only to potentials which vanish for $r > a$, but also to other types of *short-range* potentials such as those that decrease exponentially with r. In contrast, for the important case of the Coulomb potential $V(r) = V_0/r$, the partial-wave series is not convergent and other methods must be used.

13.4 Applications of the partial-wave method

Scattering by a square well

In general, for a given potential, phase shifts are calculated from a numerical solution of the radial equation (13.36). In a few cases an analytical solution is possible, in particular for scattering by an attractive square well, for which the reduced potential $U(r)$ is

$$\begin{aligned} U(r) &= -U_0, \quad r < a \quad (U_0 > 0) \\ &= 0, \quad r > a. \end{aligned} \tag{13.63}$$

Inside the well ($r < a$) the radial equation (13.36) is

$$\left[\frac{d^2}{dr^2} + \frac{2}{r}\frac{d}{dr} - \frac{l(l+1)}{r^2} + K^2\right] R_l(k,r) = 0 \tag{13.64}$$

where we have set $K^2 = k^2 + U_0$. This equation is the same as (13.38), but with K^2 replacing k^2, and its regular solution (finite at the origin) is

$$R_l^I(K,r) = N_l(K) j_l(Kr), \qquad r < a \tag{13.65}$$

where $N_l(K)$ is a normalisation constant. In the exterior region, outside the well, the general solution can be written (see (13.39) and (13.43)) as

$$R_l^E(k,r) = B_l(k)[j_l(kr) - \tan\delta_l(k) n_l(kr)], \qquad r > a. \tag{13.66}$$

The solutions for $r < a$ and $r > a$ can be joined smoothly at $r = a$ by requiring that

$$R_l^I(K,a) = R_l^E(k,a) \tag{13.67a}$$

and

$$\left[\frac{dR_l^I(K,r)}{dr}\right]_{r=a} = \left[\frac{dR_l^E(k,r)}{dr}\right]_{r=a}. \tag{13.67b}$$

Because the overall normalisation of the wave function is not of interest, but only the phase shifts $\delta_l(k)$, we can divide (13.67b) by (13.67a) to eliminate the constants N_l and B_l. Solving for $\tan\delta_l(k)$ we find that

$$\tan\delta_l(k) = \frac{k j_l'(ka) j_l(Ka) - K j_l(ka) j_l'(Ka)}{k n_l'(ka) j_l(Ka) - K n_l(ka) j_l'(Ka)} \tag{13.68}$$

where $j_l'(x) = dj_l(x)/dx$ and $n_l'(x) = dn_l(x)/dx$.

Since at low energies scattering is dominated by the $l = 0$ partial wave, it is of particular interest to study its behaviour. Using (7.64a) we have

$$\tan\delta_0(k) = \frac{k \tan(Ka) - K \tan(ka)}{K + k \tan(ka)\tan(Ka)} \tag{13.69}$$

and the $l = 0$ partial cross-section is

$$\sigma_0 = \frac{4\pi}{k^2} \sin^2\delta_0 = \frac{4\pi}{k^2} \frac{1}{1 + \cot^2\delta_0(k)}. \tag{13.70}$$

Near $k = 0$, it is easily seen from (13.69) that $k\cot\delta_0$ can be expanded in the form

$$k \cot\delta_0(k) = -\frac{1}{\alpha} + \frac{1}{2} r_0 k^2 + \cdots \tag{13.71}$$

where α is the scattering length defined by (13.59) and the constant r_0 is known as the *effective range*. In the present case the scattering length is given by

$$\alpha = \left(1 - \frac{\tan\gamma}{\gamma}\right) a, \qquad \gamma = U_0^{1/2} a \tag{13.72}$$

13.4 Applications of the partial-wave method

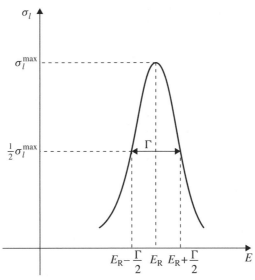

Figure 13.4 The shape of the partial cross-section σ_l near a resonance at energy $E = E_R$. The background scattering is omitted.

and the effective range can also be obtained in terms of the well depth U_0 and its range a. The expansion (13.71) is known as the *effective-range* formula; it can be shown to be valid for all short-range potentials.

If $\gamma \ll 1$ we see from (13.72) that α is finite and negative. If γ is increased, then, when $\gamma = \pi/2$, α and σ_0 become infinite and there is said to be a 'zero-energy resonance'. This condition corresponds to a well which is nearly able to support an s-wave bound state (see Section 7.4). If γ is further increased, α is again finite, until $\gamma = 3\pi/2$. At this point a second $l = 0$ state is nearly bound and the cross-section is again resonant at zero energy. In the same way, a zero-energy resonance occurs whenever γ is an odd multiple of $\pi/2$.

Resonances

In general, the lth partial cross-section

$$\sigma_l = \frac{4\pi}{k^2}(2l+1)\frac{1}{1+\cot^2 \delta_l(k)} \tag{13.73}$$

takes its maximum possible value of $4\pi(2l+1)/k^2$ if there is an energy at which $\cot \delta_l(k)$ vanishes. If this occurs as a result of $\delta_l(k)$ increasing rapidly through an odd multiple of $\pi/2$, the cross-section exhibits a narrow peak as a function of energy and there is said to be a *resonance*. Near the resonance, we can represent $\cot \delta_l(k)$ as

$$\cot \delta_l(k) = \frac{E_R - E}{\Gamma(E)/2} \tag{13.74}$$

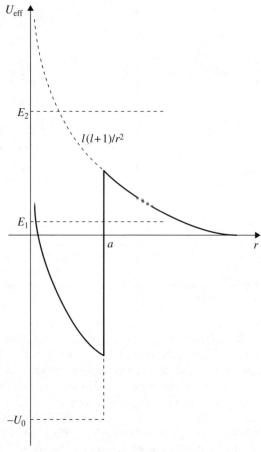

Figure 13.5 The effective reduced potential for the *l*th partial wave ($l \neq 0$) for an attractive square well.

where E_R is the resonance energy at which δ_l is an odd multiple of $\pi/2$. If $\Gamma(E)$ is a slowly varying function of the energy E, it can be approximated by $\Gamma(E_R)$, its value at the resonance energy. Using (13.73), the partial cross-section as a function of energy in the vicinity of the resonance is

$$\sigma_l(E) = \frac{4\pi}{k^2}(2l+1)\frac{\Gamma^2(E_R)/4}{(E-E_R)^2 + \Gamma^2(E_R)/4}. \tag{13.75}$$

This is known as the *Breit–Wigner resonance formula*; it is illustrated in Fig. 13.4. The shape is Lorentzian and $\Gamma \equiv \Gamma(E_R)$ is known as the *resonance width*. We note that at $E = E_R$, σ_l reaches its maximum value of $4\pi(2l+1)/k^2$. In practice, resonances are found superimposed on background scattering arising from the *l*th partial wave itself and from the other partial waves.

13.4 Applications of the partial-wave method

At the resonance energy the amplitude of the wave function within the potential well is very large, and the probability of finding the scattered particle inside the well is correspondingly high. To see how this arises consider the case of the attractive square well (13.63). The effective reduced potential in the radial equation (13.36) for the lth partial wave is

$$U_{\text{eff}} = U(r) + l(l+1)/r^2 \qquad (13.76)$$

and is illustrated in Fig. 13.5. We see that if the barrier at $r = a$ due to the repulsive centrifugal potential $l(l+1)/r^2$ were infinitely high the potential well would support genuine bound states at energies $E_1, E_2, \ldots$. In the real situation, where the barrier is finite, the bound states above the barrier, such as E_2, disappear completely, but at an energy E_R close to E_1 an incoming particle can be trapped inside the well in a state that is nearly bound, since it can only escape by the tunnel effect. Thus a resonance may be considered as a *metastable state*, whose lifetime τ, which is much longer than a typical collision time ($\sim a/v$) can be related to the resonance width Γ by using the uncertainty relation (2.84). Thus, with $\Delta t \simeq \tau$ and $\Delta E \simeq \Gamma$, we have

$$\tau \simeq \frac{\hbar}{\Gamma}. \qquad (13.77)$$

Hard-sphere potential

Another simple, but interesting example, is the 'hard-sphere' potential

$$U(r) = +\infty \qquad r < a$$
$$ = 0 \qquad r > a. \qquad (13.78)$$

Since the scattered particle cannot penetrate into the region $r < a$, the wave function in the exterior region, given by (13.66), must vanish at $r = a$, from which

$$\tan \delta_l(k) = \frac{j_l(ka)}{n_l(ka)}. \qquad (13.79)$$

Using (7.65) we see that in the low-energy limit ($ka \ll 1$)

$$\tan \delta_l(k) \simeq -\frac{(ka)^{2l+1}}{(2l+1)!!(2l-1)!!} \qquad (13.80)$$

where

$$(2l-1)!! = 1.3.5\ldots(2l-1), \qquad l > 0$$
$$ = 1, \qquad l = 0. \qquad (13.81)$$

Hence $|\tan \delta_l|$ quickly decreases as l increases. As a result, the low-energy scattering is always dominated by the s-wave ($l = 0$). Since $j_0(x) = x^{-1} \sin x$ and $n_0(x) = -x^{-1} \cos x$, we see from (13.79) that

$$\delta_0 = -ka. \qquad (13.82)$$

As $k \to 0$, the differential cross-section is isotropic and given by $d\sigma/d\Omega = a^2$, and the total cross-section at zero energy becomes

$$\sigma_{\text{tot}} \xrightarrow[k \to 0]{} 4\pi a^2 \tag{13.83}$$

which is four times the classical value πa^2.

At high energies ($ka \gg 1$) we can use the asymptotic formulae (7.66) to obtain from (13.79) approximate expressions for the phase shifts. We find in this way that

$$\delta_l \simeq \frac{l\pi}{2} - ka \tag{13.84}$$

so that

$$\sigma_{\text{tot}} \sim \frac{4\pi}{k^2} \sum_{l=0}^{l_{\text{max}}} (2l+1) \sin^2\left(\frac{l\pi}{2} - ka\right) \tag{13.85}$$

Taking $l_{\text{max}} = ka$, in accordance with our general discussion of partial waves, and pairing successive terms in (13.85), we have

$$\sigma_{\text{tot}} \simeq \frac{4\pi}{k^2}\left[\left\{\sin^2(ka) + \sin^2\left(\frac{\pi}{2} - ka\right)\right\}\right.$$
$$\left. + 2\left\{\sin^2\left(\frac{\pi}{2} - ka\right) + \sin^2(\pi - ka)\right\} + \cdots\right]$$
$$\simeq \frac{4\pi}{k^2} \sum_{l=0}^{ka} (l) \simeq 2\pi a^2. \tag{13.86}$$

It is interesting to compare this result with that obtained from classical mechanics. In classical mechanics, for a central potential the angle of scattering θ is determined by the impact parameter b (see Fig. 13.3). The number of particles scattered per unit time with scattering angles between θ and $\theta + d\theta$, is equal to the number of incident particles per unit time having impact parameters between b and $b + db$, or in other words the number of incident particles per unit time crossing a ring of area $2\pi b\,db$. This number is equal to $F 2\pi b\,db$, where F is the incident flux. The number of particles scattered into the element of solid angle $d\Omega = 2\pi \sin\theta\,d\theta$ per unit time is therefore equal to $N d\Omega = F 2\pi b\,db$, and from the definition (13.1) the classical differential cross-section is

$$\frac{d\sigma_{\text{cl}}}{d\Omega} = \frac{b}{\sin\theta}\left|\frac{db}{d\theta}\right|. \tag{13.87}$$

For elastic scattering by a hard sphere we obtain (see Fig. 13.6)

$$b(\theta) = a \sin \beta = a \sin\left(\frac{\pi - \theta}{2}\right) \tag{13.88}$$

and hence

$$\left|\frac{db(\theta)}{d\theta}\right| = \frac{a}{2} \sin\left(\frac{\theta}{2}\right). \tag{13.89}$$

13.4 Applications of the partial-wave method

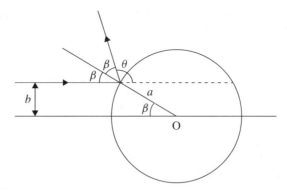

Figure 13.6 Classical scattering from a hard sphere.

Using (13.87)–(13.89) the classical differential cross-section is

$$\frac{d\sigma_{cl}}{d\Omega} = \frac{a^2}{4} \qquad (13.90)$$

and is seen to be independent of the scattering angle. The total classical cross-section is equal to πa^2, which is just the total area of the incident beam that is intercepted by the target. Looking back at (13.86), we see that the high-energy quantum mechanical result is equal to twice the classical value. This is at first sight surprising since in the limit $ka \gg 1$ (i.e. when the de Broglie wavelength $\lambda = 2\pi/k$ is small compared with the size of the target) we might expect that classical mechanics should apply. However, because the 'hard-sphere' potential has a sudden discontinuity at $r = a$ the scattering can never be treated classically. A detailed analysis of this problem shows that at high energies ($ka \gg 1$) half of the total quantum mechanical cross-section arises from 'diffraction' scattering which is produced by interference between the incident wave and the outgoing wave, and occurs within a narrow peak in the forward direction, of angular width given approximately by $1/ka$. We note that as ka becomes very large, diffraction scattering occurs into such small angles that the diffracted particles cannot be distinguished from the undeflected particles, with the consequence that (macroscopic) billiard balls of radius a do behave in the manner predicted by classical mechanics.

Nucleon–nucleon scattering

As a final example of the partial-wave method, we shall consider briefly the scattering of neutrons by protons at low energies. We have already seen in Chapter 10 that there is a single bound state of the neutron and proton of binding energy $E_B = 2.22$ MeV, called the deuteron, which is in the triplet ($S = 1$) spin state. There is no bound state in the singlet ($S = 0$) state. The existence of an electric quadrupole moment shows that the deuteron cannot be entirely in a zero-orbital angular momentum state, so that the nuclear force cannot entirely be a central force. However, the non-central component is small and the low-energy properties of the nucleon–nucleon system can be well described by a central potential of approximate range 2×10^{-15} m. Given this range, we expect (Problem 13.10) that only scattering in the ($l = 0$) s-wave is important for centre-of-mass energies E below about 10 MeV. The shape of the potential well is not determined by a knowledge of the binding energy E_B because for a given shape and range the well depth can be adjusted to provide a particular value for E_B. For example, if the range is taken to be 1.93×10^{-15} m, a square well of this range with a depth of 38.5 MeV will produce one bound state having the correct binding energy $E_B = 2.22$ MeV, but this value can also be obtained by using potentials of appropriate depth having Gaussian, exponential, ..., shapes. It might be thought that a knowledge of the low-energy scattering cross-section could determine the well shape, but we shall now see that this is not the case.

Consider the $l = 0$ radial equation in the centre-of-mass frame of reference for a neutron–proton system in the triplet state interacting via the potential well $^3V(r)$:

$$\left[\frac{d^2}{dr^2} + \frac{2}{r}\frac{d}{dr} - {}^3U(r) + k^2\right] R_0(k, r) = 0 \tag{13.91}$$

where $k^2 = 2\mu E/\hbar^2$,

$$^3U(r) = \frac{2\mu}{\hbar^2} {}^3V(r) \tag{13.92}$$

and the reduced mass μ is defined in terms of the proton and neutron masses, M_p and M_n, by

$$\mu = \frac{M_p M_n}{M_p + M_n} \simeq \frac{1}{2} M_p. \tag{13.93}$$

It is convenient to work in terms of the function $y(k, r) = r R_0(k, r)$ which satisfies the equation

$$\left[\frac{d^2}{dr^2} - {}^3U(r) + k^2\right] y(k, r) = 0 \tag{13.94}$$

subject to the boundary conditions

$$y(k, 0) = 0 \tag{13.95a}$$

and

$$y(k, r) \underset{r \to \infty}{\to} \bar{y}(k, r) = \cot[{}^3\delta_0(k)] \sin kr + \cos kr \tag{13.95b}$$

where $^3\delta_0(k)$ is the s-wave ($l = 0$) phase shift in the triplet ($S = 1$) state. We also recall that the radial equation for the triplet *bound state* function $Y(r)$ introduced in Section 10.7 is (see (10.136))

$$\left[\frac{d^2}{dr^2} - {}^3U(r) - \lambda^2\right] Y(r) = 0 \tag{13.96}$$

where $\lambda^2 = 2\mu E_B/\hbar^2$. Moreover, the function $Y(r)$ satisfies the boundary conditions

$$Y(0) = 0 \tag{13.97a}$$

and

$$Y(r) \underset{r\to\infty}{\to} \bar{Y}(r) = C_2 \exp(-\lambda r) \tag{13.97b}$$

where C_2 is a normalisation constant and $\lambda^{-1} = 4.32 \times 10^{-15}$ m.

Multiplying (13.94) from the left by $Y(r)$ and (13.96) from the left by $y(k, r)$ and subtracting the resulting equations, we obtain

$$Y(r)\frac{d^2}{dr^2}y(k,r) - y(k,r)\frac{d^2}{dr^2}Y(r) = -(k^2 + \lambda^2)y(k,r)Y(r). \tag{13.98}$$

Integrating this equation with respect to r from $r = 0$ to $r = R$ gives

$$\left[Y(r)\frac{d}{dr}y(k,r) - y(k,r)\frac{d}{dr}Y(r)\right]_0^R = -(k^2 + \lambda^2)\int_0^R y(k,r)Y(r)dr. \tag{13.99}$$

The asymptotic forms $\bar{y}(k, r)$ and $\bar{Y}(r)$ satisfy equations similar to (13.94) and (13.96), but with $^3U(r) = 0$. Following the same procedure leading to (13.99) we obtain the equation

$$\left[\bar{Y}(r)\frac{d}{dr}\bar{y}(k,r) - \bar{y}(k,r)\frac{d}{dr}\bar{Y}(r)\right]_0^R = -(k^2 + \lambda^2)\int_0^R \bar{y}(k,r)\bar{Y}(r)dr. \tag{13.100}$$

We note that both equations (13.99) and (13.100) are independent of the normalisation adopted for the functions $y(k, r)$ and $Y(r)$. Hence there is no loss of generality in setting $C_2 = 1$ in (13.97b), as we shall do in what follows. On subtracting (13.100) from (13.99) and letting R tend to infinity we find by using (13.95) and (13.97) that

$$k\cot[^3\delta_0(k)] + \lambda = -(k^2 + \lambda^2)\int_0^\infty [y(k,r)Y(r) - \bar{y}(k,r)\bar{Y}(r)]dr. \tag{13.101}$$

Whatever the shape of the potential well, for $r > a$ the product $y(k, r)Y(r)$ rapidly approaches its asymptotic form $\bar{y}(k, r)\bar{Y}(r)$. It follows that the integrand on the right-hand side of (13.101) is only large in the region in which the potential is large. At energies small compared with the absolute value of the well depth, the integral is nearly independent of k and can be approximated by its value at $k = 0$. This shows that $k\cot[^3\delta_0(k)]$ satisfies the effective range formula (13.71) where the triplet scattering length $^3\alpha$ and effective range $^3r_0$ are given by

$$^3\alpha = [\lambda - \frac{1}{2}\lambda^2\,^3r_0]^{-1} \tag{13.102}$$

and

$$^3r_0 = -2\int_0^\infty [y(0,r)Y(r) - \bar{y}(0,r)\bar{Y}(r)]dr. \tag{13.103}$$

It is clear that measurements of the low-energy triplet ($S = 1$) scattering cross-section determine only the two constants $^3\alpha$ and $^3r_0$ and do not determine the shape of the potential well. Furthermore, no additional information can be obtained from a knowledge of the bound state energy, since this is also determined by $^3\alpha$ and $^3r_0$. As $(\lambda^2\, ^3r_0) < \lambda$, the total cross-section at low energies in the triplet state can be approximated by setting $^3r_0 = 0$ in (13.71) and (13.102). This is known as the zero-range approximation:

$$^3\sigma(k) = \frac{4\pi}{k^2}\frac{1}{1+\cot^2[^3\delta_0(k)]} \simeq \frac{4\pi}{k^2+\lambda^2} = \frac{\hbar^2}{2\mu}\frac{4\pi}{E+E_B}. \qquad (13.104)$$

Since $E_B = 2.22$ MeV, the cross-section at zero energy is predicted to be $^3\sigma(0) \simeq 2 \times 10^{-28}$ m^2. This is much smaller than the observed cross-section. However, the experiments are performed with unpolarised beams which contain an equal population of the three $S = 1$ and the single $S = 0$ spin states. As there are four states in all, 3/4 of the collisions will occur in the triplet state, and 1/4 in the singlet state. The observed cross-section σ is therefore

$$\sigma(k) = \tfrac{1}{4}[3\,^3\sigma(k) + {}^1\sigma(k)] \qquad (13.105)$$

where $^3\sigma$ and $^1\sigma$ are the cross-sections in the triplet and single states, respectively. The singlet cross-section is also determined by an effective range formula and at low energies the zero-range approximation to this cross-section is

$$^1\sigma(k) \simeq \frac{4\pi}{k^2 + (^1\alpha)^{-2}} = \frac{\hbar^2}{2\mu}\frac{4\pi}{E+|^1E|} \qquad (13.106)$$

where $^1\alpha$ is the singlet scattering length and $|^1E| = \hbar^2/[2\mu(^1\alpha)^2]$. The value of the total cross-section at zero energy is $\sigma(0) \simeq 20 \times 10^{-28}$ m^2, from which $|^1\alpha| \simeq 24 \times 10^{-15}$ m and the corresponding energy $|^1E| \simeq 70$ keV. The scattering data alone cannot determine the sign of ^1E, and hence cannot determine whether there is a singlet bound state or not. As in fact there is no bound singlet state (which is known by studying the properties of the deuteron), the energy ^1E is positive and corresponds to what is called a *virtual bound state*, that is a state of near zero energy which is nearly, but not quite bound.

Further information about the low energy nucleon–nucleon system can be obtained by studying proton–proton scattering. In this case the scattering formulae must be modified to allow for the long-range Coulomb interaction. Since the two protons are identical fermions, the $l = 0$ partial wave, which is spatially symmetric, must correspond to a singlet (antisymmetric) spin state and low-energy scattering does not depend on the interaction in the triplet state. Details of the analysis are described in texts on nuclear physics.

13.5 The integral equation of potential scattering

So far we have defined the stationary scattering wave function $\psi_{\mathbf{k}}(\mathbf{r})$ as a solution of the Schrödinger equation (13.21) satisfying the boundary condition (13.28). In this section, we shall show that $\psi_{\mathbf{k}}(\mathbf{r})$ is also the solution of an equivalent *integral equation*

13.5 The integral equation of potential scattering

which incorporates the boundary condition. We shall also obtain an important *integral representation* of the scattering amplitude.

Let us return to the Schrödinger equation (13.21), which we rewrite as

$$(\nabla^2 + k^2)\psi(\mathbf{r}) = U(\mathbf{r})\psi(\mathbf{r}). \tag{13.107}$$

The general solution of this equation can be written as

$$\psi(\mathbf{r}) = \phi(\mathbf{r}) + \int G_0(k, \mathbf{r}, \mathbf{r}')U(\mathbf{r}')\psi(\mathbf{r}')d\mathbf{r}', \tag{13.108}$$

where $\phi(\mathbf{r})$ is a solution of the homogeneous (free particle) Schrödinger equation

$$(\nabla^2 + k^2)\phi(\mathbf{r}) = 0 \tag{13.109}$$

and $G_0(k, \mathbf{r}, \mathbf{r}')$ is a *Green's function* corresponding to the operator ∇^2 and the number k, namely

$$(\nabla^2 + k^2)G_0(k, \mathbf{r}, \mathbf{r}') = \delta(\mathbf{r} - \mathbf{r}'). \tag{13.110}$$

Looking at the asymptotic boundary condition (13.28), we see that for the scattering problem we are considering the function $\phi(\mathbf{r})$ is just the incident plane wave (13.25). We shall denote this particular solution of the homogeneous equation by $\phi_\mathbf{k}(\mathbf{r})$, so that

$$\phi_\mathbf{k}(\mathbf{r}) = e^{i\mathbf{k}\cdot\mathbf{r}}. \tag{13.111}$$

The equation (13.108) is an *integral equation* for $\psi(\mathbf{r})$. With the function $\phi(\mathbf{r})$ given by (13.111), we may write this equation as

$$\psi_\mathbf{k}(\mathbf{r}) = \phi_\mathbf{k}(\mathbf{r}) + \int G_0(k, \mathbf{r}, \mathbf{r}')U(\mathbf{r}')\psi_\mathbf{k}(\mathbf{r}')d\mathbf{r}'. \tag{13.112}$$

As in (13.28), we have added the subscript $\mathbf{k}$ on the wave function $\psi(\mathbf{r})$ to indicate that this is the scattering wave function corresponding to the particular incident plane wave $\exp(i\mathbf{k}\cdot\mathbf{r})$.

The Green's function

In order to determine the Green's function $G_0(k, \mathbf{r}, \mathbf{r}')$, we first use the Fourier integral representation (A.18) of the Dirac delta function to write

$$\delta(\mathbf{r} - \mathbf{r}') = (2\pi)^{-3}\int \exp[i\mathbf{K}\cdot(\mathbf{r} - \mathbf{r}')]d\mathbf{K}. \tag{13.113}$$

We also write $G_0(k, \mathbf{r}, \mathbf{r}')$ in the form

$$G_0(k, \mathbf{r}, \mathbf{r}') = (2\pi)^{-3}\int \exp(i\mathbf{K}\cdot\mathbf{r})g_0(k, \mathbf{K}, \mathbf{r}')d\mathbf{K}. \tag{13.114}$$

Substituting (13.113) and (13.114) in (13.110), we find that

$$g_0(k, \mathbf{K}, \mathbf{r}') = \frac{\exp(-i\mathbf{K}\cdot\mathbf{r}')}{k^2 - K^2} \tag{13.115}$$

giving

$$G_0(k, \mathbf{r}, \mathbf{r}') = -(2\pi)^{-3} \int \frac{\exp[i\mathbf{K}.(\mathbf{r} - \mathbf{r}')]}{K^2 - k^2} d\mathbf{K}. \tag{13.116}$$

The integrand in (13.116) has poles at $K = \pm k$, so that a well-defined prescription is required to avoid these singularities and give a meaning to the integral. This may be done by using the boundary condition (13.28). Comparing (13.28) with (13.108), we see that the Green's function $G_0(k, \mathbf{r}, \mathbf{r}')$ must be determined in such a way that it leads to a scattered wave function which for large r exhibits an *outgoing* spherical wave behaviour of the form $r^{-1} \exp(ikr)$.

Let us set $\mathbf{R} = \mathbf{r} - \mathbf{r}'$, take the z-axis to be along the vector $\mathbf{R}$, and denote by (K, Θ, Φ) the spherical coordinates of the vector $\mathbf{K}$. We then have from (13.116)

$$G_0(k, R) = -(2\pi)^{-3} \int_0^\infty dK\, K^2 \int_0^\pi d\Theta \sin\Theta \int_0^{2\pi} d\Phi \frac{\exp(iKR\cos\Theta)}{K^2 - k^2} \tag{13.117}$$

After performing the angular integrations, we find that

$$G_0(k, R) = -\frac{1}{4\pi^2 R} \int_{-\infty}^{+\infty} \frac{K \sin(KR)}{K^2 - k^2} dK \tag{13.118}$$

where we have used the fact that the integrand is an even function of K, so that the integral can be extended from $-\infty$ to $+\infty$. We may also write (13.118) as

$$G_0(k, R) = \frac{i}{16\pi^2 R}(I_1 - I_2), \tag{13.119}$$

where

$$I_1 = \int_{-\infty}^{+\infty} e^{iKR} \left(\frac{1}{K - k} + \frac{1}{K + k}\right) dK \tag{13.120}$$

and

$$I_2 = \int_{-\infty}^{+\infty} e^{-iKR} \left(\frac{1}{K - k} + \frac{1}{K + k}\right) dK \tag{13.121}$$

We can give a meaning to the integrals I_1 and I_2 by regarding them as contour integrals in the complex K-plane. Suppose, for example, that we avoid the poles at $K = \pm k$ by choosing the path P shown in Fig. 13.7(a). The integral I_1 can then be evaluated by writing

$$I_1 = \oint_C e^{iKR} \left(\frac{1}{K - k} + \frac{1}{K + k}\right) dK \tag{13.122}$$

where the contour C consists of the path P plus an infinite semi-circle C_1 in the upper-half K-plane (see Fig. 13.7(b)). Since $\exp(iKR)$ vanishes on C_1, the contribution to I_1 from the infinite semicircle C_1 is equal to zero, and the integral (13.122) is equal to

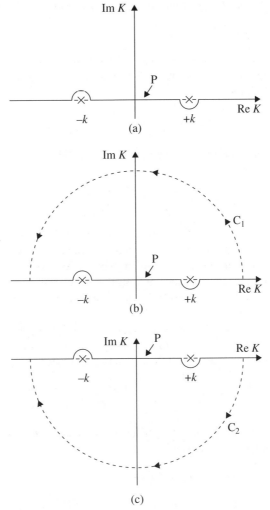

Figure 13.7 (a) The path P avoiding the poles at $K = \pm k$. (b) The contour (P + C_1) for calculating the integral I_1. (c) The contour (P + C_2) for calculating the integral I_2.

its value along the path P. Using the Cauchy theorem[2], we then obtain for the integral I_1, evaluated along the path P, the value

$$I_1 = 2\pi i \exp(ikR). \tag{13.123}$$

The integral I_2 can be evaluated in a similar way by closing the contour with an infinite semicircle C_2 in the lower-half K-plane as shown in Fig. 13.7(c). Using again

---

[2] See for example Byron and Fuller (1969).

the Cauchy theorem, we find that I_2, evaluated along the path P, is given by

$$I_2 = -2\pi i \exp(ikR). \tag{13.124}$$

Hence, going back to (13.119) and using the results (13.123) and (13.124), we see that by choosing the path P to avoid the poles at $K = \pm k$, we obtain the Green's function

$$G_0(k, R) = -\frac{1}{4\pi} \frac{\exp(ikR)}{R}. \tag{13.125}$$

Since $\mathbf{R} = \mathbf{r} - \mathbf{r}'$, we may also write this result in terms of the original variables $\mathbf{r}$ and $\mathbf{r}'$ as

$$G_0(k, \mathbf{r}, \mathbf{r}') = -\frac{1}{4\pi} \frac{\exp(ik|\mathbf{r} - \mathbf{r}'|)}{|\mathbf{r} - \mathbf{r}'|} \tag{13.126}$$

and we note that this expression exhibits the required purely outgoing spherical wave behaviour of the form $r^{-1} \exp(ikr)$ when $r \to \infty$. It is easy to verify that any other choice of integration contour which avoids the poles at $K = \pm k$ in a way different from the path P shown in Fig. 13.7(a) leads to an *incoming* spherical wave behaviour of the form $r^{-1} \exp(-ikr)$ in addition to or in place of the outgoing wave behaviour obtained above.

We remark that the integration prescription which consists in choosing the path P is equivalent to keeping the integration path along the real axis, and shifting the two poles at $K = \pm k$ off the real axis as indicated in Fig. 13.8. The poles are now located at

$$K = \pm(k + i\epsilon'), \qquad \epsilon' \to 0^+ \tag{13.127}$$

where the symbol 0^+ means that ϵ' tends to zero while remaining positive. To first order in ϵ', we have

$$K^2 = k^2 + i\epsilon, \qquad \epsilon \to 0^+ \tag{13.128}$$

where $\epsilon = 2k\epsilon'$. Using (13.116) and (13.128), we may then write for the Green's function $G_0(k, \mathbf{r}, \mathbf{r}')$ the integral representation

$$G_0(k, \mathbf{r}, \mathbf{r}') = -(2\pi)^{-3} \lim_{\epsilon \to 0^+} \int \frac{\exp[i\mathbf{K}\cdot(\mathbf{r} - \mathbf{r}')]}{K^2 - k^2 - i\epsilon} d\mathbf{K}. \tag{13.129}$$

The integration prescription on the right of this equation being equivalent to the choice of the path P of Fig. 13.7(a) leads evidently to the correct result (13.126) for the Green's function $G_0(k, \mathbf{r}, \mathbf{r}')$.

Using (13.111) and the Green's function $G_0(k, \mathbf{r}, \mathbf{r}')$ given by (13.126), the final form of the integral equation (13.112) is

$$\psi_{\mathbf{k}}(\mathbf{r}) = e^{i\mathbf{k}\cdot\mathbf{r}} - \frac{1}{4\pi} \int \frac{\exp(ik|\mathbf{r} - \mathbf{r}'|)}{|\mathbf{r} - \mathbf{r}'|} U(\mathbf{r}')\psi_{\mathbf{k}}(\mathbf{r}')d\mathbf{r}'. \tag{13.130}$$

13.5 The integral equation of potential scattering

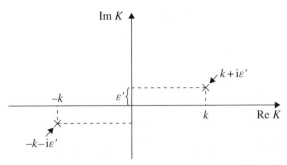

Figure 13.8 Displacement of the poles in the complex K-plane. The pole at $K = k$, after receiving a small positive imaginary part ε', has moved slightly off the real axis to the position $k + i\varepsilon'$. The pole at $K = -k$ has received a negative imaginary part $-\varepsilon'$ and has moved to the position $-k - i\varepsilon'$.

This integral equation is known as the *Lippmann–Schwinger equation* of potential scattering. It is equivalent to the Schrödinger equation (13.21) *plus* the boundary condition (13.28) which is taken care of through the Green's function $G_0(k, \mathbf{r}, \mathbf{r}')$.

Integral representation of the scattering amplitude. The transition matrix element

Let us examine more closely the asymptotic behaviour of the wave function $\psi_\mathbf{k}(\mathbf{r})$ satisfying the integral equation (13.130). We first note that for $r \to \infty$ and r' finite (so that $r' \ll r$), we have

$$|\mathbf{r} - \mathbf{r}'| \underset{r \to \infty}{\to} r - \hat{\mathbf{r}}.\mathbf{r}' + \cdots \tag{13.131}$$

and therefore

$$\frac{\exp(ik|\mathbf{r} - \mathbf{r}'|)}{|\mathbf{r} - \mathbf{r}'|} \underset{r \to \infty}{\to} \frac{\exp(ikr)}{r} \exp(-i\mathbf{k}'.\mathbf{r}') + \cdots \tag{13.132}$$

where terms of higher order in $1/r$ have been neglected. In (13.132) we have introduced the final wave vector $\mathbf{k}' = k\hat{\mathbf{r}}$, which points in the direction of the scattered particle, and has spherical polar coordinates (k, θ, ϕ). Thus, from (13.130) and (13.132), we have

$$\psi_\mathbf{k}(\mathbf{r}) \underset{r \to \infty}{\to} e^{i\mathbf{k}.\mathbf{r}} - \frac{1}{4\pi} \frac{e^{ikr}}{r} \int \exp(-i\mathbf{k}'.\mathbf{r}') U(\mathbf{r}') \psi_\mathbf{k}(\mathbf{r}') d\mathbf{r}'. \tag{13.133}$$

Upon comparison with (13.28), we see that $\psi_\mathbf{k}(\mathbf{r})$ exhibits the desired asymptotic behaviour, and we obtain for the scattering amplitude the *integral representation*

$$f(k, \theta, \phi) = -\frac{1}{4\pi} \int \exp(-i\mathbf{k}'.\mathbf{r}) U(\mathbf{r}) \psi_\mathbf{k}(\mathbf{r}) d\mathbf{r}$$

$$= -\frac{1}{4\pi} \langle \phi_{\mathbf{k}'} | U | \psi_\mathbf{k} \rangle, \tag{13.134}$$

where we have introduced the plane wave

$$\phi_{\mathbf{k}'}(\mathbf{r}) = e^{i\mathbf{k}'\cdot\mathbf{r}} \tag{13.135}$$

corresponding to the final wave vector $\mathbf{k}'$ of the scattered particle. In terms of the potential $V(\mathbf{r}) = (\hbar^2/2m)U(\mathbf{r})$, we may also write the scattering amplitude in the form

$$f = -\frac{m}{2\pi\hbar^2}\langle\phi_{\mathbf{k}'}|V|\psi_{\mathbf{k}}\rangle \tag{13.136}$$

The *transition matrix element* $T_{\mathbf{k}'\mathbf{k}}$ is defined as

$$T_{\mathbf{k}'\mathbf{k}} = \langle\phi_{\mathbf{k}'}|V|\psi_{\mathbf{k}}\rangle \tag{13.137}$$

so that

$$f = -\frac{m}{2\pi\hbar^2}T_{\mathbf{k}'\mathbf{k}} \tag{13.138}$$

and the differential cross-section is given by

$$\frac{d\sigma}{d\Omega} = |f|^2 = (2\pi)^{-2}m^2\hbar^{-4}|T_{\mathbf{k}'\mathbf{k}}|^2. \tag{13.139}$$

Partial wave analysis of the Lippmann–Schwinger equation

The integral equation (13.130) can also be analysed in partial waves. Assuming that we are dealing with a central potential, we can expand the scattering wave function $\psi_{\mathbf{k}}$ in Legendre polynomials as in (13.34), where $R_l(k,r)$ are the radial functions. Let us choose for these functions the normalisation (see (13.39) and (13.43))

$$R_l(k,r) \underset{r\to\infty}{\to} j_l(kr) - \tan\delta_l(k)n_l(kr)$$

$$\underset{r\to\infty}{\to} \frac{1}{kr}\left[\sin\left(kr - \frac{l\pi}{2}\right) + \tan\delta_l(k)\cos\left(kr - \frac{l\pi}{2}\right)\right]. \tag{13.140}$$

It can then be shown (Problem 13.11) that each radial function $R_l(k,r)$ satisfies the radial integral equation

$$R_l(k,r) = j_l(kr) + \int_0^\infty G_l(k,r,r')U(r')R_l(k,r')r'^2 dr'. \tag{13.141}$$

In this equation $G_l(k,r,r')$ is a radial Green's function given by

$$G_l(k,r,r') = kj_l(kr_<)n_l(kr_>), \tag{13.142}$$

where $r_<$ and $r_>$ denote respectively the lesser and the greater of r and r'. By analysing the behaviour of the radial equation (13.141) as $r \to \infty$ and comparing with (13.140), one obtains (Problem 13.11) the integral representation

$$\tan\delta_l(k) = -k\int_0^\infty j_l(kr)U(r)R_l(k,r)r^2 dr, \tag{13.143}$$

where we recall that $R_l(k,r)$ is normalised according to (13.140).

13.6 The Born approximation

The partial-wave method for calculating cross-sections is not always very convenient, for example if the number of partial wave is large. Fortunately, if the potential is weak, or the energy high, perturbation theory can be used. In this section, we shall concentrate our attention on a perturbative method called the *Born series*, which consists of expanding the scattering wave function, or the scattering amplitude, in powers of the interaction potential. In particular, we shall study in detail the first term of this expansion, known as the (first) Born approximation. A generalised version of this approximation will be used in Section 13.8 to study collisions involving composite systems.

The Born series

Let us attempt to solve the integral equation (13.130), starting from the incident plane wave $\phi_\mathbf{k}(\mathbf{r})$ as our 'zero-order' approximation. We obtain in this way the sequence of functions

$$\psi_0(\mathbf{r}) = \phi_\mathbf{k}(\mathbf{r}) = e^{i\mathbf{k}\cdot\mathbf{r}}, \tag{13.144a}$$

$$\psi_1(\mathbf{r}) = \phi_\mathbf{k}(\mathbf{r}) + \int G_0(k, \mathbf{r}, \mathbf{r}')U(\mathbf{r}')\psi_0(\mathbf{r}')d\mathbf{r}', \tag{13.144b}$$

$$\vdots$$

$$\psi_n(\mathbf{r}) = \phi_\mathbf{k}(\mathbf{r}) + \int G_0(k, \mathbf{r}, \mathbf{r}')U(\mathbf{r}')\psi_{n-1}(\mathbf{r}')d\mathbf{r}', \tag{13.144c}$$

where $G_0(k, \mathbf{r}, \mathbf{r}')$ is the Green's function (13.126).

Assuming for the moment that this sequence converges towards the exact wave function $\psi_\mathbf{k}$, we may write for $\psi_\mathbf{k}$ the *Born series*

$$\psi_\mathbf{k}(\mathbf{r}) = \phi_\mathbf{k}(\mathbf{r}) + \int G_0(k, \mathbf{r}, \mathbf{r}')U(\mathbf{r}')\phi_\mathbf{k}(\mathbf{r}')d\mathbf{r}'$$
$$+ \int G_0(k, \mathbf{r}, \mathbf{r}')U(\mathbf{r}')G_0(k, \mathbf{r}', \mathbf{r}'')U(\mathbf{r}'')\phi_\mathbf{k}(\mathbf{r}'')d\mathbf{r}'d\mathbf{r}'' + \cdots \tag{13.145}$$

which is clearly a perturbative expansion in powers of the interaction potential. Upon substitution of the series (13.145) into the integral representation (13.134) of the scattering amplitude, we obtain the *Born series for the scattering amplitude*. That is

$$f = -\frac{1}{4\pi}\langle\phi_{\mathbf{k}'}|U + UG_0U + UG_0UG_0U + \cdots|\phi_\mathbf{k}\rangle \tag{13.146}$$

The first term of this series, namely

$$f^B = -\frac{1}{4\pi}\langle\phi_{\mathbf{k}'}|U|\phi_\mathbf{k}\rangle \tag{13.147}$$

is called the *(first) Born approximation to the scattering amplitude*.

A similar Born series may evidently be written for the transition matrix element $T_{\mathbf{k'k}}$. Using (13.137) and (13.144a), we see that the (first) Born approximation to $T_{\mathbf{k'k}}$ is given by

$$T^B_{\mathbf{k'k}} = \langle \phi_{\mathbf{k'}} | V | \phi_{\mathbf{k}} \rangle \tag{13.148}$$

and from (13.138) it follows that $f^B = -m(2\pi\hbar^2)^{-1} T^B_{\mathbf{k'k}}$.

If the potential is central, and we use the partial wave approach, we may solve the radial integral equations (13.141) by iteration, starting from the 'zero-order' approximation $R_l^{(0)}(k, r) = j_l(kr)$. Upon substitution in (13.143) we then generate a Born series for $\tan \delta_l$, whose first term, called the (first) Born approximation to $\tan \delta_l$, is given by

$$\tan \delta_l^B(k) = -k \int_0^\infty [j_l(kr)]^2 U(r) r^2 \, dr. \tag{13.149}$$

The problem of the convergence of the Born series is a difficult one, which lies outside the scope of this book. A crude sufficient condition of convergence may be obtained by requiring that the time τ_1 spent by the particle in the 'range' a of the potential should be small with respect to a 'characteristic' time τ_2 necessary for the potential to have a significant effect. The time τ_1 is given approximately by $\tau_1 \simeq a/v$, where v is the magnitude of the incident velocity of the particle. On the other hand, if $|V_0|$ denotes a typical strength of the potential (which may be attractive or repulsive) and $|U_0| = 2m|V_0|/\hbar^2$ is the corresponding strength of the reduced potential, we may take $\tau_2 \simeq \hbar/|V_0| = 2m/(\hbar|U_0|)$. If we require that $\tau_1 \ll \tau_2$, we must therefore have

$$\frac{|V_0|a}{\hbar v} = \frac{|U_0|a}{2k} \ll 1 \tag{13.150}$$

The first Born approximation

Let us now study in more detail the first Born approximation. Using (13.147), (13.111) and (13.135), the first Born approximation for the scattering amplitude is given by

$$\begin{aligned} f^B &= -\frac{1}{4\pi} \int \exp(-i\mathbf{k'}\cdot\mathbf{r}) U(\mathbf{r}) \exp(i\mathbf{k}\cdot\mathbf{r}) d\mathbf{r} \\ &= -\frac{1}{4\pi} \int \exp(i\mathbf{\Delta}\cdot\mathbf{r}) U(\mathbf{r}) d\mathbf{r}, \end{aligned} \tag{13.151}$$

where in the second line we have introduced the wave vector transfer

$$\mathbf{\Delta} = \mathbf{k} - \mathbf{k'}, \tag{13.152}$$

$\hbar\mathbf{\Delta}$ being the momentum transfer. Since for the elastic scattering case considered here $k = k'$ and $\mathbf{k}\cdot\mathbf{k'} = k^2 \cos\theta$, the magnitude of $\mathbf{\Delta}$ is

$$\Delta = 2k \sin\frac{\theta}{2}, \tag{13.153}$$

where θ is the angle of scattering. From (13.148) we see that the corresponding first Born approximation for the transition matrix element is given by

$$T^B_{\mathbf{k'k}} = \int \exp(-i\mathbf{k'}\cdot\mathbf{r})V(\mathbf{r})\exp(i\mathbf{k}\cdot\mathbf{r})d\mathbf{r}$$

$$= \int \exp(i\mathbf{\Delta}\cdot\mathbf{r})V(\mathbf{r})d\mathbf{r}. \tag{13.154}$$

It is apparent from (13.151) and (15.154) that the first Born quantities f^B and $T^B_{\mathbf{k'k}}$ are proportional to the Fourier transform of the potential. We also note that the first Born differential cross-section

$$\frac{d\sigma^B}{d\Omega} = |f^B|^2 \tag{13.155}$$

remains unchanged when the sign of the potential is reversed.

Let us consider the particular case of *central* potentials. The angular integrations in (13.151) are then readily performed. Taking $\mathbf{\Delta}$ as our polar axis and denoting by (α, β) the polar angles of $\mathbf{r}$ with respect to that axis, we have

$$f^B(\Delta) = -\frac{1}{4\pi}\int_0^\infty dr\, r^2 U(r) \int_0^\pi d\alpha\, \sin\alpha \int_0^{2\pi} d\beta\, \exp(i\Delta r \cos\alpha)$$

$$= -\frac{1}{2}\int_0^\infty dr\, r^2 U(r) \int_{-1}^{+1} d(\cos\alpha)\, \exp(i\Delta r \cos\alpha)$$

$$= -\frac{1}{\Delta}\int_0^\infty r \sin(\Delta r) U(r) dr. \tag{13.156}$$

We see that this quantity is real and depends on k (i.e. on the energy) and on the scattering angle θ only via the magnitude Δ of the wave-vector transfer. The corresponding total cross-section in the first Born approximation is given by

$$\sigma^B_{\text{tot}}(k) = 2\pi \int_0^\pi |f^B(\Delta)|^2 \sin\theta\, d\theta$$

$$= \frac{2\pi}{k^2}\int_0^{2k} |f^B(\Delta)|^2 \Delta\, d\Delta \tag{13.157}$$

where we have used the fact that $\sin\theta\, d\theta = \Delta d\Delta/k^2$ (see (13.153)). It is worth noting from (13.157) that

$$\lim_{k\to\infty}[k^2 \sigma^B_{\text{tot}}(k)] = 2\pi \int_0^\infty |f^B(\Delta)|^2 \Delta\, d\Delta. \tag{13.158}$$

Since the right-hand side of (13.158) is independent of k, it follows that σ^B_{tot} is proportional to k^{-2} as k becomes large. Because $E = \hbar^2 k^2/2m$ we may also write

$$\sigma^B_{\text{tot}} \underset{E\to\infty}{\sim} AE^{-1} \tag{13.159}$$

where A is a constant. Thus σ^B_{tot} falls off like E^{-1} at high energies.

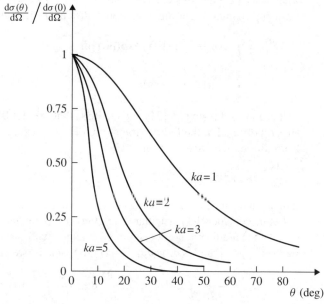

Figure 13.9 The differential cross-section in the first Born approximation for scattering by the Yukawa potential (13.160), for various values of ka.

As an illustration of the foregoing discussion, we consider the Yukawa (or 'screened-Coulomb') reduced potential

$$U(r) = U_0 \frac{e^{-\alpha r}}{r} = U_0 \frac{e^{-r/a}}{r}, \qquad a = \alpha^{-1}. \tag{13.160}$$

The evaluation of the integral on the right of (13.156) is then straightforward, and yields

$$f^B = -\frac{U_0}{\alpha^2 + \Delta^2}. \tag{13.161}$$

The corresponding first Born differential cross-section

$$\frac{d\sigma^B}{d\Omega} = \frac{U_0^2}{(\alpha^2 + \Delta^2)^2} \tag{13.162}$$

is plotted in Fig. 13.9 as a function of the scattering angle θ, for various values of $ka = k/\alpha$. We see that for large values of ka the differential cross-section (13.162) is essentially concentrated within a forward cone of angular aperture $\delta\theta \simeq (ka)^{-1}$. This behaviour is a direct consequence of (13.151) and of the fact that the Fourier transform of a function $U(r)$, which is negligible for $r \gtrsim a$, is appreciable only for values of Δ such that $\Delta \lesssim a^{-1}$, corresponding to scattering angles $\theta \lesssim (ka)^{-1}$ (see (13.153)).

The Coulomb potential

The asymptotic form (13.28) of the scattering wave function applies to potentials decreasing faster than r^{-1} as $r \to \infty$ and therefore not to the Coulomb potential acting between charges q_A and q_B:

$$V_c(r) = \frac{q_A q_B}{4\pi \varepsilon_0} \frac{1}{r}. \tag{13.163}$$

Nevertheless, one can obtain the scattering amplitude and differential cross-section for this potential in first Born approximation by letting α tend to zero in the results derived above for the Yukawa potential. Writing the reduced Coulomb potential corresponding to (13.163) as

$$U_c(r) = \frac{U_0}{r}, \qquad U_0 = \frac{2m}{\hbar^2} \frac{q_A q_B}{4\pi \varepsilon_0} \tag{13.164}$$

the first Born Coulomb scattering amplitude, obtained by taking the limit $\alpha \to 0$ in (13.161), is

$$f_c^B = -\frac{U_0}{\Delta^2}. \tag{13.165}$$

The corresponding first Born differential cross-section for Coulomb scattering is given by

$$\frac{d\sigma_c^B}{d\Omega} = \frac{U_0^2}{\Delta^4} = \left(\frac{\gamma}{2k}\right)^2 \frac{1}{\sin^4(\theta/2)}$$

$$= \left(\frac{q_A q_B}{4\pi \varepsilon_0}\right)^2 \frac{1}{16 E^2 \sin^4(\theta/2)} \tag{13.166}$$

where

$$\gamma = \frac{q_A q_B}{(4\pi \varepsilon_0) \hbar v} = \frac{U_0}{2k}. \tag{13.167}$$

The result (13.166) is identical to the formula that Rutherford obtained in 1911 by using classical mechanics.

It is a remarkable feature of Coulomb scattering that an *exact* quantum mechanical treatment of the Coulomb potential[3] yields the *same* result (13.166) for the differential cross-section. However, the exact Coulomb scattering amplitude f_c differs from the first Born result (13.165) by a phase factor. It is found that

$$f_c = -\frac{\gamma}{2k \sin^2(\theta/2)} \frac{\Gamma(1+i\gamma)}{\Gamma(1-i\gamma)} \exp\{-i\gamma \log[\sin^2(\theta/2)]\} \tag{13.168}$$

where Γ denotes the gamma function.

The Rutherford formula (13.166) for scattering by a Coulomb potential exhibits other remarkable features. Indeed, the differential cross-section (13.166) does not

---

[3] See Bransden (1983) or Joachain (1983).

depend on the *sign* of the potential. This feature is obvious within the framework of the first Born approximation, but it is striking if we recollect that the Rutherford formula is in fact an exact result. Moreover, since the energy E and the scattering angle θ enter into separate factors, the Rutherford differential cross-section (13.166) is scaled at *all* angles by the factor $(q_A q_B/16\pi\varepsilon_0 E)^2$, so that the angular distribution is independent of the energy. We also note that at fixed θ the differential cross-section is proportional to E^{-2}. Finally, the differential cross-section (13.166) is infinite in the forward direction ($\theta = 0$) where it diverges like θ^{-4}. Consequently, the total cross-section is not defined for pure Coulomb scattering. However, when considering real scattering processes, we must remember that all Coulomb potentials are modified at large distances because of the screening effect of other charges. As a result of this screening, the *quantum mechanical*[4] differential cross-section becomes finite in the forward direction, and the corresponding total cross-section is then defined.

13.7 Collisions between identical particles

Collisions between identical particles are particularly interesting as a direct illustration of the fundamental differences between classical and quantum mechanics. We shall examine first the elastic scattering of two identical spinless bosons and then analyse elastic collisions between two identical spin-1/2 fermions.

Scattering of two identical spinless bosons

Let us consider the elastic scattering of two identical bosons of mass m. For simplicity we shall consider only the case of spinless bosons. We work in the centre-of-mass system, in which the time-independent Schrödinger equation is

$$\left[-\frac{\hbar^2}{2\mu}\nabla^2 + V(\mathbf{r})\right]\psi(\mathbf{r}) = E\psi(\mathbf{r}) \tag{13.169}$$

where $\mu = m/2$ is the reduced mass and $\mathbf{r} = \mathbf{r}_1 - \mathbf{r}_2$ is the relative position vector of the two colliding particles. The situation in the centre-of-mass system is illustrated in Fig. 13.10. Two identical particles 1 and 2 approach one another, moving parallel to the z-axis in opposite directions. After an elastic collision the velocity of each particle is changed in direction but remains unchanged in magnitude. A detector counts the particles scattered into the direction characterised by the polar angles (θ, ϕ). Since the particles 1 and 2 are identical, there is no way of deciding whether a particle recorded by the detector results from a collision event in which particle 1 is scattered in the direction (θ, ϕ) (see Fig. 13.10(a)), or from a collision process in which particle 2 is scattered in that direction, so that particle 1 is scattered in the opposite direction $(\pi - \theta, \phi + \pi)$ (see Fig. 13.10(b)).

---

[4] It is worth noting that in classical mechanics the total cross-section does not exist for any potential that does not strictly vanish beyond a certain distance.

13.7 Collisions between identical particles

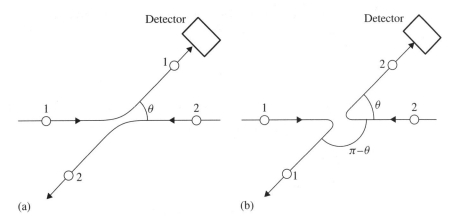

Figure 13.10 The scattering of identical particles in the centre-of-mass frame.

In classical mechanics the differential cross-section for scattering in the direction (θ, ϕ) would simply be the sum of the differential cross-sections for observation of particle 1 and particle 2 in that direction. If the same were to be true in quantum mechanics, we would obtain for the differential cross-section the 'classical' result

$$\frac{d\sigma_{cl}}{d\Omega} = |f(\theta, \phi)|^2 + |f(\pi - \theta, \phi + \pi)|^2 \tag{13.170}$$

where $f(\theta, \phi)$, the centre of mass amplitude for scattering in the direction (θ, ϕ), is related to the asymptotic behaviour of the solution $\psi_{\mathbf{k}}(\mathbf{r})$ of (13.169) satisfying the usual boundary condition

$$\psi_{\mathbf{k}}(\mathbf{r}) \underset{r \to \infty}{\to} e^{i\mathbf{k} \cdot \mathbf{r}} + f(\theta, \phi) \frac{e^{ikr}}{r} \tag{13.171}$$

and we have omitted the k-dependence of the scattering amplitude for notational simplicity. However, we shall now show that the expression (13.170) for the differential cross-section is incorrect. Indeed, we have seen in Chapter 10 that wave functions describing systems of identical particles must be properly symmetrised with respect to permutations of identical particles. In particular, a wave function describing a system of identical bosons must be completely symmetric. Thus, in the case of two identical spinless bosons, the wave function must be *symmetric* under the interchange of the spatial coordinates of the two particles. Now the interchange $\mathbf{r}_1 \leftrightarrow \mathbf{r}_2$ corresponds to replacing the relative position vector $\mathbf{r}$ by $-\mathbf{r}$, which in polar coordinates corresponds to (r, θ, ϕ) being replaced by $(r, \pi - \theta, \phi + \pi)$. The wave function $\psi_{\mathbf{k}}(\mathbf{r})$ satisfying the boundary condition (13.171) does not have the required symmetry, but the symmetric combination

$$\psi_{+}(\mathbf{r}) = \psi_{\mathbf{k}}(\mathbf{r}) + \psi_{\mathbf{k}}(-\mathbf{r}) \tag{13.172}$$

is also a solution of the Schrödinger equation (13.169) and does have the required symmetry: $\psi_+(-\mathbf{r}) = \psi_+(\mathbf{r})$. Using (13.171), the asymptotic form of $\psi_+(\mathbf{r})$ is seen to be

$$\psi_+(\mathbf{r}) \underset{r\to\infty}{\to} [e^{i\mathbf{k}\cdot\mathbf{r}} + e^{-i\mathbf{k}\cdot\mathbf{r}}] + [f(\theta,\phi) + f(\pi-\theta,\phi+\pi)]\frac{e^{ikr}}{r}. \quad (13.173)$$

The amplitude of the spherically outgoing wave is the symmetric amplitude

$$f_+(\theta,\phi) = f(\theta,\phi) + f(\pi-\theta,\phi+\pi) \quad (13.174)$$

so that the differential cross-section is

$$\frac{d\sigma}{d\Omega} = |f(\theta,\phi) + f(\pi-\theta,\phi+\pi)|^2 \quad (13.175)$$

a result which we can write in the form

$$\frac{d\sigma}{d\Omega} = |f(\theta,\phi)|^2 + |f(\pi-\theta,\phi+\pi)|^2 + 2\mathrm{Re}[f(\theta,\phi)f^*(\pi-\theta,\phi+\pi)]. \quad (13.176)$$

It is important to note that this formula differs from the 'classical' result (13.170) by the presence of the third term on the right, which arises from the *interference* between the amplitudes $f(\theta,\phi)$ and $f(\pi-\theta,\phi+\pi)$. We also remark that the total cross-section

$$\sigma_{\mathrm{tot}} = \int |f(\theta,\phi) + f(\pi-\theta,\phi+\pi)|^2 d\Omega \quad (13.177)$$

is equal to twice the number of particles removed from the incident beam per unit time and unit incident flux.

In the simple case for which the interaction potential is central, the scattering amplitude is independent of the azimuthal angle ϕ. The differential cross-section (13.175) then reduces to

$$\frac{d\sigma}{d\Omega} = |f(\theta) + f(\pi-\theta)|^2 \quad (13.178)$$

or

$$\frac{d\sigma}{d\Omega} = |f(\theta)|^2 + |f(\pi-\theta)|^2 + 2\mathrm{Re}[f(\theta)f^*(\pi-\theta)] \quad (13.179)$$

so that the scattering is symmetric about the angle $\theta = \pi/2$ in the centre-of-mass system. Since

$$P_l[\cos(\pi-\theta)] = (-1)^l P_l(\cos\theta) \quad (13.180)$$

it is clear that the partial-wave expansion of the symmetrised scattering amplitude $f_+(\theta) = f(\theta) + f(\pi-\theta)$ contains only *even* values of l. Moreover, at $\theta = \pi/2$ we note that the quantum mechanical differential cross-section (13.178) is equal to

$$\frac{d\sigma(\theta=\pi/2)}{d\Omega} = 4|f(\theta=\pi/2)|^2 \quad (13.181)$$

and hence is four times as big as if the two colliding particles were distinguishable, and twice as big as the 'classical' result

$$\frac{d\sigma_{cl}(\theta = \pi/2)}{d\Omega} = 2|f(\theta = \pi/2)|^2 \tag{13.182}$$

following from (13.170). Furthermore, if there is only s-wave ($l = 0$) scattering, so that the scattering amplitude f is isotropic, we see from (13.178) that two colliding spinless bosons have a differential cross-section four times as big as if they were distinguishable particles, and twice as big as the classical result (13.170).

The non-classical effects due to the interference term in (13.179) can be illustrated in a particularly simple and striking way for the case of *Coulomb scattering*. Let us consider two identical spinless bosons of charge Ze interacting only through Coulomb forces. This is the case for example in the scattering of two identical spinless nuclei (e.g. the scattering of two alpha particles or two ^{12}C nuclei) at colliding energies which are low enough so that the nuclear forces between the two colliding particles can be neglected due to the presence of the Coulomb barrier. Equation (13.178) now reads

$$\frac{d\sigma}{d\Omega} = |f_c(\theta) + f_c(\pi - \theta)|^2 \tag{13.183}$$

where $f_c(\theta)$ is the Coulomb scattering amplitude (13.168). Therefore, we have

$$\frac{d\sigma}{d\Omega} = \left(\frac{\gamma}{2k}\right)^2 \left| \frac{\exp\{-2i\gamma \log[\sin(\theta/2)]\}}{\sin^2(\theta/2)} + \frac{\exp\{-2i\gamma \log[\cos(\theta/2)]\}}{\cos^2(\theta/2)} \right|^2 \tag{13.184}$$

or

$$\frac{d\sigma}{d\Omega} = \left(\frac{\gamma}{2k}\right)^2 \Big[\operatorname{cosec}^4(\theta/2) + \sec^4(\theta/2)$$

$$+ 2\operatorname{cosec}^2(\theta/2) \sec^2(\theta/2) \cos\{2\gamma \log[\tan(\theta/2)]\} \Big] \tag{13.185}$$

where (see (13.167))

$$\gamma = \frac{(Ze)^2}{(4\pi\varepsilon_0)\hbar v}. \tag{13.186}$$

The result (13.185) is called the *Mott formula for the Coulomb scattering of two identical spinless bosons*. The corresponding classical calculation would only yield the first two terms on the right of (13.185) while the Rutherford formula (13.166), which applies to the Coulomb scattering of two *distinct* particles, is obtained by retaining only the first term on the right of (13.185). We remark that the interference term (the third term on the right of (13.185)) oscillates increasingly about zero as γ becomes larger and (or) one departs from the angle $\theta = \pi/2$. In the limit $\gamma \gg 1$, the differential cross-section (13.185) averaged over a small solid angle $\delta\Omega$ tends towards the classical cross-section. We also verify that at $\theta = \pi/2$ the actual (quantum mechanical) cross-section (13.185) is twice as big as the classical one.

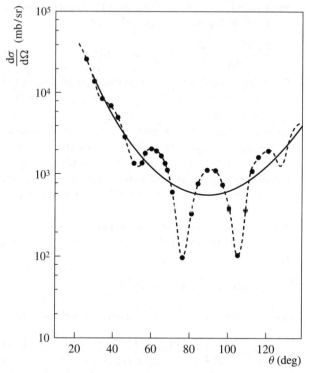

Figure 13.11 The differential cross-section (in mb/sr, where 1 mb ≡ 1 millibarn = 10^{-31} m^2), corresponding to the elastic scattering of two ^{12}C nuclei having a centre-of-mass energy of 5 MeV. The dashed curve represents the results obtained by using the Mott formula (13.185). The solid curve shows the corresponding 'classical' differential cross-section obtained by omitting the interference term in (13.185). The dots are the experimental data of D. A. Bromley, J. A. Kuehner and E. Almqvist.

As an example, let us consider the elastic scattering of two spinless ^{12}C nuclei having a centre-of-mass kinetic energy of 5 MeV. The corresponding relative velocity of the two colliding particles is $v \simeq 1.3 \times 10^7$ m s^{-1}, and since $Z = 6$ for a carbon nucleus we have $\gamma \simeq 6$. As seen from Fig. 13.11, the experimental data are in excellent agreement with the Mott formula (13.185), and demonstrate very clearly the presence of the non-classical interference term.

Scattering of two identical spin-1/2 fermions

The scattering of identical fermions is more difficult to analyse than that of spinless bosons because of the complications due to the spin. For simplicity, we shall only consider the case of two identical spin-1/2 fermions interacting through *central* forces. Since the interaction is in general different in the singlet ($S = 0$) and triplet ($S = 1$) spin states of the two fermions, we shall start from two (unsymmetrised) scattering amplitudes $f_s(\theta)$ and $f_t(\theta)$ corresponding respectively to the singlet and triplet cases.

The full wave function describing a system of two identical spin-1/2 fermions must be *antisymmetric* in the interchange of the two particles, i.e. when all their coordinates (spatial and spin) are interchanged. Now, if the system is in a *singlet* spin state ($S = 0$), the spin part of the wave function is given by (6.301) and is antisymmetric. Hence the corresponding spatial part of the wave function must be symmetric in the interchange of the position vectors $\mathbf{r}_1$ and $\mathbf{r}_2$ of the two particles. As a result, the symmetrised singlet scattering amplitude is

$$f_{s+} = f_s(\theta) + f_s(\pi - \theta) \tag{13.187}$$

and the differential cross-section in the singlet spin state is

$$\frac{d\sigma_s}{d\Omega} = |f_s(\theta) + f_s(\pi - \theta)|^2. \tag{13.188}$$

If, on the other hand, the two spin-1/2 fermions are in a triplet spin state ($S = 1$) the corresponding three spin functions (6.302) are symmetric in the interchange of the spin coordinates of the two particles. The spatial part of the wave function must therefore be antisymmetric in the interchange of the position vectors $\mathbf{r}_1$ and $\mathbf{r}_2$, so that the symmetrised triplet scattering amplitude is given by

$$f_{t-} = f_t(\theta) - f_t(\pi - \theta) \tag{13.189}$$

and the differential cross-section in the triplet spin state is

$$\frac{d\sigma_t}{d\Omega} = |f_t(\theta) - f_t(\pi - \theta)|^2. \tag{13.190}$$

If the 'incident' and 'target' particles are unpolarised (i.e. their spins are randomly orientated), the probability of obtaining triplet states is three times that of singlet states, so that the differential cross-section is given by

$$\frac{d\sigma}{d\Omega} = \frac{1}{4}\frac{d\sigma_s}{d\Omega} + \frac{3}{4}\frac{d\sigma_t}{d\Omega}$$

$$= \frac{1}{4}|f_s(\theta) + f_s(\pi - \theta)|^2 + \frac{3}{4}|f_t(\theta) - f_t(\pi - \theta)|^2. \tag{13.191}$$

For the particular case of *spin-independent* central interactions, where

$$f_s(\theta) = f_t(\theta) = f(\theta) \tag{13.192}$$

we find from (13.191) that

$$\frac{d\sigma}{d\Omega} = |f(\theta)|^2 + |f(\pi - \theta)|^2 - \text{Re}[f(\theta)f^*(\pi - \theta)]. \tag{13.193}$$

We note that this formula differs from the 'classical' result by the presence of the third term on the right, which again is an *interference* term. We also remark that at $\theta = \pi/2$ the quantum mechanical differential cross-section (13.193) is given by

$$\frac{d\sigma(\theta = \pi/2)}{d\Omega} = |f(\theta = \pi/2)|^2 \tag{13.194}$$

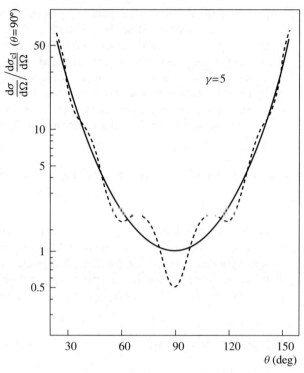

Figure 13.12 The differential cross-section $d\sigma/d\Omega$ corresponding to the scattering of two spin-$\frac{1}{2}$ fermions interacting through a Coulomb potential, divided by $d\sigma_{cl}(\theta = 90°)/d\Omega$ for the value $\gamma = (Ze)^2/[(4\pi\varepsilon_0)\hbar v] = 5$. The dashed curve represents the results obtained by using the Mott formula (13.195). The solid curve shows the corresponding 'classical' results, obtained by omitting the interference term in (13.195).

and hence is equal to one-half of the classical result $d\sigma_{cl}(\theta = \pi/2)/d\Omega = 2|f(\theta = \pi/2)|^2$. Furthermore, if there is only s-wave ($l = 0$) scattering, the differential cross-section (13.193) is four times smaller than the corresponding one (given by (13.179)) for the scattering of two identical spinless bosons.

As an illustration of equation (13.193), let us consider the case of *Coulomb scattering* of two identical spin-1/2 fermions of charge Ze, for example electron–electron scattering ($Z = -1$) or low-energy proton–proton scattering ($Z = +1$). Using the Coulomb scattering amplitude (13.168), we obtain from (13.193) the *Mott formula for the Coulomb scattering of two identical spin-1/2 fermions*:

$$\frac{d\sigma}{d\Omega} = \left(\frac{\gamma}{2k}\right)^2 \Big[\operatorname{cosec}^4(\theta/2) + \sec^4(\theta/2)$$

$$- \operatorname{cosec}^2(\theta/2) \sec^2(\theta/2) \cos\{2\gamma \log[\tan(\theta/2)]\}\Big] \quad (13.195)$$

where γ is given by (13.186). As in the case of equation (13.185), the corresponding 'classical' calculation would only yield the two first terms on the right of (13.195), and the interference term oscillates increasingly about zero as γ increases and (or) one departs from the scattering angle $\theta = \pi/2$. In contrast with the boson–boson case, however, there is a *minimum* in the differential cross-section (13.195) for spin-1/2 identical fermion–fermion scattering at $\theta = \pi/2$. At this value of the scattering angle the actual (quantum mechanical) differential cross-section (13.195) is indeed smaller than the 'classical' one by a factor of two, as we have seen above. The comparison between the classical and quantum mechanical differential cross-sections for the Coulomb scattering of two identical spin-1/2 fermions is illustrated in Fig. 13.12 for the case $\gamma = 5$.

13.8 Collisions involving composite systems

The general theory of collisions involving composite systems, for example collisions between two nuclei, is beyond the scope of this book. However, it is rather easy to obtain the appropriate generalisation of the 'potential scattering' results derived in Section 13.6, if one works within the framework of the *first Born approximation*, i.e. to *first order* in the interaction between the projectile and the target. This can be done in various ways: we shall use here the Golden Rule of perturbation theory, derived in Section 9.2.

In order to keep the discussion simple, we consider a particular example: the collision of a fast (but non-relativistic) positron with a hydrogen atom initially in the ground state 1s (characterised by the quantum numbers $n = 1, l = 0$ and $m = 0$). We want to calculate the cross-sections for:

(1) the *elastic scattering* process

$$e^+ + H(100) \rightarrow e^+ + H(100) \tag{13.196}$$

in which the positron is scattered and the hydrogen atom remains in the same (100) state;

(2) *inelastic* (excitation) *collisions* in which the scattering of the positron is accompanied by the excitation of the hydrogen atom to some *other* state characterised by the quantum numbers n, l and m (with $n \neq 1$)

$$e^+ + H(100) \rightarrow e^+ + H(nlm). \tag{13.197}$$

Let us denote respectively by $\mathbf{r}_1$ and $\mathbf{r}_2$ the coordinates of the incident positron and the target electron with respect to the hydrogen nucleus (which is massive enough so that we can neglect its motion). The unperturbed Hamiltonian is the sum of the kinetic energy operator of the positron and the Hamiltonian of the hydrogen atom

$$H_0 = -\frac{\hbar^2}{2m}\nabla_1^2 - \frac{\hbar^2}{2m}\nabla_2^2 - \frac{e^2}{(4\pi\varepsilon_0)r_2} \tag{13.198}$$

where m is the electron (positron) mass. The unperturbed wave functions are eigenfunctions of H_0. In particular, the *initial* unperturbed wave function is

$$\Phi_a = (2\pi)^{-3/2} \exp(i\mathbf{k}\cdot\mathbf{r}_1)\psi_{100}(r_2) \tag{13.199}$$

where $\mathbf{k}$ is the wave vector of the incident positron, the plane wave has been 'normalised' to a Dirac delta function, and ψ_{100} denotes the hydrogen atom ground state wave function. The total energy E_a corresponding to the initial state (13.199) is the sum of the kinetic energy $\hbar^2 k^2/2m$ of the incident positron and the energy E_1 of the bound electron in the ground state ($n = 1$) of the hydrogen atom:

$$E_a = \frac{\hbar^2 k^2}{2m} + E_1. \tag{13.200}$$

The *final unperturbed wave function* is given in the case of an elastic transition by

$$\Phi_b = (2\pi)^{-3/2} \exp(i\mathbf{k}'\cdot\mathbf{r}_1)\psi_{100}(r_2) \tag{13.201}$$

where $\mathbf{k}'$ is the wave vector of the scattered positron. For an inelastic (excitation) transition, we have instead

$$\Phi_b = (2\pi)^{-3/2} \exp(i\mathbf{k}'\cdot\mathbf{r}_1)\psi_{nlm}(\mathbf{r}_2) \tag{13.202}$$

where ψ_{nlm} is the wave function of a hydrogen atom in the state characterised by the quantum numbers (nlm), with $n \neq 1$. In both cases (13.201) and (13.202) the total energy E_b corresponding to the state Φ_b is given by

$$E_b = \frac{\hbar^2 k'^2}{2m} + E_n, \tag{13.203}$$

E_n being the bound-state energy of the hydrogen atom in a state of principal quantum number n. Energy conservation requires that $E_a = E_b$, or in other words

$$\frac{\hbar^2 k^2}{2m} + E_1 = \frac{\hbar^2 k'^2}{2m} + E_n. \tag{13.204}$$

Thus for an elastic collision we have $k = k'$. On the other hand, for an excitation process, we see from (13.204) that

$$k' = \left[k^2 - \frac{2m}{\hbar^2}(E_n - E_1)\right]^{1/2}$$

$$= \left[k^2 - \frac{1}{a_0^2}\left(1 - \frac{1}{n^2}\right)\right]^{1/2}, \quad n \neq 1 \tag{13.205}$$

where we have used (7.114) with $Z = 1$ and $\mu = m$, and a_0 is the Bohr radius (1.66). We remark that for fast incident positrons one has $ka_0 \gg 1$. For example, if the incident positrons have an energy of 500 eV one has $ka_0 \simeq 6$.

The perturbation H' is the interaction (which we shall continue to call V) between the projectile and the target. In the present case it is the sum of the Coulomb interactions between the incident positron and the hydrogen nucleus, and between the positron and the bound electron. That is

$$V(\mathbf{r}_1, \mathbf{r}_2) = \frac{e^2}{4\pi\varepsilon_0}\left(\frac{1}{r_1} - \frac{1}{r_{12}}\right) \tag{13.206}$$

where $r_{12} = |\mathbf{r}_1 - \mathbf{r}_2|$. The matrix element H'_{ba} appearing in the Golden Rule formula (9.57) is therefore given by

$$H'_{ba} = \langle \Phi_b | V | \Phi_a \rangle = (2\pi)^{-3} T^B_{ba} \tag{13.207}$$

where the first Born transition matrix element

$$T^B_{ba} = \int e^{-i\mathbf{k}'\cdot\mathbf{r}_1} \psi^*_{nlm}(\mathbf{r}_2) V(\mathbf{r}_1, \mathbf{r}_2) e^{i\mathbf{k}\cdot\mathbf{r}_1} \psi_{100}(\mathbf{r}_2) d\mathbf{r}_1 d\mathbf{r}_2 \tag{13.208}$$

is the obvious generalisation of the potential scattering first Born transition matrix element (13.154). Using the explicit expression (13.206) of the interaction potential and introducing the wave vector transfer $\mathbf{\Delta} = \mathbf{k} - \mathbf{k}'$, we can also rewrite (13.208) as

$$T^B_{ba} = \frac{e^2}{4\pi\varepsilon_0} \int \exp(i\mathbf{\Delta}\cdot\mathbf{r}_1) \psi^*_{nlm}(\mathbf{r}_2) \left(\frac{1}{r_1} - \frac{1}{r_{12}}\right) \psi_{100}(\mathbf{r}_2) d\mathbf{r}_1 d\mathbf{r}_2. \tag{13.209}$$

In order to obtain the first Born differential cross-section, we first calculate the first-order transition rate given by the Golden Rule formula (9.57). The density of final states $\rho_b(E_b)$ can be found in a way similar to that discussed in Section 11.7. With the 'normalisation' of plane waves adopted in (13.201) and (13.202), the number of states whose wave vector $\mathbf{k}'$ lies within the solid angle $d\Omega$ and having an energy in the range $(E_b, E_b + dE_b)$ is

$$\rho_b(E_b) d\Omega dE_b = d\mathbf{k}' = k'^2 dk' d\Omega \tag{13.210}$$

so that $\rho_b(E_b) = k'^2 dk'/dE_b = mk'/\hbar^2$. Using (9.57) and (13.207), the first-order transition rate corresponding to positrons scattered within an element of solid angle $d\Omega$ is thus given by

$$\begin{aligned} dW_{ba} &= \frac{2\pi}{\hbar} |H'_{ba}|^2 \rho_b(E_b) d\Omega \\ &= \frac{(2\pi)^{-5}}{\hbar} |T^B_{ba}|^2 \frac{mk'}{\hbar^2} d\Omega. \end{aligned} \tag{13.211}$$

The first Born differential cross-section is now obtained upon dividing dW_{ba} by the element of solid angle $d\Omega$, and by the incident flux F of positrons. With the 'normalisation' adopted for the incident plane wave in (13.199), we have $F = (2\pi)^{-3} v$, where $v = \hbar k/m$ is the incident positron velocity. Thus the first Born differential cross-section for either elastic or inelastic positron–atomic hydrogen scattering is

$$\frac{d\sigma^B_{ba}}{d\Omega} = (2\pi)^{-2} m^2 \hbar^{-4} \frac{k'}{k} |T^B_{ba}|^2 \tag{13.212}$$

where T^B_{ba} is given by (13.209).

Elastic scattering

We shall now illustrate the above formulae for a few typical processes, beginning with *elastic scattering*. In this case $k = k'$, so that the magnitude of the wave vector transfer is given as in potential scattering by the expression $\Delta = 2k \sin(\theta/2)$, where θ is the scattering angle. The transition matrix element (13.209) reduces to

$$T_{\text{el}}^{\text{B}} = \frac{e^2}{4\pi\varepsilon_0} \int \exp(i\boldsymbol{\Delta}\cdot\mathbf{r}_1)|\psi_{100}(r_2)|^2 \left(\frac{1}{r_1} - \frac{1}{r_{12}}\right) d\mathbf{r}_1 d\mathbf{r}_2$$

$$= \int \exp(i\boldsymbol{\Delta}\cdot\mathbf{r}_1) V_{\text{st}}(\mathbf{r}_1) d\mathbf{r}_1 \qquad (13.213)$$

where

$$V_{\text{st}}(\mathbf{r}_1) = \langle \psi_{100}|V|\psi_{100}\rangle$$

$$= \frac{e^2}{4\pi\varepsilon_0} \int |\psi_{100}(r_2)|^2 \left(\frac{1}{r_1} - \frac{1}{r_{12}}\right) d\mathbf{r}_2 \qquad (13.214)$$

is the average or *static* potential felt by the incident positron in the field of the ground-state hydrogen atom. We see from (13.213) that T_{el}^{B} is proportional to the Fourier transform of V_{st}. Using the explicit form of $\psi_{100}(r_2)$ (see Table 7.1) and expanding $1/r_{12}$ in spherical harmonics according to (10.78), one finds (Problem 13.20) that the static potential is given by

$$V_{\text{st}}(r_1) = \frac{e^2}{(4\pi\varepsilon_0)a_0}\left(1 + \frac{a_0}{r_1}\right)\exp(-2r_1/a_0). \qquad (13.215)$$

This potential is central, repulsive and of short range.

Upon substituting the expression (13.215) for $V_{\text{st}}(r_1)$ in (13.213) and using the methods developed in Section 13.5 for potential scattering in the first Born approximation, one obtains (Problem 13.21)

$$T_{\text{el}}^{\text{B}} = 4\pi a_0^2 \frac{e^2}{4\pi\varepsilon_0} \frac{(\Delta a_0)^2 + 8}{[(\Delta a_0)^2 + 4]^2} \qquad (13.216)$$

and from (13.212) the corresponding elastic first Born differential cross-section is found to be

$$\frac{d\sigma_{\text{el}}^{\text{B}}}{d\Omega} = 4a_0^2 \frac{[(\Delta a_0)^2 + 8]^2}{[(\Delta a_0)^2 + 4]^4}. \qquad (13.217)$$

We remark that in the forward direction ($\Delta = 0$) this cross-section has the constant value a_0^2, and falls off like $(\Delta a_0)^{-4}$ for large values of Δa_0. The differential cross-section (13.217) is illustrated in Fig. 13.13 for an incident positron energy of 500 eV.

13.8 Collisions involving composite systems

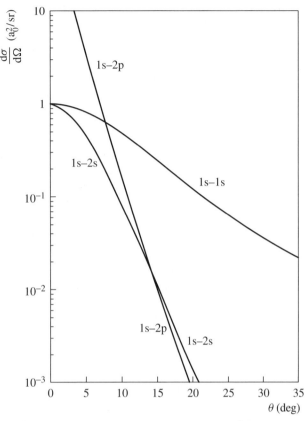

Figure 13.13 The first Born differential cross-sections (in units of a_0^2/sr) for (a) elastic scattering of positrons by the ground state of atomic hydrogen, (b) the excitation of the 2s level and (c) the excitation of the 2p level by positron impact as a function of the scattering angle θ. The energy of the incident positrons is 500 eV.

The total first Born cross-section for elastic scattering is given by

$$\sigma_{\text{tot}}^{B}(\text{el}) = 2\pi \int_0^\pi \frac{d\sigma_{\text{el}}^{B}}{d\Omega} \sin\theta d\theta$$

$$= \frac{2\pi}{k^2} \int_0^{2k} \frac{d\sigma_{\text{el}}^{B}}{d\Omega} \Delta d\Delta \qquad (13.218)$$

where we have used the fact that the differential cross-section (13.217) only depends on the magnitude $\Delta = 2k\sin(\theta/2)$ of the wave vector transfer. The integral on the right of (13.218) is readily performed and yields the result

$$\sigma_{\text{tot}}^{B}(\text{el}) = \pi a_0^2 \frac{7(ka_0)^4 + 18(ka_0)^2 + 12}{3[1 + (ka_0)^2]^3}. \qquad (13.219)$$

At high energies (so that $ka_0 \gg 1$) the leading term of $\sigma_{\text{tot}}^B(\text{el})$ is given by

$$\sigma_{\text{tot}}^B(\text{el}) \simeq \frac{7\pi}{3} k^{-2}. \tag{13.220}$$

Thus, if $E = \hbar^2 k^2/2m$ is the energy of the incident positron, we see that $\sigma_{\text{tot}}^B(\text{el})$ is proportional to E^{-1} when E is large, as in the potential scattering case (see (13.159)).

Inelastic collisions

Let us now consider inelastic collisions of the type (13.197). In this case the final target state (ψ_{nlm}) differs from the initial one (ψ_{100}). Since target wave functions corresponding to different states are orthogonal, the term $(1/r_1)$ arising from the positron–nucleus interaction does not contribute to the integral on the right of (13.209). Thus, in first Born approximation, the inelastic process (13.197) can only occur because of the interaction between the incident positron and the bound electron, and the first Born transition matrix element for inelastic scattering reduces to

$$T_{ba}^B = -\frac{e^2}{4\pi\varepsilon_0} \int \exp(i\boldsymbol{\Delta}\cdot\mathbf{r}_1) \psi_{nlm}^*(\mathbf{r}_2) \frac{1}{r_{12}} \psi_{100}(r_2) d\mathbf{r}_1 d\mathbf{r}_2. \tag{13.221}$$

We also note that for excitation scattering the magnitude of the wave vector transfer is such that

$$\Delta^2 = k^2 + k'^2 - 2kk'\cos\theta \tag{13.222}$$

where θ is the scattering angle, and k' is related to k by (13.205). In contrast to elastic collisions, where $0 \leqslant \Delta \leqslant 2k$, in the case of excitation scattering Δ varies from a non-zero minimum value $\Delta_{\min} = k - k'$ (reached for $\theta = 0$) to the maximum value $\Delta_{\max} = k + k'$ (reached for $\theta = \pi$).

The integral over the variable $\mathbf{r}_1$ in (13.221) can be performed by using the *Bethe integral*

$$\int \frac{\exp(i\boldsymbol{\Delta}\cdot\mathbf{r}_1)}{r_{12}} d\mathbf{r}_1 = \exp(i\boldsymbol{\Delta}\cdot\mathbf{r}_2) \int \frac{\exp(i\boldsymbol{\Delta}\cdot\mathbf{r}_{12})}{r_{12}} d\mathbf{r}_{12}$$

$$= \exp(i\boldsymbol{\Delta}\cdot\mathbf{r}_2) \lim_{\alpha\to 0} \int \exp(i\boldsymbol{\Delta}\cdot\mathbf{r}_{12}) \frac{\exp(-\alpha r_{12})}{r_{12}} d\mathbf{r}_{12}$$

$$= \frac{4\pi}{\Delta^2} \exp(i\boldsymbol{\Delta}\cdot\mathbf{r}_2) \tag{13.223}$$

so that (13.221) becomes

$$T_{ba}^B = -\frac{e^2}{4\pi\varepsilon_0} \frac{4\pi}{\Delta^2} \int \psi_{nlm}^*(\mathbf{r}_2) \exp(i\boldsymbol{\Delta}\cdot\mathbf{r}_2) \psi_{100}(r_2) d\mathbf{r}_2. \tag{13.224}$$

As a first example, let us consider the excitation of the 2s state of atomic hydrogen ($n = 2, l = 0, m = 0$) from the ground state (1s) by fast positrons. In this case the first Born transition matrix element is given by

$$T^B_{2s,1s} = -\frac{e^2}{4\pi\varepsilon_0}\frac{4\pi}{\Delta^2}\int \psi^*_{200}(r_2)\exp(i\mathbf{\Delta}\cdot\mathbf{r}_2)\psi_{100}(r_2)d\mathbf{r}_2. \quad (13.225)$$

Expanding $\exp(i\mathbf{\Delta}\cdot\mathbf{r}_2)$ in spherical harmonics (see (7.78)) and using the fact that both $\psi_{100}(r_2)$ and $\psi_{200}(r_2)$ are spherically symmetric, we have (Problem 13.22)

$$T^B_{2s,1s} = -16\pi\sqrt{2}\frac{e^2}{4\pi\varepsilon_0}a_0^2\frac{1}{[(\Delta a_0)^2 + \frac{9}{4}]^3}. \quad (13.226)$$

Substituting this result in (13.212), we find that the corresponding first Born differential cross-section is

$$\frac{d\sigma^B_{2s,1s}}{d\Omega} = \frac{k'}{k}\frac{128a_0^2}{[(\Delta a_0)^2 + \frac{9}{4}]^6} \quad (13.227)$$

where (see (13.205))

$$k' = \left(k^2 - \frac{3}{4a_0^2}\right)^{1/2}. \quad (13.228)$$

It is interesting to note that at high energies (where $k' \simeq k$), the cross-section (13.227) is nearly constant as a function of k in the forward direction. We also remark that for large values of Δa_0, the expression (13.227) falls off like $(\Delta a_0)^{-12}$, i.e. much faster than the first Born elastic differential cross-section (13.217), which decreases like $(\Delta a_0)^{-4}$ for large values of Δa_0. The much slower decrease of $d\sigma^B_{el}/d\Omega$ at large scattering angles is due to the fact that the projectile-nucleus interaction $e^2/(4\pi\varepsilon_0 r_1)$ does contribute to the first Born transition matrix element (13.209) in the elastic scattering case, while it contributes nothing for inelastic collisions. The differential cross-section (13.227) is illustrated in Fig. 13.13 for an incident positron energy of 500 eV.

The total first Born cross-section for the 1s–2s transition is given by integrating the differential cross-section (13.227) over all scattering angles. Using (13.222), we have

$$\sigma^B_{tot}(1s \to 2s) = 2\pi\frac{128a_0^2}{k^2}\int_{\Delta_{min}}^{\Delta_{max}} \frac{\Delta}{[(\Delta a_0)^2 + \frac{9}{4}]^6}d\Delta. \quad (13.229)$$

This integral can readily be evaluated analytically, but we shall only be interested here in obtaining the leading term of $\sigma^B_{tot}(1s \to 2s)$ at high energies ($ka_0 \gg 1$). To this end, we remark from (13.228) that for large values of ka_0 one has

$$k' \simeq k\left(1 - \frac{3}{8(ka_0)^2} + \cdots\right) \quad (13.230)$$

so that $\Delta_{min} = k - k' \simeq 3/(8ka_0^2)$ and $\Delta_{max} = k + k' \simeq 2k$. Since most of the contribution to the integral (13.229) comes from the region of small Δ, and because

$(\Delta_{\min}a_0)^2 \ll 9/4$, the leading term of $\sigma_{\text{tot}}^B(1s \to 2s)$ can be obtained by writing

$$\sigma_{\text{tot}}^B(1s \to 2s) \simeq 2\pi \frac{128 a_0^2}{k^2} \int_0^\infty \frac{\Delta}{[(\Delta a_0)^2 + \frac{9}{4}]^6} d\Delta. \qquad (13.231)$$

Hence

$$\sigma_{\text{tot}}^B(1s \to 2s) \simeq \frac{128\pi}{5} \left(\frac{2}{3}\right)^{10} k^{-2} \qquad (13.232)$$

and we see that this cross-section falls off like E^{-1} for large incident positron energies, as in the elastic scattering case (see (13.220)).

As a second example of inelastic collisions, we shall analyse the excitation of the $2p_m$ ($m = 0, \pm 1$) states of atomic hydrogen by fast positrons. In this case the first Born inelastic matrix element (13.224) becomes

$$T_{2p_m,1s}^B = -\frac{e^2}{4\pi\varepsilon_0} \frac{4\pi}{\Delta^2} \int \psi_{2p_m}^*(\mathbf{r}_2) \exp(i\boldsymbol{\Delta}\cdot\mathbf{r}_2) \psi_{100}(r_2) d\mathbf{r}_2. \qquad (13.233)$$

The integral on the right of this equation is easier to evaluate if one chooses the quantisation axis to lie along the wave vector transfer $\boldsymbol{\Delta}$. Upon performing the integration over the azimuthal angle ϕ_2, the factors $\exp(\pm i\phi_2)$ that appear in the wave functions $\psi_{2p_{\pm 1}}$ for the magnetic substates $m = \pm 1$ prevent those substates from being excited. The only final substate which is excited is therefore the $2p_0$ one, the corresponding first Born transition matrix element being (Problem 13.23)

$$T_{2p_0,1s}^B = -i24\pi\sqrt{2} \frac{e^2}{4\pi\varepsilon_0} a_0^2 \frac{1}{\Delta a_0 [(\Delta a_0)^2 + \frac{9}{4}]^3}. \qquad (13.234)$$

The first Born differential cross-section for the transition $1s \to 2p_0$, obtained by substituting (13.234) into (13.212), is

$$\frac{d\sigma_{2p_0,1s}^B}{d\Omega} = \frac{k'}{k} \frac{288 a_0^2}{(\Delta a_0)^2 [(\Delta a_0)^2 + \frac{9}{4}]^6}. \qquad (13.235)$$

Because of the presence of the factor $(\Delta a_0)^{-2}$, this differential cross-section exhibits at small wave vector transfers a much stronger peak than the one corresponding to the elastic differential cross-section (13.217) or the 1s–2s differential cross-section (13.227) (see Fig. 13.13). On the other hand, we see that the first Born differential cross-section (13.235) falls off like $(\Delta a_0)^{-14}$ for large Δa_0, which is much faster than the $(\Delta a_0)^{-4}$ decrease of the first Born elastic differential cross-section (13.217). As we mentioned above, the much slower decrease of $d\sigma_{\text{el}}^B/d\Omega$ at large values of Δa_0 is due to the non-vanishing contribution of the projectile–nucleus interaction term $e^2/(4\pi\varepsilon_0 r_1)$ to the first Born transition matrix element (13.209) for elastic collisions.

The total first Born cross-section for 1s–2p excitation is given by

$$\sigma_{\text{tot}}^B(1s \to 2p) = 2\pi \frac{288 a_0^2}{k^2} \int_{\Delta_{\min}}^{\Delta_{\max}} \frac{\Delta}{(\Delta a_0)^2 [(\Delta a_0)^2 + \frac{9}{4}]^6} d\Delta. \qquad (13.236)$$

As in the case of the 1s–2s transition we shall only be interested in obtaining the leading term of $\sigma_{\text{tot}}^{\text{B}}(1s \to 2p)$ at high energies ($ka_0 \gg 1$). Most of the contribution to the integral on the right of (13.236) comes from the region of small values of Δ near $\Delta_{\min} = k - k' \simeq 3/(8ka_0^2)$, so that when $ka_0 \gg 1$ one finds that (Problem 13.24)

$$\sigma_{\text{tot}}^{\text{B}}(1s \to 2p) \simeq \frac{576\pi}{k^2} \left(\frac{2}{3}\right)^{12} \log(ka_0). \tag{13.237}$$

We see that at high energies this cross-section behaves like $k^{-2} \log k$, or in other words is proportional to $E^{-1} \log E$, where $E = \hbar^2 k^2/2m$ is the projectile energy. This is in contrast with the elastic and 1s $\to$ 2s first Born cross-sections which fall off more rapidly (like E^{-1}) for large E (see (13.220) and (13.232)). The presence of the logarithmic factor on the right of (13.237) is due to the fact that the differential cross-section (13.235) behaves like Δ^{-2} at small wave vector transfers, so that the integrand on the right of (13.236) is proportional to Δ^{-1} for small Δ. It is interesting to note that this feature is in turn a consequence of the infinite range of the Coulomb interaction. As one might expect from the above discussion, the enhancement of the high-energy differential cross-section at small angles and the ($E^{-1} \log E$) fall-off of the total cross-section is not only exhibited in the 1s–2p transition, but in any 'dipole allowed' transition in which the change in the orbital angular momentum quantum number of the target hydrogen atom satisfies the selection rule $|\Delta l| = 1$. Finally, we remark that since in first Born approximation the cross-sections are independent of the sign of the interaction potential, the first Born cross-sections corresponding to fast incident *electrons* are identical to those we have obtained above for fast incident *positrons*, provided exchange effects between the incident and target electrons are neglected. Such exchange effects are in fact small at high incident energies.

Problems

13.1 Consider the scattering of a beam of particles by a potential of range a. Find the uncertainty in the angle of scattering when the particles are:

(a) protons of energy 10 MeV and $a = 2 \times 10^{-15}$ m;
(b) electrons of energy 5 eV and $a = 10^{-10}$ m; and
(c) protons of energy 10 keV and $a = 10^{-10}$ m.

13.2 Consider the elastic scattering of particles A of mass m_A by target particles B of mass m_B.

(a) Show that all the target particles emerge in the forward hemisphere in the laboratory coordinate system.
(b) If the angular distribution is spherically symmetric in the centre-of-mass system, what is the angular distribution of the target particles in the laboratory system?

13.3 Two beams of protons intersect, the angle between the beam directions being 5°. If the kinetic energy of the protons is 5 keV in one beam and is 5.1 keV in the other, calculate:

(a) the magnitude of the relative velocity of a proton in one beam with respect to a proton in the other beam; and

(b) the energy in the centre-of-mass system.

13.4 Consider a collision between two atoms A and B in which the atoms of type A are excited,
$$A + B \to A^* + B.$$
If the masses of A and B are m_A and m_B, respectively, and if the excitation energy transferred to A during the collision is Q ($Q > 0$), show that the laboratory and centre-of-mass differential cross-sections for observation of the excited atoms A^* in a given direction are related by
$$\frac{d\sigma}{d\Omega_L}(\theta_L, \phi_L) = \frac{(1 + \tau^2 + 2\tau \cos\theta)^{3/2}}{|1 + \tau \cos\theta|} \frac{d\sigma}{d\Omega}(\theta, \phi)$$
where the subscript L refers to laboratory quantities, and
$$\tau = \frac{m_A}{m_B}\left(\frac{T_i}{T_i - Q}\right)^{1/2}$$
T_i being the initial kinetic energy in the centre-of-mass system.

13.5 Prove that the asymptotic form (13.28) holds provided $rV(r) \to 0$ as $r \to \infty$.

(*Hint*: Write solutions of the radial equation (13.36) for large r in the form
$$R_l(k, r) = F_l(k, r) \exp(\pm ikr)/r$$
where F_l is a slowly varying function of r. Obtain an equation for F_l and prove that if $r|V(r)| \to 0$ as $r \to \infty$, then F_l is independent of r in the limit $r \to \infty$.)

13.6 Consider a potential of the form $V(r) = V_0/r^2$.

(a) Assuming that $V_0 > 0$, obtain the phase shifts δ_l and show that $\delta_l = -(\pi 2m V_0/\hbar^2)\{2(2l + 1)\}^{-1}$ when l is large. Discuss the angular distribution. Is the differential cross-section finite in the forward direction? Is the total cross-section finite?
(*Hint*: For given ν and large ρ one has $j_\nu(\rho) \simeq \rho^{-1} \sin(\rho - \nu\pi/2)$.)

(b) Suppose now that $V_0 < 0$. How must the treatment of (a) be modified? Show that the radial equation (13.36) has physically acceptable solutions if $(2mV_0/\hbar^2) > -1/4$.

13.7 Suppose that in an elastic scattering experiment between two spinless particles the centre-of-mass differential cross-section may be represented by an expression of the type

$$\frac{d\sigma}{d\Omega} = A + B P_1(\cos\theta) + C P_2(\cos\theta) + \cdots$$

Express the coefficients A, B and C in terms of the phase shifts δ_l.

(*Hint*: Use the orthogonality relation (6.80) of the Legendre polynomials and the recurrence relation (6.83a).)

13.8 Consider the scattering of a particle of mass m by a central potential $V(r)$ which is negligible for $r > d$.

(a) Show that

$$\tan\delta_l(k) = \frac{kj_l'(kd) - \gamma_l(k)j_l(kd)}{kn_l'(kd) - \gamma_l(k)n_l(kd)}, \tag{1}$$

where $j_l'(x) = dj_l(x)/dx$, $n_l'(x) = dn_l(x)/dx$ and

$$\gamma_l(k) = \left[\frac{dR_l^{\mathrm{I}}(k,r)/dr}{R_l^{\mathrm{I}}(k,r)}\right]_{r=d} \tag{2}$$

is the value of the logarithmic derivative of the regular, interior solution $R_l^{\mathrm{I}}(k,r)$ of the radial equation (13.36), evaluated at $r = d$.

(Hint: Use the fact that in the exterior region $r \geq d$, the solution of the radial equation (13.36) can be written in the form (13.66))

(b) Using the properties of the functions j_l and n_l (see Section 7.3), prove that

$$\tan\delta_l(k) \underset{k\to 0}{\to} \frac{(kd)^{2l+1}}{D_l} \frac{l - \hat\gamma_l d}{l+1+\hat\gamma_l d}, \tag{3}$$

where

$$D_l = (2l+1)!!(2l-1)!!, \qquad l > 0, \tag{4a}$$
$$D_0 = 1 \tag{4b}$$

and

$$\hat\gamma_l = \lim_{k\to 0}\gamma_l(k). \tag{5}$$

(c) Assuming first that $\hat\gamma_l d \neq -(l+1)$, prove that the partial wave amplitudes $f_l(k)$ exhibit the low-energy behaviour

$$f_l(k) \underset{k\to 0}{\sim} k^{2l}, \tag{6}$$

so that except for the s-wave ($l = 0$) contribution which in general tends to a non-zero constant, all partial cross sections σ_l ($l \geq 1$) vanish as k^{4l}. The scattering is therefore isotropic at very low energies. Prove also that the scattering amplitude f is given as $k \to 0$ by

$$f \underset{k \to 0}{\to} -\alpha \tag{7}$$

where α is the scattering length (13.59).

(d) Examine how the results obtained in (c) must be modified when by accident

$$\hat{\gamma}_l d = -(l+1). \tag{8}$$

(Hint: Show first that the result (7) is unchanged if the 'accident' (8) occurs for $l \geqslant 2$. Then prove that if equation (8) holds for $l = 0$, one has $\tan \delta_0(k) \sim k^{-1}$ as $k \to 0$, so that

$$f_0(k) \underset{k \to 0}{\sim} i/k \tag{9}$$

and the scattering length α diverges like k^{-2} when $k \to 0$. This singular case is often called a *zero-energy resonance*. Finally, if equation (8) holds for $l = 1$, prove that the scattering amplitude is given as $k \to 0$ by

$$f \underset{k \to 0}{\to} -\alpha + \beta \cos \theta \tag{10}$$

where α is the scattering length and β is a constant. Hence in this case the differential cross section is not isotropic when $k \to 0$.)

13.9 Find the total cross-section for low-energy (s-wave) scattering by a potential barrier such that

$$V(r) = \begin{cases} V_0(>0) & r < a \\ 0, & r > a. \end{cases}$$

Derive the 'hard-sphere' zero-energy result (13.83) as a particular case.

13.10 Assuming that the potential between a neutron and proton in the triplet ($S = 1$) state is a square well of depth 38.5 MeV and of range 1.93×10^{-15} m, calculate the scattering length and effective range. Repeat the calculation for the singlet ($S = 0$) state, taking the singlet well depth and range to be 14.3 MeV and 2.50×10^{-15} m, respectively. Find in each case the centre-of-mass energy up to which the effective range formula for $k \cot \delta_0$ is accurate to within 10%.

13.11 Obtain the radial integral equation (13.141) and the expression (13.143) for $\tan \delta_l$ by analysing the integral equation (13.130) in partial waves.

(*Hint*: Use the expansion (13.45) of the incident plane wave in partial waves, and the expansion

$$\frac{e^{ik|\mathbf{r} - \mathbf{r}'|}}{|\mathbf{r} - \mathbf{r}'|} = ik \sum_{l=0}^{\infty} (2l+1) j_l(kr_<)[j_l(kr_>) + i n_l(kr_>)] P_l(\cos \alpha)$$

where $r_<$ and $r_>$ denote respectively the lesser and the greater of r and r', and α is the angle between the vectors $\mathbf{r}$ and $\mathbf{r}'$.)

13.12 Using the first Born approximation (13.149) for $\tan \delta_l$, prove that for the square well (13.63) one has for $l \gg ka$

$$\tan \delta_l \underset{l \gg ka}{\simeq} U_0 a^2 \frac{(ka)^{2l+1}}{[(2l+1)!!]^2 (2l+3)}$$

so that

$$\frac{\delta_{l+1}}{\delta_l} \simeq \left(\frac{ka}{2l}\right)^2, \qquad l \gg ka$$

(*Hint*: Use the approximate formula (7.65a) for $j_l(kr)$.)

13.13 Use the first Born approximation (13.149) for $\tan \delta_l$ to show that

$$\tan \delta_l(k) \underset{k \to \infty}{\to} -\frac{1}{2} k^{-1} \int_0^\infty U(r) dr + \mathcal{O}(k^{-2})$$

(*Hint*: Use the asymptotic expression (7.66a) for $j_l(kr)$.)

13.14 Using the first Born approximation (13.149) for $\tan \delta_l$, obtain the $l = 0$ phase shift for scattering by

(a) the Yukawa potential $V(r) = V_0 \, r^{-1} \exp(-\alpha r)$,
(b) the 'polarisation' potential $V(r) = V_0/(r^2 + d^2)^2$, where d is a constant.

13.15 Obtain in the first Born approximation the scattering amplitude, the differential and the total cross-sections for scattering by the following potentials:

(a) Exponential potential

$$V(r) = V_0 \exp(-\alpha r)$$

(b) Gaussian potential

$$V(r) = V_0 \exp(-\alpha^2 r^2)$$

(c) Square-well potential

$$V(r) = V_0 \quad r < a$$
$$ = 0 \quad r > a.$$

(d) 'Polarisation' potential

$$V(r) = V_0/(r^2 + d^2)^2.$$

Discuss the angular distributions and compare your results with those obtained in the text for the Yukawa potential $V(r) = V_0 \exp(-\alpha r)/r$. Verify that $\sigma_{\text{tot}}^B \sim AE^{-1}$ as $E \to \infty$, and find the coefficient A in each case.

13.16 Consider the elastic scattering of a carbon nucleus $^{12}_{6}$C by an oxygen nucleus $^{16}_{8}$O at a centre-of-mass energy of 1 MeV.

(a) Use a simple classical argument to show that only the Coulomb interaction should be taken into account. Obtain the corresponding differential cross-section in the centre-of-mass system.

(b) Suppose that the pure Coulomb interaction is screened at a distance $r_0 = 10^{-10}$ m by the presence of atomic electrons, and that the effective interaction is given by

$$V(r) = \frac{q_A q_B}{(4\pi\varepsilon_0)r} \exp(-r/r_0).$$

Use the first Born approximation to obtain an estimate of the angular range over which the differential cross-section arising from the potential $V(r)$ differs from the pure Coulomb (Rutherford) result obtained in (a).

13.17 Alpha particles having a laboratory kinetic energy of 4 MeV are scattered elastically by copper nuclei $^{63}_{29}$Cu. In performing this experiment, it is observed that the pure Coulomb scattering law is obeyed even for scattering in the backward direction. Using simple classical considerations, obtain an upper limit of the radius R of the $^{63}_{29}$Cu nucleus.

13.18 Two alpha particles having a centre-of-mass energy of 1 MeV scatter elastically. Obtain and plot the Mott differential cross-section in the centre-of-mass system. Compare your results with those obtained (a) by ignoring the interference term and (b) by ignoring the fact that the two particles are identical.

13.19 Consider proton–proton elastic scattering in the centre-of-mass system.

(a) Compute and plot the Mott differential cross-section at an energy of 25 keV.
(b) Justify the neglect of nuclear forces made in (a). About what energy would you expect this approximation to fail badly?

13.20 Obtain the result (13.215).

13.21 Derive equation (13.216).

13.22 Derive equation (13.226).

13.23 Obtain the result (13.234).

13.24 Obtain the expression (13.237).

14 Quantum statistics

14.1 The density matrix   642
14.2 The density matrix for a spin-1/2 system. Polarisation   645
14.3 The equation of motion of the density matrix   654
14.4 Quantum mechanical ensembles   655
14.5 Applications to single-particle systems   661
14.6 Systems of non-interacting particles   663
14.7 The photon gas and Planck's law   667
14.8 The ideal gas   668
Problems   676

Until now we have considered quantum systems which can be described by a *single* wave function (state vector). Such systems are said to be in a *pure state*. They are prepared in a specific way, their state vector being obtained by performing a *maximal measurement* in which all values of a complete set of commuting observables have been ascertained. In this chapter we shall study quantum systems such that the measurement made on them is not maximal. These systems, whose state is incompletely known, are said to be in *mixed states*. Instead of a single wave function, a statistical mixture of wave functions must be used to describe these systems. *Quantum statistical mechanics* is the branch of physics dealing with such quantum systems in mixed states; it is the quantum analogue of classical statistical mechanics. It should be noted that statistics enters at two levels in quantum statistical mechanics: first, because of the statistical interpretation of the wave function and second, because of our incomplete knowledge of the dynamical state of the system.

We shall begin our study of quantum statistical mechanics by developing the *density matrix* formalism for handling mixed states. As an application, we shall analyse the spin properties of spin-1/2 particles. The equation of motion of the density matrix will then be derived and the equation of motion of an observable will be obtained. The density matrix will also be used to characterise various statistical ensembles. Finally, we shall obtain Bose–Einstein and Fermi–Dirac distribution laws for ensembles of identical bosons and fermions respectively, and shall discuss several applications.

14.1 The density matrix

Let us consider a system consisting of an ensemble of N sub-systems $\alpha = 1, 2, \ldots, N$. We suppose that each of these sub-systems is in a pure state and is therefore characterised by a distinct state vector $\Psi^{(\alpha)}$, which using the Dirac notation will be denoted by $|\alpha\rangle$. The state vectors $|\alpha\rangle$ are assumed to be normalised to unity, but need not be orthogonal to each other.

Next, let us choose a complete set of basis vectors $|n\rangle$, namely orthonormal eigenvectors of some complete set of operators. Since these basis states are orthonormal,

$$\langle n'|n\rangle = \delta_{n'n} \tag{14.1}$$

where the symbol $\delta_{n'n}$ should be replaced by $\delta(n - n')$ when the indices n and n' are continuous. Because the set of basis vectors is complete, we have

$$\sum_n |n\rangle\langle n| = I. \tag{14.2}$$

Let us expand the pure state $|\alpha\rangle$ in the basis states $|n\rangle$, namely

$$|\alpha\rangle = \sum_n c_n^{(\alpha)} |n\rangle. \tag{14.3}$$

Using (14.1) the coefficients $c_n^{(\alpha)}$ are given by

$$c_n^{(\alpha)} = \langle n|\alpha\rangle \tag{14.4}$$

and since the state $|\alpha\rangle$ is normalised to unity we have

$$\sum_n |c_n^{(\alpha)}|^2 = 1. \tag{14.5}$$

Consider an observable represented by an operator A. The expectation value of this operator in the pure state $|\alpha\rangle$ is

$$\begin{aligned}\langle A\rangle_\alpha &= \langle \alpha|A|\alpha\rangle = \sum_n \sum_{n'} c_{n'}^{(\alpha)*} c_n^{(\alpha)} \langle n'|A|n\rangle \\ &= \sum_n \sum_{n'} \langle n|\alpha\rangle \langle \alpha|n'\rangle \langle n'|A|n\rangle. \end{aligned} \tag{14.6}$$

The average value of A over the ensemble, called the *ensemble* (or *statistical*) *average* of A is given by

$$\langle A\rangle = \sum_{\alpha=1}^N W_\alpha \langle A\rangle_\alpha \tag{14.7}$$

where W_α is the *statistical weight* of the pure state $|\alpha\rangle$, namely the probability of finding the system in this state. The statistical weights must clearly be such that

$$0 \leqslant W_\alpha \leqslant 1 \tag{14.8}$$

and

$$\sum_{\alpha=1}^{N} W_\alpha = 1. \tag{14.9}$$

Using the result (14.6), we may write (14.7) as

$$\langle A \rangle = \sum_{\alpha=1}^{N} \sum_n \sum_{n'} W_\alpha c_{n'}^{(\alpha)*} c_n^{(\alpha)} \langle n'|A|n \rangle$$

$$= \sum_{\alpha=1}^{N} \sum_n \sum_{n'} \langle n|\alpha\rangle W_\alpha \langle\alpha|n'\rangle \langle n'|A|n\rangle. \tag{14.10}$$

Let us now introduce the *density operator (or statistical operator)*

$$\rho = \sum_{\alpha=1}^{N} |\alpha\rangle W_\alpha \langle\alpha|. \tag{14.11}$$

Taking matrix elements of the density operator between the basis states $|n\rangle$ we obtain the density matrix in the $\{n\}$ representation, whose elements are

$$\rho_{nn'} \equiv \langle n|\rho|n'\rangle = \sum_{\alpha=1}^{N} \langle n|\alpha\rangle W_\alpha \langle\alpha|n'\rangle$$

$$= \sum_{\alpha=1}^{N} W_\alpha c_{n'}^{(\alpha)*} c_n^{(\alpha)}. \tag{14.12}$$

Note that while the density operator is independent of the choice of the representation, the density matrix has a different form in different representations. Returning to (14.10) and using (14.12), we see that

$$\langle A \rangle = \sum_n \sum_{n'} \langle n|\rho|n'\rangle \langle n'|A|n\rangle$$

$$= \sum_n \langle n|\rho A|n\rangle$$

$$= \mathrm{Tr}(\rho A) \tag{14.13}$$

where the symbol Tr denotes the trace. Thus *the knowledge of the density matrix enables us to obtain the ensemble average of A*. We also remark that if we take A to be the identity operator I and use the fact that the pure states $|\alpha\rangle$ are normalised to unity, we obtain the normalisation condition

$$\mathrm{Tr}\rho = 1. \tag{14.14}$$

If the pure states $|\alpha\rangle$ are not normalised to unity then the ensemble average of A is given by

$$\langle A \rangle = \mathrm{Tr}(\rho A)/\mathrm{Tr}\rho. \tag{14.15}$$

As seen from its definition (14.12) the density matrix is Hermitian, namely

$$\langle n|\rho|n'\rangle = \langle n'|\rho|n\rangle^* \tag{14.16}$$

and hence can always be diagonalised by means of a unitary transformation. The diagonal elements of the density matrix,

$$\rho_{nn} = \langle n|\rho|n\rangle = \sum_{\alpha=1}^{N} W_\alpha |c_n^{(\alpha)}|^2 \tag{14.17}$$

have a simple physical interpretation. Indeed, the probability of finding the system in the pure state $|\alpha\rangle$ is W_α and the probability that $|\alpha\rangle$ is to be found in the state $|n\rangle$ is $|c_n^{(\alpha)}|^2$. Thus the diagonal element ρ_{nn} gives the probability of finding a member of the ensemble in the (pure) state $|n\rangle$. We also note from (14.8) and (14.17) that

$$\rho_{nn} \geqslant 0 \tag{14.18}$$

so that ρ is a positive semi-definite operator. Moreover, combining (14.18) with (14.14), we see that all diagonal elements of the density matrix must be such that

$$0 \leqslant \rho_{nn} \leqslant 1. \tag{14.19}$$

Let us choose a representation $\{k\}$ in which the density matrix is diagonal. In this representation we have

$$\rho_{kk'} = \rho_{kk}\delta_{kk'} \tag{14.20}$$

where ρ_{kk} is the fraction of the members of the ensemble in the state $|k\rangle$. Furthermore, using (14.14) and (14.19), we find that

$$\mathrm{Tr}(\rho^2) \leqslant \mathrm{Tr}\rho = 1 \tag{14.21}$$

a relation which remains valid in any representation since the trace is invariant under a unitary transformation. Note that since the density matrix is Hermitian we may write the result (14.21) in the form

$$\sum_n \sum_{n'} |\rho_{nn'}|^2 \leqslant 1. \tag{14.22}$$

Let us now consider the special case such that the system is in a particular pure state $|\lambda\rangle$. Then $W_\alpha = \delta_{\alpha\lambda}$ and we see from (14.11) that the density operator is just

$$\rho^\lambda = |\lambda\rangle\langle\lambda|. \tag{14.23}$$

This is a projection operator onto the state $|\lambda\rangle$, with

$$(\rho^\lambda)^2 = \rho^\lambda. \tag{14.24}$$

Hence, in this case the relation (14.21) becomes

$$\mathrm{Tr}(\rho^\lambda)^2 = \mathrm{Tr}\rho^\lambda = 1. \tag{14.25}$$

It is worth noting that the equation $\text{Tr}(\rho^\lambda)^2 = 1$ gives us a criterion for deciding whether a state is pure or not that is invariant under all unitary transformations. We also see from (14.10) that if the system is in a particular pure state $|\lambda\rangle$ we must have

$$\langle A \rangle = \langle \lambda | A | \lambda \rangle \tag{14.26}$$

as expected. This result can equally be deduced from (14.13), since

$$\text{Tr}(\rho^\lambda A) = \sum_n \sum_{n'} \langle n | \rho^\lambda | n' \rangle \langle n' | A | n \rangle = \langle \lambda | A | \lambda \rangle. \tag{14.27}$$

Suppose that we are using a representation $\{k\}$ in which ρ^λ is diagonal. Then the above equation will be satisfied if

$$\rho^\lambda_{kk'} = \delta_{k\lambda} \delta_{k'\lambda} \tag{14.28}$$

so that the only non-vanishing matrix element of ρ^λ is the diagonal element in the λth row and column, which is equal to one. As a result, all the eigenvalues of the pure state density operator ρ^λ are equal to zero, except one which is equal to unity. This last property is independent of the choice of the representation, and may therefore be used to characterise the density matrix of a pure state.

In the foregoing discussion we have labelled the rows and columns of the density matrix $\rho_{nn'}$ by simple indices n and n'. In general, the symbol n refers to a set of indices, some of which take on discrete values while others may vary continuously. In many cases, however, we are interested in some particular property of the system, for example the spin. We then omit the dependence of the density matrix on all other variables, keep only the relevant indices and define in that manner a *reduced density matrix*.

14.2 The density matrix for a spin-1/2 system. Polarisation

As an illustration of the general methods outlined above, we shall now consider the simple case of a spin-1/2 system, for example a beam of electrons. The possible pure states of a spin-1/2 particle can be labelled by the eigenvalues (p_x, p_y, p_z) of the three components of its momentum, and by the eigenvalues $m_s \hbar$ (with $m_s = \pm 1/2$) of its spin projection along an axis which we choose as our z-axis. We shall denote the corresponding kets by $|p_x, p_y, p_z, m_s\rangle$. Hence the elements of the density matrix are given in this case by

$$\langle n | \rho | n' \rangle = \langle p_x, p_y, p_z, m_s | \rho | p'_x, p'_y, p'_z, m'_s \rangle. \tag{14.29}$$

We note that the momentum indices are continuous, while the spin indices are discrete. Since we are only interested here in the spin properties, we shall disregard the momentum labels and focus our attention on the *reduced* density matrix $\langle m_s | \rho | m'_s \rangle$, which is a 2×2 matrix in spin space.

As an example, consider two beams of electrons. The first one contains N_a electrons prepared in the pure state $|\chi^{(a)}\rangle$ and the second N_b electrons prepared independently in the pure state $|\chi^{(b)}\rangle$. From (14.11) the density operator describing the joint beam is given by

$$\rho = W_a |\chi^{(a)}\rangle\langle\chi^{(a)}| + W_b |\chi^{(b)}\rangle\langle\chi^{(b)}| \tag{14.30}$$

where the statistical weights W_a and W_b are given by

$$W_a = \frac{N_a}{N_a + N_b}, \qquad W_b = \frac{N_b}{N_a + N_b}. \tag{14.31}$$

Let us now choose a set of two basis states $|\chi_1\rangle$ and $|\chi_2\rangle$, for example the two basic spinors $|\alpha\rangle$ and $|\beta\rangle$, defined by (6.208), and corresponding respectively to 'spin up' and 'spin down' states. Expanding $|\chi^{(a)}\rangle$ and $|\chi^{(b)}\rangle$ in terms of these basis states $|\chi_1\rangle \equiv |\alpha\rangle$ and $|\chi_2\rangle \equiv |\beta\rangle$, we have

$$|\chi^{(a)}\rangle = c_1^{(a)}|\chi_1\rangle + c_2^{(a)}|\chi_2\rangle, \tag{14.32a}$$

$$|\chi^{(b)}\rangle = c_1^{(b)}|\chi_1\rangle + c_2^{(b)}|\chi_2\rangle \tag{14.32b}$$

and we see from (14.12) that the density matrix is given in the representation $\{|\chi_i\rangle, i = 1, 2\}$ by

$$\rho = \begin{pmatrix} W_a |c_1^{(a)}|^2 + W_b |c_1^{(b)}|^2 & W_a c_1^{(a)} c_2^{(a)*} + W_b c_1^{(b)} c_2^{(b)*} \\ W_a c_1^{(a)*} c_2^{(a)} + W_b c_1^{(b)*} c_2^{(b)} & W_a |c_2^{(a)}|^2 + W_b |c_2^{(b)}|^2 \end{pmatrix}. \tag{14.33}$$

In particular, if the mixture consists of N_1 electrons prepared in the 'spin up' state $|\chi_1\rangle \equiv |\alpha\rangle$ and N_2 electrons prepared in the 'spin down' state $|\chi_2\rangle \equiv |\beta\rangle$ the joint beam will be represented by the density operator

$$\rho = W_1 |\chi_1\rangle\langle\chi_1| + W_2 |\chi_2\rangle\langle\chi_2| \tag{14.34}$$

with $W_1 = N_1/(N_1 + N_2)$ and $W_2 = N_2/(N_1 + N_2)$. Since in this case $c_1^{(a)} = c_2^{(b)} = 1$ and $c_1^{(b)} = c_2^{(a)} = 0$, we see that the density matrix (14.33) becomes

$$\rho = \begin{pmatrix} W_1 & 0 \\ 0 & W_2 \end{pmatrix} \tag{14.35}$$

and is therefore diagonal in the $\{|\chi_i\rangle\}$ representation.

Polarisation

Let ρ be a general 2×2 density matrix describing a spin-1/2 system. Because the unit 2×2 matrix I and the three Pauli spin matrices (6.243) form a complete set of 2×2 matrices, we may write the density matrix ρ as

$$\rho = a_0 I + a_x \sigma_x + a_y \sigma_y + a_z \sigma_z = a_0 I + \mathbf{a} \cdot \boldsymbol{\sigma} \tag{14.36}$$

where a_0, a_x, a_y and a_z are four complex parameters. The value of the parameter a_0 can immediately be obtained by taking the trace of both sides of this equation. Since $\text{Tr}\rho = 1$ (see (14.14)), $\text{Tr}I = 2$ and $\text{Tr}\sigma_i = 0$ (with $i = x, y, z$), we find that

$$a_0 = \tfrac{1}{2}. \tag{14.37}$$

14.2 The density matrix for a spin-1/2 system. Polarisation

To understand the meaning of the coefficients $a_i (i = x, y, z)$, let us calculate the average value of σ_i. According to (14.13), we have

$$\langle \sigma_i \rangle = \text{Tr}(\rho \sigma_i). \tag{14.38}$$

From (14.36) together with the fact that $\text{Tr}(\sigma_i \sigma_j) = 2\delta_{ij}$, we obtain

$$\langle \sigma_i \rangle = 2 a_i. \tag{14.39}$$

Using (14.37) and (14.39) we can therefore rewrite (14.36) in the form

$$\rho = \tfrac{1}{2}(I + \boldsymbol{\sigma} \cdot \mathbf{P}) \tag{14.40}$$

where we have introduced the *polarisation vector*

$$\mathbf{P} = \langle \boldsymbol{\sigma} \rangle. \tag{14.41}$$

Making use of the explicit representation (6.243) of the Pauli matrices, we may also write (14.40) in the form

$$\rho = \tfrac{1}{2} \begin{pmatrix} 1 + P_z & P_x - iP_y \\ P_x + iP_y & 1 - P_z \end{pmatrix}. \tag{14.42}$$

In order to give a simple physical interpretation to the polarisation vector $\mathbf{P}$, let us diagonalise ρ. We find in this way that

$$\rho = \tfrac{1}{2} \begin{pmatrix} 1 + P & 0 \\ 0 & 1 - P \end{pmatrix} \tag{14.43}$$

where the quantity $P = \pm|\mathbf{P}| = \pm(P_x^2 + P_y^2 + P_z^2)^{1/2}$ may be obtained from the equation

$$P^2 = 1 - 4 \det \rho \tag{14.44}$$

in which $\det \rho$ denotes the determinant of the density matrix. From (14.42) and (14.43) we see that in the representation where ρ is diagonal one has

$$P_x = P_y = 0, \qquad P = P_z \tag{14.45}$$

so that the density matrix assumes its diagonal form if one chooses the direction of the polarisation vector $\mathbf{P}$ as the z-axis of the coordinate system. Using (14.40) we remark that in this representation $\boldsymbol{\sigma} \cdot \mathbf{P}$ is also diagonal. Thus, if $|\chi \uparrow\rangle$ and $|\chi \downarrow\rangle$ denote the kets corresponding to 'spin up' and 'spin down' with $\mathbf{P}$ as the z-axis, we may write

$$\boldsymbol{\sigma} \cdot \mathbf{P} \, |\chi \uparrow\rangle = P |\chi \uparrow\rangle \tag{14.46a}$$

and

$$\boldsymbol{\sigma} \cdot \mathbf{P} \, |\chi \downarrow\rangle = -P |\chi \downarrow\rangle. \tag{14.46b}$$

Returning to (14.43), and remembering the physical interpretation of the diagonal elements of the density matrix given after (14.17), we see that the quantity $(1 + P)/2$ is the probability of finding in the mixture the pure states with 'spin up' along $\mathbf{P}$. This probability can also be expressed as $N_+/(N_+ + N_-)$, where N_+ is the number of spin measurements giving the value $\hbar/2$ in the direction **P** (the z-direction), and N_- is the number of spin measurements yielding the value $-\hbar/2$. Thus we have

$$\frac{1}{2}(1+P) = \frac{N_+}{N_+ + N_-}. \tag{14.47}$$

Similarly, the quantity $(1-P)/2$ is the probability of finding in the mixture the pure states with 'spin down' along **P** and this probability is also given by $N_-/(N_+ + N_-)$:

$$\frac{1}{2}(1-P) = \frac{N_-}{N_+ + N_-}. \tag{14.48}$$

Hence, from (14.47) or (14.48) we see that

$$P = \frac{N_+ - N_-}{N_+ + N_-} \tag{14.49}$$

is equal to the probability of finding the system in the state $|\chi\uparrow\rangle$ minus that of finding it in the state $|\chi\downarrow\rangle$. It is called the *polarisation* of the system.

We remark that when $P = 0$ the diagonal density matrix (14.43) becomes

$$\rho = \begin{pmatrix} \frac{1}{2} & 0 \\ 0 & \frac{1}{2} \end{pmatrix} \tag{14.50}$$

so that $\rho = \frac{1}{2}I$ and $\text{Tr}(\rho^2) = \frac{1}{2}$. The system is said to be completely *unpolarised* or in a completely *random state*. More generally, the diagonal form of an $N \times N$ density matrix corresponding to a completely random state is given by

$$\rho = \frac{1}{N}I. \tag{14.51}$$

On the contrary, when the system is in a *pure state* the density operator is such that $\rho^2 = \rho$ (see equation (14.24)), so that for the spin-1/2 system under consideration here, we have from (14.40)

$$\frac{1}{4}[I + 2\boldsymbol{\sigma}\cdot\mathbf{P} + (\boldsymbol{\sigma}\cdot\mathbf{P})^2] = \frac{1}{4}(I + 2\boldsymbol{\sigma}\cdot\mathbf{P} + P^2)$$
$$= \frac{1}{2}(I + \boldsymbol{\sigma}\cdot\mathbf{P}) \tag{14.52}$$

which implies that $P^2 = 1$. Thus there are two pure states, corresponding to the values $P = +1$ and $P = -1$, respectively, and such that the system is totally polarised in the direction of the polarisation vector **P** (when $P = 1$) or in the opposite direction (when $P = -1$). For such pure states with spin projection $\pm\hbar/2$ along the z-axis (taken along **P**) the density matrices are given respectively by

$$\rho = \begin{pmatrix} 1 & 0 \\ 0 & 0 \end{pmatrix} \quad \text{for } P = +1 \tag{14.53a}$$

and

$$\rho = \begin{pmatrix} 0 & 0 \\ 0 & 1 \end{pmatrix} \quad \text{for } P = -1. \tag{14.53b}$$

14.2 The density matrix for a spin-1/2 system. Polarisation

Comparing (14.50) and (14.53), we see that the density matrix for a completely unpolarised system can be written as

$$\rho_{\text{unp}} = \begin{pmatrix} \frac{1}{2} & 0 \\ 0 & \frac{1}{2} \end{pmatrix} = \frac{1}{2}\begin{pmatrix} 1 & 0 \\ 0 & 0 \end{pmatrix} + \frac{1}{2}\begin{pmatrix} 0 & 0 \\ 0 & 1 \end{pmatrix}. \tag{14.54}$$

Hence an unpolarised system of spin-1/2 particles (in which all spin directions are equally likely) can be regarded as made up of an equal mixture of two fully polarised systems, one in which all the particles have their spins aligned in the z-direction, and a second one in which an equal number of particles have their spins aligned in the opposite direction. It should be stressed, though, that the decomposition of equation (14.54) is not unique and can be made in an infinite number of ways.

We have just seen that when $P = 0$ or $|P| = 1$ we have completely unpolarised and completely polarised systems, respectively. For all intermediate values of $|P|$ such that $0 < |P| < 1$ the system is said to be *partially polarised*. In that case we have

$$\tfrac{1}{2}\text{Tr}(\rho^2) = \tfrac{1}{2}(1 + P^2) < 1. \tag{14.55}$$

Since $\text{Tr}(\rho^2)$ is an increasing function of $|P|$, $\text{Tr}(\rho^2) = 1/2$ for a completely unpolarised system and $\text{Tr}(\rho^2) = 1$ for a fully polarised system, it is natural to call $|P|$ the *degree of polarisation* of the system. Returning to the diagonal form (14.43) of the density matrix, and using (14.50) and (14.53) together with the identity

$$\rho = \frac{1}{2}\begin{pmatrix} 1+P & 0 \\ 0 & 1-P \end{pmatrix} = P\begin{pmatrix} 1 & 0 \\ 0 & 0 \end{pmatrix} + (1-P)\begin{pmatrix} \frac{1}{2} & 0 \\ 0 & \frac{1}{2} \end{pmatrix} \tag{14.56}$$

we see that a spin-1/2 system with polarisation P can be regarded as made up of a fully polarised part and an unpolarised part, mixed in the ratio $P/(1-P)$.

We conclude this analysis of spin-1/2 systems by considering the problem of the number of parameters required to specify the density matrix. From (14.42) we remark that for a spin-1/2 system in a *mixed* state the 2×2 density matrix is entirely specified by the *three* real independent parameters (P_x, P_y, P_z). This is readily understood as follows. A complex 2×2 matrix has four complex elements corresponding to eight real parameters. Since the density matrix satisfies the hermicity condition (14.16) the number of independent real parameters is reduced to four. Finally, the normalisation condition (14.14) reduces this number to three. Thus, for a spin-1/2 system in a mixed state we need *three independent measurements* to determine the density matrix and therefore the state of the system. If we now specialise to the case of *pure states*, our foregoing discussion shows that the additional constraint $P^2 = 1$ must be obeyed, so that only *two* real independent parameters remain.

The scattering of particles of spin 1/2 by particles of spin 0

As an example of the use of the density matrix and of the polarisation vector, we shall analyse the elastic scattering of particles of spin 1/2 by particles of spin 0. The analysis can be applied, for instance, to the scattering of nucleons by pions.

Taking the axis of quantisation to be the direction of incidence, which is parallel to the z-axis, the z-component of the spin of the particles of spin $1/2$ is $S_z = m_s\hbar$ with $m_s = \pm 1/2$. Let us consider scattering in which the initial spin state has $S_z = m\hbar$, and the final spin state has $S_z = m'\hbar$. The corresponding scattering amplitude will be denoted by $f_{m'm}(\theta, \phi)$, where (θ, ϕ) are the polar angles of the relative momentum $\hbar \mathbf{k}'$ of the particles in the final state. If the initial relative momentum of the particles is $\hbar \mathbf{k}$ parallel to the z-axis, then for elastic scattering

$$k = |\mathbf{k}| = |\mathbf{k}'| \quad \text{and} \quad \mathbf{k}.\mathbf{k}' = k^2 \cos\theta. \tag{14.57}$$

The amplitude $f_{m'm}(\theta, \phi)$ forms a two-dimensional matrix in spin space, since both m and m' can take on the values $\pm 1/2$. The matrix $f_{m'm}(\theta, \phi)$ is not, in general, diagonal because the interaction between the spin-0 and the spin-1/2 particles may be spin dependent. If the spin 1/2 particles are unpolarised in the initial state and no spin measurement is made in the final state, the differential cross-section is found by summing $|f_{m'm}(\theta, \phi)|^2$ over the final spin states and averaging over the initial spin states

$$\left(\frac{d\sigma}{d\Omega}\right)_{\text{unp}} = \frac{1}{2} \sum_m \sum_{m'} |f_{m'm}(\theta, \phi)|^2. \tag{14.58}$$

The form of $f_{m'm}(\theta, \phi)$ can be deduced by appealing to rotational invariance. The only vectors from which $f_{m'm}(\theta, \phi)$ can be constructed are $\mathbf{k}, \mathbf{k}'$ and $\boldsymbol{\sigma}$, where $\mathbf{S} = (\hbar/2)\boldsymbol{\sigma}$ is the spin operator of the spin-1/2 particles. From these three vectors we can construct five independent quantities invariant under rotations, namely the scalar products $\mathbf{k}.\mathbf{k}, \mathbf{k}.\mathbf{k}', \boldsymbol{\sigma}.(\mathbf{k} \times \mathbf{k}'), \boldsymbol{\sigma}.\mathbf{k}$ and $\boldsymbol{\sigma}.\mathbf{k}'$. Of these only the first three are unaltered under the transformations $\mathbf{k} \to -\mathbf{k}, \mathbf{k}' \to -\mathbf{k}'$ and $\mathbf{S} \to \mathbf{S}$, and hence conserve parity. Accordingly, for parity-conserving interactions (such as the electromagnetic and strong nuclear interactions), $f_{m'm}(\theta, \phi)$ can be written as

$$f_{m'm}(\theta, \phi) = f(\theta)\delta_{m'm} + ig(\theta)\langle m'|\boldsymbol{\sigma}|m\rangle.\hat{\mathbf{n}} \tag{14.59}$$

where $\hat{\mathbf{n}}$ is the unit vector along the normal to the plane of scattering, defined as

$$\hat{\mathbf{n}} = (\mathbf{k} \times \mathbf{k}')/|\mathbf{k} \times \mathbf{k}'|. \tag{14.60}$$

The amplitudes $f(\theta)$ and $g(\theta)$ are functions of k^2 (or the incident energy) and $\mathbf{k}.\mathbf{k}'$ only. From (14.59), we can see that $f_{m'm}(\theta, \phi)$ is the matrix element in spin space of the operator $\mathbf{f}$ defined by

$$\mathbf{f} = f(\theta)I + ig(\theta)\boldsymbol{\sigma}.\hat{\mathbf{n}} \tag{14.61}$$

Now consider scattering from an arbitrary initial spin state $|\chi\rangle$ to an arbitrary final spin state $|\chi'\rangle$, where $|\chi\rangle$ and $|\chi'\rangle$ are of the form (14.32). The differential cross-section is then given by

$$\frac{d\sigma}{d\Omega} = |\langle\chi'|\mathbf{f}|\chi\rangle|^2. \tag{14.62}$$

14.2 The density matrix for a spin-1/2 system. Polarisation

Suppose that the initial beam of spin-1/2 particles is not in a pure state but is partially polarised and is described by a density operator ρ or by a polarisation vector **P** related to ρ by (14.40). If **P** is parallel to the z-axis (longitudinal polarisation), then from (14.49)

$$P = \frac{N_+ - N_-}{N_+ + N_-} \tag{14.63}$$

where N_+ and N_- are equal to the number of measurements of S_z in the incident beam that would give the results $\hbar/2$ and $-\hbar/2$, respectively.

The density operator ρ' describing the beam of scattered particles in the direction (θ, ϕ) can be found from the operator **f**, as follows. The spin function $|\chi'\rangle$ arising from the initial pure spin state $|\chi\rangle$ can be expressed as $C\mathbf{f}|\chi\rangle$, where C is a normalisation factor. Using (14.30), the density operator ρ' is given by

$$\rho' = \sum_i W_i |C|^2 |\mathbf{f}\chi^i\rangle\langle \mathbf{f}\chi^i| \tag{14.64}$$

where the quantity W_i is the statistical weight of the pure spin state $|\chi^i\rangle$ in the initial beam. The expression (14.64) for ρ' can be rewritten in the form

$$\rho' = |C|^2 \mathbf{f} \left(\sum_i W_i |\chi^i\rangle\langle\chi^i| \right) \mathbf{f}^\dagger$$

$$= |C|^2 \mathbf{f}\rho\mathbf{f}^\dagger. \tag{14.65}$$

The condition $\text{Tr}\rho' = 1$ determines $|C|^2$ and we find finally that

$$\rho' = \frac{\mathbf{f}\rho\mathbf{f}^\dagger}{\text{Tr}(\mathbf{f}\rho\mathbf{f}^\dagger)}. \tag{14.66}$$

This relationship between the final and initial density operators is, in fact, quite general. In the present case for scattering of spin-1/2 particles by spin-0 particles it can be usefully expressed in terms of the polarisation vectors **P** and **P**' for the initial and final beams. From (14.40), (14.13) and (14.66), we find

$$\mathbf{P}' = \text{Tr}(\rho'\boldsymbol{\sigma})$$

$$= \frac{\text{Tr}(\mathbf{f}\mathbf{f}^\dagger\boldsymbol{\sigma}) + \text{Tr}(\mathbf{f}(\boldsymbol{\sigma}.\mathbf{P})\mathbf{f}^\dagger\boldsymbol{\sigma})}{\text{Tr}(\mathbf{f}\mathbf{f}^\dagger) + \text{Tr}(\mathbf{f}(\boldsymbol{\sigma}.\mathbf{P})\mathbf{f}^\dagger)}. \tag{14.67}$$

If the initial beam is unpolarised and no spin measurement is made in the final state, the differential cross-section is given by (14.58), which can be written as

$$\left(\frac{d\sigma}{d\Omega}\right)_{\text{unp}} = \frac{1}{2}\text{Tr}(\mathbf{f}\mathbf{f}^\dagger) = |f(\theta)|^2 + |g(\theta)|^2 \tag{14.68}$$

and the polarisation produced in the scattered beam in the direction (θ, ϕ) is

$$\mathbf{P}' = \frac{\text{Tr}(\mathbf{f}\mathbf{f}^\dagger\boldsymbol{\sigma})}{2(d\sigma/d\Omega)_{\text{unp}}} = S(\theta)\hat{\mathbf{n}} \tag{14.69}$$

where

$$S(\theta) = \frac{2\text{Im}\{f(\theta)g^*(\theta)\}}{(d\sigma/d\Omega)_{\text{unp}}} \tag{14.70}$$

and we have used (14.61). The function $S(\theta)$ is called the *Sherman function*.

Another interesting quantity is the differential cross-section summed over all final spin states arising from a particular pure-spin state χ^ν. From (14.62) this is

$$\begin{aligned}
\frac{d\sigma^\nu}{d\Omega} &= \sum_f |\langle \chi^f|\mathbf{f}|\chi^\nu\rangle|^2 \\
&= \sum_f \langle \chi^\nu|\mathbf{f}^\dagger|\chi^f\rangle\langle \chi^f|\mathbf{f}|\chi^\nu\rangle \\
&= \langle \chi^\nu|\mathbf{f}^\dagger\mathbf{f}|\chi^\nu\rangle.
\end{aligned} \tag{14.71}$$

If the incident beam is an incoherent mixture of the pure states χ^ν with statistical weights W_ν described by the density matrix ρ, the *average* differential cross-section summed over all final spin states becomes

$$\begin{aligned}
\frac{d\bar\sigma}{d\Omega} &= \sum_\nu W_\nu \langle \chi^\nu|\mathbf{f}^\dagger\mathbf{f}|\chi^\nu\rangle \\
&= \text{Tr}(\mathbf{f}\rho\mathbf{f}^\dagger).
\end{aligned} \tag{14.72}$$

Using (14.40) and (14.61) together with the result (6.241), we find that

$$\begin{aligned}
\frac{d\bar\sigma}{d\Omega} &= |f(\theta)|^2 + |g(\theta)|^2 + 2[\text{Im}\{f(\theta)g^*(\theta)\}](\hat{\mathbf{n}}\cdot\mathbf{P}) \\
&= \left(\frac{d\sigma}{d\Omega}\right)_{\text{unp}}[1 + S(\theta)(\hat{\mathbf{n}}\cdot\mathbf{P})]
\end{aligned} \tag{14.73}$$

where, as before, $(d\sigma/d\Omega)_{\text{unp}}$ is the differential cross-section (14.68) for an unpolarised incident beam, and $S(\theta)$ is the Sherman function (14.70). We see that the cross-section (14.73) depends not only on the energy and the scattering angle θ but also on the initial polarisation vector $\mathbf{P}$ of the spin-1/2 particles.

Double scattering

The polarisation produced by a collision can be detected by allowing the scattered beam of spin-1/2 particles to be scattered a second time by a spin-0 target. This double scattering is illustrated in Fig. 14.1. After the first collision at A the initially unpolarised beam is polarised in the direction $\hat{\mathbf{n}}$ (see (14.69)) which is directed out of the plane of the figure. This direction can be taken to be the x-axis, and for the second collision at B, the line AB can be taken to be the z-axis. The differential cross-section for scattering through angles (θ_2, ϕ_2) at the second collision is given in terms of the

14.2 The density matrix for a spin-1/2 system. Polarisation   653

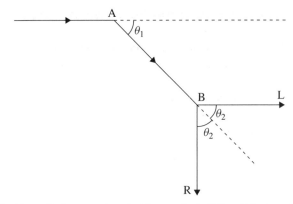

Figure 14.1 A double-scattering experiment. A beam of spin-1/2 particles is scattered by a spin-0 target at A, followed by a second scattering by a spin-0 target at B.

polarisation $P_A(\theta_1)$ produced by the first collision by

$$\frac{d\bar{\sigma}(\theta_2, \phi_2)}{d\Omega} = |f(\theta_2)|^2 + |g(\theta_2)|^2$$
$$+ 2[\text{Im}\{f(\theta_2)g^*(\theta_2)\}]n_x^B P_A(\theta_1). \quad (14.74)$$

Taking the plane of the second scattering to be the same as that of the first, $n_x^B = 1$ for scattering to the right and -1 for scattering to the left of the line AB.

The *asymmetry parameter* ε is defined as

$$\varepsilon = \frac{(d\bar{\sigma}/d\Omega)_R - (d\bar{\sigma}/d\Omega)_L}{(d\bar{\sigma}/d\Omega)_R + (d\bar{\sigma}/d\Omega)_L} \quad (14.75)$$

where $(d\bar{\sigma}/d\Omega)_{R,L}$ stands for the differential cross-section for the second scattering in the right- and left-hand directions. From (14.74), we have

$$\varepsilon = \frac{2P_A(\theta_1)\text{Im}\{f(\theta_2)g^*(\theta_2)\}}{|f(\theta_2)|^2 + |g(\theta_2)|^2}$$
$$= P_A(\theta_1) P_B(\theta_2) \quad (14.76)$$

where $P_B(\theta_2)$ is the polarisation produced by scattering at B from an initial unpolarised beam. If the spin-0 particles at B are of the same kind as those at A and if $\theta_1 = \theta_2$

$$\varepsilon = (P_A(\theta_1))^2 \quad (14.77)$$

so that a measurement of ε determines the magnitude (but not the sign) of P_A.

14.3 The equation of motion of the density matrix

Let us assume that we are working in the Schrödinger picture and that at time $t = t_0$ a certain statistical mixture of states is represented by the density operator

$$\rho(t_0) = \sum_{\alpha=1}^{N} |\alpha(t_0)\rangle W_\alpha \langle \alpha(t_0)|. \tag{14.78}$$

In Section 5.7, we have seen that pure states $|\alpha\rangle$ vary in time according to the equation

$$|\alpha(t)\rangle = U(t, t_0)|\alpha(t_0)\rangle \tag{14.79a}$$

while for the adjoint states we have

$$\langle \alpha(t)| = \langle \alpha(t_0)|U^\dagger(t, t_0) \tag{14.79b}$$

where $U(t, t_0)$ is the evolution operator. As a result, the density operator is also a function of time:

$$\rho(t) = \sum_{\alpha=1}^{N} |\alpha(t)\rangle W_\alpha \langle \alpha(t)|$$

$$= \sum_{\alpha=1}^{N} U(t, t_0)|\alpha(t_0)\rangle W_\alpha \langle \alpha(t_0)|U^\dagger(t, t_0)$$

$$= U(t, t_0)\left[\sum_{\alpha=1}^{N} |\alpha(t_0)\rangle W_\alpha \langle \alpha(t_0)|\right] U^\dagger(t, t_0). \tag{14.80}$$

Using (14.78) and assuming that the statistical weights W_α are constants, we find for the density operator the evolution equation

$$\rho(t) = U(t, t_0)\rho(t_0)U^\dagger(t, t_0) \tag{14.81}$$

which shows that $\rho(t)$ can be obtained from $\rho(t_0)$ by performing the unitary transformation $U(t, t_0)$. It should be noted that if the Hamiltonian H is time independent then from (5.243) we have $U(t, t_0) = \exp[-iH(t - t_0)/\hbar]$ and

$$\rho(t) = e^{-iH(t-t_0)/\hbar} \rho(t_0) e^{iH(t-t_0)/\hbar}. \tag{14.82}$$

Let us now differentiate (14.81) with respect to t, using the equation (5.236) satisfied by the evolution operator $U(t, t_0)$

$$i\hbar \frac{\partial}{\partial t} U(t, t_0) = HU(t, t_0) \tag{14.83a}$$

and the corresponding equation for the adjoint operator $U^\dagger(t, t_0)$

$$-i\hbar \frac{\partial}{\partial t} U^\dagger(t, t_0) = U^\dagger(t, t_0) H. \tag{14.83b}$$

Thus we have

$$i\hbar \frac{\partial}{\partial t}\rho(t) = \left[i\hbar \frac{\partial}{\partial t}U(t,t_0)\right]\rho(t_0)U^\dagger(t,t_0) + U(t,t_0)\rho(t_0)\left[i\hbar \frac{\partial}{\partial t}U^\dagger(t,t_0)\right]$$
$$= HU(t,t_0)\rho(t_0)U^\dagger(t,t_0) - U(t,t_0)\rho(t_0)U^\dagger(t,t_0)H \quad (14.84)$$

and by using (14.81), we obtain for $\rho(t)$ the equation of motion

$$i\hbar \frac{\partial}{\partial t}\rho(t) = [H,\rho(t)]. \quad (14.85)$$

This equation is often called the *Liouville equation* because it has the same form as the equation of motion giving the phase space probability distribution in classical statistical mechanics. It is worth noting that the Liouville equation (14.85) is somewhat similar to the Heisenberg equation of motion (5.285) for an *observable*. We recall, however, that the operators which appear in (14.85) are in the Schrödinger picture. As we saw in Section 5.8, in order to pass from the Schrödinger to the Heisenberg picture the unitary transformation $U^\dagger(t,t_0)$ must be made. As a result, we see from (14.81) that in the Heisenberg picture the density operator ρ_H is such that

$$\rho_H = \rho(t_0) \quad (14.86)$$

so that it is fixed in time and the Liouville equation (14.85) does not apply.

Returning to the Schrödinger picture, we may now readily obtain the equation of motion for the average value of an observable. Using the basic equation (14.13) and the Liouville equation (14.85), we find that for an operator A which does not explicitly depend upon the time

$$\frac{\partial}{\partial t}\text{Tr}(\rho A) = \text{Tr}\left(A \frac{\partial}{\partial t}\rho\right)$$
$$= -\frac{i}{\hbar}\text{Tr}(A[H,\rho])$$
$$= -\frac{i}{\hbar}\text{Tr}(AH\rho - A\rho H)$$
$$= -\frac{i}{\hbar}\text{Tr}([A,H]\rho) \quad (14.87)$$

where we have used the fact that the trace of a product of matrices does not depend on the order of its factors.

14.4 Quantum mechanical ensembles

In introducing the density matrix in Section 14.1, we considered an ensemble composed of N identical sub-systems, all described by the same Hamiltonian. Physical quantities of interest are averages over the ensemble and are determined by (14.13) in terms of the density matrix, the average taking into account both the statistical nature of the ensemble and the quantum mechanical average arising from the probabilistic nature of the theory. We shall see that when dealing with systems of identical particles the requirement that the total wave function should be symmetric (for bosons) or antisymmetric (for fermions) under the interchange of any pair of particles, leads to profound effects in quantum statistical mechanics.

We shall be concerned only with systems in equilibrium, for which the average physical quantities are constant, so that the density operator is time-independent:

$$\frac{\partial}{\partial t}\rho = 0. \tag{14.88}$$

For this to be the case, the density operator ρ must commute with the Hamiltonian H (see (14.85)) and H must be independent of t. It follows that ρ can be expressed as an explicit function of H

$$\rho = \rho(H) \tag{14.89}$$

and if the basis functions $|n\rangle$ are introduced which are eigenfunctions of H, both H and ρ must be diagonal

$$H_{nm} = E_n \delta_{nm}; \qquad \rho_{nm} = \rho_{nn} \delta_{nm}. \tag{14.90}$$

The probability that a sub-system chosen at random from an ensemble is found to be in the eigenstate $|n\rangle$ is determined by ρ_{nn} (see the discussion below (14.17)), and this is a function of the corresponding eigenvalue E_n. Several different kinds of ensemble are found to be useful and the precise functional dependence of ρ_{nn} on E_n depends on which ensemble is under discussion. We shall start by introducing the *micro-canonical* ensemble, which is appropriate for the description of a closed system.

The micro-canonical ensemble

In a micro-canonical ensemble each of the identical sub-systems composing the ensemble consists of a fixed number of particles $\mathcal{N}$ contained within a volume $\mathcal{V}$ and for which the energy E of each sub-system lies within a specified small interval $E_0 \leqslant E \leqslant E_0 + \Delta$, where $\Delta \ll E_0$. Subject to these conditions, the total number of *distinct* (pure) states in which a member of the ensemble can be found is denoted by Γ, which is a function of $\mathcal{N}$, $\mathcal{V}$, E_0 and Δ. We shall denote by $\mathcal{H}$ the Hamiltonian of each sub-system. The fundamental postulates are now made that:

(1) there is an equal *a priori* probability of finding the system in any one of these states;

(2) the phases of the probability amplitudes $c_n^{(\alpha)}$ (see (14.4)) which determine the wave function of each member α of the ensemble are distributed at random. This is equivalent to requiring the members of the ensemble to be non-interacting, so that the corresponding wave functions do not interfere with one another.

14.4 Quantum mechanical ensembles

As a consequence of postulate 1, in the energy representation (in which $\mathcal{H}$ is diagonal)

$$\rho_{nn} = 1/\Gamma \tag{14.91}$$

for all states in the interval $E_0 \leqslant E \leqslant E_0 + \Delta$ and

$$\rho_{nn} = 0 \tag{14.92}$$

for all other states.

The *entropy* of the system can now be defined as

$$S = k \log \Gamma \tag{14.93}$$

where k is Boltzmann's constant. This agrees with the usual thermodynamical definition, but now Γ must be calculated using quantum mechanics. It should be noted that if $\Gamma = 1$ each sub-system is in the same state. In this case, the entropy S vanishes and the system is perfectly ordered. If Γ is greater than 1 the system is disordered, S providing a measure of the disorder. It can be shown that the entropy defined by (14.93) has the familiar properties:

(1) The entropy of a closed system has its maximum value when the system is in thermodynamical equilibrium.

(2) The entropy is additive, in the sense that if a closed system is divided into two parts with entropies S_1 and S_2 respectively, then $S = S_1 + S_2$.

(3) The temperature T of the system can be defined by the relation

$$\frac{1}{T} = \frac{\partial S}{\partial E}. \tag{14.94}$$

Although the micro-canonical ensemble is of fundamental importance in describing the properties of a closed system, it is difficult to use in practice, and for this reason we shall consider two further types of ensemble introduced by J. W. Gibbs, called the *canonical* ensemble and the *grand canonical* ensemble.

The canonical ensemble

In this case the system of interest, $\mathcal{S}$, is in thermal contact with a much larger system, called the heat reservoir, $\mathcal{R}$. The combined system composed of $(\mathcal{R} + \mathcal{S})$ is closed, and its statistical properties are described by a micro-canonical ensemble. We shall suppose that $\mathcal{S}$ is composed of a fixed number of particles $\mathcal{N}_\mathcal{S}$ within a volume $\mathcal{V}_\mathcal{S}$ (see Fig. 14.2). Since $\mathcal{R}$ and $\mathcal{S}$ are in thermal contact, energy can be exchanged between them, however the interaction is supposed to be weak so that both $\mathcal{R}$ and $\mathcal{S}$ at any given time can be considered to be in definite energy eigenstates with energies $E_\mathcal{R}$ and $E_\mathcal{S}$, respectively. In equilibrium both $\mathcal{R}$ and $\mathcal{S}$ share a common temperature T. Conservation of energy requires that

$$E = E_\mathcal{R} + E_\mathcal{S} \tag{14.95}$$

where E is one of the eigenenergies of the combined system. Since the combined system is described by a micro-canonical ensemble, E is confined to the narrow range

658 Quantum statistics

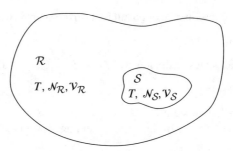

Figure 14.2 A canonical ensemble: $\mathcal{S}$ is in thermal contact with the heat reservoir $\mathcal{R}$. Energy is exchanged between $\mathcal{S}$ and $\mathcal{R}$ but $\mathcal{N}_\mathcal{S}$ and $\mathcal{V}_\mathcal{S}$ are constant.

$E_0 \leqslant E \leqslant E_0 + \Delta$ where $\Delta \ll E_0$. The basic problem is to determine the probability that at a given temperature T, the system $\mathcal{S}$ is found to be in a particular state, with energy $\mathcal{E}_\mathcal{S}$.

Let $d\Gamma(E)$ be the total number of distinct states of the combined system with energies in a range E to $E+dE$. Then, by the postulate of equal *a priori* probabilities, the probability of finding the combined system with energy in this range is

$$dP = Cd\Gamma(E) \tag{14.96}$$

if $E_0 \leqslant E \leqslant E_0 + \Delta$ and zero otherwise, where C is a constant. If $d\Gamma_\mathcal{S}$ and $d\Gamma_\mathcal{R}$ are the corresponding numbers of states in the system $\mathcal{S}$ and the reservoir $\mathcal{R}$, we have that $d\Gamma = d\Gamma_\mathcal{S} d\Gamma_\mathcal{R}$ and

$$dP = Cd\Gamma_\mathcal{S}(E_\mathcal{S})d\Gamma_\mathcal{R}(E_\mathcal{R})\delta(E - E_\mathcal{R} - E_\mathcal{S}) \tag{14.97}$$

where the delta function expresses the conservation of energy condition (14.95). We now want to find the probability dP_n that the combined system is in a state such that the system of interest $\mathcal{S}$ has a prescribed energy $\mathcal{E}_\mathcal{S} = E_n$, without regard to the state of the reservoir. This can be found by integrating dP over the allowed states of the reservoir:

$$dP_n = Cd\Gamma_\mathcal{S}(E_n) \int \delta(E - E_\mathcal{R} - E_n)d\Gamma_\mathcal{R}(E_\mathcal{R})$$
$$= Cd\Gamma_\mathcal{S}(E_n)\Delta\Gamma_\mathcal{R}(E - E_n) \tag{14.98}$$

where $\Delta\Gamma_\mathcal{R}(E - E_n)$ is the total number of states of the reservoir, subject to the conservation of energy condition. Using (14.93), the entropy of the reservoir is

$$S_\mathcal{R} = k \log \Delta\Gamma_\mathcal{R} \tag{14.99}$$

so that

$$\Delta\Gamma_\mathcal{R} = \exp(S_\mathcal{R}/k). \tag{14.100}$$

14.4 Quantum mechanical ensembles

Since the system S is small compared with the total system, $E_n \ll E$, and

$$S_\mathcal{R}(E_\mathcal{R}) = S_\mathcal{R}(E - E_n)$$
$$\simeq S_\mathcal{R}(E) - \frac{\partial S_\mathcal{R}(E)}{\partial E} E_n = S_\mathcal{R}(E) - E_n/T \quad (14.101)$$

where in the second line equation (14.94) has been employed. Combining (14.101) with (14.100) and (14.98), we find that

$$dP_n = \mathcal{A} \exp(-E_n/kT) d\Gamma_S(E_n) \quad (14.102)$$

where $\mathcal{A}$ is defined by

$$\mathcal{A} = C \exp(S_\mathcal{R}(E)/k). \quad (14.103)$$

In view of (14.102), the density matrix of the system S in the energy representation can be taken to be

$$\rho_{nm} = \mathcal{A} \exp(-E_n/kT) \delta_{nm} \quad (14.104)$$

where ρ_{nn} is the probability of finding S in the state n with energy E_n. The constant $\mathcal{A}$ can be found from the normalisation condition $\mathrm{Tr}(\rho) = 1$. Defining $Q_{\mathcal{N}_S}(T)$ by

$$Q_{\mathcal{N}_S}(T) = \mathcal{A}^{-1} \quad (14.105)$$

and taking the trace of both sides of (14.104) we find that

$$Q_{\mathcal{N}_S}(T) = \sum_n \exp(-E_n/kT) = \mathrm{Tr}[\exp(-\beta H)]. \quad (14.106)$$

where

$$\beta = 1/kT \quad (14.107)$$

and H is the Hamiltonian of the system of interest S. The quantity $Q_{\mathcal{N}_S}(T)$ is called the *partition function*. It should be emphasised that the sum over n is over all distinct states and not over eigenenergy values.

From (14.104) and (14.105) the density operator is

$$\rho = \frac{1}{Q_{\mathcal{N}_S}(T)} \exp(-\beta H) \quad (14.108)$$

and the average value of an observable A over the ensemble is (see (14.13))

$$\langle A \rangle = \mathrm{Tr}(\rho A) = \frac{1}{Q_{\mathcal{N}_S}} \mathrm{Tr}[\exp(-\beta H) A]$$
$$= \frac{\mathrm{Tr}[\exp(-\beta H) A]}{\mathrm{Tr}[\exp(-\beta H)]}. \quad (14.109)$$

The average $\bar{E}$ of the energy is found by setting $A = H$. Using (14.106) it can be expressed in the useful form

$$\bar{E} = \langle H \rangle = -\frac{1}{Q_{\mathcal{N}_S}} \frac{\partial Q_{\mathcal{N}_S}}{\partial \beta} = -\frac{\partial}{\partial \beta} (\log Q_{\mathcal{N}_S}). \quad (14.110)$$

The grand canonical ensemble

The difference between the micro-canonical and canonical ensembles is that the energy of the system is constant in the former case, while fluctuations in energy are taken into account in the latter case. In many cases, it is easier to obtain physical averages by also allowing for fluctuations in the number of particles contained in the system S. Particles can be allowed to transfer between S and the heat reservoir $\mathcal{R}$ in such a way that the total number of particles $\mathcal{N}$ in the combined system is constant (see Fig. 14.3)

$$\mathcal{N} = \mathcal{N}_\mathcal{R} + \mathcal{N}_\mathcal{S}. \tag{14.111}$$

Since the fluctuations in $\mathcal{N}_\mathcal{S}$ and $E_\mathcal{S}$ about the mean values are small for a large system, the physical results obtained using the micro-canonical, canonical and grand canonical ensembles are identical. The treatment of the grand canonical ensemble parallels that of the canonical ensemble. The difference is that the entropy of the reservoir is now a function of both $(E - E_\mathcal{S})$ and $(\mathcal{N} - \mathcal{N}_\mathcal{S})$, so that (14.101) is replaced by

$$S_\mathcal{R}(E - E_n, \mathcal{N} - \mathcal{N}_\mathcal{S}) \simeq S_\mathcal{R}(E, \mathcal{N}) - \frac{\partial S_\mathcal{R}(E, \mathcal{N})}{\partial E} E_n(\mathcal{N}_\mathcal{S}) - \frac{\partial S_\mathcal{R}(E, \mathcal{N})}{\partial \mathcal{N}} \mathcal{N}_\mathcal{S} \tag{14.112}$$

where we have written the energy eigenvalues of the system S as $E_n(\mathcal{N}_\mathcal{S})$ to emphasise that the energy levels depend on the number of particles that the system contains. The density matrix is now of the form

$$\rho_{nm} = B \exp(-\beta E_n(\mathcal{N}_\mathcal{S}) - \alpha \mathcal{N}_\mathcal{S}) \delta_{nm} \tag{14.113}$$

where B is a constant, $\alpha = (\partial S_\mathcal{R}/\partial \mathcal{N})/k$ and β is defined by (14.107). If an operator $\nu_\mathcal{S}$ is introduced with eigenvalues $N = 0, 1, 2, \ldots$, representing the number of particles in the system S, the density operator can be expressed as

$$\rho = B \exp[-\beta H(\nu_\mathcal{S}) - \alpha \nu_\mathcal{S}]. \tag{14.114}$$

The normalisation condition $\text{Tr}(\rho) = 1$ requires that if we set $Z(\alpha, T) = B^{-1}$, then

$$\frac{1}{B} = Z(\alpha, T) = \text{Tr}\{\exp[-\beta H(\nu_\mathcal{S}) - \alpha \nu_\mathcal{S}]\}. \tag{14.115}$$

The quantity Z is called the *grand partition function*. Since the eigenvalues of $\nu_\mathcal{S}$ are the integers (or zero), we can write

$$\begin{aligned} Z(\alpha, T) &= \sum_N \sum_n \exp[-\beta E_n(N) - \alpha N] \\ &= \sum_N \left\{ \sum_n \exp[-\beta E_n(N)] \right\} [\exp(-\alpha)]^N \\ &= \sum_N Q_N(T) z^N \end{aligned} \tag{14.116}$$

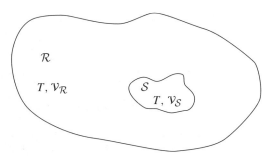

Figure 14.3 A grand canonical ensemble: $\mathcal{S}$ is in thermal contact with the heat reservoir $\mathcal{R}$. Energy is exchanged between $\mathcal{S}$ and $\mathcal{R}$ and the particle numbers fluctuate subject to $\mathcal{N_R} + \mathcal{N_S} = \mathcal{N}$ = a constant. $\mathcal{V_S}$ is constant.

where $z = \exp(-\alpha)$ is called the *fugacity*, and $Q_N(T)$ is the partition function for a system of N particles.

14.5 Applications to single-particle systems

The concept of the canonical ensemble requires that the combination of the system $\mathcal{S}$ and the reservoir $\mathcal{R}$ should possess very many states, and form a micro-canonical ensemble, subject to the condition that the total energy lies in a specified narrow band. The system $\mathcal{S}$ can be a macroscopic system, or be composed of a single particle, provided that the interaction between the particle and its neighbours is weak. We shall describe two examples of the latter type: a particle of spin-1/2 in a constant magnetic field and a particle in a box.

Spin-1/2 particle in a magnetic field

Consider a spin-1/2 particle having a magnetic moment $\mathcal{M}_s = -g\mu_B \mathbf{S}/\hbar$ subjected to a constant magnetic field of magnitude $\mathcal{B}$ in a direction parallel to the z-axis. The Hamiltonian H is (see (12.32)),

$$H = g\mu_B \mathcal{B} \sigma_z/2. \tag{14.117}$$

Using the canonical ensemble the density matrix ρ given by (14.108) is

$$\rho = \frac{\exp(-\beta H)}{\text{Tr}[\exp(-\beta H)]} \tag{14.118a}$$

and using the explicit form (6.243) of σ_z,

$$\rho = \frac{1}{\exp(-\beta g\mu_B \mathcal{B}/2) + \exp(\beta g\mu_B \mathcal{B}/2)} \begin{pmatrix} \exp(-\beta g\mu_B \mathcal{B}/2) & 0 \\ 0 & \exp(\beta g\mu_B \mathcal{B}/2) \end{pmatrix}$$

$$\tag{14.118b}$$

where, as usual, $\beta = 1/kT$.

We can now calculate the average value of the z-component of the spin $S_z = \hbar\sigma_z/2$, at the temperature T. It is

$$\langle S_z \rangle = \frac{\hbar}{2}\mathrm{Tr}(\rho\sigma_z) = \frac{\hbar}{2}\frac{\exp(-\beta g\mu_B \mathcal{B}/2) - \exp(\beta g\mu_B \mathcal{B}/2)}{\exp(-\beta g\mu_B \mathcal{B}/2) + \exp(\beta g\mu_B \mathcal{B}/2)}$$

$$= -\frac{\hbar}{2}\tanh(\beta g\mu_B \mathcal{B}/2). \tag{14.119}$$

The corresponding average value of the z-component of the magnetic moment is $\langle \mathcal{M}_z \rangle = -(g\mu_B/\hbar)\langle S_z \rangle$. For a system of N particles per unit volume, the total average magnetic moment in the z-direction at temperature T is given by $N\langle \mathcal{M}_z \rangle = -(g\mu_B/\hbar)N\langle S_z \rangle$ and using $\langle S_z \rangle$ given by (14.119), we regain equation (12.125), as we should.

Average energy of a particle in a box

As a second example, we shall work out the average energy of a spinless particle of mass μ confined to a cubic box of side L at a temperature T, again using a canonical ensemble. The eigenenergies of the system have already been obtained in Chapter 7. We found (see (7.30) and (7.31)) that

$$E_n = \frac{\hbar^2}{2\mu}\left(\frac{\pi}{L}\right)^2 n^2, \quad n^2 = n_x^2 + n_y^2 + n_z^2 \tag{14.120}$$

where each distinct state is specified by the set of three positive integers n_x, n_y, n_z. The partition function, Q, for the system is given by

$$Q = \mathrm{Tr}[\exp(-\beta H)] = \sum_{n_x,n_y,n_z} \exp(-\beta E_n) \tag{14.121}$$

where each sum is over all the positive integers. If the box is very large, the levels are closely spaced and we can express Q as the integral

$$Q = \int_0^\infty dn_x \int_0^\infty dn_y \int_0^\infty dn_z \exp\{-a(n_x^2 + n_y^2 + n_z^2)\} \tag{14.122}$$

where

$$a = \frac{\hbar^2}{2\mu}\left(\frac{\pi}{L}\right)^2 \beta. \tag{14.123}$$

Using the result

$$\int_0^\infty \exp(-ax^2)dx = \frac{1}{2}\left(\frac{\pi}{a}\right)^{1/2} \tag{14.124}$$

we find that

$$Q = L^3\left(\frac{\mu}{2\pi\hbar^2\beta}\right)^{3/2} \tag{14.125}$$

The average energy is then, using (14.110),

$$\bar{E} = \langle H \rangle = -\frac{\partial}{\partial \beta}(\log Q) = \frac{3}{2\beta} = \frac{3}{2}kT \tag{14.126}$$

in agreement with the classical principle of equipartition of energy. Note that if the box were small the sum in (14.121) could not be replaced by the integral (14.122), and the quantum and classical results would differ.

14.6 Systems of non-interacting particles

We shall now discuss the properties of systems of large numbers of non-interacting objects which are equivalent and possess the same energy levels. Three cases can be identified. In the first, known as a Maxwell–Boltzmann (MB) system, the objects can be distinguished; for example, each might be an oscillator distinguished by its position. The other two cases are those for which the objects are identical and indistinguishable particles, for which the complete wave function of the system must be either symmetric or antisymmetric under the interchange of any pair of particles. As we noted in Chapter 10, the former case applies to spin-0 or to integer-spin particles, called bosons, while the latter applies to half-odd-integer-spin particles, called fermions. Bosons are said to obey Bose–Einstein, and fermions Fermi–Dirac, statistics.

If $E_j (j = 1, 2, 3, \ldots)$ denotes an energy level of one of the particles (or objects), the total energy of the system E can be written[1]

$$E = \sum_j n_j E_j \tag{14.127}$$

where the sum runs over all the eigenenergies of a single particle, and where n_j is the number of particles with energy E_j. The total number of particles in the system is

$$N = \sum_j n_j. \tag{14.128}$$

In a Maxwell–Boltzmann or a Bose–Einstein system, any number of particles can be in each level, so that each n_j can either be zero or any positive integer, provided that (14.127) is satisfied. On the other hand, in Fermi–Dirac systems the Pauli principle allows no more than one particle to be in the same single-particle state. In this case, each n_j can only take the values 1 or 0. In calculating partition functions using (14.106) or averages by (14.109), we recall that the sums are not over eigenenergies, but are over each distinct state of the system. For this reason we need to calculate the number of distinct states g_E of the system corresponding to each level of the system with total energy E.

---

[1] For notational simplicity we shall from now on denote the energy E_S and the number of particles N_S of the system by E and N, respectively.

For a Maxwell–Boltzmann system, the total wave function is just the product of N single-particle wave functions:

$$\Psi = \phi_{E_j}(1)\phi_{E_k}(2)\ldots\phi_{E_l}(N). \tag{14.129}$$

If all the energies $E_j, E_k, \ldots, E_l$, are different, each permutation of the particles results in a different wave function, thus each energy level of the total system is $N!$-fold degenerate. However, if n_j particles with $n_j > 1$ have the same energy, interchanges between these particles do not alter the wave function. Thus the number of distinct states corresponding to a given value of the total energy of the system E is, in this case,

$$g_E = g_E^{\text{MB}} = \frac{N!}{\prod_i (n_j!)}. \tag{14.130}$$

There is only one fully symmetric and one fully antisymmetric wave function that can be built from N single-particle wave functions, so both in the Bose–Einstein and Fermi–Dirac cases the energy level E of the system is non-degenerate. To express this we set

$$g_E = g_E^{\text{BE}} = 1 \quad (\text{any } n_j) \tag{14.131}$$

for the Bose–Einstein case and

$$\begin{aligned} g_E = g_E^{\text{FD}} &= 1 \quad (n_j = 0 \text{ or } n_j = 1) \\ &= 0 \quad (\text{otherwise}) \end{aligned} \tag{14.132}$$

for the Fermi–Dirac case.

From (14.106), the partition function can be expressed, in each case, as

$$Q_N(T) = \sum_{\{n_j\}} g_E \exp\left[-\beta \sum_j n_j E_j\right], \quad \beta = 1/kT \tag{14.133}$$

where $\sum_{\{n_j\}}$ denotes a sum over every possible set of numbers n_j which is consistent with (14.128).

The average value $\langle E \rangle$ of the energy of the system can be calculated if the average value $\langle n_i \rangle$ of each of the occupation numbers n_i is specified, since

$$\langle E \rangle = \sum_i E_i \langle n_i \rangle. \tag{14.134}$$

Using a canonical ensemble, the averages $\langle n_i \rangle$ are, from (14.109) and (14.133)

$$\langle n_i \rangle = \frac{1}{Q_N(T)} \sum_{\{n_j\}} g_E \exp\left[-\beta \sum_j n_j E_j\right] n_i \tag{14.135}$$

so that

$$\langle n_i \rangle = -\frac{\partial \log Q_N(T)}{\partial (\beta E_i)}. \tag{14.136}$$

The average value $\langle n_i \rangle$ is easily calculated for the Maxwell–Boltzmann case, since taking g_E in (14.133) to be given by (14.130), we see that $Q_N(T)$ can be written as

$$Q_N(T) = \left[\sum_j \exp(-\beta E_j) \right]^N \tag{14.137}$$

and using (14.136), we find

$$\langle n_i \rangle = N \left[\sum_j \exp(-\beta E_j) \right]^{-1} \exp(-\beta E_i). \tag{14.138}$$

This result can be written as

$$\langle n_i \rangle = \exp[-(\beta E_i + \alpha)] \tag{14.139}$$

where α is a temperature-dependent constant defined by

$$\alpha = \log \left[\sum_j \exp(-\beta E_j)/N \right]. \tag{14.140}$$

The sum in (14.133) is difficult to evaluate when g_E is given by (14.131) or (14.132), and for this reason it is better to use the grand canonical ensemble to obtain $\langle n_i \rangle$ for Bose–Einstein or Fermi–Dirac systems. Averages of an operator A are again given by (14.13) with ρ expressed by (14.114). Since we are using a basis of simultaneous eigenfunctions of H and N, we have

$$\langle n_i \rangle = Z^{-1} \sum_{n_1=0}^{\infty} \sum_{n_2=0}^{\infty} \ldots g_E \exp \left\{ -\beta \sum_j [n_j E_j(N)] - \alpha \sum_j n_j \right\} n_i \tag{14.141}$$

where

$$Z = \sum_{n_1=0}^{\infty} \sum_{n_2=0}^{\infty} \ldots g_E \exp \left\{ -\beta \sum_j [n_j E_j(N)] - \alpha \sum_j n_j \right\}. \tag{14.142}$$

We notice that the grand partition function Z can be expressed as

$$Z = \sum_{n_1=0}^{\infty} \sum_{n_2=0}^{\infty} \ldots g_E \prod_j \exp\{-n_j[\beta E_j(N) + \alpha]\}$$

$$= \prod_j \sum_{n_j=0}^{\infty} g_E \exp\{-n_j[\beta E_j(N) + \alpha]\}. \tag{14.143}$$

For Bose–Einstein systems with $g_E = g_E^{\mathrm{BE}} = 1$ (see (14.131)),

$$Z^{\mathrm{BE}} = \prod_j [1 - \exp\{-\beta E_j(N) - \alpha\}]^{-1}. \tag{14.144}$$

In the case of Fermi–Dirac systems with g_E given by (14.132), such that the sum over n_j is restricted to the values 0 and 1, we find

$$Z^{\mathrm{FD}} = \prod_j [1 + \exp(-\beta E_j(N) - \alpha)]. \tag{14.145}$$

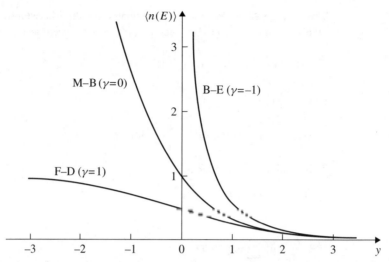

Figure 14.4 The average occupation number $\langle n(E) \rangle$ for states in an ideal gas plotted as a function of $y = \beta E + \alpha$ for the cases of Bose–Einstein ($\gamma = -1$), Maxwell–Boltzmann ($\gamma = 0$) and Fermi–Dirac ($\gamma = 1$) statistics.

In both cases, we have

$$\langle n_i \rangle = -\frac{\partial \log Z}{\partial [\beta E_i(N)]}$$

$$= \frac{1}{\exp[\beta E_i(N) + \alpha] + \gamma} \qquad (14.146)$$

where $\gamma = -1$ for Bose–Einstein and $+1$ for Fermi–Dirac systems. Comparing with (14.139), we see that if we set $\gamma = 0$ for Maxwell–Boltzmann systems, (14.146) can be used to express the average $\langle n_i \rangle$ in each of the three cases. The function

$$\langle n(E) \rangle = [\exp(\beta E + \alpha) + \gamma]^{-1}, \qquad (14.147)$$

where E varies continuously, is known as a *distribution function*.

In Fig. 14.4, $\langle n(E) \rangle$ is plotted against the parameter $y = \beta E + \alpha$ for the three cases $\gamma = 0$ (Maxwell–Boltzmann), $\gamma = 1$ (Fermi–Dirac) and $\gamma = -1$, (Bose–Einstein). We notice that for large positive values of y the three distributions become indistinguishable. In the Fermi–Dirac case $\langle n(E) \rangle$ cannot exceed unity since each individual occupation number n_i can only be 1 or 0. Since both the individual numbers n_i and the average $\langle n(E) \rangle$ must be positive or zero, only positive values of y are allowed for Bose–Einstein statistics. As $y \to 0$, $\langle n(E) \rangle \to \infty$ in the Bose–Einstein case, and this gives rise to the phenomenon of Bose–Einstein condensation discussed in Section 14.8 below.

An additional requirement is needed to determine the constant α. Although in the grand canonical ensemble the number of particles fluctuates, we can specify the average number of particles in the system, $\langle N \rangle$, and determine α through the equation

$$\langle N \rangle = \sum_j \langle n_j \rangle = \sum_j \{\exp[\beta E_j(N) + \alpha] + \gamma\}^{-1}. \tag{14.148}$$

14.7 The photon gas and Planck's law

As an application of the theory outlined in the last section, we shall re-derive Planck's law for black body radiation. The electromagnetic radiation in thermal equilibrium within an enclosure of volume V can be considered as consisting of a photon gas, each photon being a boson of zero mass, with two states of polarisation. The shape of the enclosure is immaterial, provided it is large, and will be taken to be a cube of volume $V = L^3$.

Since photons are absorbed and emitted in interactions with the walls of the enclosure, the conservation of particles expressed in (14.111) does not apply, and the entropy of the heat reservoir expressed in (14.112) is independent of N, the number of photons in the system. This has the consequence that the constant α in (14.113) must be set equal to zero. The average number of photons in each energy level at temperature T is, from (14.146) with $\alpha = 0$ and $\gamma = -1$,

$$\langle n_i \rangle = [\exp(\beta E_i) - 1]^{-1}. \tag{14.149}$$

If the volume V is large, the energy levels are closely spaced and E_i can be taken to be a continuous variable E. The number of photons $n(E, T)dE$ within the interval E to $E + dE$ at temperature T is from (14.149)

$$n(E, T)dE = D(E)[\exp(\beta E) - 1]^{-1} dE \tag{14.150}$$

where $D(E)$ is the number of photon states per unit energy interval. The energy E, wave number k, frequency ν and wavelength λ of a photon are related by

$$E = \hbar c k = 2\pi \hbar \nu = 2\pi \hbar c / \lambda. \tag{14.151}$$

Using the relationship between E and λ, the number of photon states $G(\lambda)$ within the wavelength interval $(\lambda, \lambda + d\lambda)$ can be expressed as

$$G(\lambda) = D(E)|dE/d\lambda|. \tag{14.152}$$

It follows from (14.150) that the number of photons with wavelengths between λ and $\lambda + d\lambda$ at temperature T is

$$\mathcal{N}(\lambda, T)d\lambda = G(\lambda)[\exp(\beta 2\pi \hbar c/\lambda) - 1]^{-1} d\lambda \tag{14.153}$$

and the energy density within the interval $d\lambda$ is

$$\rho(\lambda, T)d\lambda = E\mathcal{N}(\lambda, T)d\lambda/V = 2\pi \hbar c \mathcal{N}(\lambda, T)d\lambda/(\lambda V). \tag{14.154}$$

To find $D(E)$, and hence $G(\lambda)$, we note that the wave function of each photon is a plane wave which can be written as

$$\psi(x, y, z) = \exp(ik_x x + ik_y y + ik_z z). \tag{14.155}$$

If periodic boundary conditions are imposed, then (see (7.19))

$$k_x = \frac{2\pi}{L} n_x, \qquad k_y = \frac{2\pi}{L} n_y, \qquad k_z = \frac{2\pi}{L} n_z \tag{14.156}$$

where n_x, n_y, n_z are integers (positive or negative), or zero. The number of plane-wave states, up to a certain value of k, is equal to the number of cells of unit volume within a sphere of radius $k(L/2\pi)$. Recalling that each photon has two states of polarisation, the total number of photon states, n, up to a certain value k is

$$n = 2\left(\frac{L}{2\pi}\right)^3 \frac{4\pi}{3} k^3 = 2\left(\frac{1}{2\pi}\right)^3 \frac{4\pi V}{3} \left(\frac{L}{\hbar c}\right)^3 \tag{14.157}$$

from which

$$D(E) = \frac{dn}{dE} = \frac{1}{\pi^2} \frac{V E^2}{(\hbar c)^3} \tag{14.158}$$

and

$$G(\lambda) = D(E)|dE/d\lambda|$$
$$= 8\pi V/\lambda^4. \tag{14.159}$$

Using (14.153), we then have

$$\mathcal{N}(\lambda, T) = \frac{8\pi V}{\lambda^4} \frac{1}{\exp(\beta hc/\lambda) - 1} \tag{14.160}$$

and from (14.154) the spectral distribution function $\rho(\lambda, T)$ is given by

$$\rho(\lambda, T) = \frac{8\pi hc}{\lambda^5} \frac{1}{\exp(\beta hc/\lambda) - 1} \tag{14.161}$$

in agreement with Planck's law (1.13).

14.8 The ideal gas

In Section 10.3, we discussed the ground state of a system of identical non-interacting spin-1/2 particles confined within a box of volume V. In the statistical language of the present chapter, such a system forms an ideal fermion gas at zero temperature, $T = 0$. We shall now use statistical methods to describe a system of non-interacting particles confined within a cube with sides of length L and impenetrable walls at a finite temperature, treating the Maxwell–Boltzmann and boson cases as well as that of fermions.

From (14.147), the number of particles $n(E)dE$ within the energy interval E to $E + dE$ is

$$n(E)dE = D(E)[\exp(\beta E + \alpha) + \gamma]^{-1}dE, \qquad \beta = 1/kT \qquad (14.162)$$

where $D(E)$ is the number of single-particle states per unit energy interval, and where $\gamma = 0$ for Maxwell–Boltzmann, -1 for Bose–Einstein and $+1$ for Fermi–Dirac statistics. The constant α depends on the total number of particles and on the temperature.

For spin-1/2 particles, the density of states $D(E)$ has been given in (10.36). For particles of spin s there are $(2s + 1)$ spin states associated with each spatial orbital, so that, in general,

$$D(E) = (2s + 1)\frac{1}{4\pi^2}\left(\frac{2m}{\hbar^2}\right)^{3/2} V E^{1/2}. \qquad (14.163)$$

The number of particles per unit volume within the energy interval $(E, E + dE)$ is

$$n(E)\frac{dE}{V} = \frac{(2s + 1)}{4\pi^2}\left(\frac{2m}{\hbar^2}\right)^{3/2} \frac{E^{1/2}}{\exp(\beta E + \alpha) + \gamma} dE. \qquad (14.164)$$

By integrating this expression over E, the constant α can be determined in terms of the number density ρ of the particles, since

$$\rho = \frac{1}{V}\int_0^\infty n(E)dE = \frac{(2s + 1)}{4\pi^2}\left(\frac{2m}{\hbar^2}\right)^{3/2} \int_0^\infty \frac{E^{1/2}dE}{\exp(\beta E + \alpha) + \gamma}. \qquad (14.165)$$

In the case of a boson gas, this needs qualification, as we shall see in a moment. The total internal energy per unit volume, U, of the gas can be calculated in terms of α and β, using

$$U = \frac{1}{V}\int_0^\infty En(E)dE = \frac{(2s + 1)}{4\pi^2}\left(\frac{2m}{\hbar^2}\right)^{3/2} \int_0^\infty \frac{E^{3/2}dE}{\exp(\beta E + \alpha) + \gamma}. \qquad (14.166)$$

The low-density–high-temperature limit

We notice from (14.165) that if $\alpha \gg 1$, the density ρ is small. Under these circumstances $\exp(\beta E + \alpha) \gg \gamma$, so that both the distribution laws for bosons and for fermions become equal to that for the Maxwell–Boltzmann case. We conclude that quantum effects due to the identity of the particles are unimportant in systems with small particle densities. When γ can be neglected in comparison with $\exp(\beta E + \alpha)$, the integral in (14.165) is easily evaluated, and one finds

$$\rho = \frac{(2s + 1)}{4\pi^2}\left(\frac{2m}{\hbar^2}\right)^{3/2} \exp(-\alpha)\frac{\Gamma(3/2)}{\beta^{3/2}}$$

$$= (2s + 1)\left(\frac{m}{2\pi\hbar^2}\right)^{3/2} (kT)^{3/2} \exp(-\alpha) \qquad (14.167)$$

where Γ is the gamma function, and we have used the fact that $\Gamma(3/2) = (1/2)\Gamma(1/2) = (1/2)\sqrt{\pi}$. This allows the determination of the quantity α in terms of the density ρ and the temperature T. That is

$$\exp(\alpha) = z^{-1} = \frac{2s+1}{\rho}\left(\frac{m}{2\pi\hbar^2}\right)^{3/2}(kT)^{3/2} \tag{14.168}$$

where z is the fugacity.

It can be verified (see Problem 14.3) that for all gases at room temperature and pressure, z^{-1} is indeed very large compared with unity, so that Maxwell–Boltzmann statistics are always closely obeyed, and this explains the success of the classical kinetic theory of gases at room temperature and pressure.

Bose–Einstein gases

We first note from (14.147) or (14.162) that for bosons, with $\gamma = -1$, the constant α must be positive or zero, for otherwise the occupation numbers n_j could become negative. The high-density limit corresponds to small values of α and the low density limit to large values of α. We also see from (14.147) or (14.162) that for low temperatures and high densities, nearly all the particles will be in the state of lowest energy. This phenomenon is called *Bose–Einstein condensation*.

In the low-temperature, high-density limit, expressions such as (14.165) or (14.166), where E is treated as a continuous variable, must be modified because $D(0) = 0$ (see (14.163)), but in fact there is one state at $E = 0$[2]. To allow for this, it is sufficient to replace $D(E)$ by

$$\bar{D}(E) = D(E) + \delta(E) \tag{14.169}$$

where δ is the Dirac delta function. Then (14.165) is replaced by

$$\rho = \rho_0 + \frac{(2s+1)}{4\pi^2}\left(\frac{2m}{\hbar^2}\right)^{3/2}\int_0^\infty \frac{E^{1/2}dE}{\exp(\beta E + \alpha) - 1} \tag{14.170}$$

where ρ_0 is the density of particles in the ground state with $E = 0$.

As $T \to 0$, the integral term also tends to zero, so that at $T = 0$, $\rho = \rho_0$. Since for bosons any number of particles can occupy the single-particle level of lowest energy, there is no limitation on the value of the particle densities ρ_0 or ρ. The temperature T_0 below which particles are forced to condense into the ground state can be found from (14.170). Setting $x = \beta E = E/kT$, we obtain

$$\rho - \rho_0 = \frac{2s+1}{4\pi^2}\left(\frac{2mkT}{\hbar^2}\right)^{3/2}\int_0^\infty \frac{x^{1/2}dx}{\exp(x+\alpha) - 1}$$

$$= \frac{2s+1}{4\pi^2}\left(\frac{2mkT}{\hbar^2}\right)^{3/2} f_{1/2}(\alpha) \tag{14.171}$$

---

[2] It is conventional to normalise energies so that $E = 0$ represents the single-particle ground state.

where

$$f_n(\alpha) = \int_0^\infty \frac{x^n \, dx}{\exp(x+\alpha) - 1}. \tag{14.172}$$

The fraction of particles in states other than the ground state, $(\rho - \rho_0)/\rho$, is

$$\frac{\rho - \rho_0}{\rho} = \frac{2s+1}{4\pi^2 \rho}\left(\frac{2mkT}{\hbar^2}\right)^{3/2} f_{1/2}(\alpha). \tag{14.173}$$

Since $\alpha \geq 0$, the largest possible value of $f_{1/2}(\alpha)$ occurs when $\alpha = 0$ and numerically $f_{1/2}(0) = 2.315$. Defining the *critical temperature*

$$T_0(\rho) = \frac{\hbar^2}{2mk}\left(\frac{4\pi^2 \rho}{2.315(2s+1)}\right)^{2/3} \tag{14.174}$$

we find from (14.173) and (14.174) that the fraction N_0/N of particles in the ground state of the system is given by

$$\frac{N_0}{N} = \frac{\rho_0}{\rho} = 1 - \left(\frac{T}{T_0}\right)^{3/2}. \tag{14.175}$$

Thus, when the temperature T and the density ρ are such that $T < T_0(\rho)$, we see that $N_0/N > 0$, so that a finite fraction of the particles occupies the ground state level. As T is lowered and (or) ρ is raised, an increasing number of particles accumulate in the ground state, so that Bose–Einstein condensation occurs. Since the ground state has zero energy, the condensed system exerts no pressure, and is said to be 'degenerate'. It should be emphasised that the analysis following (14.169) only applies when $T < T_0(\rho)$. If $T \geq T_0(\rho)$, the number of particles in the ground state is negligible.

Using (14.174), the condition $T < T_0(\rho)$ for Bose–Einstein condensation to occur can be expressed in the form

$$\rho \lambda_T^3 > (2\pi)(2.315)(2s+1) \tag{14.176}$$

where we have introduced the *thermal de Broglie wavelength*

$$\lambda_T = \frac{h}{(2mkT)^{1/2}} \tag{14.177}$$

which is equal to the de Broglie wavelength of a particle of mass m and energy kT.

Experimental verification of Bose–Einstein condensation

Although the phenomenon of Bose–Einstein condensation of an ideal gas was predicted independently in 1924 by A. Einstein and S. N. Bose, attempts to observe it met with great difficulties for many years. These difficulties arise from the fact that at a given temperature a sufficiently large density must be attained. However, if the density is too high, the interactions between the bosons dominate the quantum statistical effects responsible for the condensation phenomenon.

Recent experimental efforts to demonstrate Bose–Einstein condensation have therefore concentrated on using *dilute* atomic gases, in which the interparticle spacing is large compared with the range of interatomic interactions. The gas atoms must of course be bosons[3]. In addition, they must be cooled to *very low* temperatures, so that their thermal de Broglie wavelength λ_T is large enough for several atoms to be contained in a volume λ_T^3.

The first observation of Bose–Einstein condensation was made in 1995 by M. H. Anderson, J. R. Ensher, M. R. Matthews, C. E. Wieman and E. A. Cornell, using a dilute gas of rubidium (^{87}Rb) atoms. The condensate fraction first appeared near a temperature of 1.7×10^{-7} K and a density of 2.5×10^{12} atoms per cubic centimetre. To obtain this result, Anderson *et al.* cooled and trapped the atoms by using methods which we now briefly describe.

The first one uses the interaction of the atoms of a gas with light beams to reduce their thermal motion. The process which is at the basis of the manipulation of atoms by light is the *radiation pressure* exerted by light on material particles. The radiation pressure was known originally to astronomers studying the deflected tails of comets. At the microscopic level, the action of light on the motion of atoms can be understood by remembering that when a photon is scattered by an atom, the net momentum transfer is given by the difference between the incident photon momentum **p** and the momentum **p'** of the scattered photon, with $p = h\nu/c$ and $p' = h\nu'/c$. Although each individual scattering process causes a very small velocity change (of the order of 1 cm/s), the accumulated momentum transfer per second can be large if the rate of scattering events is increased. This can be achieved by choosing the frequency ν of the incident light to be close to a resonance frequency $\nu_{ba} = (E_b - E_a)/h$ in the absorption spectrum of the atom.

The first experimental demonstration of the action of light on the motion of atoms was made in 1933 by R. Frisch, who observed the deflection of a beam of sodium atoms illuminated by a sodium lamp. However, the intensity of the light produced by the lamp was insufficient to manipulate the atoms in a useful way. The situation changed completely with the advent of lasers (see Section 16.3) which provide intense, highly directional and quasi-monochromatic sources of light. In 1972, H. Walther and his colleagues were able to act efficiently on the motion of atoms with a tunable laser, and between 1970 and 1975 a number of schemes were proposed to slow atoms with laser beams. In the proposal made in 1975 by T. Hänsch and A. Schawlow, atoms are cooled by using three orthogonal pairs of counterpropagating laser beams (see Fig. 14.5). The lasers are detuned below the resonance transition ($\nu < \nu_{ba}$). As a result, an atom is more likely to absorb a photon when it is moving towards the laser source because only then is the frequency Doppler-shifted towards resonance. When the atom absorbs a photon, it recoils in the direction of the laser beam, and hence slows. The subsequent re-emission of the photon by spontaneous emission leads to a recoil in random directions, so that the net effect is a deceleration of the atom. The

---

[3] A composite particle, such as an atom, is a boson if the sum of its protons, neutrons and electrons is an even number; the composite particle is a fermion if this sum is an odd number.

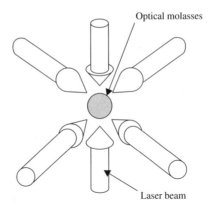

Figure 14.5 Schematic diagram of a laser arrangement producing optical molasses.

rate of these absorption–spontaneous emission cycles (fluorescence cycles) is only limited by the lifetime of the excited atomic level, and can reach 10^8 per second. Using this Doppler-cooling method, S. Chu *et al.* were able in 1985 to cool a cloud of 10^5 sodium atoms to a temperature of about 240 μK. The atoms then move as in a viscous medium, created by the laser light, called 'optical molasses'.

Even at these very low temperatures, the atoms can diffuse out of the intersection region of the molasses laser beams. In order to trap them, one can use a magneto-optical trap (MOT) which produces a restoring force due to the combination of oppositely directed circularly polarised light together with a weak magnetic field whose magnitude is zero at the centre of the trap and increases with the distance from the trap centre.

In the work of Anderson *et al.*, the rubidium atoms were first Doppler cooled and trapped. Once in the trap, the atoms were further cooled by *evaporative cooling*, which consists of selectively ejecting atoms having higher than average energy. This can be achieved by using electron spin resonance (ESR) techniques (see Section 12.4). The remaining atoms then undergo elastic collisions, which leads to thermalization at a lower temperature. In this way the temperature of the atomic cloud can be reduced to the value required for Bose–Einstein condensation to occur in the lowest energy level of the trap potential.

Following the experiment of Anderson *et al.* in 1995, Bose–Einstein condensation has been observed by a number of research groups. As an illustration, we show in Fig. 14.6 the data obtained by J. R. Ensher and colleagues in 1996 for a dilute gas of ^{87}Rb atoms. Because the gas was confined in a potential well created by the magnetic trap, the relation (14.175) is modified to read

$$\frac{N_0}{N} = 1 - \left(\frac{T}{T_0'}\right)^3 \tag{14.178}$$

where $T_0' \simeq T_0$, and $T < T_0'$. In Fig. 14.6, the onset of Bose–Einstein condensation

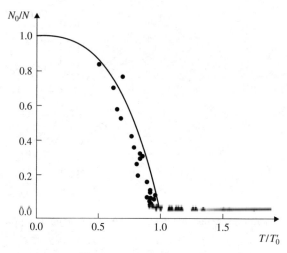

Figure 14.6 The fraction N_0/N of ^{87}Rb atoms in the ground state of the trap as a function of the scaled temperature T/T_0'. The solid curve corresponds to the theoretical prediction. The dots represent the experimental data of J. Ensher et al. At the transition, the atomic cloud contained 40 000 atoms at a temperature of 280 nK.

is clearly apparent for temperatures T less than T_0', in agreement with the theoretical result (14.178). At temperatures $T > T_0'$, the fraction of particles in the ground state is negligible, as expected from the discussion below (14.175). Improvements of the theoretical values can be obtained by taking into account the interactions between the particles.

Fermi–Dirac gases

For a fermion gas, the constant α, or the fugacity $z = \exp(-\alpha)$, is determined by specifying the density ρ and solving the implicit equation (14.165) with $\gamma = 1$,

$$\rho = \frac{(2s+1)}{4\pi^2}\left(\frac{2m}{\hbar^2}\right)^{3/2} \int_0^\infty \frac{E^{1/2}dE}{\exp(\beta E + \alpha) + 1}. \tag{14.179}$$

Since the lowest state with $E = 0$ can only be occupied by one fermion (for each spin state), for a system containing a large number of particles this state is unimportant, and in contrast to the boson case no correction need be made to (14.179). Introducing the variable $x = \beta E = E/kT$, the particle number density can be written as

$$\rho = \frac{(2s+1)}{4\pi^2}\left(\frac{2mkT}{\hbar^2}\right)^{3/2} g_{1/2}(\alpha) \tag{14.180}$$

where

$$g_n(\alpha) = \int_0^\infty \frac{x^n dx}{\exp(x+\alpha) + 1}. \tag{14.181}$$

Inspecting (14.180), we see that if $\rho/T^{3/2}$ is small α must be large and positive, in which case

$$g_{1/2}(\alpha) \simeq \exp(-\alpha) \int_0^\infty x^{1/2} \exp(-x) dx = \Gamma\left(\frac{3}{2}\right) \exp(-\alpha)$$

$$= \frac{\sqrt{\pi}}{2} \exp(-\alpha). \quad (14.182)$$

In this limit, we regain the result (14.167), and the Fermi gas behaves like a classical Maxwell–Boltzmann system. On the other hand, if $\rho/T^{3/2}$ is large, α must be negative, and large in magnitude. In this low-temperature or high-density limit, it can be shown[4] that

$$g_{1/2}(\alpha) \simeq \frac{2}{3}|\alpha|^{3/2} \quad (14.183)$$

and from (14.180)

$$\alpha \simeq -E_F/kT = -\beta E_F \quad (14.184)$$

where

$$E_F = \frac{\hbar^2}{2m}\left(\frac{2}{2s+1}\right)^{2/3} (3\pi^2\rho)^{2/3}. \quad (14.185)$$

The constant E_F is known as the Fermi energy, and for the case of spin-1/2 particles, is the same as the constant obtained in (10.39). To examine the significance of E_F we recall (14.146), which gives the average number of particles $\langle n_i \rangle$ occupying each single-particle level. At low temperatures, for fermions, using (14.184) for α, we have

$$\langle n_i \rangle = \{\exp(\beta(E_i - E_F)) + 1\}^{-1}. \quad (14.186)$$

At $T = 0$, $(\beta \to \infty)$, $\langle n_i \rangle = 1$ for all $E_i \leqslant E_F$ while $\langle n_i \rangle = 0$ for all $E_i > E_F$. This is a consequence of the Pauli exclusion principle. Since only one particle can occupy each state, in the ground state all the lowest states are occupied up to the level with $E = E_F$. For low temperatures, with $T < T_F$, where $T_F = E_F/k$ is called the *Fermi temperature*, most of the lowest energy levels remain filled. The corresponding distribution function, treating E_i as a continuous variable, has been illustrated in Fig. 10.4. In the temperature interval $0 \leqslant T \leqslant T_F$ the Fermi gas is said to be *degenerate*.

Even at $T = 0$, a Fermi gas exhibits pressure because with the exception of the $(2s+1)$ particles in the lowest energy level with $E = 0$, all the particles are in motion, in complete contrast to the boson gas, which exhibits no pressure at the absolute zero of temperature.

---

[4] See, for example, Feynman (1972) pp. 36–7.

Problems

14.1 Consider two electron beams of the same intensity, the first one being totally polarised in the z-direction, and the second one totally polarised in the x-direction.

(a) Show that the density matrix is given by

$$\rho = \begin{pmatrix} 3/4 & 1/4 \\ 1/4 & 1/4 \end{pmatrix}.$$

(*Hint*: Use the result (6.252) to obtain the normalised eigenfunction of S_x corresponding to the eigenvalue $\hbar/2$.)

(b) Prove that the degree of polarisation of the system is $|P| = 2^{-1/2}$.

14.2 Consider the elastic scattering of spin-1/2 particles by a spin-0 target. Choosing the z-axis of spin quantisation along the incident direction $\mathbf{k}$:

(a) Show that for incident spin-1/2 particles polarised parallel to $\mathbf{k}$ (longitudinal polarisation) the differential cross-section for scattering without spin flip is

$$\frac{d\sigma}{d\Omega}(\uparrow \to \uparrow) = \frac{d\sigma}{d\Omega}(\downarrow \to \downarrow) = |f(\theta)|^2$$

and that for spin-flip scattering it is

$$\frac{d\sigma}{d\Omega}(\uparrow \to \downarrow) = \frac{d\sigma}{d\Omega}(\downarrow \to \uparrow) = |g(\theta)|^2;$$

(b) Starting from the expression (14.73) for scattering into any final spin state, and denoting the Cartesian components of the unit vector $\hat{\mathbf{n}}$ (given by (14.60)) by $(-\sin\phi, \cos\phi, 0)$ show that if the incident beam of spin-1/2 particles is fully polarised in the x-direction (transverse polarisation with $\mathbf{P} = \hat{\mathbf{x}}$) then the differential cross-section (14.73) reduces to

$$\frac{d\bar{\sigma}}{d\Omega} = (|f(\theta)|^2 + |g(\theta)|^2)[1 - S(\theta)\sin\phi].$$

Compare this result with that corresponding to an incident beam of spin-1/2 particles polarised along $\hat{\mathbf{k}}$ (longitudinal polarisation).

14.3 Verify that the fugacity z of gases at room temperature and pressure satisfies $z \ll 1$.

14.4 Write down explicitly the partition function Q_N given by (14.133) for a two-particle system with three energy levels E_1, E_2 and E_3 (a) for Maxwell–Boltzmann, (b) for Bose–Einstein and (c) for Fermi–Dirac statistics.

14.5 Below 2.19 K liquid helium displays the property of *superfluidity*. Since helium atoms are bosons of spin zero, this phenomenon is believed to be connected with Bose–Einstein condensation exhibited for temperatures $T < T_0$, where T_0 is defined by equation (14.174). By evaluating T_0 explicitly, show that the result obtained is close to the critical value 2.19 K.

14.6 Evaluate the specific heat at constant volume of a Fermi gas in the low-and high-temperature limits using the expression

$$C_V = \left.\frac{\partial U}{\partial T}\right|_V.$$

Using this result, estimate the contribution made by the valence electrons in aluminium to the specific heat.

(*Hint*: Treat the electrons as a Fermi gas. Aluminium is trivalent, its density is 2.7×10^3 kg m^{-3} and its mass is 27 in atomic mass units.)

15 Relativistic quantum mechanics

15.1 The Klein–Gordon equation 679
15.2 The Dirac equation 684
15.3 Covariant formulation of the Dirac theory 690
15.4 Plane wave solutions of the Dirac equation 696
15.5 Solutions of the Dirac equation for a central potential 702
15.6 Non-relativistic limit of the Dirac equation 711
15.7 Negative-energy states. Hole theory 715
Problems 717

Until now we have based our study of quantum mechanics on the Schrödinger equation, which can be obtained from the Hamiltonian formulation of non-relativistic classical mechanics by using the correspondence rule (3.25). This equation, which is invariant under Galilean transformations (see Section 5.10), correctly describes the phenomena only if the velocities of the particles involved are small with respect to the velocity of light c. However, it is not invariant under a Lorentz change of reference frame, as required by the principle of relativity. In this chapter, we shall extend the theory to study relativistic wave equations. Two cases will be considered: the relativistic Schrödinger or Klein–Gordon equation for spinless particles, and the Dirac equation for particles of spin 1/2.

15.1 The Klein–Gordon equation

In 1926, when he was developing his non-relativistic equation, E. Schrödinger also proposed a relativistic generalisation of it, which is known as the *Schrödinger relativistic equation* or *Klein–Gordon equation*. This equation is relevant for the relativistic dynamics of spinless particles, such as π mesons. Because such particles have no internal degrees of freedom, their wave function Ψ in configuration space can only depend on the variables $\mathbf{r}$ and t.

In order to find the wave equation to be satisfied by the wave function $\Psi(\mathbf{r}, t)$, we shall proceed empirically, as we did in Chapter 3 to obtain the Schrödinger equation in the non-relativistic case. As in Chapter 3, we shall use the correspondence principle to make sure that we recover the classical laws of motion in the classical limit.

Free particle

We begin by considering a *free* particle of spin 0. The relativistic relationship between the energy E and momentum $\mathbf{p}$ of a free particle of rest mass m is

$$E = (m^2c^4 + \mathbf{p}^2c^2)^{1/2}. \tag{15.1}$$

Using the correspondence rule (see (3.25))

$$E \to E_{\mathrm{op}} = i\hbar \frac{\partial}{\partial t}, \quad \mathbf{p} \to \mathbf{p}_{\mathrm{op}} = -i\hbar \nabla \tag{15.2}$$

and operating on both sides of the equation (15.1) on a wave function $\Psi(\mathbf{r}, t)$, we obtain the relativistic analogue of (3.11):

$$i\hbar \frac{\partial}{\partial t} \Psi = (m^2c^4 - \hbar^2 c^2 \nabla^2)^{1/2} \Psi. \tag{15.3}$$

This equation has serious drawbacks. Firstly, we must interpret the square-root operator on the right-hand side. If we expand it in a power series, we obtain a differential operator of infinite order, which is very difficult to handle. Secondly, the resulting wave equation presents the unattractive feature that the space and time coordinates appear in an unsymmetrical way, so that relativistic invariance is not clearly exhibited.

To avoid these difficulties, we remove the square root by starting from the relationship

$$E^2 = m^2c^4 + \mathbf{p}^2c^2 \tag{15.4}$$

obtained by squaring the energy relation (15.1). Making the substitutions (15.2) and acting on both sides of (15.4) on a wave function Ψ, we obtain the Schrödinger relativistic equation or Klein–Gordon equation for a free particle:

$$-\hbar^2 \frac{\partial^2}{\partial t^2} \Psi = m^2 c^4 \Psi - \hbar^2 c^2 \nabla^2 \Psi. \tag{15.5}$$

It is worth noting that this is a *second-order* differential equation with respect to the time, in contrast to the (non-relativistic) Schrödinger equation, which is of *first order* in the time derivative $\partial/\partial t$.

Charged particle in an electromagnetic field

If the spinless particle has an electric charge q, and is moving in an electromagnetic field described by a vector potential $\mathbf{A}(\mathbf{r}, t)$ and a scalar potential $\phi(\mathbf{r}, t)$, we can, in analogy with the non-relativistic case (see (5.230)), make the replacements

$$E \to E - q\phi, \quad \mathbf{p} \to \mathbf{p} - q\mathbf{A}, \tag{15.6}$$

so that (15.4) is replaced by

$$(E - q\phi)^2 = m^2c^4 + c^2(\mathbf{p} - q\mathbf{A})^2. \tag{15.7}$$

15.1 The Klein–Gordon equation

Making the substitutions (15.2) and operating on both sides of (15.7) on a wave function $\Psi(\mathbf{r}, t)$, we obtain the Klein–Gordon equation for a spinless particle of charge q in an electromagnetic field:

$$\left(i\hbar \frac{\partial}{\partial t} - q\phi\right)^2 \Psi = m^2 c^4 \Psi + c^2(-i\hbar \nabla - q\mathbf{A})^2 \Psi. \tag{15.8}$$

It is interesting to investigate the non-relativistic limit of this equation. In order to do this, we introduce the new wave function $\tilde{\Psi}(\mathbf{r}, t)$, which is related to $\Psi(\mathbf{r}, t)$ by

$$\tilde{\Psi}(\mathbf{r}, t) = \Psi(\mathbf{r}, t) \exp(imc^2 t/\hbar) \tag{15.9}$$

and satisfies the equation

$$-\hbar^2 \frac{\partial^2 \tilde{\Psi}}{\partial t^2} + 2i\hbar(mc^2 - q\phi)\frac{\partial \tilde{\Psi}}{\partial t} - \left[q\phi(2mc^2 - q\phi) + i\hbar q \frac{\partial \phi}{\partial t}\right]\tilde{\Psi}$$

$$= c^2[-\hbar^2 \nabla^2 + 2i\hbar q \mathbf{A}\cdot\nabla + i\hbar q (\nabla\cdot\mathbf{A}) + q^2 \mathbf{A}^2]\tilde{\Psi}. \tag{15.10}$$

In the limit in which $|q\phi| \ll mc^2$, $|(\hbar/2mc^2)\partial\phi/\partial t| \ll |\phi|$ and $|(\hbar^2/2mc^2)\partial^2\tilde{\Psi}/\partial t^2| \ll |\hbar \partial \tilde{\Psi}/\partial t|$, this equation reduces to the non-relativistic Schrödinger equation for a spinless particle of mass m and charge q in an electromagnetic field, namely

$$i\hbar \frac{\partial \tilde{\Psi}}{\partial t} = \left[\frac{1}{2m}(-i\hbar \nabla - q\mathbf{A})^2 + q\phi\right]\tilde{\Psi}. \tag{15.11}$$

This is the equation (11.17) we used in Chapter 11.

Stationary state solutions

Let us return to the Klein–Gordon equation (15.8) and suppose that $\mathbf{A}$ and ϕ are independent of the time. We may then look for stationary state solutions of this equation, which have the form

$$\Psi(\mathbf{r}, t) = \psi(\mathbf{r}) \exp(-iEt/\hbar). \tag{15.12}$$

Substituting (15.12) into (15.8), we obtain

$$c^2(-i\hbar \nabla - q\mathbf{A})^2 \psi = \left[(E - q\phi)^2 - m^2 c^4\right]\psi. \tag{15.13}$$

In particular, if $\mathbf{A} = 0$ and ϕ is spherically symmetric, we have

$$-c^2\hbar^2 \nabla^2 \psi = [(E - q\phi(r))^2 - m^2 c^4]\psi. \tag{15.14}$$

Following the method of Section 7.2, we can separate this equation in spherical polar coordinates. As in (7.45), we write

$$\psi_{Elm}(\mathbf{r}) = R_{El}(r) Y_{lm}(\theta, \phi) \tag{15.15}$$

and obtain for the radial functions $R_{El}(r)$ the equations

$$-c^2\hbar^2\left[\left(\frac{d^2}{dr^2}+\frac{2}{r}\frac{d}{dr}\right)-\frac{l(l+1)}{r^2}\right]R_{El}(r)=[(E-q\phi)^2-m^2c^4]R_{El}(r),$$
$$l=0,1,2,\ldots \quad (15.16)$$

As in the case of the Klein–Gordon equation (15.8), it is instructive to examine the non-relativistic limit of the equations we have obtained. To do this, we write

$$E=E'+mc^2 \quad (15.17)$$

and assume that both $|E'|$ and $|q\phi|$ are small in comparison with mc^2. We may then write

$$(E-q\phi)^2-m^2c^4 = (E'+mc^2-q\phi)^2-m^2c^4$$
$$= m^2c^4\left(1+\frac{E'-q\phi}{mc^2}\right)^2-m^2c^4$$
$$\simeq m^2c^4\left[1+\frac{2}{mc^2}(E'-q\phi)+\cdots\right]-m^2c^4$$
$$\simeq 2mc^2(E'-q\phi). \quad (15.18)$$

Hence, we see that in the non-relativistic limit the equation (15.13) reduces to

$$\left[\frac{1}{2m}(-i\hbar\nabla-q\mathbf{A})^2+q\phi\right]\psi=E'\psi. \quad (15.19)$$

This is the expected result, since it can equally be obtained by starting from the non-relativistic Schrödinger equation (15.11), and assuming that $\mathbf{A}$ and ϕ are time-independent. We can then look for stationary state solutions of (15.11) of the form

$$\tilde{\Psi}(\mathbf{r},t)=\psi(\mathbf{r})\exp(-iE't/\hbar) \quad (15.20)$$

and the resulting equation for $\psi(\mathbf{r})$ is precisely (15.19).

Turning now to the particular case for which $\mathbf{A}=0$ and ϕ is spherically symmetric, we find by using (15.18) that the non-relativistic limit of the radial equations (15.16) is

$$\left[-\frac{\hbar^2}{2m}\left(\frac{d^2}{dr^2}+\frac{2}{r}\frac{d}{dr}\right)+\frac{l(l+1)\hbar^2}{2mr^2}+q\phi(r)\right]R_{El}(r)=E'R_{El}(r) \quad (15.21)$$

in agreement with the results obtained in Section 7.2.

Interpretation of the Klein–Gordon equation. Continuity equation

We now turn to the interpretation of the Klein–Gordon equation. This is non-trivial since, as pointed out above, the Klein–Gordon equation is of second order in the time derivative $\partial/\partial t$, which is different from the Schrödinger equation (3.21) upon which the probabilistic interpretation of non-relativistic quantum theory was based.

In order to simplify the discussion, we shall consider only the free-particle Klein–Gordon equation (15.5). To interpret the wave function as we did in Section 3.2 for the non-relativistic case, let us try to construct a position probability density $P(\mathbf{r},t)$ and a probability current density $\mathbf{j}(\mathbf{r},t)$ satisfying the *continuity equation* (see (3.39))

$$\frac{\partial P}{\partial t} + \nabla\cdot\mathbf{j} = 0. \tag{15.22}$$

Proceeding by analogy with the Schrödinger equation, we multiply (15.5) on the left by Ψ^*, multiply the complex conjugate equation on the left by Ψ, and take the difference of the two resulting equations. We then obtain the continuity equation (15.22) if we define the real quantities (Problem 15.1)

$$P(\mathbf{r},t) = \frac{i\hbar}{2mc^2}\left(\Psi^*\frac{\partial \Psi}{\partial t} - \Psi\frac{\partial \Psi^*}{\partial t}\right) \tag{15.23}$$

and

$$\mathbf{j}(\mathbf{r},t) = \frac{\hbar}{2mi}[\Psi^*(\nabla\Psi) - (\nabla\Psi^*)\Psi]. \tag{15.24}$$

The constants appearing in the above two equations have been determined in such a way that the expressions obtained for $P(\mathbf{r},t)$ and $\mathbf{j}(\mathbf{r},t)$ reduce to the usual ones for the Schrödinger theory in the non-relativistic limit. Indeed, we see that the expression (15.24) for $\mathbf{j}(\mathbf{r},t)$ is identical with the non-relativistic form (3.34), while the expression (15.23) for $P(\mathbf{r},t)$ is readily shown (Problem 15.2) to reduce in the non-relativistic limit to $|\Psi(\mathbf{r},t)|^2$, as given by (3.27).

Looking at (15.23), we note that $P(\mathbf{r},t)$ is not positive-definite, so that it cannot be interpreted as a position probability density. This is one of the difficulties encountered in dealing with the Klein–Gordon equation.

Another difficulty arises as follows. The free-particle Klein–Gordon equation (15.5) has plane wave solutions of the form

$$\Psi(\mathbf{r},t) = A\,\exp[i(\mathbf{k}\cdot\mathbf{r} - \omega t)], \tag{15.25}$$

where A is a constant. These are also eigenfunctions of the operators $E_{\text{op}} = i\hbar\partial/\partial t$ and $\mathbf{p}_{\text{op}} = -i\hbar\nabla$ with eigenvalues $E = \hbar\omega$ and $\mathbf{p} = \hbar\mathbf{k}$, respectively. Substituting (15.25) into (15.5), we find that

$$\hbar\omega = \pm(m^2c^4 + \hbar^2c^2\mathbf{k}^2)^{1/2}. \tag{15.26}$$

The positive and negative square roots in (15.26) correspond to an ambiguity in the sign of the energy which is also present in the classical relation (15.4). Indeed, this relation is not equivalent to (15.1), but to the more general relation

$$E = \pm(m^2c^4 + \mathbf{p}^2c^2)^{1/2} \tag{15.27}$$

which contains an additional negative-energy root. Thus, in order to obtain the 'simple' Klein–Gordon equation (15.5), we have introduced extraneous negative-energy solutions. As a result, the energy spectrum is no longer bounded from below, and it appears that arbitrary large amounts of energy could be extracted from the

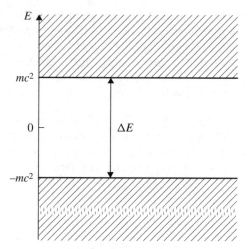

Figure 15.1 The energy gap $\Delta E = 2mc^2$ between the positive and negative continua of a free particle of mass m described by the Klein–Gordon equation (15.5).

system if an external perturbation allows it to make a transition between the positive and negative energy states. For a free particle initially at rest, this would happen if the external perturbation allowed it to jump over the energy gap $\Delta E = 2mc^2$ between the positive and negative energy continua (see Fig. 15.1).

These difficulties seemed at the time so severe that the Klein–Gordon equation fell for a while into disrepute, until W. Pauli and V. Weisskopf reinterpreted the equation in 1934 as a field equation (like Maxwell's equations for the electromagnetic field) and quantised it by using the formalism of quantum field theory. The Klein–Gordon equation then becomes a relativistic wave equation for spinless particles, within the framework of a *many-particle* theory in which the negative energy states are interpreted in terms of *antiparticles*. It should be noted that the number of particles is not conserved because of the possibility of creation and destruction of particle–antiparticle pairs.

In the Dirac theory, to which we now turn our attention, we shall find it possible to define a positive-definite position probability density $P(\mathbf{r}, t)$, but the presence of negative-energy solutions will again imply that the Dirac theory must ultimately be interpreted as a many-particle theory.

15.2 The Dirac equation

In 1928, P. A. M. Dirac discovered the relativistic wave equation which bears his name while seeking a relativistic wave equation with positive definite probability density. For several years after its discovery, the Dirac equation was thought to be the only valid relativistic wave equation for particles with mass. This idea was dismissed only after Pauli and Weisskopf reinterpreted the Klein–Gordon equation as a field equation for spinless particles. Even now, however, the Dirac equation is particularly important because it describes particles of spin-1/2, such as electrons.

In the non-relativistic quantum theory, particles of spin 1/2 are described by two-component spinor wave functions which allow for the two spin states, the z-component S_z of the spin angular momentum taking on the values $m_s \hbar$, where $m_s = \pm 1/2$. In the same way, one expects that spin-1/2 particles in the relativistic theory must be represented by wave functions having at least two components. However, since all spin-1/2 particles are associated with antiparticles having the same mass and spin, but of opposite charge, we expect to need a four-component wave function. This was unknown when Dirac obtained his equation, and it was one of the great achievements of theoretical physics that Dirac was able to predict the existence of the positron (the antiparticle of the electron) from his theory.

By analogy with non-relativistic quantum mechanics, Dirac started by looking for a wave equation of the form

$$i\hbar \frac{\partial}{\partial t} \Psi = H \Psi \qquad (15.28)$$

which like the Schrödinger equation (3.21) is *linear* in $\partial/\partial t$, and not quadratic like the Klein–Gordon equation. Since in a relativistic theory the spatial coordinates ($x_1 = x, x_2 = y, x_3 = z$) of a space–time point (called an 'event') must enter on the same footing as the time coordinate $x_4 = ict$, the equation (15.28), and hence the Hamiltonian H, is expected to be linear in the space derivatives $\partial/\partial x_k$ ($k = 1, 2, 3$). The wave function $\Psi(\mathbf{r}, t)$ in (15.28) is assumed to have N components $\Psi_i(\mathbf{r}, t)$, where $i = 1, 2, \ldots N$, and hence may be written in the form of a column vector as

$$\Psi = \begin{pmatrix} \Psi_1 \\ \Psi_2 \\ \vdots \\ \Psi_N \end{pmatrix}. \qquad (15.29)$$

Free particle

Let us first consider the case of a free particle. The Hamiltonian must then be independent of $\mathbf{r}$ and t (since there are no forces) and the simplest candidate, linear in the momentum and mass terms, may be written in the form

$$\begin{aligned} H &= c\boldsymbol{\alpha} \cdot \mathbf{p}_{\text{op}} + \beta m c^2 \\ &= c \sum_{k=1}^{3} \alpha_k (p_{\text{op}})_k + \beta m c^2 \end{aligned} \qquad (15.30)$$

where $\mathbf{p}_{\text{op}} = -i\hbar \nabla$ and $(p_{\text{op}})_k$ with $k = 1, 2, 3$ denote the Cartesian components of $\mathbf{p}_{\text{op}}$. The three components ($\alpha_1, \alpha_2, \alpha_3$) of $\boldsymbol{\alpha}$ as well as the quantity β are independent of $\mathbf{r}, t, \mathbf{p}$ and E, but need not commute with each other.

Substituting (15.30) into (15.28) and remembering that $E_{op} = i\hbar\partial/\partial t$, we obtain the Dirac wave equation for a free particle:

$$(E_{op} - c\boldsymbol{\alpha}\cdot\mathbf{p}_{op} - \beta mc^2)\Psi = 0 \tag{15.31a}$$

or

$$i\hbar\frac{\partial}{\partial t}\Psi = -i\hbar c\boldsymbol{\alpha}\cdot\boldsymbol{\nabla}\Psi + \beta mc^2\Psi. \tag{15.31b}$$

Since Ψ is a column vector with N components, this equation is a matrix equation, which can be written more explicitly as the set of N coupled equations

$$i\hbar\frac{\partial}{\partial t}\Psi_i = -i\hbar c \sum_{j=1}^{N}\left(\alpha_1\frac{\partial}{\partial r_1} + \alpha_2\frac{\partial}{\partial r_2} + \alpha_3\frac{\partial}{\partial r_3}\right)_{ij}\Psi_j + \sum_{j=1}^{N}\beta_{ij}mc^2\Psi_j$$

$$= \sum_{j=1}^{N} H_{ij}\Psi_j, \quad i = 1, 2, \ldots N. \tag{15.32}$$

In order to determine the quantities $\alpha_k (k = 1, 2, 3)$ and β which appear in the above equations, we first require that any solution Ψ of the free-particle Dirac equation (15.31) must also be a solution of the free-particle Klein–Gordon equation (15.5), which we rewrite for this purpose as

$$[E_{op}^2 - \mathbf{p}_{op}^2 c^2 - m^2 c^4]\Psi = 0. \tag{15.33}$$

This requirement, based on the correspondence principle, guarantees that in the classical limit the wave packet solutions of the Dirac equation (15.31) will exhibit the correct relativistic energy–momentum relation (15.4) for a free particle. Multiplying (15.31a) on the left by the operator $(E_{op} + c\boldsymbol{\alpha}\cdot\mathbf{p}_{op} + \beta mc^2)$, we obtain the second-order equation

$$\left\{E_{op}^2 - c^2\left(\sum_{k=1}^{3}(\alpha_k)^2(p_{op})_k^2 + \sum_{k<l=1}^{N}(\alpha_k\alpha_l + \alpha_l\alpha_k)(p_{op})_k(p_{op})_l\right)\right.$$

$$\left. - mc^3\left(\sum_{k=1}^{3}(\alpha_k\beta + \beta\alpha_k)(p_{op})_k\right) - m^2c^4\beta^2\right\}\Psi = 0. \tag{15.34}$$

Comparing (15.34) with (15.33), we see that each component Ψ_i of the Dirac wave function satisfies the Klein–Gordon equation (15.33) provided that

$$(\alpha_1)^2 = (\alpha_2)^2 = (\alpha_3)^2 = \beta^2 = 1,$$
$$[\alpha_1, \alpha_2]_+ = [\alpha_2, \alpha_3]_+ = [\alpha_3, \alpha_1]_+ = 0, \tag{15.35}$$
$$[\alpha_1, \beta]_+ = [\alpha_2, \beta]_+ = [\alpha_3, \beta]_+ = 0$$

where $[A, B]_+$ denotes the anticommutator of A and B (see (6.223)). Thus, the four quantities $\alpha_1, \alpha_2, \alpha_3$ and β anticommute in pairs, and their squares are equal to unity.

Since the quantities α_k ($k = 1, 2, 3$) and β do not commute, they cannot be numbers. We are therefore led to consider $N \times N$ matrices. In order that the Hamiltonian H be an Hermitian operator, $H = H^\dagger$, the matrices α_k and β must also be Hermitian:

$$\alpha = \alpha^\dagger, \quad \beta = \beta^\dagger \tag{15.36}$$

Dirac representation for the matrices α and β.

We shall now make use of the relations (15.35) and (15.36) to construct an explicit representation of the matrices α and β. In the interest of simplicity, we shall require this representation to be of lowest rank as possible.

We first observe that since $(\alpha_k)^2 = \beta^2 = 1$, the eigenvalues of α_k and β are ± 1. It also follows from their anticommutation properties that the trace of each α_k and of β is equal to zero. Indeed, since $[\alpha_k, \beta]_+ = 0$ and $\beta^2 = 1$, we have

$$\alpha_k = -\beta \alpha_k \beta \tag{15.37}$$

and therefore

$$\text{Tr } \alpha_k = \text{Tr}(-\beta \alpha_k \beta). \tag{15.38}$$

On the other hand, using (15.37) and the fact that $\alpha_k \beta = -\beta \alpha_k$, we have

$$\alpha_k = \beta^2 \alpha_k \tag{15.39}$$

from which

$$\text{Tr } \alpha_k = \text{Tr}(\beta^2 \alpha_k) = \text{Tr}(\beta \alpha_k \beta) \tag{15.40}$$

where we have used the fact that the trace of a product of matrices does not depend on the order of its factors. Upon comparison of (15.38) and (15.40), we see that $\text{Tr } \alpha_k = 0$. Similarly, since

$$\beta = -\alpha_k \beta \alpha_k \tag{15.41}$$

we have $\text{Tr } \beta = \text{Tr}(-\alpha_k \beta \alpha_k)$. But we also have

$$\text{Tr } \beta = \text{Tr } [(\alpha_k)^2 \beta] = \text{Tr } (\alpha_k \beta \alpha_k) \tag{15.42}$$

so that $\text{Tr } \beta = 0$. Since the trace is the sum of the eigenvalues, and the eigenvalues of the matrices α_k and β are ± 1, we conclude that there are as many $+1$ as -1 eigenvalues. As a result, the matrices α_k and β must be of *even* rank.

Let us first consider the simplest possibility, namely $N = 2$. Comparing (15.35) with (6.238) and (6.239), we see that the matrices α_k and β have similar algebraic properties as the 2×2 Pauli spin matrices σ_x, σ_y and σ_z given by (6.243). However, so long as we deal with 2×2 matrices, we cannot find a representation of more than three anticommuting quantities. Indeed, because any 2×2 matrix can be expressed as a linear combination of the four linearly independent matrices $\sigma_x, \sigma_y, \sigma_y$ and I (where I is the unit 2×2 matrix), one can prove that it is not possible to find a fourth matrix that anticommutes with all three of the Pauli spin matrices σ_x, σ_y and σ_z.

We therefore have to consider the case $N = 4$. Following Dirac, we shall choose a representation in which the matrix β is diagonal. If I denotes the unit 2×2 matrix and σ_k ($k = 1, 2, 3$) are the three Pauli spin matrices ($\sigma_x, \sigma_y, \sigma_y$), then the 4×4 matrices

$$\alpha_k = \begin{pmatrix} 0 & \sigma_k \\ \sigma_k & 0 \end{pmatrix}, \qquad \beta = \begin{pmatrix} I & 0 \\ 0 & -I \end{pmatrix} \tag{15.43}$$

satisfy all the required conditions: they are Hermitian and can readily be seen (Problem 15.4) to satisfy the relations (15.35) by using the properties of the Pauli spin matrices. The particular representation (15.43), called the *Dirac representation*, is especially convenient to discuss the non-relativistic limit of the Dirac equation, as we shall see in Section 15.6. It is worth noting that the solution of equations (15.35) and (15.36) is not unique, but it can be shown that any set of matrices satisfying these conditions provides the same physical results.

As a result of the above discussion, we see that the wave function Ψ of equation (15.29) must contain four components as anticipated at the beginning of this section. It is called a *four-component spinor* and is appropriate for particles of spin 1/2. Representations with higher rank matrices correspond to particles with spin greater than 1/2.

Charged particle in an electromagnetic field

To obtain the Dirac equation for a spin-1/2 particle of charge q moving in an electromagnetic field described by a vector potential $\mathbf{A}(\mathbf{r}, t)$ and a scalar potential $\phi(\mathbf{r}, t)$, we make the usual replacements (15.6) in (15.31) and obtain

$$[(E_{\text{op}} - q\phi) - c\boldsymbol{\alpha}\cdot(\mathbf{p}_{\text{op}} - q\mathbf{A}) - \beta mc^2]\Psi = 0 \tag{15.44}$$

or

$$i\hbar \frac{\partial}{\partial t}\Psi = [-i\hbar c\boldsymbol{\alpha}\cdot\boldsymbol{\nabla} - cq\boldsymbol{\alpha}\cdot\mathbf{A} + q\phi + \beta mc^2]\Psi. \tag{15.45}$$

Upon comparison with (15.28), we see that the Dirac Hamiltonian for a spin-1/2 particle of charge q in the presence of an electromagnetic field is given by

$$H = c\boldsymbol{\alpha}\cdot(\mathbf{p}_{\text{op}} - q\mathbf{A}) + q\phi + \beta mc^2. \tag{15.46}$$

Adjoint equation; continuity equation; probability and current densities

We have seen above that the wave function Ψ may be considered as a column matrix of the form (15.29), with four components Ψ_i ($i = 1, 2, 3, 4$). We can define $\Psi^\dagger$ to be a row vector with components Ψ_i^*, namely

$$\Psi^\dagger = (\Psi_1^* \quad \Psi_2^* \quad \Psi_3^* \quad \Psi_4^*). \tag{15.47}$$

Using (15.45), (15.46) and the fact that $\boldsymbol{\alpha}$, β and $\mathbf{p}_{op}$ are Hermitian, we see that $\Psi^\dagger$ satisfies the equation

$$-i\hbar \frac{\partial}{\partial t}\Psi^\dagger = \Psi^\dagger H$$
$$= (i\hbar c \nabla - cq\mathbf{A})\Psi^\dagger . \boldsymbol{\alpha} + q\phi \Psi^\dagger + mc^2 \Psi^\dagger \beta. \tag{15.48}$$

The quantity

$$P(\mathbf{r}, t) = \Psi^\dagger \Psi = \sum_{i=1}^{4} |\Psi_i|^2 \tag{15.49}$$

is clearly positive and can be interpreted as a *position probability density*, in the same way that $|\Psi|^2$ is the position probability density for the non-relativistic Schrödinger equation (3.21). By multiplying (15.45) on the left by $\Psi^\dagger$ and (15.48) on the right by Ψ, and taking the difference of the two results, it is found that

$$\frac{\partial P}{\partial t} + \nabla . (\Psi^\dagger c \boldsymbol{\alpha} \Psi) = 0. \tag{15.50}$$

If we interpret the vector

$$\mathbf{j}(\mathbf{r}, t) = \Psi^\dagger c \boldsymbol{\alpha} \Psi \tag{15.51}$$

as a *probability current density*, the equation (15.50) takes the form of a continuity equation

$$\frac{\partial P}{\partial t} + \nabla . \mathbf{j} = 0, \tag{15.52}$$

and we see that $c\boldsymbol{\alpha}$ can be interpreted as a velocity operator.

Stationary state solutions

Let us assume that $\mathbf{A}$ and ϕ are time-independent. We may then look for stationary state solutions of the Dirac equation having the form

$$\Psi(\mathbf{r}, t) = \psi(\mathbf{r}) \exp(-iEt/\hbar) \tag{15.53}$$

where $\psi(\mathbf{r})$ is a time-independent four-component spinor:

$$\psi = \begin{pmatrix} \psi_1 \\ \psi_2 \\ \psi_3 \\ \psi_4 \end{pmatrix}. \tag{15.54}$$

From (15.45) and (15.53) we obtain for $\psi(\mathbf{r})$ the equation

$$[-i\hbar c \boldsymbol{\alpha} . \nabla - cq\boldsymbol{\alpha} . \mathbf{A} + q\phi + \beta mc^2]\psi = E\psi. \tag{15.55}$$

Let us write the four-component spinor ψ in terms of two-component spinors,

$$\psi = \begin{pmatrix} \psi_A \\ \psi_B \end{pmatrix}, \tag{15.56}$$

where

$$\psi_A = \begin{pmatrix} \psi_1 \\ \psi_2 \end{pmatrix}, \quad \psi_B = \begin{pmatrix} \psi_3 \\ \psi_4 \end{pmatrix}. \tag{15.57}$$

Using the representation (15.43) of the matrices α and β, the two-component spinors ψ_A and ψ_B are found to satisfy the two coupled equations

$$(q\phi + mc^2)\psi_A + c\boldsymbol{\sigma}\cdot(-i\hbar\boldsymbol{\nabla} - q\mathbf{A})\psi_B = E\psi_A, \tag{15.58a}$$

$$c\boldsymbol{\sigma}\cdot(-i\hbar\boldsymbol{\nabla} - q\mathbf{A})\psi_A + (q\phi - mc^2)\psi_B = E\psi_B. \tag{15.58b}$$

These equations will be used in Section 15.6 to study the non-relativistic limit of the Dirac equation.

15.3 Covariant formulation of the Dirac theory

In order to satisfy the principle of relativity, the Dirac equation and the continuity equation upon which its physical interpretation is based must keep their form when Lorentz transformations are performed.

Let us first recall that homogeneous Lorentz transformations are real, linear, homogeneous transformations of the space–time coordinates leaving invariant the norm of the intervals between the different points of space–time. This is in accordance with the physical observation that the velocity of light *in vacuo*, c, is the same in all inertial (Lorentz) reference frames. Thus, if two observers, who are in different inertial reference frames S and S', describe the same physical event with the coordinates x_μ and x'_μ, respectively, the homogeneous Lorentz transformation between the two sets of coordinates is

$$x'_\mu = \sum_{\mu=1}^{4} L_{\mu\nu} x_\nu \tag{15.59a}$$

$$= L_{\mu\nu} x_\nu \tag{15.59b}$$

where in the second line we have followed the convention of summing over repeated indices. The coefficients $L_{\mu\nu}$ depend only on the relative velocities and spatial orientations of the two reference frames S and S'. Since the spatial components x_k and $x'_k (k = 1, 2, 3)$ are real and the time components $x_4 = ict$ and $x'_4 = ict'$ are imaginary, it follows that

$$L_{ik} \text{ is real}, \quad i, k = 1, 2, 3 \tag{15.60a}$$
$$L_{i4} \text{ and } L_{4i} \text{ are imaginary}, \quad i = 1, 2, 3 \tag{15.60b}$$

and

$$L_{44} \text{ is real}. \tag{15.60c}$$

A four-vector $a_\mu (\mu = 1, 2, 3, 4)$ is a set of four quantities transforming like the coordinates x_μ, with $a_k (k = 1, 2, 3)$ being real and a_4 being imaginary. It is often convenient to denote a four-vector a_μ in the form $a_\mu \equiv (\mathbf{a}, a_4)$. For example, in this notation $x_\mu \equiv (\mathbf{x}, ict)$. The squared norm of a four-vector is defined to be

$$a_\mu a_\mu = \mathbf{a}^2 + a_4^2 = \mathbf{a}^2 - |a_4|^2 \tag{15.61}$$

where $\mathbf{a}^2 = \mathbf{a}.\mathbf{a} = (a_1)^2 + (a_2)^2 + (a_3)^2$ is the usual squared norm of the three-vector $\mathbf{a}$. In what follows we shall continue to denote by a the norm (length) of a three-vector, where $a = |\mathbf{a}|$.

Four-vectors a_μ can be classified into three categories depending on their squared norm. Those with $a_\mu a_\mu < 0$ are called *time-like*, those with $a_\mu a_\mu > 0$ are said to be *space-like* and those with $a_\mu a_\mu = 0$ are *null vectors*. This classification corresponds to the position of the four-vector with respect to the light cone

$$x_\mu x_\mu = x^2 + y^2 + z^2 - c^2 t^2 = 0. \tag{15.62}$$

In accordance with the physical observation that the velocity of light *in vacuo*, c, is the same in all Lorentz reference frames, it is sufficient to require that the norm of the four-vector x_μ be invariant under the transformation (15.59). As a result, the coefficients $L_{\mu\nu}$ must satisfy the conditions

$$L_{\mu\nu} L_{\mu\rho} = \delta_{\nu\rho}, \tag{15.63a}$$
$$L_{\nu\mu} L_{\rho\mu} = \delta_{\nu\rho} \tag{15.63b}$$

where $\delta_{\nu\rho}$ is the usual Kronecker delta symbol. It follows that the norm of any four-vector a_μ is invariant under a homogeneous Lorentz transformation. In addition, the scalar product of two four-vectors a_μ and b_μ, defined as

$$a_\mu b_\mu = \mathbf{a}.\mathbf{b} + a_4 b_4 \tag{15.64}$$

is also an invariant.

Of particular importance in what follows is the energy–momentum (or four-momentum) four-vector

$$p_\mu \equiv \left(\mathbf{p}, i\frac{E}{c}\right) \tag{15.65}$$

whose squared norm is given by

$$p_\mu p_\mu = \mathbf{p}^2 - \frac{E^2}{c^2} = -m^2 c^2 \tag{15.66}$$

so that the four-vector p_μ is time-like.

It follows from the conditions (15.63) that

$$\det |L_{\mu\nu}| = \pm 1 \tag{15.67}$$

so that for every homogeneous Lorentz transformation there exists an inverse transformation, which we can write as

$$x_\mu = L_{\nu\mu} x'_\nu. \tag{15.68}$$

Let us set $\nu = \rho = 4$ in (15.63b). We then have

$$(L_{44})^2 = 1 - \sum_{k=1}^{3}(L_{4k})^2$$

$$= 1 + \sum_{k=1}^{3}|L_{4k}|^2 \geq 1 \qquad (15.69)$$

so that either $L_{44} \geq 1$ or $L_{44} \leq -1$. If $L_{44} \geq 1$, the transformation conserves the sign of the fourth (time) component of time-like vectors, and is called an *orthochronous* Lorentz transformation. If $L_{44} \geq 1$ and in addition $\det|L_{\mu\nu}| = +1$, the transformation also conserves the sense of Cartesian systems in three-dimensional space; it is then called a *proper* Lorentz transformation. Proper Lorentz transformations can be built up by an infinite succession of infinitesimal transformations. They include ordinary three-dimensional spatial *rotations* as well as the special Lorentz transformations called *boosts*, which are transformations to coordinates in relative motion along any spatial direction. The orthochronous Lorentz transformations ($L_{44} \geq 1$) with $\det|L_{\mu\nu}| = -1$ include the *spatial reflection* or *spatial inversion*

$$x'_k = -x_k, \quad x'_4 = x_4 \qquad (15.70)$$

and the product of a spatial reflection with a proper Lorentz transformation. The homogeneous Lorentz transformations with $L_{44} \leq -1$ include the *time reflection* or *time inversion*

$$x'_k = x_k, \quad x'_4 = -x_4 \qquad (15.71)$$

and will not be needed in what follows.

Finally, we note that in addition to the homogeneous Lorentz transformations, we can consider inhomogeneous Lorentz transformations, also called Poincaré transformations, such that

$$x'_\mu = L_{\mu\nu}x_\nu + d_\mu \qquad (15.72)$$

where d_μ is a constant four-vector.

Covariant form of the Dirac equation

Let us return to the free-particle Dirac equation (15.31). Multiplying this equation on the left by $\beta/\hbar c$ and introducing the 4×4 matrices $\gamma_\mu \equiv (\gamma_1, \gamma_2, \gamma_3, \gamma_4) \equiv (\boldsymbol{\gamma}, \gamma_4)$ such that

$$\gamma_k = -i\beta\alpha_k, \quad k = 1, 2, 3 \qquad (15.73a)$$
$$\gamma_4 = \beta \qquad (15.73b)$$

we obtain the covariant form of the Dirac equation:

$$\left(\gamma_\mu \frac{\partial}{\partial x_\mu} + \frac{mc}{\hbar}\right)\Psi = 0. \qquad (15.74)$$

15.3 Covariant formulation of the Dirac theory

The properties of the matrices γ_μ are easily derived from those of the α_k and β. The fact that α_k and β are Hermitian implies that

$$\gamma_\mu^\dagger = \gamma_\mu \tag{15.75}$$

so that the matrices γ_μ are also Hermitian. From (15.35) and (15.73), we also have

$$\gamma_\mu \gamma_\nu + \gamma_\nu \gamma_\mu = 2\delta_{\mu\nu}, \quad \mu, \nu = 1, 2, 3, 4. \tag{15.76}$$

Using the Dirac representation (15.43) for α_k and β, the matrices γ_μ have the form

$$\gamma_k = \mathrm{i}\begin{pmatrix} 0 & -\sigma_k \\ \sigma_k & 0 \end{pmatrix}, \quad \gamma_4 = \begin{pmatrix} I & 0 \\ 0 & -I \end{pmatrix}. \tag{15.77}$$

Let us now generalise our treatment to the case of a spin-1/2 particle of charge q moving in an electromagnetic field described by a vector potential $\mathbf{A}(\mathbf{r}, t)$ and a scalar potential $\phi(\mathbf{r}, t)$ which form the four-vector

$$A_\mu \equiv (\mathbf{A}, \mathrm{i}\phi). \tag{15.78}$$

Multiplying both sides of (15.45) by $\beta/\hbar c$, and using (15.73), we obtain the covariant form of the Dirac equation for a charged particle in an electromagnetic field:

$$\left[\gamma_\mu\left(\frac{\partial}{\partial x_\mu} - \frac{\mathrm{i}q}{\hbar}A_\mu\right) + \frac{mc}{\hbar}\right]\Psi = 0. \tag{15.79}$$

Taking the Hermitian conjugate of this equation, we have

$$\left(\frac{\partial}{\partial x_k} + \frac{\mathrm{i}q}{\hbar}A_k\right)\Psi^\dagger \gamma_k^\dagger + \left(-\frac{\partial}{\partial x_4} - \frac{\mathrm{i}q}{\hbar}A_4\right)\Psi^\dagger \gamma_4^\dagger + \frac{mc}{\hbar}\Psi^\dagger = 0 \tag{15.80}$$

where we have used the fact that $x_4 = \mathrm{i}ct$ and $A_4 = \mathrm{i}\phi$ are purely imaginary. It is convenient to introduce the Dirac *adjoint*

$$\bar{\Psi} = \Psi^\dagger \gamma_4 \tag{15.81}$$

so that after multiplying (15.80) on the right by γ_4 and using (15.75) and (15.76), we find that $\bar{\Psi}$ satisfies the equation

$$\left(\frac{\partial}{\partial x_\mu} + \frac{\mathrm{i}q}{\hbar}A_\mu\right)\bar{\Psi}\gamma_\mu - \frac{mc}{\hbar}\bar{\Psi} = 0 \tag{15.82}$$

which is called the *adjoint equation*.

We can also write the continuity equation (15.52) in covariant form. Multiplying (15.79) on the left by $\bar{\Psi}$ and (15.82) on the right by Ψ, and adding the two resulting equations, we find that

$$\frac{\partial}{\partial x_\mu}(\bar{\Psi}\gamma_\mu \Psi) = 0. \tag{15.83}$$

Defining the probability current density four-vector

$$j_\mu = \mathrm{i}c\bar{\Psi}\gamma_\mu \Psi \tag{15.84}$$

we can rewrite (15.83) as the continuity equation

$$\frac{\partial}{\partial x_\mu} j_\mu = 0. \tag{15.85}$$

Using (15.49), (15.51), (15.81) and (15.84), we see that

$$j_\mu \equiv (\mathbf{j}, icP) \tag{15.86}$$

so that (15.85) expresses the continuity equation (15.52) in covariant form.

Lorentz invariance of the Dirac equation

We shall now prove that the Dirac equation for a free particle is invariant under Lorentz transformations. Suppose that in a reference frame S where a given event has coordinates x_μ this equation takes the form (15.74), namely

$$\left(\gamma_\mu \frac{\partial}{\partial x_\mu} + \frac{mc}{\hbar}\right)\Psi(x) = 0 \tag{15.87}$$

where the symbol x stands collectively for $(x_1, x_2, x_3, x_4) \equiv (\mathbf{r}, ict)$. In a new Lorentz frame S' the coordinates of the same event are given by x'_μ, as defined by (15.59) for homogeneous Lorentz transformations. Using the inverse transformation (15.68), we have

$$\frac{\partial}{\partial x_\mu} = L_{\nu\mu} \frac{\partial}{\partial x'_\nu}. \tag{15.88}$$

Substituting (15.88) into (15.87), we find that

$$\left(\gamma_\mu L_{\nu\mu} \frac{\partial}{\partial x'_\nu} + \frac{mc}{\hbar}\right)\Psi(x) = 0. \tag{15.89}$$

Let us now define the four-component wave function of the particle in the new Lorentz frame S' as $\Psi'(x')$, where x' stands collectively for $(x'_1, x'_2, x'_3, x'_4) \equiv (\mathbf{r}', ict')$. We assume that $\Psi'(x')$ can be obtained from $\Psi(x)$ by a transformation of the form

$$\Psi'(x') = \Lambda \Psi(x) \tag{15.90}$$

where Λ is a 4×4 matrix. We shall now show that $\Psi'(x')$ satisfies a Dirac equation of the standard form, namely

$$\left(\gamma_\nu \frac{\partial}{\partial x'_\nu} + \frac{mc}{\hbar}\right)\Psi'(x') = 0. \tag{15.91}$$

Using (15.90), and multiplying on the left by Λ^{-1}, (15.91) becomes

$$\Lambda^{-1}\left(\gamma_\nu \frac{\partial}{\partial x'_\nu} + \frac{mc}{\hbar}\right)\Lambda \Psi(x) = 0, \tag{15.92}$$

which reduces to (15.87) provided that

$$\Lambda^{-1}\gamma_\nu \Lambda = \gamma_\mu L_{\nu\mu}. \tag{15.93}$$

15.3 Covariant formulation of the Dirac theory

Therefore, if a 4×4 matrix Λ satisfying this equation can be found, the wave functions $\Psi(x)$ and $\Psi'(x')$ satisfy Dirac equations of the same form, so that the Dirac equation is invariant under homogeneous Lorentz transformations.

Let us begin by considering infinitesimal proper Lorentz transformations, for which

$$L_{\nu\mu} = \delta_{\mu\nu} + \varepsilon_{\mu\nu} \tag{15.94}$$

where $|\varepsilon_{\mu\nu}| \ll 1$. Using the relations (15.63), we find that

$$\varepsilon_{\mu\nu} = -\varepsilon_{\nu\mu}. \tag{15.95}$$

We also have from (15.93) and (15.94)

$$\Lambda^{-1}\gamma_\nu\Lambda = \gamma_\nu + \gamma_\mu \varepsilon_{\nu\mu}. \tag{15.96}$$

By using (15.76) and keeping only first-order terms in $\varepsilon_{\mu\nu}$, it is easy to check (Problem 15.5) that (15.96) is satisfied provided that

$$\Lambda = I + \frac{1}{4}\varepsilon_{\rho\sigma}\gamma_\rho\gamma_\sigma \tag{15.97a}$$

and

$$\Lambda^{-1} = I - \frac{1}{4}\varepsilon_{\rho\sigma}\gamma_\rho\gamma_\sigma \tag{15.97b}$$

where I is the unit 4×4 matrix. Finite proper Lorentz transformations can be obtained by a succession of these infinitesimal proper Lorentz transformations.

We now consider pure spatial reflections (15.70), for which

$$L_{ik} = -\delta_{ik}, \quad i,k = 1,2,3 \tag{15.98a}$$
$$L_{4k} = L_{k4} = 0 \tag{15.98b}$$

and

$$L_{44} = 1. \tag{15.98c}$$

By direct substitution into (15.93), it is found that

$$\Lambda = \eta\gamma_4 \tag{15.99}$$

where η is a constant. Since two reflections bring one back to the original coordinate system, we may take $\eta^2 = 1$, so that $\eta = \pm 1$. Thus, under a spatial reflection

$$\Psi'(-\mathbf{r}, ict) = \eta\gamma_4\Psi(\mathbf{r}, ict). \tag{15.100}$$

We may write this equation in the alternative form

$$\Psi'(\mathbf{r}, ict) = \eta\gamma_4\Psi(-\mathbf{r}, ict)$$
$$= \eta\gamma_4\mathcal{P}\Psi(\mathbf{r}, ict) \tag{15.101}$$

where $\mathcal{P}$ is the non-relativistic parity operator introduced in Section 5.10. We note that the relativsitic wave function $\Psi(\mathbf{r}, ict)$ is not an eigenfunction of $\mathcal{P}$. It is convenient for future use to introduce a relativistic parity operator defined as

$$P = \eta\gamma_4 \mathcal{P} \tag{15.102}$$

which is Hermitian and such that $P^2 = I$.

We also remark that the Dirac equation is form-invariant under space and time translations, and hence under inhomogeneous Lorentz transformations.

Finally, we note that the method employed in this section can also be used to write the Klein–Gordon equation in covariant form, and prove its invariance under Lorentz transformations.

15.4 Plane wave solutions of the Dirac equation

Let us return to the Dirac equation (15.31) for a free particle of spin-1/2. Since the Hamiltonian (15.30) is independent of $\mathbf{r}$ and t, we can seek eigenfunctions common to both the energy and momentum operators, namely plane waves of the form

$$\Psi(\mathbf{r}, t) = Au \exp[i(\mathbf{p}.\mathbf{r} - Et)/\hbar] \tag{15.103}$$

where A is a constant and u is a four-component spinor independent of the space–time coordinates x_μ of the particle. The plane waves (15.103) are eigenfunctions of the operators $E_{op} = i\hbar\partial/\partial t$ and $\mathbf{p}_{op} = -i\hbar\nabla$ with eigenvalues $E = \hbar\omega$ and $\mathbf{p} = \hbar\mathbf{k}$, respectively. Substituting (15.103) into (15.31b) gives for u the matrix equation

$$(c\boldsymbol{\alpha}.\mathbf{p} + \beta mc^2)u = Eu. \tag{15.104}$$

It is instructive to consider first the case of a particle at rest, so that $p = 0$. Denoting by $u(0)$ the corresponding four-component spinor, and using (15.43), we find that equation (15.104) then reduces to

$$\begin{pmatrix} mc^2 I & 0 \\ 0 & -mc^2 I \end{pmatrix} u(0) = Eu(0) \tag{15.105}$$

where we recall that I is the unit 2×2 matrix. This equation has the eigenvalues $E_+ = mc^2$ (occurring twice) and $E_- = -mc^2$ (occurring twice) and four linearly independent eigenvectors which are given (up to an arbitrary constant) by

$$u^{(1)}(0) = \begin{pmatrix} 1 \\ 0 \\ 0 \\ 0 \end{pmatrix}, \quad u^{(2)}(0) = \begin{pmatrix} 0 \\ 1 \\ 0 \\ 0 \end{pmatrix}, \quad u^{(3)}(0) = \begin{pmatrix} 0 \\ 0 \\ 1 \\ 0 \end{pmatrix}, \quad u^{(4)}(0) = \begin{pmatrix} 0 \\ 0 \\ 0 \\ 1 \end{pmatrix}.$$

$$\tag{15.106}$$

The first two solutions describe a spin-1/2 particle with positive energy $E_+ = mc^2$. The last two solutions, corresponding to the negative energy $E_- = -mc^2$, will be interpreted at the end of this chapter.

15.4 Plane wave solutions of the Dirac equation

Let us now consider the case $p \neq 0$. It is convenient to write the four-component spinor u in terms of two-component spinors u_A and u_B as

$$u = \begin{pmatrix} u_A \\ u_B \end{pmatrix}, \tag{15.107}$$

where

$$u_A = \begin{pmatrix} u_1 \\ u_2 \end{pmatrix}, \quad u_B = \begin{pmatrix} u_3 \\ u_4 \end{pmatrix}. \tag{15.108}$$

We then have, from (15.104) and (15.43),

$$\begin{pmatrix} mc^2 I & c\boldsymbol{\sigma}\cdot\mathbf{p} \\ c\boldsymbol{\sigma}\cdot\mathbf{p} & -mc^2 I \end{pmatrix} \begin{pmatrix} u_A \\ u_B \end{pmatrix} = E \begin{pmatrix} u_A \\ u_B \end{pmatrix} \tag{15.109}$$

so that u_A and u_B are related by

$$u_A = \frac{c\boldsymbol{\sigma}\cdot\mathbf{p}}{E - mc^2} u_B, \quad u_B = \frac{c\boldsymbol{\sigma}\cdot\mathbf{p}}{E + mc^2} u_A. \tag{15.110}$$

Using the second of these equations to eliminate u_B, and making use of (6.241), we find that

$$c^2(\boldsymbol{\sigma}\cdot\mathbf{p})^2 u_A = \mathbf{p}^2 c^2 u_A = (E^2 - m^2 c^4) u_A. \tag{15.111a}$$

Similarly, upon elimination of u_A, we have

$$c^2(\boldsymbol{\sigma}\cdot\mathbf{p})^2 u_B = \mathbf{p}^2 c^2 u_B = (E^2 - m^2 c^4) u_B. \tag{15.111b}$$

The four eigenvalues of equation (15.109) are therefore given by

$$E_+ = +(m^2 c^4 + \mathbf{p}^2 c^2)^{1/2} \quad \text{occurring twice} \tag{15.112a}$$

and

$$E_- = -(m^2 c^4 + \mathbf{p}^2 c^2)^{1/2} \quad \text{occurring twice.} \tag{15.112b}$$

This result may also be obtained by writing the four linear, homogeneous equations corresponding to (15.109). Using the expressions (6.243) of the Pauli matrices and denoting the Cartesian components of the three-vector $\mathbf{p}$ by (p_x, p_y, p_z), we obtain a set of algebraic equations for the four components u_i of u. That is,

$$\begin{aligned}
(mc^2 - E)u_1 + cp_z u_3 + c(p_x - ip_y)u_4 &= 0 \\
(mc^2 - E)u_2 + c(p_x + ip_y)u_3 - cp_z u_4 &= 0 \\
cp_z u_1 + c(p_x - ip_y)u_2 - (mc^2 + E)u_3 &= 0 \\
c(p_x + ip_y)u_1 - cp_z u_2 - (mc^2 + E)u_4 &= 0
\end{aligned} \tag{15.113}$$

These equations have a non-trivial solution only if the determinant of the coefficients is zero. This determinant has the value $(E^2 - m^2 c^4 - \mathbf{p}^2 c^2)^2$, so that we must have

$$(E^2 - m^2 c^4 - \mathbf{p}^2 c^2)^2 = 0 \tag{15.114}$$

which indeed gives the energy eigenvalues E_+ and E_- of equation (15.112). Moreover, when the condition (15.114) is satisfied, the determinant is of rank 2 (i.e. all 3×3 minors vanish). Thus there are two linearly independent solutions of equations (15.109) corresponding to the positive energy $E_+ = +(m^2c^4 + \mathbf{p}^2c^2)^{1/2}$, which describe a free spin-1/2 particle of energy E_+ and momentum $\mathbf{p}$. They can be written as

$$u^{(1)} = N \begin{pmatrix} \alpha \\ \dfrac{c\boldsymbol{\sigma}\cdot\mathbf{p}}{E_+ + mc^2}\alpha \end{pmatrix}, \quad u^{(2)} = N \begin{pmatrix} \beta \\ \dfrac{c\boldsymbol{\sigma}\cdot\mathbf{p}}{E_+ + mc^2}\beta \end{pmatrix} \qquad (15.115)$$

where α and β are the basic two component spinors (6.230), and N is a normalisation constant. The general solution u^+ corresponding to the positive energy E_+ and the momentum $\mathbf{p}$ is therefore given by a linear combination of the solutions $u^{(1)}$ and $u^{(2)}$, namely

$$u^+ = a u^{(1)} + b u^{(2)} = N \begin{pmatrix} \chi \\ \dfrac{c\boldsymbol{\sigma}\cdot\mathbf{p}}{E_+ + mc^2}\chi \end{pmatrix} \qquad (15.116)$$

where a and b are complex coefficients, and $\chi = a\alpha + b\beta$ is a general two-component spinor (see (6.216) and (6.233)). It is worth noting that the four components u_i^+ ($i = 1, 2, 3, 4$) of u^+ are not independent of each other. Indeed, we see from (15.110) that once the two-component spinor u_A is given, the other two-component spinor u_B is specified (and vice versa). Therefore, it is always possible to express the two lower components u_i^+ ($i = 3, 4$) of any solution u^+ in terms of the two upper components u_i^+ ($i = 1, 2$).

Let us now consider the solutions corresponding to the negative energy $E_- = -(m^2c^4 + \mathbf{p}^2c^2)^{1/2}$ and the momentum $\mathbf{p}$. There are two linearly independent solutions of this type, which may be written as

$$u^{(3)} = N \begin{pmatrix} -\dfrac{c\boldsymbol{\sigma}\cdot\mathbf{p}}{-E_- + mc^2}\alpha \\ \alpha \end{pmatrix}, \quad u^{(4)} = N \begin{pmatrix} -\dfrac{c\boldsymbol{\sigma}\cdot\mathbf{p}}{-E_- + mc^2}\beta \\ \beta \end{pmatrix}. \qquad (15.117)$$

The general solution u^- corresponding to the negative energy E_- and the momentum $\mathbf{p}$ is therefore a linear combination of the solutions (15.117). That is,

$$u^- = \tilde{a} u^{(3)} + \tilde{b} u^{(4)} = N \begin{pmatrix} -\dfrac{c\boldsymbol{\sigma}\cdot\mathbf{p}}{-E_- + mc^2}\tilde{\chi} \\ \tilde{\chi} \end{pmatrix} \qquad (15.118)$$

where $\tilde{a}$ and $\tilde{b}$ are complex coefficients and $\tilde{\chi} = \tilde{a}\alpha + \tilde{b}\beta$ is a general two-component spinor. As for the case of positive-energy solutions, the four components u_i^- ($i = 1, 2, 3, 4$) of u^- are not independent, the two upper components u_i^- ($i = 1, 2$) being related to the two lower components u_i^- ($i = 3, 4$).

15.4 Plane wave solutions of the Dirac equation

We also note that the four solutions $u^{(r)}(r = 1, 2, 3, 4)$ given by (15.115) and (15.117) are orthogonal:

$$u^{(r)\dagger} u^{(s)} = 0, \quad r \neq s \tag{15.119}$$

where $r, s = 1, 2, 3, 4$ and each $u^{(r)\dagger}$ is a row vector with four components $u_i^{(r)*}$, namely

$$u^{(r)\dagger} = (u_1^{(r)*} \quad u_2^{(r)*} \quad u_3^{(r)*} \quad u_4^{(r)*}). \tag{15.120}$$

Non-relativistic limit

It is also interesting to consider the non-relativistic limit of our solutions. Then $E_+ = -E_-$ is close to mc^2, and large with respect to cp. Thus, looking back at equations (15.110), we see that for the positive energy solutions the two-component spinor u_B is of order (v/c) times u_A, where $v = p/m$ is the magnitude of the (non-relativistic) velocity of the particle; u_A is the so-called *large component* and u_B the *small component* of the four-component Dirac spinor u. On the other hand, for negative-energy solutions, u_A is of order (v/c) times u_B in the non-relativistic limit; in this case u_B is the large component and u_A the small one. We also note that in the limit $p = 0$ the solutions $u^{(r)}$ given by (15.115) and (15.117) with $N = 1$ reduce to our previous result (15.106). In particular, for the two solutions (15.115) corresponding to positive energies, we see that with $N = 1$ we have in the limit $p \to 0$

$$u^{(1)} \to \begin{pmatrix} \alpha \\ 0 \end{pmatrix}, \quad u^{(2)} \to \begin{pmatrix} \beta \\ 0 \end{pmatrix} \tag{15.121}$$

so that we retrieve the basic Pauli spin functions. Similarly, for the two negative-energy solutions (15.117), we have in the limit $p \to 0$ (choosing $N = 1$)

$$u^{(3)} \to \begin{pmatrix} 0 \\ \alpha \end{pmatrix}, \quad u^{(4)} \to \begin{pmatrix} 0 \\ \beta \end{pmatrix}. \tag{15.122}$$

The physical distinction between the two solutions for each sign of the energy can therefore be understood in the non-relativistic limit by taking into account the spin of the particle.

Spin and helicity operators

Guided by the above result, let us return to the relativistic solutions $u^{(r)}$ given by (15.115) and (15.117), and analyse the two-fold degeneracy corresponding either to the positive energy $E_+ = +(m^2c^4 + \mathbf{p}^2c^2)^{1/2}$ or the negative energy $E_- = -(m^2c^4 + \mathbf{p}^2c^2)^{1/2}$. This degeneracy implies that there must be another operator which commutes with the Hamiltonian (15.30) and with the momentum operator $\mathbf{p}_{op} = -i\hbar \nabla$, whose eigenvalues can be taken to label the states. To construct this operator, let us first introduce the Dirac spin operator

$$\mathbf{S} = \frac{\hbar}{2} \mathbf{\Sigma} \tag{15.123}$$

where $\boldsymbol{\Sigma} \equiv (\Sigma_1, \Sigma_2, \Sigma_3)$, the Σ_k being 4×4 Dirac spin matrices given by

$$\Sigma_k = \begin{pmatrix} \sigma_k & 0 \\ 0 & \sigma_k \end{pmatrix}. \tag{15.124}$$

It follows from the properties of the Pauli spin matrices that the three Cartesian components of $\mathbf{S}$ satisfy the commutation relations (6.181) characteristic of angular momentum operators. Moreover, for any state Ψ,

$$\mathbf{S}^2 \Psi = s(s+1)\hbar^2 \Psi, \quad s = \frac{1}{2} \tag{15.125}$$

and the two possible eigenvalues of S_x, S_y and S_z are $\pm\hbar/2$. Using the properties of the Pauli spin matrices, it is straightforward to show (Problem 15.6) that if H is the free-particle Dirac Hamiltonian (15.30), then

$$\begin{aligned}[H, \mathbf{S}] &= \frac{\hbar}{2} [c\boldsymbol{\alpha} \cdot \mathbf{p}_{\mathrm{op}} + \beta mc^2, \boldsymbol{\Sigma}] \\ &= i\hbar c(\boldsymbol{\alpha} \times \mathbf{p}_{\mathrm{op}}) \end{aligned} \tag{15.126}$$

so that in general the spin operator (15.123) is not a constant of the motion. In fact, for any unit vector $\hat{\mathbf{n}}$, we have

$$[H, \hat{\mathbf{n}} \cdot \mathbf{S}] = -i\hbar c \boldsymbol{\alpha} \cdot (\hat{\mathbf{n}} \times \mathbf{p}_{\mathrm{op}}) \tag{15.127}$$

which shows that the eigenfunctions of H are in general not eigenfunctions of $S_n = \hat{\mathbf{n}} \cdot \mathbf{S}$. As a result, an eigenvalue of the spin operator $\mathbf{S}$ cannot be assigned in an arbitrary direction (θ, ϕ), except in the rest frame of the particle for which $p = 0$. In that frame, we see from (15.116) and (15.118) that the corresponding spin eigenfunctions for positive energies and negative energies are respectively (with $N = 1$)

$$u^+(p=0) = \begin{pmatrix} \chi \\ 0 \end{pmatrix}, \quad u^-(p=0) = \begin{pmatrix} 0 \\ \tilde{\chi} \end{pmatrix} \tag{15.128}$$

where $\chi = a\alpha + b\beta$ and $\tilde{\chi} = \tilde{a}\alpha + \tilde{b}\beta$.

Let us now return to the equation (15.127), and choose $\hat{\mathbf{n}}$ to coincide with $\hat{\mathbf{p}}$, the unit vector pointing in the direction of the momentum. We then find that

$$[H, \hat{\mathbf{p}} \cdot \mathbf{S}] = 0 \tag{15.129}$$

The operator $\hat{\mathbf{p}} \cdot \mathbf{S}$, which is the component of the spin operator in the direction of motion $\hat{\mathbf{p}}$, is called the *helicity operator*. We see from (15.129) that this operator commutes with the free-particle Dirac Hamiltonian (15.30). In addition, the operator $\hat{\mathbf{p}} \cdot \mathbf{S}$ commutes with the momentum operator $\mathbf{p}_{\mathrm{op}} = -i\hbar \boldsymbol{\nabla}$. It follows from (15.129) that the helicity operator $\hat{\mathbf{p}} \cdot \mathbf{S}$ is a constant of the motion. Its eigenvalues $\lambda\hbar$ define a quantum number λ, called the *helicity*, which in the present case can take the values $\lambda = +1/2$ (positive helicity) and $\lambda = -1/2$ (negative helicity). Hence, for a given momentum, there are two different positive-energy four-component spinors having the form (15.116) and corresponding to positive and negative helicity, respectively.

Similarly, there are two four-component spinors of the form (15.118), which correspond respectively to positive and negative helicity.

It is worth stressing that the positive-energy four-component spinors (15.115) are in general not eigenfunctions of the helicity operator $\hat{\mathbf{p}}.\mathbf{S}$. However, appropriate linear combinations of these spinors can be formed, which are eigenstates of $\hat{\mathbf{p}}.\mathbf{S}$; the same is true for the negative-energy four-component spinors (15.117). Thus, returning to (15.116), we can express the two-component spinor χ as a superposition of the two linearly independent spinors $\chi_\uparrow$ and $\chi_\downarrow$ which are eigenstates of the operator $\hat{\mathbf{p}}.\mathbf{S}$. These are (see (6.252) and (6.255))

$$\chi_\uparrow = \begin{pmatrix} \cos\frac{\theta}{2} \\ \sin\frac{\theta}{2} e^{i\phi} \end{pmatrix}, \quad \chi_\downarrow = \begin{pmatrix} \sin\frac{\theta}{2} \\ -\cos\frac{\theta}{2} e^{i\phi} \end{pmatrix} \tag{15.130}$$

where (θ, ϕ) are the polar angles of the vector $\mathbf{p}$. A four-component spinor u^+ such as (15.116) with χ given by $\chi_\uparrow$ then represents a spin-1/2 particle with its spin directed parallel to its momentum $\mathbf{p}$, that is with helicity $\lambda = +1/2$. In the same way, if in (15.116) χ is set equal to $\chi_\downarrow$, then u^+ represents a particle with its spin anti-parallel to $\mathbf{p}$, that is with helicity $\lambda = -1/2$. If $\chi = a_\uparrow \chi_\uparrow + a_\downarrow \chi_\downarrow$, then $|a_\uparrow|^2$ and $|a_\downarrow|^2$ give the relative probabilities of finding the particle with spin 'up' and spin 'down' relative to the direction of $\mathbf{p}$, that is, with helicity $+1/2$ and $-1/2$.

An interesting particular case is that for which the z-axis is chosen along the momentum $\mathbf{p}$ of the particle, so that $\mathbf{p} \equiv (0, 0, p)$. The positive energy solutions (15.115) then become

$$u^{(1)} = N \begin{pmatrix} 1 \\ 0 \\ \frac{cp}{E_+ + mc^2} \\ 0 \end{pmatrix}, \quad u^{(2)} = N \begin{pmatrix} 0 \\ 1 \\ 0 \\ -\frac{cp}{E_+ + mc^2} \end{pmatrix} \tag{15.131}$$

while the negative-energy solutions (15.117) become

$$u^{(3)} = N \begin{pmatrix} -\frac{cp}{-E_- + mc^2} \\ 0 \\ 1 \\ 0 \end{pmatrix}, \quad u^{(4)} = N \begin{pmatrix} 0 \\ \frac{cp}{-E_- + mc^2} \\ 0 \\ 1 \end{pmatrix}. \tag{15.132}$$

Furthermore, we now have

$$\hat{\mathbf{p}}.\mathbf{S}\, u^{(r)} = S_z u^{(r)} = \lambda \hbar u^{(r)} \tag{15.133}$$

with $\lambda = +1/2$ when $r = 1, 3$ and $\lambda = -1/2$ when $r = 2, 4$. Thus we see that the solutions $u^{(r)}$ are eigenfunctions of S_z. This result is a direct consequence of the discussion following equation (15.129), since in the present case S_z is just the component of the spin operator in the direction of motion.

Normalisation of the solutions

Let us now return to the four solutions $u^{(r)}$ given by (15.115) and (15.117). Each of these solutions can be normalised to unity by requiring that

$$u^{(r)\dagger}u^{(r)} = 1, \quad r = 1, 2, 3, 4 \tag{15.134}$$

It is then easily found (Problem 15.7) that the normalisation constant N is given (apart from an arbitrary complex multiplicative factor of modulus one) by

$$N = \left[1 + \frac{c^2\mathbf{p}^2}{(E_+ + mc^2)^2}\right]^{-1/2} = \left(\frac{E_+ + mc^2}{2E_+}\right)^{1/2}. \tag{15.135}$$

Combining the normalisation condition (15.135) with the orthogonality relations (15.119), we see that the solutions $u^{(r)}$ satisfy the orthonormality relations

$$u^{(r)\dagger}u^{(s)} = \delta_{rs}, \quad r, s = 1, 2, 3, 4. \tag{15.136}$$

To each of the four four-component spinors $u^{(r)}$ solutions of equation (15.104) corresponds a four-component plane-wave solution of the free-particle Dirac equation, which, using (15.103), is given by

$$\Psi^{(r)}(\mathbf{r}, t) = A\, u^{(r)} \exp[i(\mathbf{p}\cdot\mathbf{r} - Et)/\hbar], \quad r = 1, 2, 3, 4. \tag{15.137}$$

The normalisation of these plane-wave solutions can be chosen in various ways. Let us assume that the particle is localised in same large volume V. Then

$$\int_V |\Psi^{(r)}|^2 d\mathbf{r} = |A|^2 V, \quad r = 1, 2, 3, 4. \tag{15.138}$$

If we require that the solutions be normalised to unity, then we can take

$$A = V^{-1/2}. \tag{15.139}$$

However, it is often convenient to adopt a normalisation such that

$$A = \left(\frac{E_+}{Vmc^2}\right)^{1/2}, \tag{15.140}$$

in which case the normalisation condition (15.138) is Lorentz invariant (Problem 15.7).

15.5 Solutions of the Dirac equation for a central potential

Let us consider a spin-1/2 particle of mass m and charge q in a static central field such that $\mathbf{A} = 0$ and $V(r) = q\phi(r)$ is the potential energy. The Dirac Hamiltonian (15.46) is therefore given in this case by

$$H = c\boldsymbol{\alpha}\cdot\mathbf{p}_{op} + \beta mc^2 + V(r) \tag{15.141}$$

15.5 Solutions of the Dirac equation for a central potential

and the Dirac equation to be solved is

$$[c\boldsymbol{\alpha}\cdot\mathbf{p}_{op} + \beta mc^2 + V(r)]\psi(\mathbf{r}) = E\psi(\mathbf{r}) \qquad (15.142)$$

where $\psi(\mathbf{r})$ is a four-component spinor.

In order to separate variables in this equation, let us look for operators which commute with the Hamiltonian (15.141) and hence yield good quantum numbers. We saw in Section 7.2 that in the non-relativistic Schrödinger theory the Hamiltonian $H = \mathbf{p}_{op}^2/2m + V(r)$ of a spinless particle of mass m in a central field commutes with every Cartesian component of the orbital angular momentum operator $\mathbf{L} = \mathbf{r} \times \mathbf{p}_{op}$, as well as with $\mathbf{L}^2$. As a result, simultaneous eigenstates of the operators H, $\mathbf{L}^2$ and L_z exist in Schrödinger's non-relativistic theory, with eigenvalues given by E, $l(l+1)\hbar^2$ and $m_l\hbar$, respectively[1]. In Dirac's theory, however, neither the Cartesian components of $\mathbf{L}$, nor $\mathbf{L}^2$ commute with the Dirac Hamiltonian (15.141). Indeed, it is readily shown (Problem 15.8) that if H denotes this Hamiltonian

$$[H, \mathbf{L}] = -i\hbar c(\boldsymbol{\alpha} \times \mathbf{p}_{op}). \qquad (15.143)$$

Let us now consider the Dirac spin operator $\mathbf{S} = (\hbar/2)\boldsymbol{\Sigma}$ introduced in the previous section (see (15.123)). We recall that for any state Ψ one has $\mathbf{S}^2\Psi = s(s+1)\hbar^2\Psi$ with $s = 1/2$ (see (15.125)) and that the two possible eigenvalues of S_x, S_y and S_z are $\pm\hbar/2$. We also remark that any Cartesian component of $\mathbf{S}$ commutes with any Cartesian component of $\mathbf{L}$. Moreover, using the properties of the Pauli spin matrices, we have

$$[H, \mathbf{S}] = i\hbar c(\boldsymbol{\alpha} \times \mathbf{p}_{op}) \qquad (15.144)$$

where H is the Dirac Hamiltonian (15.141).

We can now define a *total angular momentum* operator $\mathbf{J}$ as the sum of the orbital and spin angular momentum operators:

$$\mathbf{J} = \mathbf{L} + \mathbf{S}. \qquad (15.145)$$

It is easily checked that the three Cartesian components of $\mathbf{J}$ satisfy the basic commutation relations (6.139) of angular momentum operators and that $[\mathbf{J}^2, \mathbf{J}] = 0$, so that $\mathbf{J}^2$ commutes with each of the components of $\mathbf{J}$. As a result, simultaneous eigenfunctions of $\mathbf{J}^2$ and one of its components (which we choose to be J_z) can be found. As usual, we denote the eigenvalues of $\mathbf{J}^2$ by $j(j+1)\hbar^2$ and those of J_z by $m_j\hbar$. Making use of the results (15.143) and (15.144), we find that

$$[H, \mathbf{J}] = 0 \qquad (15.146)$$

so that every Cartesian component of $\mathbf{J}$ commutes with the Dirac Hamiltonian (15.141). Using (5.101c), we see that $\mathbf{J}^2$ also commutes with this Hamiltonian:

$$[H, \mathbf{J}^2] = 0. \qquad (15.147)$$

---

[1] As in Sections 6.10, 8.2 and 10.5 we denote by m_l the quantum number corresponding to the operator L_z.

Since we have $[H, \mathbf{J}^2] = 0$, $[H, J_z] = 0$ and $[\mathbf{J}^2, J_z] = 0$, it follows that simultaneous eigenstates of the Dirac Hamiltonian (15.141) and of the operators $\mathbf{J}^2$ and J_z can be found, with eigenvalues given respectively by E, $j(j+1)\hbar^2$ and $m_j\hbar$.

Let us now consider the non-relativistic parity operator $\mathcal{P}$ defined by (5.336a). The Dirac Hamiltonian (15.141) does not commute with $\mathcal{P}$, since the first term on the right of (15.141) is odd under the operation $\mathbf{r} \to -\mathbf{r}$, while the second and third terms are even. However, the relativistic parity operator P introduced in (15.102) is easily seen to commute with H, as well as with $\mathbf{J}^2$ and J_z.

We also remark that the operator

$$\mathcal{K} = \beta\left(\frac{\mathbf{\Sigma}.\mathbf{L}}{\hbar} + 1\right) \tag{15.148}$$

commutes with H. To prove this, we first define the radial momentum operator

$$p_r = -i\hbar \frac{1}{r}\frac{\partial}{\partial r}r = \frac{1}{r}(\mathbf{r}.\mathbf{p}_{\mathrm{op}} - i\hbar) \tag{15.149}$$

and the 'radial velocity' operator

$$\alpha_r = \frac{1}{r}(\boldsymbol{\alpha}.\mathbf{r}) \tag{15.150}$$

both of which are Hermitian. Using (15.43) and (15.124), together with (6.241), we have

$$(\boldsymbol{\alpha}.\mathbf{r})(\boldsymbol{\alpha}.\mathbf{p}_{\mathrm{op}}) = (\mathbf{\Sigma}.\mathbf{r})(\mathbf{\Sigma}.\mathbf{p})$$
$$= \mathbf{r}.\mathbf{p}_{\mathrm{op}} + i\mathbf{\Sigma}.\mathbf{L}$$
$$= rp_r + i(\mathbf{\Sigma}.\mathbf{L} + \hbar). \tag{15.151}$$

Multiplying on the left by $r^{-1}\alpha_r$ and using the fact that $\alpha_r^2 = 1$, we obtain the identity

$$\boldsymbol{\alpha}.\mathbf{p}_{\mathrm{op}} = \alpha_r\left[p_r + \frac{i}{r}(\mathbf{\Sigma}.\mathbf{L} + \hbar)\right]$$
$$= \alpha_r p_r + i\hbar \frac{\alpha_r}{r}\beta\mathcal{K}. \tag{15.152}$$

This result allows us to rewrite the Hamiltonian (15.141) as

$$H = c\alpha_r p_r + i\hbar\frac{c}{r}\alpha_r\beta\mathcal{K} + \beta mc^2 + V(r). \tag{15.153}$$

It is easy to verify that the operator $\mathcal{K}$ commutes with p_r, α_r and β, so that it also commutes with H. In addition, since $\mathbf{J} = \mathbf{L} + \mathbf{S}$ and $\mathbf{S} = (\hbar/2)\mathbf{\Sigma}$, we have

$$\hbar\mathbf{\Sigma}.\mathbf{L} = 2\mathbf{S}.\mathbf{L} = \mathbf{J}^2 - \mathbf{L}^2 - \mathbf{S}^2 = \mathbf{J}^2 - \mathbf{L}^2 - \frac{3}{4}\hbar^2 \tag{15.154}$$

so that we may write

$$\hbar^2\mathcal{K} = \beta\left(\mathbf{J}^2 - \mathbf{L}^2 + \frac{\hbar^2}{4}\right). \tag{15.155}$$

15.5 Solutions of the Dirac equation for a central potential

Let us now return to the Dirac equation (15.142). Using (15.153), we can put it in the form

$$H\psi = [c\alpha_r p_r + i\hbar \frac{c}{r}\alpha_r \beta \mathcal{K} + \beta mc^2 + V(r)]\psi = E\psi. \quad (15.156)$$

As in (15.56), we express the Dirac four-component spinor ψ in terms of two-component spinors ψ_A and ψ_B. From the above discussion it follows that we can take ψ to be a similtaneous eigenfunction of the operators H, P, $\mathbf{J}^2$, J_z and $\mathcal{K}$. Hence, choosing $\eta = +1$ in (15.102), we can write

$$H \begin{pmatrix} \psi_A \\ \psi_B \end{pmatrix} = E \begin{pmatrix} \psi_A \\ \psi_B \end{pmatrix}, \qquad P \begin{pmatrix} \psi_A \\ \psi_B \end{pmatrix} = \pm \begin{pmatrix} \psi_A \\ \psi_B \end{pmatrix} \quad (15.157a)$$

$$\mathbf{J}^2 \begin{pmatrix} \psi_A \\ \psi_B \end{pmatrix} = j(j+1)\hbar^2 \begin{pmatrix} \psi_A \\ \psi_B \end{pmatrix}, \qquad J_z \begin{pmatrix} \psi_A \\ \psi_B \end{pmatrix} = m_j\hbar \begin{pmatrix} \psi_A \\ \psi_B \end{pmatrix} \quad (15.157b)$$

$$\mathcal{K} \begin{pmatrix} \psi_A \\ \psi_B \end{pmatrix} = \kappa \begin{pmatrix} \psi_A \\ \psi_B \end{pmatrix} \quad (15.157c)$$

In terms of the non-relativistic parity operator $\mathcal{P}$, we have

$$\mathcal{P} \begin{pmatrix} \psi_A \\ \psi_B \end{pmatrix} = \pm \begin{pmatrix} \psi_A \\ -\psi_B \end{pmatrix}. \quad (15.158)$$

We found in Section 6.10 that the simultaneous eigenfunctions of the operators $\mathbf{L}^2$, $\mathbf{S}^2$, $\mathbf{J}^2$ and J_z are the spin-angle functions $\mathcal{Y}_{ls}^{jm_j}$, which have the parity $(-1)^l$. For the case of a spin-1/2 particle ($s = 1/2$), they are given by (6.297) and $l = j \pm 1/2$ with $j = 1/2, 3/2, \ldots$. It follows that ψ_A is proportional to $\mathcal{Y}_{l,\frac{1}{2}}^{j,m_j}$. Similarly, ψ_B is proportional to $\mathcal{Y}_{l',\frac{1}{2}}^{j,m_j}$ which has the parity $(-1)^{l'}$. From (15.158) we see that ψ_B has opposite (non-relativistic) parity to ψ_A. Thus, remembering that l' can only take the values $j \pm 1/2$, it follows that $l' = l \pm 1$. Therefore, if $l = j - 1/2$, then $l' = l + 1 = j + 1/2$ while if $l = j + 1/2$, then $l' = l - 1 = j - 1/2$. Moreover, from (15.155) and (15.157c) we have

$$\left(\mathbf{J}^2 - \mathbf{L}^2 + \frac{\hbar^2}{4}\right)\psi_A = \kappa\hbar^2\psi_A, \qquad \left(\mathbf{J}^2 - \mathbf{L}^2 + \frac{\hbar^2}{4}\right)\psi_B = -\kappa\hbar^2\psi_B. \quad (15.159)$$

As a result, ψ_A and ψ_B are eigenfunctions of $\mathbf{J}^2 - \mathbf{L}^2 + \hbar^2/4$ with eigenvalues $\kappa\hbar^2$ and $-\kappa\hbar^2$, respectively, where

$$\kappa = j(j+1) - l(l+1) + \frac{1}{4} = \left(j + \frac{1}{2}\right)^2 - l(l+1). \quad (15.160)$$

Hence

$$\kappa = \begin{cases} l+1, & j = l + \frac{1}{2} \\ -l, & j = l - \frac{1}{2} \end{cases} \quad (15.161)$$

and we see that κ takes on all positive and negative integer values except zero. We also observe that the value of κ determines both j and l, since

$$j = |\kappa| - \frac{1}{2} \tag{15.162}$$

and

$$l = \begin{cases} \kappa - 1, & \kappa > 0 \\ -\kappa, & \kappa < 0 \end{cases}. \tag{15.163}$$

In terms of the usual spectroscopic notation the states labelled by $\kappa = +1, -1, +2, -2, \ldots$ correspond to $s_{1/2}, p_{1/2}, p_{3/2}, d_{3/2}, \ldots$ states.

We may therefore write the solutions of (15.156) in the form

$$\psi_{E\kappa m_j} = r^{-1} \begin{pmatrix} P_{E\kappa}(r) \mathcal{Y}^{j,m_j}_{l,\frac{1}{2}} \\ iQ_{E\kappa}(r) \mathcal{Y}^{j,m_j}_{l',\frac{1}{2}} \end{pmatrix}. \tag{15.164}$$

In this equation $P_{E\kappa}(r)$ and $Q_{E\kappa}(r)$ are radial functions, and the factor i in front of $Q_{E\kappa}(r)$ has been introduced in order to obtain real radial equations for $P_{E\kappa}(r)$ and $Q_{E\kappa}(r)$.

Substituting (15.164) into (15.156) and making use of the identities

$$(\boldsymbol{\sigma}\cdot\hat{\mathbf{r}})\mathcal{Y}^{j,m_j}_{l,\frac{1}{2}} = -\mathcal{Y}^{j,m_j}_{l',\frac{1}{2}}, \tag{15.165a}$$

$$(\boldsymbol{\sigma}\cdot\hat{\mathbf{r}})\mathcal{Y}^{j,m_j}_{l',\frac{1}{2}} = -\mathcal{Y}^{j,m_j}_{l,\frac{1}{2}} \tag{15.165b}$$

we find that the radial functions $P_{E\kappa}(r)$ and $Q_{E\kappa}(r)$ satisfy the coupled first-order differential equations

$$\left[\frac{d}{dr} - \frac{\kappa}{r}\right] P_{E\kappa}(r) = \frac{E + mc^2 - V(r)}{\hbar c} Q_{E\kappa}(r), \tag{15.166a}$$

$$\left[\frac{d}{dr} + \frac{\kappa}{r}\right] Q_{E\kappa}(r) = -\frac{E - mc^2 - V(r)}{\hbar c} P_{E\kappa}(r), \tag{15.166b}$$

These coupled equations play the role of the radial Schrödinger equation (7.52) in the non-relativistic theory.

By eliminating the function $Q_{E\kappa}(r)$, between the two equations (15.166), we can find a single second-order equation for $P_{E\kappa}(r)$, which has the form

$$\left[\frac{d^2}{dr^2} - \frac{A'}{A}\frac{d}{dr} + \left(AB + \frac{A'}{A}\frac{\kappa}{r} - \frac{\kappa(\kappa-1)}{r^2}\right)\right] P_{E\kappa}(r) = 0 \tag{15.167}$$

where

$$A(r) = \frac{E + mc^2 - V(r)}{\hbar c}, \qquad A' = \frac{dA}{dr}, \tag{15.168a}$$

$$B(r) = \frac{E - mc^2 - V(r)}{\hbar c}. \tag{15.168b}$$

Free spherical waves

As a first example, let us consider the case of a free particle, for which $V = 0$. The equation (15.167) then reduces to

$$\left[\frac{d^2}{dr^2} - \frac{l(l+1)}{r^2} + k^2\right] P_{El}(r) = 0 \tag{15.169}$$

where we have used the facts that $\kappa(\kappa - 1) = l(l + 1)$, and

$$\frac{E^2 - m^2c^4}{\hbar^2 c^2} = \mathbf{k}^2 = k^2, \tag{15.170}$$

The equation (15.169) is identical in form to the free-particle non-relativistic equation (7.60). Its regular solution (vanishing at $r = 0$) is given apart from an arbitrary multiplicative constant by

$$P_{El}(r) = r j_l(kr) \tag{15.171}$$

where j_l is a spherical Bessel function.

The corresponding function $Q_{E\kappa}(r)$ can be obtained from $P_{E\kappa}(r)$ by using (15.166a), namely

$$Q_{E\kappa}(r) = \frac{\hbar c}{E + mc^2} \left[\frac{d}{dr} - \frac{\kappa}{r}\right] P_{E\kappa}(r). \tag{15.172}$$

For the case $\kappa = l + 1$ we have

$$Q_{E\kappa}(r) = -\frac{\hbar kc}{E + mc^2} r j_{l+1}(kr) \tag{15.173}$$

while for $\kappa = -l$, we find that

$$Q_{E\kappa}(r) = \frac{\hbar kc}{E + mc^2} r j_{l-1}(kr). \tag{15.174}$$

In obtaining these results, we have used the relations

$$j'_l(\rho) = \frac{1}{\rho} j_l(\rho) - j_{l+1}(\rho)$$

$$= -\frac{l+1}{\rho} j_l(\rho) + j_{l-1}(\rho) \tag{15.175}$$

satisfied by the spherical Bessel functions j_l.

In summary, for any value of the energy E outside the energy gap $(-mc^2, mc^2)$, there exist free spherical wave solutions of the Dirac equation (15.156). When $\kappa = l + 1$, we have $j = l + 1/2$, and the free spherical wave is given by

$$\psi_{E\kappa m_j} = C \begin{pmatrix} j_l(kr) \mathcal{Y}^{j,m_j}_{l,\frac{1}{2}} \\ -i \frac{\hbar kc}{E + mc^2} j_{l+1}(kr) \mathcal{Y}^{j,m_j}_{l+1,\frac{1}{2}} \end{pmatrix}. \tag{15.176a}$$

where C is a constant. When $\kappa = -l$, we have $j = l - 1/2$, and the free spherical wave is

$$\psi_{E\kappa m_j} = C \begin{pmatrix} j_l(kr) \mathcal{Y}^{j,m_j}_{l,\frac{1}{2}} \\ -i \dfrac{\hbar k c}{E + mc^2} j_{l-1}(kr) \mathcal{Y}^{j,m_j}_{l-1,\frac{1}{2}} \end{pmatrix}. \tag{15.176b}$$

The hydrogenic atom

As a second example, we shall consider the bound states of an electron, of mass m and charge $q = -e$, in the Coulomb field of a nucleus of charge Ze, which we assume to be infinitely heavy. In this case

$$V(r) = -\frac{Ze^2}{(4\pi\varepsilon_0)r} \tag{15.177}$$

and the two coupled equations (15.166) become

$$\left[\frac{d}{dr} - \frac{\kappa}{r}\right] P_{E\kappa}(r) = \left[\frac{mc}{\hbar}\left(1 + \frac{E}{mc^2}\right) + \frac{Z\alpha}{r}\right] Q_{E\kappa}(r), \tag{15.178a}$$

$$\left[\frac{d}{dr} + \frac{\kappa}{r}\right] Q_{E\kappa}(r) = \left[\frac{mc}{\hbar}\left(1 - \frac{E}{mc^2}\right) - \frac{Z\alpha}{r}\right] P_{E\kappa}(r), \tag{15.178b}$$

where $\alpha = e^2/(4\pi\varepsilon_0 \hbar c)$ is the fine structure constant (1.68).

To solve these equations, one can follow a procedure similar to that of Section 7.5. We first put

$$\rho = \nu r \tag{15.179}$$

where

$$\nu = \frac{mc}{\hbar}\left(1 - \frac{E^2}{m^2c^4}\right)^{1/2} \tag{15.180}$$

so that the coupled equations (15.178) can be written as

$$\left[\frac{d}{d\rho} - \frac{\kappa}{\rho}\right] P_{E\kappa}(\rho) = \left[\frac{mc}{\nu\hbar}\left(1 + \frac{E}{mc^2}\right) + \frac{Z\alpha}{\rho}\right] Q_{E\kappa}(\rho) \tag{15.181a}$$

$$\left[\frac{d}{d\rho} + \frac{\kappa}{\rho}\right] Q_{E\kappa}(\rho) = \left[\frac{mc}{\nu\hbar}\left(1 - \frac{E}{mc^2}\right) - \frac{Z\alpha}{\rho}\right] P_{E\kappa}(\rho). \tag{15.181b}$$

Let us first examine the asymptotic behaviour of $P_{E\kappa}(\rho)$ and $Q_{E\kappa}(\rho)$. For large ρ, the equations (15.181) become

$$\frac{d}{d\rho} P_{E\kappa}(\rho) = \left[\frac{mc}{\nu\hbar}\left(1 + \frac{E}{mc^2}\right)\right] Q_{E\kappa}(\rho), \tag{15.182a}$$

$$\frac{d}{d\rho} Q_{E\kappa}(\rho) = \left[\frac{mc}{\nu\hbar}\left(1 - \frac{E}{mc^2}\right)\right] P_{E\kappa}(\rho). \tag{15.182b}$$

15.5 Solutions of the Dirac equation for a central potential

Since we are interested in bound states, we require that both $P_{E\kappa}(\rho)$ and $Q_{E\kappa}(\rho)$ vanish at infinity. In this case, the asymptotic equations (15.182) are satisfied by taking (Problem 15.9)

$$P_{E\kappa}(\rho) = a_1 e^{-\rho}, \tag{15.183a}$$

$$Q_{E\kappa}(\rho) = a_2 e^{-\rho} \tag{15.183b}$$

where the ratio a_1/a_2 is given by

$$\frac{a_1}{a_2} = -\left(\frac{1+E/mc^2}{1-E/mc^2}\right)^{1/2}. \tag{15.184}$$

We can therefore look for solutions of the coupled equations (15.181) having the form

$$P_{E\kappa}(\rho) = N\left(1 + \frac{E}{mc^2}\right)^{1/2} e^{-\rho} f(\rho) \tag{15.185a}$$

$$Q_{E\kappa}(\rho) = -N\left(1 - \frac{E}{mc^2}\right)^{1/2} e^{-\rho} g(\rho) \tag{15.185b}$$

where N is a normalisation constant and the functions $f(\rho)$ and $g(\rho)$ are required to tend to unity when $\rho \to \infty$. We now write series expansions for $f(\rho)$ and $g(\rho)$ in the form

$$f(\rho) = \rho^s \sum_{k=0}^{\infty} c_k \rho^k, \qquad c_0 \neq 0, \tag{15.186a}$$

$$g(\rho) = \rho^s \sum_{k=0}^{\infty} d_k \rho^k, \qquad d_0 \neq 0. \tag{15.186b}$$

Substituting the expressions (15.185) and (15.186) into the equations (15.181) and equating the coefficients of each power of ρ, one obtains a sequence of equations, the first of which yields for s the relation

$$s = \pm(\kappa^2 - Z^2\alpha^2)^{1/2}. \tag{15.187}$$

In order that the radial functions $P_{E\kappa}$ and $Q_{E\kappa}$ be regular at the origin (that is, $P_{E\kappa}(0) = Q_{E\kappa}(0) = 0$), we must take the positive square root[2] in (15.187), so that $s = +(\kappa^2 - Z^2\alpha^2)^{1/2}$.

The remaining equations of the sequence provide recurrence relations for the coefficients $c_1, d_1, \ldots c_n, d_n, \ldots$. To satisfy the asymptotic boundary conditions that $P_{E\kappa}(\rho)$ and $Q_{E\kappa}(\rho)$ vanish at infinity, the series (15.186) for $f(\rho)$ and $g(\rho)$ must

---

[2] For this treatment to be valid, we must have $Z^2\alpha^2 < \kappa^2$, which is always the case if $Z < 137$.

terminate. As in the non-relativistic case, it is found that this is only possible for certain values of E, which are the Dirac energy eigenvalues

$$E^D_{nj} = mc^2 \left[1 + \left(\frac{Z\alpha}{n - j - 1/2 + [(j+1/2)^2 - Z^2\alpha^2]^{1/2}} \right)^2 \right]^{-1/2}. \quad (15.188)$$

By expanding this result in powers of $(Z\alpha)^2$, one obtains

$$E^D_{nj} = mc^2 \left[1 - \frac{(Z\alpha)^2}{2n^2} - \frac{(Z\alpha)^4}{2n^4} \left(\frac{n}{j+1/2} - \frac{3}{4} \right) + \cdots \right]. \quad (15.189)$$

Subtracting the rest energy mc^2 from this result, we obtain the energy values

$$E_{nj} = E^D_{nj} - mc^2$$

$$= E^{(0)}_n \left[1 + \frac{(Z\alpha)^2}{n^2} \left(\frac{n}{j+1/2} - \frac{3}{4} \right) + \cdots \right] \quad (15.190)$$

where $E^{(0)}_n$ are the non-relativistic (Schrödinger) energy eigenvalues (7.114) with $\mu = m$. Upon comparison of (15.190) with the expression (8.115) obtained in Section 8.2, we see that the two results agree if terms of higher order in $(Z\alpha)^2$ are neglected in (15.190). As explained in the discussion following equation (8.115), a non-relativistic energy level $E^{(0)}_n$, which depends only on the principal quantum number n, gives rise to *fine structure splitting* into n different Dirac levels E_{nj} (with $j = 1/2, 3/2, \ldots, n - 1/2$) when relativistic effects are taken into account. It should be emphasized that in Dirac theory two levels having the same values of the quantum numbers n and j but with $l = j \pm 1/2$ (such as the $2s_{1/2}$ and $2p_{1/2}$ levels) are still degenerate.

Comparison with experiment

Many spectroscopic studies of the fine structure of atomic hydrogen and hydrogenic ions (in particular He^+) were made to test the Dirac theory, but no definite conclusion had been reached by 1940. Although there was some evidence strongly supporting the theory, the measurements performed by W. V. Houston in 1937 and R. C. Williams in 1938 were interpreted in 1938 by S. Pasternack as indicating that the $2s_{1/2}$ and $2p_{1/2}$ levels did not coincide exactly, but that there existed a slight upward shift of the $2s_{1/2}$ level with respect to the $2p_{1/2}$ level of about 0.03 cm^{-1}. However, the experimental attempts to obtain accurate information about the fine structure of hydrogenic atoms were frustrated by the broadening of the spectral lines, due mainly to the Doppler effect. In fact, other spectroscopists disagreed with the results of Houston and Williams, and found no discrepancy with the Dirac theory.

The question was settled in 1947 by W. E. Lamb and R. C. Retherford. Using microwave techniques to induce a radio-frequency transition between the $2s_{1/2}$ and $2p_{1/2}$ levels, they demonstrated in a decisive way the existence of an energy difference between these two levels, called the *Lamb shift* (see Section 8.2). The need to explain the Lamb shift was a key factor which stimulated the development of quantum electrodynamics (QED), perhaps the most successful of all physical theories.

15.6 Non-relativistic limit of the Dirac equation

In this section we shall investigate the non-relativistic limit of the Dirac theory, restricting our attention to stationary problems for which $\mathbf{A}$ and ϕ are time-independent. Our starting point is therefore the system of two equations (15.58) for stationary states. In order to examine its non-relativistic limit, we write, as in (15.17)

$$E = E' + mc^2. \tag{15.191}$$

Substituting (15.191) into (15.58), we find that the two-component spinors $\psi_A(\mathbf{r})$ and $\psi_B(\mathbf{r})$ satisfy the system of coupled equations

$$q\phi\psi_A + c\boldsymbol{\sigma}\cdot(-i\hbar\boldsymbol{\nabla} - q\mathbf{A})\psi_B = E'\psi_A, \tag{15.192a}$$

$$c\boldsymbol{\sigma}\cdot(-i\hbar\boldsymbol{\nabla} - q\mathbf{A})\psi_A + q\phi\psi_B = (E' + 2mc^2)\psi_B. \tag{15.192b}$$

This pair of equations is still exact. Solving (15.192b) for ψ_B, we obtain

$$\psi_B = \frac{1}{E' + 2mc^2 - q\phi} c\boldsymbol{\sigma}\cdot(-i\hbar\boldsymbol{\nabla} - q\mathbf{A})\psi_A. \tag{15.193}$$

In the non-relativistic limit both $|E'|$ and $|q\phi|$ are small in comparison with mc^2. We may then write approximately

$$\psi_B = \frac{1}{2mc}\boldsymbol{\sigma}\cdot(-i\hbar\boldsymbol{\nabla} - q\mathbf{A})\psi_A \tag{15.194}$$

and we see that ψ_B is smaller than ψ_A by a factor of order p/mc (that is, v/c, where v is the magnitude of the velocity). The two-component spinors ψ_A and ψ_B are known in this case as the *large* and *small components* of the four-component spinor ψ.

The Pauli equation and the magnetic moment of the electron

Substituting (15.194) into (15.192a), we find that

$$\frac{1}{2m}[\boldsymbol{\sigma}\cdot(-i\hbar\boldsymbol{\nabla} - q\mathbf{A})]^2\psi_A + q\phi\psi_A = E'\psi_A. \tag{15.195}$$

Using the identity (6.241) satisfied by the Pauli spin matrices, we can write the first term on the left-hand side of (15.195) as

$$\frac{1}{2m}[\boldsymbol{\sigma}\cdot(-i\hbar\boldsymbol{\nabla} - q\mathbf{A})]^2\psi_A = \frac{1}{2m}(-i\hbar\boldsymbol{\nabla} - q\mathbf{A})^2\psi_A - \frac{q\hbar}{2m}\boldsymbol{\sigma}\cdot(\boldsymbol{\nabla}\times\mathbf{A})\psi_A. \tag{15.196}$$

Now $\boldsymbol{\nabla}\times\mathbf{A} = \mathcal{B}$, where $\mathcal{B}$ is the magnetic field, so that equation (15.195) becomes

$$\left[\frac{1}{2m}(-i\hbar\boldsymbol{\nabla} - q\mathbf{A})^2 - \frac{q\hbar}{2m}(\boldsymbol{\sigma}\cdot\mathcal{B}) + q\phi\right]\psi_A = E'\psi_A. \tag{15.197}$$

This equation is known as the *Pauli equation*. It differs from the corresponding non-relativistic form (15.19) of the Klein–Gordon equation for spinless particles in predicting an interaction between the external magnetic field $\mathcal{B}$ and the Pauli spin operator $\mathbf{S} = (\hbar/2)\boldsymbol{\sigma}$ of the particle. We emphasise that the Pauli equation (15.197) is an equation for a two-component spinor wave function.

To apply the Pauli equation to the case of an electron, we put $q = -e$. We then have

$$\left[\frac{1}{2m}(-i\hbar\boldsymbol{\nabla} + e\mathbf{A})^2 + \frac{e\hbar}{2m}(\boldsymbol{\sigma}.\mathcal{B}) - e\phi\right]\psi_A = E'\psi_A. \tag{15.198}$$

The term $e\hbar(\boldsymbol{\sigma}.\mathcal{B})/2m$ on the left of this equation corresponds to an interaction $-\mathcal{M}_s.\mathcal{B}$ between the magnetic field $\mathcal{B}$ and an intrinsic *magnetic moment* $\mathcal{M}_s$ of the electron, due to its spin, with

$$\mathcal{M}_s = -\frac{e\hbar}{2m}\boldsymbol{\sigma} = -\mu_B\boldsymbol{\sigma} = -\frac{e}{m}\mathbf{S} \tag{15.199}$$

where $\mu_B = e\hbar/2m$ is the Bohr magneton (1.84). We may also write the above equation as

$$\mathcal{M}_s = -g_s\mu_B\mathbf{S}/\hbar = -g_s\frac{e}{2m}\mathbf{S} \tag{15.200}$$

where the spin gyromagnetic ratio g_s has the value $g_s = 2$. We see that the Dirac theory not only predicts the existence of an intrinsic magnetic moment $\mathcal{M}_s$ for the electron, but it also predicts for it the value $\mathcal{M}_s = -(e/m)\mathbf{S}$ (corresponding to $g_s = 2$), in very good agreement with experiment[3].

In the particular case such that the magnetic field $\mathcal{B}$ is uniform, the vector potential $\mathbf{A}$ can be written as $\mathbf{A} = (\mathcal{B} \times \mathbf{r})/2$ and the scalar potential ϕ can be taken to be $\phi = 0$ (see (12.64)). The Pauli equation (15.197) then reduces to equation (12.69), which we studied in Section 12.2.

Higher order corrections for a spin-1/2 particle in a central potential. Fine structure of hydrogenic atoms

We have shown above that to lowest order in v/c, the Dirac theory is equivalent to the two-component Pauli theory. We shall now investigate higher order corrections (of order v^2/c^2) for the case of a spin-1/2 particle of charge q in a central potential $V(r)$.

Our starting point is again the pair of equations (15.192), in which we set $\mathbf{A} = 0$ and $q\phi(r) = V(r)$. We then obtain the system of two coupled equations

$$V\psi_A + c\boldsymbol{\sigma}.(-i\hbar\boldsymbol{\nabla})\psi_B = E'\psi_A, \tag{15.201a}$$

$$c\boldsymbol{\sigma}.(-i\hbar\boldsymbol{\nabla})\psi_A + V\psi_B = (E' + 2mc^2)\psi_B. \tag{15.201b}$$

---

[3] In fact, the electron has an 'anomalous' magnetic moment, which differs slightly from the value $\mathcal{M} = -(e/m)\mathbf{S}$ corresponding to $g_s = 2$. The quantity $a = (g_s - 2)/2$ has been measured with tremendous accuracy by R. S. Van Dyck, Jr, P. B. Schwinberg and H. G. Dehmelt in 1987 to give the value $a = (1.159\,652\,188\,4 \pm 0.000\,000\,004\,3) \times 10^{-3}$. The present theoretical result, calculated by using quantum electrodynamics, is $(1.159\,652\,2 \pm 0.000\,000\,2) \times 10^{-3}$, the main source of inaccuracy being relative uncertainties of the order of 10^{-7} in the value of the fine structure constant α.

15.6 Non-relativistic limit of the Dirac equation

Solving (15.201b) for ψ_B, we have

$$\psi_B = \frac{1}{E' + 2mc^2 - V} c\boldsymbol{\sigma}\cdot(-i\hbar\boldsymbol{\nabla})\psi_A. \tag{15.202}$$

Substituting this expression for ψ_B into (15.201a), we obtain

$$-c^2\hbar^2(\boldsymbol{\sigma}\cdot\boldsymbol{\nabla})\frac{1}{E' + 2mc^2 - V}(\boldsymbol{\sigma}\cdot\boldsymbol{\nabla})\psi_A + V\psi_A = E'\psi_A \tag{15.203}$$

and we note that no approximation has been made so far.

Let us now expand $(E' + 2mc^2 - V)^{-1}$ in powers of $(E' - V)/2mc^2$. Keeping the lowest order terms, we have

$$(E' + 2mc^2 - V)^{-1} \simeq \frac{1}{2mc^2}\left[1 - \frac{E' - V}{2mc^2}\right] \tag{15.204}$$

so that equation (15.203) becomes

$$-\frac{\hbar^2}{2m}\left[1 - \frac{E' - V}{2mc^2}\right](\boldsymbol{\sigma}\cdot\boldsymbol{\nabla})^2\psi_A - \frac{\hbar^2}{4m^2c^2}(\boldsymbol{\sigma}\cdot\boldsymbol{\nabla}V)(\boldsymbol{\sigma}\cdot\boldsymbol{\nabla}\psi_A) + V\psi_A = E'\psi_A. \tag{15.205}$$

Now, using the identity (6.241), we have

$$(\boldsymbol{\sigma}\cdot\boldsymbol{\nabla})^2 = \nabla^2 \tag{15.206}$$

and

$$(\boldsymbol{\sigma}\cdot\boldsymbol{\nabla}V)(\boldsymbol{\sigma}\cdot\boldsymbol{\nabla}\psi_A) = (\boldsymbol{\nabla}V)\cdot(\boldsymbol{\nabla}\psi_A) + i\boldsymbol{\sigma}\cdot[(\boldsymbol{\nabla}V)\times(\boldsymbol{\nabla}\psi_A)]. \tag{15.207}$$

Moreover, since $V(r)$ is spherically symmetric,

$$\boldsymbol{\nabla}V = \frac{dV}{dr}\hat{\mathbf{r}}, \tag{15.208a}$$

$$(\boldsymbol{\nabla}V)\cdot(\boldsymbol{\nabla}\psi_A) = \frac{dV}{dr}\frac{\partial\psi_A}{\partial r} \tag{15.208b}$$

and

$$i\boldsymbol{\sigma}\cdot[(\boldsymbol{\nabla}V)\times(\boldsymbol{\nabla}\psi_A)] = -\frac{2}{\hbar^2}\frac{1}{r}\frac{dV}{dr}\mathbf{L}\cdot\mathbf{S}\,\psi_A \tag{15.208c}$$

where we have used the facts that the orbital angular momentum operator is given by $\mathbf{L} = \mathbf{r}\times(-i\hbar\boldsymbol{\nabla})$, and that the Pauli spin operator is $\mathbf{S} = (\hbar/2)\boldsymbol{\sigma}$.

Using (15.206)–(15.208), we can rewrite the equation (15.205) in the form

$$\left[-\frac{\hbar^2}{2m}\nabla^2 + V(r) + \frac{\hbar^2}{2m}\frac{E' - V(r)}{2mc^2}\nabla^2 + \frac{1}{2m^2c^2}\frac{1}{r}\frac{dV}{dr}\mathbf{L}\cdot\mathbf{S} - \frac{\hbar^2}{4m^2c^2}\frac{dV}{dr}\frac{\partial}{\partial r}\right]\psi_A$$
$$= E'\psi_A. \tag{15.209}$$

The first and the second terms on the left-hand side of this equation constitute the non-relativistic Hamiltonian of a particle of mass m in a central potential $V(r)$. Since $\mathbf{p}_{\mathrm{op}} = -i\hbar\nabla$ and $E' - V(r) \simeq \mathbf{p}_{\mathrm{op}}^2/2m$, the third term can be written as

$$\frac{\hbar^2}{2m}\frac{E' - V(r)}{2mc^2}\nabla^2 \simeq -\frac{\mathbf{p}_{\mathrm{op}}^4}{8m^3c^2} \tag{15.210}$$

and is easily seen to be a *relativistic correction* (of order v^2/c^2) to the non-relativistic *kinetic energy* operator $-(\hbar^2/2m)\nabla^2 \equiv \mathbf{p}_{\mathrm{op}}^2/2m$ (see Problem 8.10). The fourth term is the *spin–orbit interaction*, which is readily shown to be of order v^2/c^2 times the potential energy $V(r)$.

The last term on the left of (15.209) is a relativistic correction (of order v^2/c^2) to the potential energy which gives rise to some difficulty because it is non-Hermitian. The origin of this difficulty is that if the original Dirac four-component spinor $\psi(\mathbf{r})$ is normalised to unity, namely

$$\int \psi^\dagger(\mathbf{r})\psi(\mathbf{r})\mathrm{d}\mathbf{r} = 1, \tag{15.211}$$

then from (15.56) we have

$$\int [\psi_A^\dagger(\mathbf{r})\psi_A(\mathbf{r}) + \psi_B^\dagger(\mathbf{r})\psi_B(\mathbf{r})]\mathrm{d}\mathbf{r} = 1 \tag{15.212}$$

so that the 'large' two-component spinor ψ_A only satisfies approximately the normalisation condition

$$\int \psi_A^\dagger(\mathbf{r})\psi_A(\mathbf{r})\mathrm{d}\mathbf{r} = 1. \tag{15.213}$$

C. G. Darwin has shown that the correct normalisation (15.213) of ψ_A can be obtained by replacing the last term on the left of (15.209) by the symmetrical combination

$$\frac{1}{2}\left[\left(-\frac{\hbar^2}{4m^2c^2}\frac{\mathrm{d}V}{\mathrm{d}r}\frac{\partial}{\partial r}\right) + \left(-\frac{\hbar^2}{4m^2c^2}\frac{\mathrm{d}V}{\mathrm{d}r}\frac{\partial}{\partial r}\right)^\dagger\right]$$

$$= \frac{1}{2}\left[\left(-\frac{\hbar^2}{4m^2c^2}\frac{\mathrm{d}V}{\mathrm{d}r}\frac{\partial}{\partial r}\right) + \left(-\frac{\hbar^2}{4m^2c^2}\frac{\overleftarrow{\partial}}{\partial r}\frac{\mathrm{d}V}{\mathrm{d}r}\right)\right]$$

$$= \frac{\hbar^2}{8m^2c^2}\nabla^2 V(r). \tag{15.214}$$

where the arrow indicates operation on the left. The operator in (15.214) is clearly Hermitian and is called the *Darwin term*.

Using the above results, we find that the wave equation for a spin-1/2 particle in a central potential $V(r)$, including relativistic corrections of order v^2/c^2, is

$$H\psi_A(\mathbf{r}) = E'\psi_A(\mathbf{r}) \tag{15.215}$$

where $\psi_A(\mathbf{r})$ is a two-component spinor and the Hamiltonian is given by

$$H = \frac{\mathbf{p}^2}{2m} + V(r) - \frac{\mathbf{p}^4}{8m^3c^2} + \frac{1}{2m^2c^2}\frac{1}{r}\frac{dV}{dr}\mathbf{L}\cdot\mathbf{S} + \frac{\hbar^2}{8m^2c^2}\nabla^2 V(r) \quad (15.216)$$

where we have omitted the subscript on the momentum operator for notational simplicity.

In particular, for hydrogenic atoms the central potential is the Coulomb potential $V(r) = -Ze^2/(4\pi\varepsilon_0 r)$, and we have

$$\nabla^2 V(r) = 4\pi\left(\frac{Ze^2}{4\pi\varepsilon_0}\right)\delta(\mathbf{r}) \quad (15.217)$$

so that the Darwin term (15.214) is then given by

$$\frac{\hbar^2}{8m^2c^2}\nabla^2 V(r) = \frac{\pi\hbar^2}{2m^2c^2}\left(\frac{Ze^2}{4\pi\varepsilon_0}\right)\delta(\mathbf{r}). \quad (15.218)$$

Since this term acts only at the origin and the hydrogenic wave functions vanish at $r = 0$ when $l \neq 0$, it follows that the Darwin term gives a relativistic shift only to the energy levels of the s states, for which $l = 0$. Thus, the wave equation for one-electron atoms (ions) including relativistic corrections through order v^2/c^2 is given by (15.215), where

$$H = \frac{\mathbf{p}^2}{2m} + V(r) - \frac{\mathbf{p}^4}{8m^3c^2} + \frac{1}{2m^2c^2}\frac{1}{r}\frac{dV}{dr}\mathbf{L}\cdot\mathbf{S} + \frac{\pi\hbar^2}{2m^2c^2}\left(\frac{Ze^2}{4\pi\varepsilon_0}\right)\delta(\mathbf{r}) \quad (15.219)$$

with $V(r) = -Ze^2/(4\pi\varepsilon_0 r)$. The Hamiltonian (15.219) is the starting point of our discussion of the fine structure of hydrogenic atoms in Section 8.2.

15.7 Negative-energy states. Hole theory

We have seen that both the Klein–Gordon and Dirac relativistic wave equations admit negative-energy solutions which are not realised in nature. This gives rise to an apparently fatal difficulty, since an external electromagnetic perturbation will cause a charged particle (for example an electron) with energy greater than its rest mass energy mc^2 to make a radiative transition to a state of negative energy less than mc^2.

In 1930, Dirac proposed a way out of this difficulty for electrons by formulating his *hole theory*. He suggested that all the negative-energy electron states in the universe are full. Since, according to the Pauli exclusion principle, only one electron can occupy each state, transitions from positive to negative energy states are then forbidden, unless there is some mechanism to empty negative-energy states. In this picture, the vacuum consists of an infinite sea of electrons filling all the negative-energy states, and no electrons in positive-energy states.

It is possible for a negative-energy electron to interact with radiation of energy greater than $2mc^2$, thereby making a transition to a state of positive energy larger than mc^2. This transition leaves behind a *hole* in the sea of negative-energy electrons.

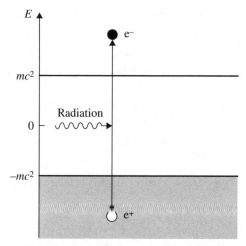

Figure 15.2 Illustration of electron–positron pair production. A negative-energy electron e⁻ interacts with radiation and makes a transition to a positive energy state, leaving behind a hole (a positron e⁺) in the sea of negative-energy states.

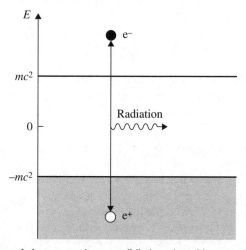

Figure 15.3 Illustration of electron–positron annihilation. A positive-energy electron e⁻ annihilates with a hole (a positron e⁺) in the sea of negative-energy states, emitting radiation.

Measured with respect to the vacuum, this hole appears to have positive charge (+e) and positive energy; it can therefore be interpreted as a *positron*. It is in this way that Dirac predicted the existence of positrons, which were subsequently discovered in 1933 by C. D. Anderson. The reaction in which radiation of energy greater than $2mc^2$ produces a positive-energy electron e⁻ together with a positive-energy positron e⁺ is called *pair production*; it is illustrated in Fig. 15.2. The reverse process, in which a positive-energy electron e⁻ falls into a negative-energy hole, emitting radiation, corresponds to electron–positron *annihilation* (see Fig. 15.3). Both electron–positron pair production and annihilation have been observed and studied.

The Dirac hole theory is asymmetric with respect to electrons and positrons. However, one can equally start from a Dirac equation for positive-energy positrons. The vacuum would in this case consist of an infinite sea of filled negative-energy positron states, with no positron in positive energy states. A hole in the sea of negative-energy positron states would then correspond to a positive-energy electron.

In spite of its success in predicting the existence of positrons, the hole theory leaves many questions unanswered. In particular, no account is taken of the interactions of the particles in the sea of negative-energy states. In fact, the hole theory is no longer the one-particle theory that we started with, as it involves an infinite number of particles. It has been superseded by quantum field theory, in which the Klein–Gordon and Dirac equations are interpreted as field equations.

Problems

15.1 Starting from the Klein–Gordon equation (15.5), obtain the continuity equation (15.22), where $P(\mathbf{r}, t)$ is given by (15.23) and $\mathbf{j}(\mathbf{r}, t)$ by (15.24).

15.2 Using the phase transformation (15.9), show that the expression of $P(\mathbf{r}, t)$ given by (15.23) reduces to $|\Psi(\mathbf{r}, t)|^2$ in the non-relativistic limit.

15.3 Using the radial Klein–Gordon equation (15.16), find the energy levels for a spinless particle of mass m and charge q moving in the Coulomb field of an infinitely heavy nucleus of charge Ze.

(*Hint*: Define the quantities

$$\beta = \frac{2}{\hbar c}(m^2 c^4 - E^2)^{1/2},$$

$$\gamma = \frac{2Z\alpha E}{\hbar c \beta},$$

$$\rho = \beta r$$

and

$$l'(l' + 1) = l(l + 1) - Z^2 \alpha^2$$

where $\alpha = e^2/(4\pi\varepsilon_0 \hbar c)$ is the fine structure constant. Then, show that the radial Klein–Gordon equation (15.16) can be written in the form of equation (7.98), namely

$$\left[\frac{d^2}{d\rho^2} - \frac{l'(l'+1)}{\rho^2} + \frac{\gamma}{\rho} - \frac{1}{4}\right] u_{El}(\rho) = 0$$

where $u_{El}(\rho) = \rho R_{El}(\rho)$.)

15.4 Prove that the matrices α_k and β given by (15.43) in the Dirac representation satisfy the relations (15.35).

15.5 Using the properties of the γ matrices and the fact that $\varepsilon_{\mu\nu} = -\varepsilon_{\nu\mu}$, prove that equation (15.96) is satisfied to first order in $\varepsilon_{\mu\nu}$ provided that Λ and Λ^{-1} are given by (15.97a) and (15.97b), respectively.

15.6 Let H be the Dirac Hamiltonian (15.30) for a free particle of spin 1/2 and $\mathbf{S} = (\hbar/2)\mathbf{\Sigma}$ be the Dirac spin operator (15.123). Prove that

$$[H, \mathbf{S}] = i\hbar c(\boldsymbol{\alpha} \times \mathbf{p}_{\text{op}}).$$

15.7 (a) Using the normalisation condition (15.134), obtain the expression (15.135) for the normalisation constant N.

(b) Show that if the normalisation condition (15.140) is adopted, then the right-hand side of (15.138) is a Lorentz invariant.

15.8 Let H be the Dirac Hamiltonian (15.141) for a spin-1/2 particle in a central potential $V(r)$. Prove that

$$[H, \mathbf{L}] = -i\hbar c(\boldsymbol{\alpha} \times \mathbf{p}_{\text{op}})$$

where $\mathbf{L} = \mathbf{r} \times \mathbf{p}_{\text{op}}$ is the orbital angular momentum operator.

15.9 Show that the asymptotic coupled equations (15.182) are satisfied by taking

$$P_{E\kappa}(\rho) = a_1 e^{-\rho}, \qquad Q_{E\kappa}(\rho) = a_2 e^{-\rho}$$

where the ratio a_1/a_2 is given by (15.184).

15.10 Starting from the Dirac equation (15.79) for a spin-1/2 particle of mass m and charge q in an electromagnetic field, one can write the corresponding equation for a particle of the same mass, but opposite charge $-q$, namely

$$\left[\gamma_\mu\left(\frac{\partial}{\partial x_\mu} + \frac{iq}{\hbar}A_\mu\right) + \frac{mc}{\hbar}\right]\Psi_c = 0,$$

where Ψ_c is called the charge conjugate wave function. Show that

$$\Psi_c = \gamma_2 \Psi^* = \gamma_2 K \Psi$$

where K is the operator of complex conjugation introduced in Section 5.10. The operator $C = \gamma_2 K$ is called the *charge conjugation operator* for particles of spin 1/2.

16 Further applications of quantum mechanics

16.1 The van der Waals interaction 719
16.2 Electrons in solids 723
16.3 Masers and lasers 735
16.4 The decay of K-mesons 746
16.5 Positronium and charmonium 753

In this chapter a number of interesting applications of quantum mechanics, chosen to illustrate the wide variety of problems that can be solved, are discussed. The first concerns the long-range interaction between neutral atoms, known as the van der Waals interaction, which is an example of the use of second-order perturbation theory. This is followed by a short account of the properties of electron motion in crystals drawing on the theory of motion in a periodic potential introduced in Chapter 4. Next we discuss some features of masers and lasers, which illustrate the application of stimulated emission studied in Chapter 11. The chapter concludes with two examples drawn from elementary particle physics. The first, a direct illustration of the superposition principle, is concerned with the properties of neutral K-mesons, and the second shows how the mass spectrum of a class of mesons can be analysed in terms of a bound 'charmed' quark and anti-quark system.

16.1 The van der Waals interaction

In this section we shall discuss the forces that act between two neutral atoms. For simplicity, we take the example of two hydrogen atoms, the first composed of a proton situated at A and electron 1, and the second of a proton situated at B and electron 2. The coordinate system is shown in Fig. 16.1. The internuclear distance AB is R, and is taken as the z-axis. The distances of electron 1 from A and B are denoted by r_{1A}, r_{1B} and, similarly, r_{2A} and r_{2B} denote the distances of electron 2 from A and B, respectively. The distance between the two electrons is r_{12}. The Hamiltonian for the system in the adiabatic approximation, in which the internuclear distance R is fixed, is

$$H = H_A + H_B + V(1, 2) \tag{16.1}$$

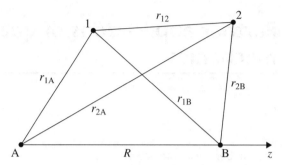

Figure 16.1 A coordinate system for calculating the long-range interaction between two hydrogen atoms.

where H_A is the Hamiltonian of the atom $(A + e_1^-)$. H_B is that for the atom $(B + e_2^-)$ and $V(1, 2)$ is the interaction energy between the two atoms. We have

$$H_A = -\frac{\hbar^2}{2m}\nabla^2_{r_{1A}} - \left(\frac{e^2}{4\pi\varepsilon_0}\right)\frac{1}{r_{1A}} \tag{16.2a}$$

$$H_B = -\frac{\hbar^2}{2m}\nabla^2_{r_{2B}} - \left(\frac{e^2}{4\pi\varepsilon_0}\right)\frac{1}{r_{2B}} \tag{16.2b}$$

and

$$V(1, 2) = \left(\frac{e^2}{4\pi\varepsilon_0}\right)\left(\frac{1}{R} + \frac{1}{r_{12}} - \frac{1}{r_{1B}} - \frac{1}{r_{2A}}\right). \tag{16.2c}$$

Two remarks should be made. First, because we are interested in large distances of separation R, $V(1, 2)$ is small and can be treated as a perturbation. Second, since exchange effects arising from the identity of the two electrons are of short range, we can, for large R, ignore the antisymmetry of the wave function. Although the long-range forces are important for scattering, the binding forces considered in Chapter 10 are greater in magnitude by a factor of 10^3, which is why, even for states of large vibrational excitation, the long-range forces play no significant role in the bound-state problem.

Let the Cartesian coordinates of electron 1 with respect to A as origin be x_{1A}, y_{1A}, z_{1A}, and of electron 2 with respect to B as origin be x_{2B}, y_{2B}, z_{2B}. Then, as we are taking the z-direction to be along AB,

$$r_{12} = [(x_{2B} - x_{1A})^2 + (y_{2B} - y_{1A})^2 + (z_{2B} - z_{1A} + R)^2]^{1/2}$$
$$r_{1B} = [x_{1A}^2 + y_{1A}^2 + (z_{1A} - R)^2]^{1/2}$$
$$r_{2A} = [x_{2B}^2 + y_{2B}^2 + (z_{2B} + R)^2]^{1/2}. \tag{16.3}$$

Each of the terms $1/r_{12}$, $1/r_{1B}$ and $1/r_{2A}$ can be expanded in a Taylor series in powers of $1/R$. Thus

$$\begin{aligned}\frac{1}{r_{1B}} &= \frac{1}{R}\left[1 + \left\{\frac{x_{1A}^2 + y_{1A}^2 + z_{1A}^2 - 2Rz_{1A}}{R^2}\right\}\right]^{-1/2} \\ &= \frac{1}{R}\left[1 - \frac{1}{2}\left\{\frac{x_{1A}^2 + y_{1A}^2 + z_{1A}^2 - 2Rz_{1A}}{R^2}\right\}\right. \\ &\quad \left. + \frac{3}{8}\left\{\frac{x_{1A}^2 + y_{1A}^2 + z_{1A}^2 - 2Rz_{1A}}{R^2}\right\}^2 - \cdots\right] \\ &= \frac{1}{R}\left[1 + \frac{z_{1A}}{R} - \frac{x_{1A}^2 + y_{1A}^2 - 2z_{1A}^2}{2R^2} + \cdots\right]. \end{aligned}$$ (16.4)

In the same way, we have

$$\frac{1}{r_{2A}} = \frac{1}{R}\left[1 - \frac{z_{2B}}{R} - \frac{x_{2B}^2 + y_{2B}^2 - 2z_{2B}^2}{2R^2} + \cdots\right]$$ (16.5)

and

$$\frac{1}{r_{12}} = \frac{1}{R}\left[1 - \frac{z_{2B} - z_{1A}}{R} - \frac{(x_{2B} - x_{1A})^2 + (y_{2B} - y_{1A})^2 - 2(z_{2B} - z_{1A})^2}{2R^2} + \cdots\right].$$ (16.6)

The expansion of $V(1, 2)$ in powers of $1/R$ is obtained by combining (16.2c) with (16.4), (16.5) and (16.6). The terms in $1/R$ and $1/R^2$ are seen to cancel, so that for large R

$$V(1, 2) = \left(\frac{e^2}{4\pi\varepsilon_0}\right)\frac{1}{R^3}(x_{1A}x_{2B} + y_{1A}y_{2B} - 2z_{1A}z_{2B}).$$ (16.7)

This long-range interaction has the form of the interaction energy of two dipoles, the first composed of proton A and electron 1, with electric dipole moment $\mathbf{D}_1 = -e\mathbf{r}_{1A}$, and the second of proton B and electron 2 with electric dipole moment $\mathbf{D}_2 = -e\mathbf{r}_{2B}$. The energy of interaction is

$$V = \frac{1}{(4\pi\varepsilon_0)R^3}\left\{\mathbf{D}_1.\mathbf{D}_2 - 3\frac{(\mathbf{D}_1.\mathbf{R})(\mathbf{D}_2.\mathbf{R})}{R^2}\right\}$$ (16.8)

which on taking the z-axis along $\mathbf{R}$, reduces to (16.7).

The perturbation energy

The energy of the unperturbed system in which both hydrogen atoms are in the ground state is $2E_{1s}$, and the unperturbed wave function is $\psi_{1s}(r_{1A})\psi_{1s}(r_{2B})$, which satisfies the equation

$$(H_A + H_B - 2E_{1s})\psi_{1s}(r_{1A})\psi_{1s}(r_{2B}) = 0.$$ (16.9)

The total energy of the system, allowing for the perturbation $V(1, 2)$, can be written as

$$E = 2E_{1s} + E^{(1)} + E^{(2)} + \cdots, \tag{16.10}$$

where $E^{(n)}$ is calculated from nth-order perturbation theory, and $V(1, 2)$ is given by the expression (16.7).

The first-order correction

$$E^{(1)} = \int \psi_{1s}^*(r_{1A})\psi_{1s}^*(r_{2B})V(1, 2)\psi_{1s}(r_{1A})\psi_{1s}(r_{2B})d\mathbf{r}_{1A}d\mathbf{r}_{2B} \tag{16.11}$$

is immediately seen to vanish, because the matrix elements of the angular-dependent terms, such as x_{1A}, y_{1A}, z_{1A}, are zero when taken between spherically symmetric wave functions such as $\psi_{1s}(r_{1A})$. The first non-vanishing perturbative correction is $E^{(2)}$, which is given by (see (8.28))

$$E^{(2)} = \sum_{j \neq 0} \frac{\langle \psi_0|V|\psi_j\rangle\langle \psi_j|V|\psi_0\rangle}{E_0 - E_j} \tag{16.12}$$

where $\psi_0 = \psi_{1s}(r_{1A})\psi_{1s}(r_{2B})$, ψ_j are the wave functions of the intermediate states, and the summation runs over all discrete and continuous intermediate states, excluding the ground state. In our case the wave functions ψ_j are products of hydrogen wave functions centred on protons A and B. The energy denominator $E_0 - E_j$ is always negative. The numerator is positive and behaves like $1/R^6$, so that the long-range interaction between two hydrogen atoms is

$$E^{(2)}(R) = -\frac{C_W}{R^6} \tag{16.13}$$

where C_W is a positive constant known as the *van der Waals constant*. The same procedure can be carried through for any pair of neutral atoms, and (16.13) may be shown to give the general form of the long-range (van der Waals) interaction, although of course the quantity C_W varies from system to system.

The long-range force is always attractive, but when R becomes small, the force becomes repulsive in character, as we saw in Chapter 10. This has suggested the introduction of emperical potentials to describe atom–atom scattering. One of the most widely used of these is the Lennard-Jones potential, which has the form

$$V(R) = C\left[\frac{1}{2}\left(\frac{R_0}{R}\right)^{12} - \left(\frac{R_0}{R}\right)^6\right] \tag{16.14}$$

where C and R_0 are constants. The constant C can be related to the van der Waals constant C_W, but both C and R_0 are usually treated as empirical constants to be determined from the data on atom–atom scattering.

16.2 Electrons in solids

Solids can be divided into those that are *crystalline* and those that are *amorphous*. The subject of this section are the crystalline solids, which are distinguished by the fact that the atomic nuclei of the constituent atoms occupy equilibrium positions forming a regular lattice, which possesses a definite periodicity and symmetry. The inner shell electrons remain localised about the nuclei situated at the *lattice sites*, but the outer (valence) electrons are not localised and move freely throughout the crystal. This arises because the spacing between the ions, formed by the nuclei and the inner electrons, is comparable with or smaller than the extent of the wave functions of the valence electrons in the corresponding isolated neutral atoms.

The Born–Oppenheimer approximation can be applied, as in the theory of molecules (see Section 10.6). In this approximation, the motion of the valence electrons is determined by supposing that the positive ions formed by the nuclei and inner shell electrons remain completely fixed at the lattice sites. In fact, these ions can vibrate about the equilibrium positions, but as in the case of molecules this motion can be treated to a good approximation independently from that of the valence electrons. This idealised situation was studied in Section 4.8 for a one-dimensional array of atoms forming a linear crystal. If such a crystal is formed of N atoms with one valence electron per atom, each valence electron moves independently in an effective field due to all the other electrons and the nuclei. The Coulomb interaction between two valence electrons is of the same order of magnitude as the interaction between a valence electron and an ion fixed at a lattice site. However, the electron–electron interactions tend to give rise to a nearly constant effective potential. This constant potential is superimposed on the periodic potential due to the ions situated at the lattice sites.

As we saw in Chapter 4, each energy level of the isolated atoms gives rise to a band of energies when the atoms are brought together to form a crystal. The simplest case is that in which all the atoms are of one species, with one valence electron per atom. If the ground state is an S state (with zero orbital angular momentum) then the lowest energy level of a system of N isolated atoms is N-fold degenerate. On bringing these atoms together the interactions remove the degeneracy and the energy splits into N separate levels. Since N is very large this *energy band* is quasi-continuous. Higher energy levels of the isolated atoms give rise to further energy bands of the crystal in a similar way. In a one-dimensional crystal (see Section 4.8), we showed that the different energy bands do not overlap and are separated by energy gaps. This is not necessarily the case for three-dimensional crystals because the periodicity of the lattice structure may be different in different directions, and the energy bands overlap in some cases and not in others. In view of the Pauli principle, each of the N levels in a band can be occupied by up to two electrons, so that in the monovalent example we have been discussing, only half the available energy levels of the lowest band (the ground state) are full. On the other hand, if the atoms contribute two valence electrons per atom, then the lowest band is completely full. We shall use these ideas to explain qualitatively why some materials have much greater electrical resistivity than others.

Resistivity

The resistivity of materials varies over an enormous range from 10^{-8} Ω m for a good conductor to 10^{18} Ω m for a very good insulator. This can be explained in terms of the band structure and the number of valence electrons of the material concerned. As we have just seen, for a material such as sodium, which is monovalent with an atomic ground state of zero orbital angular momentum, there are N valence electrons in a crystal containing N atoms, and in the ground state these occupy only one-half of the N available levels. This is called the *valence band*. There are many unoccupied levels close to those which are occupied, so that the electrons can easily be excited from one level to another and under the influence of an external electric field the electrons can move freely. Such materials are good conductors. With an odd number, (greater than one) of valence electrons per atom in the ground state, the lowest band (or bands) will be full and the next highest half occupied, so that such a material is also a good conductor. In contrast, if the material has an even number of valence electrons per atom, the *valence band* is completely occupied. The resistivity of the material now depends on whether the next highest band overlaps the valence band or not. If the bands overlap, the electrons are easily excited into vacant levels of the higher band, called the *conduction band*, and the material is a conductor. Examples are calcium, magnesium and the other alkaline earths. If there is a large energy gap, the electrons cannot be excited easily and the material is an insulator. For example, diamond is an insulator with an energy gap of 5.33 eV and a resistivity of 10^{18} Ω m. In an intermediate situation such that the energy gap is small, some electrons can be excited thermally to the higher band and the material is a *semiconductor*. For instance silicon, with an energy gap of 1.14 eV is a semiconductor. The resistivity of a semiconductor is very temperature-dependent. In particular, silicon at very low temperature is a good insulator, but at room temperature it has a resistivity of about 10^3 Ω m. Germanium, another semiconductor, is also a good insulator at very low temperature, but at room temperature is a much better conductor than silicon and has a resistivity of ~ 0.4 Ω m because the energy gap (0.63 eV) is much smaller. Apart from thermal energy, radiative energy can be used to change the resistivity of a semiconductor dramatically, since electrons can be excited into the conduction band by the absorption of a photon.

The size of the energy gap between bands can be altered significantly by the presence of impurities in a crystalline material. The deliberate addition of impurities to obtain some desired characteristic is known as *doping* and is a crucial process in the manufacture of transistors and other electronic devices.

Electron dynamics

We shall now discuss the motion of an electron through a crystal lattice, starting with the one-dimensional model of a crystal discussed in Section 4.8. There we saw that an electron moving in the periodic potential due to the ions situated at the lattice sites could be described by a Bloch wave function of the form (4.199):

$$\psi_K(x) = e^{iKx} u_K(x) \tag{16.15}$$

where $u_K(x)$ is a function with the periodicity of the lattice

$$u_K(x+l) = u_K(x). \tag{16.16}$$

The Schrödinger equation satisfied by $\psi_K(x)$ is

$$\left(-\frac{\hbar^2}{2m}\frac{d^2}{dx^2} + V(x)\right)\psi_K(x) = E(K)\psi_K(x). \tag{16.17}$$

If the periodic boundary conditions (4.210) are imposed, the only allowed values of K are those given by (4.212), namely

$$K = \frac{2\pi n}{Nl}, \qquad n = 0, \pm 1, \pm 2, \ldots \tag{16.18}$$

where N is the number of atoms in the linear crystal. Since the values of K are discrete, the wave functions $\psi_K(x)$ can be normalised to unity and the wave functions corresponding to different values of K are orthogonal. Thus

$$\int_L \psi_{K'}^*(x)\psi_K(x)\,dx = \delta_{K'K} \tag{16.19}$$

where the symbol L means that the integration is over the length of the crystal. Let us now calculate the average velocity $\langle v_x \rangle$ of an electron described by $\psi_K(x)$. It is equal to the expectation value of the operator p_x/m so that

$$\begin{aligned}\langle v_x\rangle &= \frac{1}{m}\langle p_x\rangle = \frac{1}{m}\int_L \psi_K^*(x)\left(-i\hbar\frac{d}{dx}\right)\psi_K(x)\,dx\\ &= \frac{1}{m}\int_L u_K^*(x)\left(-i\hbar\frac{d}{dx} + \hbar K\right)u_K(x)\,dx\end{aligned} \tag{16.20}$$

where in the last line we have used (16.15).

To obtain information about the effect of the operator $(-i\hbar d/dx + \hbar K)$ in (16.20), we note from (16.15) and (16.17) that $u_K(x)$ satisfies the equation

$$H(K)u_K(x) = E(K)u_K(x) \tag{16.21a}$$

where

$$H(K) = \frac{1}{2m}\left(-i\hbar\frac{d}{dx} + \hbar K\right)^2 + V(x) \tag{16.21b}$$

and in view of (16.19)

$$\int_L u_{K'}^*(x)u_K(x)\,dx = \delta_{K'K}. \tag{16.22}$$

To proceed further we require a result known as the *Feynman–Hellmann theorem*. Consider a Hamiltonian $H(\lambda)$ which depends on a parameter λ, such that

$$H(\lambda)\phi_\lambda = E(\lambda)\phi_\lambda \qquad (16.23)$$

where ϕ_λ is the normalised eigenfunction corresponding to the eigenenergy $E(\lambda)$. Under an infinitesimal change in λ so that $\lambda \to \lambda + d\lambda$, the change in the Hamiltonian is equal to $(\partial H(\lambda)/\partial \lambda) d\lambda$. From first-order perturbation theory (which is exact since the change in H is infinitesimal), the corresponding change in $E(\lambda)$ is

$$\frac{dE(\lambda)}{d\lambda} d\lambda = \left\langle \phi_\lambda \left| \frac{\partial H(\lambda)}{\partial \lambda} d\lambda \right| \phi_\lambda \right\rangle. \qquad (16.24)$$

Dividing throughout by $d\lambda$, we obtain the Feynman–Hellmann theorem which states that $dE(\lambda)/d\lambda$ is equal to the expectation value of $\partial H(\lambda)/\partial \lambda$:

$$\frac{dE(\lambda)}{d\lambda} = \left\langle \phi_\lambda \left| \frac{\partial H(\lambda)}{\partial \lambda} \right| \phi_\lambda \right\rangle. \qquad (16.25)$$

This result can now be applied to the Hamiltonian (16.21b) by taking K as the parameter λ. It is possible to treat K as a continuous variable in this way since as N is very large, being of the order of 10^{23}, equation (16.18) shows that the discrete values of K are very closely spaced. Thus we have

$$\frac{dE(K)}{dK} = \frac{\hbar}{m} \int_L u_K^*(x) \left(-i\hbar \frac{d}{dx} + \hbar K \right) u_K(x) dx. \qquad (16.26)$$

Comparing (16.26) with (16.20), we obtain the result

$$\langle v_x \rangle = \frac{1}{\hbar} \frac{dE(K)}{dK}. \qquad (16.27)$$

The dependence of $E(K)$ on K can be plotted as a *dispersion curve*. Recalling that the energy is unchanged if K is changed by a multiple of $2\pi/l$, we can adopt the *reduced-zone* scheme whereby K is confined to the interval $-\pi/l$ to π/l. This zone is known as the first Brillouin zone, the second Brillouin zone covers the adjacent intervals $\pi/l \leqslant K \leqslant 2\pi/l$ *and* $-2\pi/l \leqslant K \leqslant -\pi/l$, and so on. Typical dispersion curves in this scheme are shown in Fig. 16.2, where $E_1(K), E_2(K), \ldots$, represent the lowest energy bands. Note the energy gaps between the bands, the importance of which to the electrical properties of the material has already been emphasised.

For a free particle $E(K)$ is parabolic, $E(K) = \hbar^2 K^2/2m$ so that $\langle v_x \rangle = \hbar K/m$. For small K, the band energy $E_1(K)$ is approximately quadratic in K, but near the zone boundary, at $K = \pm \pi/l$, the behaviour is quite different, producing the gap between the energy bands $E_1(K)$ and $E_2(K)$.

In a three-dimensional crystal, the Bloch wave functions can be written as

$$\psi_{\mathbf{K}}(\mathbf{r}) = e^{i\mathbf{K}\cdot\mathbf{r}} u_{\mathbf{K}}(\mathbf{r}) \qquad (16.28)$$

where $u_{\mathbf{K}}(\mathbf{r})$ is periodic in the lattice, that is

$$u_{\mathbf{K}}(\mathbf{r} + \mathbf{R}_j) = u_{\mathbf{K}}(\mathbf{r}) \qquad (16.29)$$

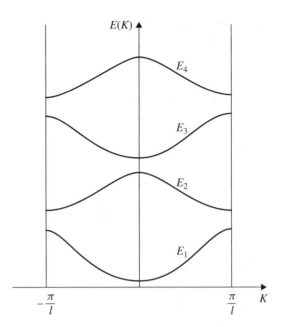

Figure 16.2 Dispersion curves for a one-dimensional crystal in the reduced zone scheme.

and $\mathbf{R}_j$ is a vector joining the origin, situated at a lattice site, to any other lattice site j. The energy bands are now functions of the three components K_x, K_y and K_z of $\mathbf{K}$. As, in general, the periodicity of a three-dimensional crystal is different in different directions, the dispersion plot of $E(K_x, K_y, K_z)$ against K_x, for example, is not necessarily the same as the plot against K_y or K_z. It is for this reason that in the three-dimensional case energy bands can overlap. Despite the difference, the average velocity can be computed as for the one-dimensional case, and it is found that

$$\langle \mathbf{v} \rangle = \frac{1}{\hbar} \nabla_{\mathbf{K}} E(\mathbf{K}). \tag{16.30}$$

It is illuminating to examine the problem of the electron velocity in a periodic lattice from another point of view. Returning to the one-dimensional case, the time-dependent Bloch wave functions can be written as

$$\Psi_K(x, t) = u_K(x) \exp[\mathrm{i}(Kx - E(K)t/\hbar)]. \tag{16.31}$$

By superposing solutions belonging to different values of K, a normalised wave packet $\Psi(x, t)$ can be constructed which represents a localised electron moving along the one-dimensional crystal

$$\Psi(x, t) = \int_{-\pi/l}^{\pi/l} a(K') \Psi_{K'}(x, t) \mathrm{d}K' \tag{16.32}$$

where K' has again been treated as a continuous variable, and the integration is over the reduced zone.

Let us take $a(K')$ to be sharply peaked about $K' = K$. In this case, expanding the exponent in $\Psi_{K'}(x,t)$ as

$$i(K'x - E(K')t/\hbar) \simeq i(Kx - E(K)t/\hbar) + i(K'-K)\left\{x - \left[\frac{dE(K')}{dK'}\right]_{K'=K} t/\hbar\right\} \quad (16.33)$$

we find that (16.32) can be written approximately as

$$\Psi(x,t) = \exp[i(Kx - E(K)t/\hbar)] \int_{-\pi/l}^{\pi/l} a(K')u_{K'}(x)$$

$$\times \exp\left[i(K' - K)\left\{x - \left[\frac{dE(K')}{dK'}\right]_{K'=K} t/\hbar\right\}\right] dK'. \quad (16.34)$$

The factor $\exp[i(Kx - E(K)t/\hbar)]$ has no effect on the probability density $|\Psi(x,t)|^2$, which depends entirely on the integral in (16.34). The value of this integral is small because of the oscillatory exponential factor in the integrand, except for values of x and t connected by the relation

$$x = \left[\frac{dE(K')}{dK'}\right]_{K'=K} t/\hbar. \quad (16.35)$$

This condition determines the motion of the centre of the wave packet, and the group velocity v_g of the packet is given by

$$v_g = \frac{1}{\hbar}\left[\frac{dE(K')}{dK'}\right]_{K'=K} \quad (16.36)$$

which agrees with (16.27). In a three-dimensional crystal the group velocity becomes

$$\mathbf{v}_g = \frac{1}{\hbar}[\nabla_{\mathbf{K}'} E(\mathbf{K}')]_{\mathbf{K}'=\mathbf{K}} \quad (16.37)$$

in agreement with (16.30).

Since it has been assumed that $a(K')$ is peaked about $K' = K$, the uncertainty principle shows that the wave packet is spread over an appreciable range of x, covering several cells of the crystal. It is only under this circumstance that v_g is constant in time and coincides with the expectation value $\langle v_x \rangle$.

External electric fields

Now we shall consider what happens if a static electric field of strength $\mathcal{E}$ is applied to a one-dimensional crystal. Provided the field varies slowly over distances comparable to the lattice interval l, by Ehrenfest's theorem (see Section 3.4) the motion of the wave packet representing an electron is classical. The additional potential energy due to the interaction of the electric field with the charge $(-e)$ of an electron is

$$W(x) = e\mathcal{E}x. \quad (16.38)$$

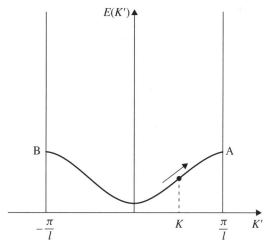

Figure 16.3 The motion of a point on a dispersion curve for a one-dimensional crystal in the presence of an electric field in the negative x-direction.

Let us first consider the effect of this potential when the first Brillouin zone contains only one electron, so that there are no restrictions arising from the Pauli exclusion principle. In this case, the rate of change of the energy of the electron must equal the rate of work done by the field, so that, classically,

$$\frac{dE(K)}{dt} = -\frac{dW(x)}{dt} = -e\mathcal{E}v_g. \tag{16.39}$$

Since

$$\frac{dE(K)}{dt} = \frac{dE(K)}{dK}\frac{dK}{dt} \tag{16.40}$$

we see from (16.36), (16.39) and (16.40) that

$$\hbar\frac{dK}{dt} = -e\mathcal{E}. \tag{16.41}$$

If the electrical field is in the negative direction, $dK/dt > 0$ and K increases steadily with time, moving along the dispersion curve (see Fig. 16.3) until the maximum value π/l is reached at point A. As the state for $K' = +\pi/l$ coincides with that for $K' = -\pi/l$ the point representing K appears at B and again moves steadily towards A. Recollecting that $E(K) = E(-K)$ and that $v_g(K) = -v_g(-K)$, we see that the electron oscillates back and forth along the x-axis.

The effective mass

In discussing the motion of an electron in an external field a useful concept is that of effective mass, which can be introduced as follows. The acceleration of the electrons

$\mathrm{d}v_\mathrm{g}/\mathrm{d}t$ can be expressed, using (16.36) as

$$a = \mathrm{d}v_\mathrm{g}/\mathrm{d}t = \frac{1}{\hbar}\left[\frac{\mathrm{d}^2 E(K')}{\mathrm{d}K'^2}\right]_{K'=K}\frac{\mathrm{d}K}{\mathrm{d}t}$$

$$= \frac{1}{\hbar^2}\left[\frac{\mathrm{d}^2 E(K')}{\mathrm{d}K'^2}\right]_{K'=K}(-e\mathcal{E}) \qquad (16.42)$$

where in the second line (16.41) has been employed. Since for a free electron moving in a field $\mathcal{E}$

$$ma = -e\mathcal{E} \qquad (16.43)$$

it is natural to rewrite (16.42) in the form

$$m^* a = -e\mathcal{E} \qquad (16.44)$$

where

$$m^* = \hbar^2 \left[\frac{\mathrm{d}^2 E(K')}{\mathrm{d}K'^2}\right]^{-1}_{K'=K} \qquad (16.45)$$

is known as the *effective mass*. Near the energy minimum at the bottom of an energy band m^* is positive, but in the upper half of the energy band m^* is negative and the electron behaves like a positively charged particle.

Electric current

Let $P(K)$ be the probability that the state of energy $E(K)$ is occupied, then the electric current carried by the electrons in an energy band is

$$I = -e \int_{-\pi/l}^{\pi/l} P(K) v_\mathrm{g}(K) \mathrm{d}K. \qquad (16.46)$$

If a band is completely full so that $P(K) = 1$ for each value of K, the current vanishes since $v_\mathrm{g}(K) = -v_\mathrm{g}(-K)$. If a band is not completely full but is filled symmetrically, as in Fig. 16.4, the current is likewise zero. However, under the influence of an electric field, the points representing the states on the dispersion curve all move in a direction opposite to the direction of the field, so that the distribution becomes asymmetric as in Fig. 16.5, and a net current flows. As we have already seen, since K values increase until reaching π/l and then reappear at $-\pi/l$ the current oscillates. This is because a perfect crystal does not obey Ohm's law and possesses no resistivity. In real crystals, the electrons interact with impurities and also are scattered by the ions, which are not completely fixed but vibrate about equilibrium positions. These interactions have the effect of reducing the kinetic energy of the electrons so that states at the top of the band remain unoccupied and the K distribution remains permanently asymmetric, with the consequence that a unidirectional current flows.

16.2 Electrons in solids

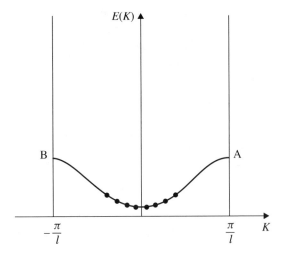

Figure 16.4 When a band is not full but is filled symmetrically no current flows.

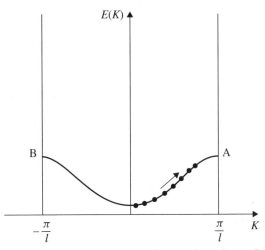

Figure 16.5 When a band is not full, under the influence of an electric field the distribution becomes asymmetric and a net current flows.

Particles and holes

We have already discussed the important effects in real three-dimensional crystals that the overlapping of bands and the size of the energy gap between bands have on the electrical conductivity. In an intrinsic semiconductor, the valence band is full at zero temperature. At finite temperatures some electrons are thermally excited into the conduction band and move under the influence of an external field as particles of effective mass m^* and charge $-e$. The valence band now possesses a few 'holes' in an otherwise full band. Since a full band carries no current, a hole in a full band must carry a current which is equal in magnitude and opposite in sign to the current carried by an electron. As the velocity of a hole and an electron are the same, this means that a hole carries a positive charge of magnitude e. Equation (16.44) for such a positive charge reads

$$m_H^* a = e\mathcal{E} \tag{16.47}$$

where m_H^* is the effective mass of a hole. The holes occur near the top of the energy band, for which the effective mass of an electron is negative. Hence a hole behaves like a particle of positive effective mass, and in general

$$m_H^* = -m^* = -\hbar^2 \left[\frac{d^2 E(K')}{dK'^2} \right]^{-1}_{K'=K}. \tag{16.48}$$

The current carried by a band containing holes is given by

$$I = e \int_{-\pi/l}^{\pi/l} [1 - P(K)] v_g(K) dK \tag{16.49}$$

where $P(K)$ is, as before, the probability that the state of energy $E(K)$ is occupied.

Quantum dots and artificial atoms

We have seen that electrons can move freely in a crystal lattice, being associated with a certain effective mass. In a recent development it has become possible to produce bound states of a number of electrons by fabricating crystals of semi-conducting material so that these electrons are confined to a small region with dimensions of the order of 10^{-6} to 10^{-7} m. These dimensions are much larger than either normal atomic dimensions or the spacing of the lattice sites (3–4×10^{-9} m). Such bound systems of from one to several hundred electrons are called *quantum dots* or *artificial atoms*. Their existence allows the quantum properties of many-electron systems of different sizes and shapes to be studied without the limitations of working with the naturally occurring elements.

A particular example which has been studied is a multilayer crystal fabricated in the form of a cylindrical rod, shown schematically in Fig. 16.6. A central conducting layer A of gallium arsenide (GaAs) is sandwiched between two insulating layers B of aluminium gallium arsenide (AlGaAs) in which about 7% of the gallium atoms have been replaced by aluminium atoms. Let us consider a single electron confined to the conducting layer and unable to penetrate the potential barrier represented by the insulating layers. Taking the axis of the crystal to be the z-direction and the conducting layer to be in the xy-plane, the potential confining the electron to the crystal can be modelled by the two-dimensional harmonic oscillator potential $V(r)$, where

$$V(r) = \frac{1}{2}\mu\omega^2 r^2 \tag{16.50}$$

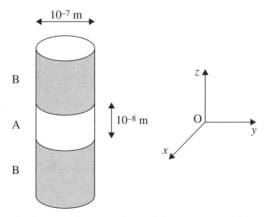

Figure 16.6 Schematic diagram of a crystal containing a quantum dot or artificial atom. A central conducting layer A of gallium arsenide is sandwiched between two insulating layers B of aluminium gallium arsenide.

with

$$r^2 = x^2 + y^2 \tag{16.51}$$

In (16.50), the region is taken to be the centre of an conducting disc, μ is an effective mass and ω is a constant. The energy levels of such a system have been worked out in problem 7.20 (c) where it was found that

$$E_n = \hbar\omega(n+1). \tag{16.52}$$

with

$$n = 2n_r + |m_l| = 0, 1, 2 \ldots. \tag{16.53}$$

Here we have written the magnetic quantum number m as m_l for future convenience. This quantum number can be a positive or negative integer, or zero, while n_r is a positive integer, or zero.

By applying an electric field further electrons can be added to the quantum dot by tunnelling through the insulating barrier and the energy required to add an additional electron can be measured. As in the case of real atoms discussed in Section 10.5, artificial atoms exhibit a shell structure. Two electrons can occupy the lowest energy state with $n_r = m_l = 0$, but because of the Pauli exclusion principle a third electron must occupy the next higher state with $n_r = 0$, $m_l = \pm 1$. Allowing for the electron spin, four electrons can occupy the second shell, so that when the first two shells are full, containing six electrons, extra energy is required to add a seventh. In the third shell, either $n_r = 0$ and $m_l = \pm 2$ or $n_r = 1$ and $m_l = 0$; thus the third shell can contain six electrons and the first three shells are full when the dot contains 12 electrons. The numbers of electrons corresponding to the filling of successive shells,

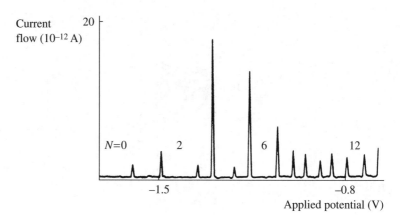

Figure 16.7 The current flow through a quantum dot as a function of the applied potential.

2, 6, 12, 20, 30 ... are known as *magic numbers* following a terminology borrowed from nuclear physics.

In a particular experimental arrangement the electric field allowing the number of electrons in the dot to be altered is produced by encircling the dot by an electrode to which is applied a negative potential. As the magnitude of this potential is increased the diameter of the dot is reduced and correspondingly the number of electrons within the dot can be decreased one by one. Since energy is required to add an extra electron to the dot, if the dot contains N electrons no current can flow until the magnitude of the applied potential is decreased to the point that $N+1$ electrons are confined. The result of a measurement of the current flow as a function of the applied potential is illustrated in Fig. 16.7. The current is zero between the peaks, which correspond to the number of electrons increasing from N to $N+1$, with $N = 1, 2, \ldots$. The potential difference between the peaks is a function of the extra energy required to add an electron to the dot. It is seen that this difference is greatest when $N = 2, 6, 12, 20, \ldots$, corresponding to the magic numbers[1].

Because the artificial atoms are so large magnetic fields of moderate strength (of the order of 1 T) produce effects that cannot be observed directly in the laboratory for real atoms since in that case field strengths of 10^6 T would be required. For a field in the direction of the z-axis the Schrödinger equation for a single electron in a dot can be obtained from (12.75) by adding the potential (16.50) and setting $k^2 = 0$, since there is no motion in the z-direction. We have

$$\left[-\frac{\hbar^2}{2\mu} \left(\frac{\partial^2}{\partial x^2} + \frac{\partial^2}{\partial y^2} \right) + \mu_B (m_l + 2m_s) \mathcal{B} + \left(\frac{e^2}{8\mu} \mathcal{B}^2 + \frac{1}{2} \mu \omega^2 \right) r^2 \right] \psi(x, y)$$

$$= E \psi(x, y) \tag{16.54}$$

---

[1] A more detailed account of experiments with quantum dots can be found in the review of Ashoori (1996).

where the mass m has been replaced by the effective mass μ. Setting $\bar{\omega} = (\omega_L^2 + \omega^2)^{1/2}$ where $\omega_L = (\mu_B/\hbar)\mathcal{B}$ is the Larmor angular frequency and proceeding as in (12.77)–(12.79), the energy levels are given by

$$E = \hbar\omega_L(m_l + 2m_s) + (n+1)\hbar\bar{\omega} \tag{16.55}$$

where n is given by (16.53) as before. The energy levels now depend on m_l as well as on $|m_l|$. This reflects the fact that the magnetic field distinguishes between the directions parallel and antiparallel to the z-axis.

16.3 Masers and lasers

In this section, we will show how the phenomenon of stimulated emission discussed in Chapter 11 can be used to construct amplifiers or generators of electromagnetic radiation. Suppose 1 and 2 are two levels, with energies $E_1, E_2 (E_2 > E_1)$, out of the many levels of a particular material. Consider a beam of electromagnetic radiation of intensity I and angular frequency $\omega = (E_2 - E_1)/\hbar$ passing through this material. The rate of change of the energy density because of absorption from the beam is

$$\frac{d\rho_a}{dt} = -N_1(\hbar\omega)W_{21} \tag{16.56}$$

where N_1 is the number of atoms in the lower energy level per unit volume and W_{21} is the transition rate per atom for absorption. Similarly, the rate of change of the energy density because of stimulated emission is

$$\frac{d\rho_s}{dt} = N_2(\hbar\omega)W_{12} \tag{16.57}$$

where N_2 is the number of atoms in the upper energy level per unit volume, and W_{12} is the transition rate per atom for stimulated emission. In Chapter 11, it was shown that W_{12} and W_{21} are equal and both are proportional to the intensity I of the incident radiation. The cross-section σ, defined as

$$\sigma = (\hbar\omega)W_{12}/I \tag{16.58}$$

is characteristic of the particular pair of levels, but is independent of the intensity of the beam radiation. In terms of σ, we can write the net rate of change of energy per unit volume traversed by the beam as

$$\frac{d\rho}{dt} = \sigma I(N_2 - N_1). \tag{16.59}$$

If the beam is of cross-sectional area A, and is travelling parallel to the z-axis, then by using (11.13) we have

$$\frac{dI}{dz} = \sigma I(N_2 - N_1). \tag{16.60}$$

We see that if $N_1 > N_2$ the incident radiation is absorbed as it traverses the material, but if $N_2 > N_1$ the radiation is amplified. Spontaneous emission will also increase the number of transitions from the upper level 2 to the lower level 1, but the corresponding transition rate is independent of the intensity I, and provided that I is sufficiently large this contribution can be ignored.

Under thermal equilibrium, we know that for non-degenerate energy levels

$$\frac{N_2}{N_1} = \exp[-(E_2 - E_1)/kT] \tag{16.61}$$

where k is Boltzmann's constant and T is the absolute temperature. Since $E_2 > E_1$ it follows that $N_2 < N_1$ and the material acts as an absorber. To achieve amplification, a population inversion must be arranged with $N_2 > N_1$, so that the substance cannot be in thermal equilibrium.

When amplification is achieved in the microwave region, we speak of a *maser* (microwave amplification by stimulated emission of radiation) and for radiation of higher frequencies we use the term *laser* (light amplification by stimulated emission of radiation).

The ammonia maser

The first maser, constructed in 1954 by C. H. Townes, J. Gordon and H. Zeiger, was based on some properties of the ammonia molecule, which we shall now describe briefly.

The inversion spectrum of NH$_3$

The ammonia molecule NH$_3$ has the form of a pyramid, whose summit is occupied by the nitrogen atom, while the base is an equilateral triangle formed by the three hydrogen atoms (see Fig. 16.8). At equilibrium, the distance NH is $d = 1.014$ Å, the distance of the nitrogen atom from the plane of the hydrogen atoms is $z_0 = 0.38$ Å and the angle α between an N–H bond and the threefold axis of symmetry of the molecule is $\alpha = 67°58'$.

There are many degrees of freedom in this system, involving electronic, vibrational and rotational motions, which result in a variety of energy levels and various quantum numbers to specify them. In this section, however, we shall assume that the NH$_3$ molecule is in its lowest electronic state and shall analyse a particular vibrational motion which is associated with the *inversion* of the molecule. To see how this comes about, let us consider one of the possible vibrational motions of the NH$_3$ molecule, analogous to the movement of an umbrella which is being opened and closed, and during which the angle α oscillates around its equilibrium position. Neglecting all other degrees of freedom, the potential energy of the system is then a function $V(z)$ of the algebraic distance z between the plane of the hydrogen atoms and the nitrogen atom. The curve $V(z)$ is sketched in Fig. 16.9(a). Because the system is symmetric with respect to the plane $z = 0$ it is clear that the potential $V(z)$ must be an even function of z. The two minima of $V(z)$ correspond to symmetrical

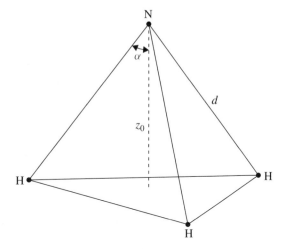

Figure 16.8 Schematic diagram of the ammonia molecule.

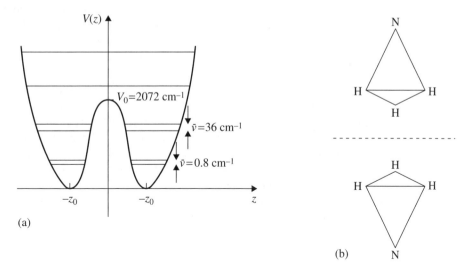

Figure 16.9 (a) The potential well for the inversion motion of ammonia illustrated in (b).

configurations of the molecule such that the nitrogen atom is located respectively above and below the plane of the hydrogen atoms [see Fig. 16.9(b)] at the equilibrium positions $z = \pm z_0 = \pm 0.38$ Å. We shall refer to these two configurations as the 'up' and 'down' configurations, respectively. The molecule can vibrate in the manner indicated above in either of the two potential wells, with the nitrogen atom on one side of the plane of the hydrogen atoms. The wave number corresponding to this vibrational motion is $\tilde{\nu} = 950$ cm^{-1}, which is in the infrared region.

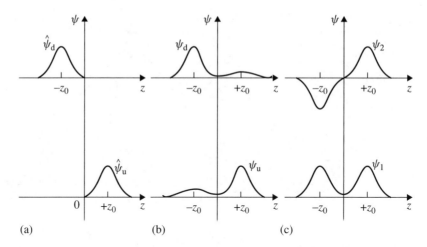

Figure 16.10 The wave functions (a) $\hat{\psi}_u$ and $\hat{\psi}_d$; (b) ψ_u and ψ_d; (c) ψ_1 and ψ_2.

As seen from Fig. 16.9(a), the potential $V(z)$ forms a barrier about $z = 0$. This barrier is due to the Coulomb repulsion between the nitrogen nucleus and the three protons. If it were of infinite height, the nitrogen atom would never be able to penetrate the plane of the hydrogen atoms and be found on the other side of this plane. However, the barrier has a finite height $V_0 = 2072$ cm^{-1}, so that there is a certain probability that the molecule will invert during the course of its vibrations, i.e. make transitions between the 'up' and 'down' configurations. It is important to emphasise that in the ground state ($v = 0$) as well as in the first excited state ($v = 1$) of the vibrational mode considered here, the energy of the molecule is lower than the potential height. As a result, the inversion of the molecule NH$_3$ in the vibrational states $v = 0$ and $v = 1$ is a classically forbidden (or hindered) motion which can only take place because of the quantum mechanical tunnel effect.

In order to understand the characteristics of this tunnelling motion let us write the one-dimensional Schrödinger equation for the motion along the z-axis, namely

$$-\frac{\hbar^2}{2m}\frac{d^2\psi(z)}{dz^2} + V(z)\psi(z) = E\psi(z) \qquad (16.62)$$

where m is an effective mass. If the potential barrier between the two wells were of infinite height, the two wells would be totally 'disconnected' and the energy spectrum would consist of the same set of energy eigenvalues in each well. Thus, each energy eigenvalue of the system would be doubly degenerate, and the eigenfunctions corresponding to a given energy would be linear combinations of the 'up' and 'down' wave functions $\hat{\psi}_u(z)$ and $\hat{\psi}_d(z)$ which vanish identically for $z \leqslant 0$ and $z \geqslant 0$, respectively. A pair of wave functions $\hat{\psi}_u$ and $\hat{\psi}_d$ is shown in Fig. 16.10(a) for the case of the lowest ($v = 0$) vibrational state.

In the actual molecule, with a finite barrier, there is a 'coupling' between the two wells which allows the inversion motion to occur. As a result, the degeneracy is removed, and the energy levels are split into doublets (see Problem 4.11). In the simple model considered here, the separation between the pair of energy levels forming a doublet depends only on the nature of the potential barrier and on the vibrational state of the molecule. As indicated in Fig. 16.9(a), the two energy levels forming the lowest ($v = 0$) doublet are separated by about 9.84×10^{-5} eV (0.8 cm^{-1}), while the next ($v = 1$) lowest pair of levels are about 4.4×10^{-3} eV (36 cm^{-1}) apart. We note that the inversion wave numbers $\tilde{v}(v = 0) = 0.8$ cm^{-1} and $\tilde{v}(v = 1) = 36$ cm^{-1} are much smaller than the wave number $\tilde{v} = 950$ cm^{-1} which corresponds to the vibrational motion, since inversion is considerably inhibited by the presence of the potential barrier.

In what follows we shall focus our attention on the lowest doublet ($v = 0$) which we shall treat as a two-level system. It is clear from the above discussion that since the potential barrier is finite we do not have rigorous 'up' and 'down' eigenfunctions $\hat{\psi}_u(z)$ and $\hat{\psi}_d(z)$ vanishing identically for $z \leqslant 0$ and $z \geqslant 0$, respectively. Instead, we define the corresponding wave functions $\psi_u(z)$ and $\psi_d(z)$ to be those for which the nitrogen atom is most probably located above or below the plane of the hydrogen atoms. These wave functions are sketched in Fig. 16.10(b). It is important to realise that because the two wells are coupled, the functions ψ_u and ψ_d are not energy eigenfunctions, and are not orthogonal to each other. Indeed, the true energy eigenfunctions must be either symmetric or antisymmetric with respect to the inversion operation $z \to -z$. In terms of normalised wave functions ψ_u and ψ_d, the normalised energy eigenfunctions are therefore given by

$$\psi_1 = \frac{1}{\sqrt{2}}(\psi_u + \psi_d) \tag{16.63a}$$

and

$$\psi_2 = \frac{1}{\sqrt{2}}(\psi_u - \psi_d). \tag{16.63b}$$

The symmetric wave function ψ_1 corresponds to the lower energy E_1 of the doublet, while the antisymmetric wave function ψ_2 corresponds to the higher energy E_2. Both functions ψ_1 and ψ_2 are shown in Fig. 16.10(c). The energy splitting of the doublet is

$$\Delta E = E_2 - E_1. \tag{16.64}$$

and we recall that its experimental value is $\Delta E \simeq 9.84 \times 10^{-5}$ eV.

Having obtained the normalised energy eigenfunctions ψ_1 and ψ_2 we may write the general time-dependent wave function of our two-level system as

$$\Psi(z, t) = c_1 \psi_1(z) e^{-iE_1 t/\hbar} + c_2 \psi_2(z) e^{-iE_2 t/\hbar} \tag{16.65}$$

where c_1 and c_2 are constants. Let us assume that at time $t = 0$ the wave function describing the system is ψ_u, so that the nitrogen atom is most probably to be found above the plane of the hydrogen atoms at that time. Using (16.63) and (16.65) we then have

$$\Psi(z, t=0) = c_1\psi_1(z) + c_2\psi_2(z) = \psi_u(z) = \frac{1}{\sqrt{2}}[\psi_1(z) + \psi_2(z)] \quad (16.66)$$

so that

$$c_1 = c_2 = \frac{1}{\sqrt{2}}. \quad (16.67)$$

Substituting (16.67) into (16.65) and using (16.64), we see that the wave function $\Psi(z, t)$ will evolve in time according to

$$\Psi(z, t) = \frac{1}{\sqrt{2}}[\psi_1(z)e^{-iE_1 t/\hbar} + \psi_2(z)e^{-iE_1 t/\hbar}e^{-i\Delta E t/\hbar}]$$

$$= \frac{1}{\sqrt{2}}[\psi_1(z) + \psi_2(z)e^{-2\pi i \nu t}]e^{-iE_1 t/\hbar} \quad (16.68)$$

where we have written $\Delta E = h\nu$. At the time $t = 1/(2\nu)$ the wave function (16.68) is given by

$$\Psi(z, t=1/(2\nu)) = \frac{1}{\sqrt{2}}[\psi_1(z) - \psi_2(z)]e^{-iE_1 t/\hbar}$$

$$= \psi_d(z)e^{-iE_1 t/\hbar}. \quad (16.69)$$

Thus

$$|\Psi(z, t=1/(2\nu))|^2 = |\psi_d(z)|^2 \quad (16.70)$$

and the nitrogen atom is most probably to be found under the plane of the hydrogen atoms at $t = 1/(2\nu)$. Since the energy difference $\Delta E = h\nu \simeq 9.84 \times 10^{-5}$ eV corresponds to a frequency $\nu \simeq 23\,800$ MHz, we see that the time required for the NH$_3$ molecule to invert is $t = 1/(2\nu) \simeq 2.1 \times 10^{-11}$ s.

The existence of the energy doublets of the ammonia molecule was first inferred from the analysis of its infrared vibrational–rotational and pure rotational spectra. However, radiative transitions between the two states forming a doublet can also occur, the corresponding lines being in the *microwave* region. In 1934, the progress made in radio-frequency techniques allowed C. E. Cleeton and N. H. Williams to observe directly a peak in the absorption spectrum of the ammonia molecule at a wavelength $\lambda \simeq 1.25$ cm, corresponding to the inversion frequency $\nu \simeq 23\,800$ MHz of the lowest doublet. The experiment of Cleeton and Williams opened the field of *microwave spectroscopy* and eventually made possible the development of the ammonia maser by C. H. Townes and his colleagues.

Population inversion

The ammonia molecule has many levels other than the doublet we have just discussed, but maser action is sought just between these two particular levels. The necessary

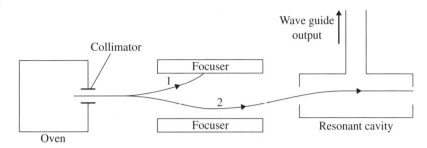

Figure 16.11 The ammonia maser. Molecules produced by the oven are collimated, selected into the upper level 2 by the inhomogeneous focusing field and then passed into the maser cavity. The whole apparatus is placed in an evacuated vessel.

coupling with the electromagnetic field takes place because the molecule possesses an electric dipole moment. In the configuration 'u' (up) this dipole moment is of magnitude D and is directed in the negative z-direction, while in the configuration 'd' (down), it is of the same magnitude but in the opposite direction.

Because the energy separation of the levels is so small a normal population in thermal equilibrium contains very nearly equal numbers of ammonia molecules in each of the energy eigenstates labelled E_1 and E_2. However, by passing a beam of ammonia molecules through an inhomogeneous electric field (see Fig. 16.11) a separation of the molecules in the two levels can be achieved, just as in a Stern–Gerlach magnet a separation is achieved between levels with different components of a magnetic dipole. To see this, consider a static electric field directed in the positive z-direction of magnitude $\mathcal{E}$. Although this field is non-uniform macroscopically, it can be considered to be uniform over distances of the order of molecular sizes. The additional interaction energy in configuration ψ_u is $D\mathcal{E}$ and in configuration ψ_d it is $-D\mathcal{E}$. This additional interaction slightly alters the eigenenergies of the E_1 and E_2 levels. Using the result (8.129) and assuming that $\Delta E \gg (D\mathcal{E})^2$, we find that the perturbed energy levels are

$$E'_1 = E_1 - (D\mathcal{E})^2/\Delta E \tag{16.71a}$$

$$E'_2 = E_2 + (D\mathcal{E})^2/\Delta E. \tag{16.71b}$$

The force on the molecule in the z-direction, F_z, depends on which of the states 1 or 2 is concerned, since

$$(F_z)_{1,2} = -\frac{\partial}{\partial z}(E'_{1,2}) = \pm 2\left(\frac{D^2}{\Delta E}\right)\mathcal{E}\left(\frac{\partial \mathcal{E}}{\partial z}\right). \tag{16.72}$$

In a uniform field the force vanishes, but in an inhomogeneous field with $\partial \mathcal{E}/\partial z > 0$ the molecules in level 1 are deflected in the positive z-direction and those in level 2 in the negative z-direction.

Having obtained a population entirely in the state of higher energy E_2, maser action occurs by stimulated emission of the transition from the level 2 to the level 1, which is reinforced by passing the beam through a cavity tuned to the required frequency. It is this method which was used in 1954 by C. H. Townes *et al.* to achieve for the first time a population inversion.

Lasers

The extension of the maser concept into the optical frequency range was proposed in 1958 by A. L. Schawlow and C. H. Townes, who considered a plane-parallel resonator such that an active material is placed between two plane mirrors (see Fig. 16.12). If one of the mirrors is partially transparent, an output beam can be obtained. If the power generated within the material by stimulated emission is greater than the sum of the power output and the power losses, then the laser acts as an oscillator and the intensity of the radiation between the mirrors increases exponentially. This increase in intensity will be finally limited by the ability of the 'pumping' mechanism producing the population inversion to keep up the number of atoms or molecules in the upper level. The laser oscillations can be started by a single photon resulting from spontaneous emission from the upper to the lower level. In applications, masers are frequently used as amplifiers, while lasers are more usually employed as oscillators generating radiation.

The characteristic properties of a beam generated by a laser are (a) monochromaticity, (b) directionality, (c) brightness (d) spatial coherence and (e) temporal coherence. The monochromaticity is a consequence of the fact that only light arising from a transition between a pair of levels is amplified. The output of a laser is a parallel beam which emerges perpendicular to the plane of the mirrors in an arrangement such as that illustrated in Fig. 16.12. This is due to the fact that only electromagnetic waves propagating in this direction will be reflected back and forth between the mirrors. This directionality also accounts for the brightness of a laser beam. The power output of other light sources is usually spread out into a large solid angle, but in a laser it is concentrated into a narrow unidirectional beam.

In stimulated emission, at each transition one photon is added to a mode of the resonator (i.e. a stationary electromagnetic field configuration which satisfies both Maxwell's equations and the boundary conditions) containing N photons. The extra photon is completely in phase with the incident photons and has the same polarisation. It follows that if laser (maser) action is initiated by a single photon, at each transition one extra photon will be produced and after N transitions, all $(N + 1)$ photons will be in phase and contribute to the same mode of the electromagnetic radiation: the laser (maser) light is said to be *coherent*. This is in contrast to other sources of light (such as a lamp filament), where the dominant process is spontaneous emission, so that the phases and polarisations associated with each photon are different. There are two independent concepts of coherence: spatial coherence and temporal coherence. If the phase difference between two points on a wave front, normal to the direction of propagation, is zero at all times, the wave is said to exhibit perfect spatial coherence. If the active material in a laser is homogeneous, the output beam exihibits spatial coherence over its whole cross-sectional area. It is in fact effectively a single plane

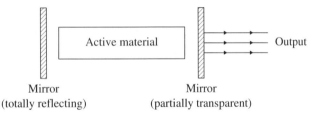

Figure 16.12 Schematic diagram of a laser in which the active material is confined between plane parallel mirrors.

wave of the form $\mathcal{E}(\mathbf{r}, t) = \mathcal{E}_0 \sin(\mathbf{k}\cdot\mathbf{r} - \omega t)$, with a single angular frequency and with all points on the wave front in phase. Temporal coherence relates to the duration of the output wave, and the coherence time is the interval over which the output is represented by the same plane wave. The stability of the laser determines the coherence time; it can be extremely long compared with the periodic time of the radiation. A detailed account of the properties of laser light can be found in the book of Svelto (1982).

We shall now briefly describe some ways of producing a population inversion, which can be used to obtain laser action. In the first of these, one looks for a three-level system (see Fig. 16.13) such that $E_1 < E_2 < E_3$, with a fast decay between levels 3 and 2 and a slow decay between 2 and 1. Incident radiation of angular frequency $\omega_{31} = (E_3 - E_1)/\hbar$ is used to raise as many atoms as possible from the level 1 to the level 3: this is known as 'pumping'. If level 3 decays rapidly to level 2, a population inversion can be obtained between levels 2 and 1. It should be noted that a population inversion cannot be obtained between levels 3 and 1, because when the number N_3 of atoms in level 3 equals the number N_1 of atoms in level 1, absorption is balanced by stimulated emission. The ruby laser constructed by T. H. Maiman in 1960 (which was the first laser to operate) provides an example of a three-level laser. A ruby is a crystalline alumina (Al_2O_3) which contains Cr^{3+} ions. These Cr^{3+} ions are excited by green light ($\lambda = 5500$ Å) to a number of closely spaced levels. Interaction with the crystal lattice de-excites these levels by a non-radiative process to a metastable level 2, which has a particularly long life of the order of 10^{-3} s. Laser action can then occur between the levels 2 and 1, resulting in red light ($\lambda = 6943$ Å).

Except in special cases, such as the ruby, it is difficult to produce a population inversion between a ground state and an excited state, because initially nearly all the atoms are likely to be in the ground state, and we have to get more than half the atoms in level 2 before a population inversion can be achieved. An easier approach is to use a four-level system (see Fig. 16.14) with $E_1 < E_2 < E_3 < E_4$, and attempt to create a population inversion between the two excited levels E_2 and E_3. We start with nearly all the atoms in the ground state 1. Level 4 is chosen so that it has a fast decay to level 3, and pumping between levels 1 and 4 then rapidly produces a population inversion between levels 2 and 3. As level 2 begins to fill up by stimulated emission at the angular frequency $\omega_{32} = (E_3 - E_2)/\hbar$, the population inversion will decrease.

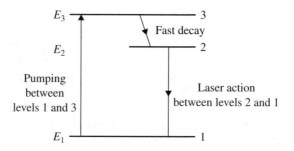

Figure 16.13 The three-level laser.

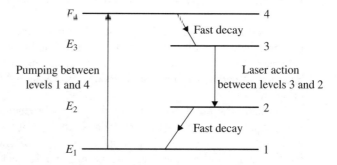

Figure 16.14 The four-level laser.

To minimise this, level 2 is chosen so that it has a fast decay to the ground state.

A gas laser presents an example of a multi-level system, which can be pumped by an electric discharge, rather than by incident radiation. An important example is the He–Ne laser, in which the active material is a mixture of helium and neon gases at low pressure. The energy levels concerned are shown in Fig. 16.15. In an electric discharge the helium atoms are raised to the 2^1S and 2^3S levels which are metastable. The ground state of neon has the configuration $(1s)^2(2s)^2(2p)^6$ and the lowest excited states have the configuration $(1s)^2(2s)^2(2p)^5(nl)$. Of these, the $nl = 4s$ and $nl = 5s$ states are coincident in energy with the 2^3S and 2^1S helium levels, respectively. As a result, in collisions between the excited helium atoms and ground-stated neon atoms, there is a high probability that neon atoms will be excited to these levels, the helium atoms reverting to the ground state. The selection rules allow transitions to the lower-lying neon 3p and 4p levels. Furthermore, the lifetimes of the 4s and 5s neon levels are of the order of 10^{-7} s, which is about ten times longer than the lifetimes of the 3p and 4p levels. The He–Ne mixture forms a 'double' laser system, which can show laser action between the 4s or 5s level and the 3p level, or between the 5s level and the 4p level. Each of the neon levels consists of several sub-levels, and out of the various possible transitions the strongest are: (a) between the 5s and 4p levels at $\lambda = 33\,900$ Å; (b) between the 5s and 3p levels at $\lambda = 6330$ Å; and (c) between the 4s

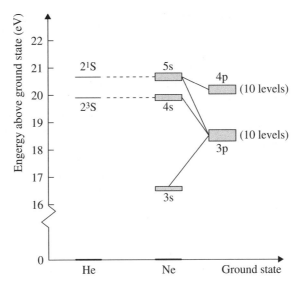

Figure 16.15 The energy levels of the He–Ne laser. The neon configuration concerned is $(1s)^2$, $(2s)^2$, $(2p)^5 nl$, each configuration giving rise to a number of levels in the shaded regions.

and 3p levels at $\lambda = 11\,500$ Å. The wavelength of the light generated in a He–Ne laser depends on the reflectivity of the mirrors between which the gas is placed. Oscillation will occur at the wavelength for which this reflectivity is a maximum.

Physics with lasers and masers

The availability of laser (or maser) light has opened up entirely new domains in physics. One of them is *laser spectroscopy*, which since the advent of widely tunable lasers around 1970 has far surpassed classical spectroscopic techniques in resolution and measurement accuracy. It has many important applications, not only in ultra-sensitive spectroscopy and metrology, but also in related areas such as laser cooling and trapping of atoms (see Section 14.8) and more generally the manipulation of atoms with light.

Another area of great interest is *quantum optics*, which covers quantum phenomena in the interaction of radiation with atoms. It therefore provides an excellent ground for testing basic quantum features. For example, a fundamental system for studying radiation coupling with matter is a single two-level atom interacting with a single mode of an electromagnetic field in a cavity. This system, which received much attention shortly after the maser was invented, appeared at first glance of be another example of a '*gedanken*' experiment. However, such a *one-atom maser* was built by H. Walther and his colleagues in 1985, and a *one-atom laser* emitting in the visible range has been realised by M. Feld *et al.* in 1994. A recent survey of the fields of quantum optics and laser spectroscopy has been given by Hänsch and Walther (1999).

If radiation fields of sufficient intensity interact with atoms or molecules, processes of higher order than the linear, single photon absorption or emission can play a significant role, and even dominate. These higher order processes, called *multiphoton processes*, correspond to the net absorption or emission of more than one photon in an atomic or molecular transition. Except for spontaneous emission (which does not require the presence of an external radiation field), the observation of multiphoton transitions requires large radiation intensities (typically in excess of 10^{10} W cm^{-2}) which, for optical and nearby frequencies, have become available through the development of the laser. Examples of multiphoton processes are the *multiphoton excitation* or *ionisation* (by the absorption of $n \geq 2$ photons) of atoms or molecules, and the *multiphoton dissociation* of molecules. An atomic or molecular system interacting with a strong laser field can also emit radiation at higher-order multiples, or harmonics, of the frequency of the laser: this process is known as *harmonic generation*. Finally, radiative collisions involving the exchange (absorption or emission) of n photons can occur in *laser-assisted atomic collisions* such as electron–atom or atom–atom collisions in the presence of a laser field. In recent years, super-intense laser fields have become available in the form of short pulses yielding intensities up to 10^{20} W cm^{-2}. These fields are strong enough to compete with the Coulomb forces in controlling the electron dynamics in atomic systems. As a result, atoms and molecules in superintense fields exhibit new properties, discovered by studying multiphoton processes. These modified properties generate new behaviour of bulk matter in intense laser fields, with wide-ranging applications such as the development of high-frequency lasers, the investigation of the properties of plasmas and condensed matter under extreme conditions of temperature and pressure (relevant for studies of inertial confinement fusion and in astrophysics) and electron acceleration to relativistic energies. A review of high-intensity laser-atom physics has been given by Joachain *et al.* (2000). The generation of super-intense laser fields and their applications is discussed by Mourou *et al.* (1998).

16.4 The decay of K-mesons

A particularly vivid illustration of the superposition principle of quantum mechanics is offered by the study of the decay of neutral K-mesons (kaons). To appreciate the results of experiments on K-meson decay, we first need to know a few facts about the interactions between elementary particles. The forces between particles can be classified into the strong, the weak, the electromagnetic and the gravitational interactions. Here we will be concerned only with the strong and weak interactions.

The strong interaction

The strong interaction is responsible for, among other things, the forces binding atomic nuclei, which we have already met in Chapters 10 and 13. Particles which participate in the strong interactions are called *hadrons*. There are two kind of hadrons: the *baryons* (such as the proton p, the neutron n, the hyperons Σ, Ξ) which have half-odd integer spin $\left(\frac{1}{2}, \frac{3}{2}, \ldots\right)$ and are therefore fermions, and the *mesons* (such as the π-mesons and the K-mesons) which have integer spin $(0, 1, \ldots)$ and hence are bosons. These strong interactions are responsible for many production and scattering reactions between the nucleons, hyperons, mesons and their excited states. Typical examples are the Σ-hyperon and the K-meson which can be produced in the collision of a π-meson (pion) and a proton

$$\pi^+ + p \to \Sigma^+ + K^+. \tag{16.73}$$

Another example of a reaction due to the strong interaction is the production of two π-mesons and a Λ^0-hyperon in the collision of a K^--meson and a proton

$$K^- + p \to \Lambda^0 + \pi^+ + \pi^-. \tag{16.74}$$

Such strong reactions conserve charge, energy, momentum and angular momentum. The rest masses and other characteristics of a few of the many particles which have been discovered are shown in Table 16.1.

When excited states are formed in a reaction, these can subsequently decay rapidly through the strong interaction. For example, in the interaction between two protons a Δ^{++} particle may be formed which subsequently decays with a lifetime of approximately 10^{-23}s to a proton and a π^+ meson

$$\begin{array}{c} p + p \to \Delta^{++} + n \\ \searrow p + \pi^+ \end{array} \tag{16.75}$$

Study of reactions of very many types shows that certain selection rules are satisfied which can be expressed by assigning two extra quantum numbers to each type of particle. These quantum numbers are like electric charge, in that their algebraic sum is conserved in any strong interaction. The first of these is the baryon number, B, which is $+1$ for the nucleons and hyperons (and their excited states) and zero for the mesons. The second is the hypercharge, Y, which is $+1$ for a nucleon, the K^+- and K^0-mesons, and is 0 for the Λ- and Σ-hyperons and for the π and η^0 mesons. Some other assignments are included in Table 16.1.

It has been discovered that for each type of particle another type exists with the same mass and spin but with charge Q, hypercharge Y and baryon number B of *opposite* sign. This second type of particle is called the *antiparticle* of the first type. For example, the antiproton $\bar{p}$ ($Q = -e$, $Y = B = -1$) is the antiparticle of the proton, and the antineutron $\bar{n}$ ($Q = 0$, $Y = B = -1$) is the antiparticle of the neutron. In strong interactions, the transition rate for a given reaction between particles is exactly the same as that for the reaction between the corresponding antiparticles. For instance, the transition rate for elastic scattering of neutrons by protons is the same (at the same energy) as that for the elastic scattering of antineutrons by antiprotons, while the transition rate for the reaction of equation (16.73) is the same as that for the reaction

$$\pi^- + \bar{p} \to \bar{\Sigma}^- + K^- \tag{16.76}$$

Table 16.1 Strongly interacting particles (ground states).

Particle	Spin (in units of $\hbar$)	Mass (in MeV/c^2)*	Q/|e|	B	Y
Nucleons					
p	$\frac{1}{2}$	938.3	1	1	1
n	$\frac{1}{2}$	939.6	0	1	1
Hyperons					
Λ	$\frac{1}{2}$	1116	0	1	0
Σ^+	$\frac{1}{2}$	1189	1	1	0
Σ^0	$\frac{1}{2}$	1192	0	1	0
Σ^-	$\frac{1}{2}$	1197	−1	1	0
Ξ^0	$\frac{1}{2}$	1315	0	1	−1
Ξ^-	$\frac{1}{2}$	1321	−1	1	−1
Ω^-	$\frac{3}{2}$	1673	−1	1	−2
Mesons					
π^+	0	139.6	1	0	0
π^0	0	135.0	0	0	0
π^-	0	139.6	−1	0	0
η^0	0	549	0	0	0
K^+	0	494	1	0	1
K^0	0	498	0	0	1
K^-	0	494	−1	0	−1
$\bar{K}^0$	0	498	0	0	−1

* 1 MeV/$c^2 = 1.78268 \times 10^{-30}$ kg.

where a bar over the letter designates the antiparticle. (Note that π^-, the antiparticle of π^+ is written without a bar, and similarly for K^-). Because of this symmetry between particles and antiparticles, it is useful to introduce an operator $\mathcal{C}$, called the *charge conjugation operator*, which converts the wave function of a particle of momentum **p** into the wave function of the corresponding antiparticle with the same momentum. In the same way, when $\mathcal{C}$ acts on the wave function of an antiparticle of momentum **p**, it yields the wave function of the corresponding particle also with momentum **p**. If A denotes a particle and $\bar{\text{A}}$ the corresponding antiparticle, then

$$\mathcal{C}|\text{A}, \mathbf{p}\rangle = \eta_c |\bar{\text{A}}, \mathbf{p}\rangle \qquad \mathcal{C}|\bar{\text{A}}, \mathbf{p}\rangle = \eta_c^*|\text{A}, \mathbf{p}\rangle \qquad (16.77)$$

where η_c is a phase factor. Clearly $C^2 = I$, since applying C twice restores the original situation. Taking η_c to be real we have $\eta_c = \pm 1$. The parameter η_c is called the *intrinsic charge conjugation parity*. By convention we shall assign $\eta_c = -1$ to the K-mesons so that (suppressing **p**), for neutral K-mesons

$$C|K^0\rangle = -|\bar{K}^0\rangle \quad \text{and} \quad C|\bar{K}^0\rangle = -|K^0\rangle. \tag{16.78}$$

In the case of charged pions π^+ and π^-, η_c is also chosen, by convention, to be equal to -1 so that

$$C|\pi^+\rangle = -|\pi^-\rangle \quad \text{and} \quad C|\pi^-\rangle = -|\pi^+\rangle. \tag{16.79}$$

On the other hand since the π^0 meson is its own antiparticle, that is $|\bar{\pi}^0\rangle \equiv |\pi^0\rangle$, we have that

$$C|\pi^0\rangle = |\pi^0\rangle. \tag{16.80}$$

The symmetry between strong interactions involving particles and antiparticles can be expressed by the condition that the operator C commutes with the strong interaction Hamiltonian H_s

$$[H_s, C] = 0. \tag{16.81}$$

In our discussion we shall need one further particle property. It can be shown that each particle can be assigned an *intrinsic parity* in such a way that a particle and the corresponding antiparticle have opposite intrinsic parity for fermions and the same intrinsic parity for bosons. Thus if $\mathcal{P}$ is the parity operator and A is a particle (or antiparticle) of momentum **p**, then

$$\mathcal{P}|A, \mathbf{p}\rangle = \eta_p |A, -\mathbf{p}\rangle, \tag{16.82}$$

where η_p, the intrinsic parity, is chosen to be $+1$ for the particle fermions p, n, Λ, ..., and -1 for the antiparticle fermions $\bar{p}$, $\bar{n}$, $\bar{\Lambda}$, ... It is found that the intrinsic parity of the π-mesons is $\eta_p = -1$ and the intrinsic parity of the K-mesons and $\bar{K}$-mesons is also -1. Thus for K^0 and $\bar{K}^0$ mesons at rest

$$\mathcal{P}|K^0\rangle = -|K^0\rangle \quad \mathcal{P}|\bar{K}^0\rangle = -|\bar{K}^0\rangle. \tag{16.83}$$

The weak interaction

Many elementary particle decays occur with transition rates which are extremely small compared with those associated with the strong nuclear interactions we have been discussing. These decays are due to the *weak* interaction. The first weak process to be investigated was the beta decay of nuclei, which takes place through the decay of the neutron

$$n \to p + e^- + \bar{\nu}_e \tag{16.84}$$

where $\bar{\nu}_e$ denotes an electron antineutrino. Weak decays into final states containing only hadrons can also take place, for example

$$\Lambda^0 \to p + \pi^-. \tag{16.85}$$

This process occurs with a lifetime of 10^{-10} s, which can be contrasted with the much faster strong decay of the Δ^{++} resonance, $\Delta^{++} \to p + \pi^+$ (see (16.75)), which takes place with a lifetime of 10^{-23} s. There is clearly a useful qualitative distinction to be made between strong and weak interactions, although at a deeper level it is thought that all interactions are connected, in the same sense that electricity and magnetism are connected by Maxwell's electromagnetic theory. A satisfactory unification of the weak nuclear interaction with electromagnetism has already been achieved in the 'standard' model of S. Weinberg and A. Salam (see, for example, the text by Halzen and Martin (1984)) and current research is concentrated on achieving a larger unification with the strong interactions and with gravitation.

In weak decays, charge and baryon number are both conserved, but hypercharge is not. For example the Λ^0 hyperon ($Y = 0$) decays to a proton ($Y = 1$) and a π^--meson ($Y = 0$) (see (16.85)) so that the hypercharge Y increases by one unit during the reaction. Similarly, Y decreases by one unit in the K^+ decay

$$K^+ \to \pi^0 + \mu^+ + \nu_\mu \tag{16.86}$$

where π^0 is a neutral π-meson, μ^+ is a positively charged muon and ν_μ is a muon-neutrino. At the same time neither the charge conjugation operator $\mathcal{C}$ nor the parity operator $\mathcal{P}$ commute with the weak Hamiltonian H_w

$$[\mathcal{C}, H_w] \neq 0, \qquad [\mathcal{P}, H_w] \neq 0. \tag{16.87}$$

However it is found that the product operator $\mathcal{CP}$ *does* commute with H_w to a very high degree of approximation (although not exactly), and this will have important consequences for our analysis of K^0 and $\bar{K}^0$ decay.

The production and decay of K^0-mesons

We now have everything we need to analyse the experiments on the production and decay of neutral K-mesons. As far as the strong interactions are concerned the K^0 and $\bar{K}^0$ particles are distinct, being distinguished by different values of the hypercharge: $Y = 1$ for K^0 and $Y = -1$ for $\bar{K}^0$. For example, in the strong interaction between a π^- meson and a proton, a K^0 meson is formed in the reaction

$$\pi^- + p \to \Lambda^0 + K^0. \tag{16.88}$$

In both the initial and final states the total hypercharge is $Y = 1$. Similarly, in the corresponding reaction between a π^+-meson and an antiproton a $\bar{K}^0$ meson is produced, the total hypercharge being $Y = -1$:

$$\pi^+ + \bar{p} \to \bar{\Lambda}^0 + \bar{K}^0. \tag{16.89}$$

16.4 The decay of K-mesons

If only strong interactions existed, the K^0 and $\bar{K}^0$ mesons would not only have exactly the same mass, but would also be stable. However because the weak interaction does not conserve hypercharge, the $|K_0\rangle$ and $|\bar{K}_0\rangle$ wave functions are coupled and the K_0, $\bar{K}_0$ particles are not energy eigenstates of the total (strong plus weak) Hamiltonian. In fact the wave functions of the K_0 and $\bar{K}_0$ particles are linear superpositions of the wave functions of the 'observed' particles: the short- and long-lived kaons, K_S^0 and K_L^0, which are identified and distinguished by different (weak) decay modes. The K_S^0 particle decays with a lifetime $\tau_1 \simeq 10^{-10}$ s mostly into a pair of π-mesons and the most important weak decay of the K_L^0 particle is into three π-mesons with a lifetime $\tau_2 \simeq 6 \times 10^{-8}$ s. Since both the K^0 and $\bar{K}^0$ particles created in the strong interaction reactions (16.88) and (16.89) are in states which are superpositions of the states representing the K_S^0 and K_L^0 particles, after creation both the two and the three π-meson decay modes are observed.

As we mentioned earlier, the weak interaction Hamiltonian H_w commutes to a high degree of approximation with the combined parity and charge conjugation operations described by the operator $\mathcal{CP}$. It is therefore useful, in a first approach, to formulate the problem in terms of $\mathcal{CP}$ eigenstates, $|K_1^0\rangle$ and $|K_2^0\rangle$. From (16.78) and (16.83) it is straightforward to check that

$$|K_1^0\rangle = \frac{1}{\sqrt{2}}(|K^0\rangle + |\bar{K}^0\rangle) \tag{16.90a}$$

$$|K_2^0\rangle = \frac{1}{\sqrt{2}}(|K^0\rangle - |\bar{K}^0\rangle) \tag{16.90b}$$

are such eigenstates, with

$$\mathcal{CP}|K_1^0\rangle = |K_1^0\rangle, \qquad \mathcal{CP}|K_2^0\rangle = -|K_2^0\rangle. \tag{16.91}$$

The transformation (16.90) being unitary, it is readily inverted to yield

$$|K^0\rangle = \frac{1}{\sqrt{2}}(|K_1^0\rangle + |K_2^0\rangle) \tag{16.92a}$$

$$|\bar{K}^0\rangle = \frac{1}{\sqrt{2}}(|K_1^0\rangle - |K_2^0\rangle). \tag{16.92b}$$

These equations show that the $|K^0\rangle$ or $|\bar{K}^0\rangle$ states created in strong interactions are equal mixtures of the $|K_1^0\rangle$ and $|K_2^0\rangle$ states more appropriate to the description of weak decays.

Since the spin of the K-meson is zero, if it decays into a two-pion state the orbital angular momentum of that state must also be zero because the pion has zero spin. A two-pion state of zero orbital angular momentum is clearly an eigenstate of $\mathcal{CP}$ belonging to the eigenvalue 1, and it follows that the K_2^0 particle cannot decay to two pions (see (16.91)). The three-pion state $\pi^+\pi^-\pi^0$ can be in eigenstates of $\mathcal{CP}$ with eigenvalues either 1 or -1, depending on the relative orbital angular momenta, which however must be consistent with a *total* orbital angular momentum of zero.

As a consequence both the K_1^0 and K_2^0 particles can make $\mathcal{CP}$ conserving transitions to three pions.

As the K_1^0 and K_2^0 particles decay at different rates, with lifetimes τ_1 and τ_2 respectively, after a time t following the instant of creation the wave functions (16.92) become

$$|\Psi_\pm(t)\rangle = \frac{1}{\sqrt{2}}\{|K_1^0\rangle \exp[-iE_1 t/\hbar - t/(2\tau_1)]$$
$$\pm |K_2^0\rangle \exp[-iE_2 t/\hbar - t/(2\tau_2)]\} \quad (16.93\text{a})$$

where

$$|\Psi_+(t=0)\rangle = |K^0\rangle, \qquad |\Psi_-(t=0)\rangle = |\bar{K}^0\rangle. \quad (16.93\text{b})$$

In the frame of reference in which the particles are at rest $E_1 = M_1 c^2$ and $E_2 = M_2 c^2$, M_1 and M_2 being the masses of the K_1^0 and K_2^0 mesons. Because only weak interactions are involved both M_1 and M_2 are extremely close to the mass M_K of the K^0 (or $\bar{K}^0$) particle and $M_1 - M_2 \ll M_K$.

Since $\tau_2 \gg \tau_1$, we see from (16.93) that after times $t \gg \tau_1$ the wave functions $|\Psi_\pm(t)\rangle$ are entirely composed of $|K_2^0\rangle$ components. This prediction can be tested experimentally since we see from (16.90) that the $|K_2^0\rangle$ wave function is an equal mixture of the $|K^0\rangle$ and $|\bar{K}^0\rangle$ wave functions. That is, if we start from reaction (16.88) which produces a pure $|K^0\rangle$ state, after the passage of time we obtain a mixture of $|K^0\rangle$ and $|\bar{K}^0\rangle$ states. Since the $\bar{K}^0$ particle has hypercharge $Y = -1$ it can produce Λ^0- and Σ-hyperons via the reactions

$$\bar{K}^0 + p \to \Lambda^0 + \pi^+$$
$$\to \Sigma^+ + \pi^0$$
$$\to \Sigma^0 + \pi^+ \quad (16.94)$$

and their interactions can be observed. Such experiments can also provide a value for the small mass difference $M_1 - M_2$ which is found to be about 10^{-14} times the mass of the K^0-meson.

$\mathcal{CP}$ violation

If the weak interaction were exactly invariant under the $\mathcal{CP}$ operation we would have $|K_L^0\rangle \equiv |K_2^0\rangle$ and the decay

$$K_L^0 \to \pi^+ + \pi^- \quad (16.95)$$

would be absolutely forbidden. However, in 1964 an experiment was performed by J. H. Christenson, J. W. Cronin, V. L. Fitch and R. Turlay in which it was found that some K_L^0 particles decay to the two-pion state with a branching ratio of about 2×10^{-3}. That is

$$\frac{W_2}{W_T} = (0.2 \pm 0.04) \times 10^{-2} \quad (16.96)$$

where W_2 is the decay rate corresponding to the reaction (16.95) and W_T is the total decay rate of the K_L^0 particle.

Two possibilities exist (they are not exclusive):

(1) The states $|K_L^0\rangle$ and $|K_S^0\rangle$ differ from the states $|K_1^0\rangle$ and $|K_2^0\rangle$, of which they are now superpositions. This violation of $\mathcal{CP}$ is usually described by introducing a parameter ε:

$$|K_S^0\rangle = \frac{|K_1^0\rangle + \varepsilon |K_2^0\rangle}{(1+|\varepsilon|^2)^{1/2}}, \qquad (16.97a)$$

$$|K_L^0\rangle = \frac{|K_2^0\rangle + \varepsilon |K_1^0\rangle}{(1+|\varepsilon|^2)^{1/2}} \qquad (16.97b)$$

(2) The state $|K_2^0\rangle$ itself can decay into two pions. This violation of $\mathcal{CP}$ is traditionally measured by a parameter ε'.

Which of these two possibilities is realised is a matter of experiment. It has been found that $|\varepsilon'/\varepsilon| = (2.2 \pm 0.3) \times 10^{-3}$, so that the parameter ε is actually directly measured by (16.96); a recent value is $|\varepsilon| = (2.28 \pm 0.02) \times 10^{-3}$.

The existence of this $\mathcal{CP}$ violating interaction means that the analysis we have presented of the decay of a neutral K-meson beam is not completely accurate. However, the analysis of the decay of a beam of K-mesons into components K_S^0, K_L^0 is similar to that discussed above and will not be given here.

Since the discovery of $\mathcal{CP}$ violation was made, searches for further decay processes showing the same behaviour (which is equivalent to a lack of invariance under time reversal) have been carried out, and some evidence has been found recently.

16.5 Positronium and charmonium

In this section we shall study the energy spectrum of two simple two-body systems which have a similar structure. The first one, *positronium*, is a bound electron–positron system and the second one, *charmonium*, is a bound system of a heavy quark and its antiquark.

Positronium

Let us first consider positronium. In the non-relativistic approximation, the Hamiltonian of this system is the hydrogenic Hamiltonian

$$H = \frac{\mathbf{p}^2}{2\mu} - \frac{e^2}{(4\pi\varepsilon_0)r} \qquad (16.98)$$

in which the reduced mass $\mu = m/2$, where m is the electron mass. The non-relativistic energy levels E_n of positronium are therefore given by (7.114) in which

$Z = 1$ and $\mu = m/2$, namely

$$E_n = -\frac{e^2}{(4\pi\varepsilon_0)a_0}\frac{1}{4n^2}, \qquad n = 1, 2, \ldots \tag{16.99}$$

where a_0 is the Bohr radius (1.66). Hence, the ionisation potential of positronium is given in the non-relativistic approximation by

$$I_p = \frac{e^2}{(4\pi\varepsilon_0)4a_0} \simeq 6.8 \text{ eV}. \tag{16.100}$$

According to (7.116) each energy level (16.99) is n^2 degenerate. If we also take into account the fact that both the electron and the positron have spin 1/2, this degeneracy is multiplied by a factor of four and hence becomes equal to $4n^2$.

Let us denote by $\mathbf{L}$ the orbital angular momentum operator (in the centre-of-mass system) and by $\mathbf{S}_1$ and $\mathbf{S}_2$ the spin angular momentum operators of the electron and the positron, respectively. The total spin operator is

$$\mathbf{S} = \mathbf{S}_1 + \mathbf{S}_2 \tag{16.101}$$

and the total angular momentum operator of the positronium system is

$$\mathbf{J} = \mathbf{L} + \mathbf{S}. \tag{16.102}$$

From our study of angular momentum in Chapter 6 we know that the allowed values of the orbital angular momentum quantum number (which we shall denote by L) are $L = 0, 1, 2, \ldots$. The allowed values of the total spin quantum number S are $S = 0$ (corresponding to spin singlets) and $S = 1$ (corresponding to spin triplets). Finally, the allowed values of the total angular momentum quantum number J are given by $J = |L - S|, |L - S| + 1, \ldots, L + S$.

The non-relativistic Hamiltonian (16.98) commutes not only with the total angular momentum $\mathbf{J}$, but also separately with the orbital angular momentum $\mathbf{L}$ and the total spin orbital angular momentum $\mathbf{S}$. The energy levels of H can therefore be labelled by the quantum numbers L and S, a term corresponding to definite values of L and S being denoted as ^{2S+1}L, with the code letters S, P, D, $\ldots$, corresponding to the values $L = 0, 1, 2, \ldots$.

When relativistic corrections are taken into account, the Hamiltonian of the positronium system no longer commutes with $\mathbf{L}$ and $\mathbf{S}$, but it still commutes with the total angular momentum $\mathbf{J}$. The energy levels (16.99) are then slightly shifted and a term ^{2S+1}L splits into a number of *fine-structure* components, characterised by the value of J, and written in the Russell–Saunders notation as $^{2S+1}L_J$. Of course, in the absence of external fields, each fine-structure term $^{2S+1}L_J$ is $(2J+1)$-fold degenerate with respect to the magnetic quantum number M_J (where $M_J\hbar$ are the eigenvalues of the operator J_z), the possible values of M_J for a given J being $M_J = -J, -J+1, \ldots, J$.

The energy levels of positronium can further be characterised by two additional 'good quantum numbers', corresponding to the parity and charge conjugation operators, respectively. Consider first the *parity* operator $\mathcal{P}$. Since the intrinsic parity of a fermion–antifermion pair is (-1) (see Section 16.4), the overall parity associated with the orbital angular momentum quantum number L is $P = (-1)^{L+1}$. Moreover, because positronium is composed of a particle (e$^-$) and its own antiparticle (e$^+$), its states are eigenstates of the *charge-conjugation operator C* introduced in Section 16.4. In order to determine the charge-conjugation parity C of a positronium state, we can consider the electron and the positron to be two states of the same particle, distinguished by the charge label. Since this particle is a fermion, it satisfies the Pauli exclusion principle, by which when two identical fermions are exchanged their wave function changes sign. In the present case, to exchange an electron with a positron we must exchange (a) the charges of the particles (C) and (b) the positions and spins of the particles. The exchange operation (b) introduces the factors $(-1)^L$ for the spatial part and $(-1)^{S+1}$ for the spin part, where we recall that the singlet spin function (corresponding to $S = 0$) is antisymmetric, while the triplet spin functions (corresponding to $S = 1$) are symmetric. Thus we have

$$(-1)^{L+S+1} C = -1 \tag{16.103}$$

so that

$$C = (-1)^{L+S} \tag{16.104}$$

showing that the positronium states are eigenstates of C with eigenvalues $C = (-1)^{L+S}$. As a result, the states of positronium can be labelled by the quantum numbers J, P and C, which are usually displayed in the form J^{PC}. It is interesting to note that for each combination of J^{PC} only one value of L and of S is allowed. The first few energy levels of positronium are shown in Fig. 16.16, where the fine-structure splittings have been greatly magnified and are not to scale.

Charmonium

In Section 10.2 we saw that the diversity of the hadrons can be explained in terms of a quark model. According to this model quarks can combine in two ways to form hadrons: a bound system of three quarks constitutes a baryon while a quark and an antiquark bound together form a meson. In the original form of the quark model introduced in 1964 by Gell-Mann and by Zweig, there were three 'flavours' of quarks denoted by 'u' (up), 'd' (down) and 's' (strange). Estimates of the rest energies of the u, d and s quarks show that they are comparable to the energy binding these quarks in a hadron. As a result, fully relativistic calculations must be performed to study these systems; such calculations are in general too complicated to be practical.

This situation was significantly modified in 1974, when a new meson, called the ψ particle, was discovered which could not be explained by the 'traditional' quark model with three flavours of quarks. However, a few years earlier, it had been suggested that there might exist a fourth 'flavour' of quark, the *charmed* quark c, which carries an extra additive quantum number–analogous to hypercharge–called charm. The mass of the charmed quark c was estimated to be much greater than that of the u, d and s quarks, so that a charmed quark c and its antiquark c̄ would form a non-relativistic bound system (cc̄). By analogy with positronium, the bound system (cc̄) is called

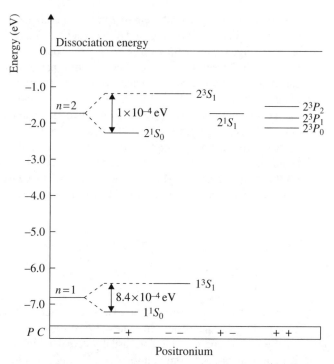

Figure 16.16 The fine structure of the $n = 1$ and $n = 2$ levels of positronium. The ordinate shows the energy of the levels in eV. The energy splitting is extremely small, being of the order of 10^{-3} eV for the $n = 1$ and 10^{-4} eV for the $n = 2$ level; it is not shown to scale.

charmonium. The ψ particle was soon identified to be a particular charmonium state (the 1^3S_1 state), and subsequently other mesons have been identified with low-lying states of charmonium (see Fig. 16.17).

Since the symmetry properties of the charmonium ($c\bar{c}$) system are identical to those discussed above for positronium, we can expect that the energy-level diagram of charmonium will resemble that of positronium. Comparison of Figures 16.17 and 16.16 shows that this is indeed the case. Of course, the energy scale is very different because the force at work in the positronium case is the electromagnetic one, while in the charmonium case it is the strong 'colour' force, 'colour' being believed to be the strong interaction's 'charge'. Looking at Figures 16.16 and 16.17, we see that the energy scale in the case of the charmonium spectrum is about 10^8 larger than that used for positronium. In particular, the ionisation potential of positronium is 6.8 eV, while charmonium becomes quasi-bound at an energy of about 630 MeV above that of the ψ particle. We also notice that the 'fine-structure' splittings are comparatively much larger in charmonium than in positronium.

The discovery of charmonium has stimulated interest in the search for still heavier quarks, such as the 'b' (bottom) quark and the 't' (top) quark (see Table 16.2). These

16.5 Positronium and charmonium

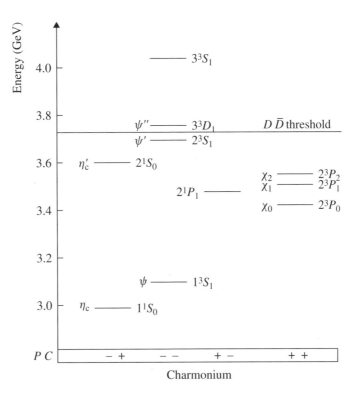

Figure 16.17 The energy levels of charmonium. The structure of the energy levels is similar to that of positronium but the splitting of the energy levels is of the order of hundreds of MeV. The levels correspond to particles which are known as the ψ, η_c and χ mesons as shown. The ordinate shows the total energy E of the levels in GeV. The corresponding particle masses are given by $M = E/c^2$. The states above the horizontal line can decay into the D- and $\bar{\text{D}}$-mesons and are only quasi-bound.

quarks, together with their corresponding antiquarks, could form bound systems $(b\bar{b})$, $(t\bar{t})$, analogous to charmonium. The word *quarkonium* is now used to denote in general a bound system made of a heavy quark and its antiquark. It should be noted that as the b and t quarks are heavier than the c quark, the non-relativistic approximation should be even more accurate to study the properties of the $(b\bar{b})$ and $(t\bar{t})$ systems. In 1977 the first $(b\bar{b})$ states were discovered and identified with Y-mesons.

Properties of the strong 'colour' force can be inferred from the quarkonium spectrum. Quantum chromodynamics–the equivalent of quantum electrodynamics for strong interactions–predicts that at *small* distances the interaction potential acting between quarks should vary inversely with r, like a Coulomb potential. On the other hand, since free quarks are not observed, the potential at large distances must be such that the quarks are 'confined' and cannot become unbound. A phenomenological

Table 16.2 The quark system. The 'flavour', name, approximate mass at rest and electric charge of the various quarks is given. All quarks have spin 1/2 and baryon number 1/3, and exist in three 'colours'.

Flavour	Symbol	Approximate mass at rest (in MeV/c^2)	Electric charge (in units of e)
up	u	2–8	2/3
down	d	5–15	−1/3
charmed	c	1000–1600	2/3
strange	s	100–300	−1/3
top	t	$(174 \pm 8) \times 10^3$	2/3
bottom	b	4100–4500	1/3

potential with these features at short and large distances is

$$V(r) = -\frac{a}{r} + br \qquad (16.105)$$

where a and b are positive constants.

17 Measurement and interpretation

17.1   Hidden variables?   759
17.2   The Einstein–Podolsky–Rosen paradox   760
17.3   Bell's theorem   762
17.4   The problem of measurement   766
17.5   Time evolution of a system. Discrete or continuous?   772

It will have become evident to the reader that quantum mechanics is a most successful theory. This has been illustrated in many applications described in this book. In fact, quantum mechanics has never been shown to fail, although some specific applications may be beyond the reach of our calculational ability. Nevertheless, quantum theory has features which seem strange compared with classical Newtonian mechanics and which some physicists have found difficult to accept. In this chapter, we shall discuss briefly some of the conceptual difficulties and experiments designed to resolve them, giving references for further study.

17.1   Hidden variables?

One of the most fundamental characteristics of quantum theory is its lack of determinism. When a single measurement of an observable A is made, the result is one of the eigenvalues a_n of A. However, unless the system is in an eigenstate of A it is impossible to predict in any particular measurement which of the eigenvalues a_n will be obtained. All that can be predicted is the probability of obtaining the eigenvalue a_n when the measurement is repeated many times on a set of identically prepared systems. This lack of determinism is quite different from anything in classical physics. It is, of course, true that many situations arise in classical physics which can only be described statistically, for example the motion of the molecules in a gas. This classical indeterminism arises merely from our lack of detailed knowledge about the positions and velocities of each molecule. It is believed that although unobservable in practice, in fact each 'classical' molecule at a given time has a well defined position and velocity, and the results of future measurements of the position and velocity of each molecule could, in principle, be determined[1]. Such considerations about

---

[1] In practice, even the smallest errors made in the initial positions and velocities will cause the future positions and velocities to be indeterminable after a short interval of time, and the behaviour of the gas becomes 'chaotic'.

classical systems have led to the supposition that quantum mechanics is an incomplete theory in that there are other variables, called '*hidden variables*', of which we are not directly aware, but which are required to determine the system completely. These hidden variables are postulated to behave in a classical deterministic manner, the apparent indeterminism exhibited by experiment arising from our lack of knowledge of the hidden sub-structure of the system studied. Thus apparently identical systems are perhaps characterised by different values of one or more hidden variables, which determine in some way which particular eigenvalues are obtained in a particular measurement.

It is of historical interest that de Broglie's original interpretation of the wave function falls into the class of hidden-variable theories. He supposed the wave function to be a physically real field propagating in space and coupled to an associated particle which has both a well defined position and momentum. The coupling between the particle and the 'pilot wave' gives rise to the observed diffraction phenomena. A deterministic theory of this type was elaborated in 1952 by D. Bohm who was able to account for the diffraction and interference shown in particle scattering, obtaining exactly the same results as those given by quantum mechanics. However, this model containing both waves and particles as separate, but connected, entities is extremely complex. To most people's minds it has even stranger features than those of quantum mechanics, and most physicists would reject it on the grounds of 'Occam's razor'[2]. Perhaps the least acceptable feature of Bohm's model, for those seeking an underlying classical mechanism to quantum theory, is its non-locality. For instance, in the analysis of the two-slit experiment (see Chapter 2) using Bohm's model there is a force acting on a particle traversing one slit which is changed instantly if the second slit is opened, or closed. Such theories, in which an action at one place is transmitted instantaneously (or at least faster than the speed of light) to alter the situation at another, are called *non-local*. As we shall see, quantum mechanics is a non-local theory, but non-locality is generally not considered to be an acceptable feature of a classical theory. It might be thought that a sufficiently ingenious hidden variable theory might be constructed, which is both deterministic and local. However, J. S. Bell in 1965 was able to lay down conditions that all deterministic local theories must satisfy. As we shall see in the next two sections these conditions are found to be violated by experiment.

17.2 The Einstein–Podolsky–Rosen paradox

The most famous physicist to question the completeness of quantum theory was A. Einstein. In 1935, in collaboration with N. Rosen and B. Podolsky, he proposed the following criteria as the basis of any acceptable theory:

(1) The quantities concerned in the theory should be 'physically real', physical

---

[2] William of Occam (1285–1349). His maxim states that 'It is vain to do with more what can be done with fewer'.

17.2 The Einstein–Podolsky–Rosen paradox

reality being defined as follows: 'If, without in any way disturbing a system, we can predict with certainty the value of a physical quantity, then there exists an element of physical reality corresponding to this physical quantity.'

(2) The theory should be local, i.e. there is no action at a distance in nature.

Einstein, Podolsky and Rosen were able to give an example of a quantum mechanical system which did not satisfy these conditions and concluded that the quantum description of nature was incomplete. Following Bohm, we shall investigate a simpler situation than that proposed by Einstein and his collaborators, but which exhibits similar features. Consider a system with total spin $S = 0$ which splits into two identical particles, 1 and 2, each of spin 1/2. When the particles are well separated let us measure the component of the spin of particle 1 parallel to some direction, which we shall define as the z-axis, Since the particle has spin 1/2, either the result $+\hbar/2$ or the result $-\hbar/2$ is obtained. Suppose that in a particular experiment the result $+\hbar/2$ is found, then since the total spin of the two-particle system is zero, a measurement of the z-component of the spin of particle 2 has to produce the result $-\hbar/2$. Subsequent measurements of the component of the spin of particle 2 parallel to the x-axis produce the results $+\hbar/2$ and $-\hbar/2$ on a fifty–fifty basis. Similarly, if the x-component of the spin of particle 1 is measured and found to be $\hbar/2$, the x-component of the spin of particle 2 must be $-\hbar/2$ and the z-component of the spin of particle 2 will be found to be $+\hbar/2$ or $-\hbar/2$ on a fifty–fifty basis. We see that the act of measuring a component of the spin of particle 1 has altered the result obtained in measuring a component of the spin of the other particle. This alteration takes place instantly no matter how far apart particles 1 and 2 may be. Thus the quantum description does not obey the conditions 1 and 2. This fact is often known as the Einstein–Podolsky–Rosen paradox. However, N. Bohr rejected the idea that the result is paradoxical, taking the view that in condition 1 physical reality can only refer to situations in which the experimental arrangement is completely specified and suggesting that this is not the case because the system is disturbed from the outset by the experimenters' decision to measure the x-component rather than the z-component of the spin of particle 1.

The quantum situation can be contrasted with what would be observed if spin were a classical variable. It would remain true that the spin components of particles 1 and 2 would be found to be equal and opposite, because the total spin is zero. However, this would be the case because the spin vectors have definite values and directions right from the beginning when the state was created, and the act of measurement on particle 1 would not change the state of particle 2 in any way. Now it might be that the quantum results could be explained by this common-cause argument. For example, there might be a classical hidden variable (or variables), the value of which was determined when the spin-zero system was created and which subsequently determined the experimental results. However, the experimental violation of Bell's theorem, which we will now discuss, shows that this explanation is in fact incorrect.

17.3 Bell's theorem

Consider again the spin-zero system of the previous paragraph, composed of two spin-1/2 particles. Choosing a particular direction z, the spin wave function is (see (6.301))

$$\chi(1,2) = \frac{1}{\sqrt{2}}[\alpha(1)\beta(2) - \beta(1)\alpha(2)]. \tag{17.1}$$

The component of the spin of particle 1 along a certain direction specified by the unit vector $\hat{\mathbf{a}}$ is $S_a(1)$ where

$$S_a(1) = \mathbf{S}(1).\hat{\mathbf{a}} = \boldsymbol{\sigma}(1).\hat{\mathbf{a}}(\hbar/2) \tag{17.2}$$

where $\boldsymbol{\sigma}(i)$ are the Pauli spin matrices for particle i. The result of a single measurement of $S_a(1)$ is either $\hbar/2$ or $-\hbar/2$. These values occur in a series of measurements on a fifty–fifty basis, so that the expectation value of $S_a(1)$ is zero. This result also follows by calculating $\langle S_a(1) \rangle$ using the wave function (17.1).

If both the components of the spin of particle 1 along $\hat{\mathbf{a}}$ and that of particle 2 along $\hat{\mathbf{b}}$ are measured jointly, the corresponding observable is $(\boldsymbol{\sigma}(1).\hat{\mathbf{a}}\ \boldsymbol{\sigma}(2).\hat{\mathbf{b}})\hbar^2/4$. The average result of a series of joint measurements of $S_a(1)$ and $S_b(2)$ is the expectation value of this operator. Denoting the expectation value of $(\boldsymbol{\sigma}(1).\hat{\mathbf{a}}\ \boldsymbol{\sigma}(2).\hat{\mathbf{b}})$ by $E(\hat{\mathbf{a}}, \hat{\mathbf{b}})$ we find using (17.1) that

$$E(\hat{\mathbf{a}}, \hat{\mathbf{b}}) = \langle \chi | \boldsymbol{\sigma}(1).\hat{\mathbf{a}}\ \boldsymbol{\sigma}(2).\hat{\mathbf{b}} | \chi \rangle = -\hat{\mathbf{a}}.\hat{\mathbf{b}}$$
$$= -\cos\phi \tag{17.3}$$

where ϕ is the angle between $\hat{\mathbf{a}}$ and $\hat{\mathbf{b}}$. If $\hat{\mathbf{a}}$ and $\hat{\mathbf{b}}$ are in the same direction, the value of $E(\hat{\mathbf{a}}.\hat{\mathbf{b}})$ is -1, which just tells us that the two spin components are both of magnitude $\hbar/2$ and of opposite sign.

Now suppose there is a hidden variable[3] λ which specifies the state of the system completely and which determines the values of the quantum variables obtained in a given experiment. The dynamical evolution of the hidden variable is considered to be subject to conditions 1 and 2 of the previous section. Each spin-zero system has a definite value of λ. When a large number of such systems are prepared identically let a fraction $p(\lambda)$ have values of the hidden variable between λ and $\lambda + d\lambda$, normalised so that

$$\int p(\lambda)d\lambda = 1, \qquad p(\lambda) \geqslant 0. \tag{17.4}$$

Let us denote the result of a measurement of the spin component $S_a(1)$ of particle 1 along a direction $\hat{\mathbf{a}}$ by $A(\hat{\mathbf{a}}, \lambda)\hbar/2$ and of a measurement of the spin component $S_b(2)$ of particle 2 along a direction $\hat{\mathbf{b}}$ by $B(\hat{\mathbf{b}}, \lambda)\hbar/2$, where $A(\hat{\mathbf{a}}, \lambda)$ and $B(\hat{\mathbf{b}}, \lambda)$ can only take the values ± 1. Since the overall spin is zero, we must have

$$A(\hat{\mathbf{a}}, \lambda) = -B(\hat{\mathbf{a}}, \lambda). \tag{17.5}$$

---

[3] There may be many such variables but this does not alter the essential nature of the discussion.

In a 'complete' theory $A(\hat{\mathbf{a}}, \lambda)$ and $B(\hat{\mathbf{b}}, \lambda)$ are entirely specified by the value of λ. In a local theory, if the measurements are made at points which are well separated, $A(\hat{\mathbf{a}}, \lambda)$ can only depend on λ and $\hat{\mathbf{a}}$ but not on $\hat{\mathbf{b}}$. Similarly, $B(\hat{\mathbf{b}}, \lambda)$ is independent of $\hat{\mathbf{a}}$, so that the result of a measurement of S_a and S_b jointly is the uncorrelated product of $A(\hat{\mathbf{a}}, \lambda)\hbar/2$ and $B(\hat{\mathbf{b}}, \lambda)\hbar/2$. The average of a series of joint measurements is $(\hbar^2/4)\mathcal{E}(\hat{\mathbf{a}}, \hat{\mathbf{b}})$, where

$$\mathcal{E}(\hat{\mathbf{a}}, \hat{\mathbf{b}}) = \int p(\lambda) A(\hat{\mathbf{a}}, \lambda) B(\hat{\mathbf{b}}, \lambda) d\lambda. \tag{17.6}$$

This is in general non-factorisable and is correlated despite the fact that the individual product $A(\hat{\mathbf{a}}, \lambda)B(\hat{\mathbf{b}}, \lambda)$ is not. However, the correlation is through the common cause, the theory remaining a local one. Note that to distinguish the averages calculated by the local hidden variable theory from the quantum result (17.3), we have employed a script $\mathcal{E}$. We notice that if $\hat{\mathbf{a}} = \hat{\mathbf{b}}$, then

$$\mathcal{E}(\hat{\mathbf{a}}, \hat{\mathbf{a}}) = E(\hat{\mathbf{a}}, \hat{\mathbf{a}}) = -1 \tag{17.7}$$

because of the conditions (17.4) and (17.5).

Now consider a joint measurement of the component of the spin of 1 along $\hat{\mathbf{a}}$ and of 2 along $\hat{\mathbf{c}}$, where $\hat{\mathbf{c}}$ is a different direction from $\hat{\mathbf{b}}$. We find that

$$\mathcal{E}(\hat{\mathbf{a}}, \hat{\mathbf{b}}) - \mathcal{E}(\hat{\mathbf{a}}, \hat{\mathbf{c}}) = \int p(\lambda)\{A(\hat{\mathbf{a}}, \lambda)B(\hat{\mathbf{b}}, \lambda) - A(\hat{\mathbf{a}}, \lambda)B(\hat{\mathbf{c}}, \lambda)\} d\lambda$$

$$= -\int p(\lambda) A(\hat{\mathbf{a}}, \lambda) A(\hat{\mathbf{b}}, \lambda)\{1 + A(\hat{\mathbf{b}}, \lambda)B(\hat{\mathbf{c}}, \lambda)\} d\lambda \tag{17.8}$$

where we have used (17.5), together with the fact that $\{A(\hat{\mathbf{b}}, \lambda)\}^2 = 1$.

The inequality

$$|\mathcal{E}(\hat{\mathbf{a}}, \hat{\mathbf{b}}) - \mathcal{E}(\hat{\mathbf{a}}, \hat{\mathbf{c}})| < \int p(\lambda)\{1 + A(\hat{\mathbf{b}}, \lambda)B(\hat{\mathbf{c}}, \lambda)\} d\lambda \tag{17.9}$$

or

$$|\mathcal{E}(\hat{\mathbf{a}}, \hat{\mathbf{b}}) - \mathcal{E}(\hat{\mathbf{a}}, \hat{\mathbf{c}})| < 1 + \mathcal{E}(\hat{\mathbf{b}}, \hat{\mathbf{c}}) \tag{17.10}$$

then follows because A only takes the values $+1$ or -1, and $p(\lambda)$ is always non-negative. This is one of a family of inequalities deduced by Bell, which must be satisfied by any real local theory. It is easy to find directions $\hat{\mathbf{a}}, \hat{\mathbf{b}}$ and $\hat{\mathbf{c}}$ for which the quantum expectation values violate (17.10), so we can conclude that *quantum mechanics is inherently a non-local theory*. For example, let the angle between $\hat{\mathbf{a}}$ and $\hat{\mathbf{c}}$ be $2\pi/3$ and let $\hat{\mathbf{b}}$ lie in the same plane as $\hat{\mathbf{a}}$ and $\hat{\mathbf{c}}$ making an angle $\pi/3$ with each of them. Then from (17.3) we see that the quantum expectation values satisfy the relations

$$|E(\hat{\mathbf{a}}, \hat{\mathbf{b}}) - E(\hat{\mathbf{a}}, \hat{\mathbf{c}})| = |-\cos(2\pi/3) + \cos(\pi/3)| = 1 \tag{17.11}$$

and

$$1 + E(\hat{\mathbf{b}}, \hat{\mathbf{c}}) = [1 - \cos(\pi/3)] = \frac{1}{2} \qquad (17.12)$$

with the consequence that

$$|E(\hat{\mathbf{a}}, \hat{\mathbf{b}}) - E(\hat{\mathbf{a}}, \hat{\mathbf{c}})| > 1 + E(\hat{\mathbf{b}}, \hat{\mathbf{c}}) \qquad (17.13)$$

which violates the inequality (17.10).

Experimental verification

Now the question arises as to whether the correlations observed experimentally agree with the quantum result and violate the locality condition, or whether Bell's inequality holding for real local theories is satisfied. Since 1972 a number of experiments of increasing accuracy have been carried out and analysed using an elaboration of the preceding theory which allows for the less than perfect efficiencies of the detectors. Rather than using particles most experiments have measured the states of polarisation of a two-photon system in which the photons propagate in opposite directions. In the experiments of A. Aspect and co-workers carried out in 1982, the $4p^2\ ^1S_0$ level of calcium is populated by two-photon excitation, using two lasers – a krypton laser at 4060 Å and a tunable dye laser at 5810 Å. This level decays back into the $4s^2\ ^1S_0$ level by cascading through the $4s4p\ ^1P_1$ level (see Fig. 17.1) with the emission of two photons of wavelengths 5513 Å and 4227 Å, respectively. Since the total angular momentum is zero in both the initial $4p^2\ ^1S_0$ and final $4s^2\ ^1S_0$ levels, the overall angular momentum of the two-photon system is also zero. As a consequence, if the photons propagate in opposite directions parallel to the z-axis, the polarisation part of the wave function is of the form

$$\psi(1, 2) = \frac{1}{\sqrt{2}}[u(1)u(2) + v(1)v(2)] \qquad (17.14)$$

where $u(i)$ represents a photon linearly polarised along the x-axis and $v(i)$ a photon linearly polarised along the y-axis. The polarisations are thus completely correlated and similar arguments apply to those we outlined for the spin one-half particles. The principle of the experiment is illustrated in Fig. 17.2. Two photons, $h\nu_1$ and $h\nu_2$, emerge in opposite directions parallel to the z-axis from the source S, and are detected by the linear polarisation analysers I and II. The analysers I and II are orientated parallel to the unit vectors $\hat{\mathbf{a}}$ and $\hat{\mathbf{b}}$, respectively, which lie in xy planes normal to the direction of propagation. These analysers record the result $+1$ if a photon is found to be linearly polarised parallel to $\hat{\mathbf{a}}$ (or $\hat{\mathbf{b}}$) and the result -1 when the polarisation is normal to $\hat{\mathbf{a}}$ (or $\hat{\mathbf{b}}$). Four coincidence rates, $N_{++}(\hat{\mathbf{a}}, \hat{\mathbf{b}})$, $N_{--}(\hat{\mathbf{a}}, \hat{\mathbf{b}})$, $N_{+-}(\hat{\mathbf{a}}, \hat{\mathbf{b}})$ and $N_{-+}(\hat{\mathbf{a}}, \hat{\mathbf{b}})$ are measured, where $N_{++}(\hat{\mathbf{a}}, \hat{\mathbf{b}})$ is the rate at which results $+1$ are obtained in I oriented parallel to $\hat{\mathbf{a}}$ and II oriented parallel to $\hat{\mathbf{b}}$ simultaneously. The rates N_{--}, N_{+-} and N_{-+} are defined similarly.

17.3 Bell's theorem   765

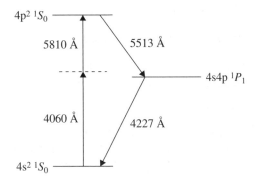

Figure 17.1 Study of two-photon correlations. The $4p^2\ ^1S_0$ level of calcium is populated by two-photon excitation. The correlations between the polarisation of two photons emitted in the decay back to the $4s^2\ ^1S_0$ level, via the 4s4p ^{1}P level, are studied in the experiment of Aspect and co-workers.

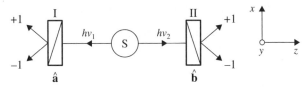

Figure 17.2 Schematic diagram of an experiment to measure the correlation of linear polarisation when two photons $h\nu_1$ and $h\nu_2$ are emitted from a source S. *(By courtesy of A. Aspect.)*

A correlation coefficient $E(\hat{\mathbf{a}}, \hat{\mathbf{b}})$ can be defined in terms of the four coincidence rates as

$$E(\hat{\mathbf{a}}, \hat{\mathbf{b}}) = \frac{N_{++}(\hat{\mathbf{a}}, \hat{\mathbf{b}}) + N_{--}(\hat{\mathbf{a}}, \hat{\mathbf{b}}) - N_{+-}(\hat{\mathbf{a}}, \hat{\mathbf{b}}) - N_{-+}(\hat{\mathbf{a}}, \hat{\mathbf{b}})}{N_{++}(\hat{\mathbf{a}}, \hat{\mathbf{b}}) + N_{--}(\hat{\mathbf{a}}, \hat{\mathbf{b}}) + N_{+-}(\hat{\mathbf{a}}, \hat{\mathbf{b}}) + N_{-+}(\hat{\mathbf{a}}, \hat{\mathbf{b}})} \tag{17.15}$$

The coefficient $E(\hat{\mathbf{a}}, \hat{\mathbf{b}})$ can be calculated using quantum mechanics by an argument rather similar to that used in obtaining the spin-1/2 result (17.3). In the present case it is found that

$$E(\hat{\mathbf{a}}, \hat{\mathbf{b}}) = \cos 2\phi \tag{17.16}$$

where ϕ is the angle between $\hat{\mathbf{a}}$ and $\hat{\mathbf{b}}$.

The corresponding correlation coefficient $\mathcal{E}(\hat{\mathbf{a}}, \hat{\mathbf{b}})$ for a real local hidden variable theory is found by following a similar reasoning to that given above for the case of spin-1/2 particles. We again introduce a hidden variable (or set of variables) λ, and a density function $p(\lambda)$ satisfying the condition (17.4). The result of a measurement of the polarisation when the analyser I is parallel to $\hat{\mathbf{a}}$ is denoted by $A(\hat{\mathbf{a}}, \lambda)$, where $A(\hat{\mathbf{a}}, \lambda)$ can take the values $+1$ or -1, while $B(\hat{\mathbf{b}}, \lambda)$ is the corresponding quantity for a measurement using the analyser II. The correlation function $\mathcal{E}(\hat{\mathbf{a}}, \hat{\mathbf{b}})$ is then of exactly the same form as (17.6), that is

$$\mathcal{E}(\hat{\mathbf{a}}, \hat{\mathbf{b}}) = \int p(\lambda) A(\hat{\mathbf{a}}, \lambda) B(\hat{\mathbf{b}}, \lambda) d\lambda. \tag{17.17}$$

Out of the several Bell's inequalities which can be deduced, the one chosen in the experiment of Aspect and co-workers is based on a quantity S defined as

$$S = \mathcal{E}(\hat{\mathbf{a}}, \hat{\mathbf{b}}) - \mathcal{E}(\hat{\mathbf{a}}, \hat{\mathbf{b}}') + \mathcal{E}(\hat{\mathbf{a}}', \hat{\mathbf{b}}) + \mathcal{E}(\hat{\mathbf{a}}', \hat{\mathbf{b}}') \tag{17.18}$$

where $\hat{\mathbf{a}}, \hat{\mathbf{a}}'$ and $\hat{\mathbf{b}}, \hat{\mathbf{b}}'$ represent two different orientations of the analysers I and II, respectively. Provided A and B can only take the values $+1$ or -1 and that (17.4) and (17.17) hold, it is not difficult to obtain the Bell inequality

$$-2 \leqslant S \leqslant 2. \tag{17.19}$$

On the other hand, if the correlations $\mathcal{E}$ in (17.18) are replaced by the corresponding quantum correlations E, defined by (17.16), $|S|$ can exceed 2 for various orientations of the four vectors $\hat{\mathbf{a}}, \hat{\mathbf{a}}', \hat{\mathbf{b}}$ and $\hat{\mathbf{b}}'$. In the 1982 experiments of A. Aspect and co-workers, the orientations were chosen such that

$$\hat{\mathbf{a}}.\hat{\mathbf{b}} = \hat{\mathbf{b}}.\hat{\mathbf{a}}' = \hat{\mathbf{a}}'.\hat{\mathbf{b}}' = \cos\phi \tag{17.20a}$$

and

$$\hat{\mathbf{a}}.\hat{\mathbf{b}}' = \cos 3\phi. \tag{17.20b}$$

The experimental values of S were then obtained for various angles ϕ between $0°$ and $90°$. The results are shown in Fig. 17.3 together with the predictions of quantum mechanics (the solid curve). The experimental results violate Bell's inequalities and are in excellent agreement with the quantum predictions. Further experiments, reviewed by Zeilinger (1999) also show a violation of Bell's inequalities. Hence it can be concluded that no local hidden variable theory underlies quantum mechanics. As we pointed out earlier, it is possible to construct non-local hidden variable theories which lead to the same experimental consequences as quantum mechanics, but there is no compelling reason to introduce such complexities.

17.4 The problem of measurement

The process of measurement differs in quantum mechanics in several respects from that in classical mechanics. For example, we have already seen that *complementary* pairs of variables exist, such as momentum and position which can only be measured simultaneously up to limits imposed by the uncertainty relations. The word measurement itself is used in two rather different senses in quantum mechanics, which it is important to distinguish. First, there are those measurements, often called *ideal measurements*, which can be repeated immediately. For example, if a beam of atoms, each of spin one-half, is passed through a Stern–Gerlach apparatus (see Chapter 1), orientated to split the beam in the z-direction, one of the two emerging

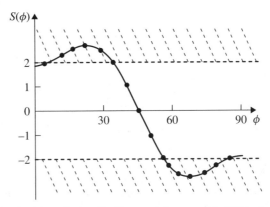

Figure 17.3 Experimental data obtained for the correlation quantity $S(\phi)$ by A. Aspect, P. Grangier and G. Roger. The solid curve represents the prediction of quantum mechanics and the shaded areas are those in which Bell's inequality (17.19) is violated. (*By courtesy of A. Aspect.*)

beams (see Fig. 17.4) contains only atoms with $S_z = +\hbar/2$ and the other only atoms with $S_z = -\hbar/2$. This measurement, which is better called *state preparation*, can be immediately repeated by passing, for instance, the $S_z = +\hbar/2$ beam through a second Stern–Gerlach apparatus, orientated in the same way as the first. If this is done, all the particles will emerge deflected in the direction corresponding to $S_z = \hbar/2$ and none in that corresponding to $S_z = -\hbar/2$. The passing of a beam of atoms through a Stern–Gerlach apparatus prepares an ensemble of atoms in a known state and is not really a measurement in the usual sense. To measure whether or not an atom has spin component $S_z = \hbar/2$ or $S_z = -\hbar/2$, it is not only necessary to pass the atom through such an apparatus but also to detect it. The act of detection usually involves the absorption of a photon, or the capture of a particle, which are not repeatable processes. Thus measurements which can be immediately repeated on the same system are the exception rather than the rule. To repeat, we will distinguish between those 'ideal measurements' in which an ensemble of objects is prepared in a reproducible state, and true measurements which involve the detection of the object together with the determination of one or more of its properties, within certain limits–for example, energy, position, momentum, and so on. The former process allows one to say something about the future behaviour of the system, while the latter gives information about its immediate past.

The relationship between the wave function Ψ, or the corresponding state vector $|\Psi\rangle$, and the results of measurements have been discussed in Chapter 5. It is all-important to recognise that a knowledge of $|\Psi\rangle$ does not allow us to predict the results of a measurement on a single object of an observable A, except in the special case for which $|\Psi\rangle$ is an eigenstate of A belonging to some eigenvalue a_n, in which case the value a_n will be obtained. Apart from this case, $|\Psi\rangle$ only determines the frequency of obtaining a particular eigenvalue when the measurement is repeated a large number of times on a set of identically prepared objects or systems (called an

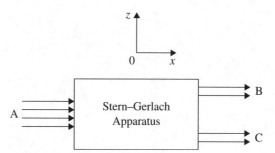

Figure 17.4 An example of state preparation. A Stern–Gerlach apparatus is orientated to split an unpolarised beam of spin one-half atoms A in the z-direction. The upper energy beam B contains only atoms with $S_z = +\hbar/2$ and the lower beam C contains only atoms with $S_z = -\hbar/2$.

ensemble of systems). It is therefore incorrect to associate a wave function with a single object (with the exception noted). On the contrary, $|\Psi\rangle$ provides us with the maximum possible information about an ensemble of identically prepared systems. It is important to be clear that the individual systems do not possess precise (but unknown) values of the variable A before a measurement is made. This would amount to the kind of hidden variable theory that we have seen to be untenable; rather a member of the ensemble cannot be said to possess a value for A (unless $|\Psi\rangle$ is an eigenvector of A) until a measurement has been performed. This view, although apparently correct, gives rise to some difficulties in interpretation, as Schrödinger pointed out in his famous cat paradox.

Schrödinger's cat

In this thought experiment, a cat is placed in a closed box together with a radioactive atom and a device which releases a deadly poison on the decay of the atomic nucleus. If the wave function of a live cat is ψ_l and that of the initial state of the atom is ϕ_i, and if the wave function of a dead cat is ψ_d and that of the final state of the atom is ϕ_f, then after a time t the wave function of the system will be the superposition

$$\Psi(t) = \psi_l\, \phi_i \exp[-t/(2\tau)] + \psi_d\, \phi_f [1 - \exp(-t/\tau)]^{1/2} \tag{17.21}$$

where τ is the lifetime of the radioactive atom. On opening the box at a certain time, the cat is clearly found to be either alive or dead. This is an act of measurement which forces the system into one of the eigenstates of being alive or being dead. It is an example of the reduction or collapse of the wave function on measurement, referred to in Chapter 2. All sorts of curious questions now arise. Before the box is opened the cat is apparently neither alive or dead, but in a superposition; what does that mean? If the cat is found to be dead, did it die on opening the box, or before? All these paradoxical questions arise in the most acute form if the wave function is held to determine the properties of a particular cat. Part of the difficulty is resolved by recognising that the wave function (17.21) refers to a large number of identically prepared cat experiments. After a time t, a certain fraction of the cats will have died and a certain fraction will be alive, as determined by $\Psi(t)$, but no prediction of the time of death of an individual cat can be made. This picturesque thought experiment is a paradigm of the problems that arise in interpreting the meaning of an entangled state such as (17.21) if a wave function is associated with a single object rather than an ensemble. This particular example should not, of course, be taken too far; there are no pure alive or dead cat states. A cat is a complex thermodynamical system at finite temperature describable (in principle!) by a density matrix with no coherence, and hence no interference, between 'alive' and 'dead' states.

An analysis of measurement

Let us now try to analyse the quantum mechanics of measurement a little more closely. We shall suppose we have an apparatus which is capable of measuring the value of an observable A. The apparatus will be taken to be macroscopic and capable of recording the result of the measurement on a piece of paper or on a magnetic tape or other medium. The result recorded in a single measurement must be one of the eigenvalues a_n of A, and for the purpose of discussion we will imagine that there is no experimental error. Let the eigenvector of A corresponding to the eigenvalue a_n be denoted by $|\phi_n\rangle$, and the set of eigenvectors $|\phi_n\rangle$, $n = 1, 2, \ldots$, be complete. The apparatus will be described by a very complicated wave function, but when presented with a system described by the eigenvector $|\phi_n\rangle$ it must record the value a_n for A, so that the apparatus must possess distinct eigenstates $|\Phi_n\rangle$, one for each n.

Before the measurement, the initial eigenvector describing the apparatus can be denoted by $|\Phi_0\rangle$, and if the system being measured is in the state $|\phi_n\rangle$, the initial state vector of the (system + apparatus) is

$$|\Psi(t)\rangle = |\phi_n\rangle \otimes |\Phi_0\rangle = |\phi_n; \Phi_0\rangle. \tag{17.22}$$

This state vector will satisfy some time-dependent Schrödinger equation, and will evolve so that after the measurement it will become

$$|\Psi(t')\rangle = |\phi_n; \Phi_n\rangle. \tag{17.23}$$

The apparatus will now be in the state with eigenvector $|\Phi_n\rangle$, and the value a_n will have been recorded, while the measured system will be described by the state vector $|\phi_n\rangle$. More generally, if the initial state of the system is described by the state vector $|\psi(t)\rangle$ having the expansion

$$|\psi(t)\rangle = \sum_n b_n |\phi_n\rangle \tag{17.24}$$

then the final state of the combined (system + apparatus) will be described by $|\Psi(t')\rangle$ where

$$|\Psi(t')\rangle = \sum_n b_n |\phi_n; \Phi_n\rangle. \tag{17.25}$$

According to our interpretation, the meaning of $|\Psi(t')\rangle$ is that in a large number of measurements on the ensemble of identically prepared systems, the fraction in which the value a_n is found is given by $|b_n|^2$, all state vectors being normalised to unity. We notice again that if we attempt to associate $|\Psi(t')\rangle$ with a single system the equation (17.25) would be difficult to interpret.

The reduction of the wave function

Instead of concentrating on the combined (system + apparatus), let us now see what happens if we work in terms of the state vector of the system only. We take the special case in which the measurements are repeatable. Before the measurement the system is described by $|\psi\rangle$ with the expansion (17.24). After making a large number of measurements on members of the system, we know that a fraction $|b_n|^2$ of these will have given the result a_n. That fraction of the system will be described after the measurements by the corresponding eigenvector $|\phi_n\rangle$. Looking at the whole ensemble, we see that after the measurements, we have a *mixed state*, a fraction $|b_1|^2$ of the system being in the pure state $|\phi_1\rangle$, $|b_2|^2$ in $|\psi_2\rangle$, and so on. This situation can be described by a density matrix ρ_{nm} which is diagonal in the basis $|\phi_n\rangle$ and is given by

$$\rho_{nm} = |b_n|^2 \delta_{nm}. \tag{17.26}$$

The change from the pure state $|\psi\rangle$ to the mixed state described by (17.26) is an example of the reduction of the wave function on measurement. This has caused confusion because there is no way in which a linear Schrödinger equation can convert a pure state into a mixed state. However, we see that this reduction is not present in the treatment of the complete physical system (measured system + apparatus) and arises because we choose to direct our attention to the measured system only – it does not correspond to any physical process in nature.

We now ask the question as to whether the two descriptions of the state of affairs after the measurement, by the complete wave function (17.25) or by the density matrix (17.26) are mutually consistent. To do this, consider some different observable C. For consistency we must obtain the same expectation value of C from either description. Starting from the mixed state we have

$$\langle C \rangle = \text{Tr}(\rho C) = \sum_n |b_n|^2 \langle \phi_n | C | \phi_n \rangle. \tag{17.27}$$

On the other hand, from the pure state (17.25), we have

$$\langle C \rangle = \langle \Psi(t') | C | \Psi(t') \rangle$$
$$= \sum_m \sum_n b_m^* b_n \langle \phi_m; \Phi_m | C | \phi_n; \Phi_n \rangle. \tag{17.28}$$

The matrix element $\langle \phi_m; \Phi_m | C | \phi_n; \Phi_n \rangle$ describes a transition in which the apparatus changes the record from a_n to a_m. However, the making of a record is essentially an irreversible process; the record is *indelible*. No process exists which will undo a measurement and delete a record, substituting another. Hence the matrix element of C must be diagonal in n. The operator C does not alter $|\Phi_n\rangle$ and (17.28) becomes equal to (17.27). There is no inconsistency between the two descriptions, and the use of the reduction of the wave function is a matter of choice. It is very useful in that a detailed description of the apparatus is not then needed.

Other views

In the previous paragraph, we have followed in part the views of L. E. Ballentine (1970) and F. J. Belinfante (1975), but other views have been expressed, often based on the application of the wave function to individual systems, rather than ensembles, and which sometimes lead to bizarre and unusual conclusions. The basic problem is that if a single system of (object + measuring apparatus) is described by the coherent entangled wave function (17.25), rather than an ensemble of such systems, then the measuring apparatus is not in a definite eigenstate with the consequence that no definite result of the measurement can be obtained. We have to explain why the pointer on a measuring apparatus takes a definite position and why no interference effects are observed.

One explanation stems from the idea of *decoherence*. A coherent system does not remain coherent for ever since the coherence is destroyed by interaction with other objects. It has been suggested that the interaction between the macroscopic measuring apparatus and the environment destroys the coherence between the different terms in (17.25) so rapidly that it can be treated as being instantaneous. This results in the pure state (17.25) reducing effectively to a statistical mixture of the states $|\phi_n; \Phi_n\rangle$ rather than a superposition of those states. The apparatus will end in one or another of the states $|\Phi_n\rangle$, which one is not known until the record is inspected.

Another approach is to go beyond conventional quantum mechanics by introducing an extra term into the Schrödinger equation so that the collapse of the wave function during a measurement occurs as a result of the dynamics. In these attempts the collapse of the wave function is treated as a physical process, in contrast to the position taken by the majority of physicists in which the collapse is a convenient shorthand allowing the dynamics of the measuring apparatus and of the measured system to be studied separately.

Many more explanations of the process of measurement have been offered. One of the most well known is the 'many-worlds' hypothesis of H. Everett. This accepts all the usual formulation of quantum mechanics including the superposition of different states of the measuring apparatus indicated by (17.25). The fact that a measurement results in a definite outcome is attributed to a relationship between the wave function (17.25) and the consciousness of the experimenter. Each term in (17.25) corresponds to a different state of awareness of the experimenter and the actual outcome results from his perception. This can be expressed by saying that when a measurement is made the world splits into many copies which cannot communicate with one another, and the outcome of the measurement belongs to one such world.

Several books describing these and other approaches to the measurement problem have been published recently and the interested reader can with advantage consult Peres (1993), or the non-technical account of Squires (1994).

17.5 Time evolution of a system. Discrete or continuous?

A feature of quantum mechanics related to the problem of measurement concerns the time evolution of a system. Consider for example an atom prepared in an excited energy eigenstate. After a certain time, the atom may make a spontaneous transition to a state of lower energy, emitting a photon. Since the final state of the atom has to be an energy eigenstate, the transition must occur instantaneously. It follows that the time evolution of a single system must consist of a series of instantaneous changes, known as *quantum jumps*. In the early days of quantum theory, this feature, which was clearly predicted by the matrix mechanics developed by Heisenberg, Born and Jordan, appeared to be in conflict with Schrödinger's wave mechanics. This is because the Schrödinger equation predicts a continuous, smooth evolution of the wave function with time. This apparent conflict was resolved when Born proposed his probabilistic interpretation of the wave function, in which the continuously evolving wave function determines the probability that a discrete change in state occurs at a given time. This resolution of the difficulty never entirely satisfied Schrödinger, who is reported to have remarked to Bohr: 'If I had known it would come to this damned jumps then I am sorry I ever got involved in quantum theory'.

At the time of the Bohr–Schrödinger debate in 1926, and until very recently, no direct observation of quantum jumps could be made because all experiments were concerned with ensemble averages over many quantum systems. New developments in experimental techniques have radically changed the situation and experiments can now be performed on single atoms. The first observations of quantum jumps were made in 1986 by scattering light from a single Ba^+ ion confined in an electromagnetic trap. A simplified version of the original idea put forward by H. G. Dehmelt in 1975 can be explained with reference to Fig. 17.5. An atom in the ground state 1 can be excited to a strongly coupled state 2 by absorbing a photon from a laser of the appropriate frequency. The fluorescent radiation due to the decay of the atom from state 2 back to state 1 can be easily observed as this is a strong transition. The atom under consideration is selected to have a third level 3, which is weakly coupled to the excited state 2, and for which the decay back to the ground state 1 is strongly inhibited. The level 3, known as the *shelving level*, has a very long lifetime, with the result that if a transition occurs from level 2 to level 3 the fluorescence is suddenly cut off. The abrupt ending of the fluorescence radiation indicates when the transition to the shelving state has taken place. After a certain period, the atom in level 3 will slowly decay by some mechanism to level 1 and the fluorescence will resume.

This method is only applicable if a single atom can be observed, for otherwise the abrupt switching off of the fluorescence would be wiped out by the radiation from other atoms. The fluorescence signal recorded from single Ba^+ ions in an experiment carried out by Th. Sauter, R. Blatt, W. Neuhauser and P. E. Toschek in 1986 is shown in Fig. 17.6. The sudden alternations between light and dark periods is clearly exhibited. The dark periods are of varying duration, and can be studied using statistical methods. The results clearly establish the existence of quantum jumps, and

17.5 Time evolution of a system. Discrete or continuous?   773

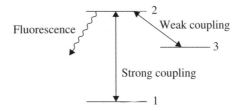

Figure 17.5 Schematic diagram of a three-level system containing a shelving state.

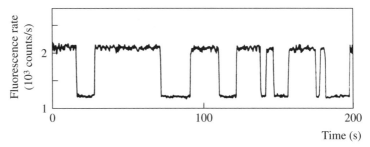

Figure 17.6 The fluorescence rate from a Ba$^+$ ion showing dark periods caused by transitions to a shelving state.

are also consistent with the smooth time evolution of probability amplitudes predicted by the Schrödinger equation. A detailed discussion of the theoretical implications of these experiments can be found in the review of Stenholm and Wilkens (1997).

A Fourier integrals and the Dirac delta function

In this appendix, we shall give a brief introduction to the Fourier integrals and the Dirac delta function, which are used in Chapter 2 and in subsequent chapters.

A.1 Fourier series

Consider a single-valued function $f(x)$ defined in the interval $-\pi \leqslant x \leqslant \pi$, and determined outside this interval by the condition $f(x + 2\pi) = f(x)$, so that $f(x)$ has the period 2π. The Fourier series corresponding to $f(x)$ is defined to be

$$f(x) = \frac{1}{2}A_0 + \sum_{n=1}^{\infty}[A_n \cos(nx) + B_n \sin(nx)]. \tag{A.1}$$

Provided $f(x)$ and $f'(x)$ are piecewise continuous in the interval $(-\pi, +\pi)$, this series converges. The constants $A_0, A_1, \ldots, A_m, \ldots$ are determined by multiplying both sides of equation (A.1) by $\cos mx$ and integrating over the interval $-\pi$ to π. We find

$$A_m = \frac{1}{\pi}\int_{-\pi}^{\pi} f(x)\cos(mx)\mathrm{d}x, \qquad m = 0, 1, \ldots. \tag{A.2}$$

Similarly, by multiplying (A.1) throughout by $\sin mx$ and integrating, we find

$$B_m = \frac{1}{\pi}\int_{-\pi}^{\pi} f(x)\sin(mx)\mathrm{d}x, \qquad m = 1, 2, \ldots. \tag{A.3}$$

Since $\cos(nx)$ and $\sin(nx)$ can be written in terms of the complex exponentials $\exp(\pm inx)$, the Fourier expansion (A.1) can be written in the alternative form

$$f(x) = (2\pi)^{-1/2}\sum_{n=-\infty}^{\infty} C_n \exp(inx) \tag{A.4}$$

where the factor $(2\pi)^{-1/2}$ has been inserted to make the formulae that will come later more symmetrical in appearance. The coefficients C_n can be found directly using the relation

$$(2\pi)^{-1}\int_{-\pi}^{\pi} \exp[i(n-m)x]\mathrm{d}x = \delta_{mn} \tag{A.5}$$

where δ_{mn} is the Kronecker delta symbol defined as

$$\begin{aligned}\delta_{mn} &= 1, \quad &\text{if } m = n \\ &= 0, \quad &\text{if } m \neq n.\end{aligned} \tag{A.6}$$

Multiplying both sides of (A.4) by $(2\pi)^{-1/2}\exp(-imx)$, integrating over x and using (A.5), we find that

$$C_m = (2\pi)^{-1/2} \int_{-\pi}^{\pi} f(x)\exp(-imx)dx \qquad (A.7)$$

where m is a positive or negative integer, or zero.

A function $f(x)$ defined in some other interval $(-L, +L)$ and periodic with period $2L$, so that $f(x+2L) = f(x)$, can also be expanded in a Fourier series by making the change of variable $x \to \pi x/L$. We have from (A.4)

$$f(x) = (2\pi)^{-1/2} \sum_{n=-\infty}^{\infty} C_n \exp(in\pi x/L) \qquad (A.8)$$

where the coefficients C_m are given by

$$C_m = L^{-1}\left(\frac{\pi}{2}\right)^{1/2} \int_{-L}^{L} f(x)\exp(-im\pi x/L)dx. \qquad (A.9)$$

A.2 Fourier transforms

Frequently, the functions with which we have to deal are not periodic, but are defined for all real values of x, $-\infty < x < \infty$. Such functions can also be expressed in terms of complex exponentials by taking the limit $L \to \infty$ in (A.8) and (A.9). As L increases in value the difference between successive terms in the series (A.8) becomes smaller and smaller, and the sum over n can be replaced by an integral:

$$f(x) = (2\pi)^{-1/2} \int_{-\infty}^{\infty} C_n \exp(in\pi x/L)dn. \qquad (A.10)$$

Introducing a new variable k by the relation

$$k = n\pi/L \qquad (A.11)$$

and defining a new function $g(k)$ by

$$g(k) = LC_n/\pi \qquad (A.12)$$

the integral (A.10) can be written as

$$f(x) = (2\pi)^{-1/2} \int_{-\infty}^{\infty} g(k)\exp(ikx)dk. \qquad (A.13)$$

By taking the limit $L \to \infty$ in (A.9) we find

$$g(k) = (2\pi)^{-1/2} \int_{-\infty}^{\infty} f(x)\exp(-ikx)dx. \qquad (A.14)$$

The integrals (A.13) and (A.14) are known as *Fourier integrals*. The function $g(k)$ is said to be the *Fourier transform* of $f(x)$ and, similarly, $f(x)$, is the Fourier transform of $g(k)$. A function $f(x)$ can only be expressed as a Fourier transform if the integrals (A.13) and (A.14) converge. This will be the case if $f(x)$ and $g(k)$ are square-integrable functions, which means that

$$\int_{-\infty}^{\infty} |f(x)|^2 dx < \infty \quad \text{and} \quad \int_{-\infty}^{\infty} |g(k)|^2 dk < \infty. \tag{A.15}$$

The Dirac delta function

By inserting $g(k)$ given by (A.14) into (A.13) we find that

$$f(x) = (2\pi)^{-1} \int_{-\infty}^{\infty} \left[\int_{-\infty}^{\infty} f(x') \exp(-ikx') dx' \right] \exp(ikx) dk \tag{A.16}$$

or

$$f(x) = \int_{-\infty}^{\infty} f(x') \delta(x - x') dx' \tag{A.17}$$

where in (A.16) the order of the integrals has been reversed and the function $\delta(x - x')$ has been defined by

$$\delta(x - x') = (2\pi)^{-1} \int_{-\infty}^{\infty} \exp[ik(x - x')] dk. \tag{A.18}$$

This procedure is certainly open to question since the integral in (A.18) does not converge! However we shall see that nevertheless a meaning can be attached to the function $\delta(x - x')$, which was introduced by P. A. M. Dirac and is known as the *Dirac delta function*. Since $f(x)$ is an arbitrary function, we note from (A.17) that $\delta(x - x')$ must be zero if $x \neq x'$, and the integral over $\delta(x - x')$ must have the value unity. Such a function does not exist, but we can consider $\delta(x - x')$ as the limit of a sequence of functions increasingly peaked about the point $x = x'$. The width of the peak must decrease and its height must increase along the sequence in such a way that the area under the peak is always unity. The result of this limiting procedure is called a *distribution* (see Lighthill 1958).

To look further into this question, let us set $x' = 0$ in (A.18) and introduce a factor $\exp(-\varepsilon|k|)$ to make the integral converge, where $\varepsilon > 0$. Writing

$$\delta_\varepsilon(x) = (2\pi)^{-1} \int_{-\infty}^{\infty} \exp(ikx - \varepsilon|k|) dk \tag{A.19}$$

we can consider $\delta(x)$ to be the limit of $\delta_\varepsilon(x)$ as $\varepsilon \to 0^+$. From (A.19) we have

$$\delta_\varepsilon(x) = (2\pi)^{-1} \left[\frac{1}{ix + \varepsilon} - \frac{1}{ix - \varepsilon} \right]$$
$$= \frac{\varepsilon/\pi}{x^2 + \varepsilon^2}. \tag{A.20}$$

We see that $\delta_\varepsilon(x)$ is exactly the kind of function that we require. As ε becomes very small $\delta_\varepsilon(x)$ also is very small, except near the point $x = 0$, where it has the large value $1/(\pi\varepsilon)$. In the limit $\varepsilon \to 0^+$, it is easy to verify that the integral over $\delta_\varepsilon(x)$ is equal to unity provided the range of integration includes the point $x = 0$, and is zero otherwise:

$$\lim_{\varepsilon \to 0^+} \int_a^b \delta_\varepsilon(x) dx \to 1, \qquad a < 0, \qquad b > 0. \tag{A.21}$$

The representation of the delta function in terms of the sequence of functions (A.20), as $\varepsilon \to 0^+$, is not unique and any other sequence of peaked functions $\delta_\varepsilon(x)$, with unit area under the peak and with $\delta_\varepsilon(x) \to 0$ as $\varepsilon \to 0^+$ for $x \neq 0$, can be used. For example the sequence of functions

$$\delta_\varepsilon(x) = (\pi \varepsilon)^{-1/2} \exp(-x^2/\varepsilon) \tag{A.22}$$

also tends to the Dirac delta function as $\varepsilon \to 0^+$. Other representations of the delta function are

$$\delta(x) = \lim_{\varepsilon \to 0^+} \frac{\sin(x/\varepsilon)}{\pi x} \tag{A.23}$$

$$\delta(x) = \lim_{\varepsilon \to 0^+} \frac{\varepsilon}{\pi x^2}[1 - \cos(x/\varepsilon)] \tag{A.24}$$

$$\delta(x) = \lim_{\varepsilon \to 0} \frac{\theta(x + \varepsilon) - \theta(x)}{\varepsilon} \tag{A.25a}$$

where in the last equation $\theta(x)$ denotes the step function

$$\theta(x) = \begin{cases} 1, & x > 0 \\ 0, & x < 0. \end{cases} \tag{A.25b}$$

The Dirac delta function is a very useful concept since by its use we can avoid explicitly representing various limiting processes and we can interchange orders of integration as in (A.16) without explicitly introducing convergence factors.

The main properties of the Dirac delta function are the following:

$$\int_a^b f(x)\delta(x - x_0)dx = f(x_0) \qquad \text{if } a < x_0 < b$$

$$= 0 \qquad \text{if } x_0 < a \quad \text{or} \quad x_0 > b \tag{A.26}$$

$$\delta(x) = \delta(-x) \tag{A.27}$$

$$x\delta(x) = 0 \tag{A.28}$$

$$\delta(ax) = \frac{1}{|a|}\delta(x), \qquad a \neq 0 \tag{A.29}$$

$$f(x)\delta(x - a) = f(a)\delta(x - a) \tag{A.30}$$

$$\int \delta(a - x)\delta(x - b)dx = \delta(a - b) \tag{A.31}$$

$$\delta[g(x)] = \sum_i \frac{1}{|g'(x_i)|}\delta(x - x_i) \tag{A.32}$$

where in the last relation x_i is a zero of $g(x)$ and $g'(x_i) \neq 0$. Particular cases of (A.32) are

$$\delta[(x-a)(x-b)] = \frac{1}{|a-b|}[\delta(x-a) + \delta(x-b)], \qquad a \neq b \tag{A.33}$$

and

$$\delta(x^2 - a^2) = \frac{1}{2|a|}[\delta(x-a) + \delta(x+a)]. \tag{A.34}$$

The derivative of the delta function

$$\delta'(x) = \frac{\mathrm{d}}{\mathrm{d}x}\delta(x) \tag{A.35}$$

can also be given a meaning, since

$$\int_a^b f(x)\delta'(x)\mathrm{d}x = [f(x)\delta(x)]_a^b - \int_a^b \delta(x)f'(x)\mathrm{d}x$$
$$= -f'(0) \tag{A.36}$$

where we have assumed that the point $x = 0$ lies in the interval (a, b). We also have

$$\int_a^b f(x)\delta'(x - x_0)\mathrm{d}x = -f'(x_0), \qquad a < x_0 < b \tag{A.37}$$

$$\delta'(x) = -\delta'(-x) \tag{A.38}$$

$$x\delta'(x) = -\delta(x) \tag{A.39}$$

$$x^2\delta'(x) = 0 \tag{A.40}$$

$$\int \delta'(a-x)\delta(x-b)\mathrm{d}x = \delta'(a-b) \tag{A.41}$$

$$\delta'(x) = \mathrm{i}\,(2\pi)^{-1} \int_{-\infty}^{+\infty} k \exp(\mathrm{i}kx)\mathrm{d}k. \tag{A.42}$$

Further properties of Fourier Transforms

Returning to equation (A.13), we have that

$$\int_{-\infty}^{\infty} |f(x)|^2 \mathrm{d}x$$
$$= (2\pi)^{-1} \int_{-\infty}^{\infty} \mathrm{d}x \left[\int_{-\infty}^{\infty} g^*(k) \exp(-\mathrm{i}kx)\mathrm{d}k \int_{-\infty}^{\infty} g(k') \exp(\mathrm{i}k'x)\mathrm{d}k' \right]$$
$$= \int_{-\infty}^{\infty} g^*(k) \left[\int_{-\infty}^{\infty} g(k')\delta(k'-k)\mathrm{d}k' \right] \mathrm{d}k$$
$$= \int_{-\infty}^{\infty} |g(k)|^2 \mathrm{d}k. \tag{A.43}$$

This result is known as *Parseval's theorem*.

The *convolution* of two functions f_1 and f_2 is defined as the integral

$$F(x) = \int_{-\infty}^{\infty} f_1(y) f_2(x - y) dy. \tag{A.44}$$

A straightforward calculation then shows that if $G(k)$ is the Fourier transform of $F(x)$ and $g_1(k)$ and $g_2(k)$ are the Fourier transforms of $f_1(x)$ and $f_2(x)$, respectively, then

$$G(k) = (2\pi)^{1/2} g_1(k) g_2(k). \tag{A.45}$$

This property is known as the *convolution theorem*.

It is easy to generalise Fourier transforms to functions in three dimensions, $f(x, y, z)$. In place of (A.13) and (A.14), we have

$$f(x, y, z) = (2\pi)^{-3/2} \int_{-\infty}^{\infty} dk_x \int_{-\infty}^{\infty} dk_y \int_{-\infty}^{\infty} dk_z \, g(k_x, k_y, k_z)$$
$$\times \exp(ik_x x + ik_y y + ik_z z) \tag{A.46}$$

and

$$g(k_x, k_y, k_z) = (2\pi)^{-3/2} \int_{-\infty}^{\infty} dx \int_{-\infty}^{\infty} dy \int_{-\infty}^{\infty} dz \, f(x, y, z)$$
$$\times \exp(-ik_x x - ik_y y - ik_z z). \tag{A.47}$$

These relations are more conveniently written in a vector notation as

$$f(\mathbf{r}) = (2\pi)^{-3/2} \int g(\mathbf{k}) \exp(i\mathbf{k}\cdot\mathbf{r}) d\mathbf{k} \tag{A.48}$$

and

$$g(\mathbf{k}) = (2\pi)^{-3/2} \int f(\mathbf{r}) \exp(-i\mathbf{k}\cdot\mathbf{r}) d\mathbf{r}. \tag{A.49}$$

The generalisation of the relation (A.18) is then

$$\delta(\mathbf{r} - \mathbf{r}') \equiv \delta(x - x')\delta(y - y')\delta(z - z')$$
$$= (2\pi)^{-3} \int \exp[i\mathbf{k}\cdot(\mathbf{r} - \mathbf{r}')] d\mathbf{k} \tag{A.50}$$

and we also note the property

$$\int f(\mathbf{r}')\delta(\mathbf{r} - \mathbf{r}') d\mathbf{r}' = f(\mathbf{r}). \tag{A.51}$$

In three dimensions Parseval's theorem (A.43) is generalised to

$$\int |f(\mathbf{r})|^2 d\mathbf{r} = \int |g(\mathbf{k})|^2 d\mathbf{k} \tag{A.52}$$

and the *convolution theorem* states that if

$$F(\mathbf{r}) = \int f_1(\mathbf{R}) f_2(\mathbf{r} - \mathbf{R}) d\mathbf{R} \tag{A.53}$$

then
$$G(\mathbf{k}) = (2\pi)^{3/2} g_1(\mathbf{k}) g_2(\mathbf{k}) \qquad \text{(A.54)}$$
where $G(\mathbf{k})$, $g_1(\mathbf{k})$ and $g_2(\mathbf{k})$ are the Fourier transforms of $F(\mathbf{r})$, $f_1(\mathbf{r})$ and $f_2(\mathbf{r})$, respectively.

B WKB connection formulae

In Chapter 8, WKB approximations are obtained for the solution of the one-dimensional Schrödinger equations in regions well separated from a classical turning point. We shall now show how a solution can be found in the vicinity of a classical turning point which will allow the WKB approximations to the left and right of the turning point to be connected. There are several methods which can be followed, but we shall only describe one of them, known as a *uniform approximation*.

The case of a left-hand barrier will be discussed with a turning point at $x = x_1$. The WKB solution is given in the region $x > x_1$ by (8.192) and in the region $x < x_1$ by (8.193). The local classical momentum $p(x)$, defined by (8.175), is zero at the turning point. For $x > x_1$, $p^2(x)$ is positive and for $x < x_1$, $p^2(x)$ is negative so that for x sufficiently close to x_1, we can approximate $p^2(x)$ by

$$p^2(x) \simeq A(x - x_1), \qquad A = \left[\frac{\mathrm{d}p^2(x)}{\mathrm{d}x}\right]_{x=x_1} \tag{B.1}$$

where the constant A is positive, $A > 0$.

Using (B.1), the Schrödinger equation

$$\left[\frac{\mathrm{d}^2}{\mathrm{d}x^2} + \hbar^{-2}p^2(x)\right]\psi(x) = 0 \tag{B.2}$$

can be approximated in the immediate vicinity of the turning point by the equation

$$\left[\frac{\mathrm{d}^2}{\mathrm{d}x^2} + A\hbar^{-2}(x - x_1)\right]\psi(x) = 0. \tag{B.3}$$

Defining a new variable q by the relation

$$q(x) = (\hbar^{-2}A)^{1/3}(x - x_1) \tag{B.4}$$

and setting

$$\Psi(q(x)) = \psi(x) \tag{B.5}$$

equation (B.3) becomes

$$\left[\frac{\mathrm{d}^2}{\mathrm{d}q^2} + q\right]\Psi(q) = 0. \tag{B.6}$$

The general solution of this equation is well known and can be expressed as a linear combination of Airy functions[1] $Ai(-q)$ and $Bi(-q)$:

$$\Psi(q) = \lambda Ai(-q) + \mu Bi(-q) \tag{B.7}$$

where λ and μ are arbitrary constants.

The function $Ai(-q)$ is bounded and has the asymptotic forms

$$Ai(-q) \sim \pi^{-1/2} q^{-1/4} \cos\left(\frac{2}{3} q^{3/2} - \frac{\pi}{4}\right), \qquad q \to \infty$$

$$\sim \frac{1}{2} \pi^{-1/2} |q|^{-1/4} \exp\left(-\frac{2}{3} |q|^{3/2}\right), \qquad q \to -\infty \tag{B.8}$$

while the linearly independent function $Bi(-q)$ is unbounded as $q \to -\infty$ and has the asymptotic forms

$$Bi(-q) \sim -\pi^{-1/2} q^{-1/4} \sin\left(\frac{2}{3} q^{3/2} - \frac{\pi}{4}\right), \qquad q \to \infty$$

$$\sim \pi^{-1/2} |q|^{-1/4} \exp\left(\frac{2}{3} |q|^{3/2}\right), \qquad q \to -\infty. \tag{B.9}$$

Having obtained in (B.7) a solution of the Schrödinger equation which is accurate near the turning point $x = x_1$ corresponding to $q = 0$, we shall now show that the original Schrödinger equation (B.2) can be *approximately* transformed to equation (B.6) over the whole interval $-\infty < x < \infty$, so that (B.7) becomes an approximate solution for all values of x. We first write down the WKB solutions of (B.6) which are

$$\Psi(q) = N_\pm (\hbar^2 |q|)^{-1/4} \exp\left[\pm \frac{i}{\hbar} \int_0^q (\hbar^2 q')^{1/2} dq'\right] \tag{B.10}$$

where N_+ and N_- are arbitrary constants. Evaluating the integral we obtain

$$\Psi(q) = N_\pm (\hbar^2 |q|)^{-1/4} \exp\left[\pm i \frac{2}{3} q^{3/2}\right]. \tag{B.11}$$

It should be noted that q is always real and that $q > 0$ for $x > x_1$ and $q < 0$ for $x < x_1$. We now compare (B.10) and (B.11) with the WKB solutions of the original Schrödinger equation, which we can write as

$$\psi(x) = N_\pm [|p^2(x)|]^{-1/4} \exp\left[\pm \frac{i}{\hbar} \int_{x_1}^x p(x') dx'\right]. \tag{B.12}$$

This comparison suggests that $\psi(x)$ can be approximated by a function

$$\bar{\psi}(x) = \left[\frac{\hbar^2 |q(x)|}{|p^2(x)|}\right]^{1/4} \Psi(q(x)) \tag{B.13}$$

---

[1] See M. Abramowitz and I. A. Stegun (1965).

where $q(x)$ is defined by the implicit relation

$$\frac{2}{3}[q(x)]^{3/2} = \frac{1}{\hbar}\int_{x_1}^{x} p(x')\mathrm{d}x', \qquad p^2 > 0 \tag{B.14a}$$

$$= \mathrm{e}^{\pm 3\pi\mathrm{i}/2}\frac{1}{\hbar}\int_{x_1}^{x} |p(x')|\mathrm{d}x', \quad p^2 < 0. \tag{B.14b}$$

The phase factor in (B.14b) has been chosen so that q is real and negative for $p^2 < 0$. We see that near $x = x_1$, the relations (B.14) reduce to (B.4), and it follows that $\bar{\psi}(x)$ is an accurate solution of the Schrödinger equation in this region. To see how accurately $\bar{\psi}(x)$ approximates $\psi(x)$ in general, we note that using (B.6) together with (B.13) and (B.14), we find that $\bar{\psi}(x)$ satisfies exactly the equation

$$\left[\frac{\mathrm{d}^2}{\mathrm{d}x^2} + \hbar^{-2}p^2(x) + \varepsilon(x)\right]\bar{\psi}(x) = 0 \tag{B.15}$$

where

$$\varepsilon(x) = -\left|\frac{\mathrm{d}q}{\mathrm{d}x}\right|^{1/2}\frac{\mathrm{d}^2}{\mathrm{d}x^2}\left\{\left|\frac{\mathrm{d}q}{\mathrm{d}x}\right|^{-1/2}\right\}. \tag{B.16}$$

It follows that $\bar{\psi}(x)$ is a good approximation to $\psi(x)$ provided that

$$\left|\frac{\hbar^2\varepsilon(x)}{p^2(x)}\right| \ll 1. \tag{B.17}$$

At the turning point $\varepsilon(x) = 0$ and the mapping of the Schrödinger equation (B.2) onto (B.6) is exact. Far from the turning point the condition (B.17) is a weaker condition than (8.186) which expresses the accuracy of the WKB approximation (B.12). It follows that $\bar{\psi}(x)$, given by (B.13) with $\Psi(q(x))$ expressed as (B.7), is a function which coincides with the WKB solutions (B.12) of the Schrödinger equation far from the turning point and is an accurate solution of the Schrödinger equation in the immediate vicinity of the turning point.

Inspecting (B.8) and (B.9), we see that to obtain the solution which is bounded to the left of the turning point we must set $\mu = 0$ in (B.7) so that $\psi(x)$ is given approximately for all values of x by

$$\psi(x) \simeq \lambda\left[\frac{\hbar^2|q(x)|}{|p^2(x)|}\right]^{1/4}\mathrm{Ai}(-q(x)). \tag{B.18}$$

Using the asymptotic form of the Airy function Ai given by (B.8), together with (B.14), immediately establishes the connection formula (8.194). In the same way, to obtain the solution asymptotic to $[p(x)]^{-1/2}\sin[(1/\hbar)\int_{x_1}^{x} p(x')\mathrm{d}x' - \pi/4]$ for $x > x_1$ which increases exponentially to the left of $x = x_1$ we set $\lambda = 0$ in (B.7) and using (B.9), we immediately obtain the connection formula (8.195).

References

Abramowitz, M. and Stegun, I. A. (1965) *Handbook of Mathematical Functions.* Dover, New York.

Ashoori, R. C. (1996) *Nature*, **379**, 413.

Ballentine, L. E. (1970) 'The Statistical Interpretation of Quantum Mechanics'. *Reviews of Modern Physics*, **42**, 358.

Belinfante, F. J. (1975) *Measurements and Time Reversal in Objective Quantum Theory.* Pergamon, Oxford.

Bell, W. W. (1968) *Special Functions for Scientists and Engineers.* Van Nostrand, London.

Bohm, D. (1951) *Quantum Theory.* Prentice Hall, New York.

Bransden, B. H. (1983) *Atomic Collision Theory*, 2nd edn. Benjamin, New York.

Bransden, B. H. and Joachain, C. J. (1983) *Physics of Atoms and Molecules.* Longman, London.

Burcham, W. E. and Jobes, M. (1995) *Nuclear and Particle Physics.* Longman, London.

Byron, F. W. and Fuller, R. W. (1969) *Mathematics of Classical and Quantum Physics.* Addison-Wesley, Reading, Massachusetts.

Dirac, P. A. M. (1958) *The Principles of Quantum Mechanics*, 4th edn. Oxford University Press, New York.

Duffin, W. J. (1968) *Advanced Electricity and Magnetism for Undergraduates.* McGraw-Hill, London.

Edmonds, A. R. (1957) *Angular Momentum in Quantum Mechanics.* Princeton University Press, Princeton, New Jersey.

Feller, W. (1970) *An Introduction to Probability Theory and its Applications*, 3rd edn. Wiley, New York.

Feynman, R. P. (1972) *Statistical Mechanics.* Benjamin, Reading, Massachusetts.

Goldstein, H. (1980) *Classical Mechanics*, 2nd edn. Addison-Wesley, Reading, Massachusetts.

Halzen, F. and Martin, A. D. (1984) *Quarks and Leptons.* Wiley, New York.

Hänsch, T. W. and Walther, H. (1999) *Reviews of Modern Physics*, **71**, S242.

Heisenberg, W. (1949) *The Physical Principles of the Quantum Theory.* Dover, New York.

Huang, K. (1998) *Quantum Field Theory*, Wiley, New York.

Jackson, J. D. (1998) *Classical Electrodynamics*, 3rd edn. Wiley, New York.

Joachain, C. J. (1983) *Quantum Collision Theory*, 3rd edn. North-Holland, Amsterdam.

Joachain, C. J., Dörr, M. and Kylstra, N. J. (2000) *Advances in Atomic, Molecular and Optical Physics*, **42**, 225.

Kibble, T. W. B. and Berkshire, F. H. (1996) *Classical Mechanics*, 4th edn. Longman, London.

Kittel, C. and Kroemer, H. (1980) *Thermal Physics*. Freeman, San Francisco.

Lighthill, M. J. (1958) *Introduction to Fourier Analysis and Generalised Functions*. Cambridge University Press, London.

Marion, J. B. and Heals, M. A. (1980) *Classical Electromagnetic Radiation*, 2nd edn. Academic Press, New York.

Mathews, J. and Walker, R. L. (1973) *Mathematical Methods of Physics*, World Student Series edn. Benjamin, New York.

Messiah, A. (1968) *Quantum Mechanics*. North-Holland, Amsterdam.

Mourou, G. A., Barty, C. P. J. and Perry, M. D. (1998) *Physics Today*, **51**, 22.

Peres, A. (1993) *Quantum Theory: Concepts and Methods*. Kluwer, Dordrecht.

Richtmyer, F. K., Kennard, E. H. and Cooper, J. N. (1969) *Introduction to Modern Physics*, 6th edn. McGraw-Hill, New York.

Rose, M. E. (1957) *Elementary Theory of Angular Momentum*. Wiley, New York.

Squires, E. J. (1994) *The Mystery of the Quantum World*. Institute of Physics, Bristol.

Stenholm, S. and Wilkens, M. (1997) *Contemporary Physics*, **38**, 257.

Svelto, O. (1982) *Principles of Lasers*, 2nd edn. Plenum, New York.

Taylor, E. F. and Wheeler, J. A. (1966) *Spacetime Physics*. Freeman, San Francisco.

Watson, G. N. (1966) *A Treatise on the Theory of Bessel Functions*. Cambridge University Press, Cambridge.

Weinberg, S. (1977) *The First Three Minutes*. Basic Books, New York.

Zeilinger, A. (1999) *Reviews of Modern Physics*, **71**, S288.

Table of fundamental constants

Quantity	Symbol	Value
Planck's constant	h	$6.626\,18 \times 10^{-34}$ J s
	$\hbar = \dfrac{h}{2\pi}$	$1.054\,59 \times 10^{-34}$ J s
Velocity of light in vacuum	c	$2.997\,92 \times 10^8$ m s^{-1}
Elementary charge (absolute value of electron charge)	e	$1.602\,19 \times 10^{-19}$ C
Permeability of free space	μ_0	$4\pi \times 10^{-7}$ H m^{-1}
		$= 1.256\,64 \times 10^{-6}$ H m^{-1}
Permittivity of free space	$\varepsilon_0 = \dfrac{1}{\mu_0 c^2}$	$8.854\,19 \times 10^{-12}$ F m^{-1}
Gravitational constant	G	6.672×10^{-11} N m^2 kg^{-2}
Fine structure constant	$\alpha = \dfrac{e^2}{4\pi \varepsilon_0 \hbar c}$	$\dfrac{1}{137.036} = 7.297\,35 \times 10^{-3}$
Avogadro's number	N_A	$6.022\,05 \times 10^{23}$ mol^{-1}
Faraday's constant	$F = N_A e$	$9.648\,46 \times 10^4$ C mol^{-1}
Boltzmann's constant	k	$1.380\,66 \times 10^{-23}$ J K^{-1}
Gas constant	$R = N_A k$	$8.314\,41$ J mol^{-1} K^{-1}
Atomic mass unit	a.m.u. $= \dfrac{1}{12} M_{12_C}$	$1.660\,57 \times 10^{-27}$ kg
Electron mass	m or m_e	$9.109\,53 \times 10^{-31}$ kg
		$= 5.485\,80 \times 10^{-4}$ a.m.u.
Proton mass	M_p	$1.672\,65 \times 10^{-27}$ kg
		$= 1.007\,276$ a.m.u.
Neutron mass	M_n	$1.674\,92 \times 10^{-27}$ kg
		$= 1.008\,665$ a.m.u.
Ratio of proton to electron mass	M_p/m	1836.15
Electron charge to mass ratio	$|e|/m$	$1.758\,80 \times 10^{11}$ C kg^{-1}
Compton wavelength of electron	$\lambda_C = \dfrac{h}{mc}$	$2.426\,31 \times 10^{-12}$ m
Classical radius of electron	$r_0 = \dfrac{e^2}{4\pi \varepsilon_0 m c^2}$	$2.817\,94 \times 10^{-15}$ m

Table of fundamental constants

Quantity	Symbol	Value
Bohr radius for atomic hydrogen (with infinite nuclear mass)	$a_0 = \dfrac{4\pi\varepsilon_0 \hbar^2}{me^2}$	$5.291\,77 \times 10^{-11}$ m
Non-relativistic ionisation potential of atomic hydrogen for infinite nuclear mass	$I_P^H(\infty) = \dfrac{e^2}{8\pi\varepsilon_0 a_0} = \dfrac{1}{2}\alpha^2 mc^2$	$2.179\,91 \times 10^{-18}$ J $= 13.6058$ eV
Rydberg's constant for infinite nuclear mass	$R(\infty) = \dfrac{me^4}{8\varepsilon_0^2 h^3 c} = \dfrac{\alpha}{4\pi a_0}$	$1.097\,37 \times 10^7$ m^{-1}
Rydberg's constant for atomic hydrogen	R_H	$1.096\,78 \times 10^7$ m^{-1}
Bohr magneton	$\mu_B = \dfrac{e\hbar}{2m}$	$9.274\,08 \times 10^{-24}$ J T^{-1}
Nuclear magneton	$\mu_N = \dfrac{e\hbar}{2M_p}$	$5.050\,82 \times 10^{-27}$ J T^{-1}
Electron magnetic moment	$\mathcal{M}_e$	$9.284\,83 \times 10^{-24}$ J T^{-1} $= 1.001\,16\,\mu_B$
Proton magnetic moment	$\mathcal{M}_p$	$1.410\,62 \times 10^{-26}$ J T^{-1} $= 2.792\,85\,\mu_N$
Neutron magnetic moment	$\mathcal{M}_n$	$-0.966\,30 \times 10^{-26}$ J T^{-1} $= -1.913\,15\,\mu_N$

Table of conversion factors

1 Å(angström) = 0.1 nm = 10^{-10} m = 10^{-8} cm

1 fm (femtometre or Fermi) = 10^{-6} nm = 10^{-15} m

λ (in Å) $\times \tilde{\nu}$ (in cm^{-1}) = 10^8 (from $\lambda\tilde{\nu} = 1$)

$a_0 = 5.29177 \times 10^{-11}$ m = 0.529177 Å

$a_0^2 = 2.80028 \times 10^{-21}$ m^2

$\pi a_0^2 = 8.79735 \times 10^{-21}$ m^2

1 Hz = 1 s^{-1}

1 electron mass (m_e) = 0.511003 MeV/c^2

1 proton mass (M_p) = 938.280 MeV/c^2

1 a.m.u. = $\dfrac{1}{12} M_{12_C}$ = 1.66057×10^{-27} kg = 931.502 MeV/c^2

1 J = 10^7 erg = 0.239 cal = 6.24146×10^{18} eV

1 cal = 4.184 J = 2.611×10^{19} eV

1 eV = 1.60219×10^{-19} J = 1.60219×10^{-12} erg

1 MeV = 1.60219×10^{-13} J = 1.60219×10^{-6} erg

1 eV corresponds to:

    a frequency of 2.41797×10^{14} Hz (from $E = h\nu$)

    a wavelength of 1.23985×10^{-6} m = 12398.5 Å (from $E = hc/\lambda$)

    a wave number of 8.06548×10^5 m^{-1} = 8065.48 cm^{-1} (from $E = hc\tilde{\nu}$)

    a temperature of 1.16045×10^4 K (from $E = kT$)

1 cm^{-1} corresponds to:

    an energy of 1.23985×10^{-4} eV

    a frequency of 2.99792×10^{10} Hz

1 atomic unit of energy = 27.2116 eV corresponds to:

    a frequency of 6.57968×10^{15} Hz

    a wavelength of 4.55633×10^{-8} m = 455.633 Å

    a wave number of 2.19475×10^7 m^{-1} = 219475 cm^{-1}

    a temperature of 3.15777×10^5 K

1 a.m.u. corresponds to an energy of 931.502 MeV = 1.49244×10^{-10} J

$kT = 8.61735 \times 10^{-5}$ eV at $T = 1$ K

$hc = 1.23985 \times 10^{-6}$ eV $\times$ m = 12398.5 eV $\times$ Å

$\hbar c = 1.97329 \times 10^{-7}$ eV $\times$ m = 1973.29 eV $\times$ Å

ΔE (in eV) $\times \Delta t$ (in s) = 6.58218×10^{-16} eV $\times$ s (from $\Delta E \Delta t = \hbar$)

Index

absorption of radiation
   absorption spectra, 20–2
   cross section for, 526
   in the dipole approximation, 524–6
action integral, 241
adiabatic approximation, 447–53
   application of, 454–5
adiabatic theorem, 452–3, 501
Aharonov-Bohm effect, 458, 567–71
Airy functions, 784
allowed transition 535
alpha decay, 421–6
ammonia inversion spectrum, 736–40
ammonia maser, 736–42
Anderson, C.D., 716
Anderson, M.H., 672
angular momentum
   addition of, 315–22
   in the Bohr model, 24
   eigenvalues of $\mathbf{J}^2$, J_z, 223–6
   general theory of, 292–9
   and rotations, 270–5, 312–15
   vector model for, 288–9
   see also orbital angular momentum,
   spin angular momentum,
   total angular momentum
anharmonic oscilltor, 383–5
annihilation, 716–17
anticommutator, 305–6
antihydrogen, 366
antineutrino, 464
antiparticle, 366, 684, 685, 747
artificial atoms *see* quantum dots
Aspect, A., 53, 764–7
atomic nucleus *see* nucleus
average value *see* expectation value

Balmer, J., 20
Balmer series, 20–21
band spectra, 291–2, 500

Barkla, C.G., 16
barrier penetration, 145, 417–20
baryon, 477, 746
baryon number, 747
basis set, 220
Baur, G., 366
Bell, J.S., 760
Bell's inequalities, 763
Bell's theorem, 762–7
Berry, M.V, 456
Berry connection, 457–8, 502
Berry curvature, 458
Berry phase, 456–8
   and the Aharonov-Bohm effect, 568
   and a rotating magnetic field, 566–7
Bessel spherical functions, 342–4
beta decay, 464–6, 749
binding energy, 27, 164
black body radiation, 1–11
   and age of the universe, 10–11
Blatt, R., 773
Bless, A.A., 19
Bloch waves, 182–4, 725
Bloch's theorem, 184
Bohm, D., 760
Bohr, N., 23, 761, 772
Bohr frequency, 23
Bohr model for one-electron atoms, 23–31
   finite nuclear mass in, 28–30
Bohr radius, 27, 29
Boltzmann, L., 3
Born, M., 56, 207, 448, 772
Born approximation
   in potential scattering, 615–20
   for scattering involving composite systems, 626–35
Born series, 615
Bose-Einstein condensation, 670–4
Bose-Einstein statistics, 474, 663–6

Bose, S.N., 671
boson, 299, 474, 663
Bothe, W., 19
boundary conditions, 157
    periodic, 187, 331
box normalisation, 115–16, 139–40, 208, 210, 332
Brackett series, 21
Bragg, W.L., 16
Breit-Wigner formula, 601
Brillouin, L., 408
Brillouin zone, 726
de Broglie, L., 38, 39, 760
de Broglie's hypothesis, 38–45
de Broglie wavelength, 40
    thermal, 671

canonical ensemble, 657–9
Carnal, O., 43
central-field approximation, 492–7
    corrections to, 497–8
centre of mass coordinates, 237–8
    for scattering, 590–2
Chambers, R.G., 569
charge conjugation, 718, 755
charmonium, 753, 755–7
Christenson, J.H., 752
Chu, S., 673
classical limit, 97–100, 149, 179–80, 256–9
Clebsch-Gordan coefficients, 317–20
Cleeton, C.E., 740
closure, 140, 206, 210
cold emission of electrons, 420–1
collapse of the wave packet, 67, 769–72
collisions, *see* scattering
colour, *see* quarks
commutators, 96, 210
    algebra of, 213
    for angular momentum, 266–8, 269–70, 273–5
    and compatibility, 211–12
    fundamental, 210
    and unitary transformations, 217
commuting observables, 211–13
    complete set of, 213
compatibility, 211–12
complementarity, 55, 76

completeness, 205
Compton, A.H., 16
Compton effect, 16–19
Compton wavelength, 19
Condon, E.U., 421
conduction, 730–2
conduction band, 187, 724
confluent hypergeometric function, 357
connection formulae for WKB approximation, 412–15, 783–5
conservation laws, 245–55
conservation of probability, 64, 85–90
constant of the motion, 235, 249
continuity equation, 89
continuous spectrum, 112, 208–10
convolution theorem, 780–1
Cornell, E.A., 672
correlation effects in atoms, 497
correspondence principle, 30, 61, 97, 179
cosmic black body radiation, 10–11
Coulomb barrier, 421–6
Coulomb gauge, 517, 519
Coulomb scattering, 619–20, 623, 626
    Mott formula for, 623, 626
covariant notation, 690–4
cross-sections
    differential, 589
    total, 589
Cronin, J.W., 752
crystal lattice, 182

Darwin, C.G., 714
Darwin term, 714
Davisson, C.J., 40
Debye, P., 11
decoherence, 771
degeneracy, 102, 204
Dehmelt, H.G., 712, 773
delta function *see* Dirac delta function
delta function normalisation, 140–1, 330
density matrix, 642–55
    equation of motion for, 654–5
    reduced, 645
    for scattering of spin 1/2 by spin 0 particles, 649–53
    for spin 1/2 system, 645–9
density of states, 210, 442, 479
density operator, 643

see density matrix
deuterium, 29, 366
deuteron, 29, 507–10
diatomic molecules
    electronic structure of, 500–4
    rotational levels, 291–2, 504–5
    vibrational levels, 504–5
dipole approximation, 520–2
Dirac, P.A.M., 30, 51, 443, 684, 777
Dirac delta function, 64, 777–9
Dirac equation, 684–92
    adjoint equation, 658–9
    central potential solutions of, 702–10
    covariant formulation of, 680–4
    for a free particle, 685–6
    for a charged particle in an electromagnetic field, 688
    for one-electron atoms, 706–10
    non-relativistic limit of, 699, 711
    plane wave solutions of, 696–702
    and hole theory, 715–16
    and Lorentz transormations, 694–5
    and negative energy states, 698, 701, 715–16
    and spin, 701
Dirac matrices α, β, 685–98, 693
    see also gamma matrices
Dirac notation, 197–9, 233
distribution function, 666
dynamical state, 193–4
dynamical variable, 94, 126, 198

effective mass, 729–30
effective range, 600–1
Ehrenfest, P., 98
Ehrenfest's theorem, 98–100, 257–8
eigenfunctions, 101, 199
    closure relation of, 206
    completeness of, 119, 205
    for the continuous spectrum, 208–10
    eigenket, 227
    of energy, 103
    expansion in, 119, 203–6
    hydrogenic, 351–63
    of L_z, 275–6
    of $\mathbf{L}^2$, 277–89
    of momentum, 135, 330
    normalisation, 104, 115–16, 208–10
    orthogonality, 115–18, 209
    for the linear harmonic oscillator, 176–8
    for spin, 299–300
    for the square well, 163–70
    see also wave functions
eigenvalue equations, 101, 199–200, 223
eigenvalues, 101
    degenerate, 101, 204–5
    for J_z, $\mathbf{J}^2$, 292–6
    for L_z, $\mathbf{L}^2$, 277–9
    for S_z, $\mathbf{S}^2$, 299
    for S_n, 310–11
    see also energy eigenvalues
eigenvectors, 223
    for spin 1, 300
    for spin 1/2, 304, 309–11
Einstein, A., 1, 9, 11, 14, 671, 760
Einstein A and B coefficients, 527–9
Einstein-Podolsky-Rosen paradox, 760–1
electric dipole moment, 521–2
    matrix elements of, 531–2
    and selection rules, 533–5
electromagnetic radiation, 516-18
    in the dipole approximation, 520–2
    energy density of, 517
electron diffraction, 40–5
electron dynamics in crystals, 724–32
electron spin resonance, 578–84
emission of radiation
    cross section for, 527
    in the dipole approximation, 527–9
    emission spectra, 20–2
    spontaneous, 515–16, 527–31
    stimulated, 516, 526–7
Endo, J., 53
energy bands, 185–8, 723–4
energy eigenvalues, 102–3
    for a free particle, 134–5, 329–31
    for a free particle in a box, 139, 156–9, 331–3
    for a hydrogenic atom, 354–7
    for an infinite square well, 158
    for an infinite well, 113–14
    for an isotropic oscillator, 333–5, 369–70
    for a linear harmonic oscillator, 174–5

for a potential well, 104–14
for a square well, 163–70, 349–51
for a three-dimensional oscillator, 333–5
in WKB approximation, 415–17
energy level spectrum
of crystals, 723–4
of helium, 485–92
of one-electron atoms, 25–9, 354–7
of a square well, 349–51
see also energy eigenvalues
energy operator, 59
ensembles, 194, 655–61
canonical, 657–9
ensemble average, 656
grand canonical, 660–1
micro-canonical, 656–7
Ensher, J.R., 672
entangled states, 472, 476, 769
entangled wave functions *see* entangled states
entropy, 657
equipartition of energy, 663
Esterman, I., 43
expectation values, 90–3, 125–6, 200
for particles with spin, 302
time variation of, 98–100, 234–5
Euler-Lagrange equation, 241
Everett, H., 772
Ezawa, H., 53

Feld, M., 745
Fermi-Dirac statistics, 474, 666, 674–5
Fermi, E., 443
Fermi energy, 420, 482, 675
Fermi gas, 478–82, 674–5
and degenerate stars, 483–5
and free-electron model, 482–3
Fermi temperature, 675
fermion, 300, 474–7
Feynman, R.P., 240
Feynman-Hellmann theorem, 726
fine stucture *see* hydrogenic atoms, positronium
Fitch, V.L., 752
flavour *see* quarks
Floquet's theorem, 183
Fock, V., 448

Fourier integrals, 776
Fourier series, 775–6
Fourier transforms, 776–81
Fowler-Nordheim formula, 421
Franck, J., 31
Franck and Hertz experiment, 31–3
free particle, 58–69, 134–41
Frisch, R., 43, 672
fugacity, 661

Galilean transformations, 255–6
gamma matrices, 692–3
gamma ray microscope, 70–1
Gamow, G., 421
Gamow factor, 424
gauge invariance, 516
Geiger, H.W., 19, 22
Gell-Mann, M., 477
Gerlach, W., 33
Germer, L.H., 41
Gibbs, J.W., 657
golden rule, 443, 445, 629
Gordon, J., 736
Goudsmit, S., 37
Grangier, P., 53, 767
grand canonical ensemble, 660–1
grand partition function, 660
Green's functions
for free particle wave equation, 609–13
for free particle radial equation, 614
group velocity, 61, 68, 136
Gurney, R.W., 421
gyromagnetic ratio, 37, 563

hadron, 304, 477, 646, 746
Hallwachs, W., 12
Hamiltonian operator, 85, 470
for particle in electromagnetic field, 519
Hamilton-Jacobi equation, 258–9
Hamilton's characterestic function, 259
Hamilton's principal function, 259
Hamilton, W.R., 259
Hankel spherical functions, 344
Hänsch, T., 672
harmonic generation, 746
harmonic oscillator
charged, in electric field, 454–5, 461–4

isotropic, 334–5, 367–70
linear, 170–82
three dimensional, 334–5
Hartree-Fock approximation, 496–7
heat capacity of metals, 482–3
Heisenberg, W., 30, 772
Heisenberg equations of motion, 239–40
Heisenberg picture, 238–9
Heisenberg uncertainty principle *see* uncertainty principle
helicity, 700–2
helium *see* two-electron atoms
Hermite equation, 172
Hermite polynomials, 175–6
Hermitian operator, 89
Hertz, G., 31
Hertz, H., 12
hidden variables, 759–67
holes, 731–2
hole theory, 715–6
Houston, W.V., 710
hydrogenic atom, 351–67
  fine structure of, 390–6, 710
  gravitational energy of, 385
  in a magnetic field, 574–6
  photoionisation of, 545–50
  relativistic theory of, 706–10
  spectrum of, 535–6
  wave functions for, 360–5
Hylleraas-Undheim theorem, 406–8
hypercharge, 747
hyperon, 746

inertial frame, 255
independent particle model, 492
identical particles, 472–7
  scattering of two, 620–6
  and symmetry of wave functions, 473–4
  see also boson, fermion
infinitesimal unitary transformations, 219
integral equation of potential scattering, 608–14
interaction picture, 239
inversion spectrum of the NH_3 molecule, 736–40

Jacobi coordinates, 471–2

Jeans, J., 5
Jeffreys, H., 408
Jönsson, C., 43
Jordan, P., 51, 772

Kawasaki, T., 53
kinetic energy operator, 84
Kirchhoff, G.R., 2, 20
Kirchhoff's law, 2
Klein-Gordon equation, 679–84
  for a charged particle in an electromagnetic field, 680–1
  for a free particle, 680
  interpretation of, 682–4
  non-relativistic limit of, 682
  and antiparticles, 684
K-meson (kaon), 477
  and $\mathcal{CP}$ violation, 752–3
  general properties of, 746–50
  production and decay, 750–2
k-normalisation, 140
Kramers, H.A., 408
Kronig-Penney potential, 182

laboratory coordinates, 590–2
Laguerre polynomial, 358
  associated, 358–9
Lamb, W.E., 396, 710
Lamb shift, 396, 710
Landau levels, 574
Landé g-factor, 576
Larmor formula, 516, 529–30
Larmor frequency, 35
laser, 735–6, 742–6
laser-assisted collisions, 746
laser spectroscopy, 745
von Laue, M., 16
Legendre function, 280–2
Legendre polynomial, 279–80
Lenard, P., 12
length gauge, 521
Lennard-Jones potential, 722
lepton, 366
Lewis, G.N., 14
lifetime
  of a hydrogenic atom, 539
  of an alpha emitter, 424–6
  of atomic levels, 74–5, 538–9

and energy widths, 74–5, 541–2
line intensity, 538
line shape, 540–4
line spectra of atoms, 20–1
line widths
  Doppler, 542–3
  natural, 541–2
  and pressure broadening, 542
Liouville, J., 408
Liouville equation, 655
Lippmann-Schwinger equation, 613
  see also integral equation of potential scattering
Lorentzian distribution, 540
Lorentz force, 521
Lorentz transformations, 690–2
Lorentz triplet, 575
Lummer, O., 4
Lyman series, 21

magnetic dipole moment
  in the Bohr model, 33–4
  of spin 1/2 particles, 563–4
  see also Stern-Gerlach experiment
magnetic interactions
  with atoms, 574–84
  with nuclei, 584
  with spin 1/2 particles, 564–7
  strong, 572–4
  weak, 564–6
magnetic resonance, 576–84
  nuclear, 584
  paramagnetic, 578–84
  Rabi apparatus for, 581–2
magnetic quantum number, 37, 276
magneton
  Bohr magneton, 34
  nuclear magneton, 563
Maiman, T.H, 743
many-electron atoms, 492–8
  and selection rules, 537
Marsden, E., 22
maser, 736–42
matrix mechanics, 51
matrix representation, 220–5
  of angular momentum operators, 296–9
  for the harmonic oscillator, 229–30
  for spin operators, 300

see also operator
Matsuda, T., 53
Matthews, M.R., 672
Maxwell, J.C., 1
Maxwell-Boltzmann statistics, 663–6
Mead, C.A., 456
measurement in quantum mechanics, 200–1, 207, 211, 767–72
  and angular momentum, 267–8
Melvill, T., 19
Mendeleev, D.İ., 496
meson, 477, 747
micro-canonical ensemble, 656–7
microscope
  Heisenberg, 70
  scanning tunnelling, 153–4
Millikan, R.A., 15
minimum uncertainty wave packet, 215–16
mixed states, 641
Mlynek, J., 43
molecular structure, 498–506
  hydrogen molecular ion, 500–6
  LCAO method, 503
  rovibrational spectrum, 499–500, 504–6
momentum operator, 59
momentum representation, 124–6, 218–19
momentum space, 63, 124–6, 196–7
Mössbauer effect, 76
Mott scattering formula, 623–6
multiphoton processes, 746
muon, 366
muonic atom, 366–7
muonium, 366

Neuhauser, W., 773
Neumann spherical functions, 342–4
neutron diffraction, 43
neutron stars, 484
Newton, I., 19
nuclear force, 506–7
nucleon, 506
nucleus
  discovery of, 22
  Fermi gas model of, 510-11

observable, 205, 211, 213
one-electron atom *see* hydrogenic atom
operator
  adjoint, 201
  anti-linear, 253
  anti-unitary, 253
  and dynamical variables, 198–201
  electric dipole, 521
  energy, 59
  evolution, 232–4, 240
  functions of, 202
  Hermitian, 89, 199–200
  inverse, 202–3
  linear, 93
  momentum, 59
  orbital angular momentum, 266–70
  parity, 249–51
  projection, 203
  in quantum mechanics, 59, 198–203
  raising and lowering, 227, 285–7, 294–5
  reflection, 249–51
  rotation, 270–5
  self-adjoint, 201
  time-reversal, 253
  translation, 246–8
  unitary, 202–3
optical molasses, 673
optical theorem, 598
orbital angular momentum
  eigenfunctions for, 275–89
  operators for, 265–70
  and polar coordinates, 268–9
  quantum number, 279
  see rotations
orthogonality, 115–18, 197, 204
  and degenerate eigenvalues, 116–18, 204
orthonormality, 204
orthonormal sets, 118
oscillation theorem, 119
oscillator strengths, 538

pair production, 716
paramagnetism, 576
parity, 159–61, 249–51, 284, 338, 365, 534
  intrinsic, 749

Parseval's theorem, 780
partial wave method, 595–9
particles and holes, 731–2
partition function, 659
Paschen-Back effect, 575–6
Paschen series, 21
Pasternack, S., 710
path integrals, 240–5
Pauli, W., 303, 477, 684
Pauli equation, 711–12
  higher order corrections to, 712–5
  and the magnetic moment of the electron, 712
Pauli exclusion principle, 477
Pauli spin matrices, 308–9
periodic boundary conditions, 139, 331
periodic system, 494–6
perturbation theory
  and van der Waals interaction, 719–22
  for a constant perturbation, 435–43
  degenerate, 386–96
  golden rule for, 442–3
  for a periodic perturbation, 443–7
  quasi-degenerate, 396–9
  in scattering, 615–17, 626–35
  time-dependent, 431–47
  time-independent, 375–99
  for a two-level system, 438–41, 446–7
Penzias, A.A., 10
perturbed oscillator, 381–5
Pfund series, 21
phase shifts, 596
phase velocity, 61, 136
photodisintegration
  of the deuteron, 550–5
  electric dipole, 550–2
  magnetic dipole, 552–5
photoelectric effect, 12–15
photoionisation, 545–50
  cross section for, 548–50
photon, 9, 12, 14
  helicity of, 544–5
  spin of, 544–5
pictures
  Heisenberg, 239–40
  interaction, 240
  Schrödinger, 238

Planck, M., 7, 174
Planck's constant, 1, 8–9, 15
Planck's spectral distribution law, 8, 668
plane waves, 58–9, 136
  expansion of, 345–6
Podolsky, B., 760
Poincaré transformations, 692
polarisation
  of spin 1/2 systems, 646–9
  vector, 517
positron, 366, 716
positronium, 366–7, 753–5
postulates of quantum mechanics, 193–6, 198–200, 205, 231
potentials
  Berry gauge, 457
  central, 335–40
  Coulomb, 351–65, 619–20, 623–4, 626
  hard sphere, 602–5
  infinite square well, 156–61
  isotropic oscillator, 367–70
  Kronig-Penney, 182
  linear harmonic oscillator, 170–82, 226–31
  periodic, 182–8
  potential barrier, 150–6
  potential step, 141–50
  potential well, 104–14
  square well, 163–70
  three-dimensional oscillator, 333–5
  three-dimensional square well, 347–51, 599–602
precession of spin
  in a magnetic field, 566
principal quantum number, 25, 354
Pringsheim, E, 4
probability amplitudes, 124, 206–7
probability current density, 89–90
  relativistic, 689
probability density
  position, 56, 66, 86, 195
  in momentum space, 68, 124–5, 196–7
  relativistic, 689
propagator, 121, 244–5
psi meson, 756
pure states, 641

quantisation

  of angular momentum, 24
  of energy, 9, 23–5, 31–3, 104–14
quantum dots, 732–5
quantum jumps, 772–4
quantum optics, 745–6
quarkonium, 757
quark, 304, 477
  bottom, 756–8
  charmed, 755, 758
  colour, 477
  down, 477, 758
  flavour, 477
  strange, 477, 758
  top, 756–8
  up, 477, 758

Rabi I. I., 591
Rabi flopping frequency, 447, 580
  molecular beam apparatus, 581–2
radial Schrödinger equation
  general properties of, 338–40
radial wave functions
  free particle, 340–4
  general properties of, 337–40
  for hydrogenic atoms, 357–62
  for the three-dimensional isotropic oscillator, 370
radiation
  absorption of, 524–6
  emission of, 526–31
radiative capture, 555
Rayleigh, Lord, 408
Rayleigh-Ritz method see variational method
Rayleigh-Schrödinger perturbation theory, 375–81
reduced mass, 28, 237
reflection coefficient, 144, 147–9, 154, 168
  in WKB approximation, 419
representation, 218
  change of, 223–5
  matrix, 220–1
  momentum representation, 218–19
  position representation, 218
resistivity, 724
resonance
  in scattering, 601–3

in two-level systems, 444, 446–7
Retherford, R.C., 396, 710
rigid rotator, 290–2
Ritz, W., 21
Ritz combination principle, 22
Roger, G., 53, 767
Röntgen, W.K., 16
Rosen, N., 760
rotating wave approximation, 446
rotations, 270–5, 312-14
   in spin space, 314–15
Rutherford, E., 22
Rutherford model of the atom, 22
Rydberg, J.R., 20
Rydberg constant, 20, 25

Salam, A., 750
Sauter, Th., 773
scanning tunnelling microscope, 153–4
scattering
   double, 652–3
   of positrons by atomic hydrogen, 626–35
   of spin 1/2 particles by spin 0 particles, 649–52
scattering amplitude, 594
scattering by composite systems, 626–35
   elastic scattering, 630–32
   inelastic scattering, 632–5
scattering length, 599
scattering of nucleons by nucleons, 605–8
scattering of identical particles, 620–6
scattering of particles
   in the Born approximation, 615–20
   in the centre of mass system, 590–2
   by a Coulomb potential, 619–20
   cross section, 589, 597–8
   experiments for, 588–90
   by a hard sphere, 602–5
   partial wave expansion for, 595–7
   by a potential, 592–626
   by a square well, 599–602
   by a Yukawa potential, 618
Schawlow, A.L.. 672, 742
Schmidt orthogonalisation, 116–17
Schrödinger, E, 30, 39, 51, 679, 772
Schrödinger equation, 83–6

   general solution for time-independent potential, 120–4
   for an atom in an electromagnetic field, 520
   for a charged particle in an electromagnetic field, 519
   for N-particles, 469–72
   for one-dimensional systems, 133–4
   in momentum space, 124–5
   relativistic, 679
   separation of variables for, 328–9, 335–7
   time-dependent, 82–5, 231
   time-independent, 100–1
   for two-body system, 236–8
   for two-electron atoms, 485–6
   see also radial Schrödinger equation
Schrödinger picture, 238–40
Schrödinger's cat paradox, 769
Schwinberg, P.B., 712
selection rules, 533–5, 537
self-consistent field, 496
semi-classical methods *see* WKB approximation
semi-conductors, 724
Sherman function, 652
Slater determinant, 476
S-matrix element, 597
Sommerfeld, A., 30, 417, 483
spectral distribution
   for a black body, 2–8
   and Planck's law, 8
   and the Rayleigh-Jeans law, 6
spectral term, 21
spherical Bessel functions, 342–4
spherical harmonics, 283–9
spin angular momentum
   addition of, 321–2
   and the Dirac equation, 699–701
   discovery of, 37–8
   properties of, 299–311
   singlet and triplet states, 321–2
   spin one-half, 303–11
   see Pauli spin matrices
   see also helicity
spinor, 314
   four-component, 688, 696–9

spin-orbit coupling, 391, 498, 714
spontaneous emission, 515, 527–31
    from H(2p), 530–1
stability of atoms, 72–3
Stark effect, 557–63
    in atomic hydrogen, 557–61
    linear, 559
    quadratic, 558
    for the rigid rotator, 562–3
state function, 194
states
    mixed, 641
    pure, 641
    stationary, 225–6
stationary phase condition, 61
stationary states, 23, 103–4, 127
statistical average, 642
statistical mechanics
    general, 655–61
    of ideal gases, 668–75
    of a particle in a box, 662–3
    and Planck's law, 666–8
    of a spin 1/2 particle in a magnetic field, 661–2
    of systems of particles, 663–6
statistical weight, 642
Stefan, J., 3
Stefan-Boltzmann law, 3
Stefan's constant, 3, 8
Stern-Gerlach experiment, 33–8
Stern, O., 33, 43
Stokes, G., 408
Stoletov, M., 12
strong interaction, 746–9
sudden approximation, 458–66
superposition principle, 57, 196
    and decay of K-mesons, 746–53
symmetry principles, 245–55

Taylor, G.I., 53
Thomas-Fermi model, 496
Thomas-Reiche-Kuhn sum rule, 538
Thomson, G.P, 40, 42
Thomson, J.J., 12, 16, 40
Thomson scattering, 16
time evolution of a system, 231–5, 772–4
time-reversal, 251–5
time translations, 253

Tonomura, A., 53
Toschek, P.E., 773
total angular momentum, 38, 311–15
    conservation of, 313
    and rotations, 312-15
Townes, C.H., 736, 740, 742
transition matrix, 614
transition rate
    for absorption, 526
    for emission, 526
translations, 244–8
transmission coefficient, 148–9, 151–5, 168–70
transmission coefficient in WKB approximation, 419
tritium, 366, 464
triton, 464
Truhlar, D., 456
tunnel effect, 152
Turlay, R., 752
two-electron atoms, 485–92
    perturbation theory of, 487–90
    and the variational method, 490–2
    wave functions of, 485–7
two-slit experiment, 43–4, 52–6, 71–2

Uhlenbeck, G.E., 37
uncertainty principle, 52, 69–76
    for time and energy, 73–4
    and zero-point energy, 159
uncertainty relations, 213–16
uncorrelated wave function, 472
uniform approximation, 783
unitary transformation, 216–19, 223–5
    infinitesimal, 219
    invariance under, 217-18

valence band, 724
Van Dyck, R.S., 712
variational method, 399–408
    for excited states, 404–8
    and minimum principle, 401
variation of constants, 432
vector addition coefficients
    see Clebsch-Gordan coefficients
vector model of angular momentum, 288–9, 298–9, 307
virial theorem, 235–6

Voigt profile, 543

van der Waals interaction, 719–22
Walther, H., 672, 745
wave functions
  continuity conditions on, 86, 113
  and dynamical states, 194–5
  entangled, 472, 476, 769
  for a free particle, 58–68, 134–41, 329–31, 340–6
  for hydrogenic atoms, 351–65
  interpretation of, 52–7, 86, 194–5, 759–60
  for the three-dimensional isotropic oscillator, 370
  in momentum space, 62–4, 67–8, 124–8, 196–7
  normalisation, 55–7, 87, 138–41, 194–5
  for particle in a box, 351–3
  for particle in a potential *see* potentials
  radial, 337
  regeneration of, 161–2
  and probabilities, 56–7
  symmetry of, 472–7
  time evolution of, 232–4
  time-independent, 100–4
  uncorrelated, 472
wave mechanics, 51
wave packets, 60–9
  of minimum uncertainty, 215–16
  in a slowly varying potential, 69
wave-particle duality, 15, 38–44, 52–6, 76
weak interaction, 749–50
Weinberg, S., 750
Weisskopf, V., 684
Wentzel, G.. 408
Wheeler. J.A., 366
white dwarf stars, 483–5
width of spectral lines, 541–4
Wieman, C E., 672
Wien, W., 5
Wien's laws, 4–5
Wigner, E, 507
Williams. N.H., 740
Williarns, R.C., 710
Wilson. C T.R.. 19
Wilson, R.W., 10

Wilson, W., 30, 417
WKB approximation, 408–26
Wronskian, 110, 184

Young, T., 43, 55

Zeeman, P., 374
Zeeman effect, 37, 374–5
Zeiger, H., 736
zero-energy resonance, 638
zero point energy, 159, 175
Zweig, G., 477